Drilling, Completion and Well Engineering for the Natural Energy Resources Extraction, Storage and Sustainable Management

Drilling, Completion and Well Engineering for the Natural Energy Resources Extraction, Storage and Sustainable Management

Editors

Mofazzal Hossain
Hisham Khaled Ben Mahmud
Md Motiur Rahman

MDPI

Basel • Beijing • Wuhan • Barcelona • Belgrade • Novi Sad • Cluj • Manchester

Editors

Mofazzal Hossain
Curtin University
Perth
Australia

Hisham Khaled Ben Mahmud
Universiti Teknologi
PETRONAS
Perak Darul Ridzuan
Malaysia

Md Motiur Rahman
Khalifa University of Science
and Technology
Abu Dhabi
United Arab Emirates

Editorial Office
MDPI
St. Alban-Anlage 66
4052 Basel, Switzerland

This is a reprint of articles from the Topic published online in the open access journals *Energies* (ISSN 1996-1073), *Minerals* (ISSN 2075-163X), and *Resources* (ISSN 2079-9276) (available at: https://www.mdpi.com/topics/drilling).

For citation purposes, cite each article independently as indicated on the article page online and as indicated below:

Lastname, A.A.; Lastname, B.B. Article Title. *Journal Name* **Year**, *Volume Number*, Page Range.

ISBN 978-3-7258-1121-2 (Hbk)
ISBN 978-3-7258-1122-9 (PDF)
doi.org/10.3390/books978-3-7258-1122-9

Contents

About the Editors

Mofazzal Hossain

Dr. Mofazzal Hossain, a Chartered Professional Engineer (FIEAust CPEng NER APEC Engineer IntPE(Aus)), holds a PhD in Petroleum Engineering from the University of New South Wales, Sydney, Australia. With over 20 years of extensive experience, his academic expertise spans multidisciplinary engineering fields, focusing on energy resource-based engineering and sciences. His areas of specialization include Drilling, Completion, and Production Engineering, Hydraulic Fracture Stimulation, Geomechanics, Wellbore Stability, Well Integrity, Numerical Modelling and Simulation Studies, particularly in Fluid Flow through Porous Media, and Well and Production Technology.

He has held various academic roles in the US, Australia, and the Middle East, including senior leadership and management positions. Dr. Hossain possesses significant experience in teaching and research, engineering education, and curriculum design within multidisciplinary engineering programs. He has supervised numerous academic research and industry-supported projects and has authored/co-authored over 90 high-quality journal and conference papers.

He has served as a discipline expert panel for Engineers Australia Accreditation and Chartered Professional Membership. He served as an organising and Technical Committee member of several SPE international conference, such as SPE IPTC, SPE-APOGCE, SPE-OTC, SPE-APDT, and Australian Energy Production Conference and Exhibition (AEPE). He also contributes to the academic community as an Editorial Board member of the international journal *Energies*, Academic Editor of *Energies*, Editor of Special Issues, Engineering subject Editor of the *Arabian Journal of Science and Engineering*, and Advisory board of the international scientific journal "*Sci*".

Hisham Khaled Ben Mahmud

Dr. Hisham Khaled Ben Mahmud is a Chartered Professional Engineer (MIEAust CPEng NER) with Engineers Australia and holds a PhD degree from Curtin University, Perth, Australia. He has gained more than 15 years of industry and academia experience in upstream and downstream. His professional career has spanned from academia and to research work on Production Optimization, Flow Assurance, Enhanced Oil Recovery, Energy Storage, and CO_2 Sequestration and Multiphase Flow through Subsea and Porous System.

He is currently working as an Associate Professor at the Department of Petroleum Engineering, Universiti Teknologi PETRONAS, Malaysia, where he has been actively involved in teaching, research, curriculum design, and consultancy. He supervised/co-supervised a number of undergraduate and postgraduate students and authored/co-authored over 40 papers published in international journals and conferences, and 5 book chapters with well-known publishers.

He is a member of the editorial board of the *International Journal of Petroleum and Petrochemical Engineering* and the *Journal of Subsurface Exploration and Exploitation*. Furthermore, he has been serving various positions in Malaysia, Australia, and China, as an industrial advisory panel, the head of department, a program coordinator, a faculty advisor, a regular technical reviewer for several leading international journals, and a technical committee member and organiser of technical sessions, short courses, and industrial trips.

Md Motiur Rahman

Dr. Md Motiur Rahman is an Associate Professor of Petroleum Engineering, Khalifa University of Science and Technology, Abu Dhabi, United Arab Emirates (UAE). Before joining Khalifa

University in 2010, he was a Senior Lecturer of Petroleum Engineering, The University of Adelaide, Australia. He was an Associate Professor of Petroleum Engineering, UAE University, during 2007–2008 and was an Assistant Professor in UAE University during 2004–2007. He has about 7 years' industry experience in oil/gas and geothermal energy.

He holds both Master of Engineering and PhD degrees in Petroleum Engineering from The University of New South Wales, Sydney, Australia. He has gained over 20 years' experience in teaching and research in academia. His teaching experiences are in the areas of drilling, completion, production, artificial lift, stimulation, and improved oil/gas recovery. His research areas are hydraulic fracturing tight and unconventional formations, acid fracturing carbonates, matrix acidizing, and enhanced/improved oil recovery from sandstones and carbonates with low salinity and surface-active polymer applications. He authored/co-authored over 100 articles in international journals and conference proceedings.

He is a life member of SPE. He served many technical committees of SPE Conferences, like ADIPEC, GOTECH, and IPTC. He presented papers in several international conferences (SPE, INTERPORE, and ASME) and co-chaired many technical sessions in conferences and workshops. He was a guest editor of a Special Issue of *Energies* and the Special Issue was called 'Advanced Research and Techniques in EOR Processes'. He is a regular technical reviewer of leading international journals.

Preface

The ever-increasing demand for energy calls for a dual approach: responsibly extracting existing resources to meet current needs, while facilitating a seamless transition to sustainable solutions for the future. Drilling and well completion processes are pivotal in achieving success in both meeting the world's growing energy demands and transitioning towards sustainable energy solutions. This applies to various sectors including exploration and extraction of oil, gas, geothermal energy, gas hydrates, deep mining, subsea mining, and underground storage of CO_2, hydrogen, or excess renewable energy. This book, comprising selected research articles from the Topic, delves into a pivotal role of drilling and well completion technologies in striking this delicate balance. It endeavours to bridge the gaps by presenting a diverse array of research focused on advancing drilling, completion, well engineering practices, and an understanding of associated areas such as reservoir characterisation, geomechanics, and rock mechanics including an overarching theme in the perspective of sustainability and management. It is particularly tailored to primary energy resource exploration and extraction sustainably, offering insights from a multidisciplinary pool of research articles covering scientific principles and research innovation. The goals are to advance drilling and well completion technology towards exploration, economic development, and sustainable extraction of primary resources while fostering a multidisciplinary approach, providing a thorough understanding of the scientific principles, critical technical aspects, and best practices necessary for the successful and responsible extraction of primary energy resources in a sustainable manner.

The articles submitted for the Topic covered very broad areas, which can be categorised into five main sub-areas: (i) Well Construction, Wellbore Stability and Drilling Automation; (ii) Wellbore Stability, Well Integrity, Geomechanics, Rock Mechanics; (iii) Drilling and Completion Fluid, Cementing and Cutting Transport; (iv) Reservoir Characterisation, Well Logging, Formation Damage and Production; and (v) Overarching Theme Including Sustainability, and Management. Accordingly, the book is organised with following 5 sub-areas:

(i) Well Construction, Wellbore Stability and Drilling Automation: Article 1–5;
(ii) Wellbore Stability, Well Integrity, Geomechanics, Rock Mechanics: Article 6–10;
(iii) Drilling and Completion Fluid, Cementing and Cutting Transport: Article 11–16;
(iv) Reservoir Characterisation, Well Logging, Formation Damage and Production: Article 17–20;
(v) Overarching Theme Including Sustainability, and Management: Article 21–25.

Mofazzal Hossain, Hisham Khaled Ben Mahmud, and Md Motiur Rahman
Editors

energies

MDPI

Article

Intelligent Action Planning for Well Construction Operations Demonstrated for Hole Cleaning Optimization and Automation

Gurtej Singh Saini, Parham Pournazari, Pradeepkumar Ashok and Eric van Oort *

Cockrell School of Engineering, The University of Texas at Austin, Austin, TX 78712, USA
* Correspondence: vanoort@austin.utexas.edu

Abstract: Reactive and biased human decision-making during well construction operations can result in problems ranging from minor inefficiencies to events that can have far-reaching negative consequences for safety, environmental compliance and cost. A system that can automatically generate an optimal action sequence from any given state to meet an operation's objectives is therefore highly desirable. Moreover, an intelligent agent capable of self-learning can offset the computation and memory costs associated with evaluating the action space, which is often vast. This paper details the development of such action planning systems by utilizing reinforcement learning techniques. The concept of self-play used by game AI engines (such as AlphaGo and AlphaZero in Google's DeepMind group) is adapted here for well construction tasks, wherein a drilling agent learns and improves from scenario simulations performed using digital twins. The first step in building such a system necessitates formulating the given well construction task as a Markov Decision Process (MDP). Planning is then accomplished using Monte Carlo tree search (MCTS), a simulation-based search technique. Simulations, based on the MCTS's tree and rollout policies, are performed in an episodic manner using a digital twin of the underlying task(s). The results of these episodic simulations are then used for policy improvement. Domain-specific heuristics are included for further policy enhancement, considered factors such as trade-offs between safety and performance, the distance to the goal state, and the feasibility of taking specific actions from specific states. We demonstrate our proposed action planning system for hole cleaning, a task which to date has proven difficult to optimize and automate. Comparing the action sequences generated by our system to real field data, it is shown that it would have resulted in significantly improved hole cleaning performance compared to the action taken in the field, as quantified by the final state reached and the long-term reward. Such intelligent sequential decision-making systems, which use heuristics and exploration–exploitation trade-offs for optimum results, are novel applications in well construction and may pave the way for the automation of tasks that until now have been exclusively controlled by humans.

Keywords: reinforcement learning; action planning; Monte Carlo tree search; digital twinning; well construction; hole cleaning; heuristic design

Citation: Saini, G.S.; Pournazari, P.; Ashok, P.; van Oort, E. Intelligent Action Planning for Well Construction Operations Demonstrated for Hole Cleaning Optimization and Automation. *Energies* **2022**, *15*, 5749. https://doi.org/10.3390/en15155749

Academic Editor: F. Pacheco Torgal

Received: 22 June 2022
Accepted: 5 August 2022
Published: 8 August 2022

1. Introduction

Well construction is the process of drilling and completing wells to extract subsurface fluids and gases such as water, brine and hydrocarbons, access geothermal energy, and store waste underground. Each well construction task has multiple co-occurring sub-processes and various systems mutually interacting, making it highly complex. Furthermore, drilling deep in the subsurface into variable geological environments adds an inherent irreducible level of complexity to the process. The highly involved and poorly predictable nature of well construction can result in operational inefficiencies or safety and environmental issues, potentially leading to non-productive time (NPT) associated with events such as lost circulation, stuck drillstring and well control incidents. Therefore, careful surveillance of process and equipment data for performance tracking and building intelligent systems for decision-making and action planning are needed.

There are many algorithms and systems in the well construction domain that currently perform process monitoring. To assist with decision-making, approaches such as case-based reasoning [1,2], performance tracking using digital twins [3–5], knowledge graphs and decision-trees [6,7], and type curve matching [8,9] are being utilized. Remote process monitoring and multidisciplinary collaboration among domain experts have been enabled by the use of real-time (RT) data streams, advanced process and system models, and machine learning techniques [10–13]. Final decisions, however, are often still based on an engineer's understanding and interpretation of the data, key performance indicators, and model outputs. This human-centered decision-making not only introduces a degree of bias [14], but can also result in reactive, short-term decisions instead of proactive long-term action planning.

By contrast, intelligent decision-making and action planning systems have resulted in overall performance improvements in various other fields. For instance, areas such as microgrid energy management [15,16], economic dispatch and large-scale power dispatch problems [17,18], electric vehicle systems management [19,20], and smart grid management [21] utilize reinforcement learning (RL) methods successfully. More recently, very high complexity adversarial board games (e.g., chess and Go), and complex RT strategy games (e.g., Alphastar, Dota, and Pac Man) have been solved using a combination of deep neural networks and tree-based search techniques [22–27].

The goal of the research presented here is to examine the applicability of these novel planning techniques in the well construction domain, and to utilize them for the development of intelligent decision-making systems (or decision-engines) for improved safety and performance (Figure 1). The proposed methodology is demonstrated here by structuring a decision-engine specifically for hole cleaning operations. Hole cleaning is the process of removing solids (cuttings, cavings, or metal shavings) from the borehole by circulating a drilling fluid in and out of the well. Removal of solids is critical in ensuring different well construction operations (such as drilling, tripping, running casing and cementing) are performed safely and efficiently. Issues such as downhole tool damage, stuck pipe situations, damage to downhole formations, and difficulty running casing and cementing can result from poor hole cleaning. Annually, hole cleaning issues cause several hundred million dollars in NPT costs [28,29]. Therefore, building a decision-engine for hole cleaning advisory and eventual automation that permits superior proactive action planning and execution is of considerable practical importance and value.

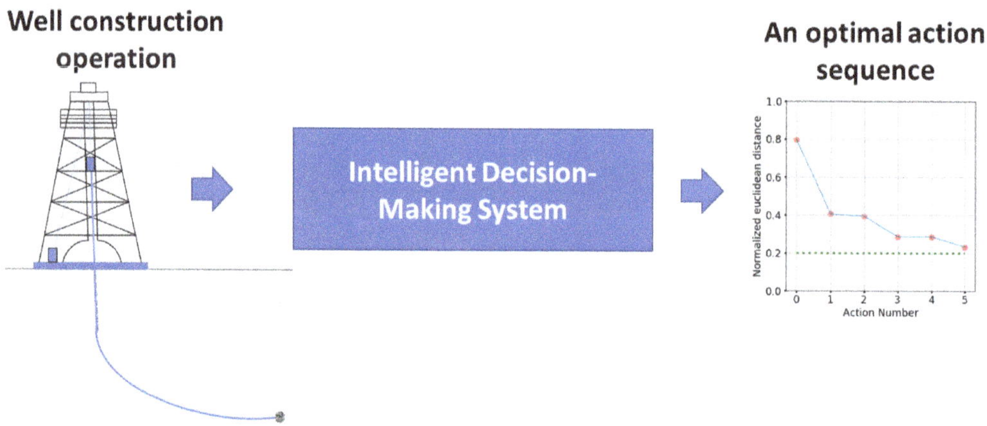

Figure 1. Schematic representation of the objective of this work: development of intelligent decision-making systems for well construction operations. The right figure shows progression of the normalized Euclidean distance of the system states (blue line) to an acceptable goal state (shown by the green dotted line, here taken to be a state at a normalized Euclidean distance of 0.2).

2. Setting Up Decision-Engines for Well Construction Operations

Planning is the process of generating an action sequence from an initial system state to a goal state to satisfy a high-level objective function. Setting up a planning system requires explicitly stating the problem objectives and constraints, defining the state- and action-space, and identifying goal states [30]. The various state-action transitions can either be fully specified in advance or be incrementally discovered (by utilizing models) as the planning proceeds. The solution of a planning problem is a policy or a strategy that suggests the sequence of actions for every successive decision-making step (called a decision epoch). The following sections provide an overview of different planning algorithms (particularly for well construction engineers unfamiliar with such algorithms, which have not yet found widespread adoption in their industry) and discuss their applicability to well construction operations.

2.1. Comparison of Planning Algorithms

Planning algorithms can be compared based on a multitude of factors, such as the search-space traversal methodology, requirement of a heuristic, ability to plan in a limited amount of time, memory requirement for storing intermediate results, computational complexity, etc. [30]. A straightforward approach to solving planning problems, however, is the exhaustive tree search, wherein all possible actions from all states are evaluated until either a goal state is reached, or the available evaluation time is exhausted. Although this method ensures finding an optimal plan, for large systems (with large state and action spaces), the associated time complexity renders this method impractical. Some other classifications of algorithms based on different search-space traversal methodologies are breadth-first search (BFS), depth-first search (DFS), and best-first search (BestFS).

BFS is a first-in, first-out (FIFO) algorithm that searches equally in all directions, i.e., all actions from a given state are evaluated first before the evaluation of the next state's actions can begin. The algorithm stops and outputs the shortest path as soon as a goal state is reached [30,31]. DFS is a more aggressive strategy, wherein, from any state, actions are chosen at random to traverse the system, until either a goal state or a dead-end state is reached. If a dead-end state is reached, the algorithm backtracks one step, then simulates any action not yet explored, i.e., this algorithm works in a last-in, first-out (LIFO) manner. The priority of such algorithms is to search deeper than to expand the search-space. A potential issue with this method is the search becoming stuck in a repetitive loop [31,32]. Neither the DFS nor the BFS utilizes an evaluation function or a heuristic to direct the search, i.e., costs associated with all actions taken from all the states are the same. Moreover, both methods require keeping track of the visited, unvisited, and dead-end states. Therefore, these algorithms run into issues when dealing with large state- or action-space systems due to high computation-time and memory requirements, with BFS more so than DFS due to its more exhaustive nature [30,33].

The BFS algorithm can be enhanced by using some form of evaluation functions or heuristics or both. Dijkstra's algorithm utilizes an evaluation function to quantify the path traversed to reach a given state after starting from the initial state. It, like the BFS, will find an optimal path (if one exists) between the initial and the goal state, but suffers from the same computational and memory issues associated with larger systems [34,35]. If a heuristic is defined to quantify the approximate cost of a state relative to its position from the goal state, this results in a greedy BestFS algorithm. The next action is chosen greedily based on this heuristic. Although greedy BestFS is usually faster than Dijkstra's algorithm, a drawback of this algorithm is the lack of exploration and backtracking, i.e., once an action is selected, it is not re-evaluated based on any new information [30,36]. This search strategy is, therefore, not guaranteed to find an optimal path. Dijkstra's algorithm supplemented by the greedy BestFS heuristic results in the A* search algorithm [30,37]. A* search is a BFS algorithm for which both an evaluation function and an admissible heuristic are defined. A* search will find an optimal path (if one exists) faster than BFS or the Dijkstra's algorithm, but it is memory-intensive since it requires storing all the visited states and keeping track of all possible and tried actions from all possible states. Another issue with A* search is the necessity for a well-defined admissible heuristic [38].

Iterative deepening search (IDS) is a variant of the DFS, especially useful in systems with high branching factors. For IDS, the DFS algorithm is continually run with increasing depth bounds until a goal state or a dead-end state is reached. During each iteration, every state encountered in the path is expanded in a BFS manner. IDS is faster than the BFS and is guaranteed to find an optimal path if one exists [39]. The iterative deepening A* (IDA*) algorithm utilizes the search technique of the IDS in conjunction with a heuristic to quantify the cost of any state relative to the goal state. IDA* is guaranteed to find the optimal solution if the heuristic is admissible [30,31]. The main advantage of iterative search algorithms (IDA and IDA*) is that they do not require much memory, since only the current path is stored. Evolutionary algorithms such as the genetic algorithm evaluate and rank multiple action sequences simultaneously using an expert-designed fitness function. The quality of the population of these action sequences is successively improved using selection, crossover, and mutation operations, until some stopping criterion is met. The intrinsic randomness in crossover and mutation steps ensures a balance between exploration and exploitation of the search space. These algorithms are the most effective if the search space is sufficiently small, or if time is not a constraint for search. Additionally, the performance of these algorithms is highly dependent on the design of the fitness function, and the crossover and mutation strategies [40,41].

For systems with a vast state- and action-space, lack of an admissible heuristic or minimal dependency on some evaluation function, memory constraints, and limited computation time availability, simulation-based search (SBS) methods are the best suited. SBS, a decision-time planning algorithm, is a type of model-based RL method suited for online action planning, starting from a given state [41]. Here, a goal-directed agent interacts (virtually) with an uncertain environment (models representing the underlying processes) based on specific policies or action plans which evolve with time. These policies are designed to balance exploration and exploitation of the search space, and every state action transition has an associated reward. These algorithms build an 'asymmetric' search tree starting from an initial or root node using a sequential BestFS strategy, as shown in Figure 2. Multiple episodes of experience (starting from the root or the current state, s_t) are simulated until either the goal state or a fixed depth bound is reached. After every episode, one or more nodes (representing a state-action transition) are added to the tree, and the values of the already present nodes are updated. SBS algorithms do not require a heuristic (i.e., they are 'aheuristic'); however, a well-defined heuristic can considerably improve the convergence time. Another useful feature of these algorithms is their 'anytime' property, i.e., the algorithm can be stopped at any time to return the best plan thus far [42].

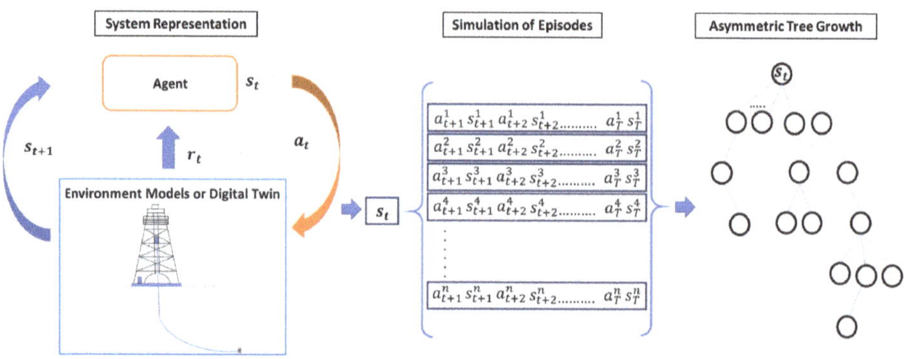

Figure 2. Asymmetric tree growth using simulation-based search algorithms. An action a_t by the agent in the environment (observed by the agent to be in a state s_t) results in an immediate reward r_t and a new observed state s_{t+1}.

Essentially, the anytime, asymmetric, and aheuristic nature of the SBS algorithms makes them a preferred candidate for large systems with memory and computational constraints, as is the case in well construction.

2.2. Utilizing SBS for Well Construction Action Planning

Well construction processes are non-holonomic since the wellbore condition at any time (state) is a function of the well's operational past, i.e., all previous wellbore conditions (past states) and previous operations (actions) [43]. The states themselves, depending on the process, can be represented by combinations of parameters such as equivalent circulation density (ECD), cuttings bed height, cuttings concentration in the flow, friction factor, and drilling dysfunction indicators. To adequately represent a process and distinguish different states, each of these parameters can take on many different values. These unique combinations of values for the many state parameters can result in a vast state-space. Similarly, various combinations of values for the many action control variables result in a vast action-space. Some action variables for drilling operations include drillstring rotation speed (RPM), weight on bit (WOB), flowrate, tripping speeds, mud properties (density and rheology), etc. [44]. Furthermore, the online or near real-time planning requirements impose constraints on computation time.

Decision-time algorithms that can plan in limited time by considering the search-space in the vicinity of the current state are therefore needed to address these requirements. SBS algorithms, such as flat Monte Carlo (MC) and Monte Carlo tree search (MCTS), have been successfully used for planning in systems with large state- and action-space (e.g., game AI engines for chess and Go, as discussed). The game AI engines utilize the MCTS in conjunction with deep neural networks (for policy and value evaluations) to identify, shortlist and sequentially simulate legal actions from a given position (state) for both players. Multiple episodes of such self-play are used to improve the policy incrementally and finally select the most promising line of play (an optimal action sequence). Likewise, the utilization of such SBS techniques for action planning during the various well construction operations can result in improved efficiency and operational performance.

2.2.1. Monte Carlo Tree Search (MCTS) Algorithm

The MCTS is an SBS method that combines MC search with an incremental tree structure. Successive MC simulations (by using a forward model of the process) are utilized for iteratively expanding the search tree and evaluating the various nodes in the tree structure. A node represents a state and contains information about its parent node, possible next actions, the number of current implementations of each action, and the average value associated with each action implementation thus far. Instead of building the entire tree, a few promising lines of play are developed further, resulting in asymmetric tree growth. The selective growth of the search tree is brought about by using two distinct action selection policies: the tree policy and the rollout policy [45,46].

Tree policies can be either greedy (i.e., focusing on exploiting what is already known), or try to balance exploring new or potentially promising paths of the search space with greedy exploitation. This balance is commonly referred to as the exploration-exploitation trade-off. One such policy can be devised by treating every state within the tree as a multi-armed bandit (MAB) problem where the Upper confidence bound (UCB1) algorithm is utilized for action selection within the tree structure [47]. The resulting algorithm is referred to as Upper confidence bound for trees (UCT). The UCT guarantees convergence to an optimal policy (given enough time) since the exploration factor (C_{exp}) not only ensures exploration of unvisited parts of the search space, but the exploration term ($\sqrt{\frac{2 \cdot \ln N(s)}{N(s,a)}}$) also gets less exploratory with increasing number of visits [48]. Equation (1) represents the basic UCT policy, where $N(s)$ is the total number of visits to the state s, $N(s,a)$ represents the number of times action a has been taken from state s. $Q(s,a)$ is the exploitation term representing the average value associated with implementing action a from state s. $Q_{\text{UCT}}(s,a)$ is the upper confidence bound or the urgency term, and the next action within the search tree is selected based on maximizing this term over the action space A_s from the state s (Equation (2)). The rollout policy, by default, is a uniform-random policy.

$$Q_{\mathrm{UCT}}(s,a) = Q(s,a) + C_{exp} \cdot \sqrt{\frac{2 \cdot \ln N(s)}{N(s,a)}} \tag{1}$$

$$a_{n+1} = \max_{a_i \in A_s}(Q_{\mathrm{UCT}}(s,a_i)) \tag{2}$$

Primarily, MC control (a model-free RL technique) is applied to simulated episodes of experiences (model-based RL) from a root node, to grow the search tree iteratively and improve the action plan by backpropagating feedback. A single MCTS iteration consists of the following four phases, as detailed in Figure 3 [47]:

- Selection from nodes already in the search tree using the tree policy;
- Expansion of the tree at the leaf node by adding one (or more) node(s);
- Simulation or rollout of actions, using the rollout policy, until the terminal condition is met (either reaching a terminal state or the end of the planning time horizon, T);
- Backpropagation or backing the rewards up the expanded tree to update the values of different state-action pairs ($Q(s, a)$) encountered during the episode.

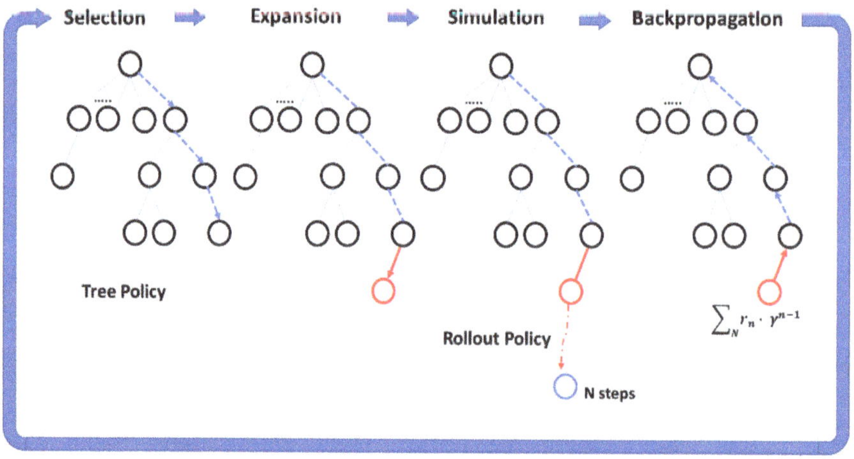

Figure 3. Schematic overview of Monte Carlo tree search algorithm phases (modified from [44]).

As the number of simulations increases over time, the tree expands, and any inherent bias is removed. Every MCTS episode commences from the root node (from which an action plan or action sequence needs to be determined), and traverses the tree based on the most recently acquired knowledge (which is incorporated in the Q values). Thus, MCTS is fundamentally a generalized policy iteration (GPI) algorithm, where the policy or action plan is iteratively evaluated and improved [41,49]. Even with naïve or vanilla policies (such as the standard UCT and uniform random), the MCTS works well and starts to move the results towards optimality.

Although the vanilla MCTS blends the generality of random sampling with the exactness of tree search, its convergence rate can be relatively low. Therefore, in practice, the two MCTS policies have been enhanced by incorporating prior-knowledge or handcrafted strategies. This has resulted in different methods, such as progressive widening, progressive bias, using prior-knowledge, and Rapid Action Value Estimation (RAVE) [47,50–52]. All these strategies require evaluation of a heuristic function, which can either be learned (using deep neural networks) or designed using domain-knowledge or can be a combination of both. The heuristic function devised for most other applications depends on factors such as the proximity of the state to the goal state, values associated with patterns, previous action values, how dangerous or safe is the state as compared to the adversary's positions, etc. [27,50,51,53].

2.2.2. Structuring Well Construction Operations as Sequential Decision-Making Systems

The goal of well construction is to drill and complete a well safely, quickly, efficiently, and economically in line with a drilling program. As previously stated, accomplishing these objectives requires efficient planning and informed decision-making at each step of every well construction process. Additionally, due to the non-holonomic property of well construction operations, any action implemented at the current time will not only affect the immediate system state but also influence the long-term evolution of the system. This necessitates the development of intelligent decision-making systems, the foundation of which is a Markov Decision Process (MDP) framework. An MDP is composed of the tuple $(\{S, A, P, R\})$ and a policy (π) that address the following elements of finite-horizon long-term sequential decision-making [30,54]:

- Identification of appropriate parameters to quantify the state (s_t) of the system, and defining the desired or goal states (s_{goal}) based on the operation's objective;
- Identification of relevant control variables or actions (a_t) that can affect the system state;
- Building the state-space (S) and the action-space (A) such that $s_t \in S$ and $a_t \in A$ for all s_t and a_t;
- Incorporating problem-specific heuristics ($\pi_{heuristic}$) to identify legitimate and promising actions from every state;
- Building models (digital twins) of the underlying processes to simulate the state transitions ($P^a_{ss'}$) brought about by different actions;
- Defining a method to quantify the various state action transitions, for instance, by using reward functions (R) to calculate long-term value functions (Q);
- Selection of a suitable action-planning technique to formulate a plan (π—the suggested sequence of actions for successive decision epochs).

For a process to be Markovian, it must satisfy the Markovian property, i.e., any transition from a given state depends only on the current state and the immediate action. In other words, the current state completely summarizes the system's operational past. As previously discussed, this is true of well construction operations; therefore, the Markovian property assumption is valid. The MCTS algorithm is particularly well-suited for well construction action planning because of its ability to handle vast search-spaces efficiently. The UCT policy assists with asymmetric tree growth along promising paths while balancing exploration and exploitation of the search space. Although MCTS is inherently aheuristic, it still permits the use of domain-knowledge-derived heuristics for speeding up the search. Moreover, the GPI property of MCTS ensures that over time better plans (action sequences) are found. Figure 4 shows the proposed structure for well construction decision-making systems.

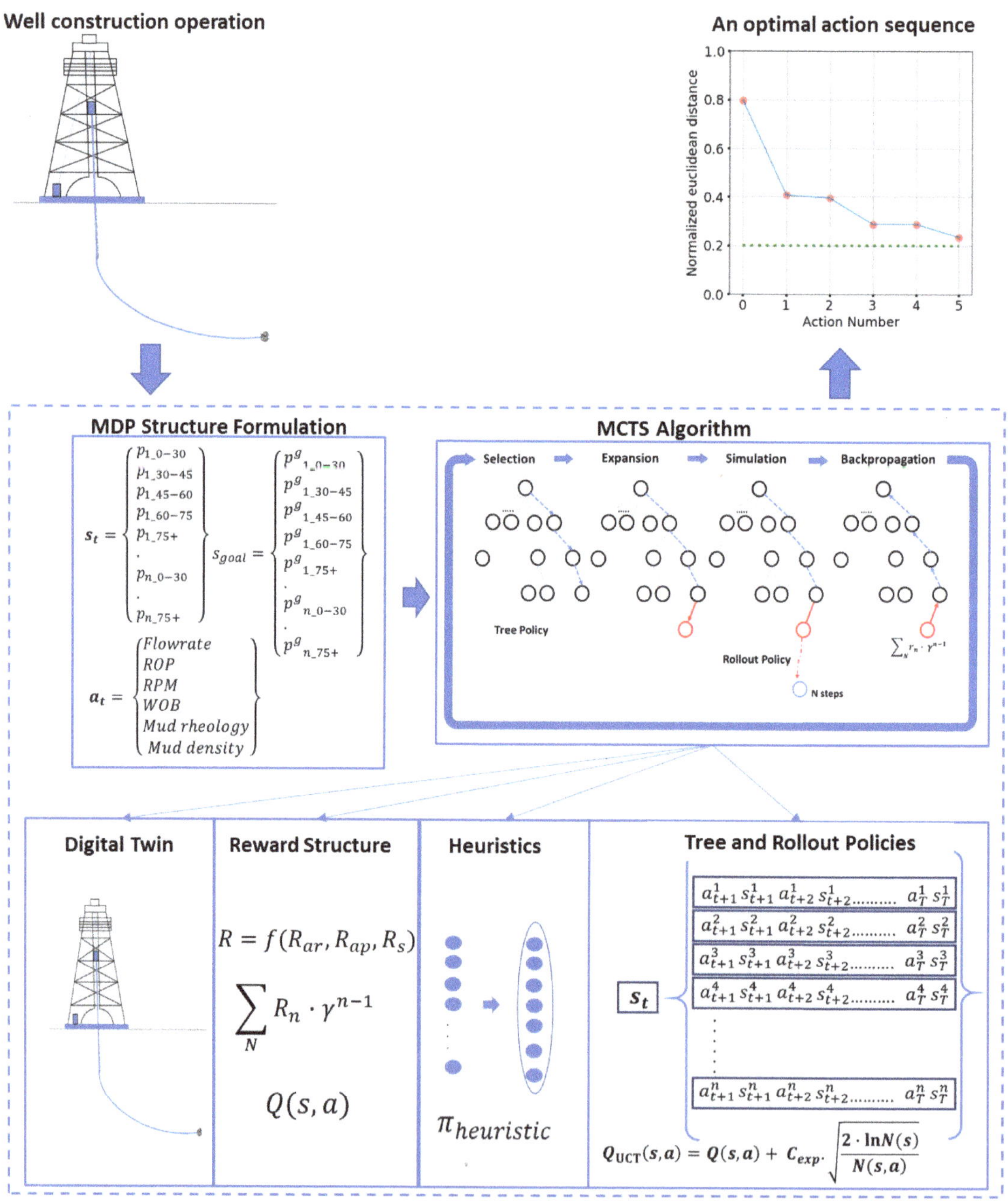

Figure 4. Structure of the proposed decision-engine. Upper right figure shows progression of the normalized Euclidean distance of the system states (blue line) to an acceptable goal state (shown by the green dotted line, here taken to be a state at a normalized Euclidean distance of 0.2).

3. Design of a Planning System for Hole Cleaning Action Planning using MCTS

The objective of hole cleaning operations is to manage:

- Height of the cuttings bed (that settles on the low side of the borehole) at low enough values to prevent issues during any subsequent stage of well construction (Figure 5a);
- Downhole ECD values to remain within a given drilling margin (Figure 5b).

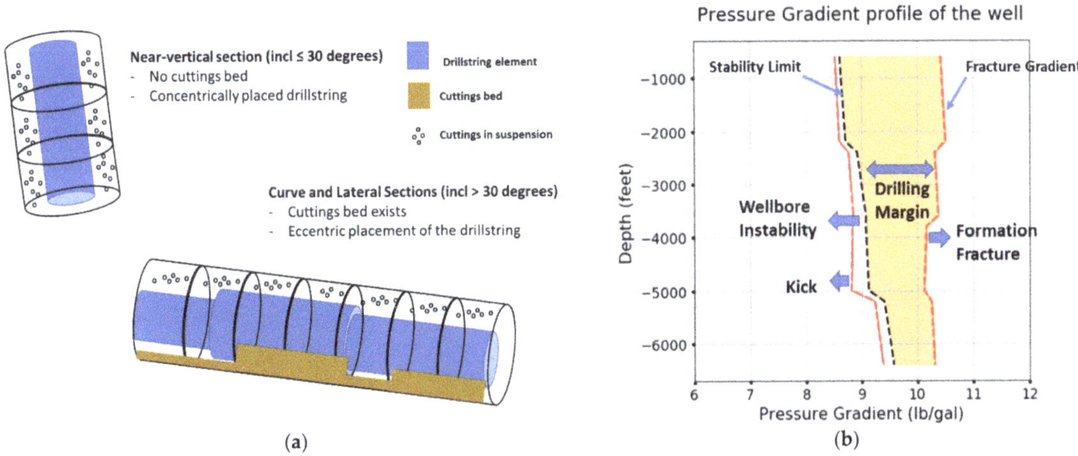

Figure 5. (**a**) Cuttings bed height in different inclination intervals for the well (**b**) Drilling margin.

ECD at any depth is the gradient of the total pressure drop at that depth, and is calculated by Equation (3). The pressure drop is the sum of the hydrostatic pressure exerted by the drilling mud (which is a function of the true vertical depth (TVD) of the well), and the total circulating frictional pressure drop in the annulus between the drillstring and the wellbore (which is a function of the measured depth (MD) along the annular space) [44]:

$$\mathrm{ECD} = \frac{P_{hydrostatic\_D_{TVD}} + P_{frictional\_pressure\_loss\_D_{MD}}}{D_{TVD} \cdot g} \tag{3}$$

The stability limit (SL) and the fracture gradient (FG) form the lower and the upper boundaries of the drilling margin. Exceeding the FG can result in fracturing the formation, which can lead to loss of the drilling mud into the formation, referred to as lost circulation events. SL is the higher value between either the pore pressure (PP) or the mud pressure required for wellbore stability. PP is the pressure exerted by the fluids (hydrocarbons or brine) in the pore spaces of formation rocks. Wellbore instability can occur when the ECD or hydrostatic mud pressure falls below the SL, and a kick, i.e., an unwanted influx of formation fluids/gases into the borehole [55] can happen if the ECD or hydrostatic mud pressure falls below the PP.

3.1. Setting Up the MDP

A directional well can be analyzed by dividing it into three distinct sections based on the inclination angles: near-vertical section, build or curve section, and horizontal section. The segment of the well with inclination angles between 0 and 30 degrees is the near-vertical section, while regions of the well with inclination angles greater than 60 degrees constitute the near-lateral horizontal section. The intermediate inclination angle segments comprise the build or the curve section of a well. This method for discretizing the state-space is proposed since the state variables' response to different actions depends to a high degree on the inclination of the well segment. Likewise, for the hole cleaning system, the cuttings transport mechanisms, and therefore the associated hole cleaning requirements, are different for the three sections. The cuttings are suspended and carried upwards by overcoming the particle slip velocity in the near-vertical section. For the horizontal section, above the angle of repose for the cuttings, a stable cuttings bed forms on the low-side of the borehole. The main requirement is to move this bed up and out of the hole continuously. In the curve section, an unstable cuttings bed forms below the angle of repose, which can avalanche back down the annulus and pack-off around the bottom hole assembly (BHA), causing stuck pipe incidents. The hole cleaning design for this section, therefore, revolves around preventing this cuttings avalanche [56,57].

3.1.1. State-Space and the Goal State

The following parameters are required for quantifying the borehole condition from a hole cleaning standpoint [58,59]:

- Height of the cuttings bed in the curve and the lateral sections of the well;
- ECD along the length of the well.

The state of the system, at any decision epoch, is represented by functional values of these parameters over the inclination intervals $\{[0,\ 30),\ [30,\ 45),\ [45,\ 60),\ [60,\ 75),\ [75+)\}$, as shown in Figure 6, and represented by Equation (4). The functional values for the cuttings bed height ($H_{inc.}$) and the ECD ($ECD_{inc.}$) parameters are whole numbers in ranges $[0,\ 4]$ and $[-3,\ 3]$, respectively. Conversion to functional values helps normalize the absolute values to specific operational thresholds, and assists in reward shaping, which is discussed in later sections. Since the cuttings bed cannot exist in the near-vertical section (inclination interval $[0,\ 30)$), H_{0-30} is not included in the definition of the state.

$$s = \begin{Bmatrix} H_{30-45} \\ H_{45-60} \\ H_{60-75} \\ H_{75+} \\ ECD_{0-30} \\ ECD_{30-45} \\ ECD_{45-60} \\ ECD_{60-75} \\ ECD_{75+} \end{Bmatrix} \tag{4}$$

Defining goal states (or desired states) is necessary to direct a decision-making agent's search of its state and action-space. For the hole cleaning system, the functional values defined for all state components are such that 0 represents the goal state for each. The method for converting the absolute parameter values to their functional values is summarized in Figure 7.

Equation (5) represents the goal state for this system.

$$s_{goal} = \begin{Bmatrix} 0 \\ 0 \\ 0 \\ 0 \\ 0 \\ 0 \\ 0 \\ 0 \\ 0 \end{Bmatrix} \tag{5}$$

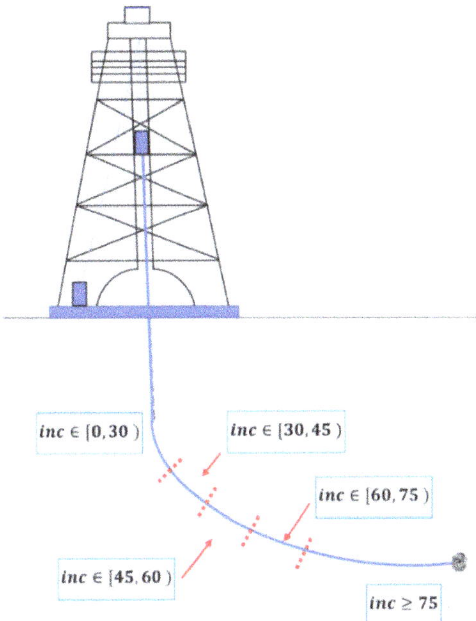

Figure 6. Strategy for defining state vector components based on wellbore inclination angles.

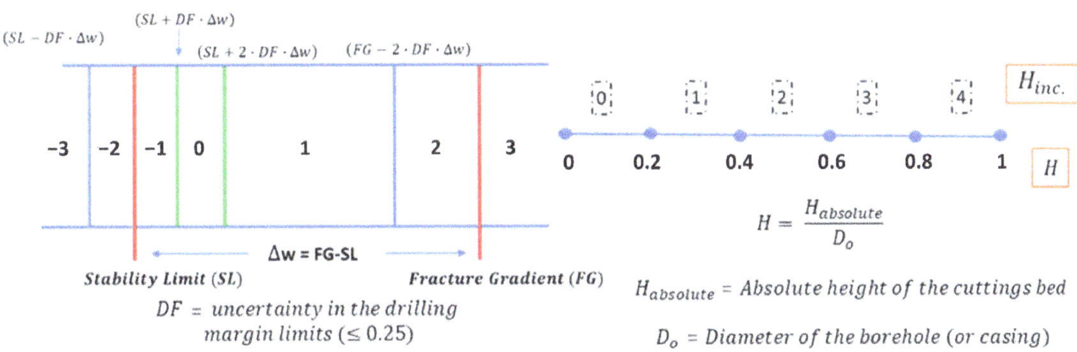

Figure 7. Functional value assignments for the state parameters.

3.1.2. Action-Space

Hole cleaning and ECD management are influenced, to varying degrees, by multiple factors, some of which can be actively controlled in real-time in the field. Factors such as drilling mud properties (particularly density and rheology), drilling parameters (such as drillstring RPM and flowrate), and rate of cuttings generation (which is a function of ROP) significantly influence hole cleaning performance, and can be actively controlled in the field [60,61]. For this research, we assume Bingham Plastic rheological behavior for the drilling mud, in which case its rheology can be characterized by a plastic viscosity (PV) and yield point (YP) [44]. The action set, as shown in Equation (6), is a combination of discrete values of these control variables. Each control variable can take on a finite number of values between some minimum and maximum thresholds that are determined by safety constraints, operational economics, and equipment and process limitations. For instance, if the flowrate has ten values in the [0, 1800] interval, its equally spaced values, in gallons per minute (GPM), are {0, 200, 400, 600, 800, 1000, 1200, 1400, 1600, 1800}.

$$a_t = \left\{ \begin{array}{c} \text{Flowrate} \\ \text{ROP} \\ \text{RPM} \\ \text{Mud density} \\ \text{Mud PV} \\ \text{Mud YP} \end{array} \right\} \quad (6)$$

3.1.3. Digital Twin

A digital twin of the hole cleaning operations was built by integrating multiple data streams with analytical implementations of the cuttings transport, hydraulics, and rig-state detection models [62]. The cuttings transport and hydraulics models estimate the height of the cuttings bed and the ECD along the well, respectively. The rig-state detection engine outputs the current operational state (such as rotary or slide drilling, tripping in or out the borehole, etc.) [63]. The purpose of this twin is twofold: first, for performance tracking by simulating different hole cleaning actions, and second, as a forward simulation model for assisting with action planning. Figure 8 illustrates the application of the twin as a forward-simulation model, where an action a_t transitions the system state from s_t, to s_{t+1}, at decision epoch t. The digital twin was designed to plan either every 5-min interval into the future, or whenever there is a change in the well operations.

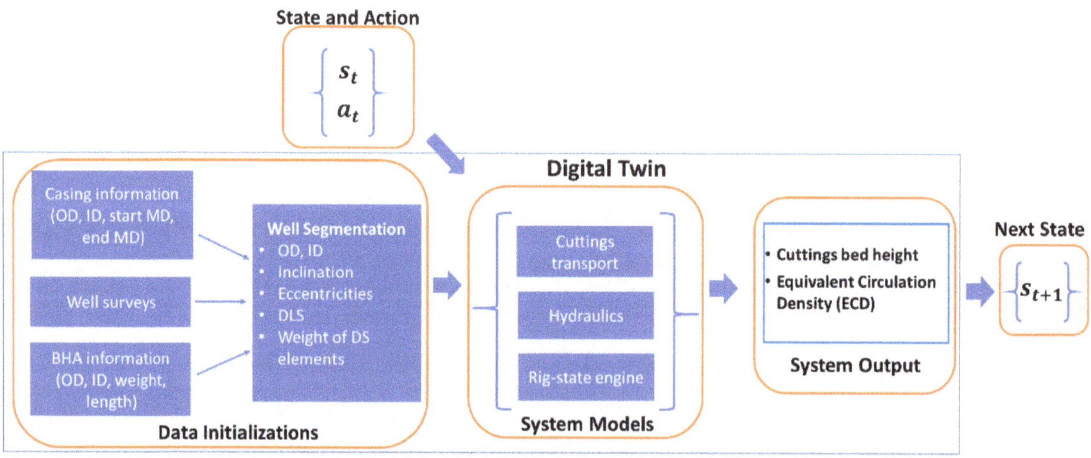

Figure 8. Application of the hole cleaning digital twin for forward-simulation.

3.1.4. Reward Function

A reward function that outputs a normalized feedback (in the [0, 1] interval) for every state-action transition is designed by considering the following:

- the reward set associated with state transition (R_s);
- the penalty set associated with action transition (R_{ap});
- the reward set associated with action values (R_{ar}).

The three reward sets are shown by Equations (7)–(9).

$$R_S = \{R_{H30-45}, R_{H45-60}, R_{H60-75}, R_{H75+}, R_{ECD0-30}, R_{ECD30-45}, R_{ECD45-60}, R_{ECD60-75}, R_{ECD75+}\} \quad (7)$$

$$R_{ap} = \left\{R_{flowrate}, R_{ROP}, R_{RPM}, R_{density}, R_{PV}, R_{YP}\right\} \quad (8)$$

$$R_{ar} = \left\{ R_{a\_flowrate}, R_{a\_ROP}, R_{a\_RPM}, R_{a\_density}, R_{a\_PV}, R_{a\_YP} \right\} \quad (9)$$

Each component of every reward set is assigned different relative weights, which are then combined to give a single number net reward (normalized in the [0, 1] range) for each: R_{s_norm}, R_{ar_norm} and R_{ap_norm}. Finally, these normalized rewards are combined in a weighted fashion, as shown by Equation (10). R_{net} is the final reward output for every state-action transition. This definition of the reward function ensures immediate feedback after every action, instead of the agent having to wait for a few decision epochs (as is the case for sparse reward functions).

$$R_{net} = \frac{W_{s\_norm} R_{s\_norm} + W_{ap\_norm} R_{ap\_norm} + W_{ar\_norm} R_{ar\_norm}}{W_{s\_norm} + W_{ap\_norm} + W_{as\_norm}} \quad (10)$$

Another advantage of this reward structure is that assigning different weights to the various reward set components provides the ability to prioritize different objectives. Prioritizing objectives, for example, would be beneficial while drilling wells that have a high probability of well control issues. In such wells, maintaining ECD within the drilling margin is more important than completely removing the cuttings bed. Similarly, there can be wells where reaching the desired state fast has a higher priority than reducing the penalty associated with taking more aggressive actions. These objectives can be managed by assigning different relative weights to the individual state or action reward components.

3.2. MCTS Setup for Hole Cleaning

Action planning with MCTS requires solving a sub-MDP starting from the current system state (root-node) in a finite amount of time. The hole cleaning digital twin is utilized as the forward-simulation model to simulate the results of different actions on the wellbore condition, and the feedback obtained using the reward function is accumulated and backpropagated. Vanilla MCTS, however, has some limitations:

- The agent is rewarded only when a terminal or a goal state is reached, i.e., sparse rewarding;
- From any state within the tree, all actions in its action-space (A_s) are evaluated by the UCT policy regardless of their practicability for the operation;
- The rollout policy is random uniform, i.e., all actions (irrespective of their feasibility) have an equal probability of being selected.

These limitations result in the requirement to conduct more simulations to expand the search tree to the extent that it can be meaningfully used for trajectory evaluation, thereby slowing down the search. To address these issues this research makes the following changes to the vanilla MCTS:

- Definition of a non-sparse reward function, such that the reward for every state-action transition during the rollout step is utilized for updating all tree nodes' Q values. This is addressed by the reward function shaping strategy, which was discussed in the previous section;
- Using a heuristic function to improve the tree policy;
- Using a heuristic function to reduce the randomness in the rollout policy.

3.2.1. Heuristic Function Development

A domain-knowledge based heuristic $\pi_{heuristic}$ is carefully crafted by balancing the following criteria:

- Safety and performance metric (A_{sp}), to prevent well control issues (such as kicks and lost circulation events) from occurring, as well as to ensure efficient cuttings bed removal and ROP maximization for optimal drilling performance;
- Performance metric, to ensure efficient cuttings bed removal and ROP maximization for optimal drilling performance;
- Sequential metric (A_{sq}), to ensure smoother or sequential changes in values of action control variables;

- Feasibility constraints (A_{fe}), to suggest realistic changes in values of the action control variables;
- Proximity metric (A_{px}), to prioritize actions based on Euclidean distance to the goal state.

Balancing safety and performance is accomplished by incorporating guidelines for well control, hole cleaning, and drilling optimization. Figure 9 is a simplistic representation of some such rules as a decision-tree. The result is a set of feasible actions A_{sp} satisfying the safety and performance requirements.

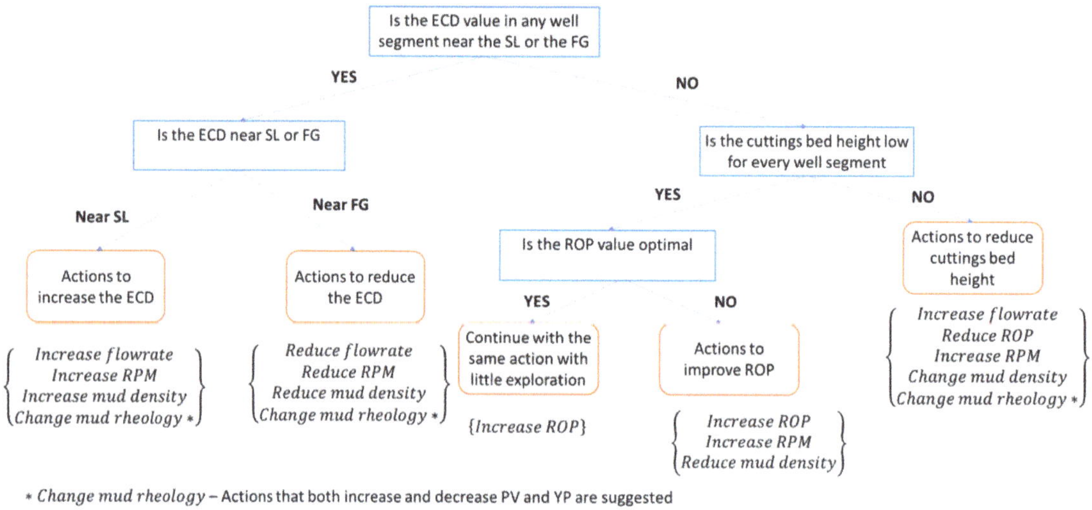

Figure 9. A simplistic representation of an action selection decision-tree for satisfying safety and performance metrics.

Similarly, to satisfy the sequential metric and the feasibility constraints, action sets A_{sq} and A_{fe} are evaluated, respectively. Figure 10 is a simple representation for estimating A_{sq} for a system with an action set containing three control variables (RPM, flowrate, and mud density). A_{sq} consists of actions in the vicinity of the most recent action. The reasoning behind such selection is to dissuade sudden changes in control variables, thereby constraining the rate of change of control variable values to safe limits. Changing mud properties (both density and rheology) is a time-consuming process; it cannot be performed in near real-time during conventional drilling or circulation operations (note that the effective density can be changed quickly in managed pressure drilling operations, where circulation takes place in a closed system where hydrostatic pressure can be quickly raised using a choke system). However, while planning for the circulation operation, adjusting mud properties is a crucial element of efficient hole cleaning. The purpose of defining A_{fe} is to incorporate such information.

To calculate the proximity metric, first, the normalized Euclidean distance for the current state (from s_{goal}), $d_{norm-euc}$, is computed using Equation (11). s_i is the ith element of the state vector s, and s_{i_max} is the maximum magnitude for the ith state vector component. The $d_{norm-euc}$ is then compared with the radius of a 'greed sphere'. This greed sphere is defined based on a normalized Euclidean distance of 0.5 from the goal state, and it is an indication of whether the states are far away from the goal state.

$$d_{norm-euc} = \frac{\sqrt{\sum_{i=i}^{N} s_i^2}}{\sqrt{\sum_{i=i}^{N} s_{i\_max}^2}} \quad (11)$$

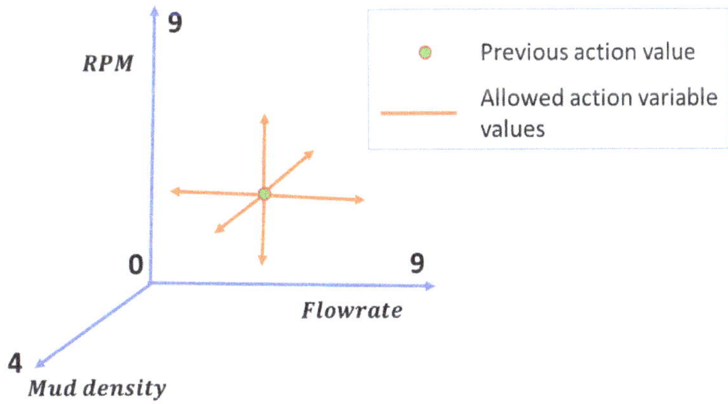

Figure 10. Method for estimating the action set associated with sequential metric. Numbers on axes are in control variable value units.

Figure 11 shows an example of this by considering a three-parameter system state (H_i, H_j and ECD_i). The near-goal states (s''_{goal}) represent the states for which the magnitudes of all state vector components are either 0 or 1.

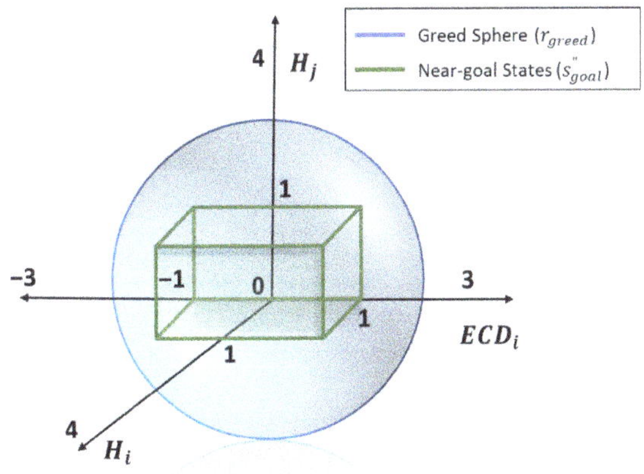

Figure 11. Method for estimating the action set associated with proximity metric.

The action selection strategy is given by Equation (12). In case the current state is at or near the goal state, the same action as the most recent action (a_{-1}) is selected. However, if the current state is further out than the greed sphere, more aggressive actions are included in the action set A_{px}. Aggressive actions represent greater magnitude changes in values of action control variables relative to the most recent action.

$$A_{px} = \begin{cases} a_{-1}, & d_{norm-euc} \leq s''_{goal} \\ a_{reg}, & s''_{goal} < d_{norm-euc} < r_{greed} \\ a_{agg}, & d_{norm-euc} \geq r_{greed} \end{cases} \quad (12)$$

Finally, the different action sets are combined, as shown in Equation (13), to output heuristic values for any action a in the action space A_s.

$$\pi_{heuristic\_a} = \begin{cases} 1, & if\ a \in A_{sp} \cap A_{sq} \cap A_{fe} \cap A_{px} \\ 0, & otherwise \end{cases} \quad (13)$$

Thus, $\pi_{heuristic}$ assigns a probability of either 0 or 1 to all the available actions in the action-space for a given state.

3.2.2. MCTS Structure

For finite-horizon action planning (T decision epochs into the future) in a limited amount of time t_{max}, the system starting in state s_0 proceeds according to the algorithm shown in Figure 12a. The root node $node_0$ corresponds to the starting state s_0, $node_{leaf}$ is the leaf node reached at the end of the selection phase in a given episode, and $node_{exp}$ is the randomly expanded node from the leaf node. The rollout phase then proceeds to plan until T epochs are reached (with root node as epoch 0), where $node_T$ is the final state (may or may not be the goal state), and G_T is the net discounted return. Equation (14) shows the calculation for G_T, where $\gamma(\leq 1)$ is the discount factor, r_k is the reward associated with the kth state-action transition, and $|exp|$ is the level for $node_{exp}$ in the tree. Finally, the backup function updates the Q value associated with all nodes from the $node_{exp}$ until $node_0$ by averaging this return value.

$$G_T = \sum_{k=|exp|}^{T} r_k.\gamma^{k-1} \tag{14}$$

```
while t_curr < t_max , do
    node_leaf ← Selection(node_0),
    node_exp ← Expansion(node_leaf),
    node_T, G_T ← Rollout(node_exp),
    Backup(node_exp, G_T)
return (actionSequence)
```

(a)

```
actionSequence
i = 0 , n = node_0
while i ≤ T , do
        actionSequence.append(bestChild(n.a))
    n ← bestChild(n)
    i ← i + 1
```

(b)

Figure 12. (**a**) MCTS algorithm (**b**) Action sequence selection method.

As previously discussed, MCTS builds an asymmetric search tree in the allocated time (t_{max}), after which an action sequence is given as output, based on the method shown in Figure 12b. The 'best child' from any node is its child-node corresponding to the action with the highest average Q value. Each node in the search tree contains the following information and can be represented by Equation (15):

- The current system state (s_t);
- The most recent action (a_{t-1});
- The action space of the node (A_{s_t});
- The total number of visits to the node so far (N_{s_t});
- All implemented actions and the resulting transitions, i.e., all child nodes ($\{a_i{:}node_i\}$);
- The total value associated with all iterations passing through the current state ($\sum_{children\ i} Q(s_t, a_i)$);
- The parent node ($node_{t-1}$);

$$node_t = \begin{pmatrix} s_t \\ a_{t-1} \\ A_{s_t} \\ N_{s_t} \\ Children : \{a_i : node_i\} \\ \sum_{children\ i} Q(s_t, a_i) \\ node_{t-1} \end{pmatrix} \quad (15)$$

The average value $(Q(s_t, a_t))$ associated with a state-action transition is calculated as the mean state-action value over all of its subsequent state's (s_{t+1}) children (a_j).

$$Q(s_t, a_t) = \frac{\sum_{children\ j} Q(s_{t+1}, a_j)}{N(s_t, a_t)} \quad (16)$$

The tree policy is modified by including the $\pi_{heuristic}$ in the exploration term of the UCT formula [23], as shown by:

$$\pi_{tree} = argmax_{a \in A_s}\left[Q(s_t, a) + \pi_{heuristic\_a} \cdot C_{exp} \cdot \sqrt{\frac{2 \cdot \ln N_{s_t}}{N_{(s_t,a)}}}\right] \quad (17)$$

Since the value of the exploitation term $(Q(s_t, a))$ is in the $[0, 1]$ range, C_{exp} is key in determining the number of simulations that would be required to make the values of the exploration $(C_{exp} \cdot \sqrt{\frac{2 \cdot \ln N_{s_t}}{N_{(s_t,a)}}})$ and the exploitation terms comparable. Figure 13 shows the plots for exploration terms calculated for different C_{exp} values as a function of the number of child node visits, given a total of 100 visits to the parent node (N_{s_t}). For the exploitation-heavy cases with C_{exp} values 0.25 and 0.5, the values of the two terms become comparable almost immediately, i.e., there is minimal exploration. On the other hand, for the exploration-heavy cases with C_{exp} values of 2 and 4, it takes a significant number of simulations to start exploiting its knowledge of the system. For the system developed in this research, a C_{exp} value of 1 is thus selected.

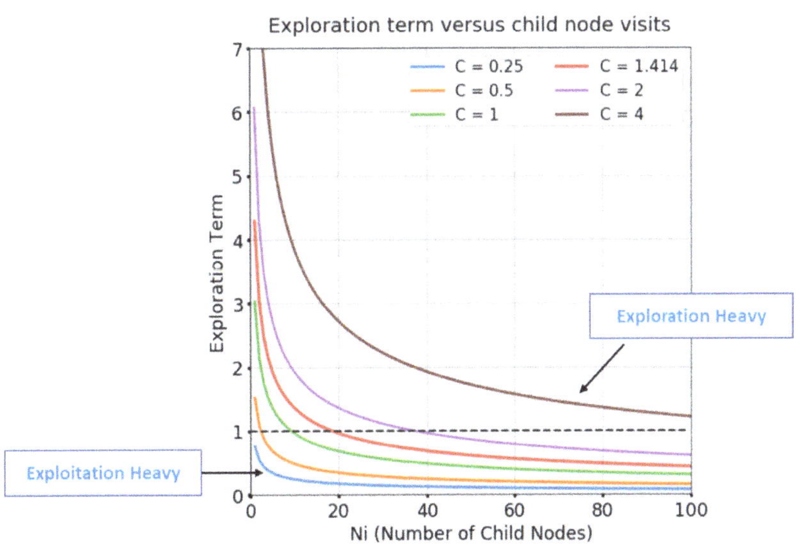

Figure 13. Calculation of exploration terms for different C_{exp} values as a function of the number of child node visits, given a total of 100 visits to the parent node.

During the rollout phase from any node, a uniform random selection policy is followed on a reduced action space (which is constructed based on Equation (18)).

$$A_{rollout} = \bigcup_{a \in A_s} \pi_{heuristic\_a} . a \tag{18}$$

4. Application of the System

The developed action planning system for hole cleaning was evaluated by performing post-mortem analysis on real-world oil well cases that exhibited hole cleaning issues. The analyses were performed by suggesting action sequences from some critical points during well construction. The final states associated with these plans were then compared with the actual well performance. A detailed investigation for one such well that exhibited issues due to insufficient hole cleaning during multiple well construction processes is presented here.

The dataset for the well included information such as:

- well trajectory, represented by directional surveys (inclination and azimuth versus the hole depth);
- well profile, represented by the BHA, casing and bit details;
- one-second surface sensor data, for directly and indirectly measuring the drilling parameters;
- mud-checks, to determine mud density and rheology among other drilling mud properties.

A hole cleaning digital twin, as discussed in Section 3.1.3 and detailed in Figure 8, was developed by integrating the various models with the relevant data sources.

4.1. Well Profile

The well had a short vertical section and a shallow kick-off point (where the well starts building inclination from the vertical) around 300 feet MD. After kick-off, the inclination angle was continuously built until the well became horizontal at about 1250 feet MD. After this, the well stayed near-horizontal until it reached its total depth (TD) of 2500 feet MD. The well profile is shown in Figure 14. This well was completed in two 'BHA runs' where each run comprised of drilling to a predetermined hole depth and subsequently, casing and cementing the hole. After the first BHA run, a surface casing of internal diameter 13.375-inch was set at a depth of 623 feet MD. Following this, in the second BHA run, a 12.25-inch hole section was drilled to well TD. Upon reaching TD, a 50-min on-bottom circulation cycle (at a flowrate of around 950 GPM and 60 RPM) was performed for hole cleaning purposes. The drillstring was then tripped out of the hole with intermittent back-reaming (at 910 GPM and 60 RPM), and finally, a 9.625-inch casing was run to TD and cemented. Poor hole cleaning negatively affected the last trip out of the hole with the drilling BHA, the run into the hole with the casing, and, ultimately, the casing cementing operation.

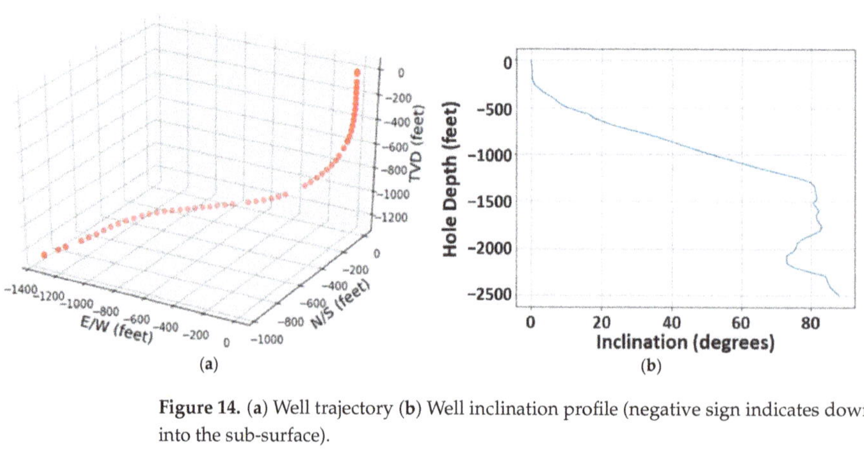

Figure 14. (**a**) Well trajectory (**b**) Well inclination profile (negative sign indicates downward depth into the sub-surface).

Calculations show that for running a 9.625-inch casing in a 12.25-inch borehole, the maximum theoretical cuttings bed height (on pulling the drillstring out of hole) should not exceed approximately 5-inches (Figure 15). This 5-inch bed height corresponds to 45.1 in^2 of cuttings in the cross-section.

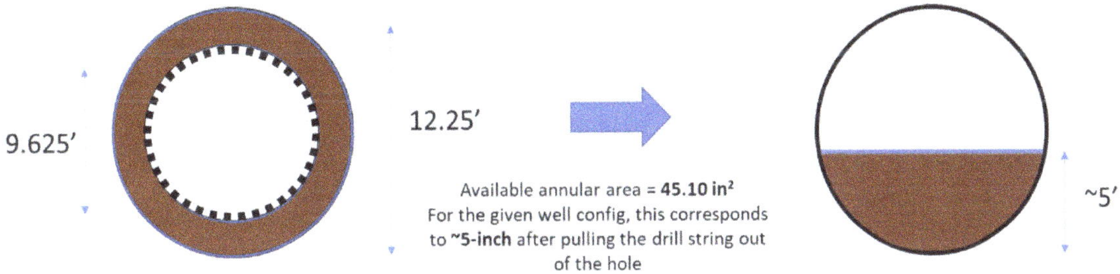

Figure 15. Theoretical maximum allowed height of the cuttings bed for the given well profile.

This cross-section of cuttings was then translated to an equivalent bed height by assuming the drillstring to be in the hole with the drill bit at TD. For the given well profile and trajectory (based on changes in outer diameters and eccentric placements of different drillstring components), this limit is represented by the red line in Figure 16a. The red shaded region corresponds to unsafe levels of cuttings bed height, while the green zone represents the goal state. Similarly, Figure 16b depicts the ECD profile (drilling window) for the well, bounded by SL and FG, with ten percent uncertainty in their values. As in Figure 16a, the green shaded zone corresponds to the goal state, while the red zones represent regions with the potential for well control issues. The zones in orange represent safe but suboptimal states.

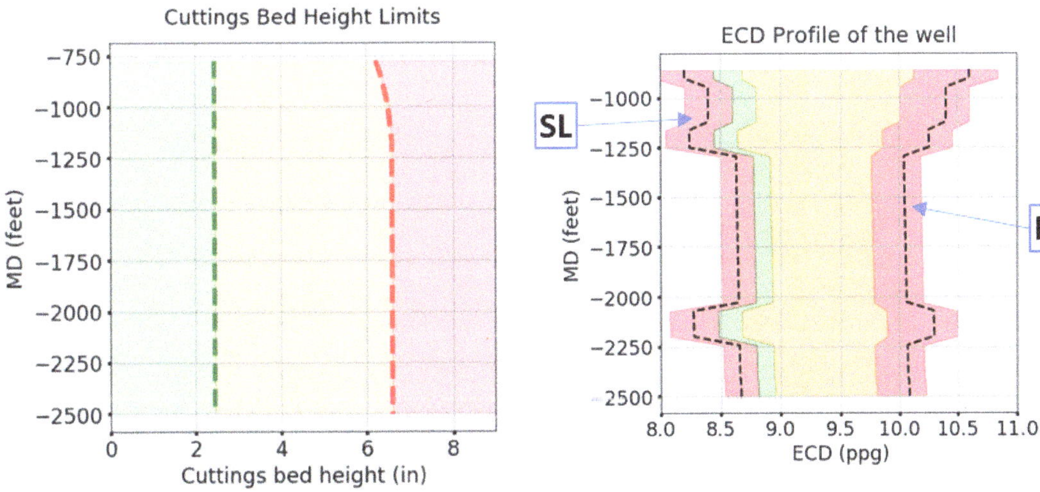

(a) (b)

Figure 16. (**a**) Equivalent bed height limits for the given well profile (**b**) ECD limits for the well considering a ten percent uncertainty in the SL and FG values (negative sign indicates downward depth into the sub-surface).

Table 1 shows the SL and FG values to define drilling margins for different sections of the well. For the near-vertical section, the SL and FG were assigned non-limiting values of 6 ppg and 18 ppg, respectively, because this interval was entirely cased while drilling the 12.25-inch section during the second BHA run.

Table 1. The SL and FG values for the different inclination intervals.

Inclination Interval	Stability Limit (ppg)	Fracture Gradient (ppg)
[0, 30)—in casing	6	18
[30, 45)	8.2	10.6
[45, 60)	8.4	10.4
[60, 75)	8.2	10.2
[75+)	8.6	10.0

4.2. *Performance Tracking and Action Planning*

The system was set up in the MDP framework by using the procedure discussed in the previous sections. Table 2 shows the number and ranges of values associated with the different control variables utilized for defining the action-space.

Table 2. Number and ranges of values for the different control variables to define the action-space.

Control Variable	Number of Discrete Values	Range of Values
Flowrate (GPM)	10	[0, 1800]
Drilling ROP (ft/h)	10	[0, 900]
Drillstring RPM (rev/min)	10	[0, 180]
Mud Density (ppg)	5	[8.5, 9.7]
Mud Plastic Viscosity (cP)	5	[7, 42]
Mud Yield Point (lbf/100 ft^2)	5	[7, 42]

The weights assigned to the different reward function components are shown in Table 3. Equal weights for all the state transitions components (W_s) assumes equal relative importance of the different state components. Similarly, the relative penalties associated with changing the different action components (W_{ap}) are also assumed to be the same. As discussed in the previous section, these weights can be tuned by the rigsite engineer to prioritize different objectives. However, for the following example, the weights for combining the normalized reward components (weights associated with R_{s_norm}, R_{ap_norm} and R_{ar_norm}) are assigned such that being in or near the goal states is prioritized over the penalty associated with changes in action variables, or the rewards due to an increased ROP. Similarly, while drilling, the weight of the reward associated with a higher ROP is more than the weight of the penalty associated with the changes in action variables. The weight of the normalized action reward component (W_{ar_norm}) has two values depending on the operation being tracked. A weight of zero is assigned to circulation operations because no new hole is being drilled (i.e., ROP is zero).

Table 3. Weight assignments for different reward function components.

$W_s = [1,1,1,1,1,1,1,1,1]$	
$W_{ap} = [1,1,1,1,1,1]$	
$W_{ar} = [0,1,0,0,0,0]$	
$W_{s\_norm} = 0.50$	
$W_{ap\_norm} = 0.20$	
Drilling	Circulation
$W_{ar\_norm} = 0.30$	$W_{ar\_norm} = 0.00$

4.2.1. Performance Tracking

State transitions were monitored, and associated rewards were calculated to track the performance of the system. The state of the system at the end of the drilling operations (during second BHA run) is represented by Equation (19), and can be visualized by Figure 17. The mud properties for drilling the last section of the well were: mud density of 9.06 ppg, PV of 11 cP, and YP of 36.5 lbf/100 ft^2.

$$s_{TD} = \left\{ \begin{array}{c} H_{30-45} \\ H_{45-60} \\ H_{60-75} \\ H_{75+} \\ ECD_{0-30} \\ ECD_{30-45} \\ ECD_{45-60} \\ ECD_{60-75} \\ ECD_{75+} \end{array} \right\} = \left\{ \begin{array}{c} 0 \\ 3 \\ 4 \\ 4 \\ 1 \\ 1 \\ 1 \\ 1 \\ 1 \end{array} \right\} \quad (19)$$

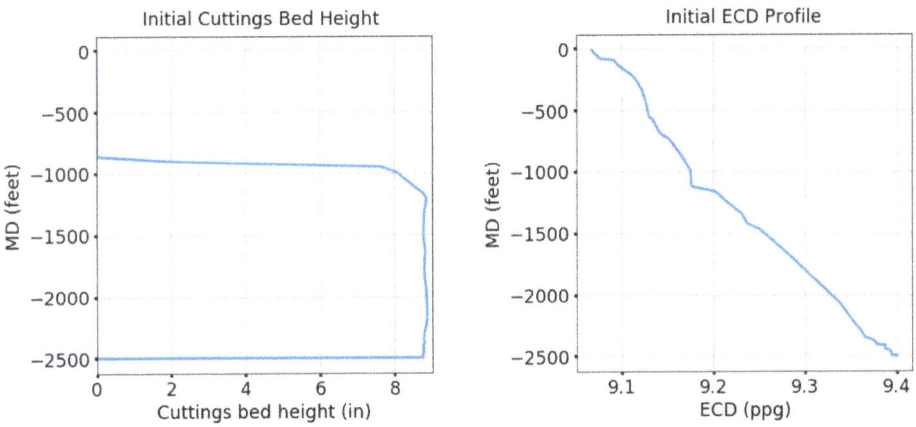

Figure 17. The state of the borehole at the end of the drilling operation during the second BHA run.

The final cuttings bed was around 9-inch, which is significantly higher than the limits specified in Figure 16a. Therefore, to ensure safe tripping operations without getting stuck, and to prepare the well for casing and cementing operations, these cuttings needed to be removed. A 50-min circulation cycle was then performed, the resulting state of which is represented by Equation (20), and Figure 18. The final cuttings bed height was still close to the allowed limit, and therefore non-optimal, explaining the issues encountered during tripping, casing, and cementing.

$$s_{TD\_circ} = \left\{ \begin{array}{c} 0 \\ 2 \\ 3 \\ 3 \\ 1 \\ 1 \\ 1 \\ 1 \\ 1 \end{array} \right\} \quad (20)$$

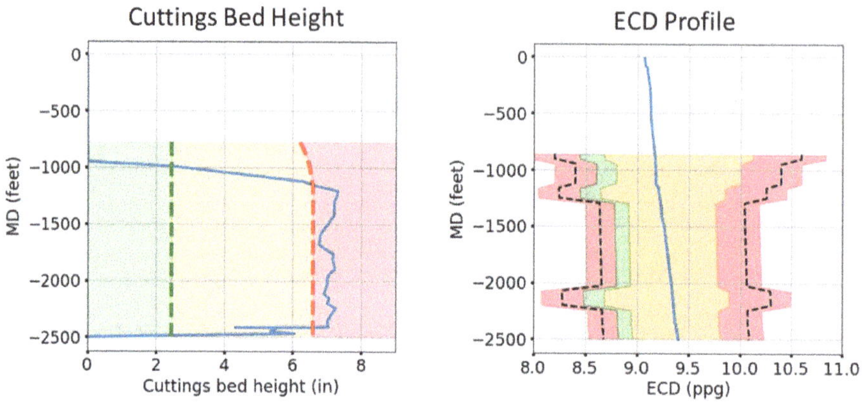

Figure 18. State of the borehole at the end of the circulation cycle.

4.2.2. Action Planning

The developed decision-making system was then used for planning the circulation cycle. The goal of the system, as previously discussed, is to reduce the cuttings bed height to safe limits while maintaining the ECD within the drilling margin (i.e., bed height and ECD need to be managed to in or around the green-shaded zones defined in Figure 16).

The following metrics are used to evaluate the different action sequences:

- The final system state;
- The average return value of the action sequence (V), which is the mean of total accumulated reward over the given trajectory that results from following the action sequence.

$$V = \frac{\sum_{t=1}^{T} R_t}{T} \tag{21}$$

For the well's actual circulation operation, the final state represented by Equation (20) was considerably far from the goal state. The V value for this action sequence was evaluated to be 0.74. The progression of the normalized Euclidean distance of the system state is shown in Figure 19. The green line at 0.2 corresponds to those states which have an absolute Euclidean distance of around two. With the definition of the state vector for this system, the theoretical maximum Euclidean distance is evaluated as 10.44, as detailed in Equation (22). For the ratio in Equation (11) to be 0.2, the maximum value for the current state's Euclidean distance can be two. For such states, the values of either four of their components have magnitude one, or one of their components has a magnitude of two, while the rest of the components are zero.

$$\sqrt{\sum_{i=i}^{9} s_{i\_max}^2} = \sqrt{4(4^2) + 5(3^2)} = 10.44 \tag{22}$$

The purpose of this 0.2 line is purely to serve as a visual aid, such that states closer to the line represent states closer to the goal state.

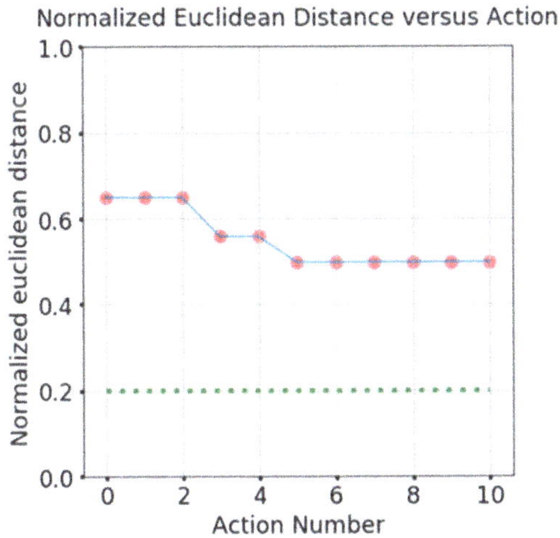

Figure 19. Progression of the normalized Euclidean distance (blue line) of the system states during the actual well circulation operation (shown by the green dotted line, here taken to be a state at a normalized Euclidean distance of 0.2).

4.2.3. Plan 1

Using the initializations defined in Tables 2 and 3, and with a C_{exp} value 1, the decision-engine was used to plan ahead for eight decision epochs (40-min) starting from the state s_{TD} (Equation (19)). Equation (23) represents the action sequence (a_{seq1}) recommended by the system. Selecting the right mud properties at the beginning of the circulation cycle is essential, since changing them is a time-consuming process. It is, therefore, highly impractical to adjust them in the middle of the circulation cycle. To this effect, the system suggested changing the mud properties (at the beginning of the circulation cycle) to mud density of 8.9 ppg, PV of 17.5 cP, and YP of 10.5 lbf/100 ft^2 from mud density 9.06 ppg, PV 11 cP, and YP 36.5 lbf/100 ft^2 initially.

$$a_{seq1} = \left\{ \begin{pmatrix} 1000 \\ 0 \\ 80 \\ 8.9 \\ 17.5 \\ 10.5 \end{pmatrix}, \begin{pmatrix} 1200 \\ 0 \\ 100 \\ 8.9 \\ 17.5 \\ 10.5 \end{pmatrix}, \begin{pmatrix} 1400 \\ 0 \\ 120 \\ 8.9 \\ 17.5 \\ 10.5 \end{pmatrix}, \begin{pmatrix} 1600 \\ 0 \\ 160 \\ 8.9 \\ 17.5 \\ 10.5 \end{pmatrix}, \begin{pmatrix} 1800 \\ 0 \\ 160 \\ 8.9 \\ 17.5 \\ 10.5 \end{pmatrix}, \begin{pmatrix} 1600 \\ 0 \\ 160 \\ 8.9 \\ 17.5 \\ 10.5 \end{pmatrix}, \begin{pmatrix} 1400 \\ 0 \\ 160 \\ 8.9 \\ 17.5 \\ 10.5 \end{pmatrix}, \begin{pmatrix} 1400 \\ 0 \\ 160 \\ 8.9 \\ 17.5 \\ 10.5 \end{pmatrix} \right\} \quad (23)$$

The predicted output state of the system is shown in Figure 20. The cuttings bed is almost entirely removed (as represented by the blue line), and the ECD values are very close to the desired regions throughout the well. Equation (24) represents the final state of the system.

$$s_{TD\_circ1} = \left\{ \begin{matrix} 0 \\ 0 \\ 0 \\ 0 \\ 1 \\ 1 \\ 1 \\ 1 \\ 1 \end{matrix} \right\} \quad (24)$$

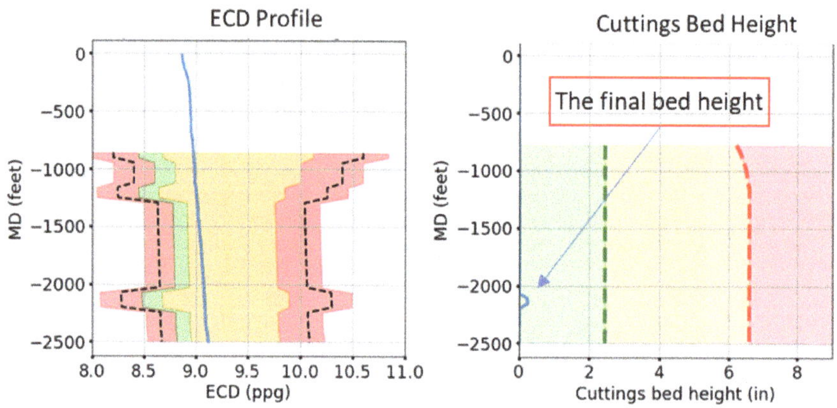

(**a**) (**b**)

Figure 20. (**a**) ECD profile (**b**) Cuttings bed height (output state) for a_{seq1}.

Figure 21 illustrates the progression of the system with actions for the sequence a_{seq1}. Figure 21a shows the reduction in the cuttings bed height with different actions, and Figure 21b represents the progression of the normalized Euclidean distance with actions.

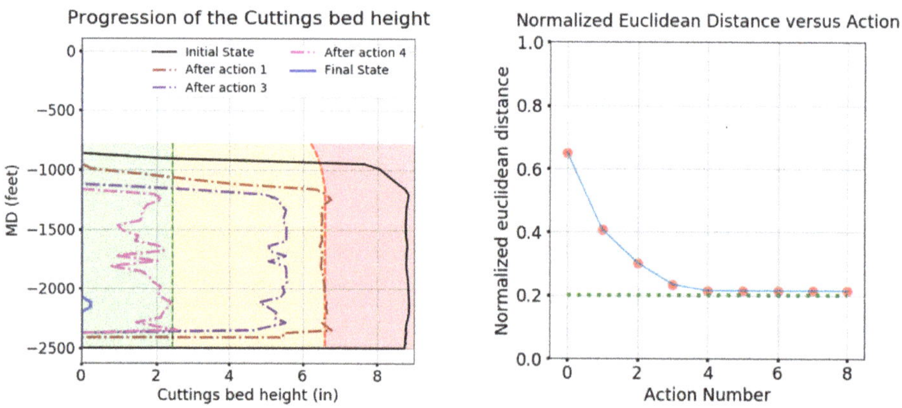

(**a**) (**b**)

Figure 21. (**a**) Progression of the cuttings bed (**b**) Progression of the normalized Euclidean distance (blue line) of the system states for a_{seq1} to an acceptable goal state (shown by the green dotted line, here taken to be a state at a normalized Euclidean distance of 0.2).

By the fourth action, the system has already moved close to the 0.2 line (by cuttings bed being reduced to the goal state), where it stays until the end. Figure 22 shows the progression of the rewards associated with this action sequence. Thus, the V value calculated for a_{seq1} is 0.82.

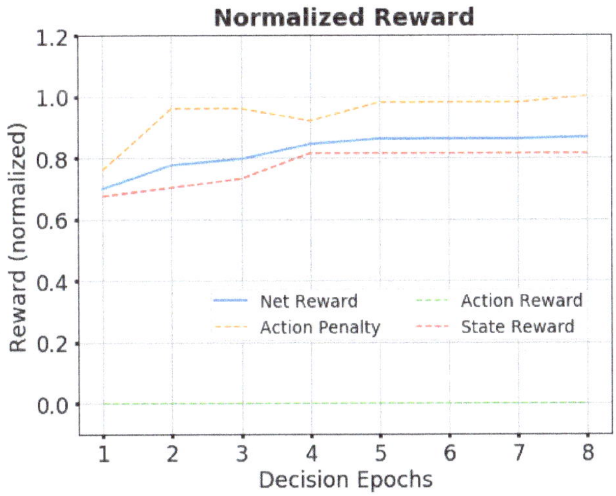

Figure 22. Progression of the rewards associated with a_{seq1}.

An interesting observation in the action sequence is the system actively trying to manage the ECD by reducing the flowrate, after the cuttings bed has been removed. Initially, the system suggests increasing the flowrate and the RPM, which helps with cuttings bed removal (as can be seen by the reduced Euclidean distance), and then later tries to reduce the ECD.

4.2.4. Plan 2

Another planning operation was performed by changing the flowrate and RPM thresholds, as shown in Table 4. No changes were made to the mud density and rheology thresholds, and the weights associated with the different reward components were also unaltered.

Table 4. Modified flowrate and RPM thresholds for varying the action-space.

Control Variable	Number of Discrete Values	Range of Values
Flowrate (GPM)	10	[0, 1500]
Drillstring RPM (rev/min)	10	[0, 150]

Equation (25) represents the action sequence (a_{seq2}) output by the planning system, over a 40-min (eight decision epochs) interval. Due to the truncated flowrate and RPM thresholds, the system suggests different mud density and rheology than for a_{seq1} to help with the cuttings removal.

$$a_{seq2} = \begin{Bmatrix} 833 \\ 0 \\ 67 \\ 9.24 \\ 10.5 \\ 10.5 \end{Bmatrix}, \begin{Bmatrix} 1000 \\ 0 \\ 83 \\ 9.24 \\ 10.5 \\ 10.5 \end{Bmatrix}, \begin{Bmatrix} 1167 \\ 0 \\ 83 \\ 9.24 \\ 10.5 \\ 10.5 \end{Bmatrix}, \begin{Bmatrix} 1333 \\ 0 \\ 100 \\ 9.24 \\ 10.5 \\ 10.5 \end{Bmatrix}, \begin{Bmatrix} 1333 \\ 0 \\ 117 \\ 9.24 \\ 10.5 \\ 10.5 \end{Bmatrix}, \begin{Bmatrix} 1500 \\ 0 \\ 133 \\ 9.24 \\ 10.5 \\ 10.5 \end{Bmatrix}, \begin{Bmatrix} 1500 \\ 0 \\ 150 \\ 9.24 \\ 10.5 \\ 10.5 \end{Bmatrix}, \begin{Bmatrix} 1333 \\ 0 \\ 150 \\ 9.24 \\ 10.5 \\ 10.5 \end{Bmatrix} \quad (25)$$

The final state of the system (s_{TD_circ2}), represented by Equation (26), is shown in Figure 23.

$$s_{TD\_circ2} = \begin{Bmatrix} 0 \\ 0 \\ 0 \\ 0 \\ 1 \\ 1 \\ 1 \\ 1 \\ 1 \end{Bmatrix} \tag{26}$$

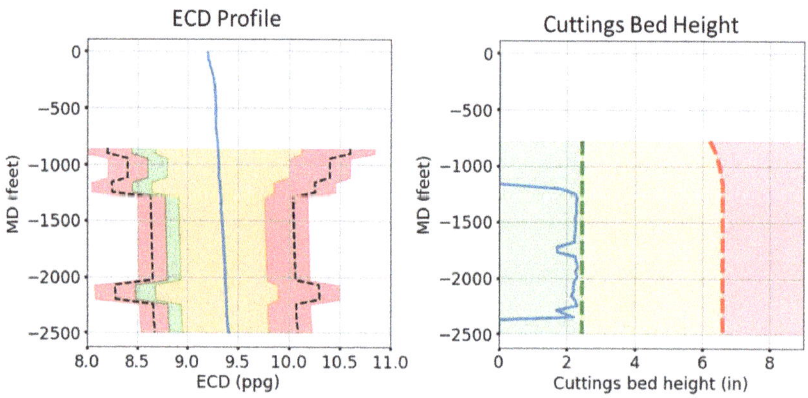

(a) (b)

Figure 23. (**a**) ECD profile (**b**) Cuttings bed height.

The normalized Euclidean distance does decrease with time, but it only reaches the 0.2 line on the seventh action (Figure 24b). An interesting observation in Figure 24a is that the cuttings bed is not entirely removed but is reduced to the goal state (green-shaded zone) values. The ECD is also in the safe-but-suboptimal region, but it is much higher than for the first case due to a higher suggested mud density.

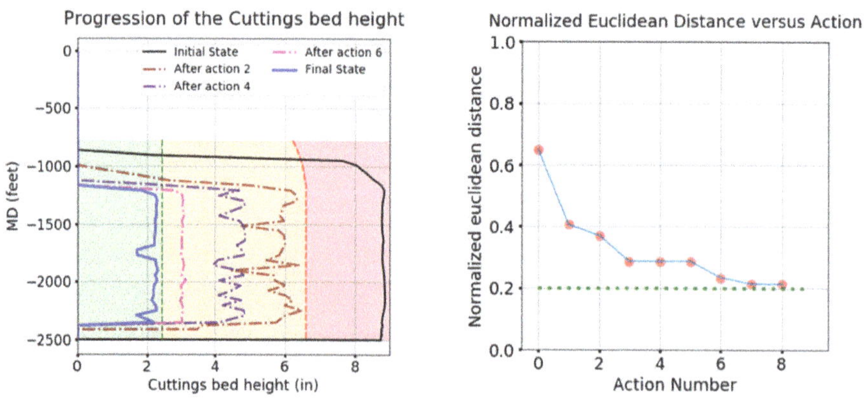

(a) (b)

Figure 24. (**a**) Progression of the cuttings bed (**b**) Progression of the normalized Euclidean distance (blue line) of the system states for a_{seq2} to an acceptable goal state (shown by the green dotted line, here taken to be a state at a normalized Euclidean distance of 0.2).

Although, the final state representations for both the plans (a_{seq1} and a_{seq2}) are the same, the average V value calculated for a_{seq2} is 0.79, which is lower than for a_{seq1}. The progression of rewards for the sequence a_{seq2} is shown in Figure 25.

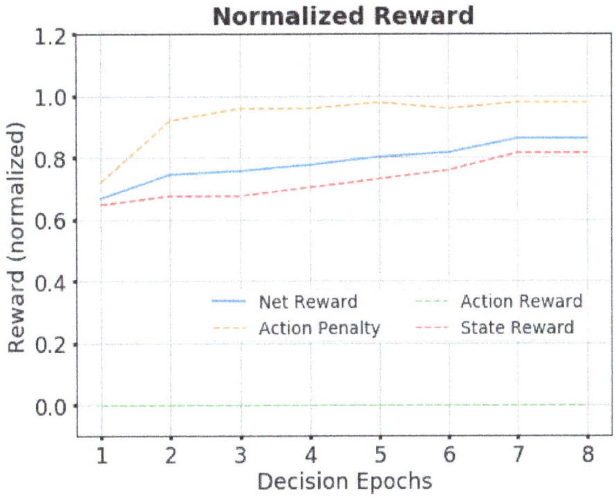

Figure 25. Progression of the rewards associated with a_{seq2}.

4.3. Discussion

For both plans a_{seq1} and a_{seq2}, the decision-engine was able to self-learn by simulating multiple episodes of experience and output action sequences that would have helped move the system towards the goal states. To summarize:

1. The non-holonomic nature of drilling operations combined with the decision-engine's long-term planning capabilities allow for more robust planning. An example of this is the system selecting appropriate mud properties at the beginning of the circulation cycle by evaluating multiple steps into the future. Both plans that were generated result in a better outcome than the actual plan generated and implemented by a human decision-maker.
2. Tuning the weights associated with the different reward components allows for prioritizing different objectives or different sections of the well over others.
3. Utilizing the domain-knowledge enriched MCTS allows for faster and more efficient planning, and well-defined heuristic functions make such planning systems implementable in the field. Exhaustively evaluating all nodes in the search tree to the eighth level (eight decision epochs or 40-min) would require over a hundred million simulations, as well as require storing the results of each state-action transition. This would be highly computationally and memory inefficient. MCTS, on the other hand, requires a number of simulations that are many orders of magnitude lower (only a few thousand in total), and all state-action transition results do not need to be stored. For the cases discussed in this paper, the planning algorithm, without any parallel processing or multi-threading on a standard laptop using a non-streamlined Python code, was able to generate these plans in under an hour.
4. Planning with higher values does result in the convergence of the state's Euclidean distance towards the 0.2-line, but it requires many more simulations. On the other hand, lower C_{exp} values introduce an element of bias depending on the order in which the nodes are added to the tree, which itself depends on the random rollout policy for MC simulations.

5. Conclusions

This paper proposes a method for the development of intelligent decision engines for well construction operations by utilizing an MDP formalism with the MCTS for action planning. This method is demonstrated by implementing a hole cleaning action planning system and comparing its performance against a human decision maker's performance. To the best of our knowledge, there are no published models or systems that address the issue of intelligent long-term sequential decision-making and action planning for the well construction domain. Such systems, however, have resulted in significant performance improvement across other domains such as game AI (such as in AlphaGo, AlphaZero), electric power distribution, dynamic resource allocation, etc. To summarize:

1. MCTS planning systems allow for a hybrid approach to managing conflicting objectives by combining the advantages of the exploration-exploitation trade-off offered by the MCTS, with domain-knowledge derived heuristics, thereby helping make better decisions.
2. A combination of the digital twin and a non-sparse reward function, with backpropagation of the episodic returns, allows the system to learn from simulated experience. A non-sparse reward structure ensures that the feedback received by the agent is frequent and meaningful, thereby speeding up the policy improvement process.
3. The underlying tree and rollout policies of the MCTS algorithm can be enhanced by using well-defined process-specific heuristics. This assists in improving the convergence rate of the system towards an optimal action sequence. For the hole cleaning system, the heuristic was designed to balance safety, performance, feasibility, and proximity constraints.
4. Utilizing such systems can aid in overall performance improvement by eliminating the need to wait on decisions, as well as suggesting optimal drilling parameters for the given wellbore condition.

Furthermore, decision engines can be developed for a multitude of other well construction applications, such as well control, drilling parameter optimization, tripping automation, cementing, etc. These decision engines can then be integrated into a rig's control system to automate monitoring, planning, and control of action variables such as drilling rate, drillstring rotation speed, tripping speed, flowrate, and mud properties. The ultimate goal is to automate important well construction tasks and optimize them by removing the (non-optimum) variability associated with subjective human-based decision-making.

Author Contributions: Conceptualization, G.S.S., P.A. and E.v.O.; methodology, G.S.S., P.P., P.A. and E.v.O.; software, G.S.S. and P.P.; validation, G.S.S., P.P. and P.A.; formal analysis, G.S.S.; investigation, G.S.S. and P.P.; resources, P.A. and E.v.O.; data curation, G.S.S. and P.A.; writing—original draft preparation, G.S.S.; writing—review and editing, P.P., P.A. and E.v.O.; visualization, G.S.S.; supervision, P.A. and E.v.O.; project administration, E.v.O.; funding acquisition, P.A. and E.v.O. All authors have read and agreed to the published version of the manuscript.

Funding: This research was funded by the RAPID Industry Affiliate Program (IAP) at the University of Texas at Austin.

Acknowledgments: The authors would like to thank the members of the Rig Automation and Performance Improvement in Drilling (RAPID) consortium at the University of Texas at Austin for their support of this research. We would also like to acknowledge our colleagues in the RAPID research group for valuable brainstorming sessions.

Conflicts of Interest: The authors declare no conflict of interest.

Unit Conversion

1 meter (m)	3.28 feet (ft)
1 meter/second (m/s)	11,811 feet/hour (ft/h)
1 psi	6894.76 Pa
1 ppg	119.83 kg/m^3
1 radian	57.2958 degrees
1 ft^3	0.02832 m^3
1 GPM	0.0000631 m^3/s
1 cP	0.001 Pa·s
1 lbf/100 ft^2	0.4788 Pa
1 lbs.	0.4536 Kg

Nomenclature

A	Action-space
A_{fe}	Action set associated with feasibility constraints (for hole cleaning heuristic)
A_{px}	Action set associated with proximity metric (for hole cleaning heuristic)
$A_{rollout}$	Reduced action space for a state during the rollout phase of the MCTS
A_s	Action space from state s
A_{s_t}	The action space of the node associated with state s_t
A_{sp}	Action set associated with safety and performance metrics (for hole cleaning heuristic)
A_{sq}	Action set associated with sequential metric (for hole cleaning heuristic)
a_{-1}	The most recently executed action
a_{agg}	Aggressive actions representing greater magnitude changes in values of action control variables relative to the most recent action
a_{reg}	Regular actions representing small changes in action control variables with respect to the most recent action
a_{seq1}	action sequence recommended by the system for the first planning case
a_{seq2}	action sequence recommended by the system for the second planning case
a_t	Action executed by the agent at time or decision epoch t
C_{exp}	The exploration factor in the UCT formula
D_{TVD}	True vertical depth for the measured depth D_{MD}
$d_{norm-euc}$	Normalized Euclidean distance for the current state
ECD	Equivalent circulation density (pounds per gallon or ppg)
$ECD_{inc.}$	Functional value of ECD in the inclination interval segment $inc.$
$\|exp\|$	The level at $node_{exp}$ is in the tree
flowrate	Rate of flow of the drilling mud through the drillstring controlled by a positive displacement reciprocating mud pump on the surface (GPM)
G_T	the net discounted return to T decision epochs
g	Acceleration due to gravity (9.81 m/s^2)
$H_{inc.}$	Functional value of the cuttings bed height in the inclination interval segment $incl.$
$incl$	Inclination angle range (degrees)
$N(s)$	Total number of visits to the state s
$N(s,a)$	Number of times action a has been taken from state s
N_{s_t}	Total number of visits to the node associated with state s_t
$node_0$	Root node corresponding to the starting state s_0
$node_{exp}$	Randomly expanded node from the leaf node during the expansion phase of the MCTS
$node_{leaf}$	Leaf node reached at the end of the selection phase of the MCTS
$node_t$	Node at decision epoch t
$P_{frictional_{pressure}\_loss\_D_{MD}}$	Frictional pressure drop in the annulus (Pa) at a measured depth D_{MD}
$P_{hydrostatic\_D_{TVD}}$	Hydrostatic Pressure (Pa) at a vertical depth of D_{TVD}
$P^a_{ss'}$	Transition probability of a system in the state s to the state s' when an agent executes action a
PV	Plastic viscosity (cP)

$Q(s,a)$	Average value associated with implementing action a from state s
$Q_{UCT}(s,a)$	The upper confidence bound or the urgency term in the UCT formula
R	Reward set
$R_{a\_density}$	Action reward component associated with the density value
$R_{a\_flowrate}$	Action reward component associated with the flowrate value
R_{a_PV}	Action reward component associated with the PV value
R_{a_ROP}	Action reward component associated with the ROP value
R_{a_RPM}	Action reward component associated with the RPM value
R_{a_YP}	Action reward component associated with the YP value
R_{ap}	Action transition-based penalty set
R_{ap_norm}	Normalized action penalty for the hole cleaning system
R_{ar}	Action value-based reward set
R_{ar_norm}	Normalized action reward for the hole cleaning system
$R_{ECDincl.}$	State reward component associated with ECD in the inclination interval $incl.$
$R_{density}$	Action Penalty component associated with changing mud density
$R_{flowrate}$	Action Penalty component associated with changing flowrate
$R_{Hincl.}$	State reward component associated with cuttings bed height in the inclination interval $incl.$
R_{net}	Net normalized reward function for the hole cleaning system
R_{PV}	Action Penalty component associated with changing mud PV
R_{ROP}	Action Penalty component associated with changing ROP
R_{RPM}	Action Penalty component associated with changing RPM
R_S	State transition-based reward set
R_{s_norm}	Normalized state reward for the hole cleaning system
R_{YP}	Action Penalty component associated with changing mud YP
ROP	Rate of penetration or drilling rate (ft/h)
RPM	Drillstring rotation speed (revs./min)
r_{greed}	Radius of the greed sphere for A_{px} evaluation
r_k	Reward associated with the kth state-action transition
S	State-space
SL	Stability limit (ppg)
s_{TD}	State of the hole cleaning system at the well TD
s_{TD_circ}	State of the wellbore after performing a circulation cycle at TD
s_{TD_circ1}	State of the wellbore after performing a circulation cycle at TD following a_{seq1}
s_{TD_circ2}	State of the wellbore after performing a circulation cycle at TD following a_{seq2}
s_{goal}	Goal or desired state for the MDP
s''_{goal}	The states near the goal state (for evaluating the A_{px} set)
s_t	State of the system at time or decision epoch t
T	Number of decision epochs to evaluate till in the future
t	Time step or decision epoch t
t_{max}	Time available for MCTS algorithm to plan
V	The average return value of an action sequence
W_{ap}	Weight set associated with the action transition penalty
W_{ap_norm}	Weight value associated with the normalized action penalty
W_{ar}	Weight set associated with the action value reward
W_{ar_norm}	Weight value associated with the normalized action reward
W_s	Weight set associated with state transition reward
W_{s_norm}	Weight value associated with the normalized state reward
WOB	Weight on bit (klbs.)
YP	Yield point (l bf/100 ft^2)
π	Policy or plan
$\pi_{heuristic\_a}$	Problem specific heuristic (probability of selecting action a)
π_{tree}	Tree policy—the action selected from a given node in the search tree during the selection phase of the MCTS
γ	Discount factor for return calculation

References

1. Sadlier, A.; Says, I.; Hanson, R. Automated Decision Support to Enhance While-Drilling Decision Making: Where Does it fit Within Drilling Automation? In Proceedings of the SPE/IADC Drilling Conference, Amsterdam, The Netherlands, 5–7 March 2013. [CrossRef]
2. Shokouhi, S.V.; Skalle, P. Enhancing decision making in critical drilling operations. In Proceedings of the SPE Middle East Oil and Gas Show and Conference, Manama, Bahrain, 15–18 March 2009; pp. 809–816. [CrossRef]
3. Mayani, M.G.; Baybolov, T.; Rommetveit, R.; Ødegaard, S.I.; Koryabkin, V.; Lakhtionov, S. Optimizing drilling wells and increasing the operation efficiency using digital twin technology. In Proceedings of the IADC/SPE International Drilling Conference and Exhibition, Galveston, TX, USA, 3–5 March 2020. [CrossRef]
4. Nadhan, D.; Mayani, M.G.; Rommetveit, R. Drilling with digital twins. In Proceedings of the IADC/SPE Asia Pacific Drilling Technology Conference and Exhibition, Bangkok, Thailand, 27–29 August 2018. [CrossRef]
5. Saini, G.S.; Ashok, P.; van Oort, E.; Isbell, M.R. Accelerating well construction using a digital twin demonstrated on unconventional well data in North America. In Proceedings of the Unconventional Resources Technology Conference, Houston, TX, USA, 23–25 July 2018. [CrossRef]
6. Danner, G.E. Using knowledge graphs to enhance drilling operations. In Proceedings of the Offshore Technology Conference, Houston, TX, USA, 4–7 May 2020. [CrossRef]
7. Miller, P.; Gouveia, J. Applying decision trees to improve decision quality in unconventional resource development. In Proceedings of the SPE Annual Technical Conference and Exhibition, Calgary, AB, Canada, 30 September–2 October 2019. [CrossRef]
8. Cordoso, J.V.; Maidla, E.E.; Idagawa, L.S. Problem detection during tripping operations in horizontal and directional wells. *SPE Drill. Complet.* **1995**, *10*, 77–83. [CrossRef]
9. Zhang, F.; Miska, S.; Yu, M.; Ozbayoglu, E.; Takach, N.; Osgouei, R.E. Is Well Clean Enough? A Fast Approach to Estimate Hole Cleaning for Directional Drilling. In Proceedings of the SPE/ICoTA Coiled Tubing & Well Intervention Conference & Exhibition, The Woodlands, TX, USA, 24–25 March 2015. [CrossRef]
10. Abbas, A.K.; Almubarak, H.; Abbas, H.; Dawood, J. Application of machine learning approach for intelligent prediction of pipe sticking. In Proceedings of the Society of Petroleum Engineers—Abu Dhabi International Petroleum Exhibition and Conference 2019 (ADIP 2019), Abu Dhabi, UAE, 11–14 November 2019. [CrossRef]
11. Da Cruz Mazzi, V.A.; Dumlao, V.C.Q.; Mourthé, A.C.L.; Elshahawi, H.; Kim, I.; Lumens, P. Machine Learning-Enabled Digital Decision Assistant for Remote Operations. In Proceedings of the Offshore Technology Conference, Houston, TX, USA, 4–7 May 2020. [CrossRef]
12. Fjellheim, R. Smart collaboration for decision making in drilling. In Proceedings of the SPE Middle East Intelligent Energy Conference and Exhibition, Manama, Bahrain, 28–30 October 2013. [CrossRef]
13. Saini, G.S.; Chan, H.C.; Ashok, P.; van Oort, E.; Behounek, M.; Thetford, T.; Shahri, M. Spider bots: Database enhancing and indexing scripts to efficiently convert raw well data into valuable knowledge. In Proceedings of the Unconventional Resources Technology Conference, Houston, TX, USA, 23–25 July 2018. [CrossRef]
14. Chan, H.C.; Lee, M.M.; Saini, G.S.; Pryor, M.; van Oort, E. Development and Validation of a Scenario-Based Drilling Simulator for Training and Evaluating Human Factors. *IEEE Trans. Hum.-Mach. Syst.* **2020**, *50*, 327–336. [CrossRef]
15. Ji, Y.; Wang, J.; Xu, J.; Fang, X.; Zhang, H. Real-Time Energy Management of a Microgrid Using Deep Reinforcement Learning. *Energies* **2019**, *12*, 2291. [CrossRef]
16. Kuznetsova, E.; Li, Y.F.; Ruiz, C.; Zio, E.; Ault, G.; Bell, K. Reinforcement learning for microgrid energy management. *Energy* **2013**, *59*, 133–146. [CrossRef]
17. Guan, J.; Tang, H.; Wang, K.; Yao, J.; Yang, S. A parallel multi-scenario learning method for near-real-time power dispatch optimization. *Energy* **2020**, *202*, 117708. [CrossRef]
18. Jasmin, E.A.; Imthias Ahamed, T.P.; Jagathy Raj, V.P. Reinforcement Learning approaches to Economic Dispatch problem. *Int. J. Electr. Power Energy Syst.* **2011**, *33*, 836–845. [CrossRef]
19. Qi, X.; Luo, Y.; Wu, G.; Boriboonsomsin, K.; Barth, M. Deep reinforcement learning enabled self-learning control for energy efficient driving. *Transp. Res. Part C Emerg. Technol.* **2019**, *99*, 67–81. [CrossRef]
20. Vandael, S.; Claessens, B.; Ernst, D.; Holvoet, T.; Deconinck, G. Reinforcement Learning of Heuristic EV Fleet Charging in a Day-Ahead Electricity Market. *IEEE Trans. Smart Grid* **2015**, *6*, 1795–1805. [CrossRef]
21. Kang, D.-J.; Park, J.H.; Yeo, S.-S. Intelligent Decision-Making System with Green Pervasive Computing for Renewable Energy Business in Electricity Markets on Smart Grid. *EURASIP J. Wirel. Commun. Netw.* **2009**, *2009*, 247483. [CrossRef]
22. Arulkumaran, K.; Cully, A.; Togelius, J. AlphaStar: An Evolutionary Computation Perspective. In Proceedings of the Genetic and Evolutionary Computation Conference Companion, Prague, Czech Republic, 13–17 July 2019; pp. 314–315. [CrossRef]
23. Moerland, T.M.; Broekens, J.; Plaat, A.; Jonker, C.M. A0C: Alpha Zero in Continuous Action Space. *arXiv* **2018**, arXiv:1805.09613.
24. Ontanon, S.; Synnaeve, G.; Uriarte, A.; Richoux, F.; Churchill, D.; Preuss, M. A survey of real-time strategy game AI research and competition in starcraft. *IEEE Trans. Comput. Intell. AI Games* **2013**, *5*, 293–311. [CrossRef]
25. Silver, D.; Sutton, R.S.; Müller, M. Temporal-difference search in computer Go. *Mach. Learn.* **2012**, *87*, 183–219. [CrossRef]
26. Silver, D.; Schrittwieser, J.; Simonyan, K.; Antonoglou, I.; Huang, A.; Guez, A.; Hubert, T.; Baker, L.; Lai, M.; Bolton, A.; et al. Mastering the game of Go without human knowledge. *Nature* **2017**, *550*, 354–359. [CrossRef] [PubMed]

27. Silver, D.; Hubert, T.; Schrittwieser, J.; Antonoglou, I.; Lai, M.; Guez, A.; Lanctot, M.; Sifre, L.; Kumaran, D.; Graepel, T.; et al. A general reinforcement learning algorithm that masters chess, shogi, and Go through self-play. *Science* **2018**, *362*, 1140–1144. [CrossRef] [PubMed]
28. Ahmed, O.S.; Aman, B.M.; Zahrani, M.A.; Ajikobi, F.I.; Aramco, S. Stuck Pipe Early Warning System Utilizing Moving Window Machine Learning Approach. In Proceedings of the Abu Dhabi International Petroleum Exhibition & Conference, Abu Dhabi, UAE, 11–14 November 2019.
29. Forshaw, M.; Becker, G.; Jena, S.; Linke, C.; Hummes, O. Automated hole cleaning monitoring: A modern holistic approach for NPT reduction. In Proceedings of the International Petroleum Technology Conference, Dhahran, Saudi Arabia, 13–15 January 2020. [CrossRef]
30. LaValle, S.M. *Planning Algorithms*; Cambridge University Press: Cambridge, UK, 2006. [CrossRef]
31. Poole, D.L.; Mackworth, A.K. *Artificial Intelligence: Foundations of Computational Agents*, 2nd ed.; Cambridge University Press: Cambridge, UK, 2017.
32. Bennett, B. Graphs: Breadth First Search vs. Depth First Search [WWW Document]. 2019. Available online: https://medium.com/javascript-in-plain-english/graphs-breadth-first-search-vs-depth-first-search-d9908c560642 (accessed on 1 July 2020).
33. Mussmann, S.; See, A. Graph Search Algorithms. 2017. Available online: http://cs.stanford.edu/people/abisee/gs.pdf (accessed on 1 June 2020).
34. Alija, A.S. Analysis of Dijkstra's and A* Algorithm to Find the Shortest Path. Ph.D. Thesis, Universiti Tun Hussein Onn Malaysia, Batu Pahat, Malaysia, 2015.
35. Reddy, H. Path Finding—Dijkstra's and A* Algorithm's. *Int. J. IT Eng.* **2013**, 1–15.
36. Bhaumik, D.; Khalifa, A.; Green, M.C.; Togelius, J. Tree Search vs Optimization Approaches for Map Generation. *arXiv* **2019**, arXiv:1903.11678.
37. Cui, X.; Shi, H. A*-based Pathfinding in Modern Computer Games. *Int. J. Comput. Sci. Netw. Secur.* **2011**, *11*, 125–130.
38. Klein, S. Attacking SameGame using Monte-Carlo Tree Search: Using Randomness as Guidance in Puzzles. Master's Thesis, KTH Royal Institute of Technology, Stockholm, Sweden, 2015.
39. Li, L.K.; Seng, K.P.; Yeong, L.S.; Ch'ng, S.I.; Li-Minn, A. The Boundary Iterative-Deepening Depth-First Search Algorithm. *Int. J. Adv. Comput. Sci. Its Appl.* **2014**, *4*, 45–50.
40. Korkmaz, M.; Durdu, A. Comparison of optimal path planning algorithms. In Proceedings of the 2018 14th International Conference on Advanced Trends in Radioelecrtronics, Telecommunications and Computer Engineering (TCSET), Slavske, Ukraine, 20–24 January 2018; pp. 255–258. [CrossRef]
41. Sutton, R.S.; Barto, A.G. *Reinforcement Learning: An Introduction*, 2nd ed.; The MIT Press: Cambridge, MA, USA, 2018.
42. Silver, D. *Reinforcement Learning and Simulation-Based Search in Computer Go*; University of Alberta: Alberta, AB, Canada, 2009.
43. Cayeux, E.; Mihai, R.; Carlsen, L.; Stokka, S. An approach to autonomous drilling. In Proceedings of the IADC/SPE International Drilling Conference and Exhibition, Galveston, TX, USA, 3–5 March 2020. [CrossRef]
44. Mitchell, R.F.; Miska, S.Z. *Fundamentals of Drilling Engineering*; Society of Petroleum Engineers: Richardson, TX, USA, 2011.
45. James, S.; Konidaris, G.; Rosman, B. An analysis of Monte Carlo tree search. In Proceedings of the 31st AAAI Conference on Artificial Intelligence, San Francisco, CA, USA, 4–9 February 2017; pp. 3576–3582.
46. Vodopivec, T.; Samothrakis, S.; Ŝter, B. On Monte Carlo Tree Search and Reinforcement Learning. *J. Artif. Intell. Res.* **2017**, *60*, 881–936. [CrossRef]
47. Browne, C.B.; Powley, E.; Whitehouse, D.; Lucas, S.M.; Cowling, P.I.; Rohlfshagen, P.; Tavener, S.; Perez, D.; Samothrakis, S.; Colton, S. A Survey of Monte Carlo Tree Search Methods. *IEEE Trans. Comput. Intell. AI Games* **2012**, *4*, 1–43. [CrossRef]
48. Kocsis, L.; Szepesvári, C.; Willemson, J. *Improved Monte-Carlo Search*; Technical. Repotrt 1; University of Tartu: Tartu, Estonia, 2006.
49. Vodopivec, T. *Monte Carlo Tree Search Strategies*; Maastricht University: Maastricht, The Netherlands, 2018.
50. Chaslot, G.; Bakkes, S.; Szitai, I.; Spronck, P. Monte-carlo tree search: A New Framework for Game AI. In Proceedings of the AAAI Conference on Artificial Intelligence and Interactive Digital Entertainment, Stanford, CA, USA, 22–24 October 2008; pp. 389–390.
51. Chaslot, G.M.J.-B.; Winands, M.H.M.; van den Herik, H.J.; Uiterwijk, J.W.H.M.; Bouzy, B. Progressive Strategies for Monte-Carlo Tree Search. *New Math. Nat. Comput.* **2008**, *4*, 343–357. [CrossRef]
52. Gelly, S.; Silver, D. Monte-Carlo tree search and rapid action value estimation in computer Go. *Artif. Intell.* **2011**, *175*, 1856–1875. [CrossRef]
53. Efroni, Y.; Dalal, G.; Scherrer, B.; Mannor, S. How to Combine Tree-Search Methods in Reinforcement Learning. In Proceedings of the AAAI Conference on Artificial Intelligence, Honolulu, HI, USA, 27 January–1 February 2019; Volume 33, pp. 3494–3501. [CrossRef]
54. Puterman, M.L. *Markov Decision Processes, Wiley Series in Probability and Statistics*; John Wiley & Sons, Inc.: Hoboken, NJ, USA, 1994. [CrossRef]
55. Bourgoyne, A.T.; Millheim, K.K.; Chenevert, M.E.; Young, F.S. *Applied Drilling Engineering*; Society of Petroleum Engineers: Richardson, TX, USA, 1986.
56. Ozbayoglu, M.E.; Saasen, A.; Sorgun, M.; Svanes, K. Effect of Pipe Rotation on Hole Cleaning for Water-Based Drilling Fluids in Horizontal and Deviated Wells. In Proceedings of the IADC/SPE Asia Pacific Drilling Technology Conference and Exhibition, Jakarta, Indonesia, 25–27 August 2008. [CrossRef]

57. Sanchez, R.A.; Azar, J.J.; Bassal, A.A.; Martins, A.L. The Effect of Drillpipe Rotation on Hole Cleaning During Directional Well Drilling. In Proceedings of the SPE/IADC Drilling Conference, Amsterdam, The Netherlands, 4–6 March 1997. [CrossRef]
58. Baldino, S.; Osgouei, R.E.; Ozbayoglu, E.; Miska, S.; Takach, N.; May, R.; Clapper, D. Cuttings settling and slip velocity evaluation in synthetic drilling fluids. In Proceedings of the Offshore Mediterranean Conference and Exhibition, Ravenna, Italy, 25–27 March 2015.
59. Cayeux, E.; Mesagan, T.; Tanripada, S.; Zidan, M.; Fjelde, K.K. Real-time evaluation of hole-cleaning conditions with a transient cuttings-transport model. *SPE Drill. Complet.* **2014**, *29*, 5–21. [CrossRef]
60. Gul, S.; van Oort, E.; Mullin, C.; Ladendorf, D. Automated Surface Measurements of Drilling Fluid Properties: Field Application in the Permian Basin. *SPE Drill. Complet.* **2020**, *35*, 525–534. [CrossRef]
61. Nazari, T.; Hareland, G.; Azar, J.J. Review of Cuttings Transport in Directional Well Drilling: Systematic Approach. In Proceedings of the SPE Western Regional Meeting, Anaheim, CA, USA, 27–29 May 2010. [CrossRef]
62. Saini, G.S.; Ashok, P.; van Oort, E. Predictive Action Planning for Hole Cleaning Optimization and Stuck Pipe Prevention Using Digital Twinning and Reinforcement Learning. In Proceedings of the IADC/SPE International Drilling Conference and Exhibition, Galveston, TX, USA, 3–5 March 2020. [CrossRef]
63. De Oliveira, G.L.; Zank, C.A.C.; De Costa, A.F.S.; Mendes, H.M.; Henriques, L.F.; Mocelin, M.R.; De Oliveira Neto, J. Offshore drilling improvement through automating the rig state detection process—Implementation process history and proven success. In Proceedings of the IADC/SPE Drilling Conference and Exhibition, Fort Worth, TX, USA, 1–3 March 2016. [CrossRef]

energies

MDPI

Article

Well Construction Action Planning and Automation through Finite-Horizon Sequential Decision-Making

Gurtej Singh Saini, Oney Erge, Pradeepkumar Ashok and Eric van Oort *

Cockrell School of Engineering, The University of Texas at Austin, Austin, TX 78712, USA
* Correspondence: vanoort@austin.utexas.edu

Abstract: Well construction operations require continuous complex decision-making and multi-step action planning. Action selection at every step demands a careful evaluation of the vast action space, while guided by long-term objectives and desired outcomes. Current human-centric decision-making introduces a degree of bias, which can result in reactive rather than proactive decisions. This can lead from minor operational inefficiencies all the way to catastrophic health and safety issues. This paper details the steps in structuring unbiased purpose-built sequential decision-making systems. Setting up such systems entails representing the operation as a Markov decision process (MDP). This requires explicitly defining states and action values, defining goal states, building a digital twin to model the process, and appropriately shaping reward functions to measure feedback. The digital twin, in conjunction with the reward function, is utilized for simulating and quantifying the different action sequences. A finite-horizon sequential decision-making system, with discrete state and action space, was set up to advise on hole cleaning during well construction. The state was quantified by the cuttings bed height and the equivalent circulation density values, and the action set was defined using a combination of controllable drilling parameters (including mud density and rheology, drillstring rotation speed, etc.). A non-sparse normalized reward structure was formulated as a function of the state and action values. Hydraulics, cuttings transport, and rig state detection models were integrated to build the hole cleaning digital twin. This system was then used for performance tracking and scenario simulations (with each scenario defined as a finite-horizon action sequence) on real-world oil wells. The different scenarios were compared by monitoring state–action transitions and the evolution of the reward with actions. This paper presents a novel method for setting up well construction operations as long-term finite-horizon sequential decision-making systems, and defines a way to quantify and compare different scenarios. The proper construction of such systems is a crucial step towards automating intelligent decision-making.

Keywords: sequential decision-making; Markov decision process; reward shaping; well construction; hole cleaning; digital twinning

Citation: Saini, G.S.; Erge, O.; Ashok, P.; van Oort, E. Well Construction Action Planning and Automation through Finite-Horizon Sequential Decision-Making. *Energies* **2022**, *15*, 5776. https://doi.org/10.3390/en15165776

Academic Editors: Mofazzal Hossain, Hisham Khaled Ben and Md Motiur Rahman

Received: 22 June 2022
Accepted: 6 August 2022
Published: 9 August 2022

1. Background and Introduction

Well construction, i.e., the process of drilling and completing wells for applications such as extracting hydrocarbons or accessing geothermal energy, is a highly technical discipline. There is irreducible complexity involved in the process while drilling through highly variable geological environments deep in the sub-surface. These environments can pose safety hazards; therefore, the various well construction operations require careful surveillance of the system variables. Currently, real-time (RT) data streams, advanced process models, and sophisticated simulation techniques are utilized for monitoring well construction operations (see e.g., [1]). Decision-making, however, is still primarily performed by humans, with little automation. The decisions are based on the understanding of the processes by the subject matter expert (e.g., engineer, or the driller out in the field) in control of the process. They are made not only based on interpretations of the model outputs, but also to a large extent on past experiences, and sometimes 'gut feelings.' The decisions are also affected by other human factors, such as situational awareness or even the physical and mental state of the decision-maker (with, e.g., fatigue playing a major role [2]). The consequences of making erroneous decisions can range from poor operational efficiency to catastrophic failures and accidents.

The overarching goal of the research presented here is the development of an automated intelligent decision-making system for improved well construction safety and performance. Developing such a system requires the identification of the necessary data and data streams, digital twinning of the underlying operation, structuring the decision-making system, and finally selecting the simulation or planning method. This paper discusses a general approach for setting up such systems by combining data and digital twins, and demonstrates this for hole cleaning operations. Hole cleaning is the process of removing solids (cuttings generated during drilling, cavings, or metal shavings) from the borehole to ensure that different well construction operations (including drilling, tripping, casing installation, and cementing) can be performed safely and efficiently. Inefficient hole cleaning can result in various problems, such as stuck pipe, downhole tool damage, formation damage, or trouble running the casing. Hole cleaning issues account for several hundred million dollars annually in lost time costs [3,4]. The literature on hole cleaning modelling is extensive; the latest developments are covered in detail by [5] and references therein. There is, however, currently no framework for setting up systems that can perform scenario analysis and action planning in well construction processes such as hole cleaning, and this is the main topic of this paper.

In other domains, the use of sophisticated decision-making systems has resulted in an overall improvement in safety and operational efficiency. Such decision-making systems, coupled with digital twins of operations or equipment, have been utilized in areas such as manufacturing [6], autonomous vehicles [7,8], and smart grid management [9,10]. Recently, high-complexity board games (e.g., go and chess) were solved using tree-based search techniques in combination with learned system models [11,12]. Complex RT strategy games, such as StarCraft and Dota, have been solved by utilizing novel reinforcement learning techniques [13,14]. Amongst these, sequential decision making is a class of algorithm that takes the dynamics of the system into consideration. The decision and actions are made in steps, and adjusted based on rewards assigned to outcomes. This approach parallels how manual decisions and actions are made with regards to hole cleaning in an actual drilling operation, which motivated its selection as the approach of choice in this paper. Further, given the need to arrive at a solution within a certain time period, the hole cleaning problem is handled as a finite-horizon sequential decision-making problem.

2. Setting up the Planning and Decision-Making Systems for Well Construction Operations

Planning is the process of generating an action sequence from an initial system state to some goal state to satisfy some high-level objective function. Multiple approaches such as forward search algorithms (A* search, greedy best-first search, Dijkstra's algorithm), exhaustive methods (tree search, policy iteration, value iteration), or simulation-based search methods (flat Monte Carlo, Monte Carlo tree search) can be used to solve planning problems [15–17]. Most of these methods utilize a combination of domain knowledge and logic with search functions, which can be further enhanced by heuristics. This paper does not cover the algorithm selection process, and the reader is referred to [18] for a detailed comparison of the approaches. The various planning algorithms were evaluated using six criteria, namely optimal solution guarantee, memory requirement, computational time requirement, necessity of a well-defined admissible function, evaluation function use, and exploration/exploitation balancing. All planning problems, however, have the following essential components [15]:

- Objective function, stating the initial and the desired (goal) states, and constraints that can influence decision-making;

- Decision epochs, or the times at which decisions need to be made. Epochs can either be explicitly represented as time intervals, or implicitly represent a sequence of actions in succession;
- State space, to describe all possible situations or scenarios (states) the system can be in, at any given decision epoch;
- Action space, to quantify all possible decisions or actions that can be utilized to manipulate states;
- Plan, or strategy that represents the sequence of actions taken at every successive decision epoch.

In effect, planning problems are sequential decision-making problems that can be solved by reinforcement learning (RL) techniques. In RL, a goal-directed learning agent interacts with an uncertain environment (either physically or virtually) based on specific policies or action plans. Every interaction is associated with immediate feedback or reward. The goal of this agent is to maximize the long-term reward. To accomplish this, the agent needs to exploit what it has already experienced and try new actions to learn from unexplored trajectories [19,20]. Figure 1 shows a schematic of this agent–environment interaction, where an action a_t by the agent in the environment (observed by the agent to be in state s_t) results in an immediate reward r_t and a new observed state s_{t+1}.

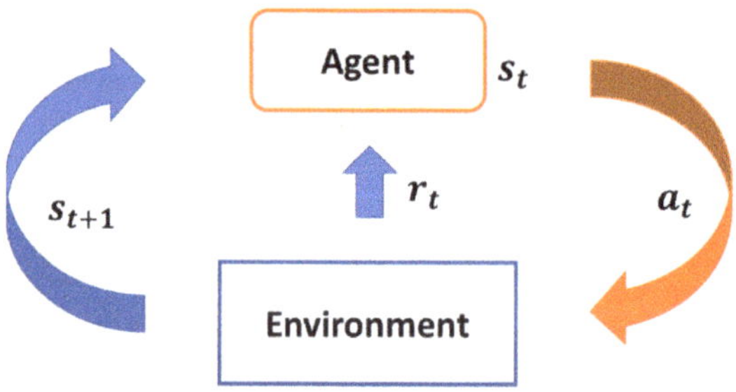

Figure 1. Agent–environment interaction in RL (modified from [20]).

Such interactions between a decision-making agent and a fully observable environment to achieve some long-term objective can be formalized in a Markov decision process (MDP) framework. A process is said to be Markovian if it follows the Markovian property, i.e., if any future outcomes depend only on the current system state and the immediate action. An MDP is defined by a tuple $(\{S, A, P, R\})$ and a policy (π) [21].

S is the state space, where s_t ($s_t \in S$) represents the state of the system, as perceived by the agent, at time t. A state is defined by a set of parameters to quantify the condition of the environment completely (fully observable). A is the action space, where a_t ($a_t \in A$) is an individual action taken by the agent at time t to manipulate the system in the state s_t. An action is a combination of different control variables that can influence the environment. P is the transition function representing the state–action transition probabilities. $P^a_{ss'}$ is the probability that a system in state s, at time t transitions to state s' at time $t+1$ on taking an action a. R is the reward function to quantify the immediate feedback associated with a state–action transition. Reward may depend either only on the final state, or on the final state and the action.

The accumulation of rewards over multiple time steps or decision epochs is the system's return. The time horizon for accumulating these rewards may be finite (fixed number of steps) or infinite, and may include a discount factor γ ($\leq$1) in the case of infinite time horizon problems. In this paper, we assume 10 finite steps and therefore assume a discount factor ($\gamma = 1$), giving every step equal weight. The policy π is the logic or set of rules used by an agent to select an action from a given state; it may be stochastic or deterministic. The goal of an agent is to find a policy that allows it to maximize its total return. There is always at least one optimal policy for an MDP that helps extract the maximum return from the system [22]. Another crucial step in MDP formulation is reward shaping, or engineering the reward function to obtain more frequent feedback on appropriate system behaviors [23]. Thus, reward shaping influences the total return, thereby affecting the system's policy.

2.1. Well Construction Sub-Processes as MDPs

Well construction is a multi-step process that requires planning and decision-making at every step of its various sub-processes. Planning necessitates identifying objectives, constraints, and required data associated with the individual sub-processes. A crucial step in developing such planning systems is setting them up properly, which requires the following elements:

- Formulating an MDP for the operation, which includes appropriately defining state and action spaces
- Defining a goal or a desired state
- Efficient shaping of the reward function
- Setting up an integrated-multi model system replicating the process (environment), i.e., building its digital twin

2.1.1. MDP Formulation

Formulating an MDP for any process requires the following [21]:

- The process should satisfy the Markovian property
- Any state defined for the process should be fully observable
- State space should be finite or countably infinite, with states defined by exhaustively incorporating all relevant parameters
- There is an explicit definition of the action space with appropriately identified control variables

For most well construction processes, the condition (state) of the wellbore at any time is a culmination of all the previous operations (actions), past conditions (past states), and state transitions. In other words, the current state is a representation of the well's operational past, and any subsequent transition depends only on this state and the immediate action. The assumption that well construction operations follow the Markovian property is, therefore, valid. The state of the system needs to be represented by all relevant parameters required to fully describe the process under consideration. The state is also continually refined based on the data received from surface or downhole sensors (at frequencies of 1 Hz or higher). This results in the state being a complete representation of the environment as perceived by the agent, i.e., the state is assumed to be fully observable. The operations or variables that can be actively controlled to bring about state transitions constitute the action.

The state and action spaces can be either discrete or continuous; however, for the work presented here, both are defined as discrete sets. Figure 2 illustrates the proposed method for discretizing the state space based on wellbore inclinations. Well inclinations in the range 0 to 30 degrees are considered to be a near-vertical section, and the regions of the well with inclination angles greater than 75 degrees are considered to be near-horizontal lateral. The intermediate inclination angle regions comprise the curve or the build section of a well. This method of discretizing the state space is proposed because the response of state variables to different actions has a high degree of dependency on the inclination of the well segment. Wellbore inclination significantly influences the cuttings transport mechanisms, which are different for near-vertical, intermediate, and lateral sections. Consequently, this affects the hole cleaning requirements. Note, however, that the state space may be discretized in other ways, such as in intervals or sections of measured depth (MD) or true vertical depth (TVD).

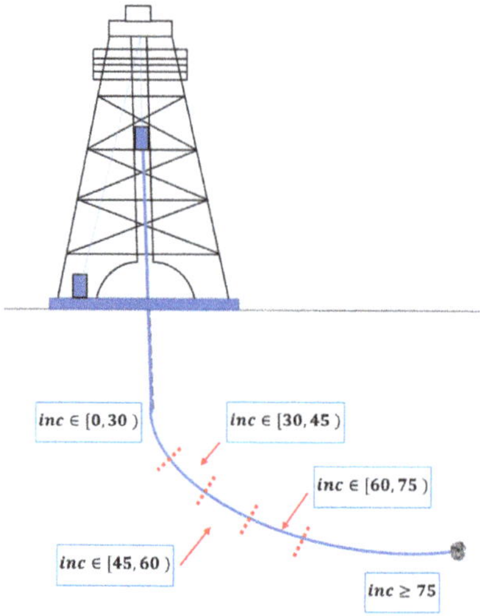

Figure 2. The proposed strategy for defining discrete state spaces based on wellbore inclination angles.

Equation (1) represents the state vector, where 'p_1' through 'p_n' are the exhaustive set of parameters required to define the system state completely. The state of the system here consists of some functional value of these parameters over the appropriate inclination intervals $\{[0,30),[30,45),[45,60),[60,75),[75+)\}$. Another point to note is that these inclination interval definitions can be adjusted depending on the requirements of the underlying process.

$$s_t = \begin{Bmatrix} p_{1\_0-30} \\ p_{1\_30-45} \\ p_{1\_45-60} \\ p_{1\_60-75} \\ p_{1\_75+} \\ \cdot \\ p_{n\_0-30} \\ \cdot \\ p_{n\_75+} \end{Bmatrix} \tag{1}$$

Similarly, the action space is constructed by different combinations of possible values of the identified control variables. For drilling operations, some such control variables are the surface drillstring rotation speed (RPM), weight on bit (WOB), drilling mud properties, flowrate, and drillstring tripping speeds. These variables can take on discrete values between specified minimum and maximum thresholds. These thresholds are dictated by safety constraints, process and equipment limitations, and operational economics. Note that in some problems, it will make more sense to break down the problem by measured depth or true vertical depth sections instead of by inclination. Additionally, the state space parameters will be dependent on the objective of the digital twin. In this paper, the parameters were largely dictated by the hole cleaning/hydraulics problem, serving as an example of how the problem can be mathematically set up for action planning.

2.1.2. Goal State

The goal or desired state, as the name suggests, refers to the subset of the state space that the drilling agent aims to achieve. The goal state is used as the reference to direct the agent's search. The desired functional values of individual goal state components are used for the construction of the overall goal state, as shown in (2).

$$s_{goal} = \begin{Bmatrix} p^{g}{}_{1\_0-30} \\ p^{g}{}_{1\_30-45} \\ p^{g}{}_{1\_45-60} \\ p^{g}{}_{1\_60-75} \\ p^{g}{}_{1\_75+} \\ \cdot \\ p^{g}{}_{n\_0-30} \\ \cdot \\ p^{g}{}_{n\_75+} \end{Bmatrix} \quad (2)$$

2.1.3. Reward Shaping

Shaping the reward function allows for rewarding or penalizing a drilling agent's behavior more frequently, instead of at sparse intervals or at the end of an episode. Frequent rewards, in turn, help with more directed and faster learning. A possible strategy for reward shaping is to provide the agent with regular feedback based on its position relative to the goal state. Another factor to consider is the contribution to the reward of the relative changes in different action control variables. For instance, if a state transition from s to s' can be achieved by two completely different actions a and a', the reward function needs to be able to recognize and quantify this difference. This is especially important, for instance, in cases where the agent suggests changing drilling RPM and flowrate with an alternative action being a change to mud rheological parameters (which may be economically and temporally more expensive).

2.1.4. Digital Twinning the Environment

For action planning, a comprehensive model or digital twin of the process needs to be constructed. This twin is then used for replicating the environment, thereby simulating multiple episodes or trajectories of experience [6,24,25]. Model-free RL techniques can then be applied to these episodes to improve the return value and, subsequently, to determine an optimal policy. Figure 3 details these steps. The Monte Carlo tree search approach was identified to be applicable to this problem and is discussed in detail in [25].

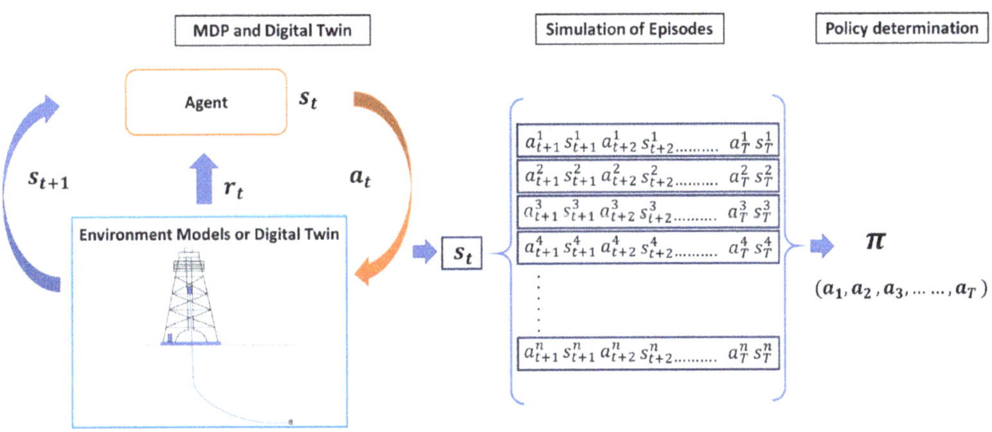

Figure 3. Application of MDP and digital twins for simulating episodes to determine the optimal policy.

3. Setting up the Hole Cleaning Decision-Making System

Here, we demonstrate step-by-step how to set up a decision-making and planning system for the hole cleaning operation.

3.1. Formulating the MDP for the Hole Cleaning System

Effective and safe hole cleaning requires keeping the cuttings bed height low enough to prevent issues at any stage of the well construction operation. Moreover, the equivalent circulation density (ECD) needs to be managed within a safety or drilling margin (Figure 4). ECD at a depth is the gradient of the sum of the hydrostatic head exerted by the drilling mud (a function of the true vertical depth (TVD) of the well) and the total circulating frictional pressure loss in the annulus between the drillstring and the wellbore, which is a function of the measured depth (MD) along the length of the well [26].

$$\mathrm{ECD} = \frac{\mathrm{P_{hydrostatic\_D_{TVD}}} + \mathrm{P_{frictional\_pressure\_loss\_D_{MD}}}}{\mathrm{D_{TVD}} \cdot \mathrm{g}} \tag{3}$$

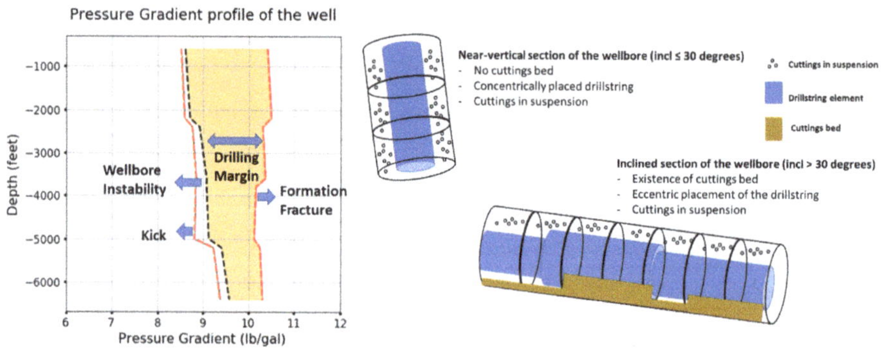

(a) (b)

Figure 4. (**a**) Drilling safety margin; (**b**) Cuttings bed distribution in different inclination segments of the well.

In Figure 4a, the lower and upper limits of the drilling margin are the stability limit (SL) and the fracture gradient (FG), respectively. SL is the higher value between the pore pressure (PP) and the mud pressure essential for maintaining wellbore stability. PP is exerted by the fluids (brine or hydrocarbons) present in the pore spaces of the formation rocks. If the ECD falls below the SL, it can cause wellbore instability and, if it falls below the PP of permeable formations with the potential to flow, an unwanted influx of formation fluids into the borehole (which is called a 'kick'). Exceeding the FG can fracture the formation and lead to mud loss. Such mud loss events are called lost circulation events [27]. It is, therefore, an objective of the system to maintain the ECD within this drilling margin at all times.

Figure 4b depicts the cuttings bed distribution in the different sections of the wellbore. In the near-vertical section of the wellbore, the principal method of suspending and carrying cuttings up hole in the mud is by overcoming the particle slip velocity, and no cuttings bed can exist. In the curve section (inclinations from 30 to 60 degrees), an unstable cuttings bed can form below the angle of repose. However, there is a high possibility that when the mud circulation stops, the cuttings avalanche back down the annulus, which can pack off around the bottom-hole assembly, causing a stuck pipe incident. For this section, the hole cleaning design requires tackling and preventing this cuttings avalanche. In the near-horizontal section (inclination angle from 60 to 90 degrees), above the angle of repose for the cuttings, a stable cuttings bed will form on the low side of the hole. The primary hole cleaning requirement is to continuously move this bed up and out of the hole (see e.g., [28,29]). Therefore, another main objective of the decision-making system is to satisfy these hole cleaning requirements.

3.1.1. State Space

The following parameters are required to quantify the condition of the borehole from the perspective of the hole [30,31]:

- Height of the cuttings bed in the curve and the lateral sections of the wellbore;
- ECD along the entire length of the wellbore.

The well can be treated as a series of interconnected control volumes, segmented based on any changes in well dimensions (e.g., changes in inner or outer diameters) or based on different survey intervals (as shown in Figure 4b). Each control volume's condition can be independently represented by absolute values of ECD and cuttings bed height. However, with the well being segmented into multiple inclination intervals (based on the strategy discussed in Figure 2), every inclination interval usually consists of many such control volumes. A functional value derived from the absolute value is calculated for each of these parameters in every control volume. These values are then averaged over the different inclination segments to obtain a single value per inclination interval for every parameter. Converting to a functional value normalizes the absolute value to specific operational thresholds, and assists in reward shaping, as discussed in later sections.

Cuttings Bed Height

The absolute value of the cuttings bed height for every control volume is normalized to its outer diameter (Figure 5). These values are then used to calculate the average normalized cuttings bed height indicator H (dimensionless) for every inclination segment of the well, as shown by Equation (4).

$$H_k^{norm} = \frac{H_k^{absolute}}{D_{o\_k}},\ H = \frac{\sum_{k=1}^{N_{seg}} H_k^{norm}}{N_{seg}} \tag{4}$$

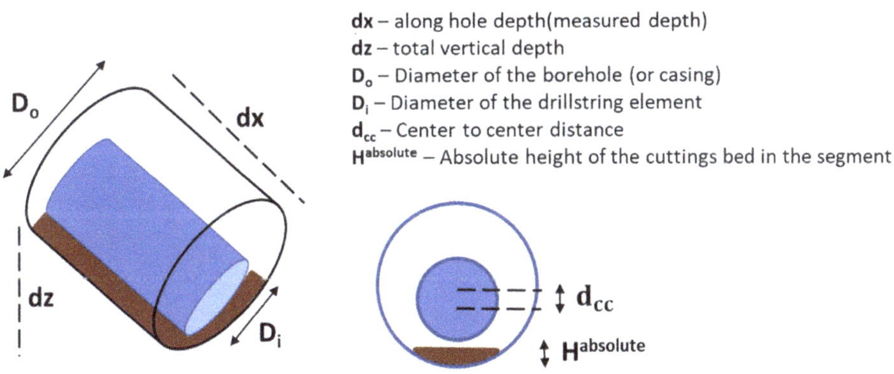

Figure 5. Absolute cuttings bed height for a single control volume element.

The functional value $H_{inc.}$ is then derived from H using Equation (5), as visualized in Figure 6.

$$H_{inc.} = \begin{cases} 0 & H \leq 0.20 \\ 1 & 0.20 < H \leq 0.40 \\ 2 & 0.40 < H \leq 0.60 \\ 3 & 0.60 < H \leq 0.80 \\ 4 & H > 0.80 \end{cases} \tag{5}$$

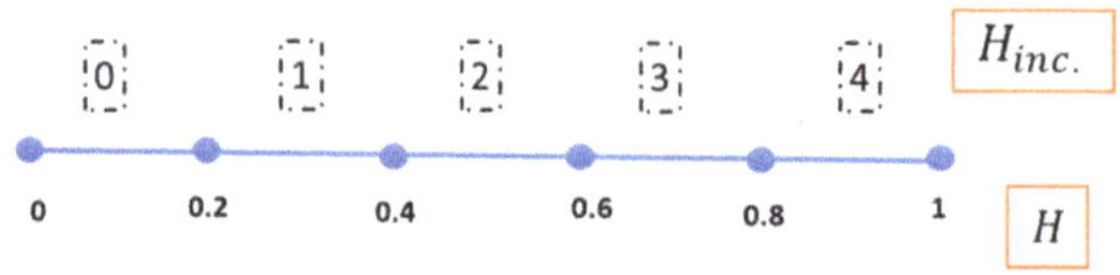

Figure 6. Functional value assignment for the cuttings bed height parameter.

The parameter $H_{inc.}$ is evaluated for all non-vertical sections, because no cuttings bed will form in the $[0, 30)$ degree inclination interval. Thus, the bed height components of the state vector are $\{H_{30-45}, H_{45-60}, H_{60-75}, H_{75+}\}$.

ECD

As previously discussed, ECD needs to be managed within the drilling margin. There is, however, some degree of uncertainty associated with its limits, which is accounted for by considering an uncertainty factor $DF(\leq 0.25)$. ECD_{avg}, the average ECD for an inclination interval, is calculated by averaging absolute ECD values over all the control volume segments in that interval (Equation (6)). Since the SL and FG values vary with depth, ECD_{avg} is calculated independently for the different intervals.

$$\text{ECD}_{avg} = \frac{\sum_{k=1}^{N_{seg}} \text{ECD}_k^{absolute}}{N_{seg}} \quad (6)$$

Using ECD_{avg}, the functional value of ECD, $\text{ECD}_{inc.}$, is calculated using Equation (7) and discussed in Figure 7.

$$\text{ECD}_{inc.} = \begin{cases} -3 & \text{ECD}_{av} \leq SL - DF{\cdot}\Delta w \\ -2 & SL - DF{\cdot}\Delta w < \text{ECD}_{av} \leq SL \\ -1 & SL < \text{ECD}_{av} \leq SL + DF{\cdot}\Delta w \\ 0 & SL + DF{\cdot}\Delta w < \text{ECD}_{av} \leq SL + 2{\cdot}DF{\cdot}\Delta w \\ 1 & SL + 2{\cdot}\, DF{\cdot}\Delta w < \text{ECD}_{av} \leq FG - 2{\cdot}DF{\cdot}\Delta w \\ 2 & FG - 2{\cdot}DF{\cdot}\Delta w < \text{ECD}_{av} \leq FG \\ 3 & \text{ECD}_{av} > FG \end{cases} \quad (7)$$

$$where \;\; \Delta w = FG - SL$$

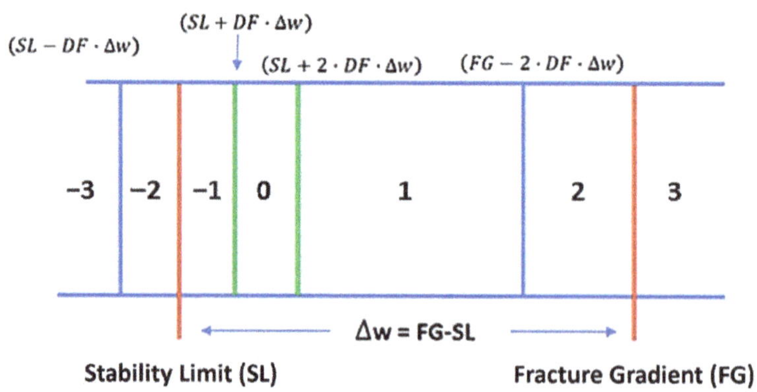

Figure 7. Functional value assignment for the ECD parameter.

Since keeping ECD within the drilling margin is essential throughout the well, the state components related to the ECD parameter, $\{ECD_{0-30}, ECD_{30-45}, ECD_{45-60}, ECD_{60-75}, ECD_{75+}\}$ are calculated for all intervals.

Equation (8) represents the complete hole cleaning state of the wellbore. In this form, every component of the state vector is represented by its functional value at every decision epoch.

$$s = \left\{\begin{array}{c} H_{30-45} \\ H_{45-60} \\ H_{60-75} \\ H_{75+} \\ \mathrm{ECD}_{0-30} \\ \mathrm{ECD}_{30-45} \\ \mathrm{ECD}_{45-60} \\ \mathrm{ECD}_{60-75} \\ \mathrm{ECD}_{75+} \end{array}\right\} \tag{8}$$

This representation of state is Markovian since it fully represents the condition of the hole cleaning system and encompasses all the information about the system's history. Any subsequent state transition depends only on the state and the action taken. As can be seen from Equation (8), the state vector has nine parameters. Each of the $H_{inc.}$ parameter can have five values while each of the $\mathrm{ECD}_{inc.}$ parameters can have seven values. This translates into a state space size of $5^4 \times 7^5$ = 10,504,375.

3.1.2. Goal State

The goal for any decision-making system is to first search the state and action space and then move towards the desired state. The functional values for all state variable components are defined such that 0 represents the desired state for each; therefore, the target goal state for the system is as shown in Equation (9).

$$s_{goal} = \left\{\begin{array}{c} 0 \\ 0 \\ 0 \\ 0 \\ 0 \\ 0 \\ 0 \\ 0 \\ 0 \end{array}\right\} \tag{9}$$

3.1.3. Actions Space

Hole cleaning, while managing the ECD within the drilling margin, is a function of (see e.g., [32–34]:

- Drilling mud properties (particularly density and viscosity);
- Cuttings properties (size and density);
- Drilling parameters such as drilling RPM and flow rate;
- Drillstring geometry and its eccentricity in the borehole;
- Rate of cuttings generation (which depends on the drilling rate);
- Borehole geometry (diameters of the open or cased hole sections along the well) and inclination angle.

Hole cleaning pills or sweeps, i.e., limited volumes of fluid with altered density and/or viscosity to aid in cuttings evacuation from the hole (mostly effective in vertical hole rather than deviated hole).

Some of these control variables affect the condition of the borehole to a greater extent than others. Moreover, some variables can be controlled more readily than others. Figure 8 presents a chart comparing the different control variables, plotted for their relative influence on hole cleaning against their ability to be actively controlled in real-time.

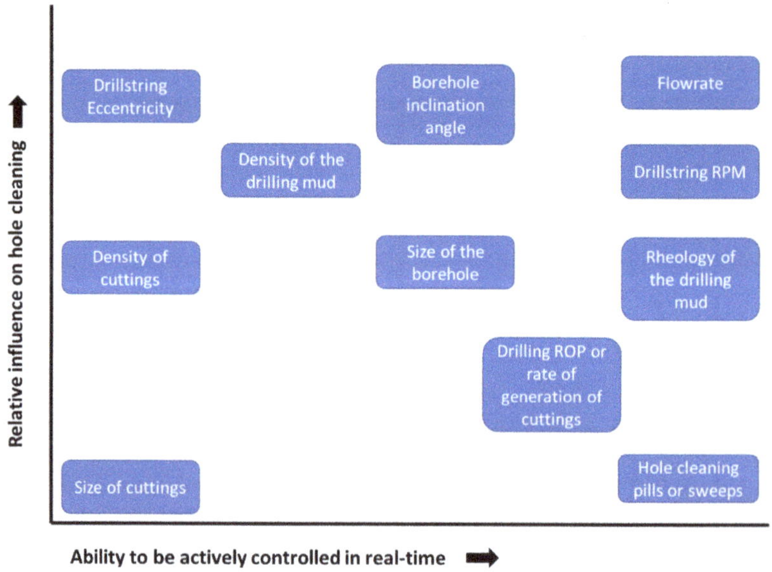

Figure 8. Variables contributing to the hole cleaning performance in a deviated hole (modified from [35]).

Thus, the key parameters that have a significant influence on the hole cleaning performance, and can be actively controlled in the field, are flow rate, RPM, mud properties (rheological parameters), and the WOB to control the rate of penetration (ROP). In the following, we will assume that the fluid behavior is that of a Bingham plastic fluid, in which case its rheology is quantified by its plastic viscosity (PV) and yield point (YP). Another critical parameter that influences the ECD is the mud density. A combination of these variables at every decision epoch constitutes an action, which is represented by Equation (10).

$$a_t = \left\{ \begin{array}{c} Flowrate \\ ROP \\ RPM \\ Mud\ density \\ Mud\ PV \\ Mud\ YP \end{array} \right\} \tag{10}$$

3.2. Digital Twin of the Environment

A digital twin was built by integrating the available well initializations (data streams such as well plans, well surveys, well geometry information, etc.) with analytical implementations of the hydraulics and cuttings transport models [18]. The hydraulics model calculates the frictional pressure losses and ECD throughout the well and utilizes sub-models presented in [32,36]. The cuttings transport model also utilizes multiple sub-models [37–43] and estimates the cuttings bed height and the cuttings concentration in the flow stream along the well. Figure 9 illustrates the use of this twin to predict the system state at the next epoch, based on the current state and immediate action. The epoch is the smallest time step of the planning problem for which an action is determined. The digital twin was designed to plan either every 5 min interval into the future or whenever there was a change in the well operations.

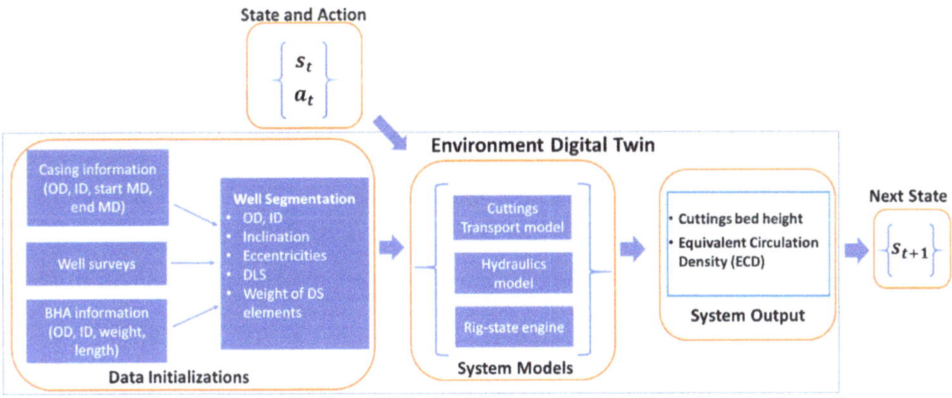

Figure 9. Digital twin of the hole cleaning system's environment (see [18] for details on the models).

3.3. Reward Function

To quantify the immediate feedback associated with state–action transitions, a reward function is defined for the hole cleaning system, which has three distinct components:

- Reward associated with state transition;
- Penalty associated with action transition;
- Reward associated with action variables.

3.3.1. Reward Associated with State Transition

Since the objective of the system is to reach the goal state, every component of the state vector tries to achieve a functional value of 0. This was used as a reference to calculate normalized reward values associated with every state vector component in the $[-1, 1]$ range. Table 1 details the functions used for these calculations.

Table 1. Reward function associated with state vector components.

Component	Reward Function	Values
H_{30-45}	$R_{H30-45} = \frac{2 - H_{30-45}}{2}$	$\{1, 0.33, -0.33, -1\}$
H_{45-60}	$R_{H45-60} = \frac{2 - H_{45-60}}{2}$	$\{1, 0.33, -0.33, -1\}$
H_{60-75}	$R_{H60-75} = \frac{2 - H_{60-75}}{2}$	$\{1, 0.33, -0.33, -1\}$
H_{75+}	$R_{H75+} = \frac{2 - H_{75+}}{2}$	$\{1, 0.33, -0.33, -1\}$
ECD_{0-30}	$R_{ECD0-30} = 1 - \frac{2}{3} \cdot \lvert E_{0-30} \rvert$	$\{-1, -0.33, 0.33, 1, 0.33, -0.33, -1\}$
ECD_{30-45}	$R_{ECD30-45} = 1 - \frac{2}{3} \cdot \lvert E_{30-45} \rvert$	$\{-1, -0.33, 0.33, 1, 0.33, -0.33, -1\}$
ECD_{45-60}	$R_{ECD45-60} = 1 - \frac{2}{3} \cdot \lvert E_{45-60} \rvert$	$\{-1, -0.33, 0.33, 1, 0.33, -0.33, -1\}$
ECD_{60-75}	$R_{ECD60-75} = 1 - \frac{2}{3} \cdot \lvert E_{60-75} \rvert$	$\{-1, -0.33, 0.33, 1, 0.33, -0.33, -1\}$
ECD_{75+}	$R_{ECD75+} = 1 - \frac{2}{3} \cdot \lvert E_{75+} \rvert$	$\{-1, -0.33, 0.33, 1, 0.33, -0.33, -1\}$

Thus, the reward function contribution of state transition is represented by the set given in Equation (11).

$$R_S = \{R_{H30-45}, R_{H45-60}, R_{H60-75}, R_{H75+}, R_{ECD0-30}, R_{ECD30-45}, R_{ECD45-60}, R_{ECD60-75}, R_{ECD75+}\} \quad (11)$$

3.3.2. Penalty Associated with Action Transition

Table 2 details the calculation of the penalty (negative reward) related to changes in action values. The purpose of these definitions is two-fold:

- To discourage the system from making extreme changes in actions, unless the reward associated with state transition offsets this penalty;
- Select the least penalizing action in case multiple actions result in the same state transition.

Table 2. Reward function associated with action transition.

Component	No. of Intervals	Reward Function	Values
Flowrate	$n_{flowrate}$	$R_{flowrate} = -\frac{\lvert \Delta N_{flowrate} \rvert}{n_{flowrate}}$	[−1, 0]
Drilling ROP	n_{ROP}	$R_{ROP} = -\frac{\lvert \Delta N_{ROP} \rvert}{n_{ROP}}$	[−1, 0]
Drillstring RPM	n_{RPM}	$R_{RPM} = -\frac{\lvert \Delta N_{RPM} \rvert}{n_{RPM}}$	[−1, 0]
Mud density	$n_{density}$	$R_{density} = -\frac{\lvert \Delta N_{density} \rvert}{n_{density}}$	[−1, 0]
Mud PV	n_{PV}	$R_{PV} = -\frac{\lvert \Delta N_{PV} \rvert}{n_{PV}}$	[−1, 0]
Mud YP	n_{YP}	$R_{YP} = -\frac{\lvert \Delta N_{YP} \rvert}{n_{YP}}$	[−1, 0]

The terms $\Delta N_{variable}$ and $n_{variable}$, respectively, are the number of interval changes between consecutive actions and the number of discrete values possible for a given control variable. Their use to calculate a penalty value is illustrated in Figure 10. Here, the action results in a jump across 3 out of a total of 10 intervals and therefore the penalty is −3/10. The action transition-based penalty set is expressed in Equation (12).

$$R_{ap} = \left\{R_{flowrate}, R_{ROP}, R_{RPM}, R_{density}, R_{PV}, R_{YP}\right\} \quad (12)$$

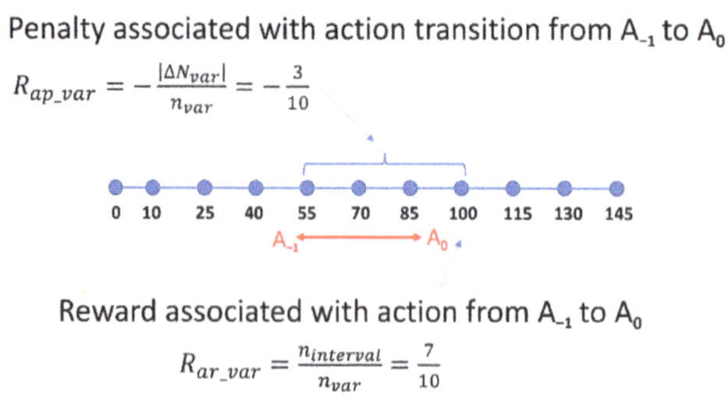

Figure 10. Example calculation of an action transition-based penalty and reward associated with action.

3.3.3. Reward Associated with Action

In the planning phase of drilling operations, the hole cleaning requirement of the system would push the ROP to zero, simply because no cuttings are generated at zero ROP leading to zero bed height and optimum ECD. However, because a critical objective of drilling is to drill a well as fast as reasonably possible (within given limits), there needs to be a positive feedback or reward associated with the ROP. Equation (13) represents this reward, which is calculated using Equation (14) as a ratio of the discrete interval number for a given ROP value to the total number of ROP intervals. This reward component is in the range $[0, 1]$. The calculation is also illustrated in Figure 10. The action results in a jump to the seventh interval and therefore the reward is $7/10$.

$$R_{ar} = \{0, R_{ROP}, 0, 0, 0, 0\} \tag{13}$$

$$R_{ROP} = \frac{n_{interval}}{n_{ROP}} \tag{14}$$

3.3.4. Calculating the Net Reward

Reward value quantifies the 'goodness' of taking some action from a given system state. Thus, the next step for the hole cleaning system is to combine the individual reward components to output a single reward value in the $[0, 1]$ range. This is accomplished by assigning different relative weights to the various components. This ability to assign different weights provides a way to prioritize different objectives. This would be advantageous in drilling wells where there is, for instance, a high risk of well control issues. In these wells, the objective of keeping the ECD within the drilling margin becomes a higher priority than completely removing the cuttings bed. Similarly, reducing the penalty associated with taking drastic actions will not be as important for certain wells as reaching the desired state quickly. Managing these objectives can be accomplished by assigning different relative weights to the individual state or action reward components.

The sets W_s, W_{ap}, and W_{ar}, respectively, are the weights associated with sets for state transition reward, action transition penalty, and the action reward. Equations (15)–(17) represent the method for combining these weights and their associated reward sets. The final values of R_{S_net}, R_{ap_net}, and R_{ar_net} are in the ranges $[-1, 1]$, $[-1, 0]$, and $[0, 1]$, respectively.

$$R_{S\_net} = \frac{\sum_i W_{si} R_{si}}{\sum_i W_{si}} \tag{15}$$

$$R_{ap\_net} = \frac{\sum_i W_{api} R_{api}}{\sum_i W_{api}} \tag{16}$$

$$R_{ar\_net} = \frac{\sum_i W_{ari} R_{ari}}{\sum_i W_{ari}} \tag{17}$$

Before further combining these three components, they are first normalized to the $[0, 1]$ range, using the method presented in Equations (18)–(20). While this normalization is not necessary, it helps to better understand prioritization of end-user choices.

$$R_{s\_norm} = \frac{R_{S\_net} + 1}{2} \tag{18}$$

$$R_{ap\_norm} = R_{ap\_net} + 1 \tag{19}$$

$$R_{ar\_norm} = R_{ar\_net} \tag{20}$$

Finally, the individual normalized rewards R_{s_norm}, R_{ap_norm}, and R_{ar_norm} are combined based on the weights W_{s_norm}, W_{ap_norm}, and W_{ar_norm} as per Equation (21). These weights define the relative importance of the individual normalized rewards and can also be tuned in real-time.

$$R_{net} = \frac{W_{s\_norm}R_{s\_norm} + W_{ap\_norm}R_{ap\_norm} + W_{ar\_norm}R_{ar\_norm}}{W_{s\_norm} + W_{ap\_norm} + W_{as\_norm}} \quad (21)$$

The above definition of the reward function ensures immediate feedback after every action, as opposed to the agent having to wait until the end of an episode (as is the case for sparse reward functions).

4. Implementation of a System as an MDP

Here, we demonstrate the developed hole cleaning decision-making system for performance tracking and action planning using a specific example. The dataset used is from an actual oil well that exhibited issues due to insufficient hole cleaning during tripping, casing, and cementing operations. This dataset included the well's directional survey data, well profile information (casing, BHA, and bit details), one-second surface sensor data, and mud check information. A digital twin of the well was developed by integrating physics-based models (cuttings transport and hydraulics with an incorporated thermal model), data-based models (rig state detection engine), and relevant raw data sources (as detailed in Figure 9).

4.1. Well Profile

The well profile and trajectory used here are from an actual drilling operation and shown in Figure 11. The well had a short vertical section with a shallow kick-off point (where the well starts building inclination angle from vertical) around 300 feet MD. The inclination angle reached 30 degrees at approximately 750 feet MD, and 75 degrees (horizontal section) around 1250 feet MD. After this, the well remained near-horizontal until it reached its total depth (TD) of 2500 feet MD. The surface casing with an internal diameter of 13.375 inch was set at a depth of 623 feet MD. Following this, a 12.25 inch hole section was drilled to well TD. Upon reaching TD, a 50 min on-bottom circulation cycle was performed (at a flow rate of 950 gallons per minute (GPM) and 60 RPM), the drillstring was then tripped out of the hole with intermittent back-reaming (at 910 GPM and 60 RPM), and a 9.625 inch casing was run to TD and cemented.

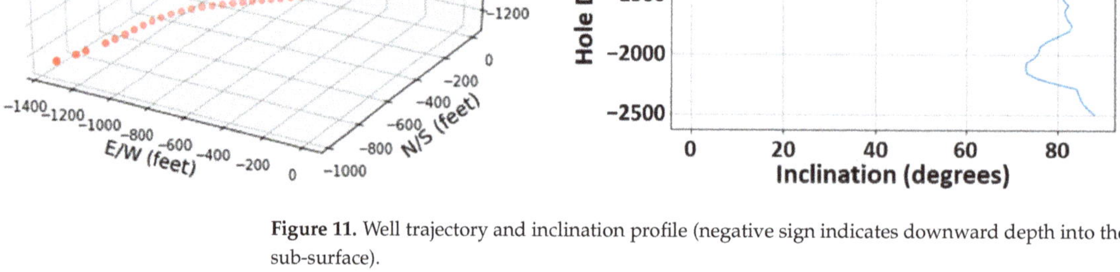

Figure 11. Well trajectory and inclination profile (negative sign indicates downward depth into the sub-surface).

The SL and FG values to define the drilling margins for the different sections of the well are shown in Table 3. For the near-vertical section, the SL and FG were assigned maximum values of 6 ppg and 18 ppg, respectively, because this interval was entirely cased while drilling the 12.25 inch section.

Table 3. The SL and FG values to define drilling margin for the different inclination intervals.

Inclination Interval	Stability Limit (ppg)	Fracture Gradient (ppg)
[0, 30)—in casing	6	18
[30, 45)	8.2	10.6
[45, 60)	8.4	10.4
[60, 75)	8.2	10.2
[75+)	8.6	10.0

To safely run the 9.625 inch casing after drilling, calculations show that the maximum theoretical cuttings bed height (on pulling the drillstring out of hole) should not exceed approximately 5 inches (Figure 12a,b). Based on the variation in the drillstring geometry (different outer diameter and eccentric placements of various drillstring components), this bed height limit was translated into an equivalent bed height [31], assuming that the drillstring was still in the hole, with the drill bit at TD (Figure 12c). The red line in Figure 13a shows the upper limit and the green line depicts the desired limit.

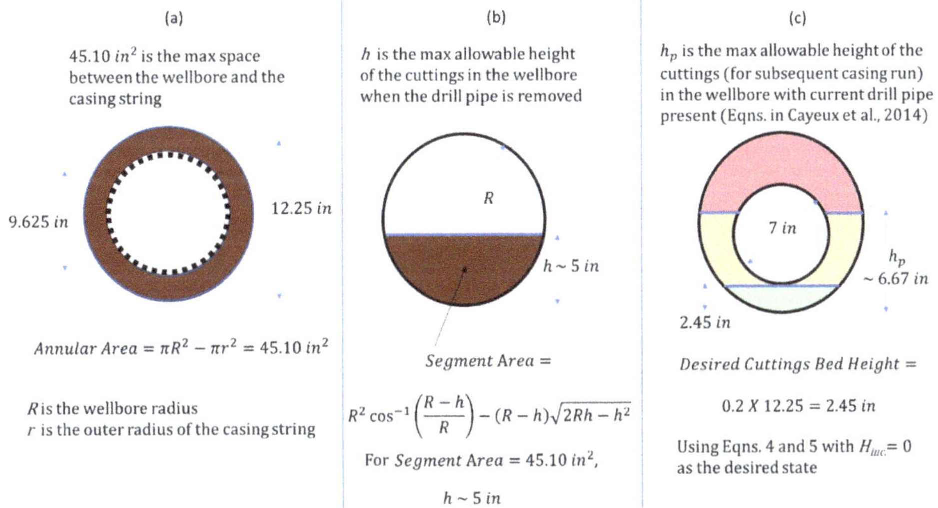

Figure 12. The theoretical limit of allowed cuttings bed height for the given well profile. (**c**) Equations in [31].

Figure 13b illustrates the drilling margin limits considering an uncertainty value (DF) of ten percent. The red-colored regions in the figure correspond to intervals with potential for well control issues; the green zone, on the other hand, is the desired ECD value (goal state). The orange region is a safe but non-optimal zone. These regions correspond to the values in Table 3 with the bands corresponding to Equation (7). Similarly, in Figure 13a the red-shaded region depicts the area above the equivalent limit height, while the green-shaded region corresponds to the desired state.

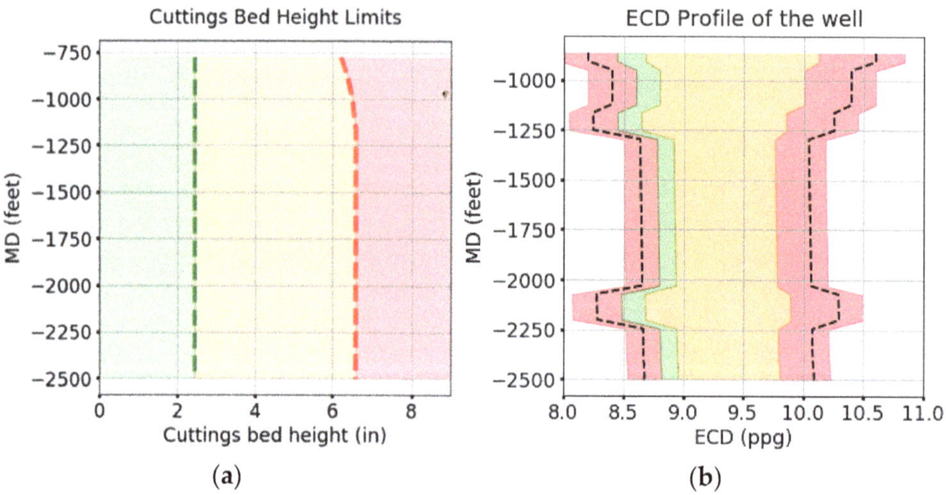

Figure 13. (**a**) Limits of the cuttings bed height profile. (**b**)The drilling margin for the well considering a ten percent uncertainty in the SL and FG limits (negative sign indicates downward depth into the sub-surface).

4.2. Performance Tracking of the System and Summary of Issues

To track the performance of the system during drilling operations, state transitions were monitored and associated rewards were calculated. State space was defined by dividing the well into five inclination-based segments, with the procedure as discussed in the previous section. The reward function was shaped based on state and action values and transitions. Determining the action space required the specification of the discrete values of the different control variables. Table 4 shows the number and range of values for the different variables.

Table 4. Value discretization of control variables.

Control Variable	Number of Discrete Values	Range of Values
Flow Rate (GPM)	10	[0, 1500]
Drilling ROP (ft/h)	10	[0, 900]
Drillstring RPM (rev/min)	10	[0, 150]
Mud Density (ppg)	5	[8.5, 9.7]
Mud Plastic Viscosity (cP)	5	[7, 42]
Mud Yield Point (lb./100ft^2)	5	[7, 42]

Table 5 shows the different weights assigned for reward calculations. For this system, the relative importance of the various state components is assumed to be the same. Similarly, relative penalties associated with altering the different action components are also assumed to be the same. Note, however, that the weights need to be determined by those who are familiar with the well, and weights used here are to provide a demonstration of the framework. The normalized reward component (W_{ar_norm}) has two values depending on the operation being tracked. Since no new hole is drilled during circulation operations, i.e., the ROP is zero; the weight is assigned a value of zero.

Table 5. Weight assignments for reward function shaping.

$W_s = [1, 1, 1, 1, 1, 1, 1, 1, 1]$	
$W_{ap} = [1, 1, 1, 1, 1, 1]$	
$W_{ar} = [0, 1, 0, 0, 0, 0]$	
$W_{s\_norm} = 0.50$	
$W_{ap\_norm} = 0.20$	
Drilling	Circulation
$W_{ar\_norm} = 0.30$	$W_{ar\_norm} = 0.00$

Figure 14 overlays the different normalized reward components calculated for decision epoch intervals of 5 min. For this system, the net reward tracks the state reward, since W_{s_norm} is significantly higher than the other weights. The reward value at the end of the drilling operation stabilizes to around 0.46. An increase in reward value to 0.68 at the end of the circulation cycle indicates an improvement in the hole condition. This improvement is also reflected in the state reward value, which increases from 0.51 to 0.59.

Figure 14. Normalized reward components versus decision epochs for the well.

Figure 15 shows the state of the system at the end of the drilling operation (12.25 inch hole section), which can be represented by Equation (22). The mud properties for drilling the last section of the well were: mud density of 9.06 ppg, PV of 11 cP, and YP of 36.5 lbf/100 ft^2. The final bed height was around 9 inches, which needed to be significantly reduced.

$$s_{TD} = \begin{Bmatrix} 0 \\ 3 \\ 4 \\ 4 \\ 1 \\ 1 \\ 1 \\ 1 \\ 1 \end{Bmatrix} \quad (22)$$

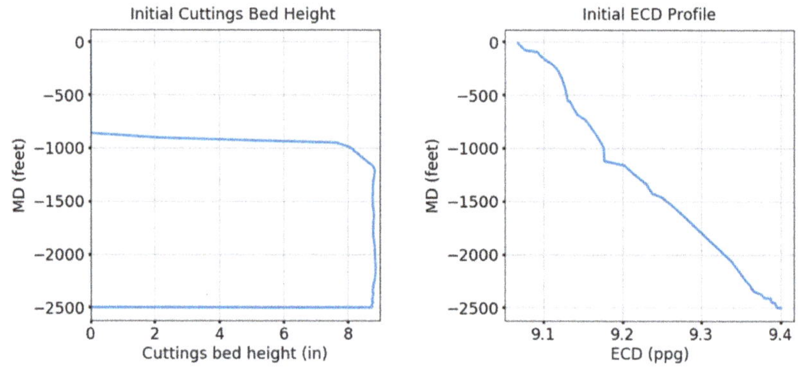

Figure 15. State of the borehole at the end of the drilling operation.

Thus, a 50 min circulation cycle to remove cuttings followed. Removal of cuttings is essential to ensure safe tripping operations without getting stuck, as well as to prepare the well for casing and cementing operations. The state of the system at the end of the circulation cycle is shown in Figure 16. As can be seen, the cuttings bed height was very close to the allowed limit, and therefore still non-optimal, leading to issues while running casing and subsequently cementing. The drilling crew did not have a process in place to evaluate various control actions; as mentioned before, this paper lays the foundational framework for that.

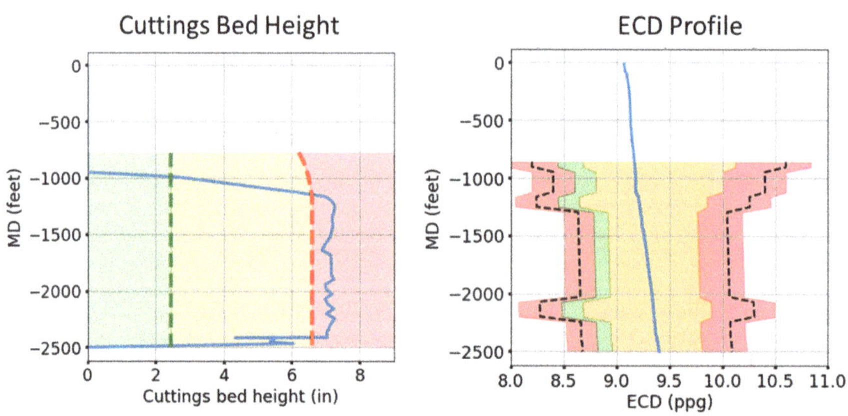

Figure 16. State of the borehole at the end of the circulation cycle.

4.3. Basic Action Planning

Here, we discuss the utilization of the hole cleaning planning system to simulate various state–action transition options. Multiple action sequences were simulated for a 50 min (10 decision epochs) circulation interval starting from the state of well at the end of the drilling operation, s_{TD} (represented by Equation (22)). The purpose of these simulations was to understand and quantify the effects of different action sequences on the hole condition, and in identifying a viable course of action. A viable action sequence would result in an improved wellbore condition, without compromising wellbore stability. Figure 17 details some of the simulated action sequences, where each action is structured in the form of Equation (10).

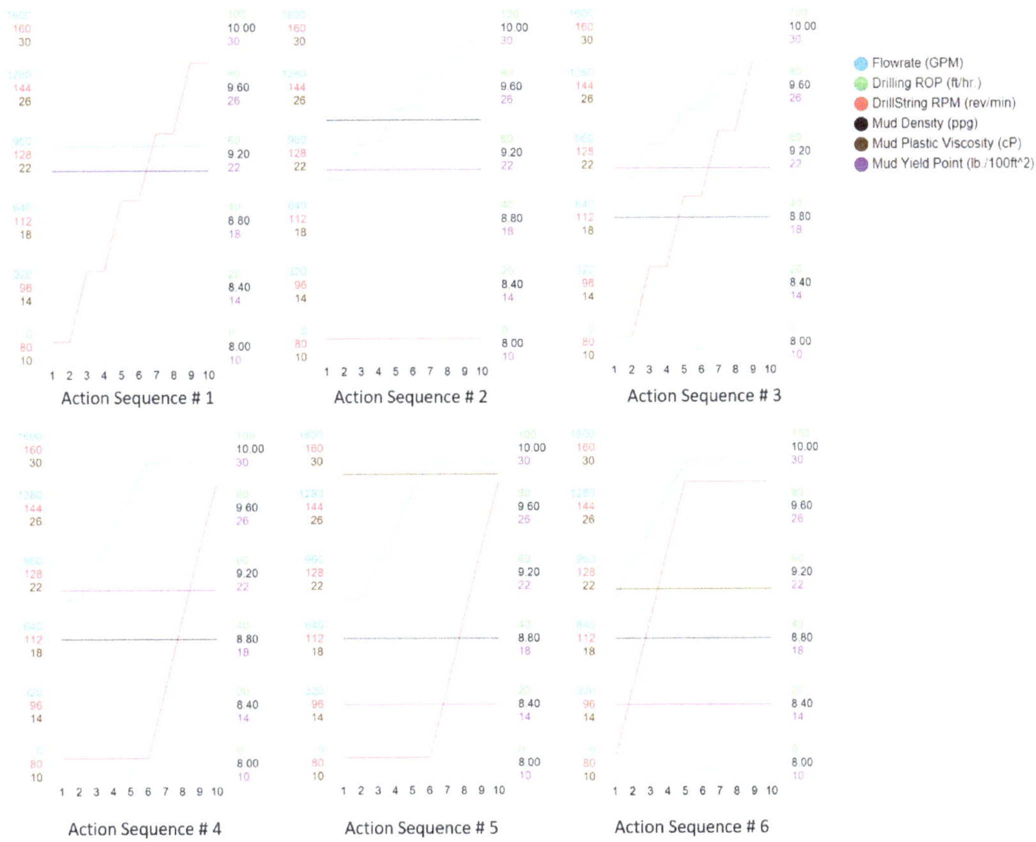

Figure 17. Six action sequences simulated in this work and discussed in detail in the text.

Changing the mud properties (density and rheology) is a time-consuming process; therefore, as it is highly impractical to change them in the middle of the circulation cycle, they are changed at the beginning of the action sequences. For the first four action sequences in Figure 17, PV value is increased to 21 cP, while the YP value is reduced to 21 lbf/100 ft^2. For the fifth action sequence, the PV and YP are changed to 28 cP and 14 lbf/100 ft^2, respectively. Finally, for the sixth sequence, PV and YP values are adjusted to 21 cP and 14 lbf/100 ft^2, respectively. For the first action sequence, the mud density remains unaltered (at 9.1 ppg); for the second sequence, the density is increased to 9.4 ppg. For the remaining four action sequences, the mud density is reduced to 8.8 ppg.

Implementing action sequence number one would have resulted in a slightly better hole condition than for the actual hole after circulation, as shown in Figure 18. Note that drillstring RPM was the only parameter that was changed (from 83 to 150 RPM) in this case. The reward obtained by the system would have stabilized at around 0.76, compared with 0.68 after the circulation cycle. Additionally, the normalized state reward would have been 0.67, as compared to 0.59 after the circulation cycle (which can be seen in Figure 14).

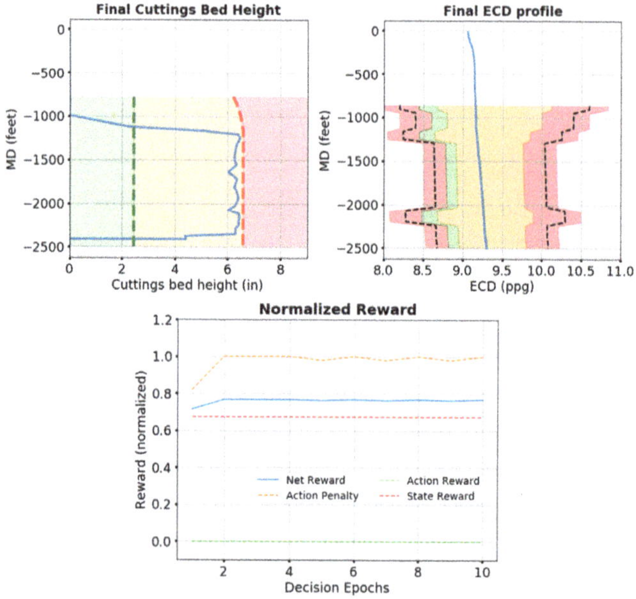

Figure 18. Predicted final system state and rewards after implementing action sequence one.

Action sequence two would have resulted in an even lower bed height; however, the predicted ECD value at greater depths nears the upper instability region, as shown in Figure 19. In this case, only the flow rate parameter is changed (from 833 to 1500 GPM) during the operation. The expected net reward for this case would have approached 0.77, and the state reward would have stabilized around 0.69.

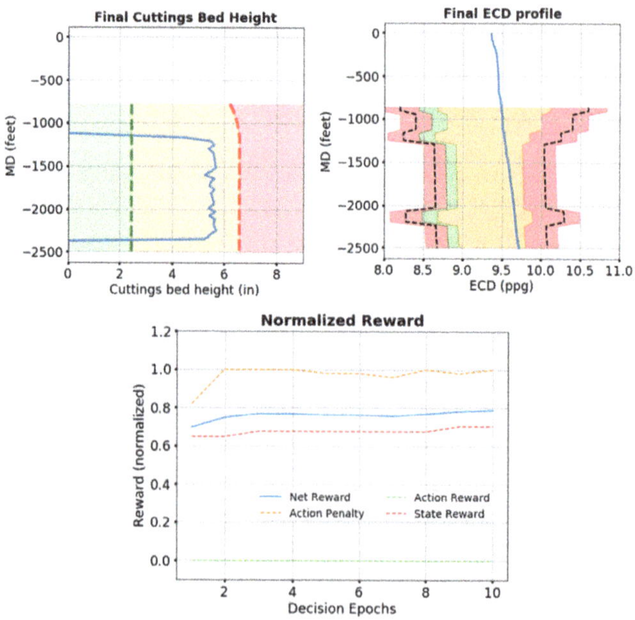

Figure 19. Predicted final system state and rewards after implementing action sequence two.

Figure 20 shows the predicted state after simulating action sequence number three. There would have been a significant reduction in the bed height (to approximately 4.5 inches), and the ECD value would be very close to the desired region. Both flow rate and drillstring RPM are changed during the operation; the flowrate varies from 833 to 1500 GPM, and the drillstring RPM from 83 to 150. The net and the state reward for this case would have been around 0.79 and 0.70, respectively.

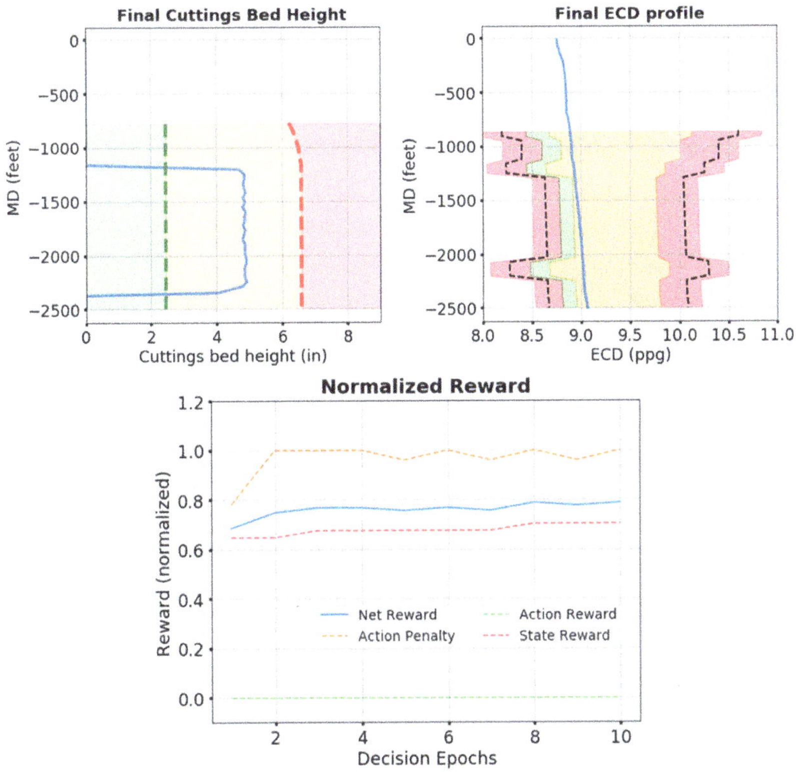

Figure 20. Predicted final system state and rewards after implementing action sequence three.

Figure 21 shows the final state of the system after simulating action sequence four. The expected net and the state reward for this case would also have been around 0.79 and 0.70, respectively. As for action sequence three, both the flow rate and the drillstring RPM are increased during the operation. The primary difference between the two sequences is the order in which the changes are suggested.

The output of the execution of action sequence five is depicted in Figure 22. This sequence would result in a substantially reduced bed height (under 3.5 inches) and an ECD value very close to the desired region. The net and the state reward values for this case are 0.82 and 0.76, respectively. In this case, both the flow rate and the RPM are increased, from 833 to 1500 GPM and 83 to 150 RPM, respectively.

Figure 23 shows the expected output of implementing action sequence six. This sequence would also result in a substantially reduced bed height (under 2.5 inches) and an ECD value very close to the desired region. The expected net and state rewards would be 0.86 and 0.81, respectively, the highest among all previous simulated trajectories. In this case, both the flow rate and the RPM are increased, from 833 to 1500 GPM and 83 to 150 RPM, respectively.

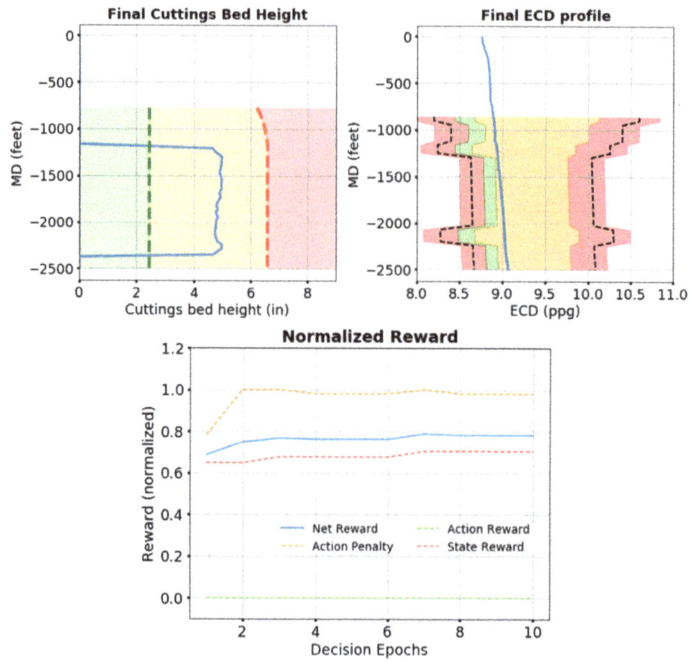

Figure 21. Predicted final system state and rewards after implementing action sequence four.

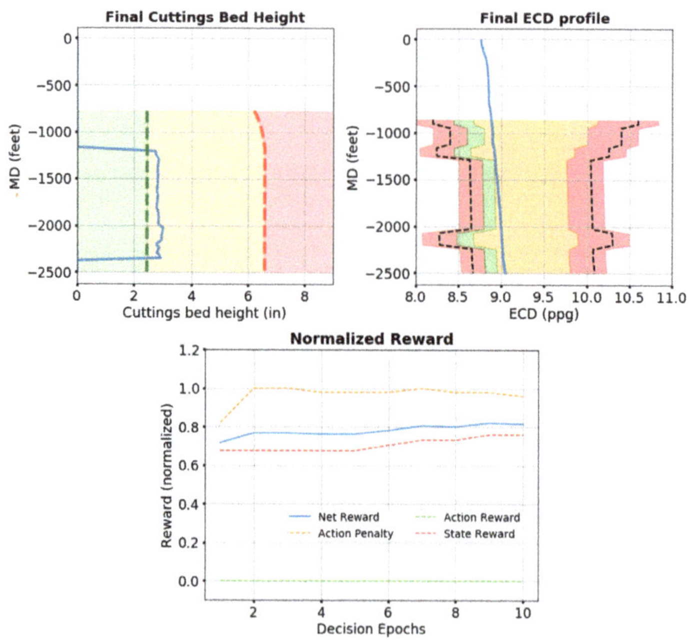

Figure 22. Predicted final system state and rewards after implementing action sequence five.

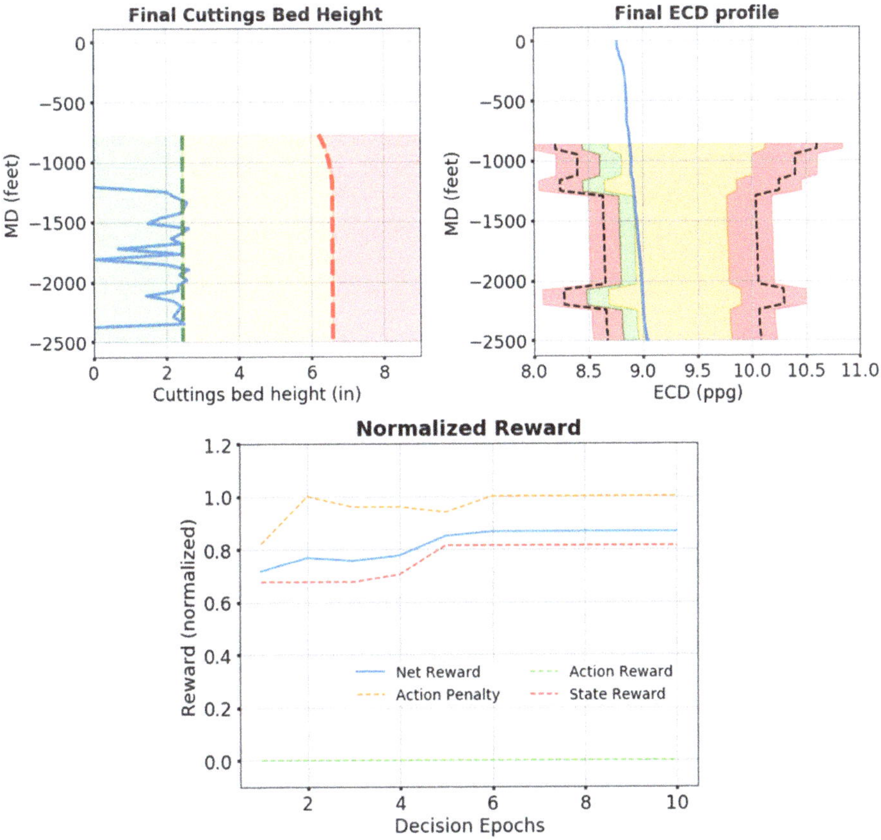

Figure 23. Predicted final system state and rewards after implementing action sequence six.

This example shown on field data clearly illustrates the potential for such a decision-making approach. Explicitly classifying the hole condition (state) and quantifying state–action transitions allows the evaluation and comparison of the different action sequences, which is a vital component in building an intelligent hole cleaning advisory system. Table 6 summarizes the net and the state rewards associated with the different action sequences. Here, action sequence number six has the best performance (as quantified by the highest final state reward) and also has the highest net reward (which depends on both the final state and the state–action transitions).

Table 6. Summary of the rewards associated with the different action sequences.

Action Sequence	Net Reward	Final State Reward
0 (Original)	0.68	0.59
1	0.76	0.68
2	0.77	0.69
3	0.79	0.70
4	0.79	0.70
5	0.82	0.76
6	0.86	0.81

5. Conclusions

This paper proposes and justifies setting up well construction operations as finite-horizon sequential decision-making systems for long-term planning. To the best of our knowledge, this is the first time a well construction operation has been structured as an MDP with carefully shaped rewards and an integrated multi-model digital twin, and subsequently utilized for evaluating action sequences. Such representation of well construction operations allows for an unbiased quantification and comparison of different scenarios. To summarize, this paper:

- Discusses the requirements and the steps in setting up such systems (i.e., formulating an MDP, defining the goal state, efficient reward shaping, and digitally twinning the underlying process) by detailing the development of a hole cleaning decision-making system.
- Discusses the importance of reward shaping for well construction operations to ensure frequent and suitable feedback, thereby facilitating effective policy design. It also demonstrates the use of a non-sparse normalized reward function designed for a hole cleaning system for performance tracking and simple action planning.
- Demonstrates the use of digital twinning for simulating various action sequences to track the state evolution and reward progression, thereby allowing ranking of the different sequences based on their long-term returns.

Furthermore, more directed search and planning methods such as simulation-based search can be deployed on these systems to enhance system performance considerably. In the longer-term, such decision engines can be incorporated into a rig's control system to help automate control of action variables such as RPM, flowrate, tripping speeds, and mud rheology. This will thereby enable the full automation of complex drilling operations such as hole cleaning. This, in turn, will avoid any human-centric biases and eliminate human mistakes in well construction operations and their negative consequences, such as stuck pipe incidents, induced well control and lost circulation incidents, etc. The outlined approach is therefore expected to have a potentially large positive impact on the efficiency, economics, and safety of future well construction operations.

Author Contributions: Conceptualization, G.S.S., P.A. and E.v.O.; methodology, G.S.S., O.E., P.A. and E.v.O.; software, G.S.S. and O.E.; validation, G.S.S., O.E. and P.A.; formal analysis, G.S.S.; investigation, G.S.S. and O.E.; resources, P.A. and E.v.O.; data curation, G.S.S. and P.A.; writing—original draft preparation, G.S.S.; writing—review and editing, O.E., P.A. and E.v.O.; visualization, G.S.S.; supervision, P.A. and E.v.O.; project administration, E.v.O.; funding acquisition, P.A. and E.v.O. All authors have read and agreed to the published version of the manuscript.

Funding: Funding was provided by the RAPID Industry Affiliate Program (IAP) at the University of Texas at Austin.

Acknowledgments: The authors would like to thank the members of the Rig Automation and Performance Improvement in Drilling (RAPID) consortium at the University of Texas at Austin for their support of this research. We would also like to acknowledge our colleagues in the research group for valuable brainstorming sessions.

Conflicts of Interest: The authors declare no conflict of interest.

Glossary

Unit conversion

1 m (m)	3.28 feet (ft)
1 meter/second (m/s)	11,811 feet/hour (ft/hr.)
1 psi	6894.76 Pa
1 ppg	119.83 kg/m^3
1 radian	57.2958 degrees
1 ft^3	0.02832 m^3
1 GPM	0.0000631 m^3/s

1 cP	0.001 Pa·s
1 lb./100ft^2	0.4788 Pa
1 lbs.	0.4536 Kg

Nomenclature

A	Action space
a_t	Action executed by the agent at time t
D_{o_k}	Outer diameter of the kth control volume segment (inches)
D_{TVD}	Total vertical depth (m)
ECD	Equivalent circulation density (pounds per gallon or ppg)
$ECD_k^{absolute}$	Absolute ECD value in the kth control volume segment (ppg)
ECD_{avg}	The average ECD value for an inclination interval (ppg)
$ECD_{inc.}$	Functional value of ECD in the inclination interval segment $inc.$
FG	Fracture gradient (ppg)
$flowrate$	Rate of flow of the drilling mud through the drillstring controlled by a positive displacement reciprocating mud pump on the surface (GPM)
GPM	Gallons per minute
g	Acceleration due to gravity (9.81 m/s^2)
H	Normalized cuttings bed height for an inclination interval
$H_{inc.}$	Functional value of the cuttings bed height in the inclination interval segment $incl.$
$H_k^{absolute}$	Absolute cuttings bed height in the kth control volume segment (inches)
H_k^{norm}	Normalized cuttings bed height in the kth control volume segment
$incl$	Inclination angle range (degrees)
N_{seg}	Number of control volume segments within an inclination interval segment
$n_{interval}$	Number of discrete values possible for a given control variable
P	Transition probability set
$P_{frictional_{pressure}\_loss\_D_{MD}}$	Frictional pressure drop in the annulus (Pa) at a measured depth H
$P_{hydrostatic\_D_{TVD}}$	Hydrostatic pressure (Pa) at a vertical depth of TVD_H
$P_{ss'}^a$	Transition probability of a system in the state s to the state $s'$when an agent executes action a
PV	Plastic viscosity (cP)
p_i	ith parameter component of the state vector
$p^g{}_i$	Goal state value of the ith parameter component of the state vector
R	Reward set
R_{ap}	Action transition-based penalty set
R_{ap_net}	Non-normalized action penalty for the hole cleaning system
R_{ap_norm}	Normalized action penalty for the hole cleaning system
R_{ar}	Action value-based reward set
R_{ar_net}	Non-normalized action reward for the hole cleaning system
R_{ar_norm}	Normalized action reward for the hole cleaning system
R_{net}	Net normalized reward function for the hole cleaning system
R_S	State transition-based reward set
R_{S_net}	Non-normalized state reward for the hole cleaning system
R_{s_norm}	Normalized state reward for the hole cleaning system
ROP	Rate of penetration or drilling rate (ft/hr.)
RPM	Drillstring rotation speed (revs. /min)
r_t	Reward received by the system at time t
S	State space
SL	Stability limit (ppg)
s_{goal}	Goal or desired state for the hole cleaning system
s_t	State of the system at time t
s_{TD}	State of the hole cleaning system at the well TD

TD	Total depth of the well (feet)
t	Time step or decision epoch
W_{ap}	Weight set associated with the action transition penalty
W_{ap_norm}	Weight value associated with the normalized action penalty
W_{ar}	Weight set associated with the action value reward
W_{ar_norm}	Weight value associated with the normalized action reward
W_s	Weight set associated with state transition reward
W_{s_norm}	Weight value associated with the normalized state reward
WOB	Weight on bit (Klbs.)
YP	Yield point (lb./100ft^2)
$\Delta N_{variable}$	Number of interval changes between consecutive actions
Δw	Difference between the FG and SL of the drilling window (ppg)
π	Policy
γ	Discount factor for return calculation

References

1. Gholami Mayani, M.; Rommetveit, R.; Oedegaard, S.I.; Svendsen, M. Drilling automated realtime monitoring using digital twin. In Proceedings of the Abu Dhabi International Petroleum Exhibition & Conference, Abu Dhabi, United Arab Emirates, 12–15 November 2018. [CrossRef]
2. Chan, H.C.; Lee, M.M.; Saini, G.S.; Pryor, M.; van Oort, E. Development and Validation of a Scenario-Based Drilling Simulator for Training and Evaluating Human Factors. *IEEE Trans. Hum.-Mach. Syst.* **2020**, *50*, 327–336. [CrossRef]
3. Ahmed, O.S.; Aman, B.M.; Zahrani, M.A.; Ajikobi, F.I.; Aramco, S. Stuck Pipe Early Warning System Utilizing Moving Window Machine Learning Approach. In Proceedings of the Abu Dhabi International Petroleum Exhibition & Conference, Abu Dhabi, United Arab Emirates, 11–14 November 2019. [CrossRef]
4. Forshaw, M.; Becker, G.; Jena, S.; Linke, C.; Hummes, O. Automated hole cleaning monitoring: A modern holistic approach for NPT reduction. In Proceedings of the International Petroleum Technology Conference 2020, IPTC 2020, Dammam, Saudi Arabia, 13–15 January 2020. [CrossRef]
5. Erge, O.; van Oort, E. Time-dependent cuttings transport modeling considering the effects of eccentricity, rotation and partial blockage in wellbore annuli. *J. Nat. Gas Sci. Eng.* **2020**, *82*, 103488. [CrossRef]
6. Kunath, M.; Winkler, H. Integrating the Digital Twin of the manufacturing system into a decision support system for improving the order management process. *Procedia CIRP* **2018**, *72*, 225–231. [CrossRef]
7. Kiran, B.R.; Sobh, I.; Talpaert, V.; Mannion, P.; Al Sallab, A.A.; Yogamani, S.; Pérez, P. Deep reinforcement learning for autonomous driving: A survey. *IEEE Trans. Intell. Transp. Syst.* **2021**, *23*, 4909–4926. [CrossRef]
8. Schwarting, W.; Alonso-Mora, J.; Rus, D. Planning and Decision-Making for Autonomous Vehicles. *Annu. Rev. Control Robot. Auton. Syst.* **2018**, *1*, 187–210. [CrossRef]
9. Kang, D.-J.; Park, J.H.; Yeo, S.-S. Intelligent Decision-Making System with Green Pervasive Computing for Renewable Energy Business in Electricity Markets on Smart Grid. *EURASIP J. Wirel. Commun. Netw.* **2009**, *2009*, 1. [CrossRef]
10. Olaf Blech, J.; Fernando, L.; Foster, K. Towards Decision Support for Smart Energy Systems based on Spatio-temporal Models. *arXiv* **2017**, arXiv:1705.03860.
11. Silver, D.; Schrittwieser, J.; Simonyan, K.; Antonoglou, I.; Huang, A.; Guez, A.; Hubert, T.; Baker, L.; Lai, M.; Bolton, A.; et al. Mastering the game of Go without human knowledge. *Nature* **2017**, *550*, 354–359. [CrossRef] [PubMed]
12. Silver, D.; Hubert, T.; Schrittwieser, J.; Antonoglou, I.; Lai, M.; Guez, A.; Lanctot, M.; Sifre, L.; Kumaran, D.; Graepel, T.; et al. A general reinforcement learning algorithm that masters chess, shogi, and Go through self-play. *Science* **2018**, *362*, 1140–1144. [CrossRef] [PubMed]
13. Arulkumaran, K.; Cully, A.; Togelius, J. AlphaStar: An Evolutionary Computation Perspective. In Proceedings of the Genetic and Evolutionary Computation Conference Companion, Prague, Czech Republic, 13–17 July 2019; pp. 314–315. [CrossRef]
14. OpenAI Five. 2018. Available online: https://openai.com/blog/openai-five/ (accessed on 1 July 2020).
15. LaValle, S.M. *Planning Algorithms*; Cambridge University Press: Cambridge, UK, 2006. [CrossRef]
16. Poole, D.L.; Mackworth, A.K. *Artificial Intelligence: Foundations of Computational Agents*, 2nd ed.; Cambridge University Press: Cambridge, UK, 2017.
17. Vodopivec, T.; Samothrakis, S.; Ŝter, B. On Monte Carlo Tree Search and Reinforcement Learning. *J. Artif. Intell. Res.* **2017**, *60*, 881–936. [CrossRef]
18. Saini, G.S. Digital Twinning of Well Construction Operations for Improved Decision-Making. Ph.D. Thesis, The University of Texas at Austin, Austin, TX, USA, 2020.
19. Silver, D.; Sutton, R.S.; Müller, M. Temporal-difference search in computer Go. *Mach. Learn.* **2012**, *87*, 183–219. [CrossRef]
20. Sutton, R.S.; Barto, A.G. *Reinforcement Learning: An Introduction*, 2nd ed.; The MIT Press: Cambridge, MA, USA, 2018.
21. Puterman, M.L. (Ed.) *Markov Decision Processes*; John Wiley & Sons, Inc.: Hoboken, NJ, USA, 1994. [CrossRef]

22. Cochran, J.J.; Cox, L.A., Jr.; Keskinocak, P.; Kharoufeh, J.P.; Smith, J.C.; Feinberg, E.A. Total Expected Discounted Reward MDPS: Existence of Optimal Policies. In *Encyclopedia of Operations Research and Management Science*; Cochran, J.J., Cox, L.A., Keskinocak, P., Kharoufeh, J.P., Smith, J.C., Eds.; Wiley: New York, NY, USA, 2011. [CrossRef]
23. Wiewiora, E. Reward Shaping. In *Encyclopedia of Machine Learning and Data Mining*; Springer: Berlin/Heidelberg, Germany, 2017; pp. 1104–1106. [CrossRef]
24. Jones, D.; Snider, C.; Nassehi, A.; Yon, J.; Hicks, B. Characterising the Digital Twin: A systematic literature review. *CIRP J. Manuf. Sci. Technol.* **2020**, *29*, 36–52. [CrossRef]
25. Saini, G.S.; Ashok, P.; van Oort, E. Predictive action planning for hole cleaning optimization and stuck pipe prevention using digital twinning and reinforcement learning. In Proceedings of the SPE/IADC Drilling Conference, Houston, TX, USA, 3 March 2020. [CrossRef]
26. Mitchell, R.F.; Miska, S.Z.; Cunha, J.; Kastor, R.; Aadnoy, B.S.; Eustes, I.A.; Sweatman, R.; Kelessidis, V.C.; Maglione, R.; Ozbayoglu, E.; et al. *Fundamentals of Drilling Engineering*; Society of Petroleum Engineers: Richardson, TX, USA, 2011.
27. Bourgoyne, A.T. *Applied Drilling Engineering*; Society of Petroleum Engineers: Richardson, TX, USA, 1986.
28. Sanchez, R.A.; Azar, J.J.; Bassal, A.A.; Martins, A.L. The Effect of Drillpipe Rotation on Hole Cleaning During Directional Well Drilling. In Proceedings of the SPE/IADC Drilling Conference, Amsterdam, The Netherlands, 4 April 1997. [CrossRef]
29. Sifferman, T.R.; Becker, T.E. Hole cleaning in full-scale inclined wellbores. *SPE Drill. Eng.* **1992**, *7*, 115–120. [CrossRef]
30. Baldino, S.; Osgouei, R.E.; Ozbayoglu, E.; Miska, S.; Takach, N.; May, R.; Clapper, D. Cuttings settling and slip velocity evaluation in synthetic drilling fluids. In Proceedings of the Offshore Mediterranean Conference and Exhibition, OMC, Ravenna, Italy, 25 March 2015.
31. Cayeux, E.; Mesagan, T.; Tanripada, S.; Zidan, M.; Fjelde, K.K. Real-time evaluation of hole-cleaning conditions with a transient cuttings-transport model. *SPE Drill. Completion* **2014**, *29*, 5–21. [CrossRef]
32. Erge, O.; Ozbayoglu, E.M.; Miska, S.Z.; Yu, M.; Takach, N.; Saasen, A.; May, R. The effects of drillstring-eccentricity, -rotation, and -buckling configurations on annular frictional pressure losses while circulating yield-power-law fluids. *SPE Drill. Completion* **2015**, *30*, 257–271. [CrossRef]
33. Gul, S.; van Oort, E.; Mullin, C.; Ladendorf, D. Automated Surface Measurements of Drilling Fluid Properties: Field Application in the Permian Basin. *SPE Drill. Completion* **2020**, *35*, 525–534. [CrossRef]
34. Saasen, A.; Løklingholm, G. The Effect of Drilling Fluid Rheological Properties on Hole Cleaning. In Proceedings of the IADC/SPE Drilling Conference, Dallas, TX, USA, 4 April 2002. [CrossRef]
35. Nazari, T.; Hareland, G.; Azar, J.J. Review of Cuttings Transport in Directional Well Drilling: Systematic Approach. In Proceedings of the SPE Western Regional Meeting, Anaheim, CA, USA, 27–29 May 2010. [CrossRef]
36. Karstad, E.; Aadnoy, B.S. Analysis of temperature measurements during drilling. In Proceedings of the SPE Annual Technical Conference and Exhibition, San Antonio, TX, USA, 5–8 October 1997. [CrossRef]
37. Bassal, A.A. The Effect of Drillpipe Rotation on Cuttings Transport in Inclined Wellbores. Ph.D. Thesis, University of Tulsa, Tulsa, OK, USA, 1995.
38. Duan, M.; Miska, S.; Yu, M.; Takach, N.; Ahmed, R.; Zettner, C. Critical conditions for effective sand-sized solids transport in horizontal and high-angle wells. *SPE Drill. Completion* **2009**, *24*, 229–238. [CrossRef]
39. Jalukar, L.S. A Study of Hole Size Effect on Critical and Subcritical Drilling Fluid Velocities in Cuttings Transport for Inclined Wellbores. Ph.D. Thesis, University of Tulsa, Tulsa, OK, USA, 1993.
40. Larsen, T.I.; Pilehvari, A.A.; Azar, J.J. Development of a new cuttings-transport model for high-angle wellbores including horizontal wells. *SPE Drill. Completion* **1997**, *12*, 129–135. [CrossRef]
41. Larsen, T.I.F. A Study of the Critical Fluid Velocity in Cuttings Transport for Inclined Wellbores. Ph.D. Thesis, University of Tulsa, Tulsa, OK, USA, 1990.
42. Naganawa, S.; Nomura, T. Simulating transient behavior of cuttings transport over whole trajectory of extended reach well. In Proceedings of the IADC/SPE Asia Pacific Drilling Technology Conference and Exhibition, Bangkok, Thailand, 11–13 October 2006. [CrossRef]
43. Rubiandini, R.S. Equation for estimating mud minimum rate for cuttings transport in an inclined-until-horizontal well. In Proceedings of the SPE/IADC Middle East Drilling Technology Conference, Abu Dhabi, United Arab Emirates, 8 November 1999. [CrossRef]

energies

MDPI

Article

Digital Twins for Real-Time Scenario Analysis during Well Construction Operations

Gurtej Singh Saini, AmirHossein Fallah, Pradeepkumar Ashok and Eric van Oort *

Cockrell School of Engineering, The University of Texas at Austin, Austin, TX 78712, USA
* Correspondence: vanoort@austin.utexas.edu

Abstract: Well construction is a complex multi-step process that requires decision-making at every step. These decisions, currently made by humans, are inadvertently influenced by past experiences and human factor issues, such as the situational awareness of the decision-maker. This human bias often results in operational inefficiencies or safety and environmental issues. While there are approaches and tools to monitor well construction operations, there are none that evaluate potential action sequences and scenarios and select the best possible sequence of actions. This paper defines a generalized iterative methodology for setting up a digital twin to address this shortcoming. Depending on its application, the objectives and constraints around the twin are formulated. The digital twin is then built using a cyclical process of defining the required outputs, identifying and integrating the necessary process models, and aggregating the required data streams. The twin is set up such that it is predictive in nature, thus enabling scenario analysis. The method is demonstrated here by setting up twinning systems for two different categories of problems. First, an integrated multi-model twin to replicate borehole cleaning operations for stuck-pipe prevention is developed and tested. Second, the creation, implementation, and testing of a twinning system for assisting with operational planning and logistics is demonstrated by considering the time it takes to drill a well to total depth (TD). These twins are also used to simulate multiple future scenarios to quantify the effects of different actions on eventual outcomes. Such systems can help improve operational performance by allowing more informed human, as well as automated, decision-making. Development of a system for well construction operations that integrates multiple sources of information with process and equipment models to quantify the system state and analyzes different scenarios by evaluating action sequences is a novel contribution of this paper. The approach presented here can be applied to the construction of digital twins for any well construction operation.

Keywords: digital twinning; well construction; drilling automation; hole cleaning; scenario analysis; well planning

Citation: Saini, G.S.; Fallah, A.; Ashok, P.; van Oort, E. Digital Twins for Real-Time Scenario Analysis during Well Construction Operations. *Energies* **2022**, *15*, 6584. https://doi.org/10.3390/en15186584

Academic Editors: F. Pacheco Torgal and Korjenic Azra

Received: 22 June 2022
Accepted: 30 August 2022
Published: 8 September 2022

1. Introduction and Background

Well construction is the process of drilling and completing a well on the subsurface for some specific purpose, such as extracting hydrocarbons, accessing geothermal energy, storing waste, or collecting rock samples. Drilling a well is typically an uncertain and high-cost operation due to the irreducible complexity of the geological environment deep below the earth's surface, leading to an inherent degree of variability and unpredictability in well construction operations. Drilling operations can pose safety hazards and therefore require careful monitoring of the equipment and process variables. Process or (and) equipment failures during any of the various steps involved in well construction can result in non-productive time (NPT) and associated trouble cost. Such failures could also lead to potentially catastrophic wellbore failures or well control issues, as was shown by the blowout on the Macondo well in the Gulf of Mexico in 2010 [1]. Monitoring, however, is only the first step to ensure safe and efficient well construction operations. Monitoring needs to be followed by appropriate actions, which today, are largely driven by the knowledge and experience possessed by the humans involved in the operation. Tools do not exist in well construction for scenario tracking and analysis in real-time, and this drastically reduces the chances of an optimum response. A major objective of this paper is detailing a process to perform real-time scenario analysis and to demonstrate the process with examples.

In general, recent advances in sensors and data processing technologies at the well construction rig site have now made access to operational and equipment data more reliable. This data, when used in conjunction with historical data and appropriate system models, can be used to improve the overall safety and efficiency of well construction. This improvement in safety and efficiency can be accomplished by developing what are referred to as 'digital twins'. Literature describes the digital twinning process as having three main components [2,3]:

- The real physical system (process or equipment) and the associated sensors for data collection;
- Virtual representation of integrated models of the system to digitally replicate its behavior to enable performance tracking and scenario analysis;
- A data-stream to exchange data and information between the real system and its twin.

Note, however, that not everyone applies the above strict definition when calling a system a digital twin. The capability for scenario analysis is missing in the digital twins published in the well construction digital twin literature, and that is the shortcoming that this paper will address. Given that no such scenario analysis digital twin exist in the well construction literature to date, we briefly introduce one from a different domain as an example. Figure 1 demonstrates the three components of digital twinning through an application for Formula One racing strategy planning and management. The physical system is the race car interacting with other competitors' race cars and the racetrack. The race car itself has over 300 sensors that stream data about different aspects of the car (such as engine and tire temperatures, racetrack conditions, etc.) in real-time to the team's control room at their headquarters [4,5]. This data is then fed into an integrated multi-model system (the digital twin) representing the car and its interaction with the environment, to run many thousands of simulations per second. These simulations analyze multiple scenarios and decide on the best race strategy going forward. Furthermore, throughout the race weekend, starting from the practice sessions on Friday until the final race on Sunday, many hundred gigabytes of data are collected. This data is used to update the understanding of the environment and the car, thereby improving the digital twin's performance for future racing events (a process known as enrichment).

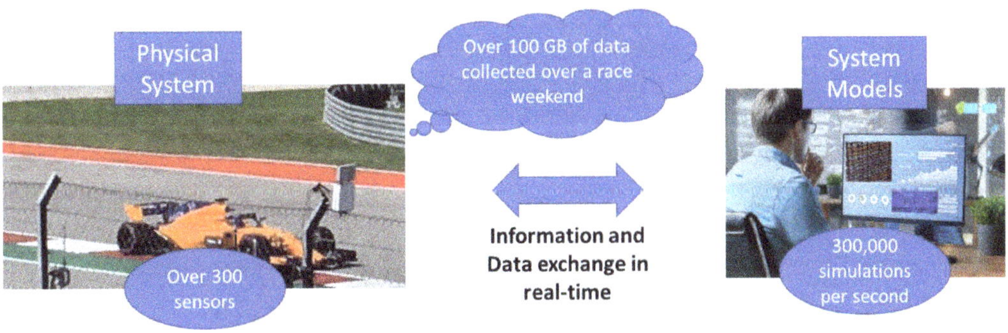

Figure 1. The three components of digital twinning demonstrated through application in auto racing strategy planning, based on information from [4].

Although the domains are entirely different, Formula One racing is actually an appropriate analog for well construction when it comes to the topic of digital twinning and pro-active scenario analysis/action planning. In both cases, we have a dynamic system (the car during racing; the rig and the well during well construction) that is outfitted with sensors collecting real-time data at various frequencies. For rigs and wells, these are the surface sensors and downhole measurements provided by an instrumented drillstring. Whereas the data in racing is used to inform the racing headquarters to take proactive actions, the rig and well data can be used for the same purpose by the rig team and/or a remote monitoring team. In both cases, the ultimate goal of digital twinning is twofold. First, the aim is performance optimization, which in racing is related to optimized speed and reduced lap time, whereas in drilling this is typically the optimized rate of penetration (ROP) and reduced time per footage drilled (e.g., a typical indicator used is days per 1000 ft drilled). Drilling inefficiently typically leads to so-called invisible lost time (ILT). Secondly, in both cases it is important to prevent unforeseen and catastrophic events. In racing, these are engine failures, tire punctures because of non-optimum tire wear management strategy, and avoidable crashes. By analogy, in drilling, one aims to prevent well control incidents and lost circulation events, as well as borehole instability and stuck pipe incidences. The latter can be a major source of non-productive time (NPT) and trouble cost on any well construction project. Excessive ILT and NPT often lead to well construction projects not meeting economic and environmental, safety, and governance (ESG) goals. Their avoidance and minimization are therefore major motivators for considering and implementing digital twinning with associated scenario analysis and action planning.

A very valuable property of digital twins is their ability to be updated in real-time (RT) based on the most recent data. This feature allows for their application in building and applying RT decision-making systems [6–8]. Apart from Formula One racing simulations and strategy planning, the concept of digital twinning has been successfully implemented in various other industries across a variety of applications. Some practical applications include twinning wind turbines for digital wind farm management [9,10], twinning gas turbines to predict failures [9,11], aerospace asset maintenance [12,13], product lifecycle management [14–16], and developing integrated models for weather forecasting [17,18].

In this paper, we first document the state of the art in well construction monitoring and support. We can call these weak digital twins, as they all lack real-time scenario analysis capabilities. Then we propose a generalized methodology for constructing a well construction digital twin that allows for scenario analysis. We subsequently illustrate the process by two applications, hole cleaning and logistical optimization, as representative examples of well construction processes.

2. Digital Twins in the Well Construction Domain

Well construction is guided by an engineering plan called the drilling program, which includes information about the current well's directional plan, the bottom hole assembly (BHA) design, subsurface geology, fluid and hydraulics plan, well control measures, casing design, and cementing plans, as well as relevant information from offset wells [19]. Apart from specifying making the hole with a drill bit attached to a hollow flexible drillstring and rotated via a motor at the surface (top drive or rotary table), the drilling program guides many other processes, such as [20–22]:

- tripping the drillstring in and out of the borehole using a drilling rig's hoisting system (drawworks, hook, traveling block, etc.)
- making or breaking connections, to add or remove discrete drillstring elements called stands using pipe handling systems (hydrarackers and hydratongs)
- directional drilling, to control the trajectory of the well using either mud motors (rotary and slide drilling operating modes) or rotary steerable systems
- circulating the drilling fluid (or drilling mud), to maintain its equivalent circulation density (ECD) within a drilling margin using the rig's hydraulics system (reciprocating positive displacement mud pumps, standpipe, etc.)
- hole cleaning, to remove the drilled cuttings from the borehole by controlling the mud flowrate, mud properties, and surface drilling parameters

- well logging, to make high-fidelity downhole measurements, such as formation evaluation logs, near-bit drilling parameter measurements, and directional surveys using measurement and logging while drilling (MLWD) tools.

Throughout the process, various drilling parameters, such as weight on bit (WOB), drillstring rotation speed (RPM), applied torque, and standpipe pressure, are measured directly or calculated using a suite of sensors installed on the drilling equipment [23]. Once a section of the well has been drilled, the next steps involve securing it by running casing and then cementing the casing in place [20].

As can be seen, well construction is a highly involved multi-step process wherein each step has multiple co-occurring sub-processes and various systems interacting with each other. Over the past few decades, much research has been conducted to manage this complexity. For instance, process modeling [24–26] and advanced data analytics [27–31] have assisted in the monitoring of well construction operations for improving its efficiency and safety. Table 1 shows the many areas within the well construction domain wherein progress has been made. While all of these approaches use models (physics or data-based) and partially satisfy the requirements for being called a digital twin, none of them have a framework for scenario analysis, which is an important component of digital twinning (as shown in the example on race strategy planning) and is essential to be able to optimize the selection of appropriate future actions. The goal of this paper is to advance the state of the art in the digital twinning of well construction operations by formulating a step by step process for building digital twins that are also capable of real-time scenario analysis.

Table 1. Current applications for improving operational efficiency and safety and in well construction.

Application Type	Value Addition	Potential Applications
Non-productive time identification (diagnostics) and prevention	• Real-time process monitoring and diagnostics • Improved and faster decision making • Event detection • Drilling optimization	• Mitigating well control issues (e.g., kicks and losses) [27,32,33] • Drilling dysfunction identification and mitigation [34,35] • Wellbore quality degradation identification and mitigation [36–38] • Hole cleaning and stuck pipe prevention [39,40]
Prognostics for equipment failure and detection	• Reduce downtime associated with operational failures • Early diagnosis of failures • Condition-based maintenance (CBM) of drilling tools and equipment • Optimizing drilling tool performances within their operating windows	• CBM of subsea blowout preventer (BOP) pipe rams [41] • Estimating the remaining useful life of downhole components (MLWD tools, mud motors, and drill pipes) [42–45] • CBM of surface equipment, such as top drive, mud pumps, drawworks, and pipe handling equipment [46–48] • CBM of artificial lift systems [49,50]
Invisible lost time evaluation and mitigation	• Improve operational efficiency for various processes • Key performance indicator (KPI) monitoring and tracking • Identification of performance gaps in different processes	• Rig crew performance monitoring and improvement [51–53] • Defining KPIs for quantifying rigs' operational performance for various auxiliary tasks [54–56]

Table 1. *Cont.*

Application Type	Value Addition	Potential Applications
Logistics and planning	• Forecasting and planning • Ability to run what-if scenarios on multiple processes • Delivery of safe, cost-effective wells • Improved well planning and more efficient oilfield development	• Predicting bit degradation and RT downhole drilling fluid properties for optimizing drilling parameters [57–59] • Replication of well construction operations such as drilling, casing, tripping, and cementing (short- and long-term operational planning) [37,60]
Training and development	• Training drilling crew on specific technology • Preparing college students • Testing new technology in a safe environment	• Replicating surface drilling operations (and equipment) and downhole process behavior [61–64] • Developing and testing drilling programs and operation manuals [65,66] • Beta-testing software before rolling them out to the field [67]

3. Digital Twinning Methodology

The methodology developed consists of the following steps:

a. The first step is to determine the objectives of the system being twinned. For instance, the goal of an 'RT drilling dysfunction mitigation twin' would be to detect signatures of dysfunctions (such as vibrations, bit balling, stick-slip), and suggest drilling parameters (such as RPM, WOB, torque, flowrate) to optimize the rate of penetration (ROP) while minimizing dysfunctions.

b. The next step is to set up the twinning system and requires a compromise between the desired and the practical outputs. This compromise depends on the system's objectives, the available data, and the computation capacity for model implementation. The following three steps are iterated repeatedly:

   1. Based on the twin's objectives, the metrics or outputs required to quantify the state of the system are identified. For the dysfunction mitigation twin, parameters such as mechanical specific energy (MSE), depth of cut, and stick-slip index, would be required to quantify the drilling performance and dysfunctions.
   2. Once these metrics are recognized, the next step is the identification and implementation of the different models (data- or physics-based, or a combination of both) to calculate them, i.e., building the digital twin. The selection of these models depends on the application type and the associated temporal and computational constraints. For instance, since online action planning applications require balancing accuracy with runtime speed, analytical model implementations may be acceptable in certain cases. On the other hand, for offline planning or more detailed post-job analysis, wherein computation time or capacity may not be a constraint, numerical models may be used.
   3. Subsequently, based on the input requirements of the various models (twin), the necessary data streams are identified and aggregated; the data is filtered and processed (data wrangling) and integrated with the twin. If the available data stream does not support the calculation of a desired output component, the required system outputs or the underlying models are tweaked accordingly.

c. The twinning system thus set up is then used for performance tracking and scenario analysis. The goal of scenario analysis is to make predictions about future system states and to understand the different driving factors and how they might influence the system behavior [68].

Figure 2 illustrates the different steps and the cyclical relationships between them. The following two sections demonstrate the application of this methodology on two important well construction tasks: one for optimizing hole cleaning operations (NPT diagnostics and mitigation) and the other for predicting and optimizing operational times to evaluate the time remaining to well total depth (logistics and planning).

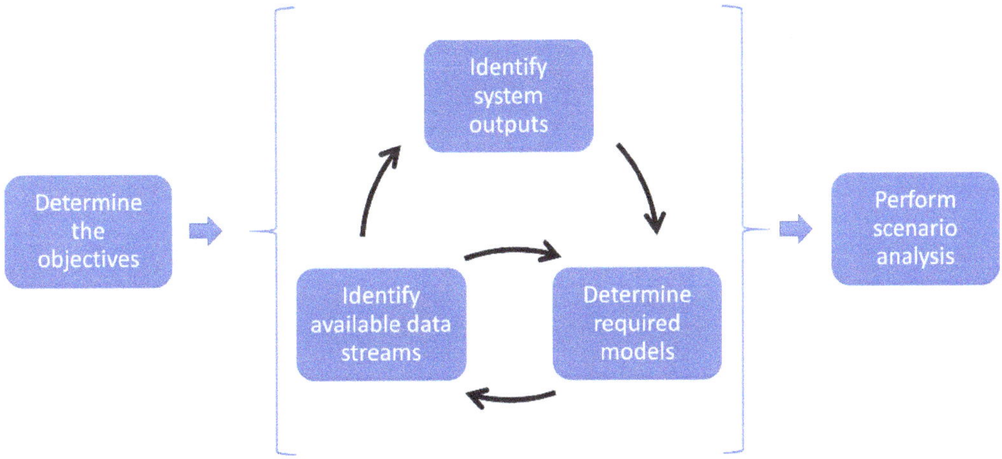

Figure 2. Digital twinning methodology.

4. Development of a Digital Twin for Hole Cleaning Advisory

In this section, we will demonstrate how the process identified in the previous section is used for developing a hole cleaning twin. Inadequate or poor hole cleaning can lead to a series of costly drilling issues such as stuck pipes, formation damage, downhole tool damage, reduction in drilling speed, difficulty tripping out of the hole, or issues while running casing [69]. Well problems such as stuck pipes and wellbore instabilities can halt drilling operations for days and cost millions of dollars in NPT [70]. Therefore, it is of value to digitally twin the hole cleaning system to understand the progression of the borehole condition in real-time and track the outcomes of different actions. Such a digital twin can serve as a baseline for developing an automated intelligent decision-making system for hole cleaning.

4.1. Determining the Objective of the Twinning System

The primary objective of hole cleaning is to remove solids from the wellbore to the extent that different operations such as drilling, tripping, casing, and cementing can be performed efficiently and safely. The solids may include cuttings, cavings, metal shavings, or cement. The following sections discuss the construction and application of a twin for hole cleaning monitoring and action planning.

4.2. Building the Digital Twin

Since a well can be many thousands of feet in depth, hole cleaning measures directed towards one interval of the well may not necessarily result in effective hole cleaning in another interval. Thus, it is essential to quantify and monitor the condition of the borehole in real-time for the entire well.

4.2.1. Identification of the System Outputs

The purpose of modeling cuttings transport is to understand the process of removal of solids (cuttings, cavings, or metal shavings) from the borehole. Removal of solids is critical to ensure that different well construction operations (such as tripping, casing, and cementing) are performed safely and efficiently. The solids in the borehole can be quantified using a combination of the cuttings bed height and the concentration of cuttings in the flow. Furthermore, not maintaining the ECD within the drilling margin can result in wellbore instability issues, such as kicks or lost circulation events. Therefore, the following metrics may be used to describe the state of the borehole for a hole cleaning system [24,71]:

- Height of the cuttings bed at different depths along the wellbore;
- Concentration of cuttings in suspension along the wellbore;
- ECD along the length of the wellbore;
- The average friction factor of the wellbore.

The mechanics of cuttings transport are different in different sections of the wellbore; they depend on wellbore inclinations and wellbore geometry. The cuttings generated during the drilling of non-vertical sections (borehole inclination angles greater than 30 degrees) can start building on the low side of the borehole, resulting in a cuttings bed of some height. More energetic cuttings that do not settle are suspended and carried up hole by the drilling mud [72]. ECD at any depth is calculated as the gradient of the total pressure drop at that depth. The total pressure drop is calculated by summing the hydrostatic head exerted by the drilling mud to the circulating frictional pressure losses in the annular space between the drillstring and the wellbore [20]. Torque and drag, respectively, are rotational and axial forces acting between the drillstring and the casing or the formation. These forces are primarily caused by a combination of the side forces and frictional forces acting on the drillstring [73]. The torque and drag forces are used to estimate a friction factor for the wellbore.

4.2.2. Determining the Required Models

The next step is to identify the models required to calculate these metrics. Analytical implementations, to balance accuracy and runtime speed, of the following models need to be adapted, integrated, and implemented to build the twin [74]:

- Hydraulics model to calculate the frictional pressure losses and ECD throughout the well;
- Torque and drag model to evaluate the friction factor of the well using rotary, slack-off, and pick-up weights;
- Cuttings transport model to estimate cuttings bed height and cuttings concentration in the flow throughout the well.

For hydraulics, an adaptation of the analytical narrow slot approximation model for yield power law (YPL) fluids proposed by Erge et al. [25] was implemented with some modifications to account for the effects of cuttings bed height and cuttings concentration in flow on the mud density. For torque and drag, an analytical soft string model was implemented based on Aadnoy et al.'s work [75,76]. For cuttings transport, a quasi-transient analytical model was derived and implemented based on a combination of concepts from work by Larsen et al. [77], Jalukar [78], Bassal [79], Duan and Miska [72], Rubiandini [80], and Naganawa et al. [81], along with further modifications to account for mass conservation and the physics of fluid flow.

4.2.3. Identification of the Data

The next step is the identification and aggregation of the required data streams to implement the above models. This data can be grouped into the following categories:

- Information about the well profile, obtained from pre-drill well plans and near real-time directional survey data;
- Details about the BHA such as the geometry and unit weights of the individual components, obtained from pre-drill well plans;
- Casing information, such as geometry and casing setting depths for different casing strings and liners, derived from wells' operational data and pre-drill well plans;

- Real-time drilling data obtained from surface and downhole sensors, such as drillstring RPM, surface torque, drilling ROP, flowrate, hookload, tripping velocity, mud rheology, downhole near-bit sensor data, and downhole pressure data.

Datasets comprising the above-mentioned field data were made available for this study by an oil and gas operator. Data from an unconventional well drilled directionally in a shale reservoir is used in the following to illustrate our digital twinning approach to hole cleaning.

Figure 3 summarizes the structure of the proposed digital twin for the hole cleaning system. Initially, well profile and well geometry information are utilized for segmenting the well into smaller discrete control volumes. These segments represent any changes in well dimensions (changes in inner or outer diameter) or different survey intervals. Solving each control volume with time and depth generates an estimate of the state of the system (hole condition quantification metrics for the well). The actual state of the system is determined by utilizing downhole tools, such as a pressure while drilling (PWD) tool for calculating ECD and near-bit MLWD tools for estimating the friction factor. The differences between the actual and the predicted state values can then be used to update the system models.

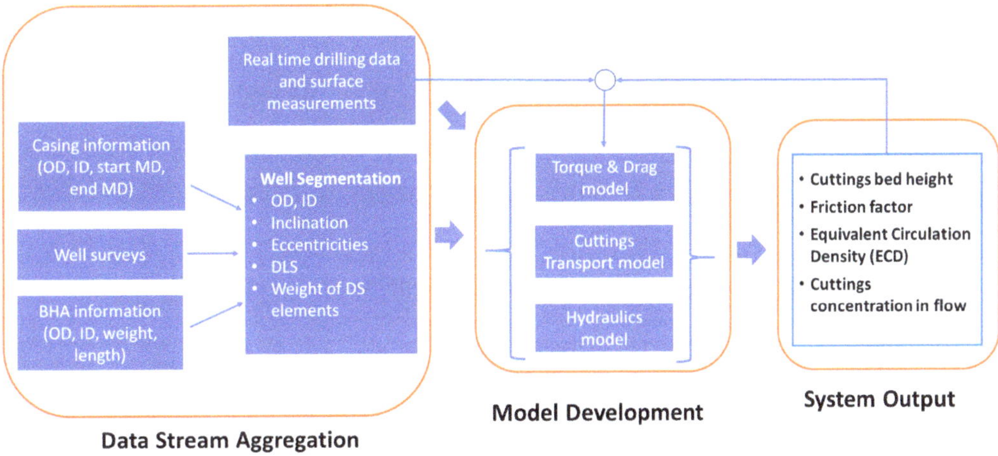

Figure 3. Final structure of the proposed digital twin for the hole cleaning system, see also [82].

4.3. Application of the Twin for Performance Tracking and Scenario Analysis

The methodology was applied for a set of wells to develop a variant of the proposed hole cleaning digital twin, as shown in Figure 4. The data stream did not include tripping information such as tripping rig states or tripping velocities; therefore, the twin's desired output was adjusted not to incorporate the friction factor. Furthermore, since the available data streams did not include downhole tool data (no PWD or near-bit MLWD data), no immediate feedback about the actual system state was possible. This resulted in a digital twin with only the cuttings bed height, ECD, and cuttings concentration in the flow as the outputs. The developed twin was used as a tool for performance tracking by continuously evaluating the state of the wellbore. This twin also offers the capability of performing scenario analysis and action planning.

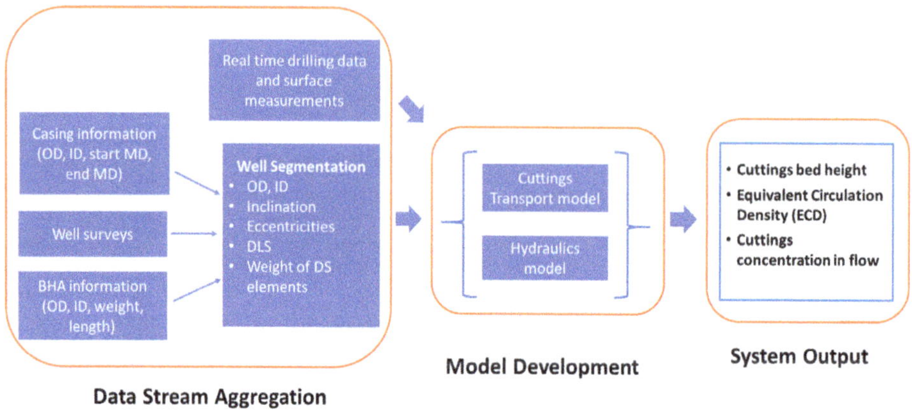

Figure 4. The developed variant of the digital twin for the hole cleaning system, see also [82].

4.3.1. Performance Tracking

The digital twin was deployed to replicate the hole cleaning performance of multiple wells. For one well, Figure 5 represents the state of the borehole after tracking its evolution from the surface in 10 min intervals. In this case, the well has been drilled to a hole depth of 10,000 feet. In the prior 10 min interval, the drilling operation involved slide drilling at an ROP of 93 feet per hour, with an average surface rotational speed of 14 RPM (slide drilling mode) and a flowrate of 598 gallons per minute (GPM). Properties of the drilling mud for this system are stated in the figure. For this system, the table (inset Figure 5) shows some of the required initializations regarding cuttings properties, thermal properties, and thermal gradients. The plots of ECD, cuttings concentration in the flow, and cuttings bed height versus hole depth are displayed in the figure. In the horizontal section at depths greater than 7500 feet, there exist cuttings beds approximately 3 to 4 inches high (in a 7.875-inch open hole). The ECD at 10,000 feet is around 13.1 ppg for drilling mud with a surface density of 12.5 ppg. The cuttings concentration in the flowstream for most of the well is under 5 percent. A relatively high cuttings concentration near the bit is due to cuttings that have not yet been deposited.

4.3.2. Scenario Analysis

The system from its current state (Figure 5) was used to simulate multiple scenarios or actions (a combination of changes in action variables, such a flowrate, RPM, and ROP) to evaluate the outcomes of each. Figure 6 describes the different actions and their predicted consequences on the hole cleaning system state. For the system at 10,000 feet, its state is represented by cuttings concentration, cuttings bed height, and ECD. Then, five different actions over the next 10 min interval were simulated using the digital twin. As expected, different actions lead to different future system states, some more desirable than the others. Some actions lead to minimal improvement in hole condition; actions 1 and 2 result in only a slight reduction in the cutting bed height and the ECD. Some actions (such as actions 3 and 4) lead to an improved hole condition by reducing the bed height while minimally affecting the ECD. On the other hand, more aggressive actions (such as action 5) result in a significantly reduced bed height, but simultaneously result in a high ECD value. Although such actions assist with cuttings removal, an increase in the ECD closing the distance to the fracture gradient can result in induced wellbore fracturing and lost circulation issues.

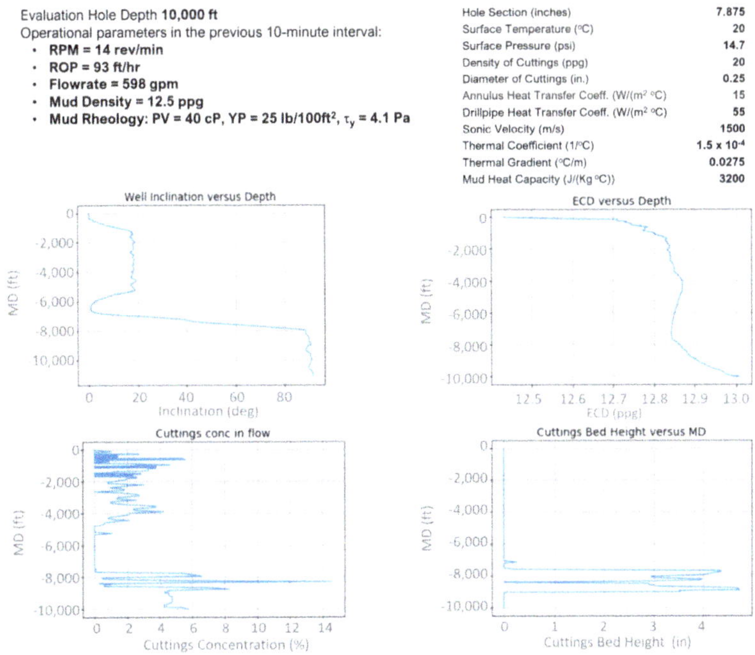

Figure 5. State of the system at a 10,000 foot hole depth after tracking its evolution from the surface.

Figure 6. Scenario analysis for predictive action planning.

This concept of a single-step scenario analysis can easily be expanded to multi-step analysis, as shown in Figure 7. This digital twin, therefore, can be used to evaluate multiple action sequences over several time steps (evaluation intervals) to find the optimal way of traversing through the system, i.e., coming up with an optimal plan (action planning). Such short- and long-term scenario analysis is only possible with a predictive model and is a prime requirement for any twin.

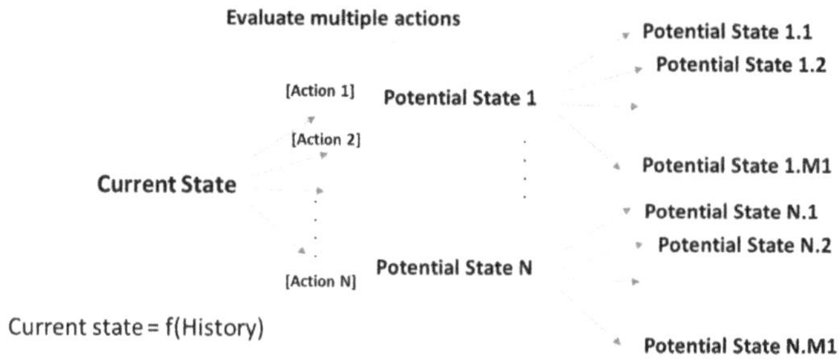

Figure 7. Simulating multiple action sequences.

5. Development of a Digital Twinning System for Optimizing Logistics and Planning

In this section, we will demonstrate how the methodology described in Section 3 is used for developing a twin for optimized logistics and well planning. During well construction, the drilling of a well section or interval is generally followed by circulation cycles for hole cleaning, then casing and cementing operations, and in some cases, by clean-out operations and casing/formation integrity tests. These steps are repeated until the objectives laid out in the drilling program are met, which includes drilling the well to its planned total depth (TD). The unpredictability in reaching TD affects the overall logistics and completion schedule of the well, which, in turn, affects the drilling program. It is therefore essential to be able to estimate and optimize the times required for the different well construction operations.

5.1. Determining the Objective of the Twinning System

This section discusses the development of a digital twin for predicting, updating, and optimizing in real-time, the time remaining to drill a section or reach well TD, referred to as 'time to TD'. This time prediction can be made using information such as:

- The operational performance of the offset (historical) wells;
- The current position of the drill bit relative to the well plan;
- The operational performance of the current well up to its present depth;
- Degradation in performance of drilling tools, equipment, or drill bit.

5.2. Building the Digital Twin

The following sections describe the different steps involved in building this digital twin.

5.2.1. Identification of the System Outputs

The first step in estimating time to TD is the identification of components that comprise this time and the factors affecting these individual components. Time to TD from the standpoint of rig activities is the sum of anticipated times spent performing the different operations (or time spent in different rig states). The following generally comprise time to TD:

- Total anticipated on-bottom drilling time ($T_{Drilling}$);
- Total time for acquiring the directional surveys ($T_{Surveys}$);
- Total expected time for making connections ($T_{Connections}$);

- Total tripping time (including times for tripping in, tripping out, making up and laying down BHAs) ($T_{Tripping}$);
- Total circulation time ($T_{Circulation}$);
- Total miscellaneous time to capture the times for all other activities and operations not considered above ($T_{Miscellaneous}$).

The anticipated on-bottom drilling time further depends on the following:

- Total distances to be drilled individually by rotary and slide drilling;
- The remaining formations to be drilled through and the approximate depths of each;
- The estimated average drilling speeds (for both slide and rotary drilling modes) through each remaining formation.

The approximation of the total connection and the total survey times requires estimating the remaining number of connections and surveys until the planned TD is reached. Similarly, metrics are designed to extrapolate the estimated circulation and tripping times until the planned TD is reached. To summarize, the time to TD can be calculated as follows:

$$T_{well\ TD} = T_{Drilling} + T_{Surveys} + T_{Connections} + T_{Tripping} + T_{Circulation} + T_{Miscellaneous} \quad (1)$$

5.2.2. Determining the Required Models

The next step is the identification of the models required to calculate the individual components. Knowledge derived from offset wells is used in combination with the current well's performance, along with the information about planned tasks, to anticipate future events and estimate the different time components. The following models need to be implemented to accomplish this:

- Rig state detection engine, to classify the real-time data into different operational categories such as on-bottom drilling (slide and rotary drilling), tripping (in and out), circulating, reaming, and making connections;
- Slide and rotary drilling predictive models, to predict relative amounts of slide and rotary drilling required until the TD, based on the well plan and the learnings from offset wells;
- Predictive data-based models, to estimate different component times by utilizing an adaptive weighed scheme for combining statistics obtained from real-time data and offset well data, as shown in Figure 8. For example, the connection times between the more recent time interval $t = t_2$ to $t = t_{current}$ would likely be given a higher weight in comparison to the connection times between the interval $t = t_1$ to $t = t_2$ when coming up with an expected connection time in the future.

In the adaptive weighted scheme, the most recently acquired data has the highest weight (W1), while historical data is assigned the lowest weight (W4). The sum of all the weights is 1, and the various times (t_1, t_2, etc.) are not fixed; instead, they are functions of the number of data points collected thus far.

5.2.3. Identification of the Data

The following data streams are required for implementing this digital twin:

- Real-time drilling data obtained from the surface sensor measurements, specifically data channels such as drillstring RPM, applied WOB, drilling ROP, flowrate, standpipe pressure, tripping speed, and surface torque and hookload measurements;
- Geologic formation top information (anticipated start and end depths of different formations);
- BHA and bit information, obtained from the well plan;
- Well trajectory information, derived from the well plan and the well survey data, quantified using azimuth and inclination angles at different depths along the well;
- Descriptive statistics, derived from offset well data, to quantify past performances for the various operations.

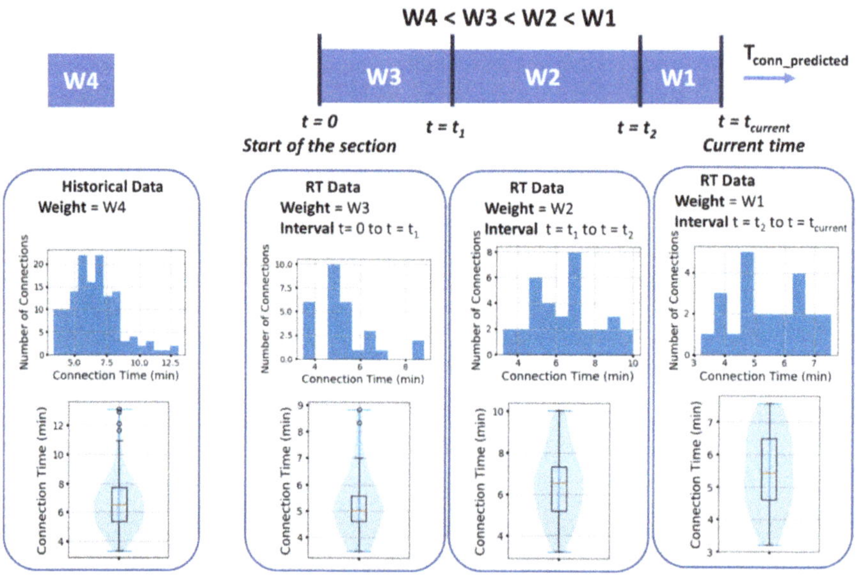

Figure 8. Adaptive weighted scheme for combining data collected at different times (for average connection time approximation).

Here again, datasets comprising the above-mentioned data were made available for this study by an oil and gas operator. Figure 9 summarizes the structure of the designed digital twin. First, the offset well data in combination with the current well's well plan, and formation top information is used to derive pre-drill or 'a priori' initializations. Subsequently, these initializations, in conjunction with RT drilling data, are fed into the multi-model digital twin. The twin then estimates the different time components that are summed together to make a time-to-TD prediction. As a new real-time data point is collected, the various statistics are re-calculated, and a new prediction is made. The differences between this prediction and the actual time are then utilized for tuning the adaptive weights of the prediction model. Therefore, the structure of the developed digital twin ensures that the system is continuously learning and updating itself based on the latest data.

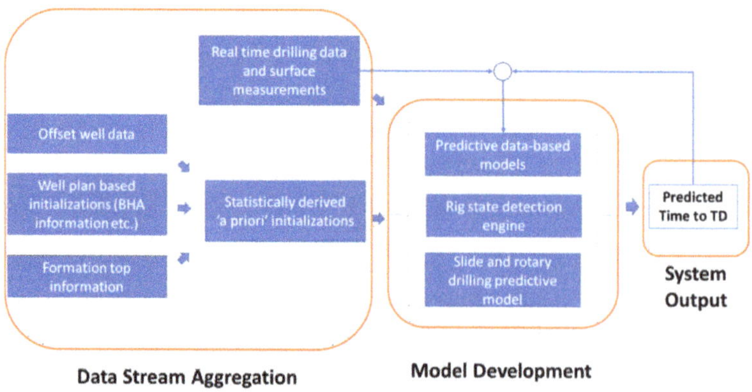

Figure 9. Final structure of the designed digital twin for making time to TD predictions, see also [82].

5.3. Application of the Twin for Performance Tracking and Scenario Analysis

The developed digital twin was implemented on a dataset comprising four wells; three were used as 'offset wells', and the fourth was treated as the 'test well'. Figure 10 illustrates a subset of the data collected from each of the offset wells to derive a priori initializations for the twin. Initially, the number of collected data points from the test well is low. This results in the times and weights associated with the adaptive weighting being tuned to give more importance to the offset well data for making predictions. However, as more real-time data is collected, the twin starts learning, thereby updating the different weights and times.

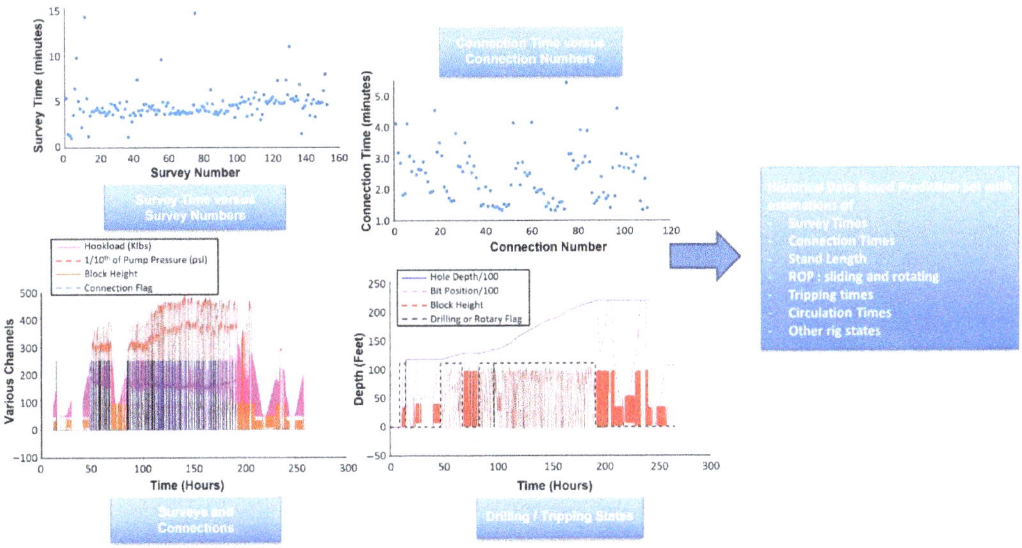

Figure 10. Subset of the data collected from each offset well for deriving a priori initializations.

5.3.1. Performance Tracking

The twin was used to make real-time predictions for the time required to reach TD for drilling the horizontal lateral section of the test well. As per the well plan, the lateral section was scheduled to be drilled with a single bit run starting at approximately 11,500 feet till the planned TD of 20,700 feet. However, unexpected tool failures mandated three bit runs (or consequently two trips in and out of the borehole to replace the failed tools). Since these trips were not planned, the initial predictions made by the twin were optimistic. However, the predictions became more realistic once the rig state engine identified the unexpected trips. Figure 11 shows the results of the predictions on plots between the predicted time versus the actual operation times. Figure 11a details the individual time predictions for different operations. Figure 11b shows the total time to TD predictions, along with the three bit runs. Another feature of the various time predictions is their ability to adapt rapidly due to:

- Different weights being assigned to different data points at different times (adaptive weighting);
- Continuous real-time re-evaluation of the slide and rotary drilling requirements.

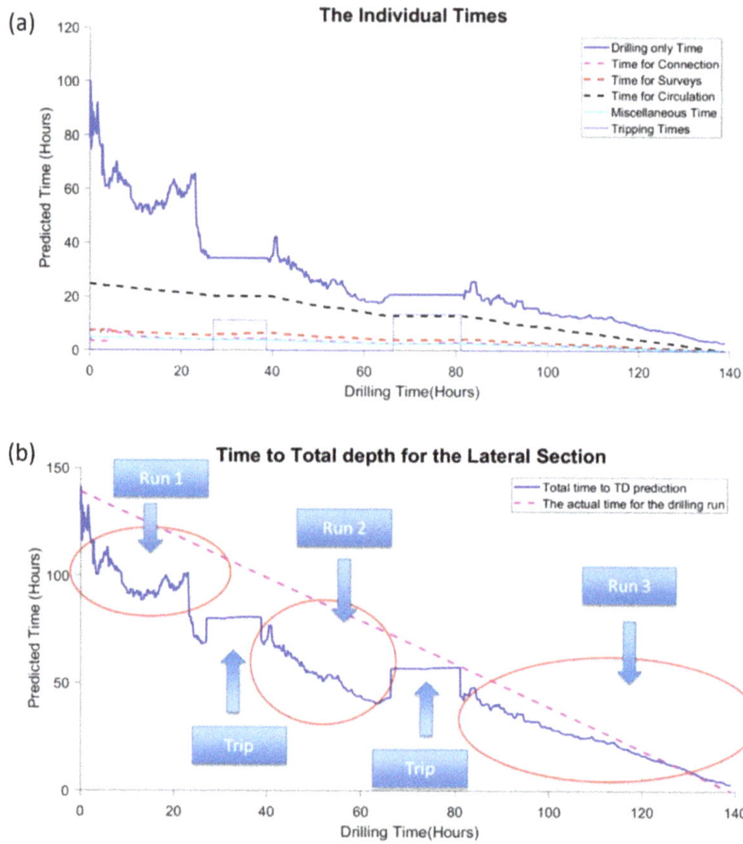

Figure 11. (**a**) Estimated individual times for different rig states versus time; (**b**) total time to TD predictions versus time.

Figure 12 is another way to visualize the predictions by overlaying the predicted and actual remaining times to TD versus the drilled hole depth. The two "time jumps" around 13,250 feet and 15,850 feet represent the trips to change the failed tools. Spikes are visible in the predicted times in both these instances, as soon as the algorithm identifies the beginning of tripping operations. Since a spike accounts for the time related to completing the tripping procedure from the given depth, it is directly proportional to the depth at which tripping starts.

5.3.2. Scenario Analysis

Similar to the hole cleaning advisory twin, this twin also permits scenario analysis to estimate the outcome of different logistical actions on the time to reach TD. Some examples of such logistical decisions include:

- Analyzing the effect of tripping out at a certain depth to change the drill bit;
- Examining the impact of different drilling parameters on the various operational times;
- Inspecting 'what-if' scenarios with varying values of time components such as tripping speeds or connection times.

Some examples of scenarios that were analyzed using basic CBM models and data are summarized in Figure 13.

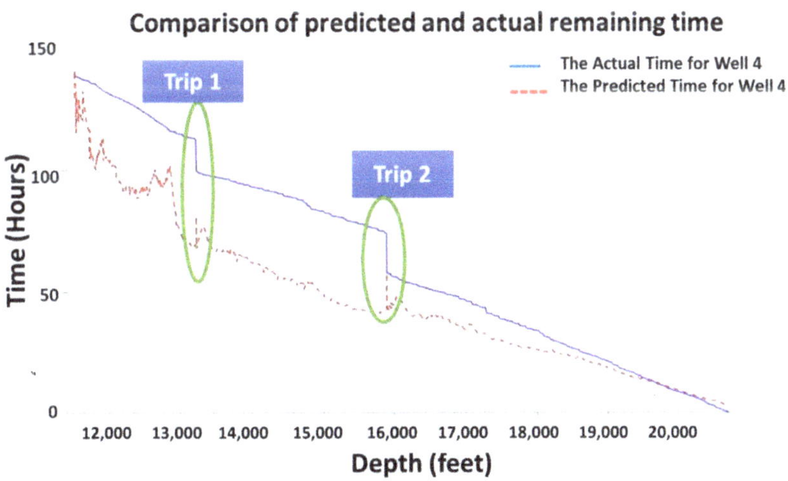

Figure 12. Comparison of the predicted and actual remaining times to TD versus drilled hole depth.

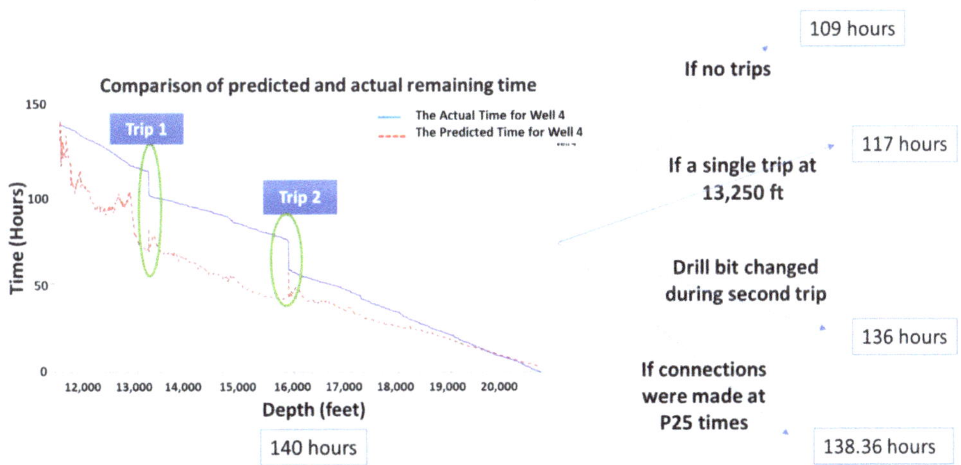

Figure 13. Analyzing post-run 'what-if' scenarios.

Observations include:

- Integrating CBM models, such as for estimating bit degradation, provides the twin with the ability to estimate the bit condition and remaining useful bit life in RT. This, in turn, allows for evaluating the benefits of continued drilling with a potentially worn bit, versus the time cost of tripping operations to change to a newer bit. The entire horizontal section (over 9000 feet) was drilled using a single drill bit. However, if the drill bit were changed (to a newer bit) during either of the trips, it would have resulted in reducing the total drilling time. Assuming linear bit degradation as a function of the depth drilled would have resulted in finishing the last section almost 4 h faster if the bit were changed during the second trip.
- Operationally, the average time for making a connection while drilling the well was 171 s, while the P25 time was 112 s (Figure 14). Making all connections throughout drilling at 112 s would have resulted in ILT reduction, with the total operation time being reduced by 1.64 h.

- This twin is a good starting point for making initial time predictions and estimations and can be further enhanced by using advanced process models. Integrating more comprehensive CBM models for, for instance, mud motor failure would not only have allowed for the prediction of 'unexpected' tool failures, thereby saving the extra downhole trips (NPT prevention), but also suggested optimal drilling parameters to prolong tool life.

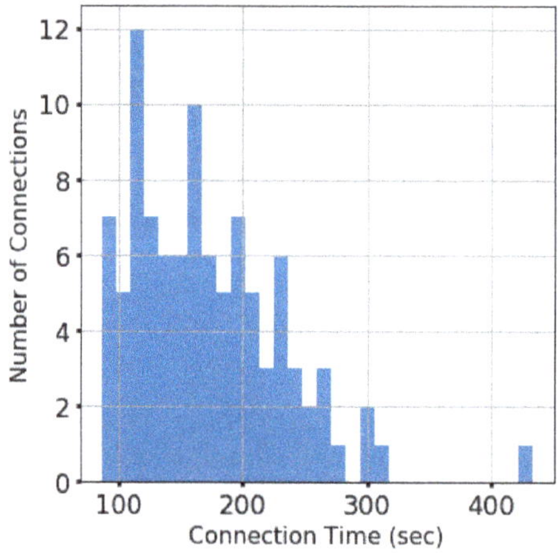

Figure 14. Variation and distribution of connection times for the drilling operation.

6. Summary and Conclusions

This paper proposes a three-step methodology (identification of the objectives, building the digital twin, and performance tracking and scenario analysis) for setting up twinning systems for well construction operations such that they can be utilized for RT performance tracking and scenario analysis. There are many models available in the drilling literature, included in conventional "weak" digital twinning approaches, to address drilling efficiency and safety issues; however, to the best of our knowledge, none of these provide a systematic framework for analyzing scenarios or action planning described in this paper. On the other hand, scenario analysis is common in other domains such as gaming artificial intelligence engines (chess and go), race strategy planning, wind farm management, etc. To enable scenario analysis in a stronger digital twinning approach for well construction, the methodology proposed in this paper addresses the following aspects of digital twinning:

- The defined cyclic process of development of the digital twin ensures that the final built twin is a compromise between the application type, the available data, and the memory and computational constraints.
- The integrated multi-model nature of the twin results in more realistic constraints around the permitted action values; for instance, as shown for the hole cleaning system, an aggressive increase in the flowrate would result in faster cuttings removal but lead to increased ECD, which could be detrimental to wellbore quality.
- Scenario design and selection is highly dependent on the twin and the available computation time. Thus, domain-knowledge can be employed to intelligently structure scenarios and build action sequences.

Due to their short- and long-term decision-making capability, digital twins that are able to perform scenario analysis can be utilized for setting up intelligent decision-making frameworks, which can be used as an advisory system for the rig teams taking improved actions but can also be coupled directly with the rig control systems to automate the control of drilling parameters such as flowrate, RPM, WOB, tripping speeds, and mud rheology. This paper provides a process for the creation of digital twins focused on scenario analysis, which, when implemented either through advisory systems or automation, will help drilling engineers and well construction managers in reducing time and cost through minimized ILT and avoided NPT events (e.g., well control, lost circulation, wellbore instability, and stuck pipes), making operations more reliable and safer by removing human bias and inconsistency.

Author Contributions: Conceptualization, G.S.S., P.A. and E.v.O.; funding acquisition, P.A. and E.v.O.; investigation, G.S.S., A.F., P.A. and E.v.O.; methodology, G.S.S., A.F., P.A. and E.v.O.; resources, P.A. and E.v.O.; software, G.S.S.; supervision, E.v.O.; visualization, P.A.; writing—original draft, G.S.S. and E.v.O.; writing—review & editing, P.A. and E.v.O. All authors have read and agreed to the published version of the manuscript.

Funding: This research was funded by the RAPID Industry Affiliate Program (IAP) at the University of Texas at Austin.

Institutional Review Board Statement: Not applicable.

Informed Consent Statement: Not applicable.

Data Availability Statement: Not applicable.

Acknowledgments: The authors would like to thank the members of the Rig Automation and Performance Improvement in Drilling (RAPID) consortium at the University of Texas at Austin for their support of this research. We especially thank Hess and Cenovus for making available the field data for this study. We would also like to acknowledge our RAPID colleagues in the research group for valuable brainstorming sessions.

Conflicts of Interest: The authors declare no conflict of interest.

Unit Conversion

1 meter (m)	3.28 feet (ft)
1 meter/second (m/s)	11,811 feet/hour (ft/h)
1 psi	6894.76 Pa
1 ppg	119.83 kg/m^3
1 radian	57.2958 degrees
1 ft^3	0.02832 m^3
1 GPM	0.0000631 m^3/s

Nomenclature

CBM	Condition-based maintenance
ECD	Equivalent circulation density (ppg)
ESG	Environmental, safety, and governance
ILT	Invisible lost time
MLWD	Measurement and logging while drilling
NPT	Non-productive time
PV	Plastic viscosity (cP)
PWD	Pressure while drilling
ROP	Rate of penetration (ft/h)
RPM	Surface rotation rate of the drillstring (revs/min)
RT	Real time
$T_{Drilling}$	Total anticipated on-bottom drilling time (h)
$T_{Circulation}$	Total circulation time (h)
$T_{Connections}$	Total expected time for making connections (h)
$T_{Miscellaneous}$	Total miscellaneous time for all other operations (h)
$T_{Surveys}$	Total time for acquiring the directional surveys (h)
$T_{Tripping}$	Total tripping time, including times for tripping in, tripping out, making up, and laying down BHAs (h)
$T_{Circulation}$	Total circulation time (h)
$T_{well\ TD}$	Total time to TD prediction (h)
WOB	Weight on bit (Klbs.)
YP	Yield point (lb./100ft^2)

References

1. Griggs, J.W. BP Gulf of Mexico Oil Spill. 2010. Available online: http://www.ktiv.com/Global/story.asp?S=13386062 (accessed on 24 April 2020).
2. Grieves, M. Digital Twin: Manufacturing Excellence through Virtual Factory Replication This Paper Introduces the Concept of a Whitepaper by Dr. Michael Grieves. *White Pap.* 2014. Available online: https://www.researchgate.net/publication/275211047_Digital_Twin_Manufacturing_Excellence_through_Virtual_Factory_Replication (accessed on 25 April 2020).
3. Grieves, M.; Vickers, J. Digital twin: Mitigating unpredictable, undesirable emergent behavior in complex systems. In *Transdisciplinary Perspectives on Complex Systems: New Findings and Approaches*; Springer International Publishing: Berlin/Heidelberg, Germany, 2016; pp. 85–113. ISBN 9783319387567.
4. Boxall, A. Inside McLaren's Secretive F1 Mission Control Room | Digital Trends. 2016. Available online: https://www.digitaltrends.com/cars/inside-mclaren-f1-racing-mission-control-room/ (accessed on 6 August 2020).
5. Breuer, H. The Future of Manufacturing | Digital Twin | Global. Available online: https://new.siemens.com/global/en/company/stories/research-technologies/digitaltwin/the-future-of-manufacturing-plm-red-bull-racing.html (accessed on 25 April 2020).
6. Jones, D.; Snider, C.; Nassehi, A.; Yon, J.; Hicks, B. Characterising the Digital Twin: A systematic literature review. *CIRP J. Manuf. Sci. Technol.* **2020**, *29*, 36–52. [CrossRef]
7. Datta, S.P.A. Emergence of Digital Twins. *arXiv* **2016**, arXiv:1610.06467.
8. Chinesta, F.; Cueto, E.; Abisset-Chavanne, E.; Duval, J.L.; Khaldi, F. El Virtual, Digital and Hybrid Twins: A New Paradigm in Data-Based Engineering and Engineered Data. *Arch. Comput. Methods Eng.* **2020**, *27*, 105–134. [CrossRef]
9. Schmidt, M.-T. Delivering a Digital Twin. *ANSYS Advant.* **2017**, *2017*, 43–45.
10. Digital Wind Farm Services & Solutions | GE Renewable Energy (n.d.). Available online: https://www.ge.com/renewableenergy/wind-energy/onshore-wind/digital-wind-farm (accessed on 25 April 2020).
11. Power Digital Solutions, G. GE Power Digital Solutions GE Digital Twin. 2016. Available online: https://www.ge.com/digital/applications/digital-twin (accessed on 6 August 2020).
12. Tuegel, E.J.; Ingraffea, A.R.; Eason, T.G.; Spottswood, S.M. Reengineering Aircraft Structural Life Prediction Using a Digital Twin. *Int. J. Aerosp. Eng.* **2011**, *2011*, 154798. [CrossRef]
13. Etong, N. Digital Twins in Commercial Aviation. 2019. Available online: https://aviation.report/whitepapers/digital-twins-in-commercial-aviation (accessed on 6 August 2020).
14. Macchi, M.; Roda, I.; Negri, E.; Fumagalli, L. Exploring the role of Digital Twin for Asset Lifecycle Management. *IFAC-PapersOnLine* **2018**, *51*, 790–795. [CrossRef]
15. Lim, K.Y.H.; Zheng, P.; Chen, C.H. A state-of-the-art survey of Digital Twin: Techniques, engineering product lifecycle management and business innovation perspectives. *J. Intell. Manuf.* **2020**, *31*, 1313–1317. [CrossRef]
16. Lu, Y.; Liu, C.; Wang, K.I.K.; Huang, H.; Xu, X. Digital Twin-driven Smart Manufacturing: Connotation, reference model, applications and research issues. *Robot. Comput. Integr. Manuf.* **2020**, *61*, 101837. [CrossRef]
17. Hoffman, R.N.; Atlas, R. Future Observing System Simulation Experiments. *Bull. Am. Meteorol. Soc.* **2016**, *97*, 1601–1616. [CrossRef]
18. Rasheed, A.; San, O.; Kvamsdal, T. Digital Twin: Values, Challenges and Enablers. *arXiv* **2019**, arXiv:1910.01719.
19. Dunn, M.D.; Payne, M.L. Design, Specification, and Implementation of Drilling Operations Database Program. In Proceedings of the SPE Annual Technical Conference and Exhibition, New Orleans, LA, USA, 5–8 October 1986. [CrossRef]
20. Mitchell, R.F.; Miska, S.Z. *Fundamentals of Drilling Engineering*; SPE Textbook Series; Society of Petroleum Engineers: Richardson, TX, USA, 2011; Volume 12.
21. Fauzi, M.; Rosli, H.W.; Sulaiman, W. Fundamentals of Petroleum Engineering Drilling Operations, n.d. Available online: http://ocw.utm.my/pluginfile.php/1080/mod_resource/content/0/Chapter_4-OCW.pdf (accessed on 6 August 2020).
22. Maidla, E.; Haci, M. Understanding Torque: The Key to Slide-Drilling Directional Wells. In Proceedings of the IADC/SPE Drilling Conference, Dallas, TX, USA, 2–4 March 2004. [CrossRef]
23. Cayeux, E.; Daireaux, B. Early Detection of Drilling Conditions Deterioration Using Real-Time Calibration of Computer Models: Field Example from North Sea Drilling Operations. In Proceedings of the SPE/IADC Drilling Conference and Exhibition, Amsterdam, The Netherlands, 17–19 March 2009. [CrossRef]
24. Cayeux, E.; Mesagan, T.; Tanripada, S.; Zidan, M.; Fjelde, K.K. Real-Time Evaluation of Hole-Cleaning Conditions with a Transient Cuttings-Transport Model. *SPE Drill. Completion* **2014**, *29*, 5–21. [CrossRef]
25. Erge, O.; Ozbayoglu, E.M.; Miska, S.Z.; Yu, M.; Takach, N.; Saasen, A.; May, R. The effects of Drillstring-Eccentricity, -Rotation, and -Buckling Configurations on Annular Frictional Pressure Losses while Circulating Yield-Power-Law Fluids. *SPE Drill. Completion* **2015**, *30*, 257–271. [CrossRef]
26. Gu, Q.; Fallah, A.; Chen, D.; Ashok, P.; van Oort, E. Higher Precision Automated Managed Pressure Drilling Control Achieved through the Addition of a Thermal Model. In Proceedings of the International Petroleum Technology Conference, Beijing, China, 26–28 March 2019. [CrossRef]
27. Isemin, I.A.; Nkundu, K.A.B.; Agwu, O.E. Utilization of Big Data Analytics and Advanced Formation Modelling for Detection of Kicks in Drilling Operations. In Proceedings of the SPE Nigeria Annual International Conference and Exhibition, Lagos, Nigeria, 5–7 August 2019. [CrossRef]

28. Al-Ghunaim, S.M.; Sadiq, B.M.; Siam, M.; Nassar, I.; Aissa, R.K. Operations Efficiency: Improved Well Planning Methodology Based on Invisible Lost Time Smart KPIs. In Proceedings of the SPE Middle East Oil & Gas Show and Conference, Manama, Bahrain, 6–9 March 2017. [CrossRef]
29. Saini, G.S.; Chan, H.C.; Ashok, P.; van Oort, E.; Isbell, M.R. Automated Large Data Processing: A Storyboarding Process to Quickly Extract Knowledge from Large Drilling Datasets. In Proceedings of the IADC/SPE Drilling Conference and Exhibition, Fort Worth, TX, USA, 6–8 March 2018. [CrossRef]
30. Coley, C. Building a Rig State Classifier Using Supervised Machine Learning to Support Invisible Lost Time Analysis. In Proceedings of the SPE/IADC International Drilling Conference and Exhibition, The Hague, The Netherlands, 5–7 March 2019. [CrossRef]
31. Okoli, P.; Vega, J.C.; Shor, R. Estimating Downhole Vibration via Machine Learning Techniques Using Only Surface Drilling Parameters. In Proceedings of the SPE Western Regional Meeting, San Jose, CA, USA, 23–26 April 2019. [CrossRef]
32. Pournazari, P.; Ashok, P.; Van Oort, E.; Unrau, S.; Lai, S. Enhanced Kick Detection with Low-Cost Rig Sensors through Automated Pattern Recognition and Real-Time Sensor Calibration. In Proceedings of the SPE Middle East Intelligent Oil and Gas Conference and Exhibition, Abu Dhabi, United Arab Emirates, 15–16 September 2015. [CrossRef]
33. Mao, Y.; Zhang, P. An Automated Kick Alarm System Based on Statistical Analysis of Real-Time Drilling Data. In Proceedings of the Abu Dhabi International Petroleum Exhibition & Conference, Abu Dhabi, United Arab Emirates, 11–14 November 2019. [CrossRef]
34. Zhao, Y.; Sun, B.; Fu, G.; Liu, S.; Xu, L.; Wang, N. Early Detection of Drillstring Washout Based on Hydraulics Model and Pattern Recognition Method. In Proceedings of the 29th International Ocean and Polar Engineering Conference, Honolulu, HI, USA, 16–21 June 2019.
35. Jeong, C.; Yu, Y.; Mansour, D.; Vesslinov, V.; Meehan, R. A Physics Model Embedded Hybrid Deep Neural Network for Drillstring Washout Detection. In Proceedings of the IADC/SPE International Drilling Conference and Exhibition, Galveston, TX, USA, 3–5 March 2020. [CrossRef]
36. Mayani, M.G.; Baybolov, T.; Rommetveit, R.; Ødegaard, S.I.; Koryabkin, V.; Lakhtionov, S. Optimizing Drilling Wells and Increasing the Operation Efficiency Using Digital Twin Technology. In Proceedings of the IADC/SPE International Drilling Conference and Exhibition, Galveston, TX, USA, 3–5 March 2020. [CrossRef]
37. Hutchinson, M.; Elliott, A.; Carlson, M.; Thornton, B.; Murray, B. Using Simulation to Prevent Non-Productive Time. In Proceedings of the SPE/IADC International Drilling Conference and Exhibition, The Hague, The Netherlands, 5–7 March 2019. [CrossRef]
38. Rommetveit, R.; Mayani, M.G.; Nabavi, J.; Helgeland, S.; Hammer, R.; Råen, J. Automatic Realtime Monitoring of Drilling Using Digital Twin Technologies Enhance Safety and Reduce Costs. In Proceedings of the Abu Dhabi International Petroleum Exhibition & Conference, Abu Dhabi, United Arab Emirates, 11–14 November 2019. [CrossRef]
39. Forshaw, M.; Becker, G.; Jena, S.; Linke, C.; Hummes, O. Automated Hole Cleaning Monitoring: A Modern Holistic Approach for NPT Reduction. In Proceedings of the International Petroleum Technology Conference, Dhahran, Saudi Arabia, 13–15 January 2020. [CrossRef]
40. Al Hindi, K.; Boushehri, M.; Pingle, D.; Kumar, R.; Tiwari, A.; Wasnik, A.; Estarabadi, J.; Shoeibi, A.; Martocchia, A.; Regan, M. Quantifying Hole Cleaning in Realtime Optimizes Drilling Performance and Demonstarbly Reduces NPT and ILT in a Complex Multilateral Well. In Proceedings of the SPWLA 2nd Asia Pacific Technical Symposium, Bogor, Java, Indonesia, 7–8 November 2018.
41. Mutlu, M.; Wassar, T.; Franchek, M.A.; Omrani, A.E.; Gutierrez, J.A. Real-Time Condition and Performance Monitoring of a Subsea Blowout Preventer Pipe Ram. *SPE Drill. Completion* **2018**, *33*, 50–62. [CrossRef]
42. You, M.; Salama, W.; Alasow, A. Technology Lifecycle Management Innovation–Prognostic Health Monitoring on Downhole Drilling Equipment. In Proceedings of the International Petroleum Technology Conference, Dhahran, Saudi Arabia, 13–15 January 2020. [CrossRef]
43. Carter-Journet, K.; Kale, A.; Zhang, D.; Pradeep, E.; Falgout, T.; Heuermann-Kuehn, L. Estimating Probability of Failure for Drilling Tools with Life Prediction. In Proceedings of the SPE Asia Pacific Oil & Gas Conference and Exhibition, Adelaide, Australia, 14–16 October 2014. [CrossRef]
44. Reckmann, H.; Jogi, P.; Kpetehoto, F.T.; Chandrasekaran, S.; Macpherson, J.D. MWD Failure Rates Due to Drilling Dynamics. In Proceedings of the IADC/SPE Drilling Conference and Exhibition, New Orleans, LA, USA, 2–4 February 2010. [CrossRef]
45. Lines, L.A.; Mauldin, C.L.; Hill, J.W.; Aiello, R.A. Advanced Drilling Dynamics Sensor Allows Real-Time Drilling Optimization, Damage Prevention and Condition Monitoring of RSS and LWD BHAs. In Proceedings of the SPE Annual Technical Conference and Exhibition, Amsterdam, The Netherlands, 27–29 October 2014. [CrossRef]
46. Pournazari, P.; Adams, D.; Ashok, P.; van Oort, E.; Holliday, K. Real-Time Health Monitoring of Top Drives for Offshore Operations Using Physics Based Models and New Sensor Technology. In Proceedings of the SPE Deepwater Drilling and Completions Conference, Galveston, TX, USA, 14–15 September 2016. [CrossRef]
47. Kyllingstad, A.; Nessjoen, P.J. Condition Based Maintenance: A New Early Leak Detection System for Mud Pumps. In Proceedings of the SPE/IADC Drilling Conference and Exhibition, Amsterdam, The Netherlands, 1–3 March 2011. [CrossRef]
48. Johnson, W.; Rao, S. Using Model-Based Design for Digital Twins to Predict the Real-Time Performance of Surface Equipment. In Proceedings of the IADC/SPE International Drilling Conference and Exhibition, Galveston, TX, USA, 3–5 March 2020. [CrossRef]

49. Liu, Y.; Yao, K.-T.; Liu, S.; Raghavendra, C.S.; Lenz, T.L.; Olabinjo, L.; Seren, F.B.; Seddighrad, S.; Dinesh Babu, C.G. Failure Prediction for Rod Pump Artificial Lift Systems. In Proceedings of the SPE Western Regional Meeting, Anaheim, CA, USA, 27–29 May 2010. [CrossRef]
50. Guo, D.; Raghavendra, C.S.; Yao, K.T.; Harding, M.; Anvar, A.; Patel, A. Data Driven Approach to Failure Prediction for Electrical Submersible Pump Systems. In Proceedings of the SPE Western Regional Meeting, Garden Grove, CA, USA, 27–30 April 2015. [CrossRef]
51. Ouahrani, L.; Abdul Haris, A.N.; Suluru, S.; Chiha, A.; Al Fakih, A. Invisible Lost Time Measurement and Reduction Contributes to Optimizing Total Well Time by Improving ROP and Reducing Flat Time. In Proceedings of the SPE Kingdom of Saudi Arabia Annual Technical Symposium and Exhibition, Dammam, Saudi Arabia, 23–26 April 2018. [CrossRef]
52. Contreras, W.; Al-hamrany, A.; Wuhaimed, A.; Aramco, S.; Mathew, B.; Albassam, B. Tripping Performance Analysis and Quantifying Invisible Lost Time. In Proceedings of the International Petroleum Technology Conference, Dhahran, Saudi Arabia, 13–15 January 2020. [CrossRef]
53. El Afifi, S.; Albassam, B.; Fahmy, F.A. Enhance The Drilling & Tripping Performance on Automated Rigs with Fully Automated Performance Measurement. In Proceedings of the SPE Middle East Intelligent Oil and Gas Conference and Exhibition, Abu Dhabi, United Arab Emirates, 15–16 September 2015. [CrossRef]
54. Mittal, M.; Lakhanpal, V.; Samuel, R. Measuring Microinvisible Loss MILT Time in Drilling Operations. In Proceedings of the IADC/SPE International Drilling Conference and Exhibition, Galveston, TX, USA, 3–5 March 2020. [CrossRef]
55. Lakhanpal, V.; Samuel, R. A New Approach to Harness Data for Measuring Invisible Lost Time in Drilling Operations. In Proceedings of the SPE Annual Technical Conference and Exhibition, San Antonio, TX, USA, 9–11 October 2017. [CrossRef]
56. De Oliveira, G.L.; Zank, C.A.C.; De Costa, A.F.S.; Mendes, H.M.; Henriques, L.F.; Mocelin, M.R.; De Oliveira Neto, J. Offshore Drilling Improvement through Automating the Rig State Detection Process—Implementation Process History and Proven Success. In Proceedings of the IADC/SPE Drilling Conference and Exhibition, Fort Worth, TX, USA, 1–3 March 2016. [CrossRef]
57. Millan, E.; Ringer, M. A New Workflow for Estimating Bit Wear and Monitoring Drilling Efficiency in Real Time During Drilling Operations. In Proceedings of the IADC/SPE Drilling Conference and Exhibition, Fort Worth, TX, USA, 6–8 March 2018. [CrossRef]
58. Liu, Y.; Kibbey, J.; Bai, Y.; Wu, X. Real-Time Bit Wear Monitoring and Prediction Using Surface Mechanics Data Analytics: A Step Toward Digitization Through Agile Development. In Proceedings of the IADC/SPE Drilling Conference and Exhibition, Fort Worth, TX, USA, 6–8 March 2018. [CrossRef]
59. Shirangi, M.G.; Ettehadi, R.; Aragall, R.; Furlong, E.; May, R.; Dahl, T. Digital Twins for Drilling Fluids: Advances and Opportunities. In Proceedings of the IADC/SPE International Drilling Conference and Exhibition, Galveston, TX, USA, 3–5 March 2020. [CrossRef]
60. Nadhan, D.; Mayani, M.G.; Rommetveit, R. Drilling with Digital Twins. In Proceedings of the IADC/SPE Asia Pacific Drilling Technology Conference and Exhibition, Bangkok, Thailand, 27–29 August 2018. [CrossRef]
61. Crichton, M.T.; Moffat, S.; Crichton, L. Developing a Team Behavioural Marker Framework Using Observations of Simulator-Based Exercises to Improve Team Effectiveness: A Drilling Team Case Study. *Simul. Gaming* **2017**, *48*, 299–313. [CrossRef]
62. Drilling and Well Control Simulators | Drilling Systems, (n.d.). Available online: https://www.drillingsystems.com/simulators/drilling-and-well-control/ (accessed on 24 April 2020).
63. Shirkavand, F.; Hareland, G.; Olson, W. The design and development of a drilling simulator for planning and optimizing under-balanced drilling operations. *J. Can. Pet. Technol.* **2010**, *49*, 68–73. [CrossRef]
64. Chan, H.C.; Lee, M.M.; Saini, G.S.; Pryor, M.; van Oort, E. Development and Validation of a Scenario-Based Drilling Simulator for Training and Evaluating Human Factors. *IEEE Trans. Human-Machine Syst.* **2020**, *50*, 327–336. [CrossRef]
65. Howell, R.; McKenzie, B.; Leslie, C.; Poletzky, I. Advanced Drilling Simulators for Well Control Training: Bringing Together People, Procedures and New Technology. In Proceedings of the SPE/IADC International Drilling Conference and Exhibition, The Hague, The Netherlands, 5–7 March 2019. [CrossRef]
66. Höhn, P.; Freifer, R.; Oppelt, J. Realistic Scenario Drilling Simulation for Deep and Critical Wells. In Proceedings of the Offshore Mediterranean Conference and Exhibition, Ravenna, Italy, 27–29 March 2019.
67. Kelessidis, V.C.; Ahmed, S.; Koulidis, A. An Improved Drilling Simulator for Operations, Research and Training. In Proceedings of the SPE Middle East Oil & Gas Show and Conference, Manama, Bahrain, 8–11 March 2015. [CrossRef]
68. Minchev, Z.B.; Shalamanov, V.M. Scenario Generation and Assessment Framework Solution in Support of the Comprehensive Approach. In Proceedings of the SAS-081 Symposium on "Analytical Support to Defence Transformation", Sofia, Bulgaria, 26–28 April 2010.
69. Zhang, F.; Miska, S.; Yu, M.; Ozbayoglu, E.; Takach, N.; Osgouei, R.E. Is Well Clean Enough? A Fast Approach to Estimate Hole Cleaning for Directional Drilling. In Proceedings of the SPE/ICoTA Coiled Tubing & Well Intervention Conference & Exhibition, The Woodlands, TX, USA, 24–25 March 2015. [CrossRef]
70. Aldea, C.; Iyoho, A.W.; Zamora, M. Hole Cleaning: The Achilles' Heel of Drilling Performance. In Proceedings of the AADE National Technical Conference and Exhibition AADE-05-NTCE-29, Houston, TX, USA, 5–7 April 2005.
71. Ozbayoglu, M.E.; Saasen, A.; Sorgun, M.; Svanes, K. Effect of Pipe Rotation on Hole Cleaning for Water-Based Drilling Fluids in Horizontal and Deviated Wells. In Proceedings of the IADC/SPE Asia Pacific Drilling Technology Conference and Exhibition, Jakarta, Indonesia, 25–27 August 2008. [CrossRef]

72. Duan, M.; Miska, S.; Yu, M.; Takach, N.; Ahmed, R. Critical Conditions for Effective Sand-Sized-Solids Transport in Horizontal and High-Angle Wells. *SPE Drill. Completion* **2009**, *24*, 229–238. [CrossRef]
73. Mirhaj, S.A.; Kaarstad, E.; Aadnoy, B.S. Torque a Drag Modeling; Soft-String versus Stiff-String Models. In Proceedings of the SPE/IADC Middle East Drilling Technology Conference and Exhibition, Abu Dhabi, United Arab Emirates, 26–28 January 2016. [CrossRef]
74. Saini, G.S.; Ashok, P.; van Oort, E. Predictive Action Planning for Hole Cleaning Optimization and Stuck Pipe Prevention Using Digital Twinning and Reinforcement Learning. In Proceedings of the IADC/SPE International Drilling Conference and Exhibition, Galveston, TX, USA, 3–5 March 2020. [CrossRef]
75. Aadnoy, B.S.; Fazaelizadeh, M.; Hareland, G. A 3D analytical model for wellbore friction. *J. Can. Pet. Technol.* **2010**, *49*, 25–36. [CrossRef]
76. Fazaelizadeh, M. Real Time Torque and Drag Analysis during Directional Drilling. Unpublished Doctoral Thesis, University of Calgary, Calgary, AB, Canada, 2013. Available online: http://hdl.handle.net/11023/564 (accessed on 9 July 2020).
77. Larsen, T.I.; Pilehvari, A.A.; Azar, J.J. Development of a new cuttings-transport model for high-angle wellbores including horizontal wells. *SPE Drill. Complet.* **1997**, *12*, 129–134. [CrossRef]
78. Jalukar, L.S. *Study of Hole Size Effect on Critical and Subcritical Drilling Fluid Velocities in Cuttings Transport for Inclined Wellbores*; University of Tulsa: Tulsa, OK, USA, 1993.
79. Bassal, A.A. *The effect of Drillpipe Rotation on Cuttings Transport in Inclined Wellbores*; University of Tulsa: Tulsa, OK, USA, 1995.
80. Rubiandini, R.S. Equation for Estimating Mud Minimum Rate for Cuttings Transport in an Inclined-Until-Horizontal Well. In Proceedings of the SPE/IADC Middle East Drilling Technology Conference, Abu Dhabi, United Arab Emirates, 8–10 November 1999. [CrossRef]
81. Naganawa, S.; Nomura, T. Simulating Transient Behavior of Cuttings Transport over Whole Trajectory of Extended Reach Well. In Proceedings of the IADC/SPE Asia Pacific Drilling Technology Conference and Exhibition, Bangkok, Thailand, 13–15 November 2006. [CrossRef]
82. Saini, G.S. Digital Twinning of Well Construction Operations for Improved Decision-Making. Ph.D. Thesis, The University of Texas at Austin, Austin, TX, USA, 2020.

energies

Article

A Nonlinear Dynamic Model for Characterizing the Downhole Motions of the Sidetracking Tool in a Multilateral Well

Xiuxing Zhu [1], Weixia Zhou [1], Yujian Lei [1], Peng Jia [1], Shifeng Xue [1,*], Bo Zhou [1,*] and Yuanbo Xia [2]

[1] College of Pipeline and Civil Engineering, China University of Petroleum (East China), Qingdao 266580, China
[2] CNPC Tianjin Bo-Xing Engineering Science &Technology Co., Ltd., Tianjin 300451, China
* Correspondence: sfeng@upc.edu.cn (S.X.); zhoubo@upc.edu.cn (B.Z.)

Abstract: It is of practical interest to investigate the mechanical behaviors of a sidetracking tool system and to describe the sidetracking tool's vibration mechanical response, as this can provide an important basis for evaluating and optimizing the tool structure and effectively controlling the profile of the sidetracking window. In this article, three nonlinear dynamic models with ten, six, and two degrees of freedom, respectively, are established using the Lagrange method to characterize the behavior of the sidetracking tool. It should be noted that in these models, the axial, lateral, and torsional vibration of the tool system are fully coupled. The process of the sidetracking tool mills in the casing-pipe wall is divide into three typical stages, i.e., the window mill, pilot mill, and watermelon mill grinding the casing, respectively. The dynamic response of the three stages is studied to more effectively analyze the influence of the sidetracking tool vibration deformation on the window width. The Runge–Kutta method, which is easy to implement, is applied to solve the supposed nonlinear dynamic model, and some useful findings are as follows. The effects of sidetracking tool vibrations at different stages on window widening size are illustrated quantitatively. The vibration trajectory pattern of the sidetracking tool is different from that of the conventional drilling tool due to the influence of the whipstock, which changes from the general whirling motion pattern to the X reciprocating pattern, and the vibration amplitude decreases. Due to the influence of the tool's lateral amplitude, the window profile is widened. The widened window size of the window mill and the pilot mill are 3.30 mm and 2.74 mm, respectively, and the extended window size of the watermelon mill is 0.07 mm, while the maximum window width formed by the sidetracking tool is 374.34 mm. This work proposes, for the first time, the coupled vibration model of the sidetracking tool system, which is helpful to better understand the nonlinear dynamic effects of the sidetracking tool, laying the foundation for the optimization of the sidetracking parameters.

Keywords: window milling; sidetracking tool; nonlinear dynamics; vibration; window profile

Citation: Zhu, X.; Zhou, W.; Lei, Y.; Jia, P.; Xue, S.; Zhou, B.; Xia, Y. A Nonlinear Dynamic Model for Characterizing the Downhole Motions of the Sidetracking Tool in a Multilateral Well. *Energies* **2023**, *16*, 588. https://doi.org/10.3390/en16020588

Academic Editors: Mofazzal Hossain, Md Motiur Rahman and Hisham Khaled Ben Mahmud

Received: 24 November 2022
Revised: 28 December 2022
Accepted: 31 December 2022
Published: 4 January 2023

1. Introduction

The casing window sidetracking technology is widely used in the recovery of residual oil in aged oil fields, especially in multilateral wells [1,2]. Its aim is to cut out a slanting and smooth window at the designated position on the casing wall of the original well hole by using sidetracking tools along the whipstock [3,4]. The huge expansion of side-tracking technology has facilitated the application of an increasing number of sidetracking systems, in which tool systems have had a significant impact on the success of sidetracking. A bottom-hole sidetracking tool system typically consists of a window mill, a flexible joint, and a watermelon mill. The tool diameter is generally smaller than the casing diameter, providing a gap for the tool to deform and move due to weight on bit and torque [5]. Thus, the motion of a sidetracking tool is a multi-coupled nonlinear motion that includes rotations around its geometric axis, roll, and rotational motion around the drilling axis, etc. These complex motions increase the uncertainty of the tool motion and make it more difficult to predict the profile of the sidetracking window.

The vibration of the sidetracking tool system during drilling is an unavoidable phenomenon due to the combined effects of weight on bit, torque, and reaction forces on the casing wall. This phenomenon consists of three different types of vibrations, namely longitudinal, transverse, and torsional. The longitudinal vibration is the most harmful [6–8]. Tool vibrations can cause tool system failure, accelerate casing wear, and reduce device safety. It is worth noting that the most intense vibration occurs at the bottom of the well, so it is extremely essential to analyze the vibration characteristics of the bottom hole assembly [9].

Regarding the reduction of unnecessary losses caused by tool vibration during sidetracking, there are numerous previous articles in the literature that have addressed the stick-slip phenomenon in drill string vibration, and these studies have proposed methods for its control. S. H. Choi et al. [10] gave a consistent derivation of a set of governing differential equations and proposed a different method to correctly calculate the axial load effect of the Timoshenko beam. However, the coupling between the bending and torsional vibrations due to mass eccentricity is not considered in this work. H. Ahmadian et al. [11] adopted the Lagrange method to obtain the control equation of drill string motion considering the coupling of axial, lateral, and torsional vibrations and proposed a fully coupled model, calculated the contact force between the drill collar and borehole wall, and studied its behavior. K. Nandakumar et al. [12] established a fully coupled two degree of freedom model, which consisted of two coupled delay differential equations, assuming a state-dependent delay and viscous damping for both axial and torsional motions. They conducted a detailed linear stability analysis of the obtained mathematical model. This work provides a reference for subsequent research on drill string vibration. Luciano P.P. de Moraes et al. [13] studied the coupled vibration of a drill string considering the four degrees of freedom non-smooth model, with parametric studies of bit bounce, stick-slip, rotation, and combination effects. Dou Xie et al. [14] developed a dynamic model of the drill string in horizontal wells with six degrees of freedom, considering longitudinal, lateral, and torsional motions. State-dependent time delays were introduced during the cutting process to couple the axial and torsional motions of the bit. Mohammadzadeh Maziar et al. studied the fully coupled nonlinear vibrations drill string, which is a composite drill string consisting of orthogonal anisotropic layers, using the Lagrange method and finite element methods and investigated the analytical capability of the dynamic model [15]. Then, they also studied the ability of the composite drill string, with different configurations, to track and transmit the fully coupled nonlinear vibration [16]. All the above studies have led to advances in the study of vibrational properties of drill strings. However, if these theories are to be applied to the sidetracking aspect, further analysis is needed to continue to consider the effects of the whipstock and milling tools.

In addition, the drill string vibrations have been investigated for control and energy conversion. Wang Peng et al. [17] developed a unified mathematical model considering the two mechanisms of static and dynamic friction conversion and friction decomposition. The model was solved by the second-order finite difference method to obtain the variations of weight on bit, drilling speed, and positive displacement motor tool face under three different sliding drilling methods. Tian Jialin et al. established the multi degrees of freedom torsional vibration model and the nonlinear dynamic model of the drill string, which is a new torsional vibration tool. They studied the viscosity reduction mechanism combined with the actual working conditions in the drilling process [18]. Then, they established the drill string axial-torsional-lateral coupled vibration model. The coupled vibration sliding mode PI controller was designed to control and analyze the drill string system [19]. Ritto Thiago G. et al. [20] proposed an active vibration control method, which considered a simplified two-DOF drill string torsion model. Through experiments, Xu Yuqiang et al. [21] studied the characteristics of downhole bit load and drill string longitudinal vibration under different conditions. Significantly, they proposed an energy conversion efficiency analysis method converting the drill string vibration to spring potential energy. Sassi Sadok et al. [22] conducted an experimental study on the suppression of drill string vibration using a new drill string design. More recently, AmirNoabahar

Sadeghi et al. [23] subdivided the drill string into smaller torsional segments and modeled the whole system dynamically. Mathematically by modeling stick-slip vibrations using an extracted torsion model, three structures have been developed and proposed to reduce and actively control stick-slip vibrations, namely, speed control, bit-weight control, and increasing damping at the bottom of the drill string.

In the process of milling window, the behavior of the bottom hole assembly (BHA) affects the deformation and geometry of the milling tool, and then affects the movement of the BHA. Therefore, the mechanical analysis of BHA is also extremely crucial. Sarker Mejbahul et al. [24] presented a model to analyze the dynamics of a horizontal oil well bottom hole assembly. The model can predict how axial and torsional bit–rock reactions are propagated to the surface and the role that lateral vibrations near the bit play in exciting those vibrations, as well as the stressing components in the bottom-hole-assembly. Liu Yongsheng et al. [25] developed a nonlinear dynamic model with four degrees of freedom (DOE) to characterize the behavior of the drill string in a deviated well. This study provided a theoretical understanding of the drill string nonlinear motion in non-vertical wells. Wang Jin et al. [26] analyzed the dynamics model of a rotary steerable system (RSS) with a single stabilizer and flexible joint, and simplified the drill collar and joint to obtain a suitable BHA mechanical analysis model and a finite element analysis model. However, these studies do not take into account the interaction between the drill string and the casing. Thus, the results for the mechanical behavior of the BHA cannot be reflected in the window profile model.

The above research results mainly focus on the vibration of the drilling tool in the drilling process, while there are few studies on the vibration deformation of the sidetracking tools. Particularly, there are few reports on the model considering the influence of the sidetracking tool vibration on the window shape under the action of construction parameters such as weight on bit. Therefore, studying the force of the sidetracking tool in the drilling process and establishing the vibration models in different drilling stages is essential. The second section of this paper depicts the 2 -DOF system considered in the analyses, as well as the control strategies employed. The third section presents the numerical results and, finally, the concluding remarks are presented in the last section.

2. Dynamic Model of Sidetracking Tool

As shown in Figure 1, the physical model of the sidetracking tool has three key parts: the watermelon mill, the pilot mill, and the window mill [27]. These three parts have different operating conditions, so their dynamic behavior will be discussed separately.

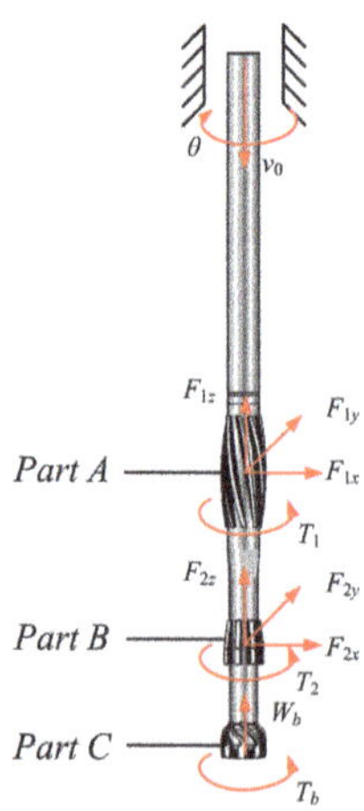

Figure 1. Physical model of the sidetracking tool.

2.1. Model Simplification of the Sidetracking Tool

For the ideal sidetracking tool considered, the assumptions are as follows: (1) The drill hole is straight, and its diameter is equal to the drill bit diameter. (2) The damping effect of the drilling fluid is neglected [28]. Based on these assumptions, a lumped mass model is proposed, which consists of three parts, as shown in Figure 1. In this system, the inertial characteristics of the watermelon mill, pilot mill, and window mill are represented by Part A, B, and C, respectively. These three parts are connected with each other through the elastic section string.

According to the physical model of the sidetracking tool, the dynamics of Parts A and B are described by four degrees of freedom, i.e., one longitudinal, one torsional, and two transverse, respectively. The dynamic characteristics of Part C are described by only two degrees of freedom, longitudinal and torsional, because its lateral motion is limited by the whipstock. This system is driven from the top with constant axial velocity v_0 and constant angular velocity $\varphi_{t} = \theta$. In Figure 2, the subscripts 1 and 2 of each physical quantity represent Part A and Part B, respectively. The subscripts x and y represent the lateral direction, respectively, and z represents the axial direction. W_b represents the weight on bit. T_b represents torque, and T_i (i = 1~2) represents friction torque. F represents the contact force.

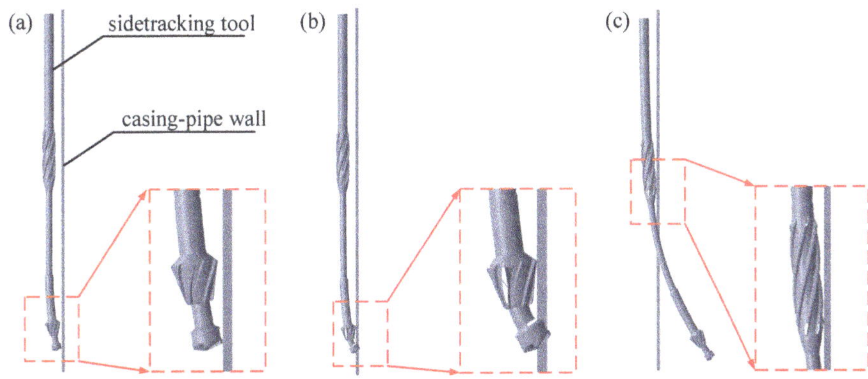

Figure 2. Three typical stages of the sidetracking process: (**a**) the first stage, (**b**) the second stage, and (**c**) the third stage.

The sidetracking process can be divided into three typical stages, as shown in Figure 2. The sidetracking tool is lowered into the casing until the window mill touches the whipstock, which can be considered as the first stage. The second stage occurs when it continues drilling until the pilot mill contacts the whipstock. In the third stage, both the window and pilot mills are included the formation, and the watermelon mill contacts the whipstock to mill the casing wall for the last time.

2.2. Motion Equation of the Sidetracking Tool

In order to describe the dynamic model, the coordinate system (x-y-z-φ) is introduced. Thus, the spatial positions of Parts A, B, and C can be determined by the coordinates (x_1-y_1-z_1-φ_1), (x_2-y_2-z_2-φ_2), and (z_3-φ_3). Parts A, B, and C all rotate around their geometric centers, (x,y) denoting their lateral displacements, and z is their longitudinal displacements. The contact between the sidetracking tool and the casing wall is modeled as linear elasticity. In order to simulate the plane motion of sidetracking tool, linear eccentricity e associated with Part A and Part B is introduced [29]. The mass of Part A is m_1 and the moment of inertia is J_1. The mass of Part B is m_2 and the moment of inertia is J_2. The mass of Part C is m_3 and the moment of inertia is J_3.

2.2.1. The First Stage

During the first stage of the sidetracking process, the windowed shoe section was run, reaching the miter section to begin milling the casing. Then, the kinetic energy of the drill string system can be written as

$$T = \frac{1}{2}m_1(\dot{x}_1{}^2 + \dot{y}_1{}^2 + \dot{z}_1{}^2) + \frac{1}{2}J_1\dot{\varphi}_1{}^2 + \frac{1}{2}m_2(\dot{x}_2{}^2 + \dot{y}_2{}^2 + \dot{z}_2{}^2) + \frac{1}{2}J_2\dot{\varphi}_2{}^2 + \frac{1}{2}m_3\dot{z}_3{}^2 + \frac{1}{2}J_3\dot{\varphi}_3{}^2 \tag{1}$$

where ($\dot{}$) denotes a time derivative, x_1 and y_1 are the lateral displacements of Part A, and z_1 and φ_1 are the axial and angular displacement, respectively. x_2 and y_2 are the lateral displacement of Part B, and z_2 and φ_2 are the axial and angular displacements, respectively. z_3 and φ_3 are the axial and angular displacements of Part C, respectively.

Euler–Bernoulli beams are used to describe the potential energy V. Let $w(z,t)$ and $\varphi(z,t)$ be the axial and torsional displacements of the sidetracking tool system, respectively, and $u(z,t)$ and $v(z,t)$ be the lateral displacements at $z \in [0, L]$. It is assumed that the deformation is linearly elastic, so V can be constructed as

$$V = \frac{1}{2}\int_0^L \left\{EI\left[(u'')^2 + (v'')^2\right] + EA(w')^2 + 2GI(\theta')^2\right\}dz \tag{2}$$

where, A and I refer to the cross-sectional area and moment of inertia of the string, respectively, and the specific expression is

$$A = \frac{\pi}{4}\left(D_o{}^2 - D_i{}^2\right) \tag{3}$$

and

$$I = \frac{\pi}{64}\left(D_o{}^4 - D_i{}^4\right) \tag{4}$$

respectively.

The system is divided into three parts according to the simplified model shown in Figure 3. Therefore, the boundary conditions related to the axial motion of the above system are

$$\begin{cases} w|_{z=0} = v_0 t \\ w|_{z=\frac{L}{2}} = z_1 \\ w|_{z=\frac{3L}{4}} = z_2 \\ w|_{z=L} = z_3 \end{cases} \tag{5}$$

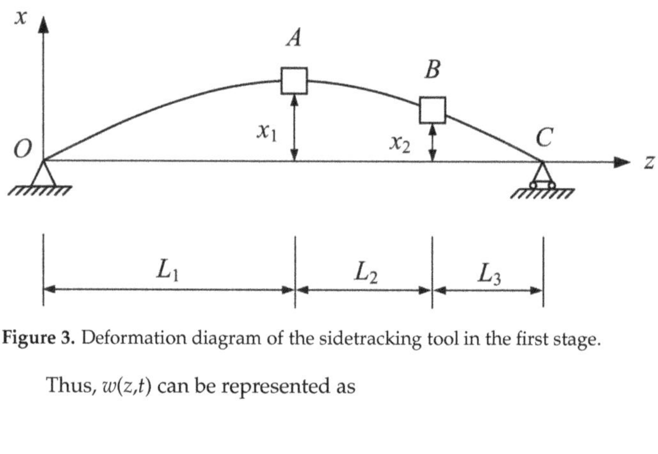

Figure 3. Deformation diagram of the sidetracking tool in the first stage.

Thus, $w(z,t)$ can be represented as

$$w(z,t)=\begin{cases}\frac{z_1-v_0t}{L/2}z+v_0t, & 0\le z\le \frac{L}{2}\\ \frac{z_2-z_1}{L/4}z-2z_2+3z_1, & \frac{L}{2}<z\le\frac{3L}{4}\\ \frac{z_3-z_2}{L/4}z-3z_3+4z_2, & \frac{3L}{4}<z\le L\end{cases} \tag{6}$$

while the boundary conditions related to the lateral motion are

$$\begin{cases}\theta|_{z=0}=\varphi_t\\ \theta|_{z=\frac{L}{2}}=\varphi_1\\ \theta|_{z=\frac{3L}{4}}=\varphi_2\\ \theta|_{z=L}=\varphi_3\end{cases} \tag{7}$$

Therefore, $\theta(z,t)$ can be represented as

$$\theta(z,t)=\begin{cases}\frac{\varphi_1-\varphi_t}{L/2}z+\varphi_t, & 0\le z\le\frac{L}{2}\\ \frac{\varphi_2-\varphi_1}{L/4}z-2\varphi_2+3\varphi_1, & \frac{L}{2}<z\le\frac{3L}{4}\\ \frac{\varphi_3-\varphi_2}{L/4}z-3\varphi_3+4\varphi_2 & ,\frac{3L}{4}<z\le L\end{cases} \tag{8}$$

In addition, the bending deflection of the sidetracking tool system in the x-direction is shown in Figure 3. The deflection equation and the angle equation of the system in the first stage can be obtained from the shear force and bending moment equations of the system and the boundary conditions of A, B, C and O. Since the bending in the y-direction is symmetric with that in the x-direction, the deformation equations of the window system in the x and y directions can be expressed as

$$u(z,t)=\begin{cases}\frac{2z}{23L^3}\left[(51x_1-24x_2)L^2-(112x_1-96x_2)z^2\right] & ,0\le z\le\frac{L}{2}\\ -\frac{1}{23L^3}\left[(144x_1-176x_2)L^3\right](-928z^3x_1+1216z^3x_2+1728Lz^2x_1-966L^2zx_1-2112Lz^2x_2+1104L^2zx_2) & ,\frac{L}{2}<z\le\frac{3L}{4}\\ \frac{2L-2z}{23L^3}(225L^2x_1-344L^2x_2+240z^2x_1-416z^2x_2-480Lzx_1+832Lzx_2) & ,\frac{3L}{4}<z\le L\end{cases} \tag{9}$$

and

$$v(z,t)=\begin{cases}\frac{2z}{23L^3}\left[(51y_1-24y_2)L^2-(112y_1-96y_2)z^2\right] & ,0\le z\le\frac{L}{2}\\ -\frac{1}{23L^3}\left[(144y_1-176y_2)L^3-(966y_1-1104y_2)L^2z+(1728y_1-2112y_2)Lz^2-(928y_1-1216y_2)z^3\right] & ,\frac{L}{2}<z\le\frac{3L}{4}\\ \frac{2L-2z}{23L^3}\left[(225y_1-344y_2)L^2-(480y_1+832y_2)Lz+(240y_1-416y_2)z^2\right] & ,\frac{3L}{4}<z\le L\end{cases} \tag{10}$$

respectively.

Substituting Equations (6)–(10) into potential energy expression (2), the expression of potential energy can be obtained as

$$\begin{aligned}V=\;&\frac{EA}{L}\left[2(z_2-z_1)^2+2(z_3-z_2)^2+(z_1-v_0t)^2\right]+\frac{GI}{L}\left[2(\varphi_1-\varphi_t)^2+4(\varphi_2-\varphi_1)^2+4(\varphi_3-\varphi_2)^2\right]\\ &+\frac{EI}{2L^3}\left[\begin{array}{l}\frac{3}{1058}(224x_1-192x_2)^2+\frac{192}{529}(211{x_1}^2-572x_1x_2+508{x_2}^2+211{y_1}^2-572y_1y_2+508{y_2}^2)\\ +\frac{3}{1058}(224y_1-192y_2)^2+\frac{192}{529}(225{x_1}^2-780x_1x_2+676{x_2}^2+225{y_1}^2-780y_1y_2+676{y_2}^2)\end{array}\right]\end{aligned} \tag{11}$$

In order to describe the viscous damping effect of this sidetracking tool system, Rayleigh dissipation function D is introduced here

$$D=\frac{1}{2}\left[c_a\left({\dot{z}_1}^2+{\dot{z}_2}^2+{\dot{z}_3}^2\right)+c_b\left({\dot{x}_1}^2+{\dot{y}_1}^2+{\dot{x}_2}^2+{\dot{y}_2}^2\right)+c_t\left({\dot{\varphi}_1}^2+{\dot{\varphi}_2}^2+{\dot{\varphi}_3}^2\right)\right] \tag{12}$$

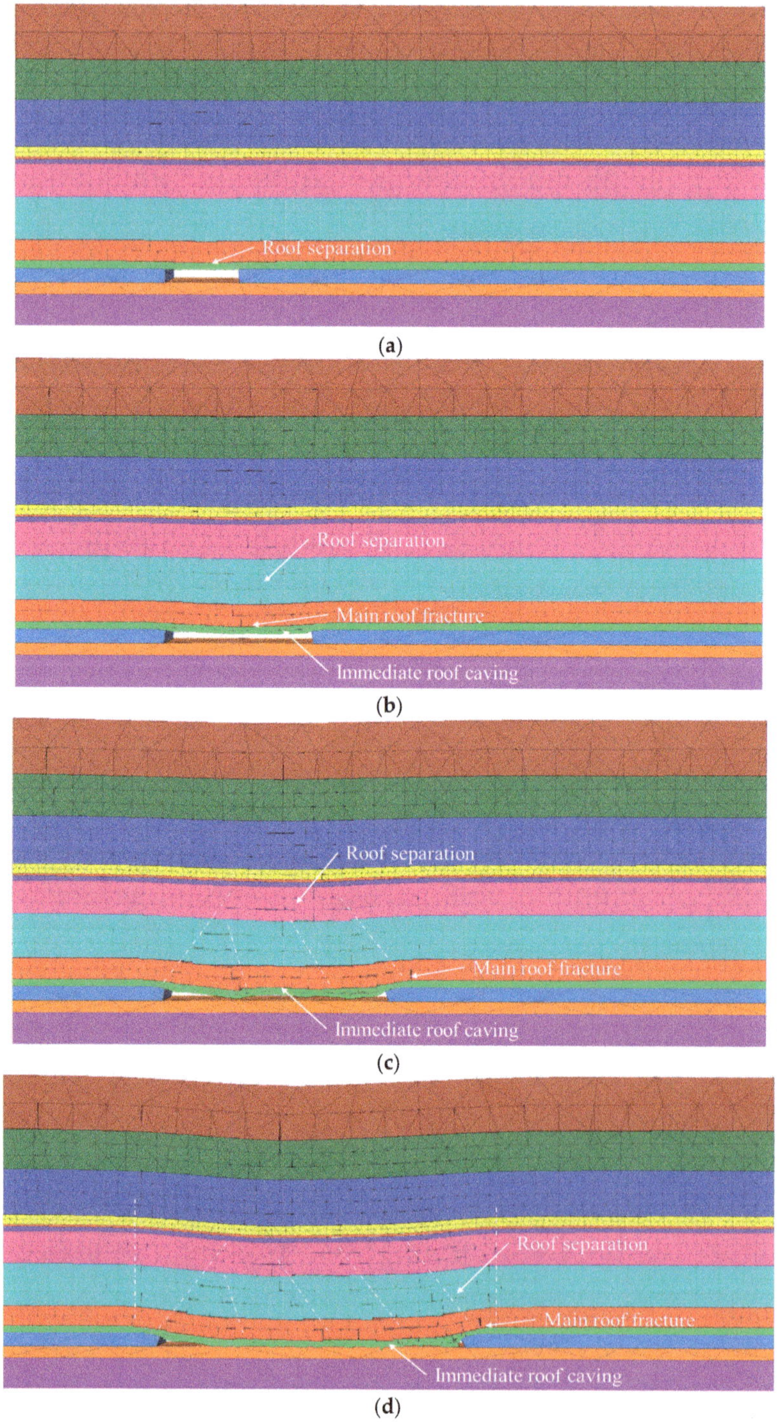

Figure 4. *Cont.*

The Lagrange equation is

$$\frac{d}{dt}\left(\frac{\partial T}{\partial \dot{q}_j}\right)-\frac{\partial T}{\partial q_j}+\frac{\partial U}{\partial q_j}+\frac{\partial D}{\partial \dot{q}_j}=Q_j \tag{13}$$

where q_j is the generalized coordinate.

Substituting T, V, and D into the Lagrange equation can obtain

$$[M]\{\ddot{x}\}+[C]\{\dot{x}\}+[K]\{x\}=\{F\} \tag{14}$$

where $[M]$, $[C]$, and $[K]$ represent the mass matrix, damping matrix, and stiffness matrix, respectively, and $\{F\}$ is the column vector of the force; the expressions are

$$[M]=\begin{bmatrix} m_1 &&&&&&&&& \\ & m_2 &&&&&&&& \\ && m_1 &&&&&&& \\ &&& m_2 &&&&&& \\ &&&& m_1 &&&&& \\ &&&&& m_2 &&&& \\ &&&&&& m_3 &&& \\ &&&&&&& J_1 && \\ &&&&&&&& J_2 & \\ &&&&&&&&& J_3 \end{bmatrix} \tag{15}$$

$$[C]=\begin{bmatrix} c_b &&&&&&&&& \\ & c_b &&&&&&&& \\ && c_b &&&&&&& \\ &&& c_b &&&&&& \\ &&&& c_a &&&&& \\ &&&&& c_a &&&& \\ &&&&&& c_a &&& \\ &&&&&&& c_t && \\ &&&&&&&& c_t & \\ &&&&&&&&& c_t \end{bmatrix} \tag{16}$$

$$[K]=\begin{bmatrix} 9k_r & -11k_r &&&&&&&& \\ -11k_r & 16k_r &&&&&&&& \\ && 9k_r & -11k_r &&&&&& \\ && -11k_r & 16k_r &&&&&& \\ &&&& 3k_w & -2k_w &&&& \\ &&&& -2k_w & 4k_w & -2k_w &&& \\ &&&&& -2k_w & 2k_w &&& \\ &&&&&&& 3k_\varphi & -2k_\varphi & \\ &&&&&&& -2k_\varphi & 4k_\varphi & -2k_\varphi \\ &&&&&&&& -2k_\varphi & 2k_\varphi \end{bmatrix} \tag{17}$$

and

$$\{F\} = \begin{Bmatrix} F_{1x} + m_1 e_1 \dot{\varphi}_1{}^2 \cos\varphi_1 \\ F_{2x} + m_2 e_2 \dot{\varphi}_2{}^2 \cos\varphi_2 \\ F_{1y} + m_1 e_1 \dot{\varphi}_1{}^2 \sin\varphi_1 \\ F_{2y} + m_2 e_2 \dot{\varphi}_2{}^2 \sin\varphi_2 \\ k_w \cdot v_0 t + m_1 g - sign(\dot{z}_1)\mu_r F_{n1} \\ m_2 g - sign(\dot{z}_2)\mu_r F_{n2} \\ -W_b - F_b \sin\alpha \\ k_\varphi \cdot \varphi_t - T_1 \\ -T_2 \\ -T_b \end{Bmatrix} \tag{18}$$

respectively.

In the above formulas, k_r, k_w, and k_φ are expressed as

$$k_r = \frac{33EI}{L^3}, k_w = \frac{2EA}{L}, k_\varphi = \frac{4GI}{L} \tag{19}$$

respectively.

2.2.2. The Second Stage

In the second stage of the sidetracking process, the window mill has been milled out into the formation, and then, the pilot mill begins milling the casing. The kinetic energy and potential energy of the sidetracking tool system are

$$T = \frac{1}{2}m_1(\dot{x}_1{}^2 + \dot{y}_1{}^2 + \dot{z}_1{}^2) + \frac{1}{2}J_1\dot{\varphi}_1{}^2 + \frac{1}{2}m_2\dot{z}_2{}^2 + \frac{1}{2}J_2\dot{\varphi}_2{}^2 \tag{20}$$

and

$$V = \frac{1}{2}\int_0^L \left\{EI\left[(u'')^2 + (v'')^2\right] + EA(w')^2 + 2GI(\theta')^2\right\}dz \tag{21}$$

respectively, where, ($\dot{}$) denotes a time derivative.

The system is divided into two parts, according to the simplified model shown in Figure 4. Therefore, the boundary conditions related to the axial and lateral motions of the above system are

$$\begin{cases} w|_{z=0} = v_0 t \\ w|_{z=\frac{L}{2}} = z_1 \\ w|_{z=L} = z_2 \end{cases} \tag{22}$$

and

$$\begin{cases} \theta|_{z=0} = \varphi_t \\ \theta|_{z=\frac{L}{2}} = \varphi_1 \\ \theta|_{z=L} = \varphi_2 \end{cases} \tag{23}$$

respectively.

Thus, the $w(z,t)$, $\theta(z,t)$, $u(z,t)$ and $v(z,t)$ are

$$w(z,t) = \begin{cases} \frac{z_1 - v_0 t}{L/2} z + v_0 t, & 0 \le z \le \frac{L}{2} \\ \frac{z_2 - z_1}{L/2} z - z_2 + 2z_1, & \frac{L}{2} < z \le L \end{cases} \tag{24}$$

$$\theta(z,t) = \begin{cases} \frac{\varphi_1 - \varphi_t}{L/2} z + \varphi_t, & 0 \le z \le \frac{L}{2} \\ \frac{\varphi_2 - \varphi_1}{L/2} z - \varphi_2 + 2\varphi_1, & \frac{L}{2} < z \le L \end{cases} \tag{25}$$

$$u(z,t)=\begin{cases}\frac{x_1}{L^3}\left(3L^2z-4z^3\right) & ,0\le z\le\frac{L}{2}\\ \frac{x_1}{L^3}\left(3L^2(L-z)-4(L-z)^3\right) & ,\frac{L}{2}<z\le L\end{cases} \tag{26}$$

and

$$v(z,t)=\begin{cases}\frac{y_1}{L^3}\left(3L^2z-4z^3\right) & ,0\le z\le\frac{L}{2}\\ \frac{y_1}{L^3}\left(3L^2(L-z)-4(L-z)^3\right) & ,\frac{L}{2}<z\le L\end{cases} \tag{27}$$

respectively.

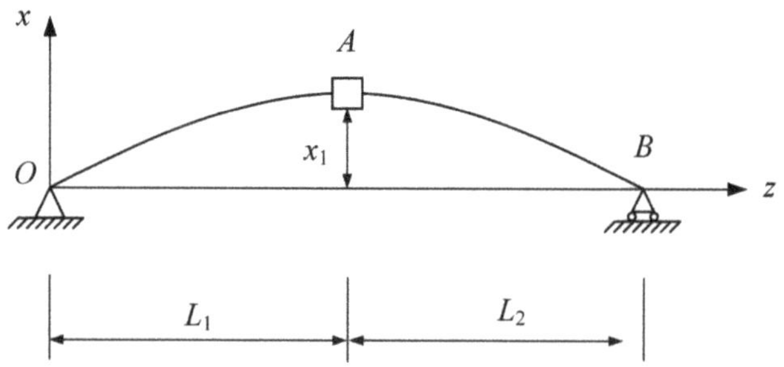

Figure 4. Deformation diagram of the sidetracking tool in the second stage.

Substituting Equations (24)–(27) into potential energy expression (21), the expression of potential energy can be obtained as

$$V=\frac{EA}{L}\left[(z_2-z_1)^2+(z_1-v_0t)^2\right]+\frac{GI}{L}\left[(\varphi_1-\varphi_t)^2+(\varphi_2-\varphi_1)^2\right]+\frac{24EI}{L^3}\left({x_1}^2+{y_1}^2\right) \tag{28}$$

The Rayleigh dissipation function D is

$$D=\frac{1}{2}\left[c_a\left({\dot{z}_1}^2+{\dot{z}_2}^2\right)+c_b\left({\dot{x}_1}^2+{\dot{y}_1}^2\right)+c_t\left({\dot{\varphi}_1}^2+{\dot{\varphi}_2}^2\right)\right] \tag{29}$$

Substituting T, V, and D into the Lagrange equation, the result is

$$[M]\{\ddot{x}\}+[C]\{\dot{x}\}+[K]\{x\}=\{F\} \tag{30}$$

where $[M]$, $[C]$, and $[K]$ represent the mass matrix, damping matrix, and stiffness matrix, respectively, and $\{F\}$ is the column vector of the force, and their expressions are

$$[M]=\begin{bmatrix}m_1 &&&&& \\ & m_1 &&&& \\ && m_1 &&& \\ &&& m_2 && \\ &&&& J_1 & \\ &&&&& J_2\end{bmatrix} \tag{31}$$

$$[C]=\begin{bmatrix}c_b &&&&& \\ & c_b &&&& \\ && c_a &&& \\ &&& c_a && \\ &&&& c_t & \\ &&&&& c_t\end{bmatrix} \tag{32}$$

$$[K] = \begin{bmatrix} k_r & & & & & \\ & k_r & & & & \\ & & 2k_w & -k_w & & \\ & & -k_w & k_w & & \\ & & & & 2k_\varphi & -k_\varphi \\ & & & & -k_\varphi & k_\varphi \end{bmatrix} \tag{33}$$

and

$$\{F\} = \begin{Bmatrix} F_{1x} + m_1 e_1 \dot{\varphi}_1{}^2 \cos \varphi_1 \\ F_{1y} + m_1 e_1 \dot{\varphi}_1{}^2 \sin \varphi_1 \\ K_w \cdot v_0 t + m_1 g - sign(\dot{z}_1) \mu_r F_{n1} \\ -W_b - F_b \sin \alpha \\ K_\varphi \cdot \varphi_t - T_1 \\ -T_b \end{Bmatrix} \tag{34}$$

respectively. In the above formulas, k_r, k_w, and k_φ are expressed as

$$k_r = \frac{48EI}{L^3}, k_w = \frac{2EA}{L} \text{ and } k_\varphi = \frac{2GI}{L} \tag{35}$$

respectively.

2.2.3. The Third Stage

In the third stage, the window mill and pilot mill have finished milling the casing and entered the formation, and the watermelon mill begins milling. At this time, the sidetracking tool system can be regarded as an axial force link element, as shown in Figure 5, and the kinetic energy and potential energy of the system are

$$T = \frac{1}{2} m_1 \dot{z}_1{}^2 + \frac{1}{2} J_1 \dot{\varphi}_1{}^2 \tag{36}$$

and

$$V = \frac{1}{2} \int_0^L \left\{ EI \left[(u'')^2 + (v'')^2 \right] + EA (w')^2 + 2GI (\theta')^2 \right\} dz \tag{37}$$

respectively.

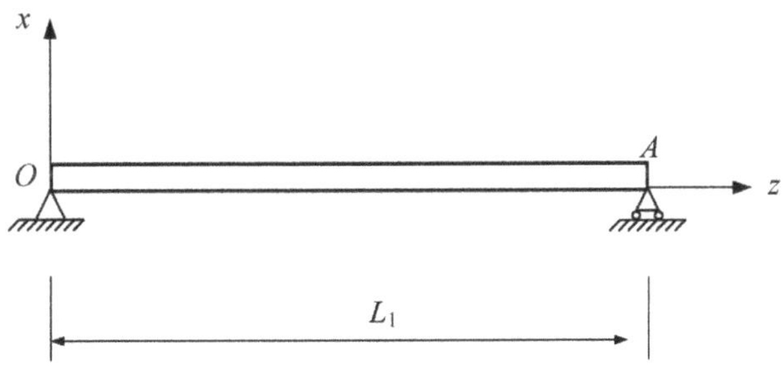

Figure 5. Deformation diagram of the sidetracking tool in the third stage.

Then, the vibration equation of this system is

$$[M]\{\ddot{x}\} + [C]\{\dot{x}\} + [K]\{x\} = \{F\} \tag{38}$$

where, $[M]$, $[C]$, and $[K]$ represent the mass matrix, damping matrix, and stiffness matrix, respectively, and $\{F\}$ is the column vector of the force; their expressions are

$$[M] = \begin{bmatrix} m_1 & \\ & m_1 \end{bmatrix} \tag{39}$$

$$[C] = \begin{bmatrix} c_a & \\ & c_t \end{bmatrix} \tag{40}$$

$$[K] = \begin{bmatrix} k_w & -k_w \\ -k_w & k_w \end{bmatrix} \tag{41}$$

and

$$\{F\} = \begin{Bmatrix} -W_b - F_b \sin\alpha \\ -T_b \end{Bmatrix} \tag{42}$$

respectively.

In the above formulas, k_w is expressed as

$$k_w = \frac{EA}{L} \tag{43}$$

2.3. Contact Model

The sidetracking tool moves within the casing, and its radial displacement should not exceed the clearance between the window mill and the casing inner wall. The radial displacement of the window mill center from the casing center is marked as ρ_i. The radial gap between the window mill and the inner casing is marked as β. The distance of the window mill into the casing wall is marked as δ_i. Here, a judgment parameter λ is introduced to judge whether the mills contact the casing wall. When $\lambda_i = 0$, it indicates that there is no contact between them, while when $\lambda_i = 1$, contact occurs. Their expressions are

$$\rho_i = \sqrt{x_i{}^2 + y_i{}^2} \tag{44}$$

$$\beta = \frac{D_c - D_{out}}{2} \tag{45}$$

$$\delta_i = \rho_i - \beta \tag{46}$$

and

$$\lambda_i = \begin{cases} 0, & \delta_i \le 0 \\ 1, & \delta_i > 0 \end{cases} \tag{47}$$

respectively.

Based on the previous simplifying assumptions, the contact can be described by a linear elastic model. The contact force acting on the sidetracking tool can be divided into two parts, including the normal force F_n and the tangential force F_t due to friction [30], which can be defined as

$$F_{ni} = \lambda_i(k_p\delta_i + c_p v_{ni}) \tag{48}$$

and

$$F_{ti} = \mu_r\left[\tanh(2v_{ri}) + \frac{2v_{ri}}{1 + 4v_{ri}2}\right]F_{ni} \tag{49}$$

respectively. The normal velocity is

$$v_{ni} = \dot{x}_i \cdot \frac{y_i}{\sqrt{x_i{}^2 + y_i{}^2}} + \dot{y}_i \cdot \frac{x_i}{\sqrt{x_i{}^2 + y_i{}^2}} \tag{50}$$

If contact occurs, the motion of the sidetracking tool with respect to the casing can be described as pure rolling or sliding. To determine the presence of sliding, the relative velocity between the two faces is marked as

$$v_{ri} = -\dot{x}_i \cdot \frac{x_i}{\sqrt{x_i^2 + y_i^2}} + \dot{y}_i \cdot \frac{y_i}{\sqrt{x_i^2 + y_i^2}} + \frac{D_{out}}{2}\dot{\varphi}_i \quad (51)$$

When $v_r \neq 0$, it means there is only relative sliding, while when $v_r = 0$, it means that there is no relative sliding, but only pure rolling. By superimposing the contact force components along their respective vectors in the x and y directions, the external forces from the casing wall acting on the mill by are calculated as

$$\begin{aligned} F_{xi} &= -F_{ni} \cdot \frac{y_i}{\sqrt{x_i^2+y_i^2}} + F_{ti} \cdot \frac{x_i}{\sqrt{x_i^2+y_i^2}} \\ F_{yi} &= -F_{ni} \cdot \frac{x_i}{\sqrt{x_i^2+y_i^2}} - F_{ti} \cdot \frac{y_i}{\sqrt{x_i^2+y_i^2}} \end{aligned} \quad (52)$$

The friction moment on the sidetracking tool is

$$T_i = \frac{1}{2}F_{ti}D_{out} \quad (53)$$

The weight on bit and drill bit torque during sidetracking can be defined as

$$Wob = \begin{cases} k_c(z_3 - s_0 \sin(n_b\varphi)), z_3 < s_0 \sin(n_b\varphi) \\ 0 \qquad\qquad\qquad\qquad\quad , z_3 \geq s_0 \sin(n_b\varphi) \end{cases} \quad (54)$$

and

$$Tob = Wob\left[\frac{2}{3}r_{out}f(\dot{\varphi}_3) + \varsigma\sqrt{r_{out}\delta_c}\right] \quad (55)$$

respectively, wherein the depth of cut per revolution and the average rate of penetration are

$$\delta_c = \frac{2\pi ROP}{\omega_d} \quad (56)$$

and

$$ROP = c_1 W_0 \sqrt{\omega_d} + c_2 \quad (57)$$

respectively. A continuous function f is used to express the influence of bit speed on bit torque [31], and its expression is

$$f(\dot{\varphi}_3) = \frac{2}{\pi}a\tan(\varepsilon\dot{\varphi}_3)(\frac{\mu_e - \mu_d}{1 + \tau|\dot{\varphi}_3|} + \mu_d) \quad (58)$$

3. Simulation of the Sidetracking Tool Motion Process

A numerical procedure is used to solve the obtained equations system, which can be solved using the Runge–Kutta method. Some system parameters involved in the calculations and their specific values are listed in Table 1, which represent typical situations in sidetracking operations.

3.1. The First Stage

Analyzing the curve diagram of displacement with respect to time variation can better describe the movement of the sidetracking tool. Figures 6 and 7 both show the historical curve of time displacement within the test time. Figure 6 shows the change diagram of lateral displacement with respect to time of Parts A and B, and Figure 7 shows the time variation of the angular displacements of Parts A, B, and C. In the above two figures, x_1 and x_2 represent the lateral displacement of Parts A and B, respectively, and φ_1, φ_2, and φ_3 represent the angular displacement of Parts A, B, and C, respectively. Figure 6 only discusses the lateral displacement in the x-direction because the bending in the y-direction is symmetrical with the bending in the x-direction.

Table 1. System parameters values used for numerical study.

Quantities	Variable	Value	Units
Unbalanced mass on BHA	m_b	10	kg
Tool external diameter	D_{out}	215.9	mm
Tool internal diameter	D_{in}	86.1	mm
Casing internal diameter	D_c	244.5	mm
Linear eccentricity	e	8	mm
Elastic modulus	E	2.1×10^{11}	Pa
Shear modulus	G	7.7×10^{10}	Pa
Density	ρ	7800	kg/m^3
Distance between restraint end and watermelon mill	L_1	640	mm
Distance between pilot mill and watermelon mill	L_2	320	mm
Distance between the window mill and the pilot mill	L_3	320	mm
Whipstock angle	α	2	°
Static friction coefficient	μ_e	0.35	
Dynamic friction coefficient	μ_d	0.1	
Initial weight on bit	W_0	30×10^3	N
Supporting force of whipstock	F_{ob}	200×10^3	N
Initial speed	θ_1	3.14	rad/s
Initial angle	φ_t	10	°
Contact stiffness of tool and casing	k_p	1×10^8	N/m
Contact stiffness of window mill	k_c	25×10^3	
Damping of tool and casing	c_p	5×10^3	N·s/m
Steepness parameter	χ	0.5	
Empirical constants	c_1	1.35×10^{-8}	
Empirical constants	c_2	1.9×10^{-4}	
Forced parameters related to borehole irregularity	s_0	0.001	
Dimensionless coefficient	ζ	0.1	
Bit coefficient	n_b	3	

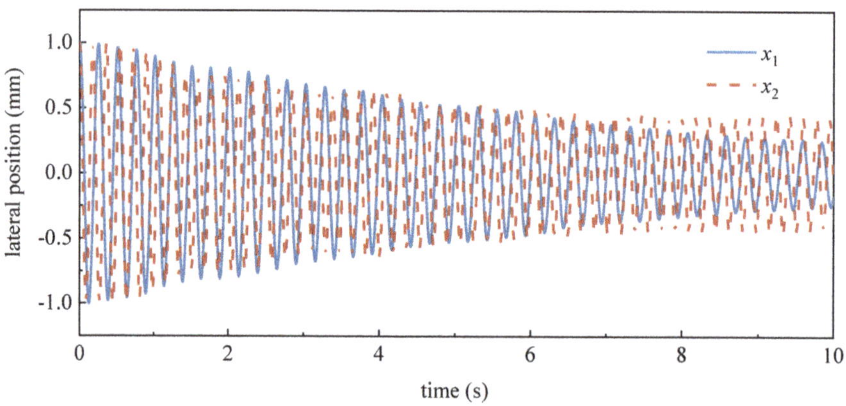

Figure 6. Schematic diagram of lateral displacements x_1 and x_2 with respect to t.

The length of the sidetracking tool analyzed in this paper is 1280 mm, and within this measurement range, only a small range of vibration occurs. In Figure 6, the amplitude of the window mill and the pilot mill at the initial time in the x direction is relatively large, and the amplitude gradually decreases with the passage of time. This phenomenon indicates that the window mill rolls continuously in the axial direction. The amplitude of φ_3 is always larger than φ_1 and φ_2 in the measured range. In the first stage of sidetracking, the window mill just begins to contact the whipstock, and in addition to the friction of casing wall, the system is also subjected to the lateral support of the whipstock, while the movements of the other two parts are still in the casing, so the dynamic response of Part C is more obvious and larger than that of the other two parts.

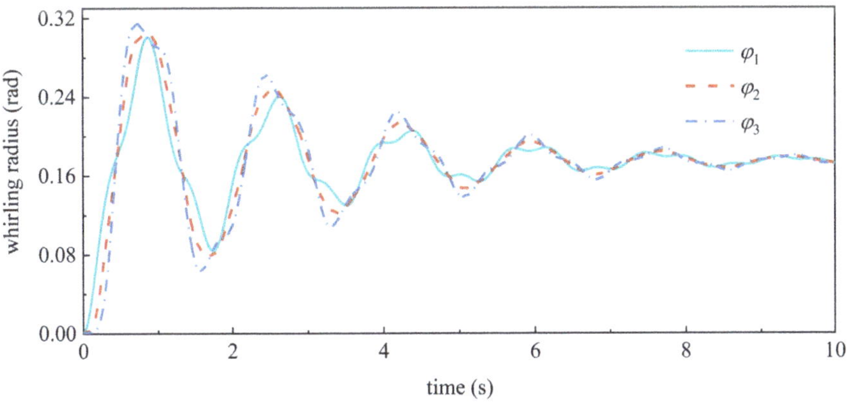

Figure 7. Schematic diagram of angular displacements φ_1, φ_2 and φ_3 with respect to *t*.

Figure 8 shows the angular velocity variation of the three Parts A, B, and C with respect to time in the first stage. The rotation direction varies continuously between forward and backward during the time period covered by the simulation, and all three curves are faster in the first half of the period. This model predicts the contact between the sidetracking tool and the casing, but the sidetracking tool is short, and its vibration amplitude is smaller than the gap between the shoe and the casing wall. The vibration is finally stabilized as a limit ring behavior due to the presence of dissipative effects such as friction.

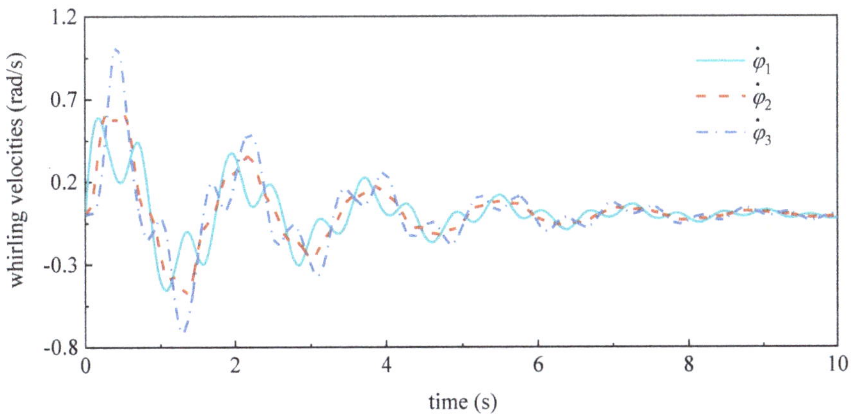

Figure 8. Schematic diagram of angular velocity of Parts A, B and C with respect to t, respectively.

Figure 9 shows the lateral motion trajectory of Part C. It can be seen from the figure that the window mill can only roll back and forth in the direction parallel to the surface of the whipstock. It is worth noting that the vibration trajectory shape changes from the vortex shape to the X reciprocating shape, which is very different from the behavior of conventional drilling tools. Its vibration amplitude is smaller than the gap between the window mill and the casing wall, and there is no random collision with the casing wall. The maximum lateral displacement is 3.30 mm.

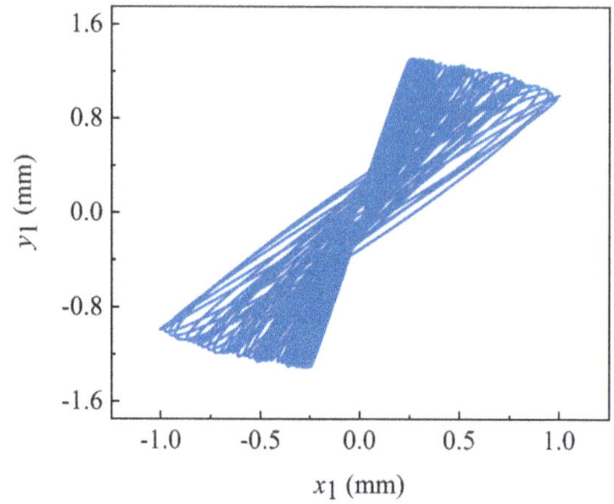

Figure 9. The movement trajectory of the window mill.

3.2. The Second Stage

During the second stage of the sidetrack process, the window mill is milled into the formation, and the pilot mill begins to mill the casing. The research objects in this stage are Parts A and B.

Figures 10 and 11 both show the historical curve of time-displacement in the test time. Figure 10 shows the lateral displacement change diagram of Part A with respect to time, and Figure 11 shows the angular displacement change diagram of Parts A and B with respect to time. In the above two figures, x_1 and y_1 represent the lateral displacement of Part A, respectively, and φ_1 and φ_2 represent the angular displacement of Parts A and B, respectively.

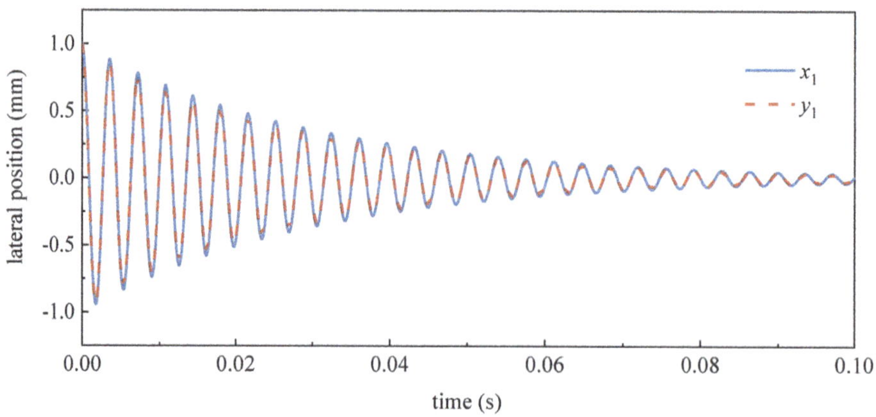

Figure 10. Schematic diagram of lateral displacements x_1 and y_1 with respect to t.

From Figure 10, it can be seen that the amplitude of the pilot mill is larger in the lateral direction at the initial moment, and it gradually decreases with time. This phenomenon indicates that the pilot mill continuously tumbles in the axial direction. The angular displacement curve (Figure 11) shows obvious vibration in the first half of the study period, and gradually levels off in the second half. In addition, φ_2 is always larger than φ_1. Figure 12 shows the angular speed variation of the two Parts A, B with respect to time in the second stage. The angular speed curve shown in Figure 12 has obvious vibration in the first half of the study period, and gradually level off in the second half. In addition, the angular speed of Part B is always larger than the angular speed of Part A.

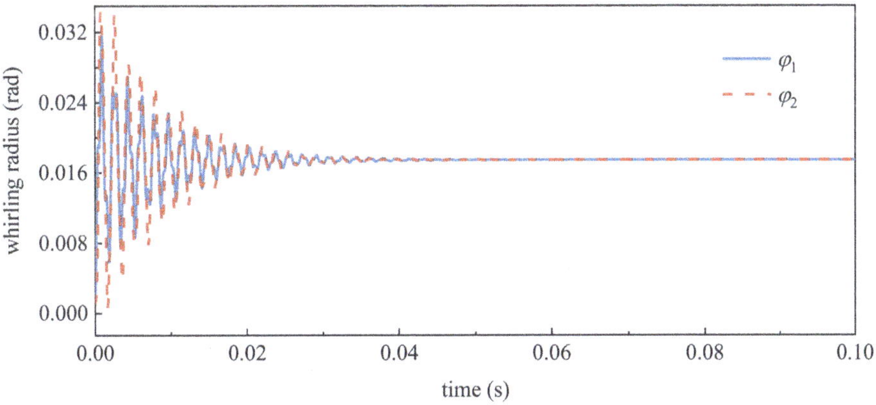

Figure 11. Schematic diagram of angular displacements φ_1 and φ_2 with respect to t.

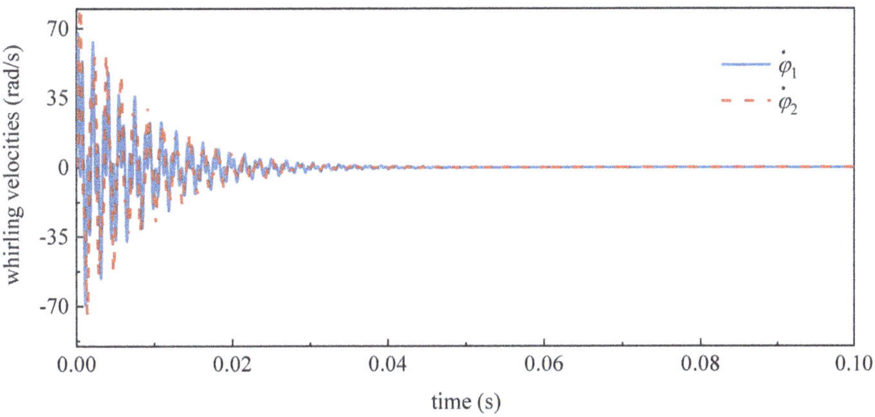

Figure 12. Schematic diagram of angular velocity with respect to t.

In the second stage of the open window, the window mill has entered the formation, and the pilot mill is in contact with the whipstock; thus, it is subject to the frictional force of the casing wall and the lateral support force of the whipstock. Therefore, the dynamic response of Part B is more obvious compared with the other two parts, and the amplitude is larger. Over time, the motion of the sidetracking tools all gradually stabilized.

The lateral movement trajectory of Part B is shown in Figure 13. It can be seen that the pilot mill rolls back and forth in the direction parallel to the surface of the whipstock. The pilot mill is located in the middle of the window mill and the watermelon mill in the axial direction. Both of its ends are subjected to tension and restriction, so its motion range is smaller than that of the window mill, which is in the first stage. The maximum lateral displacement at this moment is 2.74 mm.

3.3. The Third Stage

In the third stage, the window and pilot mills have entered the formation, and the watermelon mill is in the window position. In this case, the sidetracking tool system can be considered as a cantilever beam structure, with a fully restrained upper end and a lower end subject to the lateral forces provided by the whipstock and the weight on bit generated during the drilling process. Numerical simulation is used to calculate the lateral displacement of the tool system under the action of weight on bit and lateral force. The specific steps are as follows.

1. The upper end of the tool system is fully constrained, and the lateral force in the positive Z-axis is applied to the ground sole circle at the lower end. The lateral force is 23.3 kN.
2. Subsequently, the upper end of the sidetracking tool is fully constrained, which has been deformed at this moment, and a WOB of 5 t and a torque of 130 kN·m are applied to its lower surface.
3. The lateral displacements of the sidetracking tool under the weight on bit, torque, and lateral force are shown in Figure 14.

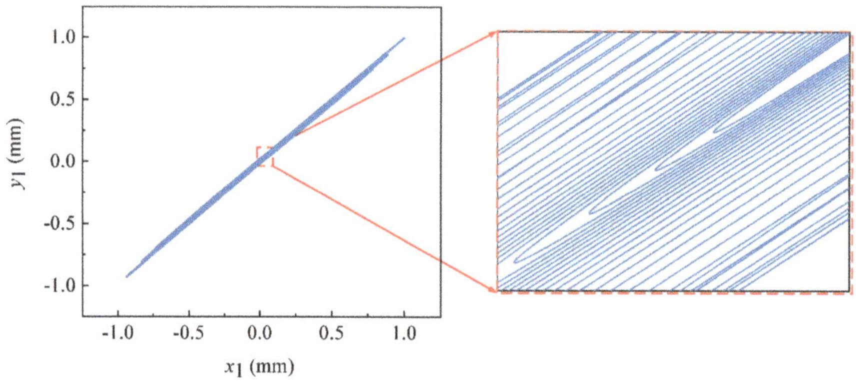

Figure 13. The movement trajectory of the pilot mill.

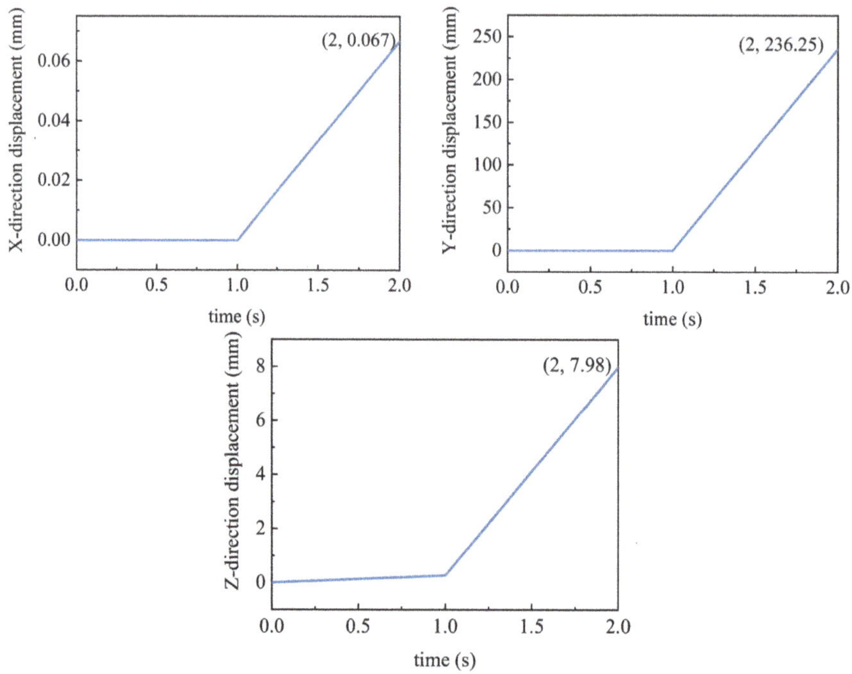

Figure 14. Displacement of BHA.

In Figure 14, the first half of the curve indicates that the tool has not yet touched the whipstock and is only subject to weight on bit and torque, producing only a small lateral displacement, which is negligible. While the latter half indicates that the tool has contacted the whipstock. The lateral force of the whipstock acts on the tool y-direction, so the displacement in this direction is the largest. None of the external forces on the tool had a large effect on the displacement in the x-direction. From the results of the numerical simulation, it can be seen that the maximum lateral displacement in the x-direction of the sidetracking tool is 0.067 mm.

3.4. Case Analysis

Through the above analysis, the window width, which is derived from the action of an arbitrary size casing and the mills, can be calculated. A longitudinal profile of the casing is taken from the centerline of the casing. In this case, the plane is perpendicular to the casing wall, and the Cartesian coordinate system is established, as shown in Figure 15.

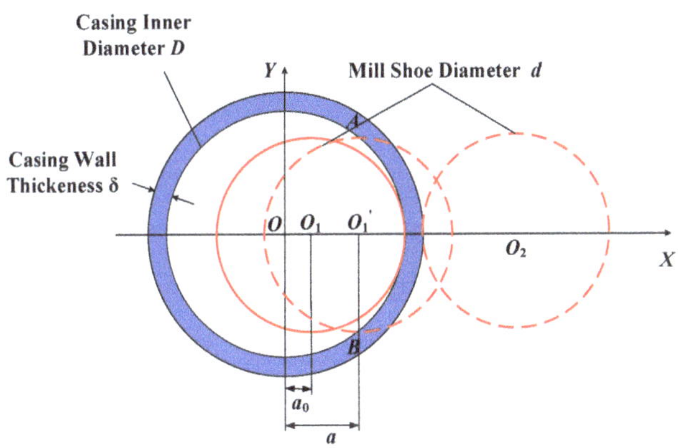

Figure 15. Schematic to model the lateral motions of window mill.

The central point of casing on this plane coincides with the central point of the window mill, and this point is set as the origin of the coordinates O. The forward direction of the window mill is taken as the X axis, and the direction perpendicular to the X axis is taken as the Y axis. The casing inner diameter, marked as D, represents the thickness, marked as δ, and the outer diameter is marked as $D + \delta$. The outer diameter of the window mill is marked as d. γ represents the bevel angle of the whipstock, and a represents the distance between the center of the casing and the center of the window mill. The window width produced by the window mill on the casing can be regarded as the length between the intersection points of two circles on the XY plane, as shown in Figure 15. The derived theoretical formula for the window width is

$$W = 2\sqrt{\frac{D^2 + d^2}{8} - \frac{\left(\frac{D-d}{2} + x\right)^2}{4} - \frac{\left(D^2 - d^2\right)^2}{64\left(\frac{D-d}{2} + x\right)^2}}, x \in [0, d + \delta] \tag{59}$$

where x is the distance between the tool cross section center and the casing cross section center. When $D = 244.48$ mm, $d = 184.15$ mm, and $\delta = 13.84$ mm, the ideal window width calculated by Equation (59) is $W = 368.23$ mm.

The schematic diagram of the window result formed by the sidetracking tool milling casing wall under stress is shown in Figure 16, where curve (1) represents the theoretical window profile calculated by Equation (59). Curve (2) represents the window profile formed by the window mill, and the maximum lateral displacement generated is 3.30 mm. Curve (3) is the window profile formed by the pilot mill, and the maximum lateral displacement generated by it is 2.74 mm. Curve (4) is the profile formed by watermelon mill, whose maximum lateral displacement is 0.07 mm. Therefore, the maximum width of the final window is 374.34 mm.

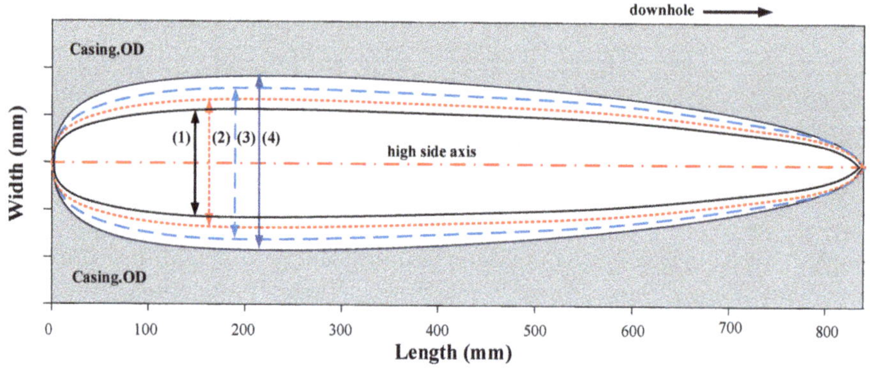

Figure 16. Schematic diagram of the window profile.

4. Conclusions

A study of the coupled axial, lateral, and torsional vibrations of an actively controlled sidetracking tool has been presented and some dynamic models have been proposed, including a ten degrees of freedom, six degrees of freedom, and a two degree of freedom dynamics model, which can be used to analyze the effect of the sidetracking tool system vibration on the casing window profile.

1. The sidetracking tool is taken as the research focus during window opening, and the simplified physical model is established including Parts A, B, and C. The vibration model of the sidetracking tool is obtained according to the Lagrange equation, which can be solved by the Runge–Kutta numerical method. This model can be used to analyze the force and motion process of different mills during the sidetracking.
2. The dynamic model is numerically solved by Runge–Kutta method, and the dynamic responses of different parts of the sidetracking tool at different stages are obtained. The analysis shows that regarding the drilling of the sidetracking tool, the vibration amplitude of the mills colliding with the whipstock is the largest. Its displacement and angular velocity are larger than those of the other parts. The vibration trajectory pattern of the sidetracking tool is different from that of the conventional drilling tool due to the influence of the whipstock, which changes from the general whirling motion pattern to the X reciprocating pattern, and the vibration amplitude decreases.
3. The sidetracking tool analyzed in this paper does not include the drill pipe, but only the mills. Within this measurement range, a small range of vibration occurs, which is smaller than the gap between the mill and the casing wall.
4. The window profile is widened by the sidetracking tool vibration deformation, and the extension of the window width by the window mill and the pilot mill are 3.30 mm and 2.74 mm, respectively. The extension of the window width by the watermelon mill is 0.07 mm, so the maximum window width formed by the sidetracking tool is 374.34 mm.

During the casing sidetracking process, the actual window shape will also be affected by parameters such as mill shape, casing material, and formation hardness. In the next study, more influencing factors will be considered to further improve the accuracy and applicability of window opening tools.

Author Contributions: Conceptualization, X.Z. and B.Z.; methodology, X.Z.; software, W.Z.; formal analysis, W.Z.; investigation, W.Z., Y.L. and P.J.; data curation, Y.L. and P.J.; writing—original draft preparation, X.Z., W.Z., Y.L. and P.J.; writing—review and editing, S.X., B.Z. and Y.X.; visualization, X.Z. and W.Z.; supervision, S.X. and B.Z.; funding acquisition, S.X. All authors have read and agreed to the published version of the manuscript.

Funding: This research was funded by [the National Key Research and Development Program of China] grant number [No. 2019YFC1509204] and [the Independent Innovation Research Program of China University of Petroleum (East China)] grant number [No.27RA2215005].

Conflicts of Interest: The authors declare no conflict of interest.

References

1. Kong, C.; Zhang, D.; Liang, Z.; Cai, R. Mechanical analysis of drill bit during dual-casing sidetracking. *J. Pet. Sci. Eng.* **2017**, *159*, 970–976. [CrossRef]
2. Yang, Y.; Yang, Y.; Ren, H.; Qi, Q.; Chen, X. Research on the working mechanism of the PDC drill bit in compound drilling. *J. Pet. Sci. Eng.* **2020**, *185*, 106647. [CrossRef]
3. Yuan, Y.; Qi, Z.; Yan, W.; Zhang, J.; Chen, C. Old well sidetracking selection standards: Sulige low-permeability gas field, China. *J. Pet. Explor. Prod. Technol.* **2020**, *10*, 1699–1709. [CrossRef]
4. Han, Y.; Ai, Z.; Kuang, Y.; Liu, J.; Yang, W. Nonlinear dynamic modeling of drill string in horizontal well—A geometrically exact approach. *J. Pet. Sci. Eng.* **2018**, *172*, 1133–1152. [CrossRef]
5. Feng, T.; Bakshi, S.; Gu, Q.; Yan, Z.; Chen, D. Design optimization of bottom-hole assembly to reduce drilling vibration. *J. Pet. Sci. Eng.* **2019**, *179*, 921–929. [CrossRef]
6. Divenyi, S.; Savi, M.A.; Franca, L.F.P.; Weber, H.I. Nonlinear dynamics and chaos in systems with discontinuous support. *Shock. Vib.* **2006**, *13*, 315–326. [CrossRef]
7. Richard, T.; Germay, C.; Detournay, E. A simplified model to explore the root cause of stick–slip vibrations in drilling systems with drag bits. *J. Sound Vib.* **2007**, *305*, 432–456. [CrossRef]
8. Yigit, A.; Christoforou, A. Coupled torsional and bending vibrations of actively controlled drillstrings. *J. Sound Vib.* **2000**, *234*, 67–83. [CrossRef]
9. Germay, C.; Denoël, V.; Detournay, E. Multiple mode analysis of the self-excited vibrations of rotary drilling systems. *J. Sound Vib.* **2009**, *325*, 362–381. [CrossRef]
10. Choi, S.H.; Pierre, C.; Ulsoy, A.G. Consistent modeling of rotating timoshenko shafts subject to axial loads. *J. Vib. Acoust.* **1992**, *114*, 249–259. [CrossRef]
11. Ahmadian, H.; Nazari, S.; Jalali, H. Drill string vibration modeling including coupling effects. *Int. J. Ind. Eng. Prod. Res.* **2007**, *18*, 59–66. Available online: http://ijiepr.iust.ac.ir/article-1-30-fa.html (accessed on 1 October 2022).
12. Nandakumar, K.; Wiercigroch, M. Stability analysis of a state dependent delayed, coupled two DOF model of drill-stringvibration. *J. Sound Vib.* **2013**, *332*, 2575–2592. [CrossRef]
13. De Moraes, L.P.; Savi, M.A. Drill-string vibration analysis considering an axial-torsional-lateral nonsmooth model. *J. Sound Vib.* **2018**, *438*, 220–237. [CrossRef]
14. Xie, D.; Huang, Z.; Ma, Y.; Vaziri, V.; Kapitaniak, M.; Wiercigroch, M. Nonlinear dynamics of lump mass model of drill-string in horizontal well. *Int. J. Mech. Sci.* **2020**, *174*, 105450. [CrossRef]
15. Mohammadzadeh, M.; Shahgholi, M.; Arbabtafti, M.; Yang, J. Vibration analysis of the fully coupled nonlinear finite element model of composite drill strings. *Arch. Appl. Mech.* **2020**, *90*, 1373–1398. [CrossRef]
16. Mohammadzadeh, M.; Arbabtafti, M.; Shahgholi, M.; Yang, J. Nonlinear vibrations of composite drill strings considering drill string-wellbore contact and bit-rock interaction. *Arch. Appl. Mech.* **2022**, *92*, 2569–2592. [CrossRef]
17. Wang, P.; Ni, H.; Wang, X.; Wang, R. Modelling the load transfer and tool surface for friction reduction drilling by vibrating drill-string. *J. Pet. Sci. Eng.* **2018**, *164*, 333–343. [CrossRef]
18. Tian, J.; Li, G.; Dai, L.; Yang, L.; He, H.; Hu, S. Torsional vibrations and nonlinear dynamic characteristics of drill strings and stick-slip reduction mechanism. *J. Comput. Nonlinear Dyn.* **2019**, *14*, 4043564. [CrossRef]
19. Tian, J.; Liu, Y.; Xiong, J.; Wen, F.; Liu, G. Analysis of drill string's vibration characteristics based on sliding mode-PI controller. *J. Vib. Eng. Technol.* **2022**, *10*, 919–927. [CrossRef]
20. Ritto, T.G.; Ghandchi-Tehrani, M. Active control of stick-slip torsional vibrations in drill-strings. *J. Vib. Control* **2018**, *25*, 194–202. [CrossRef]
21. Xu, Y.; Zhang, H.; Guan, Z. Dynamic characteristics of downhole bit load and analysis of conversion efficiency of drill string vibration energy. *Energies* **2021**, *14*, 229. [CrossRef]
22. Sassi, S.; Renno, J.; Zhou, H.; Baz, A. Experimental investigation of the vibration control of nonrotating periodic drill strings. *J. Vib. Acoust.* **2021**, *143*, 4049942. [CrossRef]
23. Sadeghi, A.N.; Arıkan, K.B.; Özbek, M.E. Modelling and controlling of drill string stick slip vibrations in an oil well drilling rig. *J. Pet. Sci. Eng.* **2022**, *216*, 110759. [CrossRef]

24. Sarker, M.; Rideout, D.G.; Butt, S.D. Dynamic model for 3D motions of a horizontal oilwell BHA with wellbore stick-slip whirl interaction. *J. Pet. Sci. Eng.* **2017**, *157*, 482–506. [CrossRef]
25. Liu, Y.; Gao, D. A nonlinear dynamic model for characterizing downhole motions of drill-string in a deviated well. *J. Nat. Gas Sci. Eng.* **2017**, *38*, 466–474. [CrossRef]
26. Wang, J.; Xue, Q.; Li, L.; Huang, L.; Li, F.; Liu, B. Influence of flex-sub on mechanical properties of rotary steerable drilling system. *Mech. Sci.* **2020**, *11*, 285–297. [CrossRef]
27. Navarro-López, E.M. An alternative characterization of bit-sticking phenomena in a multi-degree-of-freedom controlled drillstring. *Nonlinear Anal. Real World Appl.* **2009**, *10*, 3162–3174. [CrossRef]
28. Tian, J.; Yang, Y.; Yang, L. Vibration characteristics analysis and experimental study of horizontal drill string with wellbore random friction force. *Arch. Appl. Mech.* **2017**, *87*, 1439–1451. [CrossRef]
29. Liao, C.-M.; Vlajic, N.; Karki, H.; Balachandran, B. Parametric studies on drill-string motions. *Int. J. Mech. Sci.* **2012**, *54*, 260–268. [CrossRef]
30. Caijin, Y.; Zaibin, C.; Wei, J.; Shiquan, J.; Gexue, R. A Multibody dynamic model of drillstring for torque and drag analysis. *J. Offshore Mech. Arct. Eng.* **2015**, *137*, 31403. [CrossRef]
31. Christoforou, A.; Yigit, A. Fully coupled vibrations of actively controlled drillstrings. *J. Sound Vib.* **2003**, *267*, 1029–1045. [CrossRef]

energies

Article

Design and Experimental Research of a Wellhead Overflow Monitoring System for Open-Circuit Drilling of Natural Gas Hydrate

Chao Zhong [1,2,*], Jing'an Lu [1,2] and Dongju Kang [1,2]

[1] Guangzhou Marine Geological Survey, China Geological Survey, Guangzhou 511458, China
[2] National Engineering Research Center of Gas Hydrate Exploration and Development, Guangzhou 511458, China
* Correspondence: zcjf_999@126.com

Abstract: Natural gas hydrate is easy to decompose and leak due to the changes in temperature and pressure during drilling, which causes safety accidents. Early monitoring of wellhead overflow is a practical and effective measure to prevent overflow blowouts and other accidents. Herein, a wellhead methane monitoring system for the open-circuit drilling of marine natural gas was designed. The system consisted of an overwater acoustic reception part and an underwater self-contained methane monitoring part, matching the construction environment of marine natural gas hydrate exploitation. Compared with the existing gas logging technology (measurement while drilling), the monitoring and early warning of wellhead methane content were realized at all stages of drilling, casing running, cementing, completion and fracturing in the process of natural gas hydrate exploitation. System communication and data acquisition tests were completed at different water depths through sea trials, which verified the effectiveness of the system design. The research results provide important theoretical and technical implications for promoting the development of early spill monitoring technology at the wellhead of open-circuit drilling for marine gas hydrates.

Keywords: natural gas hydrate; open-circuit drilling; wellhead monitoring; methane concentration; trial

Citation: Zhong, C.; Lu, J.; Kang, D. Design and Experimental Research of a Wellhead Overflow Monitoring System for Open-Circuit Drilling of Natural Gas Hydrate. *Energies* **2022**, *15*, 9606. https://doi.org/10.3390/en15249606

Academic Editors: Mofazzal Hossain, Md Motiur Rahman and Hisham Khaled Ben Mahmud

Received: 27 October 2022
Accepted: 14 December 2022
Published: 18 December 2022

1. Introduction

Natural gas hydrates remain stable under the high-pressure and low-temperature environment of the seafloor [1]. During the drilling process, the changes in temperature and pressure easily lead to the decomposition of hydrates [2]. When the natural gas overflow into the wellbore is not detected in time and controlled effectively, more natural gas overflow enters the wellbore and expands continuously during the transportation process. The rapid expansion of natural gas overflow triggers borehole dilation, collapse, gas precipitation (gas protrusion phenomenon) and even well blowout approaching the wellhead, among which uncontrolled well blowout is a catastrophic accident with huge losses in the process of oil and gas resource exploration and development [3,4]. Early overflow monitoring is one of the main technical tools to achieve well blowout prevention in oil and gas wells [5].

Due to the complex downhole environment, the conventional platform detection method is delayed in detection results, resulting in a significant reduction in well control response time, which cannot meet the requirements of early monitoring of hydrate drilling overflow. Moreover, the volume of the gas invading annular drilling fluid for the first time is very small under the influence of a deep-water high-pressure environment, having little impact on the drilling fluid. During the continuous upward return of gas with drilling fluid, the volume of gas rapidly expands in the presence of decompression, especially when approaching the wellhead. Therefore, there is not enough time to judge the degree of gas invasion according to wellhead parameters [6,7]. It is of great significance to develop potential disaster control measures in time, promote the safe development of marine natural gas hydrates and protect the marine ecological environment to carry out research on the technology of open-circuit drilling wellhead overflow and methane monitoring of marine natural gas hydrates.

2. Current Status of Research on Monitoring Methods

2.1. Monitoring Method Based on Downhole Measurement Parameters

The downhole logging parameter monitoring method uses relevant parameters collected from the downhole logging tool for leakage monitoring. According to different downhole measurement tools, it can be divided into APWD (annular pressure while drilling), LWD (logging while drilling), ultrasonic monitoring and spectral monitoring. This monitoring method realizes the early detection of overflow and reflects the real overflow situation at the bottom of the well, but requires downhole follow-up measurement tools. Due to the complexity of the development process, the cost of the method is higher than that of the surface parameter monitoring method. In addition, downhole follow-up measurement tools still face problems such as low temperature/pressure tolerance and slow data transmission [8].

In the annular pressure monitoring while drilling method, the downhole annular pressure, temperature, and density measurements are obtained in real time by annular pressure while drilling tools. for the method achieves the detection of the changes in real-time data and the early monitoring of overflow as it occurs [9,10]. Charlez et al. obtained pressure and temperature data by a follow-hole annular pressure measurement tool to validate and optimize the hydrodynamic parameter model in real time, which ensured that the calculated value of bottom-hole pressure was closer to the actual value [11]. However, a follow-hole measurement tool was employed in the method, resulting in higher costs.

The LWD method monitors the conductivity, resistivity, salinity and other characteristics of drilling fluid at the bottom of the well with logging while drilling tools and then analyzes whether overflow occurs. Considering the high logging cost and slow data transmission, resistivity, natural gamma and acoustic sensors are usually used to monitor overflow, and experienced engineers are equipped for observation and analysis [12]. Okoli et al. identified the change in formation lithology with the natural gamma logging tool while drilling and analyzed the possibility of formation overflow. However, the accuracy of overflow warning is low and the cost is high by simply analyzing the change in formation lithology, which is due to the phenomenon of negative pressure at the bottom hole [13].

The ultrasonic while drilling method monitors the change in drilling fluid flow in the annular drill collar by installing ultrasonic sensors on the collar joint. The method is usually used together with temperature and pressure sensors to realize the real-time monitoring of overflow. Professor Chen Ping and others installed ultrasonic, temperature, pressure and other sensors at the drill collar joints to monitor the acoustic transit time, temperature and pressure at the bottom of the well, but there were problems of low pressure and temperature, slow data transmission and high cost [14].

Spectral monitoring while drilling is employed to analyze the composition of drilling fluid at the bottom of the well and observe the content change of hydrocarbon components in the drilling fluid to determine the occurrence of overflow. This technology has been successfully used to analyze offshore oil spills, but the MWD (measurement while drilling) tool is still in the development stage [15].

2.2. Comprehensive Monitoring Method

The comprehensive monitoring method collects the real-time data of surface sensors and downhole MWD tools a comprehensive logging instrument and then analyzes the real-time data changes in these logging parameters to monitor the overflow. This method fully combines the advantages of the surface parameter monitoring method and downhole

MWD parameter monitoring method, which can obtain logging parameters under various working conditions, comprehensively analyze the changes in logging parameters, and carry out the early warning of overflow. This method exhibits the best monitoring effect, but it requires expensive downhole MWD tools and a high-precision flowmeter, leading to a high monitoring cost [16–18]. Combined with the various characteristics of comprehensive logging parameters, the overflow model is established by using the dynamic neural network method. The overflow model shows an excellent monitoring effect of overflow, but it requires a high-precision flowmeter and downhole MWD tools, as well as a large amount of logging data from adjacent wells, which is costly. Nayeem et al. monitored bottom-hole pressure, temperature, wellhead flow and other data using pressure and flow sensors and detected overflow through a threshold method. However, monitoring costs are high in the presence of expensive sensors. To improve the accuracy of overflow monitoring, Andia et al. proposed a Bayesian method for overflow monitoring, in which small changes in comprehensive logging data were identified and a large amount of logging data for training before use was required, which is a heavy workload [19–21].

2.3. Methane Monitoring Based on Seabed Interface

There is no professional wellhead methane monitoring technology for marine gas hydrate trial recovery projects. Currently, the known subsea boundary layer methane monitoring system [22] and ROV are usually used.

The methane sensors on the subsea submersible markers are combined with acoustic communicators to send methane data to the shipboard receiving system, realizing methane measurement at the boundary layer of the seafloor. Due to the large size and mass of the subsea submersible observation system, the method may not be able to accurately cast to the hydrate test wellhead from the surface to the seafloor, which may not achieve the expected monitoring effect.

ROVs equipped with high-precision methane sensors can submerge in the wellhead near the mudline and monitor the submerged wellhead during the surface drilling stage of the seafloor by combining the real-time camera function of ROVs, further realizing overflow monitoring during the drilling process. As shown in Figure 1.

Figure 1. Subsea boundary layer methane monitoring methods: (**a**) ROV with methane sensor and (**b**) subsea submersible observing system.

To meet the safety development of unconventional oil–gas, it is necessary to pursue a lower detection limit and faster detection speed, realize the timeliness of gas analysis and logging information synchronization and develop high accuracy, miniaturization and intelligence in equipment manufacturing. To address the above problems, a methane monitoring system at the wellhead of offshore natural gas open-hole drilling was designed for the construction environment of offshore natural gas hydrate extraction, including an above-water acoustic communication receiving part and a subsea self-contained methane monitoring part [23]. Compared with the existing MWD methods, the as-designed system monitored the methane content at the wellhead at various stages (such as drilling, casing running, cementing, completion and fracturing) in the process of natural gas hydrate exploitation and effectively warned of the potential risks (including overflow and methane leakage that may occur during the trial production of marine natural gas hydrate).

3. System Principle and Design Scheme

3.1. System Structure

The wellhead methane monitoring system of open-circuit drilling for offshore natural gas includes the underwater acoustic communication receiving part and the underwater self-contained methane monitoring part, which is generally composed of a logging system, central control system, and communication system. The wellhead methane gas overflow status and content were monitored by placing an underwater self-contained methane monitoring system around the wellhead and combining acoustic communication technology, and the related technical parameters are listed in Table 1. As shown in Figures 2 and 3.

1. Logging system

The logging system consists of a wellhead suction filtration device and a measurement system. Among them, the wellhead suction filtration device is the core part of the front to prevent the drilling mud fluid from clogging the detection module. The measurement system mainly uses a self-contained methane monitoring system, which is composed of a sensor for measuring methane gas and a suction pump for detecting the concentration of gas carried by the drilling mud return fluid. The test gas of the system is mainly methane, and other gas sensors can be added later.

2. Control system

The central control system is composed of two parts, including a data storage and transmission system and a power supply unit, which are integrated into the battery compartment. The data storage and transmission system are responsible for controlling the data acquisition sequence to complete the data storage and transmission, and the power supply unit with a long-term service life provides the required power for the whole device.

3. Communication system

The communication system includes two modules: one is the communication module, which receives and processes underwater data; the second is the data display module, which mainly matches the communication interface for time calibration and outputs the data processing results in a graphical manner.

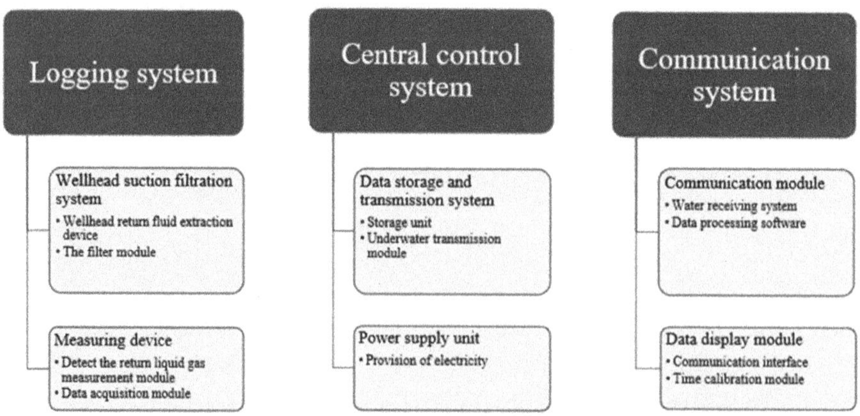

Figure 2. System architecture diagram.

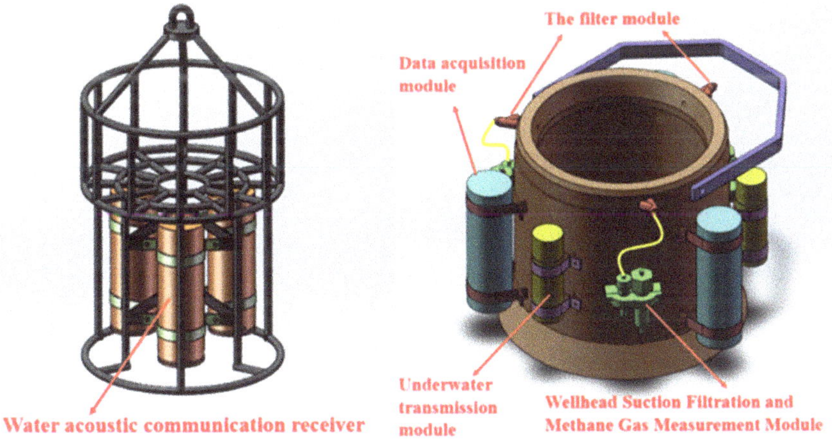

Figure 3. Schematic diagram of the system.

Table 1. Main technical indicators of the system.

Project	Performance Index
measuring range	1~500 nmol/L
working temperature	−5~55 °C
communication frequency band	21–31 Khz
working water depth	4000 m

3.2. *Working Mode and Technical Route*

The technical route is shown in Figure 4. The upper computer software was used to receive the data from the lower computer (underwater measurement system), including real-time data acquisition and historical data acquisition, which were controlled and sent by the upper computer. Specifically, the surface acoustic communicator was first put into the water through the offshore platform, and the host computer first sent acoustic instructions to wake up the underwater acoustic communicator through the surface acoustic communicator. The real-time collection or view history data command was sent to the underwater acoustic communication machine, and then the underwater acoustic communication machine was transported to the lower computer. The lower computer received the reading instructions for the real-time collection or viewing of historical data and then sent the relevant data to the underwater acoustic communication machine, and the surface acoustic communication machine and the upper computer were responsible for receiving and processing the data.

3.3. *Design of Suction Filtration Device*

The suction filtration device was used to filter the mud liquid at the mining wellhead to prevent blocking the methane measurement module. The methane measurement module included a methane sensor installed in the suction filtration device. The application of this device improved the detection effect to a certain extent. The specific working principle is as follows.

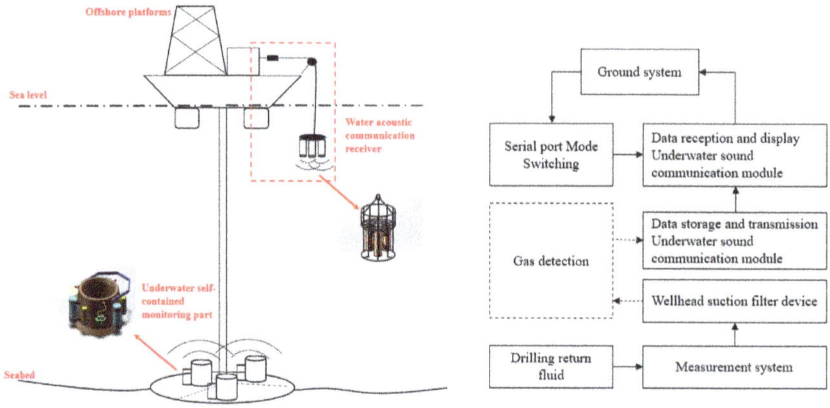

Figure 4. System working diagram and technical route.

The suction pump and the first motor were installed in the upper channel to pump out the filtered mud liquid, and the upper channel was also equipped with a one-way valve to prevent external seawater from entering the channel. After the suction pump was turned on, the mud liquid entered the upper channel through the filter, and then it was sucked into the front end of the sensor located in the flange. The methane in the mud liquid was detected by the sensor, and then the after-detected mud liquid was discharged through the upper channel. The flushing pump and the second motor were installed in the lower channel to draw seawater into the lower channel for flushing. After filtering, the seawater was filtered and injected into the front of the methane sensor in the flange by the flushing pump through the one-way valve in the lower channel, and then the seawater was discharged to complete the backflushing of the screen. The pipelines and mud screen in the lower channel were eventually cleaned. Schematic diagram of suction filtration device design is shown Figure 5.

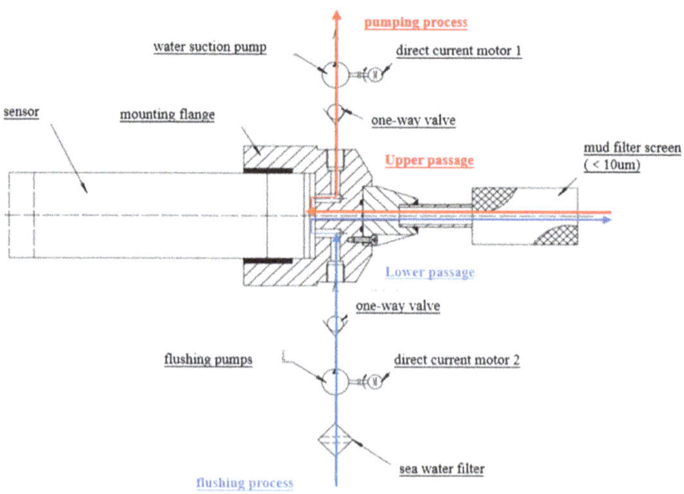

Figure 5. Schematic diagram of suction filtration device design.

4. System Sea Trial

After the preliminary lake test, the system was carried on a survey ship for sea testing to verify the system stability and data receiving reliability in a real environment.

The sea trial area is located in the northern part of the South China Sea, with a water depth of approximately 1300 m, matching the actual water depth environment for hydrate production.

Sea Trial Process

(1) Deck commissioning.

Underwater methane monitoring: three sets of underwater equipment (two in service and one for standby) were all installed in the wellhead basket, which was designed according to the actual wellhead structure and working environment.

Underwater acoustic communication: Three acoustic communication devices (two for use and one for standby) were connected to the computer upper computer software through a 200 m communication cable and powered by the built-in battery. The underwater part and the above-water part need to be paired with a pairing serial number of 18,264–18,187, 18,260–18,265, 18,263–18,259.

(2) The communication sampling frequency at a water depth of 500 m was set as 3 s, and the sampling time was 2–3 min. The test results showed that the communication was normal and the data were successfully received.As shown in Figures 6 and 7.

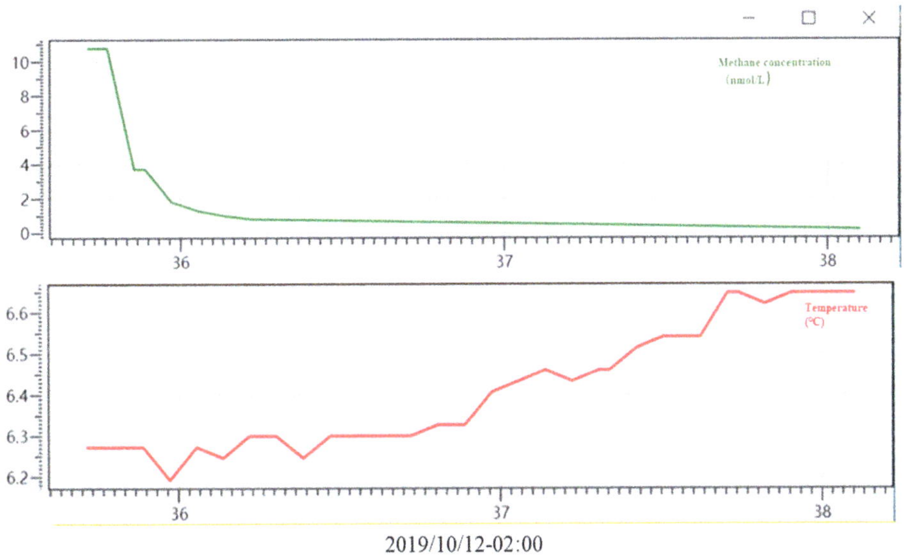

Figure 6. Test data at a water depth of 500 m (18,264–18,187 sound communication module paired successfully).

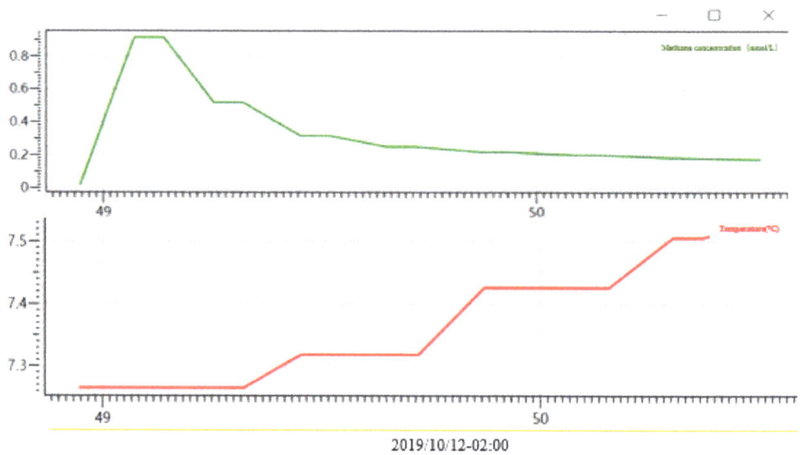

Figure 7. Test data at a water depth of 500 m (18,260–18,265 sound communication module paired successfully).

(3) The operating frequency was 30 s at a water depth of 1250 m, and the test time was 10–20 min. The test results showed that the communication was normal, and the data were successfully received. As shown in Figures 8–10.

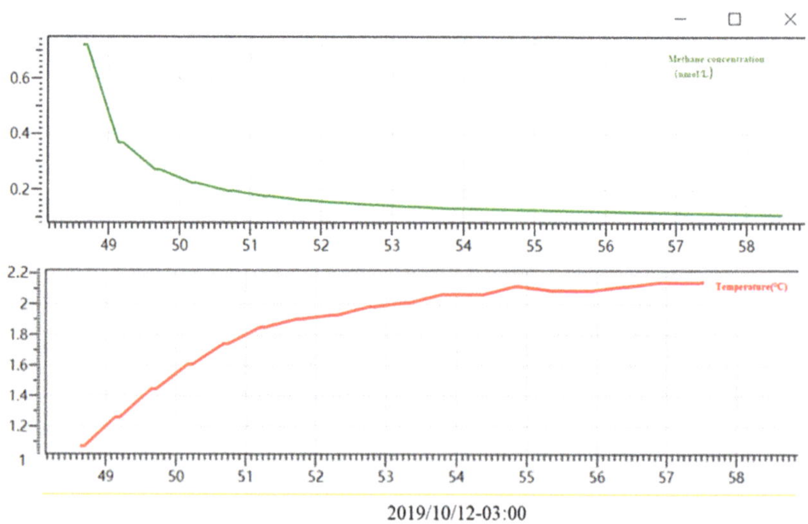

Figure 8. Test data at a water depth of 1250 m (18,264–18,187 sound communication module paired successfully).

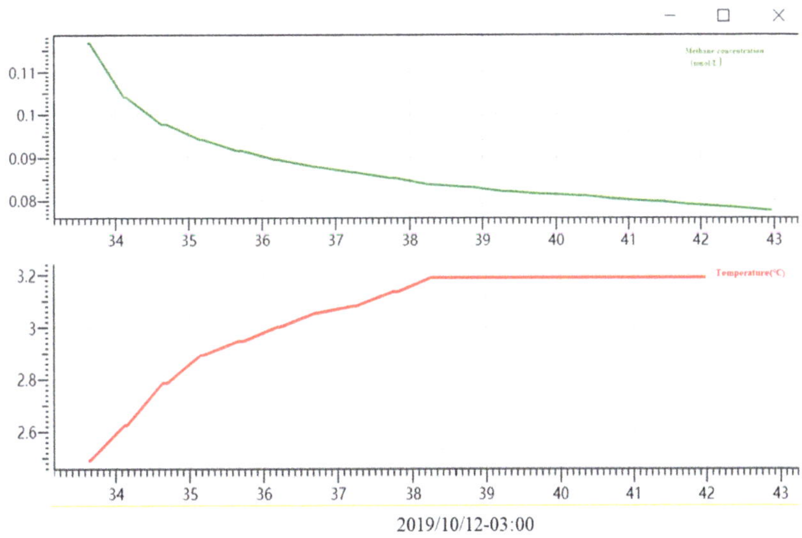

Figure 9. Test data at a water depth of 1250 m (18,260–18,265 sound communication module paired successfully).

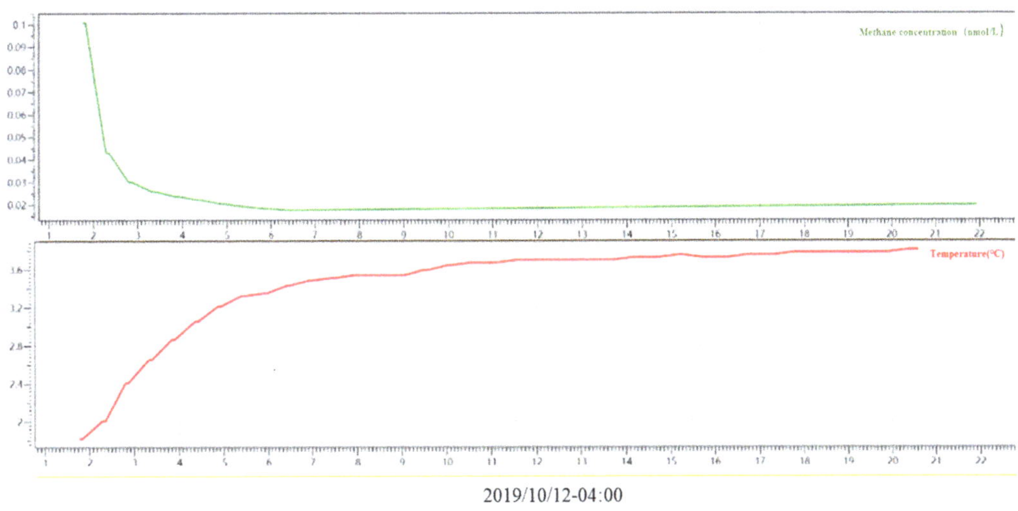

Figure 10. Test data at a water depth of 1250 m (18,263–18,259 sound communication module paired successfully).

The deep-sea environment is subject to less interference, and the sensor was preheated during the system-lowering process, which achieved the optimal state of sensor use. Therefore, the data received by the sensor tended to be flat as the depth increased from 500 m to 1250 m.

5. Results and Discussion

(1) A wellhead overflow monitoring system for the open-circuit drilling of offshore natural gas was designed, which was mainly used to monitor the methane content of the drilling fluid return fluid and surrounding areas. The as-designed system is composed of

an underwater acoustic communication receiving part and an underwater self-contained methane monitoring part. In actual use, the underwater self-contained methane monitoring system was arranged around the mining wellhead to monitor the overflow status and content of methane gas at the wellhead in real time. Compared with the current gas logging technology (measurement while drilling) used in the oil and gas exploration drilling stage, the system realized the monitoring of methane content in the process of drilling, casing running, cementing, completion and fracturing of marine natural gas hydrate when it was placed at the production wellhead.

(2) The system included a wellhead suction filter device with forward and back flushing circuits, which avoided the blockage of the sensor detection module by mud return fluid and improved the detection effect to a certain extent. In addition, the sensor needs to be preheated to achieve optimal performance.

(3) This system updated the method of monitoring methane in the drilling overflow of marine gas hydrate. The drilling mud return fluid was pumped into the measurement system to detect the methane content, and the collected data were transmitted back to the shore-based system through acoustic communication, which effectively avoided the loss of methane in the mud return fluid in the process of collection and transportation, thus enhancing the monitoring accuracy of the system to a certain extent.

(4) The stability of the system and the reliability of data reception under a real environment were further verified through 500 m and 1250 m sea trials, which provided technical reserves for wellhead overflow monitoring and early warning in the process of gas hydrate exploitation.

Author Contributions: Analysis and investigation, C.Z.; writing—review and editing, C.Z.; methodology, C.Z. and J.L.; writing—original draft preparation, D.K. All authors have read and agreed to the published version of the manuscript.

Funding: This research was jointly supported by the Key-Area Research and Development Program of Guangdong Province (2020B1111030003) and the Geological Survey Projects of China Geological Survey (DD20221700, DD20221706).

Conflicts of Interest: The authors declare no conflict of interest.

References

1. Fang, Y.X.; Li, M.B.; Jin, X.L.; Shentu, H.G. Character analysis of the marine gas hydrate stability zone. *Mar. Geol. Quat. Geol.* **2001**, *21*, 103–106.
2. Li, J.; Zhang, C.; Wu, Z.; Zhang, G.; Gao, Q.; Liu, K.; Cheng, X. Can low hydration heat cement prevent the decomposition of natural gas hydrate under high pressure and low temperature? *J. Nat. Gas Sci. Eng.* **2022**, *97*, 104347. [CrossRef]
3. Rakibul, I.; Faisal, K. Real time risk analysis of kick detection: Testing and Validation. *Reliab. Eng. Syst. Saf.* **2017**, *161*, 25–37.
4. Ali, K.V. Early kick detection and well control decision-making for managed pressure drilling automation. *J. Nat. Gas Sci. Eng.* **2015**, *27*, 354–366.
5. Wei, L.; Yang, X.; Xi, C.; Hong, C.; Yumin, L.; Xueqing, L. Research and application of downhole blowout prevention system while drilling: A review. *J. Pet. Sci. Eng.* **2019**, *188*, 106882. [CrossRef]
6. Chen, P.; Ma, T.S. Research status of early monitoring technology for deepwater drilling overflow. *Acta Pet. Sin.* **2014**, *35*, 602–612.
7. Liu, Z.; Ma, Q.; Cai, B.; Liu, Y.; Zheng, C. Risk assessment on deepwater drilling well control based on dynamic bayesian network. *Proc. Saf. Environ. Prot.* **2021**, *149*, 643–654. [CrossRef]
8. Morley, J.; Heidler, R. Field testing of an new nuclear magnetic resonance logging while drilling tool. In Proceedings of the SPE Annual Technical Conference and Exhibition, San Antonio, TX, USA, 29 September–2 October 2002.
9. Krishna, S.; Ridha, S.; Campbell, S.; Ilyas, S.U.; Dzulkarnain, I.; Abdurrahman, M. Experimental evaluation of surge/swab pressure in varying annular eccentricities using non-newtonian fluid under couette-poiseuille flow for drilling applications. *J. Pet. Sci. Eng.* **2021**, *206*, 108982. [CrossRef]
10. Gandelman, R.A.; Martins, A.L.; Teixeira, G.T.; Waldmann, A.T.A.; Rezende, M.S.C.; Aragao, A.F.L. A comprehensive methodology to avoid and remediate drilling problems by real-time pwd data interpretation. In Proceedings of the SPE Annual Technical Conference and Exhibition, New Orleans, LA, USA, 4–7 October 2009.
11. Charlez, P.A.; Easton, M.; Morrice, G.; Tardy, P. Validation of advanced hydraulic modeling using pwd data. In Proceedings of the Offshore Technology Conference, Houston, TX, USA, 4–7 May 1998.

12. Harris, M.; Byrd, D.M.; Archibald, M.; Naupari, C.E.; Bittar, M.S.; Chemali, R.E. Real-time decisions with improved confidence using azimuthal deep resistivity and at-bit gamma imaging while drilling. In Proceedings of the SPE Offshore Europe Oil and Gas Conference and Exhibition, Aberdeen, UK, 8–11 September 2009.
13. Okoli, U.; Schrader, H. Using LWD Tools to Enhance Drilling Operations: A Case Study. In Proceedings of the Nigeria Annual International Conference and Exhibition, Lagos, Nigeria, 6–8 August 2012.
14. Zhou, Q.; Zhao, H.; Zhan, H.; Zhang, H.; Lu, C. The application of ultrasonic based on doppler effect used in early kick detection for deep water drilling. In Proceedings of the International Conference on Communications, Budapest, Hungary, 9–13 June 2013.
15. Zhou, K.; Chen, G. *Analysis of Oil Spill Hyperspectral Data Based on Spectral Characteristics; "The Belt and Road" Strategy and Marine Scientific and Technological Innovation*; Oceanographic Society of China: Beijing, China, 2015.
16. Yongshou, D.A.I.; Weijie, Y.U.E.; Weifeng, S.U.N. Online monitoring and warning system for early kick foreboding on three high wells. *J. China Univ. Pet. (Ed. Nat. Sci.)* **2015**, *39*, 188–194.
17. Li, J.W.; Liu, G.H.; Li, J. Study on kick detection methods during U-tubing effect in dual gradient drilling. *Drill. Prod. Technol.* **2016**, *39*, 19–22.
18. Wu, X.; Wan, F.; Chen, Z.; Han, L.; Li, Z. Drilling and completion technologies for deep carbonate rocks in the sichuan basin: Practices and prospects. *Nat. Gas Ind. B* **2020**, *7*, 547–556. [CrossRef]
19. Nayeem, A.A.; Venkatesan, R.; Khan, F. Monitoring of down-hole parameters for early kick detection. *J. Loss Prev. Proc. Ind.* **2015**, *40*, 43–54. [CrossRef]
20. Kamyab, M.; Shadizadeh, S.R.; Jazayeri-rad, H.; Dinarvand, N. Early kick detection using real time data analysis with dynamic neural network: A case study in iranian oil fields. In Proceedings of the SPE Nigeria Annual International Conference and Exhibition, Tinapa, Calabar, Nigeria, 31 July–7 August 2010.
21. Yuan, J.L.; Fang, B.T.; Xing, X.S.; Gen, L.J.; Yin, Z.; Wang, Y. Real-time early warning of drilling overflow based on naïve Bayes algorithm. *Oil Drill. Prod. Technol.* **2021**, *43*, 455–460. (In Chinese)
22. Dong, Y.F.; Luo, W.Z.; Liang, Q.Y.; Qiu, H.L.; Ren, C.; Yan, R.; Lin, J.Q. A developed bottom-supported submersible buoyant system and its testing application to a natural gas hydrate area. *Mar. Geol. Quat. Geol.* **2017**, *37*, 195–203. (In Chinese)
23. Zhong, C.; Liang, Q.Y.; Guo, B.B.; Wu, X.M.; Dong, Y.F.; Yang, L.; Xiao, X.; Zhang, T.T.; Su, D.Y.; Wu, X.Y. An Open-Circuit Drilling Wellhead Methane Logging System. Chinese Patent ZL202011242750.1, 17 June 2022. (In Chinese)

energies

MDPI

Article

Study of Wellbore Instability and Collapse Mechanism for a Layered Continental Shale Oil Horizontal Well

Shibin Li [1], Kai Liang [1,*], Changhao Wang [1], Yao Wang [1], Yuxuan Jiao [1], Xiaoxing Zhu [1] and Chunhua Wang [2]

[1] School of Petroleum Engineering, Northeast Petroleum University, Daqing 163318, China; lishibin@nepu.edu.cn (S.L.); chwang88@nepu.edu.cn (C.W.); wangyao@stu.nepu.edu.cn (Y.W.); Pursuit@stu.nepu.edu.cn (Y.J.); zhuxiaoxing@stu.nepu.edu.cn (X.Z.)

[2] Daqing Oilfield Drilling Engineering Company, Daqing 163413, China; wangchunhua001@cnpc.com.cn

* Correspondence: liangkai@stu.nepu.edu.cn

Abstract: The shale oil horizontal wells in the Songliao Basin are affected by a lack of mature theories, technologies and experiences in the direction of wellbore stability. Wellbore collapse may occur, and in severe cases, the wellbore may be scrapped, resulting in huge economic losses. Therefore, aiming at addressing the above problems, rock mechanics experiments were carried out. Based on the theories of elasticity and rock mechanics, this paper considers not only the influence of the bedding plane, but also the influence of hydration on the strength weakening of the shale body and the bedding plane. The analysis shows that no matter under which in situ stress mechanism, the wellbore in the vertical well section is the most stable, and when the inclination angle is approximately 45°, the wellbore is most likely to be unstable. Changes in water content do not affect the most stable or unstable regions. Under the same conditions, the equivalent density of collapse pressure increases with the increase in water content. In addition, field examples are also analyzed to verify the accuracy of this model, which can provide a theoretical and technical basis for the safe construction of continental shale oil horizontal wells.

Keywords: wellbore stability; weak plane; continental shale; collapse pressure; hydration

Citation: Li, S.; Liang, K.; Wang, C.; Wang, Y.; Jiao, Y.; Zhu, X.; Wang, C. Study of Wellbore Instability and Collapse Mechanism for a Layered Continental Shale Oil Horizontal Well. *Energies* **2022**, *15*, 4538. https://doi.org/10.3390/en15134538

Academic Editors: Mofazzal Hossain, Md Motiur Rahman and Hisham Khaled Ben Mahmud

Received: 6 June 2022
Accepted: 20 June 2022
Published: 21 June 2022

1. Introduction

With the increasing importance of unconventional oil and gas resources in the global energy structure, shale oil has become a hot area of energy exploration and development [1–3]. The development of continental shale oil in China is still in its infancy. The shale oil horizontal wells in Block G in the Songliao Basin are affected by a lack of mature theories, technologies and experiences in the direction of wellbore stability [4–8]. The continental shale oil in the northern Songliao Basin has huge potential resources and has become an important strategic replacement resource in Daqing. In 2020, the daily oil and gas equivalent of Well A2 reached more than 39 tons, showing a promising prospect for exploration and development [9–13]. The Q1 and Q2 groups in block G develop large sections of shale with extremely well-developed bedding. The mineral components are mainly dolomite, calcite and quartz, which are hard and brittle shale, and the average content of clay minerals is over 40%. Therefore, during the drilling process, the shale is prone to hydration and expansion. Downhole accidents such as wellbore collapse and block drop may occur, and in severe cases, the wellbore may be scrapped, resulting in huge economic losses. The wellbore instability problem of shale oil horizontal wells is one of the technical bottlenecks restricting the safe and efficient development of shale oil [14–17].

Experts and scholars at home and abroad have carried out a large number of experiments and theoretical research on the direction of wellbore stability, and have achieved rich results. Jaeger et al. [18] originally proposed the strength failure criterion of rock weak plane. On this basis, Liu et al. [19,20] established a collapse pressure prediction model considering the influence of weak plane based on the theory of continuum mechanics and stress coordinate system transformation, and analyzed the effects of weak plane and occurrence on wellbore stability. Shuai et al. [21,22] combined the wellbore stress model and the weak plane failure criterion under the condition of deviated wells, established the formation collapse pressure model of highly deviated wells, and analyzed the factors affecting wellbore stability. The above studies only consider the effect of weak plane on wellbore stability, and do not consider the influence of hydration on wellbore stability. Aadnoy [23] established a collapse pressure model that considered both bedding occurrence and shale strength weakening, and analyzed the factors affecting the collapse pressure distribution. Combined with the weak plane strength criterion, Ma et al. [24–29] established an analysis model for the wellbore stability of layered shale horizontal wells, and analyzed the effects of bedding occurrence and water content on the collapse pressure of horizontal wells. However, the current research on shale wellbore stability in China mainly focuses on the marine shale of Longmaxi Formation in Sichuan, and the wellbore instability of continental shale in the north of Songliao Basin has not been studied. At present, for the problem of wellbore stability of continental shale in the north of Songliao Basin, only the corresponding relationship between logging parameters and wellbore rock mechanical properties has been preliminarily established. In the field, the drilling fluid density is roughly adjusted according to the occurrence of wellbore collapse [30,31]. At present, how to establish the wellbore stability model of layered continental shale under different working conditions is the bottleneck restricting further research on the wellbore stability of shale drilling in this block.

Therefore, in view of the above problems, this paper carried out rock mechanics experimental research, based on the obtained rock mechanics data, in-depth analysis of the wellbore instability mechanism and main controlling factors of the Gulong shale layer, and a clear algorithm for obtaining the formation collapse pressure. Therefore, aiming at addressing the above problems, rock mechanics experiments were carried out, including triaxial compressive strength experiments and tensile strength experiments. On the basis of rock mechanics data, the mechanism of wellbore instability and its influencing factors in Gulong shale are deeply analyzed, and the algorithm for calculating formation collapse pressure is determined. Based on the theories of elasticity and rock mechanics, this paper considers not only the influence of the bedding plane, but also the influence of hydration on the strength weakening of rock body and the bedding plane. A prediction model for the wellbore collapse pressure of a horizontal well in layered continental shale oil is established, the mechanism of wellbore instability of a horizontal well in layered shale is studied, and the influence laws of different in situ stress conditions, wellbore trajectory, water content, bedding plane occurrence and other factors on collapse pressure are analyzed. In addition, field examples are also analyzed to verify the accuracy of this model, which can provide a theoretical and technical basis for the safe construction of continental shale oil horizontal wells [32–34].

2. Rock Mechanics Experiments of Layered Shale

2.1. Experimental Core Preparation

Since the shale is prone to hydration expansion when encountering water, it cannot be cut and cored used in conventional processing methods, so the full-size core is processed by using the NC wire cutting method. According to the coring method in Figure 1 [35], the 0, 30°, 45°, 60°, 90° direction cores are processed, respectively, and both ends of the sample are cut fat and polished to make the length diameter ratio of the shale sample approximately 2.

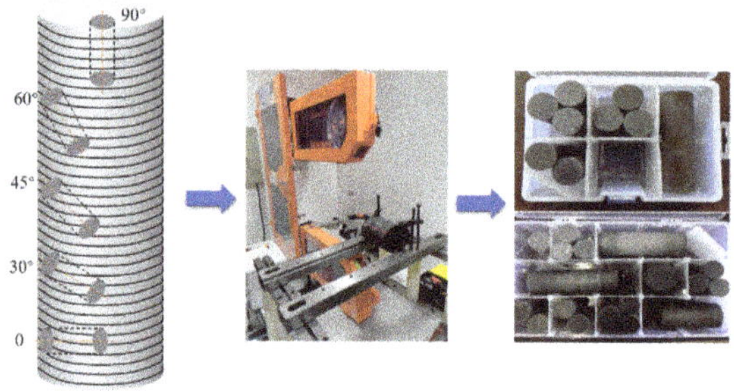

Figure 1. Preparation of standard rock samples by NC wire cutting.

2.2. Triaxial Compressive Strength Test

According to the regulations, the standard specimen is a cylinder with a diameter of 2.5 cm and a height of 5.0 cm, and the allowable variation range is 4.8~5.2 cm. During axial loading, the strain rate should be controlled within 1×10^{-5} mm/s. When the pressure is loaded in two directions, parallel to the bedding and perpendicular to the bedding, the fracture patterns of the rock are different. When the pressure is loaded parallel to the bedding direction, the shale will open many times along the bedding fractures, forming multiple groups of fragments along the bedding plane direction. When the pressure is loaded perpendicular to the bedding direction, shear cracks are first generated. When the shear cracks are connected with the micro-cracks along the bedding, large volume fractures will be formed. The practical experience of on-site horizontal well drilling shows that the wellbore collapse is the most serious in the build-up section with an inclination angle of 30°~75°. Therefore, the angle between the direction of loading pressure and bedding (approximately equal to the deviation angle in actual drilling) is divided into three areas through laboratory experimental research. As shown in Figure 2, 0°~30° is divided into layered split zone. 30°~75° is divided into failure zone of weak plane and 75°~90° is divided into failure zone of the shale body.

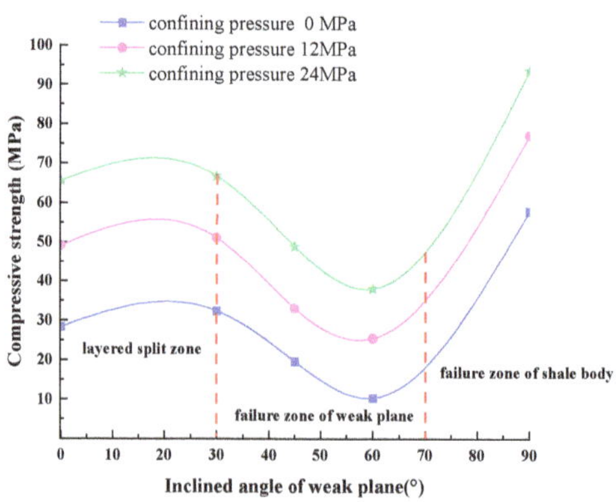

Figure 2. The relationship between the triaxial compressive strength and the angle between the loading pressure direction and the bedding direction.

According to the fracture shape of shale in Figure 3, when the pressure is loaded along the parallel bedding direction, the bedding fractures will be opened many times, resulting in tensile failure. Therefore, when the shale is loaded with pressure in the horizontal direction, it is hard and brittle, forming multiple groups of fragments. When the shale is loaded with pressure perpendicular to the bedding direction, it needs to overcome the shear strength of the shale body and shows a certain plastic deformation. It is caused by the closure and expansion of micro-fractures under the action of loading pressure, which needs to overcome the shear strength of the shale body. The following three conclusions are drawn: (1) When the angle between the loading pressure direction and the bedding is 0°~30°, the splitting failure mainly occurs along the bedding plane. (2) When the angle between the loading pressure direction and the bedding is 30°~75°, the shear slip mainly occurs along the bedding plane. When the angle is 50°~60°, the compressive strength of the shale body is the minimum. (3) When the angle between the loading pressure direction and the bedding is 75°~90°, the shear failure of the shale body mainly occurs, and the compressive strength of the shale body is the maximum at this time.

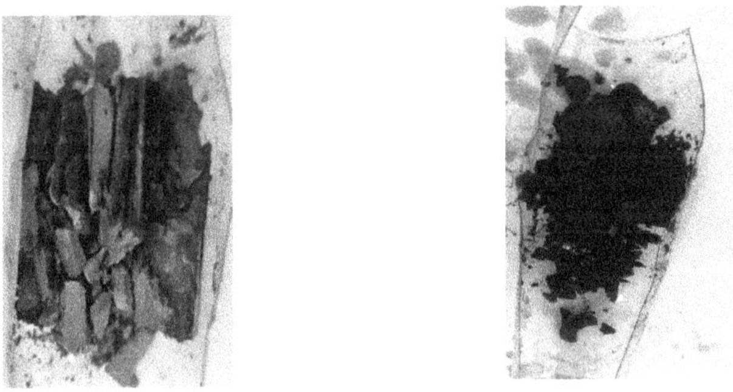

(a) (b)

Figure 3. Photos of shale destroyed in triaxial compression experiment: (**a**) photo of shale being destroyed when loaded parallel to the bedding direction; (**b**) photo of shale being destroyed when loaded perpendicular to the bedding direction.

2.3. Tensile Strength Test of Shale

The tensile strength test is carried out using an indirect test method, the Brazilian split tensile test method. The rock sample is processed into a cylindrical specimen with a diameter of approximately 38 mm and a length of approximately 19 mm, basically maintaining a diameter to thickness ratio of 2:1. The shale is more prone to tensile failure when loaded parallel to the bedding direction than when loaded perpendicular to the bedding direction. When the included angle with the bedding direction is 45°, the shale undergoes shear slip, and the shear strength of the bedding plane is much lower than the tensile strength of the shale body. Therefore, the continental shale in the Songliao Basin has well-developed bedding and low shear strength, which leads to the easy opening of fractures along the bedding direction, which is an important factor affecting the stability of the wellbore.

In the same way as the triaxial compressive strength test, according to the collapse and fracture characteristics of the wellbore, and according to the angle between the loading pressure direction and the bedding direction (well inclination angle), the tensile strength can be divided into three areas: tensile failure zone of bedding, shear slip failure zone and tensile failure zone of the shale body, as shown in Figure 4.

When the inclination angle is 0°~30°, the formation fracture belongs to the tensile failure of the bedding plane, and the collapse belongs to the shear failure of the shale body. The fracture pressure and collapse pressure are both low, and the drilling fluid density should not be too high. When the well inclination is 30°~60°, the shear strength of shale is much lower than the tensile strength, and the drilling fluid density window is the narrowest, so precise pressure control should be ensured. When the inclination angle is 60°~90°, the formation fracture belongs to the tensile failure of the shale body, and the collapse belongs to the bedding splitting or shear failure. Both the collapse pressure and the fracture pressure increase, and the drilling fluid density should be appropriately increased. The following three conclusions are drawn: (1) When the angle between the loading pressure direction and the bedding direction is 0°~30°, tensile failure mainly occurs along the bedding plane. (2) When the angle between the loading pressure direction and the bedding direction is 30°~60°, shear slip failure mainly occurs along the bedding plane. When the angle between the loading pressure direction and the bedding direction is 45°, the tensile strength is the minimum. (3) When the angle between the loading pressure direction and the bedding direction is 60°~90°, the tensile failure of the shale body mainly occurs.

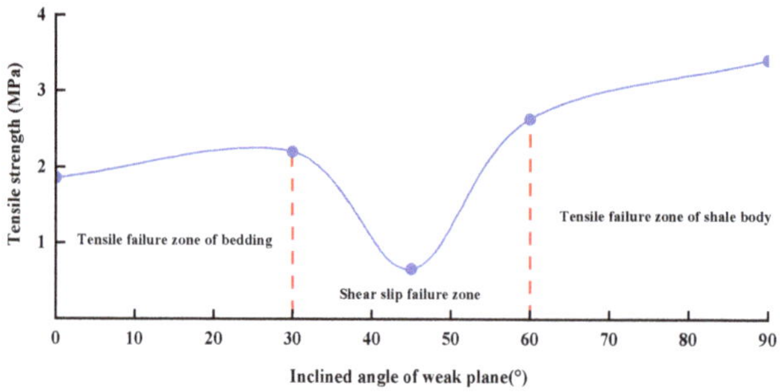

Figure 4. The relationship between the tensile strength and the angle between the loading pressure direction and the bedding direction.

3. Prediction Model of the Collapse Pressure of a Horizontal Well in Layered Shale

3.1. Borehole Stress Considering Hydration Strain

The stress–strain (hydration strain) equilibrium equation of rock around the well under hydration is [36]:

$$\frac{d\sigma_{rr}}{dr} + \frac{\sigma_{rr} - \sigma_{\theta\theta}}{r} = 0 \tag{1}$$

where σ_{rr} and $\sigma_{\theta\theta}$ are the radial stress and circumferential stress under the cylindrical coordinate system of the wellbore, respectively, Pa; r is the radius, m.

The radial stress on the wellbore is equal to the pressure of the drilling fluid column in the well, that is, the inner boundary condition is $\sigma_{rr}|_{r=R} = p_{\mathrm{m}}$. The radial stress at infinity from the wellbore is equal to the in-situ stress, that is, the outer boundary condition is $\sigma_{rr}|_{r\to\infty} = S$.

Substitute the internal and external boundary conditions into the stress-strain equilibrium equation, and the general solution of the stress-strain equilibrium equation (that is, the radial displacement u) is finally obtained as follows:

$$u = Ar + B/r \tag{2}$$

$$A = \frac{(1-2v)(1+v)}{E}\frac{(\sigma_{xx}+\sigma_{yy})}{2} \tag{3}$$

$$B = \frac{(1+v)}{2E}\left[\frac{(\sigma_{xx}+\sigma_{yy})}{2} - p_{\mathrm{m}}\right]R^2 \tag{4}$$

where σ_{xx} and σ_{yy} are the principal stresses in the X-axis and Y-axis directions under the borehole rectangular coordinate system, Pa; u is radial displacement, m; p_{m} is the effective liquid column pressure in the well, Pa; E is the elastic modulus when the formation water content is f_{w}, Pa; v is the Poisson's ratio when the formation water content is f_{w}; R is the borehole radius, m.

According to the plane strain geometric equation of the axisymmetric wellbore, the radial and circumferential hydration strains generated by hydration can be obtained:

$$\begin{cases} \varepsilon_{rr} = \frac{du}{dr} = A - \frac{B}{r^2} \\ \varepsilon_{\theta\theta} = \frac{u}{r} = A + \frac{B}{r^2} \end{cases} \tag{5}$$

where ε_{rr} is the radial strain; is the tangential $\varepsilon_{\theta\theta}$ strain.

The vertical hydration strain is:

$$\varepsilon_V = K_1 \Delta f_w + K_2 \Delta f_w^2 \tag{6}$$

where: $K_1 = 0.0708$; $K_2 = 11.08$; ε_V is the vertical hydration strain; Δf_w is the increment of water content; $\Delta f_w = f_w - f_{wi}$; Δf_{wi} is the original formation water content.

Finally, the hydration stress of the layered shale in the plane two-dimensional spatial cylindrical coordinate system is obtained as follows:

$$\begin{cases} \sigma_{rr}^{hy} = \frac{E}{(1-2v)(1+v)}[(1-v)\varepsilon_{rr} + v\varepsilon_{\theta\theta} - (m+v)\varepsilon_V] \\ \sigma_{\theta\theta}^{hy} = \frac{E}{(1-2v)(1+v)}[(1-v)\varepsilon_{\theta\theta} + v\varepsilon_{rr} - (m+v)\varepsilon_V] \\ \sigma_{zz}^{hy} = \frac{E}{(1-2v)(1+v)}[v\varepsilon_{rr} + v\varepsilon_{\theta\theta} - (1-v+2mv)\varepsilon_V] \end{cases} \tag{7}$$

where σ_{rr}^{hy}, $\sigma_{\theta\theta}^{hy}$ and σ_{zz}^{hy} are respectively the radial stress, circumferential stress and axial stress in the cylindrical coordinate system of the wellbore when the bedding plane is considered, Pa; m is the anisotropic ratio.

3.2. Wellbore Circumferential Stress Produced by In Situ Stress

As shown in Figure 5, the principal stress and shear stress in the cylindrical coordinate system around the borehole are obtained by coordinate transformation as:

$$\begin{cases} \sigma_{rr} = \frac{\sigma_{xx}+\sigma_{yy}}{2}\left(1-\frac{R^2}{r^2}\right) + \frac{\sigma_{xx}-\sigma_{yy}}{2}\left(1+\frac{3R^4}{r^4}-\frac{4R^2}{r^2}\right)\cos 2\theta \\ +\tau_{xy}\left(1+\frac{3R^4}{r^4}-\frac{4R^2}{r^2}\right)\sin 2\theta + \frac{R^2}{r^2}p_m \\ \sigma_{\theta\theta} = \frac{\sigma_{xx}+\sigma_{yy}}{2}\left(1+\frac{R^2}{r^2}\right) - \frac{\sigma_{xx}-\sigma_{yy}}{2}\left(1+\frac{3R^4}{r^4}\right)\cos 2\theta \\ -\tau_{xy}\left(1+\frac{3R^4}{r^4}\right)\sin 2\theta - \frac{R^2}{r^2}p_m \\ \sigma_{zz} = \sigma_{zz} - 2v(\sigma_{xx}-\sigma_{yy})\left(\frac{R}{r}\right)^2\cos 2\theta \\ \tau_{r\theta} = \left[\frac{(\sigma_{yy}-\sigma_{xx})\sin 2\theta}{2} + \tau_{xy}\cos 2\theta\right]\left(1-\frac{3R^4}{r^4}+\frac{2R^2}{r^2}\right) \\ \tau_{\theta z} = \tau_{yz}\left(1+\frac{R^2}{r^2}\right)\cos\theta - \tau_{xz}\left(1+\frac{R^2}{r^2}\right)\sin\theta \\ \tau_{rz} = \tau_{xz}\left(1-\frac{R^2}{r^2}\right)\cos\theta + \tau_{yz}\left(1-\frac{R^2}{r^2}\right)\sin\theta \end{cases} \tag{8}$$

where σ_{zz} is the axial stress under the cylindrical coordinate system of the wellbore, Pa; $\tau_{r\theta}$, $\tau_{\theta z}$ and τ_{rz} are the tangential stresses in three planes under the cylindrical coordinate system of the wellbore, Pa; θ is the azimuth along the wellbore circumference, (°).

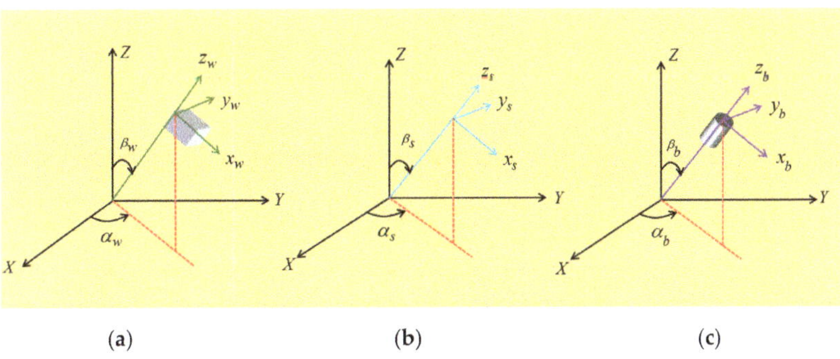

Figure 5. Reference coordinate system: (**a**) stratigraphic rectangular coordinate system; (**b**) in situ stress rectangular coordinate system; (**c**) borehole rectangular coordinate system.

3.3. Total Effective Stress of Wellbore

The total effective stress of the wellbore consists of two parts, one is the hydration stress generated by the hydration strain of shale, and the other is the stress generated by the in situ stress. The calculation formula is as follows:

$$\begin{cases} \sigma_{rr}^{t} = \sigma_{rr} + \sigma_{rr}^{hy} \\ \sigma_{\theta\theta}^{t} = \sigma_{\theta\theta} + \sigma_{\theta\theta}^{hy} \\ \sigma_{zz}^{t} = \sigma_{zz} + \sigma_{zz}^{hy} \\ \tau_{r\theta}^{t} = \tau_{r\theta} \\ \tau_{qz}^{t} = \tau_{\theta z} \\ \tau_{rz}^{t} = \tau_{rz} \end{cases} \quad (9)$$

When $r = R$, the total effective stress on the shaft wall can be obtained:

$$\begin{cases} \sigma_{rr}^{t} = \frac{E}{(1-2v)(1+v)}[(1-v)\varepsilon_{rr} + v\varepsilon_{\theta\theta} - (m+v)\varepsilon_{v}] + p_{m} - \eta p_{p} \\ \sigma_{\theta\theta}^{t} = \frac{E}{(1-2v)(1+v)}[(1-v)\varepsilon_{\theta\theta} + v\varepsilon_{rr} - (m+v)\varepsilon_{v}] + \sigma_{xx}(1-2\cos\theta) + \\ \sigma_{yy}(1+2\cos 2\theta) - 4\tau_{xy}\sin 2\theta - p_{m} - \eta p_{p} \\ \sigma_{zz}^{t} = \frac{E}{(1-2v)(1+v)}[v\varepsilon_{rr} + v\varepsilon_{\theta\theta} - (1-v+2mv)\varepsilon_{v}] + \sigma_{zz} - \\ v[2(\sigma_{xx} - \tau_{xy})\cos 2\theta + 4\tau_{xy}\sin 2\theta] - \eta p_{p} \\ \tau_{\theta z}^{t} = 2\tau_{yz}\cos\theta - 2\tau_{xz}\sin\theta \\ \tau_{r\theta}^{t} = 0 \\ \tau_{rz}^{t} = 0 \end{cases} \quad (10)$$

where σ_{rr}^{t}, $\sigma_{\theta\theta}^{t}$, σ_{zz}^{t}, $\tau_{r\theta}^{t}$, $\tau_{\theta z}^{t}$ and τ_{rz}^{t} are the total effective stress components of the wellbore after considering the hydration stress, Pa; η is the effective stress coefficient; p_p is the formation pore pressure, Pa.

According to formula (8), σ_{rr}^{t} is one of the principal stresses, that is, $\sigma_r = \sigma_{rr}^{t}$. The calculation formula of the other two principal stresses is:

$$\begin{cases} \sigma_a = \frac{\sigma_{zz}^{t} + \sigma_{\theta\theta}^{t}}{2} + \sqrt{\left(\frac{\sigma_{zz}^{t} - \sigma_{\theta\theta}^{t}}{2}\right)^2 + \left(\tau_{\theta z}^{t}\right)^2} \\ \sigma_b = \frac{\sigma_{zz}^{t} + \sigma_{\theta\theta}^{t}}{2} - \sqrt{\left(\frac{\sigma_{zz}^{t} - \sigma_{\theta\theta}^{t}}{2}\right)^2 + \left(\tau_{\theta z}^{t}\right)^2} \end{cases} \quad (11)$$

Then, the maximum and minimum principal stresses are:

$$\begin{cases} \sigma_1 = \max(\sigma_a, \sigma_b, \sigma_r) \\ \sigma_3 = \min(\sigma_a, \sigma_b, \sigma_r) \end{cases} \quad (12)$$

where σ_a, σ_b and σ_r are the principal stress on the wellbore, Pa; σ_1 and σ_3 are the maximum and minimum principal stresses, respectively, Pa.

The maximum principal stress and the minimum principal stress on the wellbore of a shale oil horizontal well can be obtained through the above model, and the collapse pressure can be obtained by combining the Mohr–Coulomb weak plane strength criterion [37].

3.4. Determination of Collapse Pressure

The layered continental shale formation has a set of approximately parallel bedding planes, and the strength of the bedding plane is lower than that of the shale body, which is also called the weak plane. Jaeger proposed the single weak plane strength theory, which is a generalization of the Mohr–Coulomb failure criterion, which can describe the shear failure of a rock mass with one or a group of parallel weak planes, as shown in Figure 6. In this paper, the strength theory of single weak surface is used as the criterion of wellbore instability.

$$\sigma_1 - \sigma_3 = \frac{2(C_W + \sigma_3 \tan \varphi_W)}{(1 - \tan \varphi_W \cot \beta) \sin 2\beta} \tag{13}$$

$$\beta = \arccos \frac{\boldsymbol{n} \cdot \boldsymbol{N}}{|\boldsymbol{n}||\boldsymbol{N}|} \tag{14}$$

$$\mathbf{n} = \sin i_W \cos \alpha_W \mathbf{i} + \sin i_W \sin \alpha_W \mathbf{j} + \cos \alpha_W \mathbf{k} \tag{15}$$

$$\mathbf{N} = \begin{cases} \sin \gamma \mathbf{j} + \cos \gamma \mathbf{k} & \sigma_{zz}^t \geqslant \sigma_{\theta\theta}^t \\ -\cos \gamma \mathbf{j} + \sin \gamma \mathbf{k} & \sigma_{zz}^t < \sigma_{\theta\theta}^t \end{cases} \tag{16}$$

$$\gamma = \begin{cases} \frac{\pi}{4} & \sigma_{zz}^t = \sigma_{\theta\theta}^t \\ 0.5\arctan\left|\frac{2\tau_{\theta z}^t}{\sigma_{\theta\theta}^t - \sigma_{zz}^t}\right| & \sigma_{zz}^t \neq \sigma_{\theta\theta}^t \end{cases} \tag{17}$$

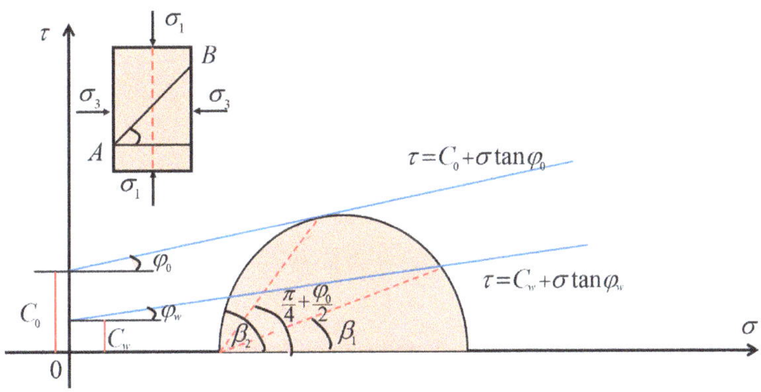

Figure 6. Schematic diagram of single weak plane strength theory [38].

The lower limit β_1 and upper limit β_2 of β when the weak plane is broken are:

$$\begin{cases} \beta_1 = 0.5\varphi_W + 0.5\arcsin\left[\frac{(\sigma_1 + \sigma_3 + 2C_W \cot \varphi_W) \sin \varphi_W}{\sigma_1 - \sigma_3}\right] \\ \beta_2 = \frac{\pi}{2} + \varphi_W - \beta_1 \end{cases} \tag{18}$$

From the Mohr stress circle, it can be seen that when a, the failure of the weak surface causes the instability of the wellbore. When a or a, the weak surface is stable, at this time the rock body is damaged and the borehole wall is unstable. The basis for determining the failure of rock body is as follows:

From the Mohr stress circle, it can be seen that under the condition $\beta_1 \leq \beta \leq \beta_2$, the failure of the weak plane causes the instability of the wellbore. Under the condition $0 < \beta < \beta_1$ or $\beta_2 < \beta < \pi/2$, the weak plane is stable, and the shale body failure causes the instability of the wellbore. The basis for determining the failure of shale body is as follows:

$$\sigma_1 - \sigma_3 = \frac{2\cos \varphi}{1 - \sin \varphi} + \sigma_3 \frac{1 + \sin \varphi}{1 - \sin \varphi} \tag{19}$$

where β is the angle between the normal of the weak plane and the maximum principal stress, (°); C_W is the cohesion of weak plane, MPa; φ_W; is the friction angle of weak plane, (°); a_W is the bedding plane inclination, (°); i_W is the dip angle of bedding plane, (°).

Based on the mechanical parameter test data of shale core under different water content, the variation law of shale mechanical characteristic parameters with water content is obtained:

$$\begin{cases} E = E_0 e^{-11}\sqrt{f_w - f_{w_i}} \\ v = v_0 + 1.3 f_w \\ C = C_0 - 58.8(f_w - f_{w_i}) \\ \varphi = \varphi_0 - 187.5(f_w - f_{wi_i}) \end{cases} \tag{20}$$

where E_0 is the initial elastic modulus, Pa; v_0 is the initial Poisson's ratio; C_0 is the initial cohesion, Pa; φ_0 is the original internal friction angle, (°); v is the Poisson's ratio when the formation water content is f_w; C is the cohesion when the formation water content is f_w, Pa; φ is the internal friction angle when the formation water content is f_w, (°).

3.5. Solution of Collapse Pressure Model

In this paper, Matlab is used to compile a program to calculate the collapse pressure by the iterative method, which is used to predict the collapse pressure of shale oil horizontal wells under the condition of mechanical–chemical coupling considering the influence of the bedding plane. The solution flow of the calculation program is shown in Figure 7.

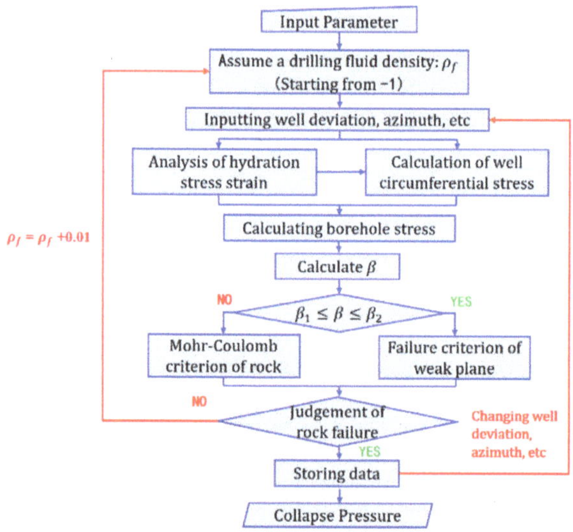

Figure 7. Solution flow of collapse pressure.

4. Analysis of Influencing Factors of Wellbore Instability in Layered Shale

For the layered shale in the Songliao Basin, the collapse pressure is closely related to the in situ stress, wellbore trajectory, rock mechanical parameters, hydration characteristics and other factors. In order to analyze the influencing factors of wellbore instability of continental shale in Songliao Basin, based on the measured data, the effects of different in situ stress conditions, different water content and different number of structural planes on the collapse pressure are calculated, respectively. The basic parameters used are shown in Table 1:

Table 1. Basic stratigraphic parameters of continental shale in Songliao Basin.

Serial Number	Parameters	Numerical Value
1	vertical depth/m	2422.0
2	overburden pressure/MPa	74.03
3	horizontal maximum in situ stress/MPa	64.06
4	horizontal minimum in situ stress/MPa	54.23
5	pore pressure/MPa	21.62
6	Biot's coefficient	0.79
7	initial Poisson ratio	0.21
8	water content of original stratum	0.032
9	water content of wellbore	0.052
10	initial cohesion of shale body/MPa	16.52
11	initial internal friction angle of shale body/°	32.73
12	cohesion of weak plane/MPa	3.28
13	internal friction angle of weak plane/°	23.87
14	dip angle of weak plane/°	3
15	tendency of weak plane/°	137
16	anisotropic ratio of formation	0.71
17	well deviation angle/°	0°~90°
18	azimuth/°	0°~360°

4.1. Influence of In Situ Stress Conditions on Collapse Pressure

Under the condition of keeping the three main in situ stresses unchanged, three different geostress conditions ($\sigma_V > \sigma_H > \sigma_h$; $\sigma_H > \sigma_V > \sigma_h$ and $\sigma_H > \sigma_h > \sigma_V$) are used to analyze the variation of equivalent density of collapse pressure with well deviation and azimuth.

Figure 8 shows the variation law of collapse pressure equivalent density with well inclination and azimuth under normal fault condition ($\sigma_V > \sigma_H > \sigma_h$). Under the conditions of this study, the collapse density can reach up to 1.41 g/cm^3 and the lowest is 0.83 g/cm^3. When the azimuth angle is constant, the collapse pressure shows an increasing trend with the increase in the inclination angle. The azimuth angles of 0° and 180° are the safest for drilling.

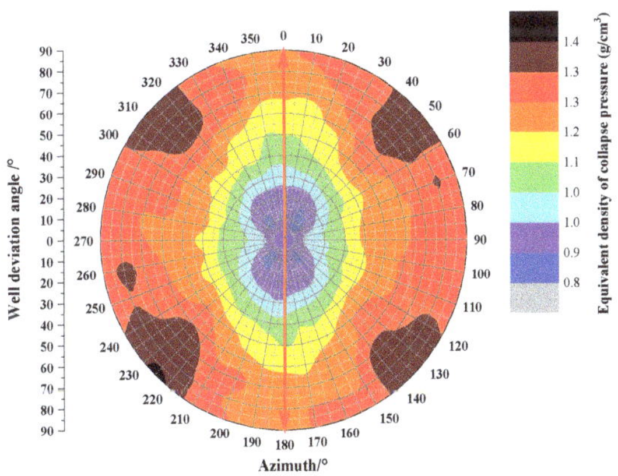

Figure 8. Under normal fault condition ($\sigma_V > \sigma_H > \sigma_h$).

Figure 9 shows the variation law of collapse pressure equivalent density with well inclination and azimuth under strike-slip fault condition ($\sigma_H > \sigma_V > \sigma_h$). The maximum collapse density can reach 1.38 g/cm^3 and the minimum is 0.91 g/cm^3 under the condition of this study. The collapse density values at azimuth angles of 50°, 130°, 230° and 330° corresponding to well deviation angles of 0~90° are low, so it is relatively safe to drill along this direction.

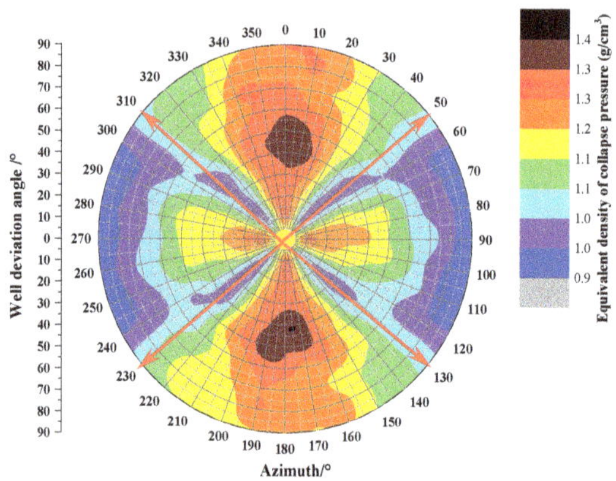

Figure 9. Under strike-slip fault condition ($\sigma_H > \sigma_V > \sigma_h$).

Figure 10 shows the variation law of collapse pressure equivalent density with well inclination and azimuth under reverse fault condition ($\sigma_H > \sigma_h > \sigma_V$). The maximum collapse density can reach 1.40 g/cm^3 and the minimum is 0.95 g/cm^3. The collapse density values at azimuth angles of 50°~130° and 230°~310° corresponding to well deviation angles of 0~90° are low, and drilling along this direction is the safest under the condition of this study.

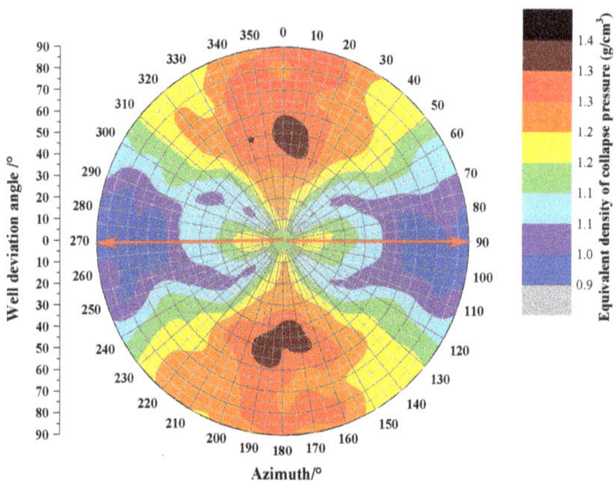

Figure 10. Under reverse fault condition ($\sigma_H > \sigma_h > \sigma_V$).

The analysis shows that no matter under which in situ stress mechanism, the wellbore in the vertical well section is the most stable, and when the inclination angle is approximately 45°, the wellbore is most likely to be unstable.

4.2. *Influence of Water Content on Collapse Pressure*

The variation law of collapse pressure equivalent density with well inclination and azimuth under different water content is analyzed, and the results are shown in Figure 11. It can be seen from the analysis that the most stable region and the easily unstable region of the wellbore hardly change with the increase in water content. The reason may be that under the given calculation conditions, the weakening effect of increasing water content on rock mechanical parameters is greater than the change of hydration expansion stress, resulting in the most stable region and the most unstable region of the wellbore hardly changing with the change of water content. The equivalent density of collapse pressure corresponding to different well inclinations and azimuths increases with the increase in water content. The reason may be that the weakening degree of hydration stress and hydration on rock mechanical parameters increases with the increase in water content. Finally, the equivalent density of collapse pressure increases with the increase in water content.

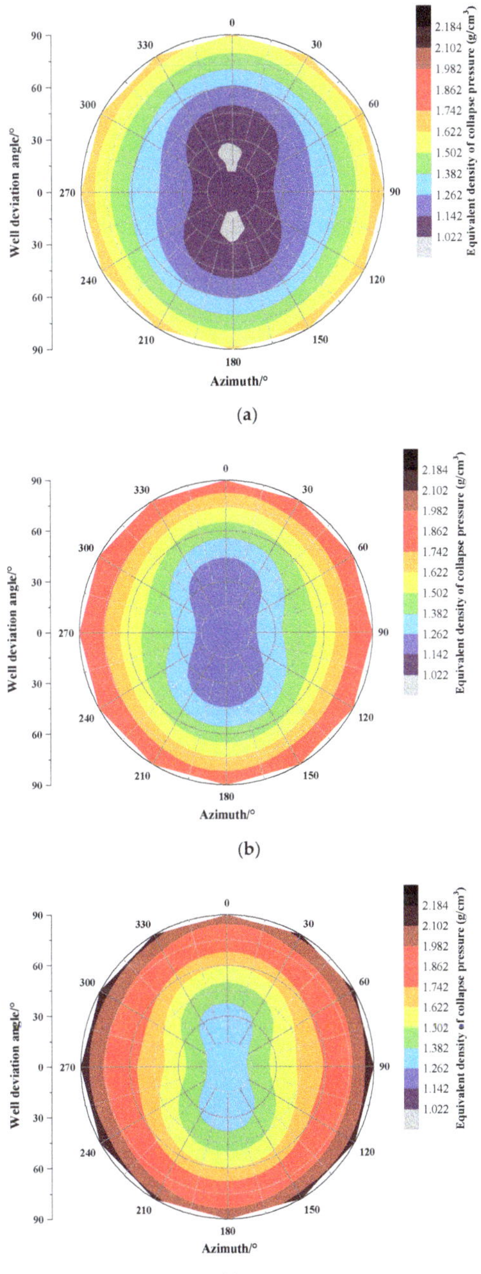

(**a**)

(**b**)

(**c**)

Figure 11. Collapse pressure distribution under different water content: (**a**) water content is 5%; (**b**) water content is 10%; (**c**) water content is 15%.

Therefore, when the water-based drilling fluid is used to drill into the layered shale formation, the hydration strain has a relatively large influence on the collapse pressure. When designing the density of the drilling fluid, not only the influence of hydration on the rock mechanical property parameters, but also the rock mechanical property parameters should be considered. The effect of hydration strain on collapse pressure needs to be considered at the same time.

4.3. Influence of the Number of Structural Planes on Collapse Pressure

The structural plane refers to the differentiation plane and discontinuity plane of various substances in the rock, such as fault, bedding, joint and schistosity. Five parameters such as occurrence, shape, filling characteristics, ductility and density can be used to quantitatively characterize the size, direction, type and development degree of structural planes. The numerical simulation mainly simulates the influence of the number of structural planes on the collapse pressure distribution when they are 0, 1 and 3, respectively.

As shown in Figure 12, as long as the number of structural planes is greater than or equal to 1, the equivalent density of the overall collapse pressure will increase. When there is no structural plane, the variation range of collapse density is 0.86~1.38 g/cm^3; When there are 1~3 structural planes, the variation range of collapse density is 0.93~1.73 g/cm^3, and the increase range is 0.17~0.35 g/cm^3. It can be seen that the existence of a structural plane will greatly increase the risk of wellbore collapse.

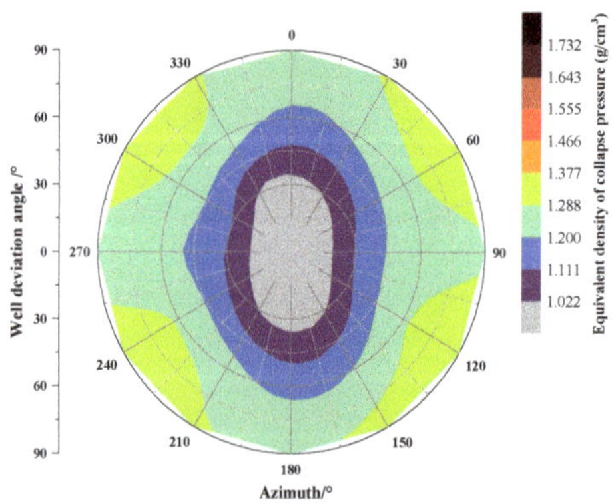

(a)

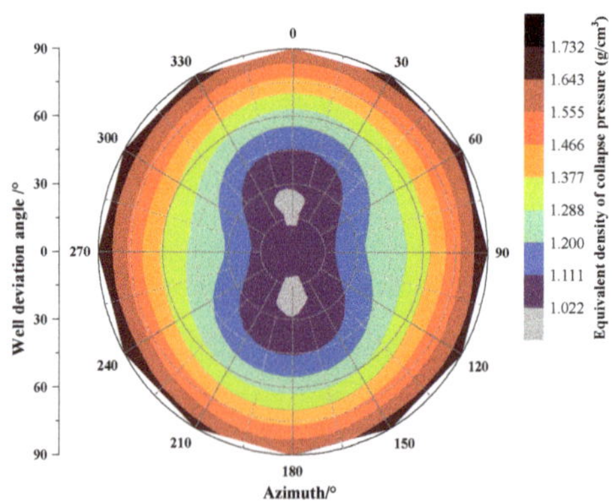

(b)

Figure 12. *Cont.*

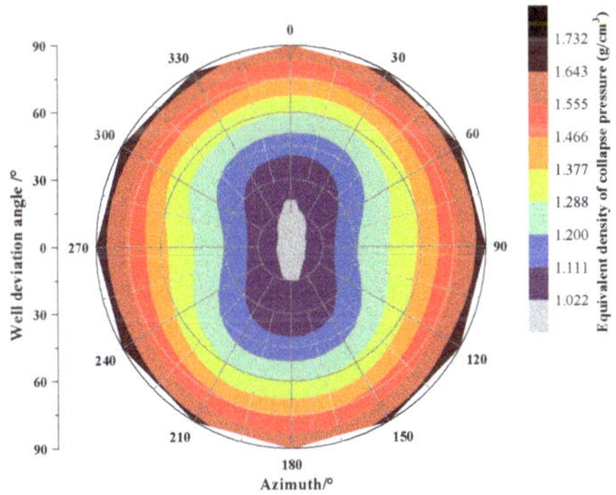

(c)

Figure 12. Collapse pressure distribution under different number of structural planes: (**a**) no structural plane; (**b**) one structural plane; (**c**) three structural planes.

5. Oilfield Case Analysis

A shale oil horizontal well G5 in the northern Songliao Basin was blocked and stuck continuously during the first trip of the third trip through the build-up section. After the calculation and correction of the well trajectory, hydration and the structural plane, the equivalent density of the minimum collapse pressure of the second drilling trip increased from 1.45 g/cm^3 to 1.51 g/cm^3. Before the completion of drilling, the density calculated according to the annular pressure consumption gradually increased to 1.53 g/cm^3. There was no collapse and block falling during the later construction, as shown in Table 2.

Table 2. Adjustment of drilling fluid density.

Well Name	Whether to Use the Model Correction of this Paper	Drilling Situation
A1	NO	The wellbore is unstable, collapses, and the well is closed for sidetracking.
Y6	NO	The wellbore peels off and blocks, the tripping is not smooth, and the drilling speed is slow.
G3	YES	The drilling operation is smooth and the drilling period is less than 15 days.

In the later stage of the G test area, after other platform wells and surrounding horizontal wells entered the Group Q, the drilling fluid density all increased to more than 1.60 g/cm^3, and the downhole complexity was greatly reduced. From 2020 to 2021, the collapse pressure model of horizontal wells in layered shale established in this paper will be widely used in the two horizontal well test areas. The drilling cycle of horizontal wells will be greatly reduced and the average mechanical drilling speed will be significantly increased. Among them, well G3 has the best effect in 2021. The shortest drilling cycle of the third section is 13.77 days, and the average mechanical drilling speed is increased to 33.53 m/h, as shown in Figure 13.

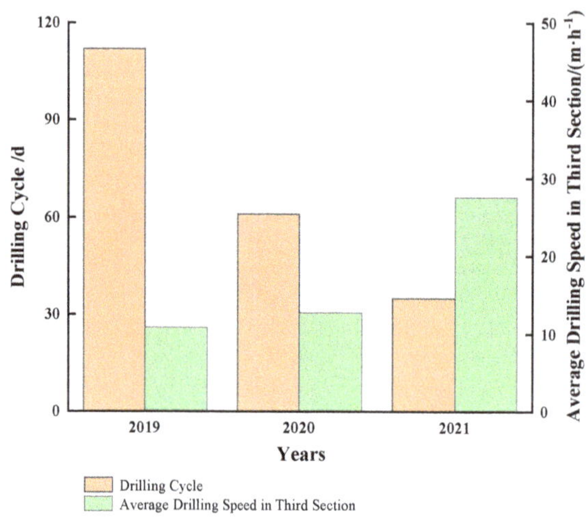

Figure 13. Annual comparisons of drilling period and ROP.

6. Conclusions

A prediction model for the wellbore collapse pressure of a horizontal well in layered continental shale oil is established, the mechanism of wellbore instability of a horizontal well in layered shale is studied, and the influence laws of different in situ stress conditions, wellbore trajectory, water content, bedding plane occurrence and other factors on collapse pressure are analyzed. Based on the calculation results and analyses of the case in the real field, the following conclusions can be drawn:

1. The analysis shows that no matter under which in situ stress mechanism, the wellbore in the vertical well section is the most stable, and when the inclination angle is approximately 45°, the wellbore is most likely to be unstable.
2. Changes in water content do not affect the most stable or unstable regions. Under the same conditions, the equivalent density of collapse pressure increases with the increase in water content.
3. Since water-based drilling fluids are used to drill into layered shale formations, the hydration strain has a greater impact on the equivalent density of collapse pressure. Therefore, it is suggested that the influence of hydration on rock mechanical parameters and the influence of hydration expansion stress should be considered when designing the density of drilling fluid. During the drilling process, it is necessary to strictly control the filtration loss of the drilling fluid and improve the inhibition of the drilling fluid to reduce the water content of the formation, reduce the hydration strain, and ensure the stability of the wellbore.
4. The results of this study are basically consistent with the previous research conclusions. However, the research results of this paper are more suitable for a layered continental shale oil horizontal well in the establishment of theoretical model and the final conclusion, and are demonstrated with field examples.

Author Contributions: Writing—review and editing, S.L.; writing—original draft preparation, K.L.; conceptualization, C.W. (Changhao Wang); resources, Y.W.; data curation, Y.J. and X.Z.; investigation, C.W. (Chunhua Wang). All authors have read and agreed to the published version of the manuscript.

Funding: This research was supported by the National natural science foundation of China (Grant No. 51874098).

Institutional Review Board Statement: Not applicable.

Informed Consent Statement: Not applicable.

Data Availability Statement: Not applicable.

Conflicts of Interest: On behalf of all the co-authors, the corresponding author states that there is no conflict of interest.

References

1. Wen, C.; Wei, L. The Wellbore Stability Study in Bedding Shale Formation on the Condition of Plasticity. *Chem. Technol. Fuels Oil* **2022**, *58*, 220–231.
2. Ewy, R.T. Wellbore Stability Predictions by Use of a Modified Lade Criterion. *SPE Drill. Completion* **1999**, *14*, 85–91. [CrossRef]
3. Bo, K.H.; Jin, Y.; Lu, Y.H.; Liu, H.T.; Zhu, J.Z. A Quantitative Evaluation Method of Anti-Sloughing Drilling Fluid Inhibition for Deep Mudstone. *Energies* **2022**, *15*, 1226. [CrossRef]
4. Ding, L.Q.; Wang, Z.Q.; Lv, J.G.; Wang, Y.; Liu, B.L. A New Model for Real-Time Prediction of Wellbore Stability Considering Elastic and Strength Anisotropy of Bedding Formation. *Energies* **2021**, *15*, 251. [CrossRef]
5. Zhang, F.; Liu, H.B.; Cui, S.; Meng, Y.F.; Wang, J.J. Influence of the Weakening Effect of Drilling Fluid on Wellbore Stability in Anisotropic Shale Formation. *Front. Phys.-Lausanne* **2021**, *9*, 541. [CrossRef]
6. Liu, Z.Y.; Pan, Z.J.; Li, S.B.; Zhang, L.G.; Wang, F.S.; Han, L.L.; Zhang, J.; Ma, Y.Y.; Li, H.; Li, W. Study on the effect of cemented natural fractures on hydraulic fracture propagation in volcanic reservoirs. *Energy* **2022**, *241*, 122845. [CrossRef]
7. Cao, G.; Cheng, Q.; Liu, Y.; Bu, R.; Zhang, N.; Wang, P. Influencing Factors of Surfactant Stripping Crude Oil and Spontaneous Imbibition Mechanism of Surfactants in a Tight Reservoir. *ACS Omega* **2022**, *7*, 19010–19020. [CrossRef]
8. Han, L.L.; Li, X.Z.; Guo, W.; Ju, W.; Cui, Y.; Liu, Z.Y.; Qian, C.; Shen, W.J. Characteristics and Dominant Factors for Natural Fractures in Deep Shale Gas Reservoirs: A Case Study of the Wufeng Longmaxi Formations in Luzhou Block, Southern China. *Lithosphere* **2022**, *2022*, 9662175. [CrossRef]
9. Li, X.R.; Zhang, C.F.; Feng, Y.C.; Wei, Y.R.; Chen, X.P.; Weng, H.Y.; Deng, J.G. An integrated geomechanics approach to evaluate and manage wellbore stability in a deep graben formation in Tarim Basin. *J. Pet. Sci. Eng.* **2022**, *208*, 109391. [CrossRef]
10. Gao, C.; Miska, S.; Yu, M.J.; Dokhani, V.; Ozbayoglu, E.; Takach, N. Experimental and numerical analysis of effective enhancement of wellbore stability in shales with nanoparticles. *J. Nat. Gas Sci. Eng.* **2021**, *95*, 104197. [CrossRef]
11. Deng, J.; Liu, W.; Yu, B.; Tan, Q.; Yang, L. Pore pressure and stress distribution analysis around an inclined wellbore in a transversely isotropic formation based on the fully coupled chemo-thermo-poroelastic theory. *J. Nat. Gas Sci. Eng.* **2017**, *40*, 24–37.
12. Epelle, E.I.; Gerogiorgis, D.I. A review of technological advances and open challenges for oil and gas drilling systems engineering. *AIChE J.* **2020**, *66*, e16842. [CrossRef]
13. Tavallali, M.S.; Karimi, I.A.; Baxendale, D. Process systems engineering perspective on the planning and development of oil fields. *AIChE J.* **2016**, *62*, 2586–2604. [CrossRef]
14. Asaka, M.; Holt, R.M. Anisotropic wellbore stability analysis: Impact on failure prediction. *Rock Mech. Rock Eng.* **2020**, *54*, 583–605. [CrossRef]
15. Ding, Y.; Liu, X.J.; Luo, P.Y. The analytical model for horizontal wellbore stability in anisotropic shale reservoir. *Geotech. Geol. Eng.* **2020**, *38*, 5109–5126. [CrossRef]
16. Ibrahim, A. A review of mathematical modelling approaches to tackling wellbore instability in shale formations. *J. Nat. Gas Sci. Eng.* **2021**, *89*, 103870. [CrossRef]
17. Kang, Q.; Chen, M.; Yan, J.; Zhang, F. Stability model of borehole wall during the well test after acidizing treatment of sandstone reservoirs. *Pet. Explor. Dev.* **2011**, *38*, 589–593.
18. Jaeger, J.C. Shear failure of anistropic rocks. *Geol. Mag.* **1960**, *97*, 65–72. [CrossRef]
19. Liu, X.J.; Zeng, W.; Liang, L.X.; Lei, M. Wellbore stability analysis for horizontal wells in shale formations. *J. Nat. Gas Sci. Eng.* **2016**, *31*, 1–8. [CrossRef]
20. Liu, J.; Yang, Z.; Sun, J.S.; Dai, Z.; You, Q. Experimental investigation on hydration mechanism of sichuan shale (China). *J. Pet. Sci. Eng.* **2021**, *201*, 108421. [CrossRef]
21. Shuai, H.; Yang, C.H.; Zhang, B.P.; Guo, Y.T.; Wang, L.; Wei, Y.L. Experimental research on anisotropic properties of shale. *Rock Soil Mech.* **2015**, *36*, 609–616.
22. Ma, Y.Y.; Li, S.B.; Zhang, L.G.; Liu, S.Z.; Liu, Z.Y.; Li, H.; Shi, E.X. Study on the effect of well layout schemes and fracture parameters on the heat extraction performance of enhanced geothermal system in fractured reservoir. *Energy* **2020**, *202*, 117811. [CrossRef]
23. Aadnoy, B. Modeling of the stability of highly inclined boreholes in anisotropic rock formations. *SPE Drill. Eng.* **1988**, *3*, 259–268. [CrossRef]
24. Ma, T.S.; Chen, P. A wellbore stability analysis model with chemical-mechanical coupling for shale gas reservoirs. *J. Nat. Gas Sci. Eng.* **2015**, *26*, 72–98. [CrossRef]
25. Ma, T.S.; Chen, P.; Zhang, Q.B.; Zhao, J. A novel collapse pressure model with mechanical-chemical coupling in shale gas formations with multi-weakness planes. *J. Nat. Gas Sci. Eng.* **2016**, *36*, 1151–1177. [CrossRef]
26. Ma, T.S.; Liu, Y.; Chen, P.; Wu, B.; Fu, J.; Guo, Z. Fracture-initiation pressure prediction for transversely isotropic formations. *J. Pet. Sci. Eng.* **2019**, *176*, 821–835. [CrossRef]

27. Ma, T.S.; Zou, J.; Chen, P.; Liu, H. Investigation on the influence coupling drilling fluid and formation boundary on acoustic wave propagation in drill string. *Geomech. Geophys. Geo-Energy Geo-Resour.* **2020**, *6*, 35. [CrossRef]
28. Peng, N.; Ma, T.S.; Chen, P.; Liu, Y. Pore pressure evaluation of formation testing while drilling under supercharged conditions. *J. Petrol. Sci. Eng.* **2021**, *203*, 108689. [CrossRef]
29. Epelle, E.I.; Gerogiorgis, D.I. Optimal rate allocation for production and injection wells in an oil and gas field for enhanced profitability. *AICHE J.* **2019**, *65*, e16592. [CrossRef]
30. Liang, C.; Chen, M.; Jin, Y.; Lu, Y. Wellbore stability model for shale gas reservoir considering the coupling of multi-weakness planes and porous flow. *J. Nat. Gas Sci. Eng.* **2014**, *21*, 364–378. [CrossRef]
31. Lee, B.H. Borehole breakouts and compaction bands in two high porosity sandstones. *Int. J. Rock Mech. Min.* **2004**, *41*, 287–301.
32. Epelle, E.I.; Gerogiorgis, D.I. A Multiperiod Optimisation Approach to Enhance Oil Field Productivity during Secondary Petroleum Production. *Comput. Aided Chem. Eng.* **2019**, *46*, 1651–1656.
33. McConnell, D.R.; Zhang, Z.; Boswell, R. Review of progress in evaluating gas hydrate drilling hazards. *Mar. Petrol. Geol.* **2012**, *34*, 209–223. [CrossRef]
34. Epelle, E.I.; Gerogiorgis, D.I. A multiparametric CFD analysis of multiphase annular flows for oil and gas drilling applications. *Comput. Chem. Eng.* **2017**, *106*, 645–661. [CrossRef]
35. Li, S.B.; Liang, K.; Wang, C.H.; Jiao, Y.X.; Liu, H.G.; Wang, C.H. Study on the mechanism of anisotropic wellbore instability in continental shale in Songliao Basin. *J. Petrol. Explor. Prod. Technol.* **2022**. [CrossRef]
36. Zhi, G.; Chen, M.; Yan, J.; Shuai, Y.; Du, X. Experimental study of brittleness anisotropy of shale in triaxial compression. *J. Nat. Gas Sci. Eng.* **2016**, *36*, 510–518.
37. Zeynali, M.E. Mechanical and physico-chemical aspects of wellbore stability during drilling operations. *J. Pet. Sci. Eng.* **2012**, *82–83*, 120–124. [CrossRef]
38. Chen, P.; Ma, T.S.; Xia, H. A collapse pressure prediction model of horizontal shale gas wells with multiple weak planes. *Nat. Gas Ind.* **2015**, *2*, 101–107. [CrossRef]

energies

MDPI

Article

Study on the Coupling Effect of Stress Field and Gas Field in Surrounding Rock of Stope and Gas Migration Law

Shizhe Li [1,2] and Zhaofeng Wang [1,*]

1 College of Safety Science and Engineering, Henan Polytechnic University, Jiaozuo 454150, China; 1710992686@qq.com
2 Sinopec Fuel Oil Sales Co., Ltd., Fujian Branch, Fuzhou 350000, China
* Correspondence: wzf3988@163.com

Abstract: In the process of working face mining, the permeability of the coal seam and the crack evolution characteristics of overlying strata are very important for efficient gas drainage. In this study, the distribution characteristics of the stress field and crack field in the working face and their relations are analyzed mainly by 3DEC numerical simulation. Furthermore, combined with the on-site measurement of coal seam stress, gas pressure, and gas seepage in front of the working face and the gas seepage in overlying strata before and after mining, the coupling effect of stress field and gas field and the law of gas migration and distribution in the working face are deeply explored. The results show that the changing trend of gas seepage and gas pressure is controlled by the stress change of the working face, and with the increase of stress, gas pressure and gas seepage also increase. The peak position of gas pressure is the farthest from the coal wall, about 22.5~25 m, followed by the peak of stress and gas seepage. When the permeability of coal and rock mass increases, the gas seepage increases and the gas pressure decreases. The coal seam stress and gas seepage in the working face and gas seepage in the overlying strata fracture zone along the tailgate side are generally greater than those on the headgate side, but the gas pressure is the opposite. Mining cracks and strata separation provide a good channel and space for gas migration and accumulation. Along the strike and tendency of the working face, gas is mainly concentrated in the overlying strata crack space above the separation zone and the roof and overlying strata crack space on the side of the tailgate, respectively. Based on this, the directional borehole gas drainage technology and borehole layout scheme in the fractured zone are put forward, which effectively reduce the gas concentration in the working face by 30~36%.

Keywords: stress field of working face; crack field; gas field; gas migration; gas drainage

Citation: Li, S.; Wang, Z. Study on the Coupling Effect of Stress Field and Gas Field in Surrounding Rock of Stope and Gas Migration Law. *Energies* **2023**, *16*, 6672. https://doi.org/10.3390/en16186672

Academic Editors: Mofazzal Hossain, Hisham Khaled Ben Mahmud and Md Motiur Rahman

Received: 2 August 2023
Revised: 9 September 2023
Accepted: 15 September 2023
Published: 18 September 2023

1. Introduction

Coal mine gas disasters mainly include coal and rock gas dynamic disasters, gas explosions, gas emissions, gas suffocation, and so on. This kind of disaster seriously threatens the safety of production, and it is also the focus of many scholars. Over the years, various coal-mining countries have done a lot of research on the mechanism and prevention technology (Gas drainage) of gas disasters and also achieved some research results [1–9]. According to the existing research results, it can be known that the stress redistribution of surrounding rock caused by mining has great influences on the development of cracks in the roof and overlying strata, gas desorption, and seepage in coal seams. In the process of coal mining, the mining stress changes, and the overlying strata move and fracture under the influence of mining, accompanied by the development and evolution of mining cracks [10–12], and then the gas migrates in the cracks [13]. Therefore, it is the foundation and key to studying the stress distribution characteristics and crack evolution laws of the surrounding rock.

For the stress distribution of mining surrounding rock and the evolution law of overlying strata cracks, Professor Qian and Professor Li [14–16] put forward such overlying strata shapes as "O" rings, elliptical throwing belts, and rectangular terrace belts with rounded corners of mining cracks, which provided a theoretical basis for the study of overlying strata in stope. Lin et al. used 3DEC numerical simulation software to analyze the distribution law of stress and displacement of mining overlying strata and study the evolution characteristics of pressure relief gas storage areas [17]. Xu et al. comprehensively use physical similarity simulation and 3DEC numerical simulation to explore the failure characteristics, stress distribution characteristics, and crack evolution laws of overlying strata. Furthermore, the evolution characteristics of coal body cracks in the working face under the influence of mining are studied [18]. Wang et al. used physical similarity simulation and PFC numerical simulation to study the failure characteristics, crack development, and porosity change characteristics of overlying strata so as to determine the development height of the gas-conducting fractured zone of overlying strata [19]. Li et al. comprehensively used physical similarity simulation and UDEC numerical simulation to study the failure characteristics, crack evolution, and pressure relief characteristics of roofs and overlying strata in protective layer mining. It also shows that protective layer mining can effectively reduce gas pressure and is beneficial to gas release [20].

For the coupling relationship between mining stress field and gas field and the law of gas migration, Zhao et al. established the coupling model of coal seam stress field, damage field, gas diffusion field, and seepage field and carried out numerical simulations on the coal seam mining process [21]. At the same time, the evolution laws of gas pressure, coal seam stress, and permeability under different conditions are analyzed. Zhang et al. studied the development height of fractured zones in overlying strata by physical experiment and 3DEC numerical simulation and verified it by the gas drainage method. Affected by mining, the stope stress and gas migration are constantly changing, and the stable overlying strata crack is an effective channel for gas migration to ensure efficient gas extraction [22]. Zhang et al. analyzed the principle of high-level directional borehole gas drainage technology, explored the influence of fractured zones in goaf on gas drainage effect, and made it clear that the gas drainage amount is closely related to the evolution law of overlying strata cracks [23]. In order to study the characteristics of gas migration in anisotropic coal seams, Nian et al. established the Coupled Anisotropic Dual-Porosity Model and analyzed the influence of permeability anisotropy and working face stress on gas pressure, gas drainage volume, and effective drainage area [24]. At present, scholars have comprehensively used a variety of methods or monitoring equipment to effectively obtain the evolution laws of overlying strata crack fields, stress fields, displacement fields, deformation, gas seepage, and gas pressure, thus guiding the design and application of directional boreholes in gas drainage [25–27]. In addition, according to the dynamic change of gas drainage concentration, it can also be directly used to judge the stress distribution characteristics of the working face [28].

Previous research work has achieved fruitful results in the crack evolution and failure movement of overlying strata induced by coal seam mining, forming a recognized theoretical system. However, the coupling relationship between mining stress fields and gas fields and the law of gas migration are complex and still need further study. In this paper, the stress distribution characteristics of stope, the evolution law of overlying strata cracks, and the evolution characteristics of the pressure relief gas accumulation area are studied by using 3DEC numerical simulation software. Combined with the on-site measurement of coal seam stress, gas pressure, and gas seepage in front of the working face and the gas seepage in overlying strata before and after mining, the coupling effect of stress field and gas field and the law of gas migration and distribution in the working face are deeply explored. Based on this, the gas drainage technology and directional borehole layout scheme in the fractured zone are put forward, and the experimental working face is designed and applied in engineering. This study further enriches the relevant theoretical system of gas migration and distribution under the coupling effect of multiple fields in the working face and provides a reference for gas drainage technology for pressure relief. So as to accelerate the realization of safe and efficient coal and gas co-mining in high-gas mines.

2. Distribution Characteristics of Mining Stress in Working Face

Mining will cause stress redistribution in the surrounding rock of the stope. With the advance of the working face, the roof collapses, and the gangue behind the goaf is gradually compacted. The "horizontal three zones" as shown in Figure 1a are formed in the working face and mainly involve the coal wall abutment pressure zone (I), the roof separation zone (II), and the recompression zone (III). At the same time, roof caving and overlying strata subsidence will also form "vertical three zones", namely the caving zone, fractured zone, and continuous bending zone [29,30]. With the disturbance of mining, when the stress reaches the secondary balance, the stress-stable area (A), the stress-concentration area (B), and the stress-relaxation area (C), as shown in Figure 1b, are respectively formed in the front and back of the working face. The coal wall in front of the working face supports most of the load of the fractured zone above the working face and its overlying strata, which causes the advance abutment pressure in front of the working face to be much greater than that behind the working face. At the same time, the working face keeps advancing, and the coal wall is also in the process of dynamically moving in the advancing direction, so the abutment pressure before and after the working face is in a moving state. Because the fractured zone has formed an arch-shaped balanced structure supported by the coal wall and the caving gangue in the goaf, the working face is in the range of the stress reduction zone. With the working face advancing forward, roof caving, and overlying strata subsidence, the gangue in the goaf behind the working face is compacted again, and its stress will further increase. When the gangue in goaf is compacted to a stable state, its stress will also be in a stable state.

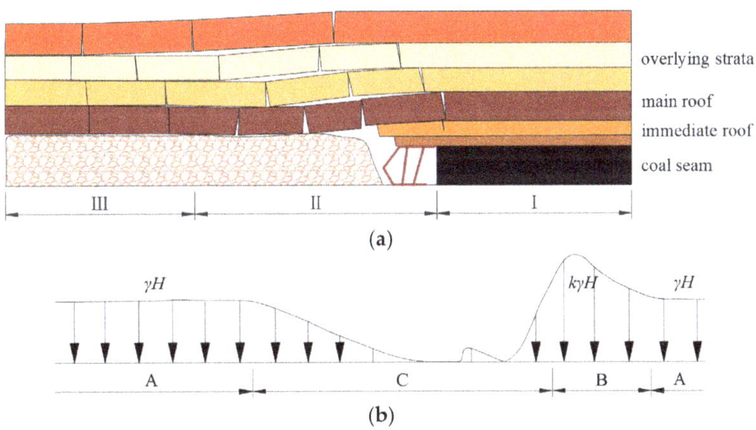

Figure 1. Schematic diagram of the abutment pressure of the working face (**a**) transverse three-zone, (**b**) abutment pressure.

3. Numerical Simulation of Failure Field and Stress Field in Working Face

3.1. Numerical Model Establishment

At present, the main coal seam in a mine in Shanxi is 9# coal seam. The dip angle of the coal seam is about 2°, and it is a flat seam with a thickness of 2.53~4.28 m. The coal seam is relatively stable, with a simple structure and no special geological structure. This study is based on working face 9102, which is close to the goaf of working face 9101. The length of the working face is 180 m along the inclined direction, and it is mined forward along the strike. The average mining height of the working face is 3.8 m, and the buried depth is about 300 m, as shown in the specific coal seam histogram (Figure 2). The roadway layout with goaf-side entry retaining without coal pillars is adopted in this working face. The mining technology is longwall retreating fully-mechanized coal mining, and the roof is managed by the full caving method. Physical and mechanical parameters and joint mechanical parameters of coal and rock mass in the working face are shown in Tables 1 and 2.

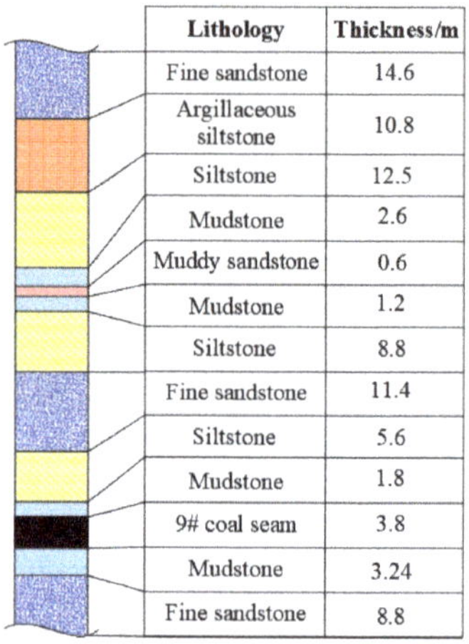

Figure 2. Coal seam histogram of working face.

Table 1. Physical and mechanical parameters of coal and rock mass.

Lithology	Unit Weight kN/m^3	Average Thick-ness/m	Elastic Modu-lus/GPa	Poisson Ratio	Cohesion/MPa	Internal Friction Angle/°	Tensile Strength/MPa
Fine sandstone	2.54	14.6	18	0.29	5.2	39	7.2
Argillaceous siltstone	2.5	10.8	16	0.15	4.6	36	6.2
Siltstone	2.46	12.5	16.6	0.17	5.5	38.7	7.15
Mudstone	2.37	2.6	15.6	0.18	2.2	40	7.05
Muddy sandstone	2.32	0.6	7.35	0.16	2.99	26.2	1.1
Mudstone	2.35	1.2	16	0.17	2.2	25	1.5
Siltstone	2.44	8.8	18.6	0.17	3.5	32	2.8
Fine sandstone	2.54	11.4	18	0.29	5.2	39	7.2
Siltstone	2.44	5.6	18.6	0.17	3.5	32	2.8
Mudstone	2.1	1.8	13	0.21	2.2	35	3.15
9# coal seam	1.35	3.8	1.7	0.32	1.8	20	0.8
Mudstone	2.1	3.24	13	0.21	2.1	35	3.18
Fine sandstone	2.54	8.8	18	0.29	5.2	34	2.2

Table 2. Joint mechanical parameters of coal and rock mass.

Lithology	Elastic Modulus/GPa	Normal Stiffness/GPa	Tangential Stiffness/GPa
Fine sandstone	2.34	4.7	3.6
Argillaceous siltstone	1.62	3.22	2.39
Siltstone	1.54	3.01	2.28
Mudstone	1.07	2.41	1.96
Muddy sandstone	1.26	2.52	2.1
Mudstone	1.07	2.41	1.96
Siltstone	1.54	3.01	2.28
Fine sandstone	2.34	4.7	3.6
Siltstone	1.54	3.01	2.28
Mudstone	1.07	2.41	1.96
9# coal seam	0.33	1.25	1.02
Mudstone	1.07	2.41	1.96
Fine sandstone	2.34	4.7	3.6

According to the conditions of working face 9102, the numerical model as shown in Figure 3 is established by using 3DEC numerical simulation software. The length (X direction), width (Y direction), and height (Z direction) of the model are 200 m, 10 m, and 85 m, respectively. 40-m coal pillars are set at both ends of the model to avoid boundary effects, and 120-m coal seam mining is simulated by advancing along the X direction. The boundary of the model is constrained by a fixed displacement, and the displacement and velocity are set to zero. At the same time, in order to simulate the rock mass load above the model to the surface, the equivalent vertical stress of 7.5 MPa is applied above the model. The elastic-plastic constitutive model based on the Mohr-Coulomb strength criterion is adopted for each stratum [17,18].

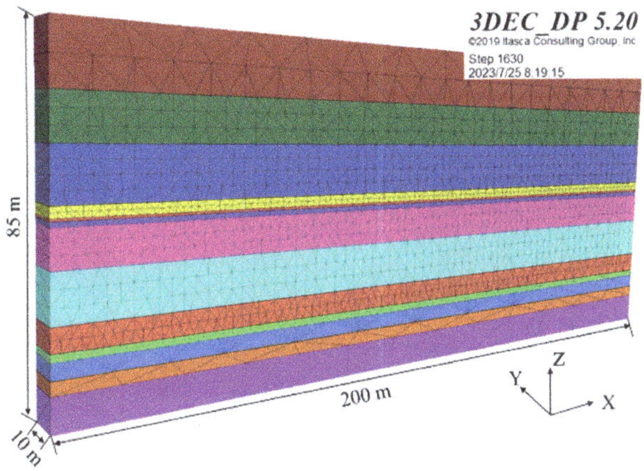

Figure 3. Numerical model.

3.2. *Failure Characteristics of Roof and Overlying Strata*

The failure characteristics of the roof and overlying strata under different advancing lengths of working face are shown in Figure 4. When the working face advances 20 m forward, the immediate roof and the main roof only produce local separation phenomena, and the roof has not obviously failed at this time. When the working face advances 40 m, the immediate roof begins to be slightly failed and caving, and the main roof also fractures in the middle of the goaf. At the same time, an obvious separation occurred on the upper roof. When the working face advances 60 m, the immediate roof caves in a large range, and the main roof behind the working face also appears to have fracture instability. The main roof above the coal wall in the working face fractures, forming a stable masonry beam structure. The phenomenon of roof separation continues to spread to the upper roof. When the working face advances 80 m, the main roof periodically fractures with the working face advancing forward. At this time, the phenomenon of roof separation is intensified, accompanied by the subsidence of overlying strata. In addition, the roof and overlying strata vertically above the coal wall in front of the working face also failed, which was caused by the large-scale subsidence of the overlying strata. When the working face advances 100 m and 120 m, the failure of the roof and overlying strata is basically the same. With the continuous advance of the working face, the roof caving continues to extend to the overlying strata. At this time, the phenomenon of the roof separation zone (II) above the coal wall of the working face is more obvious.

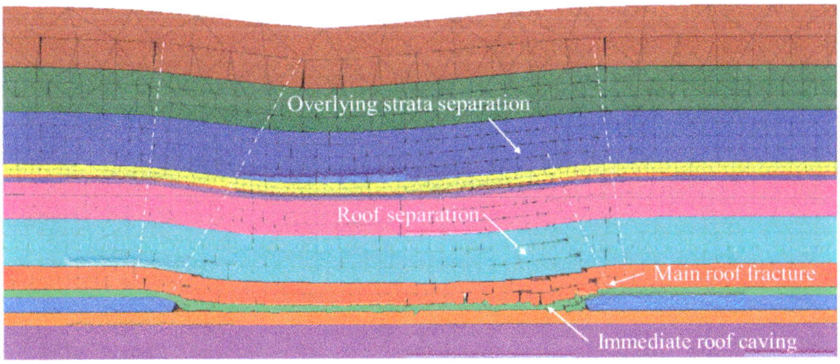

(e)

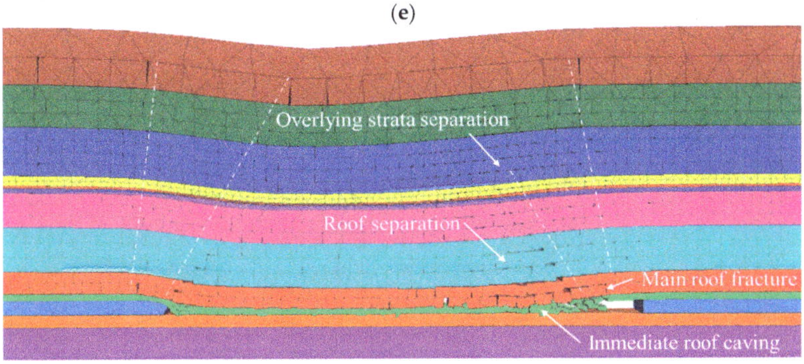

(f)

Figure 4. Failure characteristics of roof and overlying strata, (**a**) Advance 20 m, (**b**) Advance 40 m, (**c**) Advance 60 m, (**d**) Advance 80 m, (**e**) Advance 100 m, (**f**) Advance 120 m.

3.3. Evolution Law of Cracks in Roof and Overlying Strata

Figure 5 shows the development of cracks in the roof and overlying strata under different advancing lengths of the working face, and the statistics of the crack development height and the number of vertical cracks are shown in Figure 6. When the working face advances 20 m forward, the cracks at the roof position above the goaf begin to develop. When the working face advances 40 m, the roof cracks continue to develop and extend to the overlying strata. At this time, the development of cracks is concentrated in the middle of the goaf, and most of them are horizontal cracks; vertical through cracks are bred only at the immediate roof and main roof. When the working face advances 60 m, cracks in the roof and overlying strata develop continuously, and the distribution of cracks occupies the whole area of roof collapse in Figure 4c. At this time, the development of cracks is still concentrated in the middle of the goaf, and most of them are horizontal cracks. The vertical cracks are not only developed in the middle of the roof but are also mainly distributed in the position of the roof in front of the working face. When the working face advances 80 m, cracks in the roof and overlying strata above the goaf continue to develop. At the same time, a large number of vertical cracks are developing in the roof above the overlying strata in the middle of the goaf and the coal wall in front. On the whole, the cracks in the roof and overlying strata in front of the working face are concentrated. When the working face advances between 100 m and 120 m, due to the subsidence and compaction of overlying strata, the cracks in the caving roof and overlying strata in the back and middle goaf of the working face are gradually closed. At this time, cracks are mainly distributed in the roof and overlying strata in front of the working face and continue to develop to a height of about 60 m. At the same time, with the advancement of the working face, the vertical cracks in the roof and overlying strata continue to develop, and the working face, goaf, roof, and overlying strata continue to communicate.

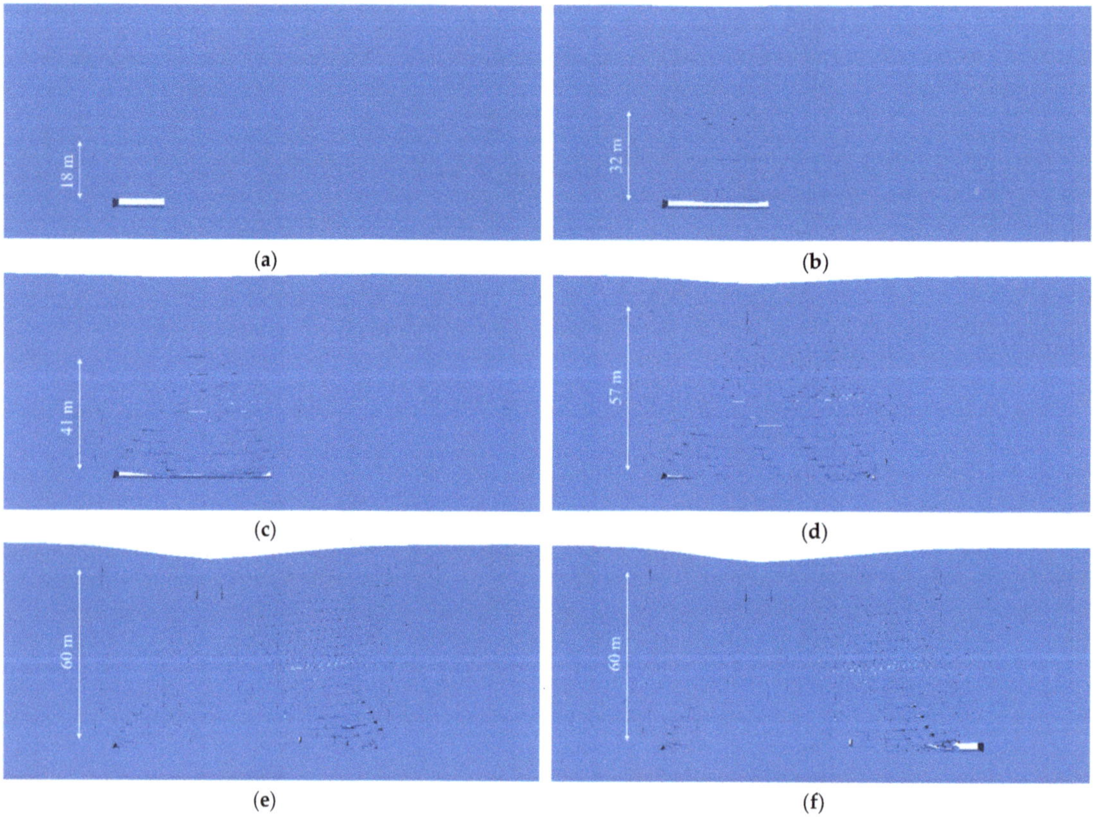

Figure 5. Development of cracks in the roof and overlying strata, (**a**) Advance 20 m, (**b**) Advance 40 m, (**c**) Advance 60 m, (**d**) Advance 80 m, (**e**) Advance 100 m, (**f**) Advance 120 m.

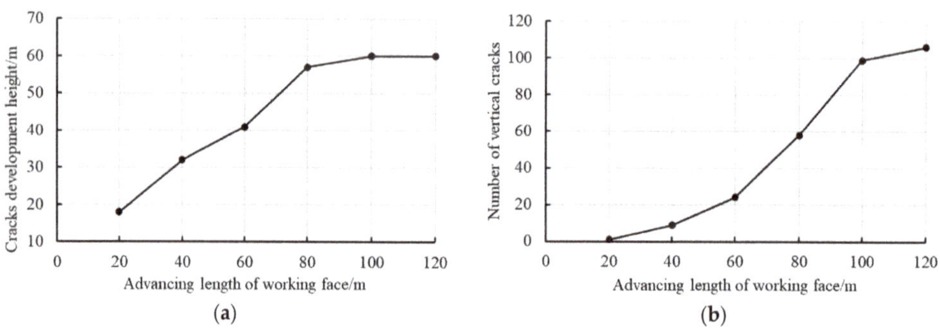

Figure 6. Statistics of crack development under different advancing lengths in the working face: (**a**) Cracks development height; (**b**) Number of vertical cracks.

3.4. Distribution Characteristics of Stress Field in Working Face

The vertical stress distribution characteristics of the working face under different advancing lengths are shown in Figure 7. And vertical stress monitoring is carried out in the range of 40 m away from the coal wall in front of the working face, as shown in Figure 8. When the working face advances 20 m forward, the stress is released due to the separation crack of the roof above the goaf. However, there is a small degree of stress concentration at the coal pillar behind the working face and the coal wall in front. As the working face advances to 40 m, 60 m, or even 80 m, the stress peak position at the coal wall in front of it also moves forward. At the same time, the peak value of stress is increasing, and the range of stress-relaxation area is roughly consistent with the distribution range of cracks in the roof and overlying strata, and it is expanding. When the working face advances between 100 m and 120 m, the peak stress increases and remains basically stable. At this time, due to the large-scale destruction, subsidence, and compaction of the roof and overlying strata, the stress in the central goaf of the working face is stabilized again and transmitted downward. Therefore, the stress in the central part of the goaf increases relative to the primary rock stress and even produces a small degree of stress increase. This phenomenon is consistent with the distribution characteristics of roof and overlying strata cracks (opening-closing) in the working face. In addition, the monitoring results of the abutment pressure of the coal wall in front of the working face in Figure 8a are consistent with the theory in Figure 1b, indicating the reliability of the numerical simulation results. With the advance of the working face, the peak stress position in front of the working face is about 12.5 m away from the coal wall. The peak stress is about 1.56~1.97 times the primary rock stress.

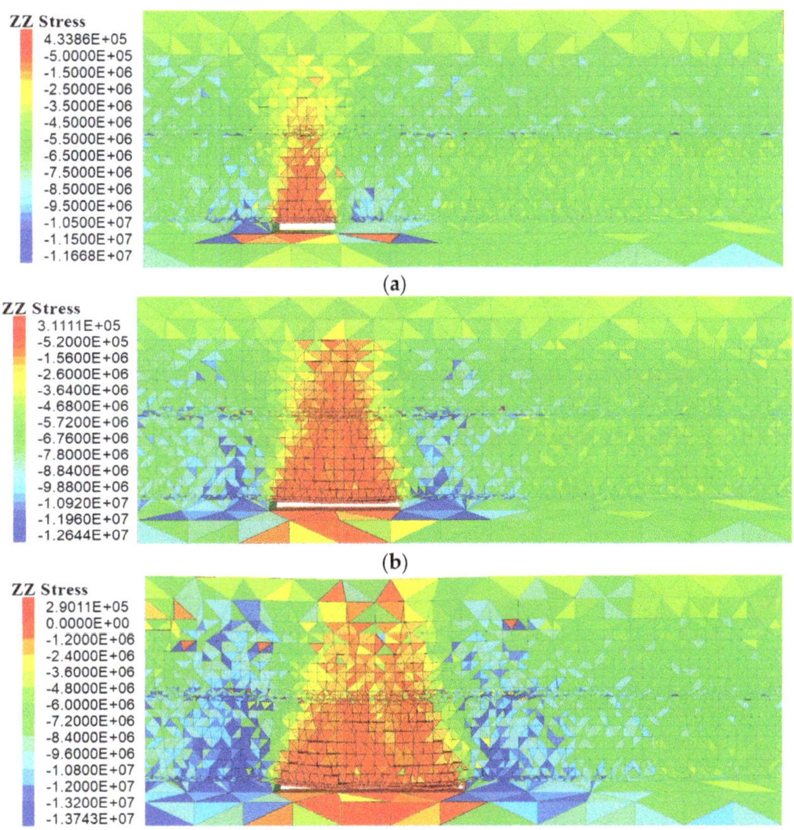

(a)

(b)

(c)

Figure 7. *Cont.*

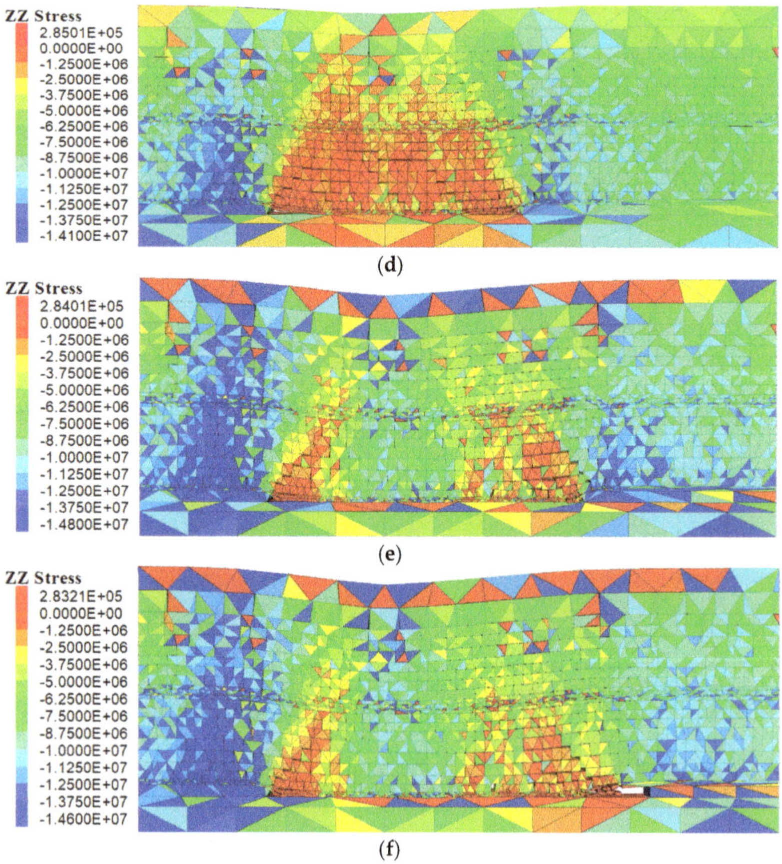

Figure 7. Vertical stress distribution characteristics of the working face, (**a**) Advance 20 m, (**b**) Advance 40 m, (**c**) Advance 60 m, (**d**) Advance 80 m, (**e**) Advance 100 m, (**f**) Advance 120 m.

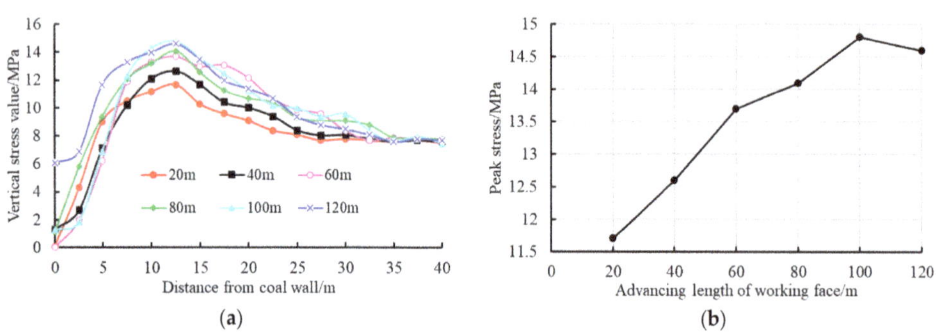

Figure 8. Evolution characteristics of vertical stress in front of the working face: (**a**) Vertical stress curve; (**b**) Peak stress.

Based on the numerical simulation study of the failure field and stress field of the working face, it can be seen that with the working face moving forward, the failure, cracks, and stress of the roof and overlying strata evolve synchronously, and their distribution ranges are basically the same. Especially when the working face reaches critical mining, the overall crack development height and stress peak value are basically stable. Among them, vertical cracks develop rapidly and penetrate through the goaf and overlying strata. At this time, the cracks in the middle of the goaf are gradually compacted and closed, and the stress is gradually increased. In this process, the crack experienced the state of "opening-closing", and the stress experienced the state of "relaxation-concentration-stable". Based on the above research, it lays the foundation for further exploring the influence of stress fields on gas distribution laws and directional borehole gas drainage technology in fractured zones.

4. Analysis of Coupling Effect between Stress Field and Gas Field

In order to further explore the coupling relationship between stress field and gas field and the influence of the law of stress field on gas pressure and permeability, 20 m deep on-site boreholes are constructed in the coal body of working face 9102 through tailgate and headgate, and four sets of survey lines are arranged. A borehole stress gauge, gas pressure gauge, and flowmeter are installed in the tailgate and headgate, respectively, and the distribution of abutment pressure, gas pressure, and gas seepage within 70 m from the coal wall of the working face is tested. Two sets of survey lines are arranged at the sides of the tailgate and headgate, respectively, and the distance between each set of survey lines is 10 m. There are 19 measuring points in each group of survey lines, and the distance between the monitoring points is 5 m. Among them, the monitoring points within 30 m of the coal wall are densely arranged with a spacing of 2.5 m. The site monitoring points on the working face are arranged as shown in Figure 9.

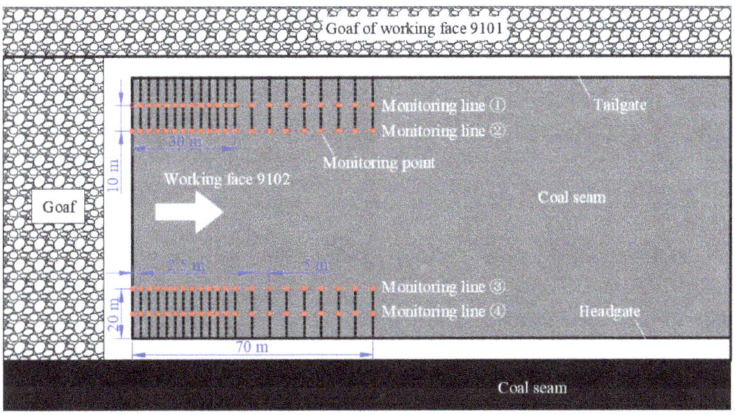

Figure 9. Schematic diagram of measuring points in coal body of working face.

The on-site monitoring results of abutment pressure distribution, gas pressure distribution, and gas seepage in front of the working face are shown in Figure 10. The gas seepage and gas pressure in the working face 9102 are closely related to the stress distribution in the working face. The maximum peak stress of the coal body on the side of the tailgate can reach 17.9 MPa, which is about 15 m away from the coal wall of the working face. The area with a larger stress value is 7.5~22.5 m. The maximum peak stress of the coal body at the side of the headgate can reach 15.3 MPa, which is about 12.5 m away from the coal wall of the working face. The area with a larger stress value is 10~20 m. The peak stress and influence range of the coal body on the side of the tailgate are larger than those on the side of the headgate. The distribution trend of gas seepage in the advancing direction of the working face is basically consistent with the stress distribution, but the peak position is different. The maximum gas seepage value is generally within the range of 10~12.5 m from the working face. Affected by the goaf of the working face 9101, the maximum gas seepage of the coal body at the side of the headgate is about 1.9 m^3/min, and that at the side of the tailgate is about 2.4 m^3/min. The gas seepage from the coal body at the side of the tailgate is generally greater than that at the side of the headgate. The distribution law of gas pressure in the advancing direction of the working face is basically the same as that of stress, and the peak area of gas pressure is ahead of the peak area of stress. Affected by the goaf of the working face 9101, the maximum gas pressure of the coal body at the side of the headgate is about 0.55 MPa, and that at the side of the tailgate is about 0.25 MPa. The gas pressure of the coal body at the side of the tailgate is generally lower than that at the side of the headgate. In addition, compared with Figure 8a, the distribution trend of abutment pressure in front of the working face obtained by numerical simulation is basically consistent with the on-site measurement results, which can be compared and verified.

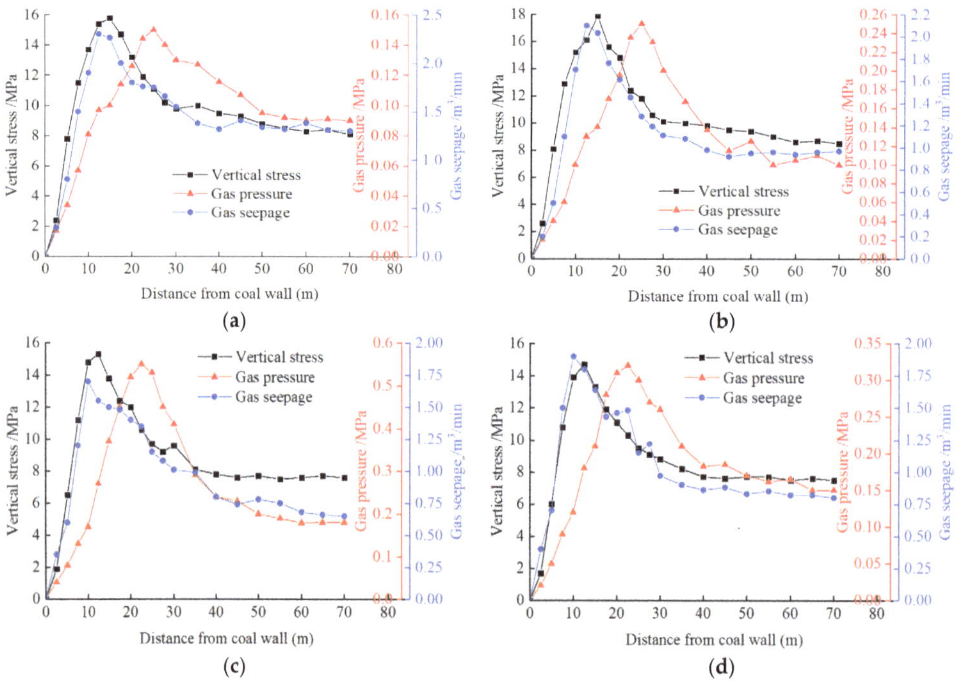

Figure 10. Variation curve of working face stress and gas pressure and seepage, (**a**) Monitoring line ① (**b**) Monitoring line ②, (**c**) Monitoring line ③ (**d**) Monitoring line ④.

According to the above analysis, it can be known that stress disturbance is the main factor affecting gas seepage and gas pressure, and the changing trend of gas seepage and gas pressure is controlled by the stress change of the working face. With the increase in stress in the advancing direction of the working face, the gas pressure and gas seepage also increase. There is a positive correlation between gas fields and stress fields. Under the action of high stress near the working face, the pores and cracks of coal and rock mass further develop [18,31,32], providing a large space for gas to migrate. With the permeability of coal and rock mass increasing, the gas seepage increases, and the gas pressure decreases at this time. The peak position of gas pressure is the farthest from the coal wall, about 22.5~25 m, followed by the peak stress and the peak gas seepage. Influenced by the mining of the previous working face, the influence range of stress value and high stress of the coal body in the working face on the side of the tailgate is generally larger than that in the working face on the side of the headgate, and the gas pressure of the coal body in the working face on the side of the headgate is generally larger than that in the working face on the side of the tailgate, but the gas seepage is the opposite. The gas pressure, gas seepage, and permeability of coal seams are all controlled by the development and change of stress. For reasonable control of coal and rock gas dynamic disasters, the first condition is to control the gas pressure of the coal body and fully ensure the discharge of displaced gas.

5. Gas Migration Law and Drainage Engineering Test in Working Face

5.1. Gas Migration and Accumulation Law in Roof and Overlying Strata Failure Field

In order to further explore the gas accumulation and migration in the roof and overlying strata of the goaf, two groups of roof boreholes (d1, d2, d3, and d4) were respectively arranged in the tailgate and headgate of the working face in 1902 at a distance of 100 m from the coal wall, and the borehole inclination depth was 100 m. The included angle between each borehole and the working face in the horizontal direction is 70°. That is to say, along the axis direction of the mining roadway, it is inclined to the coal seam by 30°. In addition, the included angle between boreholes d1 and d3 and the strike direction of the coal seam is 20°, and the included angle between boreholes d2 and d4 and the strike direction of the coal seam is 45°. The specific borehole layout is shown in Figure 11. Firstly, monitoring the gas seepage of roof boreholes with different depths before coal mining and, secondly, monitoring the gas seepage of roof boreholes with different depths after coal mining.

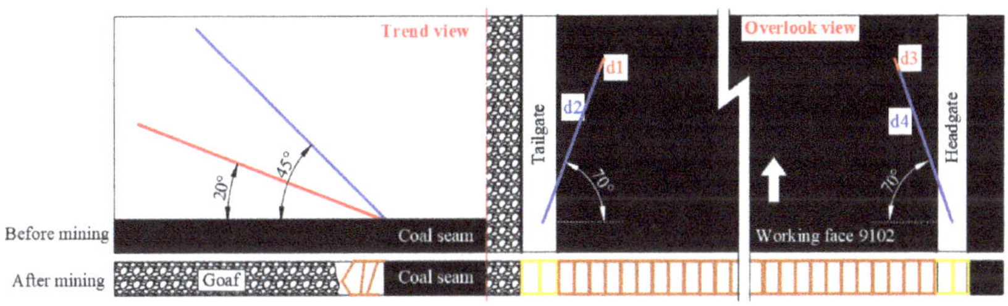

Figure 11. Schematic diagram of roof boreholes layout.

The monitoring results for roof gas seepage are shown in Figure 12. Before the working face is mined to the borehole position, the trend of gas seepage in the roof and overlying strata is basically the same, and the gas seepage is small. With the increase in borehole depth, gas seepage decreases. This is because the roof and overlying strata are less disturbed, and the degree of crack development is low when the roadway is excavated in the working face. When the working face is mined to the borehole position, the gas seepage in each borehole is large. At the same time, there is a big difference between the gas seepage of roof boreholes before and after mining in the working face, and the gas seepage increases gradually with the increase in borehole depth in the fractured zone. This is because, under the influence of mining, the overlying strata cracks are highly developed, which provides favorable channels and spaces for gas migration and accumulation.

After mining in the working face, the larger gas seepage of boreholes d1 and d2 in the tailgate is 2.9~3.5 m^3/min and 3.2~4.2 m^3/min, respectively, and the larger gas seepage of boreholes d3 and d4 in the headgate is 1.8~2.1 m^3/min and 2.1~2.9 m^3/min, respectively. This shows that the gas seepage from the roof and overlying strata failure field on the side of the tailgate is greater than that on the side of the headgate. This is due to the influence of the return air flow in the working face, and there is gas accumulation in the roof and overlying strata of the tailgate side. According to the boreholes d1 and d3, the gas seepage is 1.8~3.5 m^3/min and 0.9~2.1 m^3/min, respectively, and the overall gas seepage is increasing. After conversion, the vertical height of the deepest part of the two groups of boreholes is 34 m, and the horizontal length goes deep into the goaf at 94 m (the vertical distance from the roadway wall is 47 m). Therefore, it can be known that the boreholes d1 and d3 are completely in the fractured zone of the goaf in the vertical direction. According to the boreholes d2 and d4, the gas seepage is 1~4.2 m^3/min and 0.3~2.9 m^3/min, respectively. With the increase in borehole depth, the gas seepage first increases, then stabilizes, and finally decreases. After conversion, the vertical height of the deepest part of the two groups of boreholes is 70.7 m, and the horizontal length is 70.7 m deep into the goaf (the vertical distance from the roadway wall is 35.35 m). However, the instantaneous reduction positions of gas seepage in boreholes d2 and d4 are at borehole depths of 90 m and 80 m, respectively, so it can be judged that cracks in the roof and overlying strata developed before this position. At the same time, there are still some cracks at the depths of 95 m and 85 m. The crack development degree is poor in the range of 95~100 m depth in borehole d2 and 85~100 m depth in borehole d4. Therefore, the corresponding vertical heights of boreholes d2 and d4 at this position are calculated to be 67.2 m and 60.1 m, respectively, which is the development height of the fractured zone, and the monitoring and judgment results are basically consistent with the numerical simulation results.

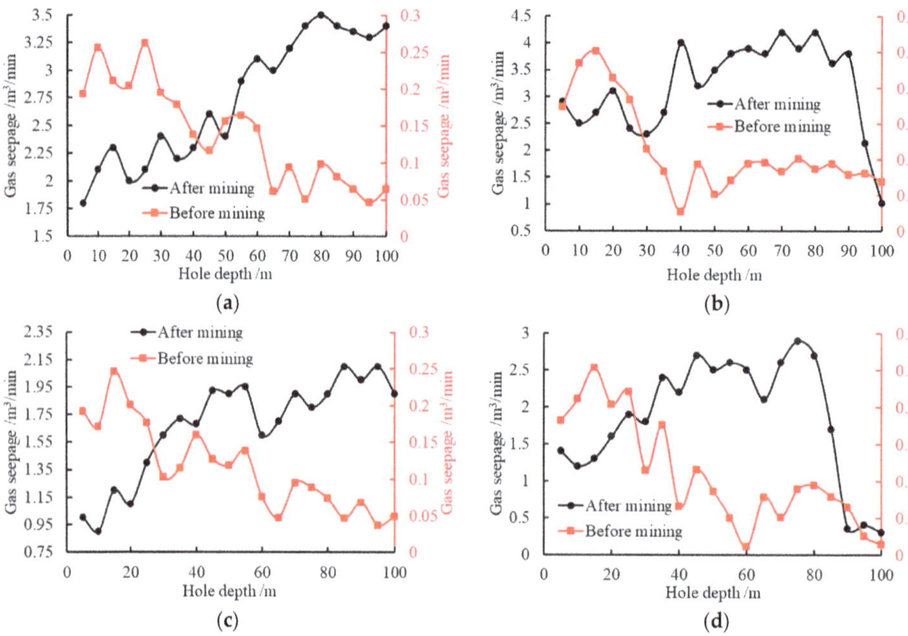

Figure 12. Gas seepage of roof and overlying strata, (**a**) Borehole d1, (**b**) Borehole d2, (**c**) Borehole d3, (**d**) Borehole d4.

According to the gas seepage monitoring of the roof and overlying strata in the goaf in Figure 12, combined with the numerical simulation results of roof failure and overlying strata crack distribution (Figures 4 and 5), the gas distribution law and migration path of the overlying strata fractured zone are analyzed along the working face advancing direction and the working face layout direction, respectively. According to the monitoring of gas seepage in the roof borehole, the roof was not disturbed and failed before coal mining, and the gas seepage was small. It can be considered that the gas distribution content of the roof above the solid coal body in the working face is extremely low, and the gas distribution is mainly the failed roof and overlying strata above the goaf in the working face. Under the influence of mining, the roof and overlying strata form separation layers and cracks, which provide a good channel and space for gas migration. As we all know, the density of gas is lower than that of air. When a large amount of gas accumulates in the stop, its pressure rises. There is a certain gas concentration and pressure difference in the crack channel of the rock stratum, so the gas will float upward along the crack in the vertical direction under the action of pneumatic power [33].

Along the advancing direction of the working face, the gas migration path in the fractured zone is shown in Figure 13. The gas is first released by the coal wall abutment pressure zone (I) in front of the working face and then mainly migrates to the roof and overlying strata above it along the crack channel of the roof separation zone (II). Among them, when the gas migrates to the large separation layer or dislocation space of the rock stratum, it will form a certain degree of accumulation phenomenon here. At the same time, the gas will also migrate to the recompression zone (III) behind the working face in the horizontal direction, but the gas migration degree is relatively small because of the recompression of the cracks in the goaf. Finally, the gas forms a stable accumulation at the highest positions A and B in the fractured zone, and the gas content is huge. However, as the working face moves forward, the coal wall abutment pressure zone (I), the roof separation zone (II), and the recompression zone (III) move forward continuously, so the main gas accumulation positions A and B are also moving forward. At the same time, it is clear that with the mining of the working face and the accumulation of time, the gas content at position A will be higher than that at position B.

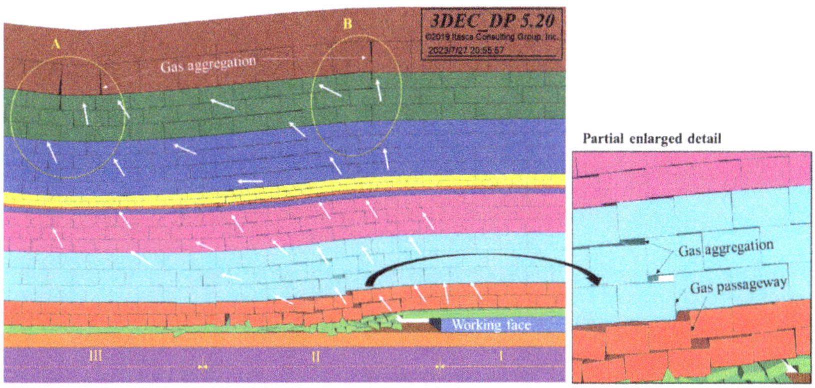

Figure 13. Gas migration path in front of working face.

Along the layout direction of the working face, the distribution of cracks in the roof and overlying strata is shown in Figure 14. The overall development height of the cracks is about 61 m, and the cracks are mainly developed in horizontal separation. Basically, the upward penetration in the inclined direction is realized. Combined with the coal seam gas seepage and roof gas seepage monitored from the tailgate and the headgate in Figures 10 and 12, the gas migration path in the fractured zone is analyzed as shown in Figure 15. The gas is first released from the coal wall abutment pressure zone (I) in front of the working face, diffused along the tailgate and the headgate of the working face, and then migrated from the crack channel of the goaf to the roof and overlying strata above it. Among them, when the gas migrates to the large separation layer or dislocation space of the rock stratum, it will form a certain degree of accumulation. Finally, the gas forms a stable accumulation at the highest positions C and D in the fractured zone, and the gas content is huge. Among them, position C is the upper corner of the working face, and the gas content at this position is higher than that at position D.

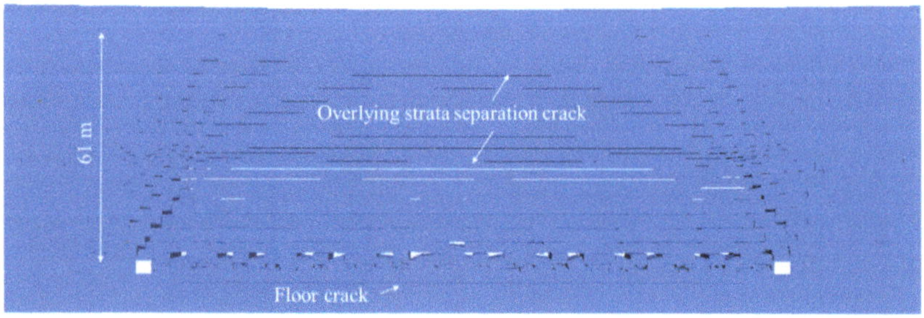

Figure 14. Cracks distribution characteristics of the roof and overlying strata in the layout direction of working face.

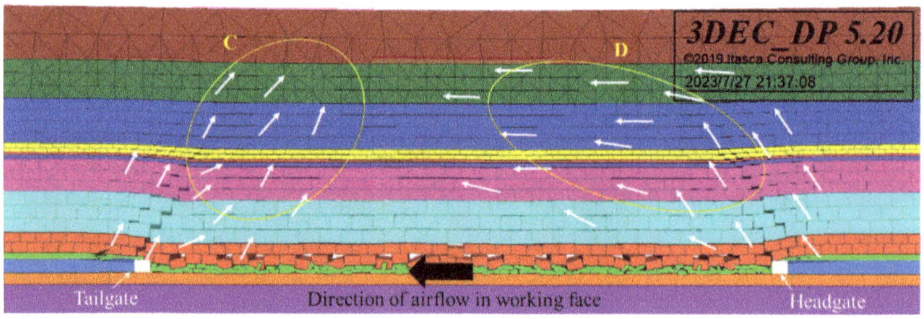

Figure 15. Gas migration path in layout direction of working face.

5.2. Test Study on Gas Drainage Technology by Directional Borehole in Fractured Zone

Based on the analysis of the law of gas accumulation and migration in the roof and overlying strata fractured zone, firstly, the development height of the fractured zone can reach 60~67.2 m, and secondly, the gas distribution and accumulation are mainly in the roof and overlying strata crack on the tailgate side. In this paper, aiming at working face 9102, not mined coal seam, gas drainage boreholes with a depth of 150 m are arranged at positions about 30 m and 45 m away from the vertical height of coal seam from the coal body side of the tailgate and headgate, respectively, as on-site tests. The direction of boreholes points to the back of the working face, and the designed dip angle of each borehole is about 30°. The specific arrangement scheme is shown in Figure 16.

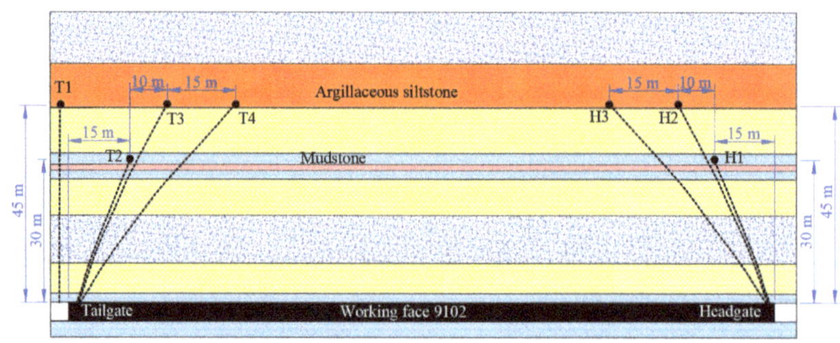

Figure 16. Schematic layout of the test boreholes.

According to the gas drainage boreholes arranged in Figure 16, the relationship between gas drainage volume and time is obtained, as shown in Figure 17. As the working face moves forward, the gas drainage amount of borehole T1 shows a decreasing trend as a whole. At the same time, before the working face advances 25 m, the gas drainage amount of borehole T1 is the largest. This phenomenon is due to the fact that the borehole T1 is close to the goaf of the previous working face, and the gas extracted at this position includes not only the working face but also the goaf of the previous working face. Moreover, due to the undeveloped cracks in the borehole T1, the gas extracted is mainly from the goaf of the previous working face. The gas drainage boreholes T2, T3, T4, H1, H2, and H3 are advancing with the working face, and the overall gas drainage volume shows a trend of increasing first and then decreasing. It shows that each borehole is affected by periodic roof caving, and the variation law of rock cracks near the borehole is that the primary cracks develop first and the secondary cracks are derived in large quantities. These cracks gradually penetrated to form a gas flow channel, and then some cracks were closed due to the influence of mining, which reduced the gas drainage amount in the borehole. Among them, the crack closure has the greatest influence on the gas drainage quantity of boreholes T2, T3, and H1. In addition, the gas drainage amount of boreholes T2, T3, and T4 on the tailgate side is larger in turn, and the gas drainage amount of boreholes H1, H2, and H3 on the headgate side is larger in turn. At the same time, the gas drainage amount at the side of the tailgate is greater than that at the side of the headgate. This phenomenon is basically consistent with the law of gas migration and distribution in fractured zones.

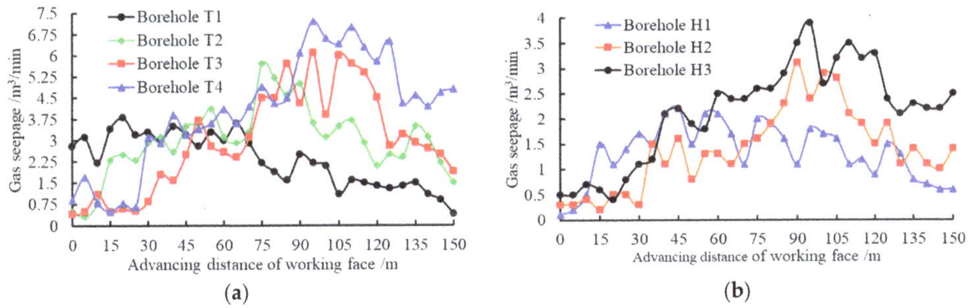

Figure 17. Gas drainage from boreholes in fractured zones: (**a**) Borehole on the side of the tailgate; (**b**) Borehole on the side of the headgate.

In order to evaluate the effect of gas drainage by boreholes in fractured zones, the gas concentration in the upper corner and tailgate during the process of the working face advancing 150 m when gas drainage is carried out in fractured zones and no gas drainage is compared, as shown in Figure 18. When gas drainage boreholes are not arranged in the fractured zone, the larger gas concentrations in the upper corner and tailgate are about 0.35~0.5% and 0.3~0.4%, respectively. There is obvious gas accumulation in the upper corner. When the gas drainage boreholes as shown in Figure 16 are arranged, the gas concentrations in the upper corner and tailgate are about 0.14~0.32% and 0.19~0.28% respectively, and the gas concentrations are reduced by 30~36%. After gas drainage by borehole in the fractured zone, the gas concentration is low and the drainage effect is remarkable. Thus, it shows that the directional borehole in the fractured zone is reasonable, which can effectively reduce the gas concentration in the upper corner of the working face and the tailgate and provide a guarantee for safe mining in the mine.

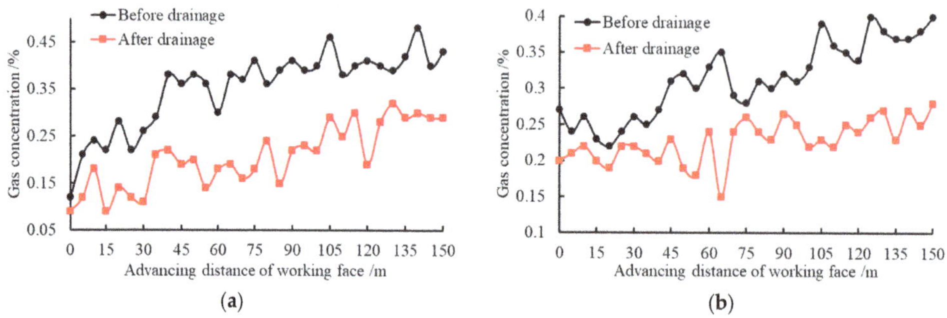

Figure 18. Comparison of concentration before and after gas drainage, (**a**) Upper corner, (**b**) Tailgate.

6. Conclusions

Through the combination of 3DEC numerical simulation and on-site monitoring, the distribution characteristics of stress field, fracture field, gas pressure, and gas seepage in the working face are deeply explored; the coupling effect of stress field and gas field in the working face and its influence on gas migration and distribution are revealed; and effective gas drainage technology in the fracture zone is put forward. The main conclusions are as follows:

(1) With the working face advancing forward, the roof cracks experienced the process of "opening-closing", and the stress experienced the process of "relaxation-concentration-stable". When the height of crack development and the peak stress of the roof of the working face reach a stable state, the height of crack development is about 60~67.2 m and the peak stress is about 1.56~1.97 times the original rock stress.

(2) The changing trend of gas seepage and gas pressure is controlled by the stress change of the working face, which has a typical coupling effect. With the increase in stress in the advancing direction of the working face, the gas pressure and gas seepage also increase. The peak position of gas pressure is the farthest from the coal wall (22.5~25 m) ahead of the working face, followed by the peak of stress (12.5~15 m) and the peak of gas seepage (10~12.5 m).

(3) There is a certain correlation between coal seam permeability and gas seepage. Under the action of high stress near the working face, the pores and cracks in the coal and rock mass further develop. The permeability of coal and rock mass increases, the gas seepage increases, and the gas pressure decreases at this time.

(4) The stress value and the influence range of high stress on the coal body in the working face on the tailgate side are generally larger than those in the working face on the headgate side, but the gas pressure is opposite. Generally, the gas seepage of the coal body in the working face and the fractured zone above the tailgate side is higher than that on the headgate side.

(5) Mining cracks and strata separation spaces provide a good channel and space for gas migration and accumulation. Along the advancing direction of the working face, the gas is mainly concentrated in the crack space above the roof separation zone, and the accumulation position moves forward. Along the layout direction of the working face, there is gas accumulation in the roof of the tailgate side and the crack space of overlying strata.

(6) Based on the law of gas migration and distribution, the gas drainage technology of directional boreholes in fractured zones is put forward. The overall gas drainage effect is remarkable, which effectively reduces the gas concentration of 30~36% in the upper corner of the working face and the tailgate.

Author Contributions: S.L.: Formal analysis, software, data curation, writing—original draft preparation. Z.W.: Conceptualization, methodology, resources, writing—review and editing. All authors have read and agreed to the published version of the manuscript.

Funding: This research received no external funding.

Data Availability Statement: Not applicable.

Conflicts of Interest: The authors declare no conflict of interest.

References

1. Jia, Z.; Ye, Q.; Li, H. Damage Assessment of Roadway Wall Caused by Dynamic and Static Load Action of Gas Explosion. *Processes* **2023**, *11*, 580. [CrossRef]
2. Waloski, R.; Korzeniowski, W.; Bołoz, Ł.; Rączka, W. Identification of Rock Mass Critical Discontinuities While Borehole Drilling. *Energies* **2021**, *14*, 2748. [CrossRef]
3. Shu, L.Y.; Wang, K.; Liu, Z.S.; Zhao, W.; Zhu, N.N.; Lei, Y. A novel physical model of coal and gas outbursts mechanism: Insights into the process and initiation criterion of outbursts. *Fuel* **2022**, *323*, 124305. [CrossRef]
4. Petkovic, M.; Chen, Y.; Gamrath, I.; Gotzes, U.; Hadjidimitrou, N.S.; Zittel, J.; Xu, X.F.; Koch, T. A hybrid approach for high precision prediction of gas flows. *Energy Syst.* **2021**, *13*, 1–26. [CrossRef]
5. Lei, Y.; Cheng, Y.P.; Wang, L.; Ren, T.; Li, Q.Y. Potential infrasonic tremors in coal seam systems: Implications for the prediction of coal and gas outbursts. *Fuel* **2022**, *326*, 125000. [CrossRef]
6. Ziętek, B.; Banasiewicz, A.; Zimroz, R.; Szrek, J.; Gola, S. A Portable Environmental Data-Monitoring System for Air Hazard Evaluation in Deep Underground Mines. *Energies* **2020**, *13*, 6331. [CrossRef]
7. Cheng, L.; Ge, Z.; Chen, J.; Ding, H.; Zou, L.; Li, K. A Sequential Approach for Integrated Coal and Gas Mining of Closely-Spaced Outburst Coal Seams: Results from a Case Study Including Mine Safety Improvements and Greenhouse Gas Reductions. *Energies* **2018**, *11*, 3023. [CrossRef]
8. Lalik, K.; Dominik, I.; Skrzypkowski, K.; Korzeniowski, W.; Zagórski, K. Self-Excited Acoustical Measurement System for Rock Mass Stress Mapping. *Sensors* **2021**, *21*, 6749. [CrossRef] [PubMed]
9. Szlązak, N.; Swolkień, J.; Kamiński, P. Design of Coal Seam Exploitation in Methane Hazard Conditions: A Case Study. *Energies* **2023**, *16*, 365. [CrossRef]
10. Xiong, Y.; Kong, D.Z.; Wen, Z.J.; Wu, G.Y.; Liu, Q.Z. Analysis of coal face stability of lower coal seam under repeated mining in close coal seams group. *Sci. Rep.* **2022**, *12*, 509. [CrossRef]
11. Li, X.B.; Ji, D.L.; Han, P.H.; Li, Q.S.; Zhao, H.B.; He, F.L. Study of water-conducting fractured zone development law and assessment method in longwall mining of shallow coal seam. *Sci. Rep.* **2022**, *12*, 7994. [CrossRef] [PubMed]
12. Zhang, F.; Zhang, J. Research on Joint Protection Layers and Gas Prevention Technology in Outburst Coal Seams. *Sustainability* **2022**, *14*, 8859. [CrossRef]
13. Wang, G.; Fan, C.; Xu, H.; Liu, X.; Wang, R. Determination of Long Horizontal Borehole Height in Roofs and Its Application to Gas Drainage. *Energies* **2018**, *11*, 2647. [CrossRef]
14. Qian, M.G.; Xu, J.L. Study on the "O–SHAPE" circle distribution characteristics of mining-induced fractures in the overlaying strata. *J. China Coal Soc.* **1998**, *23*, 20–23.
15. Li, S.G.; Lin, H.F.; Zhao, P.X.; Xiao, P.; Pan, H.Y. Mining fissure elliptic paraboloid zone and dynamic evolution of coal and methane mining. *J. China Coal Soc.* **2014**, *39*, 1455–1462.
16. Lin, H.F.; Li, S.G.; Cheng, L.H.; Wang, H.S. Experimental analysis of dynamic evolution model of fracture zone caused by overburden mining. *J. Min. Saf. Eng.* **2011**, *28*, 298–303.
17. Lin, H.F.; Wang, X.; Xu, P.Y.; Kong, X.G.; Shuang, H.Q.; Zhao, P.X. Evolution characteristics analysis and engineering application of pressure-relieved gas reservoir in extra-thick coal seam mining. *Coal Sci. Technol.* **2023**, *51*, 173–182. [CrossRef]
18. Xu, X.; Kong, D.Z.; Xiong, Y.; Chen, F. Evolution Regularity and Control Technology of Coal Face Rupture Induced by the movement of Overlying Strata Under the Influence of Mining. *Min. Metall. Explor.* **2023**, *40*, 435–452. [CrossRef]
19. Wang, G.; Wu, M.M.; Wang, R.; Xu, H.; Song, X. Height of the mining-induced fractured zone above a coal face. *Eng. Geol.* **2017**, *216*, 140–152. [CrossRef]
20. Li, L.; Wu, G.; Liu, Q. Study on Overburden Movement and Fissure Evolution Law of Protective Layer Mining in Shallow Coal Seam. *Energies* **2022**, *15*, 1831. [CrossRef]
21. Zhao, Y.; Lin, B.Q.; Liu, T.; Zheng, Y.N.; Kong, J.; Li, Q.Z.; Song, H.R. Mechanism of multifield coupling-induced outburst in mining-disturbed coal seam. *Fuel* **2020**, *272*, 117716. [CrossRef]
22. Zhang, B.C.; Liang, Y.P.; Sun, H.T.; Wang, K.Q.; Zou, Q.L.; Dai, J.H. Evolution of mining-induced fractured zone height above a mined panel in longwall coal mining. *Arab. J. Geosci.* **2022**, *15*, 6. [CrossRef]
23. Zhang, F.; Wang, G.; Wang, B. Study and Application of High-Level Directional Extraction Borehole Based on Mining Fracture Evolution Law of Overburden Strata. *Sustainability* **2023**, *15*, 2806. [CrossRef]
24. Nian, F.; Ju, F.; Zheng, C.; Wu, H.; Cheng, X. Effects of Coal Permeability Anisotropy on Gas Extraction Performance. *Processes* **2023**, *11*, 1408. [CrossRef]

25. Zhang, C.L.; Xu, J.; Peng, S.J.; Li, Q.X.; Yan, F.Z.; Chen, Y.X. Dynamic behavior of gas pressure and optimization of borehole length in stress relaxation zone during coalbed methane production. *Fuel* **2018**, *233*, 816–824. [CrossRef]
26. Zhang, D.Y.; Zhang, M.B.; Wang, L.K.; Zeng, R.P.; Wang, J.B. Determination and Application of Reasonable Levels for Highly Directional Long Boreholes in Deep Outburst Coal Seams. *Geofluids* **2022**, *2022*, 3621921. [CrossRef]
27. Liu, C.F.; Zhang, L.; Li, Y.C.; Liu, F.G.; Martyushev, D.A.; Yang, Y.F. Effects of microfractures on permeability in carbonate rocks based on digital core technology. *Adv. Geo-Energy Res.* **2022**, *6*, 86–90. [CrossRef]
28. Dzhioeva, A.K.; Brigida, V.S. Spatial non-linearity of methane release dynamics in underground boreholes for sustainable mining. *J. Min. Inst.* **2020**, *245*, 522–530. [CrossRef]
29. Zhang, P.S.; Sun, B.Y. Distribution characteristics of the advance abutment pressure in a deep stope. *J. Geophys. Eng.* **2020**, *17*, 686–699. [CrossRef]
30. Xie, J.L.; Xu, J.L.; Wang, F. Mining-induced stress distribution of the working face in a kilometer-deep coal mine—A case study in Tangshan coal mine. *J. Geophys. Eng.* **2018**, *15*, 2060–2070. [CrossRef]
31. Xiong, Y.; Yang, S.L.; Kong, D.Z.; Song, G.F.; Ma, Z.Q.; Zuo, Y.J. Analysis on early warning of coal sample failure based on crack development law and strain evolution characteristics. *Eng. Fail. Anal.* **2023**, *148*, 107170. [CrossRef]
32. Izadi, G.; Wang, S.G.; Elsworth, D.; Liu, J.S.; Wu, Y.; Pone, D. Permeability evolution of fluid-infiltrated coal containing discrete fractures. *Int. J. Coal Geol.* **2010**, *85*, 202–211. [CrossRef]
33. Li, Z.; Wang, L.; Ding, K.; Ren, B.; Wang, S.; Jiang, C.; Pan, Z. Study on Fracture and Seepage Evolution Law of Stope Covered by Thin Bedrock under Mining Influence. *Minerals* **2022**, *12*, 375. [CrossRef]

energies MDPI

Article

Wellbore Stabilization Technology of "Fluid-Solid-Chemical Coupling" in Continental Shale Oil—A Case Study of Shale Oil in Block GL

Xin Ai [1,2,*] and Mian Chen [1]

1 College of Petroleum Engineering, China University of Petroleum (Beijing), Beijing 102249, China
2 Daqing Oil Field Company Limited, Daqing 163000, China
* Correspondence: 276175542@qq.com

Abstract: During the oil shale drilling in Group Q of Block GL, the shale is prone to hydration, deterioration, sidewall exfoliation and frequent collapse, which affects the efficient exploration and development of shale oil. In order to reveal the mechanism of wellbore instability in the shale formation, the tectonic characteristics of shale are studied by combining microscopic and macroscopic methods, which identifies three key factors of physics, chemistry and mechanics about wellbore instability. Based on the analysis of earth stress and rock mechanics parameters, the experiment has established the prediction model of "fluid-solid-chemical coupling" collapse pressure of shale formation in Group Q, and calculated the safe drilling fluid density window for the horizontal wells of shale oil in Block GL by the prediction model. The main factors of wellbore instability in shale formation, deterioration characteristics and high density of oil-based drilling fluid on the well site is combined. The targeted plugging anti-sloughing and strong wetting agent were developed by using laser particle size instruments and pressure transmission experiments. A high thixotropy and strong plugging oil-based drilling fluid system is formed. The research results have been successfully applied in 16 horizontal wells of the shale oil test platform. The excellent rheological property, reliable plugging and bearing capacity and outstanding wellbore stabilization effect provides technical support for high-quality and efficient exploration and development of shale oil in Block GL. The "fluid-solid-chemical coupling" wellbore stabilization technology of continental shale oil formed in this study can provide a reference for the exploration and development of similar types of shale oil and gas reservoirs.

Keywords: shale oil; wellbore stability; "fluid-solid-chemical" coupling model; oil-based drilling fluid

Citation: Ai, X.; Chen, M. Wellbore Stabilization Technology of "Fluid-Solid-Chemical Coupling" in Continental Shale Oil—A Case Study of Shale Oil in Block GL. *Energies* **2022**, *15*, 6962. https://doi.org/10.3390/en15196962

Academic Editor: Reza Rezaee

Received: 6 July 2022
Accepted: 19 September 2022
Published: 22 September 2022

1. Introduction

In 2020, the shale in Group Q of Block GL achieved a breakthrough in oil production, and became an important strategic replacement resource of the oilfield [1]. During the drilling of several horizontal wells, taking well GYYP1 as an example, prominent problems such as the wellbore spalling and collapsing, the sidetracking of the abandoned wellbore, the loss of the reconstruction length during fracturing, workover for the casing change in the problematic stratum during production, and the instability wellbore. The concrete manifestation is that the thin shale is easily hydrated and exfoliated along the stratification crack. When the drilling fluid density is low, it is not enough to support the wellbore and when the drilling fluid density is high, it is easy to leak back and induce the wellbore instability, which seriously restricts the shale oil exploration and development process in Block GL. Compared with shale mainly deposited in marine or saline lake basins, oil shale of in Block GL has better oil generation quality [1–5], a higher proportion of shale thickness of foliation type and striation type [6,7]. The oil shale in Block GL has undeveloped structural fractures but highly developed horizontal foliation fractures, and has higher clay mineral content and a high degree of evolution [2]. These differences in rock fabric characteristics [8] make the wellbore instability mechanism [4] and mechanical properties [2,4] in the drilling process quite different from shale oil in other blocks, and the related theories and conventional technologies on wellbore stability cannot be directly applied [1–7]. Therefore, it is essential to study the mechanism and methods of wellbore instability in continental shale formation in Block GL to provide theoretical support for drilling fluid system construction, drilling design optimization and drilling technology matching.

To solve the problem of wellbore instability in shale formation, this paper presents the idea of "fluid-solid-chemical coupling" technology, which combines the study of wellbore stability mechanics with respect to drilling fluid chemistry, and carry out the analysis of formation mechanics-chemical coupling. In fact, mechanics-chemical coupling is the interaction of fluid, solid and chemical effects. Therefore, at first, it is necessary to make clear the key factors which cause wellbore instability in shale formation in Block GL, and provide the theoretical and experimental basis for the establishment of "fluid-solid-chemical coupling" model. Then, the accurately calculation of the collapse pressure changes caused by factors such as hydration of drilling fluid by the model needs to be undertaken so as to further guide the optimizing of drilling fluid, improve the drilling fluid system, and achieve wellbore stability in shale formation.

2. Causes of Wellbore Instability

Through the analysis of shale fabric characteristics, the key influencing factors of shale wellbore instability in Block GL are identified, which mainly include the following three aspects:

(1) Physical factors. Shale in Block GL is deposited in the quiet condition of the water environment, with a mudstone thickness ratio of more than 95%, and the thickness of the single layer is less than 0.2 m with the sheet structure [9]. The mudstone is mixed with thin siltstone, crust limestone and dolomite striations with poor porosity and oil content. Influenced by the overpressure of organic hydrocarbon generation and tectonic stress release, horizontal stratification fractures are extremely developed [10] with the density of 1000~3000 fractures/m (shown in Figure 1). The main types of pores are organic pores and intergranular pores of clay minerals [1,5,8], and the pore sizes are between 10 and 30 nm, with a total porosity of 6–9%. The overburden horizontal permeability is less than 0.02 mD, and the vertical permeability is one to two orders of magnitude lower than the horizontal permeability. When drilling the horizontal wells, the weak horizontal stratification section formed by the extremely developed fracture-pore nanostructures provides a channel for drilling fluid intrusion [11,12]. If the solid particles in the drilling fluid cannot effectively block the micro-fractures, it will lead to the rapid hydration and deterioration of the shale, which will aggravate the probability of wellbore instability [13].

(2) Chemical factors. The shale formation in Block GL has a high content of clay, with an average of 35.6%. It has a high degree of clay mineral evolution, which mainly consists of illite, accounting for more than 70% of clay minerals, and the silica is precipitated during transformation [1]. XRD analysis of mineral composition characteristics [14] shows high brittle mineral content, which is over 42%. Brittle mineral mainly consists of quartz with the average content of 32.5%, and the brittleness index of elastic parameter method is 33–42%. The bearing capacity of the formation is weak, and the compressive strength of rock is 28 MPa–101 MPa and it easily form cracks, a potential factors of wellbore instability. The shale is rich in organic matter, and the type of organic matter is mainly type I, with high total organic carbon, of which the content is 1.4–4.5% (TOC) [5,15,16]. The TOC in Q1 stratum is 1.8–4.5%, and the TOC in Q2 stratum is 1.4–2.2%. The maturity (Ro) is more than 1.2%, and the free hydrocarbon content (S_1) is more than 8 mg/g. The gas ratio is more than 50%, and the oil saturation is more than 60%. The crude oil density is less than 0.8 g/cm^3, and the viscosity is less than 10 cp. When the contact angle is greater than 145, the wettability is oil-wet, and the invasion of water-based drilling fluid easily hydrates the surface, which causes interlayer microcracks to open, expand and extend, and leads to

the reduction of rock strength. The invasion of oil-based drilling fluid dissolves the light components in organic matter, changes the chemical characteristics of the rock surface, reduces the cohesion of rock, promotes the release of rock stress and destroys the stability of wellbore.

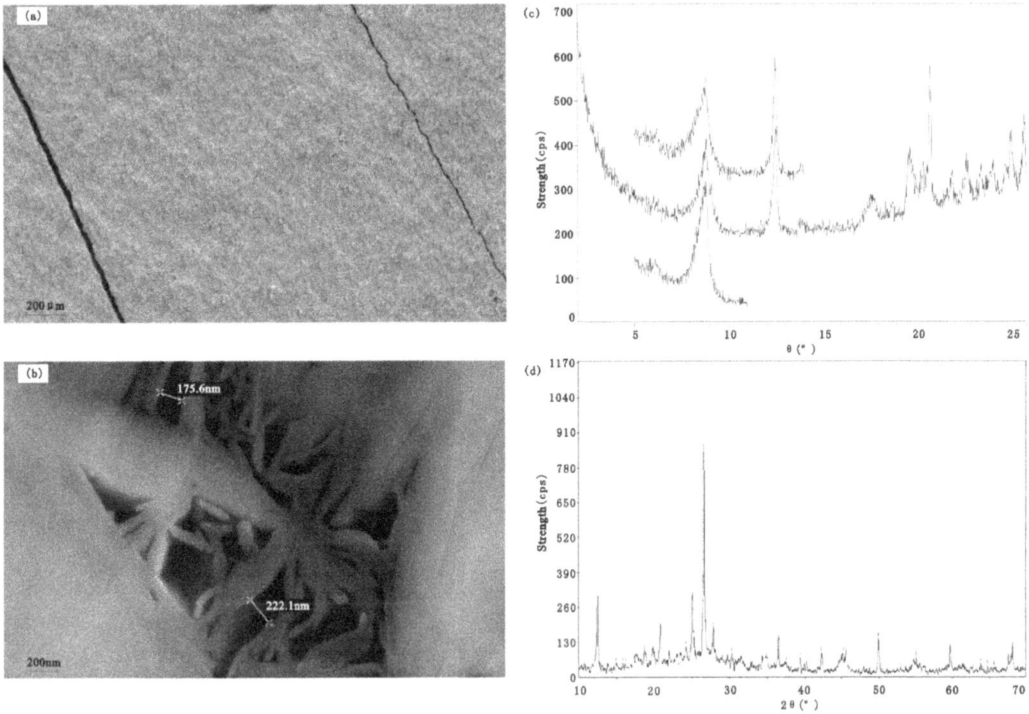

Figure 1. Geological characteristics analysis of shale in Block GL. (**a**) Scanning electron microscope picture for the crack of the shale laminated structure; (**b**) Scanning electron microscope picture for the intergranular void of clay; (**c**) Map for drill core by X-ray diffractometry of the clay; (**d**) Map for whole rock by X-ray diffractometry of the clay.

(3) Mechanical factors. The shale formation is over-pressured with the pressure coefficient greater than 1.4, and the underground physical property is better than that of the ground, which indicates that the stress release will destroy the seepage capacity of the reservoir, so the pressure coefficient of drilling fluid should be greater than that of the formation. This block is rich in illite, which is arranged along the stratification orientation, and its hydration is mainly controlled by the hydration force, which is equivalent to the repulsion force. The K^+ in the interlayer can absorb up to 20 water molecules, and the generating expansion and disintegration structure may destroy the crystal layer, which belongs to the expansion type with high stress and little strain [17]. With the influence of mineral fabric, the weak surface of shale laminated structure, and under the action of pressure difference [18] and chemical potential difference [19] during drilling, the drilling fluid filtrate quickly intrudes into the stratum along the developed micro-cracks and pores. This phenomenon changes the pore pressure near the wellbore and the stress state of the rock [20]. It also causes the partial hydration of shale, the uneven distribution of hydration swelling pressure [21,22], and mechanical damage [23]. It leads to the significant reduction of shale cohesion and internal friction angle and weakening of the rock strength [24].

3. The Safe Density Window of Drilling Fluid

During the drilling process, the collapse pressure and fracture pressure will change with the influence of hydration of the drilling fluid, the angle of inclination, the azimuth angle and many other conditions [11,13,24,25], so the safe density window of the drilling fluid should be accurately calculated under various factors, which makes the drilling fluid produce enough support for the wellbore and realize the wellbore stability.

3.1. A Prediction Model of Collapse Pressure

3.1.1. Analysis of In-Situ Stress in the Block

Using twelve lengths of drill cores from Q1, Q2 and Q3 stratums from three wells, including Well G693-66-S68 in the laboratory to carry out the paleomagnetic directional experiment combined with differential strain analysis, the magnitude and direction of in situ stress in Group Q are obtained. By using the Kaiser effect measurement method of acoustic emission [10], the drill cores of three wells including Well GX7091 in Q1 and Q2 and Q3 stratums are processed for a compression test, and the acoustic emission signals are received. The stress value at the time of increasing acoustic emission activity is the maximum stress on the rock sample, and the results are shown in Table 1. According to the paleomagnetic directional experiment combined with differential strain and acoustic emission measurement experimental results, the maximum horizontal geostress gradient of Group Q of Block GL is 0.020 MPa/m, the minimum horizontal geostress gradient is 0.017 MPa/m and the vertical principal stress gradient is 0.025 mpa/m. The maximum horizontal geostress direction in Group Q of Block GL is between N100 $\pm$ 10° (shown in Table 1).

Table 1. Experimental results of earth stress in Block GL.

Method	Well	Stratum	Depth (m)	Vertical Stress Gradient (MPa)	Maximum Horizontal Geostress Gradient (MPa)	Minimum Horizontal Geostress Gradient (MPa)	Angle with Marking Line (°)	Marking Line Direction (°)	The Maximum Horizontal Geostress Direction (°)
Paleomagnetic directional experiment combined with differential strain analysis	G693-66-S68	Q1	2319.54–2319.67	58.7	43.1	34.8	121°	N219° E	N98° E
		Q1	2325.45–2325.60	54.2	44.2	37.0	86°	N207° E	N121° E
		Q2, Q3	2277.56–2277.71	54.9	42.8	37.4	−84°	N49° E	N133° E
		Q2, Q3	2234.85–2235.01	54.7	43.6	36.0	35°	N163° E	N128° E
Kaiser effect measurement method of acoustic emission	GX7091	Q1	2247.98–2248.10	56.2	42.7	36.6	−66°	N30° E	N96° E
		Q1	2269.75–2269.87	57.1	42.0	37.0	92°	N204° E	N112° E
		Q1	2260.94–2261.07	56.8	42.7	36.2	93°	N197° E	N104° E

3.1.2. Rock Mechanics Parameter Section

In order to establish the prediction model of "three pressures" in Group Q, that is, formation pressure, collapse pressure and fracture pressure, triaxial pressure tests with different drill core development, different confining pressure conditions and different loading directions are carried out to obtain the stress-strain curve of the core, and the elastic modulus, Poisson's ratio and strength of the drill core are calculated. According to the Mohr-Coulomb criterion, the molar envelope circle of stratum rocks is drawn, and the internal friction angle and cohesion of shale are obtained, which is shown in Table 2. The cohesion and internal friction angle under the direction of vertical foliation is the maximum, and the cohesion under the direction of foliation of 60° is the minimum. Using the prediction model for the time difference of P-wave and S-wave, density, natural gamma and other well logging data of well GY2HC [26], the section of rock mechanics parameters is obtained, and the mechanical parameters including the elastic modulus and compressive strength of rock are obtained by laboratory triaxial compression experiment. The section of rock mechanics parameters is calibrated, and the empirical coefficient in the prediction model is obtained, and the longitudinal distribution section of rock mechanics parameters in this block is established, as shown in Figure 2.

Table 2. Test of rock mechanical property parameters of dry shale samples in Block GL.

Core Number	Stratum	Coring Depth (m)	Loading Direction	Cohesion (MPa)	Internal Friction Angle (°)	Internal Friction Coefficient
12	Q1	2275.17–2275.31	Parallel foliation	11.386	12.95	0.230
25	Q1	2317.79–2317.91	Vertical foliation	22.464	13.06	0.232
28	Q1	2305.75–2305.90	Foliation of 30°	15.698	7.57	0.133
39	Q1	2302.31–2302.47	Foliation of 45°	8.338	6.39	0.112
29	Q1	2345.27–2345.39	Foliation of 60°	4.775	4.40	0.077

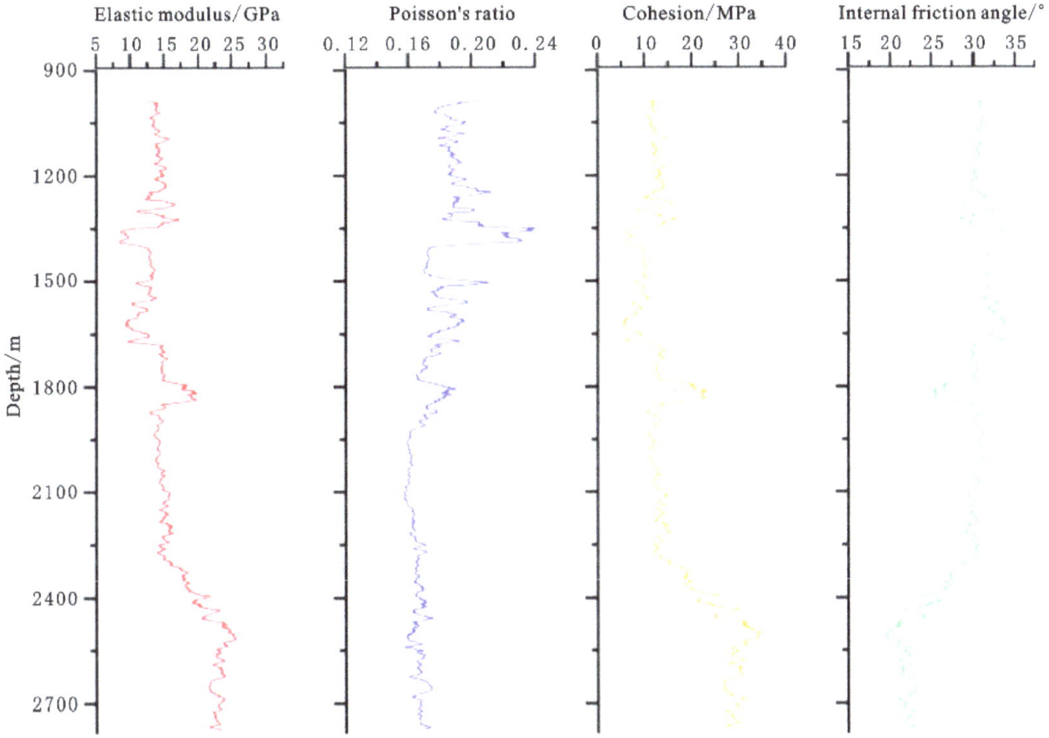

Figure 2. Rock mechanics parameter profile of Well GY2HC.

3.1.3. A "Fluid-Solid-Chemical" Coupling Collapse Pressure Prediction Model

Multi-field coupling is widely used in the drilling engineering field, as well as other engineering communities [27–29], to accurately capture the coupling effects among different physical fields. In the drilling engineering field, with the drilling bit penetrating the formation and drilling fluid being injected, it will form the complex interactions among the fluid seepage field, solid stress field and the chemical potential field in underground rock. If the mud density is too low, shear failure will occur on the wellbore wall, leading to wellbore collapse or hole shrinkage. Collapse pressure refers to the lower limit of drill fluid density to prevent wellbore collapse or hole shrinkage, and it is necessary to make the stress distribution of the strata around the wellbore clear for the calculation of collapse pressure. The special rock fabric characteristics of Block GL make it more prone to hydration and collapse than the conventional formation in the drilling process, so it is necessary to establish the corresponding characterization model. After spudding the wellbore, the hydraulic action and chemical action of drilling fluid will change the pore pressure of the formation around the well, and the essence of wellbore stability analysis is to study the physical and chemical action among the fluid seepage field, solid stress field and chemical potential field. That is, by adopting the method of fluid-solid-chemical coupling, a fluid-solid coupling model with chemical potential gradient as the chemical driving force and hydraulic pressure difference as the hydraulic driving force is comprehensively established [13,18,19,21,25,30–33]. Combined with the results of in-situ stress analysis and rock mechanics parameters, as well as clearly defining the variation rule of wellbore strength with the soaking time of the drilling fluid, we establish the prediction model of "fluid-solid-chemical" coupling collapse pressure in the shale formation of Group Q.

The finite element method in Comsol is used to calculate the coupling of multiple physical fields. Taking different well types as the analysis objects, through the relationship between water content test and rock mechanics parameters, and introducing rock mechanics parameters with different water contents, the stress model of the shaft lining is solved. On this basis, the liquid column pressure corresponding to the failure collapse of the shaft wall is analyzed by the Moore-Coulomb criterion, which is taken as the collapse pressure calculation result. The calculation result parameters under different angles of inclination and azimuth angles are extracted, and the final collapse pressure equivalent density window is obtained by interpolation with the drawing software Origin.

(1) Fluid-chemical coupling model

The numerical model is adopted, and its interior is a 2 m × 2 m × 2 m angle of inclination and the well borehole wall stress model with parametric scanning of azimuth angle. The model considers the coupling of fluid-solid-solid. The external is a large area of 10 m × 10 m × 10 m, which is used for in situ stress balance to eliminate stress concentration. The model adopts finer mesh processing, in which the stress calculation model is obtained by the boundary layer and ultra-fine division for the well borehole wall at the stress concentration position and the stratum with the structural plane.

The hydraulic pressure difference and the chemical potential difference are the two main factors that cause pore pressure change [24]. The volume flow through the unit body can be obtained by transforming the Darcy formula:

$$J = \frac{nK(\Delta P + I_m \Delta\theta)}{\mu \Delta x} \quad (1)$$

In Formula (1), J is the volume flow through the unit, cm^3/s; n is the cross-sectional area of the unit body, m^2; K is permeability, μm^2; ΔP is hydraulic pressure difference, MPa; I_m is membrane efficiency, dimensionless; μ is the viscosity of the fluid, mPa·s; θ is the chemical potential, J/mol; Δx is the unit length, cm.

The distribution equation of chemical potential [22] is as follows, which can be calculated according to the concentration of inorganic salt ions:

$$\frac{\partial \theta}{\partial t_1} = D\left(\frac{\partial^2 \theta}{\partial x^2} + \frac{\partial^2 \theta}{\partial y^2}\right) \quad (2)$$

In Formula (2), x and y are diffusion distance components of chemical potential, cm; t_1 is time, s; D is the chemical potential diffusion coefficient, cm^2/s.

(2) Fluid-solid coupling model

The rock unit is subjected to various stresses, including the extrusion load and pore pressure from surrounding rocks, and there are various forms of the strain caused by stress. According to the source of the strain, it can be divided into the average strain of dry porous media, the volume strain caused by pore pressure and the volume strain caused by tensile stress. Combined with the effective stress theory of Biot [34], considering the average strain of dry porous media, volume strain caused by pore pressure and tensile stress, the constitutive equation of solid mechanics under fluid-solid coupling is obtained.

By considering these strains, the stress distribution of the formation around the well can be accurately calculated:

$$\sigma_{ij} = \left[\lambda e_s - P_p(1-\beta)\right]\delta_{ij} + 2G\varepsilon_{ij} \tag{3}$$

In Formula (3), σ_{ij} is the stress tensor, MPa; λ is lame constant of stratum, dimensionless; e_s is the volume of solid skeleton, m^3; P_p is pore pressure, MPa; β is Biot constant, dimensionless; δ_{ij} is Kronecker symbol, dimensionless; G is formation shear modulus, MPa; and ε_{ij} is the strain tensor, dimensionless.

The stress balance equation with displacement component u_i and pore pressure P_p under fluid-solid coupling:

$$G\nabla^2 u_i - \nabla P_p(1-\beta) + \nabla e_s(\lambda + G) = 0 \tag{4}$$

In Formula (4), u_i is displacement component of solid skeleton, m.

The change of water content is expressed as:

$$\varsigma = \Phi d_i = \alpha e_s + \frac{\Phi}{K_f}(\Delta P + I_m \Delta\theta) \tag{5}$$

In Formula (5), ζ is the change of water content, %; Φ is porosity, %; d_i is the relative displacement of pore fluid of the solid skeleton, m; α is a Biot coefficient, $\alpha = 1 - \beta$, dimensionless; e_s is the volume of solid skeleton, m^3; K_f is the compressibility of pore fluid, 1/Pa; ΔP is hydraulic pressure difference, MPa; $\Delta\theta$ is the chemical potential difference, J/mol; I_m is membrane efficiency, dimensionless.

In an infinite axisymmetric area, such as around the wellbore, the control equation of pore pressure is obtained by combining the fluid-solid coupling model with the fluid-solid coupling model [24]:

$$\frac{\partial P_p}{\partial t} = K_1(I_m \nabla^2\theta + \nabla^2 P_p) \tag{6}$$

In Formula (6), K_1 is the dispersion coefficient, which is equivalent to c in the equation of Rice-Cleary [35], dimensionless. The specific expression is:

$$K_1 = \frac{K}{\left(\frac{\Phi}{K_f} + \frac{\alpha^2}{2G+\lambda}\right)\mu} \tag{7}$$

(3) The change rule model of stratum strength with the soaking time of drilling fluid.

From the mechanical point of view, wellbore collapse is caused by shear failure, which accounts for the stress on the rocks around the wellbore exceeding the strength of the rocks. Generally speaking, wellbore collapse occurs at the horizontal minimum geostress position of the wellbore wall. In the study of wellbore stability mechanics [34,36–38], the Mohr-Coulomb strength criterion expressed by principal stresses σ_1 and σ_3 is:

$$C = \frac{\sigma_1}{2}\left[(f^2+1)^{\frac{1}{2}} - f\right] - \frac{\sigma_3}{2}\left[(f^2+1)^{\frac{1}{2}} + f\right] \tag{8}$$

In Formula (8), σ_1 and σ_3 are the maximum and minimum principal stress around the wellbore, MPa; C is cohesion, MPa; f is the internal friction coefficient of rock, $f = \tan\varphi$, dimensionless; φ is the internal friction angle, °.

When shale comes into contact with drilling fluid, driven by physical and chemical actions such as chemical potential and capillary force, the drilling fluid filtrate will invade the rock, making the rock strength gradually decrease with the increase of soaking time. When the drilling fluid filtrate reaches saturation in the rock, shale stops absorbing water, and the reduction of formation strength slows down at this time [36–41]. Based on this, the variation rule of shale formation strength changed with soaking time of drilling fluid:

$$C(t) = C_0 - A \cdot \exp\left(-\frac{B}{t}\right)$$

$$\varphi(t) = \varphi_0 - E \cdot \exp\left(-\frac{F}{t}\right) \tag{9}$$

In Formula (9), $C(t)$ is the formation cohesion at time t after the wellbore is spudded in, Mpa; $\varphi(t)$ is the friction angle in the formation at time t after the wellbore is spudded in, °; C_0 is the formation cohesion when the wellbore is just spudded in, Mpa; φ_0 is the friction angle in the formation when the wellbore is just spudded in, °; t is the time after the wellbore hole is spudded in, day; A is the fitting constant of shale cohesion, dimensionless; B is the fitting constant of shale internal friction angle, dimensionless; E is the fitting constant of parameters such as dynamic viscosity of drilling fluid, dimensionless; F is the fitting constant of physical and chemical parameters of drilling fluid, dimensionless.

(4) "Fluid-solid-chemical" coupling collapse pressure model

Considering pore pressure and the chemical potential difference of drilling fluid, when the diameter of the wellbore is much smaller than the hole axis, it can be regarded as a plane strain problem, that is, the axial stress around the hole is 0 Mpa. For the surrounding rock of shale sidewall, the stress distribution around the hole is:

$$\sigma_r = P_m + I_m \Delta\theta_p - \alpha P_p$$

$$\sigma_\theta = \sigma_H(1 - 2\cos 2\theta_1) + \sigma_h(1 + 2\cos 2\theta_1) - (P_m + I_m\Delta\theta_p) - \alpha P_p \tag{10}$$

In Formula (10), σ_r is the radial stress around the well, MPa; σ_θ is circumferential stress, MPa; σ_H is the maximum horizontal geostress, MPa; σ_h is the minimum horizontal geostress, MPa; P_m is the drilling fluid column pressure, MPa; and θ_1 is the round angle of the wellbore, °.

According to the pore pressure, the stress distribution of wall rock under the action of chemical potential difference of drilling fluid, and the influence of shale formation strength with soaking time of drilling fluid, the calculation formula of "fluid-solid-chemical" coupling collapse pressure of shale stratum of Qingshankou:

$$P_c = P_m = \frac{2\sigma_h f_1(t) - (f_1(t) - f_2(t))\alpha P_p - 2C(t) - (f_1(t) + f_2(t))I_m\Delta\theta_p}{f_1(t) + f_2(t)} \tag{11}$$

In Formula (11), $f_1(t)$ and $f_2(t)$ are the internal friction coefficients of rocks at time t after the wellbore is spudded in, dimensionless. Specific Expression:

$$f_1(t) = \left(1 + \tan^2(\varphi(t))\right)^{1/2} - \tan(\varphi(t))$$

$$f_2(t) = \left(1 + \tan^2(\varphi(t))\right)^{1/2} + \tan(\varphi(t)) \tag{12}$$

By combining the above multi-field coupling equation with the actual wellbore trajectory, and integrating the influence rule of water content and the soaking time of drilling fluid on rock strength, the variation rule of shale collapse pressure with wellbore drilling time passing can be simulated and predicted. It can effectively predict when the wellbore wall may collapse and the severity of the wellbore wall collapse and calculate the wellbore wall collapse period.

3.2. Prediction of Three-Pressure Section

Combined with the laboratory shale strength deterioration law experiment with drilling fluid soaking time passing, the prediction results of shale wellbore wall collapse pressure changes with different wellbore trajectories, dip angles and water absorption of stratum rocks in the target formation are calculated by using the "fluid-solid-chemical" coupling collapse pressure prediction model, shown in Figure 3. From the calculation results, it can be seen that with the deterioration of shale by "water absorption", the collapse pressure gradually increases. When the shale does not deteriorate, the collapse density changes from 1.02 g/cm^3 to 1.73 g/cm^3. When the water content of shale reservoir increases, the shale strength value changes, and the equivalent density of wellbore collapse pressure increases rapidly. When the water content is 10% and 15%, the equivalent density respectively increases by 0.15 g/cm^3–0.22 g/cm^3 and 0.31 g/cm^3–0.45 g/cm^3, and the risk of wellbore wall instability increases continuously. The change rule of formation collapse pressure with azimuth and angle of inclination shows that when drilling directional wells, the collapse pressure is the lowest in the direction of minimum geostress, and the risk of wellbore instability is the lowest; while the collapse pressure is the highest in the direction of horizontal maximum geostress, and the risk of wellbore instability is the highest. With the increase of angle of inclination, the collapse pressure tends to increase gradually, the collapse pressure of horizontal wells is the highest and the risk of wellbore instability is the greatest.

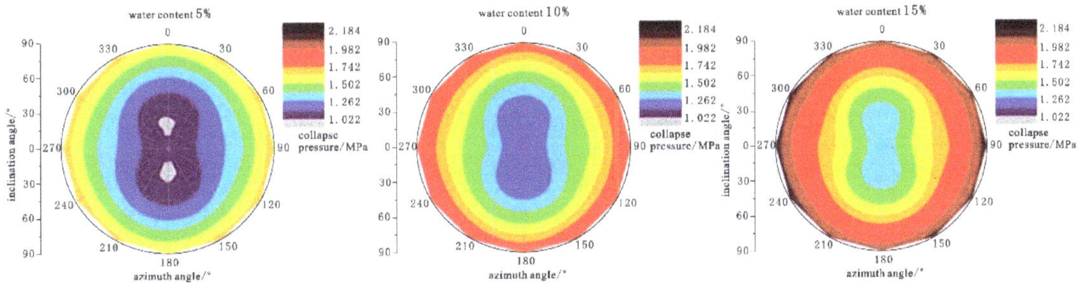

Figure 3. Nephogram of collapse pressure distribution under different well types and water content.

Combined with the previous research results and considering the relationship between the stratification strength of stratum and soaking time, the structural section deterioration and three-pressure section correction of the stratum soaked by drilling fluid are carried out, and the correction results of collapse pressure under the actual bottom hole conditions are obtained. The safe drill fluid density windows of four wells including well GY1-Q1-H3, and two vertical wells, well GY20 and GY40 are shown in Figure 4. With the increase of soaking time, the maximum collapse pressure equivalent density respectively increases from 1.33 g/cm^3 and 1.36 g/cm^3 to 1.46 g/cm^3 and 1.47 g/cm^3, and the minimum equivalent density of fracture pressure is 1.81 g/cm^3. After correction of deviation, azimuth and the deterioration of the two horizontal wells, Gy-Q1-H3 and GY2HC, the maximum equivalent density of collapse pressure respectively increases from 1.35 g/cm^3 and 1.34 g/cm^3 to 1.66 g/cm^3 and 1.65 g/cm^3, and the minimum value of fracture pressure is 1.85 g/cm^3. Therefore, when the oil-based drilling fluid is used and the bottom-hole rock carrying meets the requirements, the recommended drilling fluid density for vertical wells is 1.46 g/cm^3 to 1.51 g/cm^3, and that for horizontal wells is from 1.66 g/cm^3 to 1.71 g/cm^3, which can meet the requirements of the physical support of the wellbore wall.

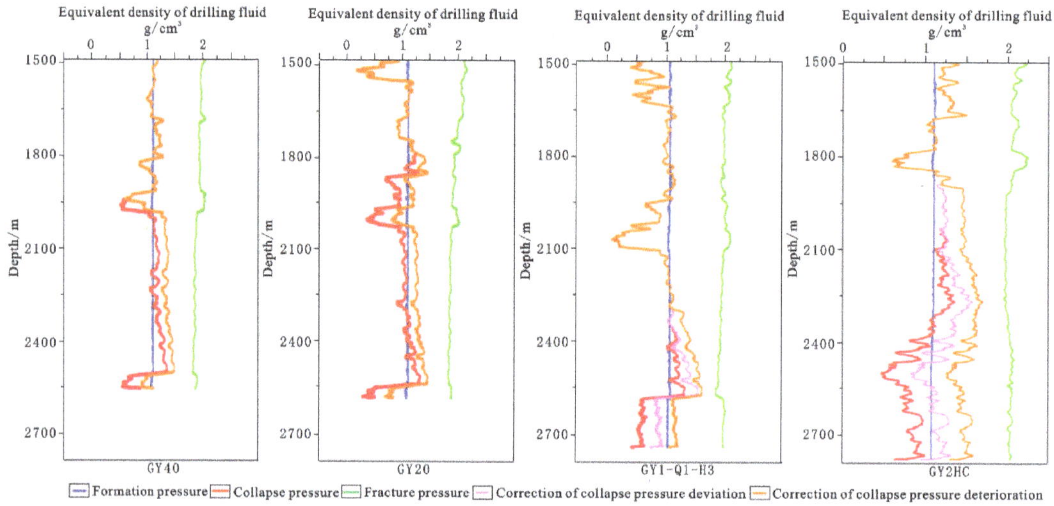

Figure 4. Prediction of safe density window of drilling fluid.

4. Development of Drilling Fluid System

4.1. Analysis of the Characteristics of Shale Deterioration

The particularity of the continental shale lithology in Block GL determines the complexity of its interaction with the drilling fluid system. In order to clarify the deterioration characteristics of shale caused by different fluids, the drill cores of the four wells including Well GY6HC from Group Q are selected. The rolling recovery, linear expansion, acoustic test, tensile strength and triaxial compression experiments are carried out in different media such as clear water, water-based and oil-based drilling fluid. The results show that the average rolling recovery rate of shale in clean water is 96.39%, it is 98.33% in water based drilling fluid, and it is 99.52% in oil-based drilling fluid. The linear expansion rate in clear water is 2.87%, while in water-based drilling fluid it is 1.47%, and in oil-based drilling fluid it is 0.05%. Compared with oil-based drilling fluid, the time difference of shale acoustic waves in water-based drilling fluid changes more greatly, and the change range in horizontal stratification direction is smaller than that in the vertical direction. The vertical tensile strength decreases by 15% after being soaked by the oil-based drilling fluid, and decreases by 60% after being soaked by the water-based drilling fluid. When the shale is soaked in oil-based medium, its cohesion decreases by 0.6 MPa and its internal friction angle decreases by 6.1° compared with that in clean water, which indicates that its hydration cracking effect in oil-based drilling fluid is weak, its expansion rate is low, and basically no expansion occurs. When the water phase medium comes into contact with the shale, the rock strength will be weakened, which has a strong weakening effect on the shale.

Dynamic changes of mechanical parameters such as triaxial compressive strength, elastic modulus and Poisson's ratio of rock are an important basis for judging shale hydration and deterioration. According to the Mohr-Coulomb failure criterion, a TAW2000 microcomputer-controlled triaxial stress testing machine is used indoors, and shale cores are soaked with water-based and oil-based drilling fluids to carry out the triaxial mechanical tests for soaking the shale cores for different times [2]. As shown in Table 3, with the increase of soaking time of the water-based drilling fluid, the strength of rock loaded in the horizontal stratification direction decreases rapidly. After soaking for two days, the triaxial strength decreases by 34%, resulting in foliation splitting and showing brittleness. In the vertical direction, the loading rock produces shear cracks connected with stratification cracks, and the fissure of the interlayer starts to crack, which shows a certain plasticity in general. It has little effect on the triaxial compressive strength, that is, it has little effect on shale matrix strength, and vertical loading overcomes the shear strength of the matrix. After soaking in oil-based drilling fluid for 10 days, the triaxial strength of rock decreased by 19% when loaded horizontally along the stratification direction. When loaded vertically along the stratification direction, the triaxial strength decreased by 17%, and the overall strength is high, which shows that the original anisotropy still exists [32]. After 10 days, the degree of rock deterioration gradually tends to be stable with the soaking time passing [42]. Compared with water-based drilling fluid, oil-based drilling fluid has less influence on shale rock mechanical parameters in Block GL.

Table 3. Triaxial test results of drilling fluid soaking.

Drilling Fluid	Coring Direction	Depth (m)	Soaking Time (d)	Rock Mechanics Parameter		
				Triaxial Compressive Strength (MPa)	Elastic Modulus (MPa)	Poisson's Ratio
Water-based	Horizontal	2326.61–2326.79	0	65.658	12,199.4	0.172
			2	43.326	7155.2	0.195
	Vertical	2317.79–2317.91	0	93.678	6567.3	0.178
			2	81.321	10,347.7	0.213
Oil-based	Horizontal	2326.61–2326.79	0	65.658	12,199.4	0.172
			10	53.262	12,589.5	0.174
	Vertical	2317.79–2317.91	0	93.678	6567.3	0.178
			10	77.450	13,453.3	0.145

4.2. Development of High Thixotropic Plugging Oil-Based Drilling Fluid System

This block is characterized with brittle shale, a high content of clay, and mainly consists of illite. It has strong surface hydration, and developed micro-nano pores and fractures, which may easily lead to the deterioration of shale in drilling fluid. It is difficult to carry cuttings in long horizontal sections, which puts forward strict requirements for the inhibition, plugging and rheology of the drilling fluid [33].

According to the development characteristics of nano-scale pores and fractures in shale stratum, the oil-soluble formula is optimized on the basis of the original brine drilling fluid system. By using a laser particle size analyzer and a pressure transmission experiment [24,43], the lipophilic nano-polymer and micro-plugging agent FTTS, of which particle size matches the pore size of the formation, are selected as the film-forming plugging materials of the system. The particle size distribution is shown in Figure 5. According to the pressure transmission experiment, after two plugging materials are added to the oil-based drilling fluid, the pressure transmission time of the drilling fluid in the core is increased by 27.8 times and 2.9 times compared with clear water and oil-based drilling fluid. The addition of a film-forming plugging agent forms a dense polymer film on the formation surface. It improves the toughness of filter cake, effectively delays the transmission of liquid column pressure in shale, and effectively blocks nano-scale pores and joints. It also improves the chemical plugging and anti-collapse ability of the drilling fluid system.

In order to meet the mechanical stability effect of horizontal wells in shale stratum, the design density of drilling fluid is more than 1.65 g/cm^3, and a large amount of barite powder and micro-nano materials in the system will greatly increase the viscosity and shear force of the system. At the same time, the horizontal section of shale oil horizontal wells is more than 2000 m long, and the grinding and dispersion of cuttings during long drilling periods will further increase the viscosity and shear force of the system. In order to reduce the force between solid particles, a strong polar wetting agent is added to the original system formula to reduce the interfacial tension, improve the wettability of the solid surface, improve the dispersion degree and optimize the rheology of the drilling fluid system, which is shown in Table 4. The addition of a wetting agent not only reduces the viscosity of the system, but also has no obvious influence on the shear force of the system. The addition can effectively improve the influence of weighting materials and the cuttings dispersion on the rheological properties of the drilling fluid system. It could realize the low viscosity and high shear of the high-density oil-based drilling fluid system and control the rheological properties of the drilling fluid in the operation process.

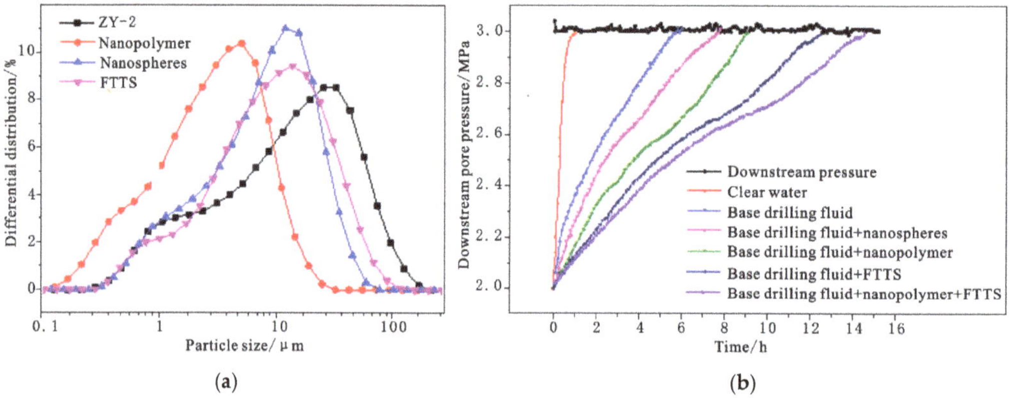

Figure 5. Optimization test of plugging materials ((**a**) Particle size distribution of plugging materials; (**b**) Pressure transmission evaluation of oil-based drilling fluid).

Table 4. Effect of wetting agent on drilling fluid rheology.

Formula	Apparent Viscosity/mPa·s			Static Shear Force/Pa		Plastic Viscosity/mPa·s	Dynamic Shear Force/Pa	Demulsification Voltage/V
	Φ600/Φ300	Φ200/Φ100	Φ6/Φ3	Initial Shear Force	Final Shear Force			
Base drilling fluid	28/18	14/10	4/3	2	3	10	4	699
Base drilling fluid + 2% wetting agent	26/17	13/9	4/3	2	3	9	4	702
Weighted drilling fluid	86/55	42/29	10/9	5	9	31	12	912
Weighted drilling fluid + 2% wetting agent	77/50	37/26	9/8	4	8	27	11.5	997

Finally, a gemini quaternary ammonium salt emulsifier which can realize special phase behavior is formed, and a low viscosity and high shear water-in-oil drilling fluid system with low apparent viscosity, low plastic viscosity and reasonable dynamic shear force is formed with high efficiency organic soil. The system can resist temperatures of 150 °C, and its density can reach 1.75 g/cm^3. Its demulsification voltage is more than 800 V, and high temperature and high pressure filtration loss is less than 3.0 mL.

5. Wellsite Test

Sixteen shale oil horizontal wells including well SL-Q1 on platforms in Block GL has been completed with the oil-based drilling fluid system under high thixotropy and strong plugging, which is developed in laboratory.

It is found that the cleaning problem of cuttings bed in the long horizontal section is prominent in the operation. During the spudding-in period of the completion operation for horizontal wells, the phenomenon of pressure hold-up occurred from time to time, and it is easy to hold back and leak the formation, which leads to the wellbore instability. The main reason is that the rheological property of drilling fluid is not controlled enough.

When the viscosity shear is high, the flow resistance increases and the ECD increases. By further optimizing the drilling fluid performance in real time, the annular pressure loss is reduced. During the second spudding-in, it has low viscosity, low water loss, the funnel viscosity is 40–50 s, and the API water loss is less than 4 mL, which reduces the probability of cuttings sticking to the wellbore wall. During the third spudding-in, the reading of the six-speed meter is adjusted to eight at six revolutions, and it is increased to 10 before landing. The dynamic-plastic ratio is increased to 0.3 Pa/mPa·s–0.4 Pa/mPa·s so as to improve the rock-carrying effect and fully purify the wellbore. At the same time, according to the change of downhole ECD, the drilling fluid density is appropriately adjusted to ensure that the downhole ECD is less than the leakage pressure and prevent the occurrence of lost circulation.

In view of the instability caused by scraping against the wellbore during operation. According to the characteristics of shale falling, the periodic tripping is cancelled. Timely circular reaming is carried out according to the change of underground ECD, with a large displacement of 34 L/s–40 L/s and a high rotation speed of 70 rpm–80 rpm. On the premise of ensuring the stability of the wellbore, the drilling operation parameters are optimized, such as low rotation speed in the mud-making stratum and high rotation speed in the expansive stratum in the second spudding-in. Adjusting the rotation speed according to the screw angle in the third spudding-in, reducing the rotation speed in the reaming stage, and adjusting the displacement in real time according to the wellbore condition. Meanwhile, elaborate operations such as controlling the spudding-in speed, starting and stopping the pump step by step, and weighting the drilling fluid in multiple cycles are adopted to reduce the downhole pressure fluctuation and ensure a smooth and controlled operation.

Relying on the reasonable drilling fluid density design, the excellent plugging and anti-collapse ability of the system, timely and effective drilling fluid adjustments on the wellsite and a smooth operation, there are no complicated situations such as wellbore collapse and sticking during the operation process, which effectively ensured the wellbore stability and wellbore cleaning. The depths of the shale oil horizontal wells on the test platform are from 4791 to 4875 m, the vertical depths are 2452–2592 m, the horizontal lengths are 2032 to 2506 m, and the design of drilling fluid density is 1.65 to 1.73 g/cm^3. During the wellsite test, the plastic viscosity and apparent viscosity of the oil-based drilling fluid are reasonable, the viscosity at six turns is high, the dynamic shear force is reasonable, the thixotropy is strong, and the cuttings return is normal. The demulsification voltage is more than 800 V, the emulsification stability is good, and there is no oil-water stratification phenomenon. There is no sticking display and wellbore wall peeling, and the tripping operation is smooth. After the well completion, there is no sticking display in drifting, well logging is smooth, and there is no sticking display during casing running. The cementing operation is normal, and the ability of stabilizing the wellbore wall is strong. The drilling and completion period is shortened from 113 days to an average of less than 30 days after the test. The successful application of sixteen test wells verifies that the "fluid-solid-chemical coupling" wellbore stabilization technology of continental shale oil can meet the requirements of safe and fast drilling of shale oil horizontal wells in Block GL.

6. Conclusions

(1) The influencing factors of wellbore instability in shale formation in Block GL are clear and definite. The average clay mineral content of shale stratum in this block is 35.6%, the average content of brittle mineral is 32.5%, the foliation density is 1000–2500 pieces/m, the TOC value is 1.4–4.5%, and the content of the organic matter is rich. Under the action of pressure difference and chemical potential difference, drilling fluid filtrate will quickly invade the stratum along the pores and fractures, and the water phase will easily cause the hydration of the illite surface, resulting in the opening, expansion and extension of interlayer micro-fractures, which result in a decrease in rock strength. The oil phase dissolves the organic matter, reduces the cohesion of the rock, and promotes the stress

release of the rock. Under the comprehensive action of physical, chemical and mechanical factors, the shale oil horizontal wells frequently suffer from wellbore instability.

(2) A theoretical model of "fluid-solid-chemical" coupling collapse pressure prediction is established. Based on the well logging interpretation data and triaxial stress experiments after soaking in different fluid media, and with full consideration of hydraulic pressure difference and chemical potential difference, the prediction method of "fluid-solid-chemical" coupling collapse pressure is established, and the change rule of formation collapse pressure under different parameters such as angle of inclination, azimuth and water content is defined. The safe drilling fluid density window of shale oil horizontal wells in this block is given, and the drilling fluid density of the target interval of vertical wells is recommended to be from 1.46 g/cm^3 to 1.51 g/cm^3, and that of the horizontal well interval is recommended to be from 1.66 g/cm^3 to 1.71 g/cm^3, which can meet the requirements.

(3) A targeted oil-based drilling fluid system has been constructed. According to the main factors affecting the wellbore instability of shale formation in Block GL and the characteristics of high density of oil-based drilling fluid in wellsite operation, a targeted plugging and anti-sloughing agent and a strong polar wetting agent are selected, and a high thixotropic and strong plugging oil-based drilling fluid system is formed. The system has excellent rheological properties under high density, the plugging pressure-bearing capacity is more than 5 MPa, the demulsification voltage is more than 800 V, and the comprehensive performance is stable. The wellsite test of sixteen horizontal wells in shale oil test platforms has been successful. Relying on reasonable drilling fluid density, excellent plugging and anti-collapse ability and timely and effective drilling fluid adjustment on the wellsite, the effect of stabilizing wellbore is outstanding, which provides technical support for the efficient exploration and development of shale oil.

Author Contributions: Conceptualization, X.A.; Data curation, X.A.; Formal analysis, X.A.; Funding acquisition, X.A.; Investigation, X.A.; Project administration, X.A.; Resources, X.A.; Methodology, M.C.; Supervision, Writing—review & editing, M.C. All authors have read and agreed to the published version of the manuscript.

Funding: The Major Science and Technology Project of PetroChina Company Limited-"Research on Exploration and Development Theory and Key Technologies of Shale Oil in Daqing Gulong" (2021ZZ10).

Conflicts of Interest: The authors declare no conflict of interest.

References

1. Sun, L.; Liu, H.; He, W.; He, W.; Li, G.; Zhang, S.; Zhu, R.; Jin, X.; Meng, S.; Jiang, H. Analysis of major scientific issues and research paths of Daqing Gulong shale oil. *Pet. Explor. Dev.* **2021**, *48*, 527–540. [CrossRef]
2. Liu, H.; Cui, S.; Zhu, D.; Li, Y.; Zhang, L. Study on microstructure and mechanical properties of hard and brittle shale. *J. Undergr. Space Eng.* **2019**, *15*, 34–39.
3. Wang, G. *Hydration Characteristics of Hard and Brittle Shale and Its Influence on Wellbore Collapse Pressure*; Southwest Petroleum University: Chengdu, China, 2017.
4. Liu, H.; Cui, S.; Meng, Y.; Li, Z.; Yu, X.; Sun, H.; Zhou, Y.; Luo, Y. Rock Mechanics and Wellbore Stability of Deep Shale During Drilling and Completion Processes. *J. Pet. Sci. Eng.* **2021**, *205*, 108882. [CrossRef]
5. He, W.; Meng, Q.; Zhang, J. Main control factors and classification evaluation of Gulong shale oil enrichment in Songliao Basin. *Daqing Pet. Geol. Dev.* **2021**, *40*, 1–12.
6. Song, M.; Liu, H.; Wang, Y.; Li, Y. Understanding and exploration practice of Paleogene shale oil enrichment in Jiyang Depression. *Pet. Explor. Dev.* **2020**, *47*, 225–235. [CrossRef]
7. Yao, J.; Zhao, Y.; Liu, G.; Qi, Y.; Li, Y.; Luo, A.; Zhang, X. Reservoir-forming conditions and key technologies of exploration and development in Qingcheng Oilfield. *Acta Pet. Sin.* **2020**, *41*, 777–795.
8. Liu, B.; Shi, J.; Fu, X.; Lyu, Y.; Sun, X.; Gong, L.; Bai, Y. Lithofacies characteristics of continental shale rock series and shale oil enrichment conditions—A case study of organic-rich shale in one section of Qingshankou Formation of Cretaceous in Gulong Depression, Songliao Basin. *Pet. Explor. Dev.* **2018**, *45*, 828–838. [CrossRef]
9. Liu, G. *Petroleum Geology*; Petroleum Industry Press: Beijing, China, 2018; pp. 158–162.
10. Chen, Z.; Meng, Q.; Wan, T.; Jia, Q.; Zhang, T. Numerical simulation of tectonic stress field in Gulong Depression of Songliao Basin by elastic-plastic incremental method. *Earth Sci. Front.* **2002**, *9*, 483–492.

11. Zhou, J.; He, S.; Tang, M.; Huang, Z.; Chen, Y.; Chi, J.; Zhu, Y.; Yuan, P. Analysis of wellbore stability considering the effects of bedding planes and anisotropic seepage during drilling horizontal wells in the laminated formation. *J. Pet. Sci. Eng.* **2018**, *170*, 507–524. [CrossRef]
12. Hu, S.; Bai, B.; Tao, S.; Bian, C.; Zhang, T.; Chen, Y.; Liang, X.; Wang, L.; Zhu, R.; Jia, J.; et al. Heterogeneous geological conditions and differential enrichment characteristics of high maturity shale oil in China. *Pet. Explor. Dev.* **2022**, *49*, 224–237. [CrossRef]
13. Chen, P.; Ma, T.; Xia, H. Prediction Model of Shale Horizontal Well Collapse with Multiple Weak Faces. *Nat. Gas Ind.* **2014**, *34*, 87–93.
14. Wang, B.; Sun, J.; Shen, F.; Li, W.; Zhang, W. Mechanism of wellbore instability in continental shale gas horizontal sections and its water-based drilling fluid countermeasures. *Nat. Gas Ind. B* **2020**, *7*, 680–688.
15. Zhang, G.; Zhang, J.; Zhao, Y.; Lv, J.; Cheng, X.; Yang, H.; Lu, G.; Ma, M. Main control factors for the enrichment of interbedded shale oil in the two section of Qingshankou Formation in Qijia area of northern Songliao Basin. *Pet. Geol. Dev.* **2019**, *38*, 143–150.
16. Wang, Y.; Liang, J.; Zhang, J.; Zhao, B.; Zhao, Y.; Liu, X.; Xia, D. Resource potential and exploration direction of shale oil in Gulong of Songliao Basin. *Pet. Geol. Dev.* **2020**, *39*, 15.
17. Chenevert, M.E. Shale alteration by water adsorption. *J. Pet. Technol.* **1970**, *22*, 141–148. [CrossRef]
18. Van Oort, E. On the physical and chemical stability of shales. *J. Pet. Eng.* **2003**, *38*, 213–235. [CrossRef]
19. Hale, A.H.; Mody, F.K.; Salisbury, D.P. Influence of chemical potential on wellbore stability. *SPE Drill. Completion* **1993**, *8*, 207–216. [CrossRef]
20. Deng, F. *Study on the Law of Drilling Fluid Invading Shale Formation and Its Influence on Wellbore Stability*; Southwest Petroleum University: Chengdu, China, 2019.
21. Chen, G.; Ewy, R.T.; Yu, M. Analytic solutions with ionic flow for a pressure transmission test on shale. *J. Pet. Sci. Eng.* **2010**, *72*, 158–165. [CrossRef]
22. Oort, E.V.; Hale, A.H.; Mody, F.K.; Roy, S. Critical parameters in modeling the chemical aspects of borehole stability in shales and in designing improved water based shale drilling fluids. *J. Pet. Technol.* **1994**, 171–186.
23. Xiang, C.; Chen, J.; Yang, G. Experiment on strength characteristics of brittle shale under drilling fluid immersion. *Fault Block Oil Gas Field* **2018**, *25*, 803–806.
24. Wang, J. *Study on the Mechanochemical Coupling Model of Shale Wellbore Stability*; China University of Petroleum: Beijing, China, 2007.
25. Lu, Y.; Chen, M.; Yuan, J.; Jin, Y.; Teng, X. Instability mechanism of inclined shaft in anisotropic stratum. *Acta Pet. Sin.* **2013**, *34*, 563–568.
26. Zhang, W.; He, S.; Guo, Q. Prediction method and prospect of seismic data pressure. *Adv. Geophys.* **2005**, *20*, 814–817.
27. Keyes, D.E.; Mcinnes, L.C.; Woodward, C.; Gropp, W.; Myra, E.; Pernice, M.; Bell, J.; Brown, J.; Clo, A.; Connors, J.; et al. Multiphysics Simulations Challenges and opportunities. In Proceedings of the IEEE International Conference on High Performance Computing, Data, and Analytics, Bengaluru, India, 18–21 December 2013.
28. Zhang, H.; Guo, J.; Lu, J.; Li, F.; Xu, Y.; Downar, T.J. An Assessment of Coupling Algorithms in HTR Simulator TINTE. *Nucl. Sci. Eng. J. Am. Nucl. Soc.* **2018**, *190*, 287–309. [CrossRef]
29. Fernández, M.A.; Moubachir, M. A Newton method using exact jacobians for solving fluid–structure coupling. *Comput. Struct.* **2005**, *83*, 127–142. [CrossRef]
30. Roatei, S. On the variational formulation of a chemically active shale constitutive model. *Proc. Rom. Acad. Ser. A* **2010**, *11*, 116–124.
31. Sarout, J.; Detournay, E. Chemoporoelastic analysis and experimental validation of the pore pressure transmission test for reactive shales. *Int. J. Rock Mech. Min. Sci.* **2011**, *48*, 759–772. [CrossRef]
32. Li, J.; Liu, G.; Chen, M. A new stress model of rock around the wellbore in orthotropic stratum. *J. Rock Mech. Eng.* **2011**, *30*, 2481–2485.
33. Zhang, S.; Wang, H.; Qiu, Z.; Cao, W.; Huang, H.; Chen, Z. Calculation method of safety density window of fluid-solid-chemical coupling in shale wellbore wall. *Pet. Explor. Dev.* **2019**, *46*, 1197–1205.
34. Biot, M.A. Theory of Elasticity and Consolidation for a Porous Anisotropic Solid. *J. Appl. Phys.* **1955**, *26*, 182–185. [CrossRef]
35. Rice, J.R.; Cleary, M.P. Some Basic Stress Diffusion Solutions for Fluid-Saturated Elastic Porous Media with Compressible Constituents. *Rev. Geophys. Space Phys.* **1976**, *14*, 227–241. [CrossRef]
36. Xiao, Z. *Study and Application of Fluid-Solid-Chemical Coupling Model of Hard and Brittle Shale Wellbore Wall Stability*; Yangtze University: Wuhan, China, 2020.
37. Ma, T.; Zhang, Y.; Qiu, Y.; Liu, Y.; Chen, P. Risk assessment method of inclined shaft instability based on reliability theory. *Acta Pet. Sin.* **2021**, *42*, 1486–1498.
38. Xiao, Z.; Jia, S.; Qi, X.; Dai, Y.; Lv, F.; Jia, L.; Wen, C. Discussion on progressive failure effect of hard and brittle shale wellbore wall under fluid-solid-chemical coupling condition. *J. Cent. South Univ. Nat. Sci. Ed.* **2019**, *50*, 2464–2480.
39. Ma, T. *Study on the Mechanism of Wellbore Collapse and Instability of Shale Gas Horizontal Wells*; Southwest Petroleum University: Chengdu, China, 2015.
40. Ding, L.; Wang, Z.; Lu, J.; Sun, Y. Stability model of inclined wellbore wall in layered strata based on Mogi-Coulomb strength criterion of surrounding rock ontology. *J. Rock Mech. Eng.* **2017**, *36*, 622–632.
41. Yan, C.; Deng, J.; Wei, B.; Tan, Q.; Deng, F.; Zhu, H.; Hu, L.; Chen, Z. Study on wellbore collapse pressure of shale gas reservoir. *J. Rock Mech. Eng.* **2013**, *32*, 1595–1602.

42. Guang, X.; Chen, Z.; Liu, X.; Wang, Y.; Si, N. Study on wellbore stability of rational shale formation under triaxial stress. *Sci. Technol. Eng.* **2015**, *15*, 54–57.
43. Chen, J. Evaluation method of drilling fluid wellbore stability for hard and brittle shale. *Oilfield Chem.* **2018**, *35*, 151–156.

energies

MDPI

Review

Ultrasonic-Assisted Rock-Breaking Technology and Oil and Gas Drilling Applications: A Review

Jinyu Feng [1], Tie Yan [1,*], Yang Cao [2] and Shihui Sun [1]

[1] School of Petroleum Engineering, Northeast Petroleum University, Daqing 163318, China
[2] Daqing Oilfield Limited Company, Daqing 163712, China
* Correspondence: yant@nepu.edu.cn

Abstract: High-efficiency rock-breaking is a problem that has long been studied in the oil- and gas-drilling industry. The successful use of ultrasonic technology in related fields has prompted us to study how to introduce ultrasonic technology into rock-breaking in oil and gas drilling. This paper introduces and discusses the successful cases of ultrasonic breaking technology in related fields, summarizes the three basic forms of ultrasonic action on rocks, namely, resonance, impact and cavitation, expounds the factors and laws that affect ultrasonic-assisted rock-breaking, and summarizes the research results reported in recent years. It is believed that, at present, the application of ultrasonic-assisted rock-breaking technology in the oil- and gas-drilling industry still faces some problems and challenges: first, the downhole high-temperature and high-pressure conditions will affect the effect of ultrasonic-assisted rock-breaking, and the related mechanisms and research are not clear; second, the impact of circulating media on ultrasonic-assisted rock-breaking is not clear; third, the problem of ultrasonic propagation and utilization in the downhole has not been well-solved; fourth, the stability of drilling tools and circulating media caused by high-frequency characteristics has not been well-solved. Therefore, it is suggested to increase research on the mechanism of ultrasonic-assisted rock-breaking with oil- and gas-drilling characteristics and the transmission and utilization of downhole ultrasonic energy in the future, and increase the development of supporting products to support the application of this technology in the oil and gas industry.

Keywords: oil and gas drilling; efficient rock-breaking; rock-breaking in deep well; ultrasonic; resonance-enhanced drilling; percussion rock-breaking; ultrasonic cavitation

Citation: Feng, J.; Yan, T.; Cao, Y.; Sun, S. Ultrasonic-Assisted Rock-Breaking Technology and Oil and Gas Drilling Applications: A Review. *Energies* **2022**, *15*, 8394. https://doi.org/10.3390/en15228394

Academic Editors: Mofazzal Hossain, Md Motiur Rahman and Hisham Khaled Ben Mahmud

Received: 4 October 2022
Accepted: 8 November 2022
Published: 10 November 2022

1. Introduction

The deep and ultra-deep oil and gas resources are rich, widely distributed, and occur in various types of reservoirs. In recent years, important breakthroughs have been made in terms of exploration [1]. As of 2008, China had cumulative, proven oil geological reserves exceeding 3×10^{10} barrels; cumulative production exceeded 4.2×10^9 barrels, the cumulative proven natural gas geological reserves at nearly 5×10^{12} m^3, and cumulative natural gas production exceeded 4.3×10^{11} m^3. A series of deep and ultra-deep large oil and gas fields were successively discovered in the Tarim Basin and the southern margin of the Junggar Basin in Northwest China and the Sichuan Basin in Southwest China. Therefore, the exploitation of deep oil and gas resources is particularly important [2,3]. Due to its high density, high hardness and poor drillability, it is very difficult to drill and break rock. Efficient, deep well-drilling and rock-breaking methods have become a key technology for oil and natural gas exploitation.

Ultrasound is a soundwave with a frequency of greater than 20 kHz, which has the characteristics of a high frequency, short wavelength, long propagation distance, concentrated direction, etc. Its penetrating ability is strong, and it can produce ultrasonic cavitation in liquids [4]. Relevant research results have been widely used in medical, manufacturing, aerospace and other fields, and achieved good results and high economic value. The successful experience and research results regarding ultrasonic application in the above fields also provide a good reference for the oil- and gas-drilling industry, and provide new ideas for the development of efficient rock-breaking methods.

By reviewing the developmental history of ultrasonic-assisted crushing technology, this paper expounds the technical characteristics and evolution of ultrasonic-assisted crushing and the technical basis and advantages of ultrasonic-assisted crushing in the oil and gas industries. Additionally, it summarizes the basic action forms and influences of ultrasonic-assisted rock-breaking. The research status of related factors to the ultrasonic-assisted rock-breaking effect is summarized, and the progress of ultrasonic-assisted rock-breaking laboratory test research in recent years is also summarized. The authors analyze the deficiencies in the current research on ultrasonic-assisted rock-breaking technology and the technical challenges faced by oil- and gas-drilling field applications, and they put forward the future research prospects on ultrasonic-assisted rock-breaking technology, which is to be realized in oil and gas drilling as soon as possible.

2. Development History of Ultrasonic-Assisted Crushing Technology

An ultrasonic wave is a type of mechanical wave that generally produces mechanical effects, thermal effects and electromagnetic effects in the process of propagation in solid media, and cavitation effects in the process of propagation in liquid media [5–7]. With good directionality, it is widely used in auxiliary machining and planetary drilling, as well as oil- and gas-layer plug removal and auxiliary production stimulation.

2.1. Development History of Ultrasonic-Aided Machining Technology

Ultrasonic-assisted processing technology is a processing method that uses ultrasonic high-frequency vibration to make abrasives impact, polish, and hydraulically impact the surface of the workpiece at a high frequency, so that the material is gradually broken and peeled off. This is used for high-hardness and high-brittleness materials that are difficult to process using traditional methods, such as drilling [8].

The original ultrasonic machining-drilling technology (USM) [8–10] only uses the vibration characteristics of an ultrasonic wave to transfer the mechanical energy of the ultrasonic wave to the abrasives between the machining gaps, so that the abrasives have a high-frequency impact on the processed materials. In this processing method, due to the small amplitude of the tool head, the output energy cannot be used efficiently. When processing materials such as ceramics and other hard rocks, the tool head wears to a high degree, and other tool parts are also easy to wear, so the processing accuracy and efficiency are reduced, and large-scale promotion is difficult. With the further development of this technology, Rotary Ultrasonic Machining (RUM) [11] emerged; in this, in addition to the axial vibration, the tool head also performs a high-speed circumferential rotational motion. This machining method improves work efficiency and can even perform a quick machining complex of three-dimensional contours. In recent years, a more advanced multi-frequency ultrasonic processing technology has been developed [12], which mainly draws on the planetary drilling sampler technology, and a free mass is connected to the connecting rod between the driver and the tool head. The mass block can reciprocate and irregularly strike the driver, which drives the top of the drill pipe to impact the contact surface between the material and the tool head, speeding up the processing speed and improving the processing accuracy.

2.2. Development History of Ultrasonic-Assisted Extraterrestrial Planet Drilling and Sampling Technology

Soil and rock research on extraterrestrial planets is crucial to understanding the origin and development of the universe. The installations on alien probes launched by humans are used to obtain soil and rock samples from extraterrestrial planets. However, compared with the earth, alien planets often have the characteristics of weak gravity and low environmental gravity [13]; therefore, it is necessary to electrically drive the ultrasonic generator to generate the vibration that is needed to assist to complete the sampling work of the drill. According to the working mode and principle, existing extraterrestrial planetary ultrasonic drills can be divided into three types: impact ultrasonic drills, rotary-impact combined ultrasonic drills and rotary-impact integrated ultrasonic drills.

The impact ultrasonic drill was first developed for obtaining the soil and rock of outer planets by the National Aeronautics and Space Administration. Scholars from the University of Paderborn in Germany [14,15] transformed the disk-shaped free block into a spherical one, effectively solving the problem of free block swinging, and carried out the high-frequency shock process of the free block. Nonlinear dynamics research was carried out, and this kind of impact ultrasonic drill was considered to have a very good effect. The impact ultrasonic driller developed by the Russian Bisk Ultrasonic Technology Center and the Institute of Space Research of the Russian Academy of Sciences [16] can drill into the simulated lunar soil at a maximum drilling rate of 25 mm/min. The United States used this drill on the end effector of the Mars rover Spirit, which was launched to obtain soil and rock samples on Mars [17].

The principle of the rotary-impact ultrasonic drill is to combine the longitudinal impact of the impact drill with the rotary motion of the traditional electric drill and, at the same time, realize the purpose of chip removal through the rotation of the drilling tool, to improve drilling efficiency. Honeybee Robotics [18,19] developed an ultrasonic drill based on this mechanism. In indoor tests, it can drill to a depth of 2 m on limestone samples. A depth of about 3 m was drilled for the rock of 40 MPa.

The rotary-impact coupling-integrated ultrasonic drill was arranged with multiple spiral grooves on the horn of the drill. This structure enabled the horn to convert part of the longitudinal vibration into torsional vibration while the horn is longitudinally vibrating, increasing the impact force. Scholars from Harbin Institute of Technology [20] made use of this principle to produce a prototype of a rotary-impact, coupled, integrated ultrasonic drill. When working, the small-diameter end of the horn longitudinally vibrates to transfer the force to the drilling tool to achieve impact motion. At the same time, the cover plate adopts a V-shaped structure to realize torsional vibration and the coupled motion of rotation and impact. The test results show that the drilling effect of the impact-coupled ultrasonic drill is obviously better than that of the impact-type ultrasonic drill [21].

2.3. *Development History of Ultrasonic-Assisted Enhancements of Oil and Gas Production*

In the development of oil and gas fields, the oil and gas layers are prone to plugging due to the perennial injection and production, which reduces the oil and gas recovery rate. With the help of an ultrasonic wave, the blocked oil and gas layers can be unblocked, and the permeability of the oil and gas layers can be improved to achieve the purpose of auxiliary production. When the ultrasonic wave propagates in the rock medium, the impurities blocked in the rock will vibrate, which will destroy the cohesion between the blocking particles and the reservoir rock, and increase the number of microcracks in the oil and gas reservoir [22]. Ultrasonic cavitation can subject the asphalt, colloid, waxy and other macromolecular compounds in crude oil to shearing and crushing. In addition, the thermal effect of ultrasonic waves in the rock medium reduces the viscosity and accelerates the flow rate of the produced fluid, thereby improving the oil layer's permeability and increasing oil and gas reservoir production. This technology has the characteristics of a simple process, convenient construction, strong layered processing ability, low cost and quick effect [23–25]. In the 1960s, the United States first carried out research on ultrasonic-assisted oil-well plug removal and production enhancement technology [26,27], and China also started research in the field of auxiliary oil and gas field development, such as ultrasonic plug removal and production enhancement technology, at the end of the 20th century [22,28,29]. With the successful research and development of high-power ultrasonic equipment, and the development and application of new materials, this technology has been tested in the United States, Russia, China's Daqing, and Shengli oilfield blocks, and has achieved good results. Table 1 compares production before and after the use of this technology in different oilfield blocks in China and abroad. The table shows that the output of each block increased to varying degrees after the ultrasonic-assisted production increase measures were adopted. Figure 1 is a bar chart comparing the production growth in each block before and after using this technology. The application of ultrasonic-assisted production stimulation technology in the Daqing Oilfield block has increased the production by about 3.5 times, which proves that ultrasonic technology can effectively assist oilfield production and efficiency.

Table 1. Compare initial production and after ultrasonic-assisted production in different locations.

Location	Frequency (kHz)	Power (kW)	Initial Production (Barrels/d)	Production after Ultrasonic Assistance (Barrels/d)
Demkinskoe Oil Field, Russia [30]	19	9	11.17	13.46
Zhongyuan Oilfield, China [28]	-	-	25.16	38.48
Shengli Oilfield, China [28]	-	-	42.92	68.08
Samara region, Russia [26]	15–30	10	85.10	137.64
Oilfields in northern Shaanxi, China [22]	5–30	>10	4.51	9.25
Yanchang Oilfield, China [29]	22	10	53.28	111.00
Western Siberia region, Russia [26]	15–30	10	29.00	61.57
Samotlor Oil Field, Russia [27]	10	18–27	23.46	56.39
Daqing Oilfield, China [28]	-	-	29.60	103.60

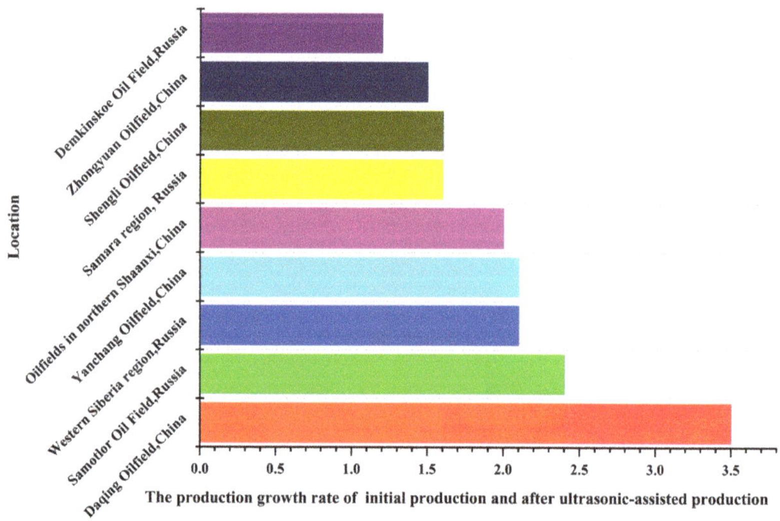

Figure 1. The production growth rate of initial production and ultrasonic-assisted production in different locations.

3. Research Progress on the Mechanism of Ultrasonic-Assisted Rock-Breaking Technology

Ultrasonic is a high-frequency soundwave with a frequency above 20 kHz, which has high-energy characteristics and can transmit energy in the form of a mechanical wave in the medium [31]. When it acts on the rock medium, it can cause two-phase or multi-phase media to interact. This can have thermal, electromagnetic and mechanical effects on the rock medium and can also lead to some chemical changes.

3.1. Effect Forms

At present, it is generally accepted in academic circles that rock-crushing mainly depends on the mechanical effects of the ultrasonic wave in the medium propagation, which include resonance effects, impact effects and cavitation effects [32].

3.1.1. Resonance Effect

When the ultrasonic wave acts on the rock, the rock is forced to vibrate. When the frequency of the rock's forced vibration is the same as the natural vibration frequency of the rock itself, the rock will have a resonance effect. At this time, the amplitude of the crystal particles inside the rock is the largest. Then, the crystals inside the rock are prone to fracture, resulting in cracks inside the rock. In this case, the rock is easily damaged [33–35]. With the help of the transducer, the electrical signal in the ultrasonic generator is converted to mechanical vibrations, and the vibration frequency of the ultrasonic generator is adjusted to be consistent with the natural frequency of the rock in the formation. Therefore, high-efficiency rock crushing can be achieved in engineering [36].

Since the 1980s, ultrasonic resonance has been applied in the field of engineering crushing. The United States and other western developed countries [37] have successively developed PB400 and RB500 ultrasonic-excitation resonance crushers according to the characteristics of ultrasonic resonance. These are widely used in road pavement construction and have achieved good results. In terms of resonance rock-breaking, researchers from the University of Aberdeen in the United Kingdom demonstrated the feasibility of resonance rock-breaking in indoor physical tests for the first time, and believed that, under resonance conditions, the rock-breaking speed was 10 times that of ordinary rock-drilling [38,39]. In recent years, scholars in China and abroad [40–45] have explored the mechanism of resonance rock-breaking, the expansion process of rock cracks and the distribution of cracks and other rock-crushing processes under resonance, and provided a theoretical model of resonance rock-breaking. The means of resonance enhancement, drilling and rock-breaking efficiency were evaluated under different resonance frequencies, amplitudes, action times, static load pressures and other conditions. A variety of resonance rock-breaking tools were designed for engineering operations, mining development, oil and gas drilling and other fields.

3.1.2. Impact Effect

Impact effect is a common action method in engineering. It mainly uses the force-applying object to generate a huge impact load in an instant, causing the force-bearing object to produce a rapid change in the state of motion over a short period of time, so that the force-bearing object produces deformation and changes in properties' dynamics [46]. The rock fracture process under the effect of impact load is very complicated. Due to the original, natural cracks and defects in the rock, under the huge impact load at the moment of external force, the stress distribution inside the rock becomes extremely uneven. Stress concentration occurs in some local locations. When a huge impact load occurs in a short time, the stress generated at the tip of the tiny crack in the rock is very large. When the stress at the tip reaches the critical value of expansion, the crack rapidly expands. There will be a certain bifurcation phenomenon during the integration surface, and the crack penetration phenomenon will appear when the crack meets the crack. When many cracks start to penetrate, macro-cracks will appear, and the rock will locally break. The periodic high-frequency impacts will lead to rock fatigue, which is more conducive to damage and fragmentation of the rock [47,48].

In order to better analyze and describe the dynamic characteristics of rocks under impact, scholars have constructed macro- and micro-impact loading models of rock media with the help of rock mechanics theory, and introduced discrete element ideas to their studies to simulate impact under discontinuous media conditions. The damage and fracture process of the rock under the shock-wave transmission process, the dynamic response law of the rock under shock action, the crack initiation conditions of the cracked rock, and the mechanical change law of the rock under strong dynamic shock have been studied [49–51] to obtain a better understanding of the mechanism of impact, and the characteristics of change and influencing factors of rocks, during the action process.

3.1.3. Cavitation Effect

Cavitation is a special effect caused by ultrasonic vibration in the liquid medium. Since the ultrasonic wave has an alternating cycle of positive and negative pressure, if the liquid medium contains dissolved gas nuclei in the positive pressure phase, the ultrasonic wave squeezes the liquid molecules, and the gas nuclei shrink accordingly. The molecular density of the liquid phase is sparse, the density of the liquid decreases, and the gas core expands [52]. The violent vibrations of the ultrasonic wave cause the bubbles in the liquid medium to be in a constant dynamic process of expansion and contraction. When the sound pressure is large enough, the bubbles in the liquid phase collapse [53]. The minimum sound intensity at which ultrasonic cavitation occurs is called the cavitation threshold. The magnitude of this parameter is related to the liquid-phase physical parameters of the ultrasonic wave. The cavitation threshold represents the difficulty of ultrasonic cavitation in the liquid medium [54].

The micro-shockwave caused by the repeated expansion and contraction of the cavitation bubble has a high-frequency impact on the rock surface and its collapse produces a micro-jet effect, which damages the rock surface. Studies have shown that the shockwave pressure generated by the rapid collapse of cavitation bubbles can reach 200–350 MPa within 10^{-3} s, and the micro-jets generated by this collapse can reach speeds of over 10^3 m/s [55,56].

The cavitation effect is affected by the viscosity of the liquid phase, the gas content and the size of the contained particles. The higher the viscosity of the liquid, the slower the collapse of the cavitation bubble, and the smaller the damage effect. The test results show [53] that there is a critical value for the gas content in the liquid phase. When the gas content is less than this threshold, the gas content is higher, and more damage is caused by cavitation. The addition of solid particles in the liquid phase can also effectively enhance the effects of ultrasonic cavitation. Experiments show that, under ultrasonic cavitation [57], the addition of silica particles to water can lead to surface roughness in sandstone and shale rock samples. Compared with pure water, the surface roughness of rock samples is increased by 20% and 400%, respectively.

3.2. *Influencing Factors*

The process of ultrasonic-assisted rock-breaking is affected by various forms, and the influencing factors. The factors affecting the rock-breaking effect can be roughly divided into ultrasonic parameters and engineering parameters. The ultrasonic parameters mainly include frequency, amplitude, power, etc., while the engineering parameters mainly include static load pressure and action time.

3.2.1. Ultrasonic Vibration Frequency

Ultrasonic vibration frequency is an important parameter in ultrasonic-assisted rock-breaking, indicating the speed of ultrasonic vibration. The high-frequency cyclic load causes the rock to fatigue, and the cracks inside the rock gradually increase and expand, which reduces the strength of the rock. The faster the cyclic loading frequency, the faster the rock fatigue damage occurs. When the fatigue failure limit is reached, the strength rapidly decreases and the rock suffers macroscopic damage [58]. Xu [59] conducted an experimental study on the characteristics of rock damage in various types of rock samples under different cyclic loads using the SHPB experimental system. Damage also increases with the increasing frequency of cyclic loading. Scholars from the University of Aberdeen in the United Kingdom [38,39] believed, through experimental research, that there is an optimal range of vibration frequency in the process of rock-breaking, and when the vibration frequency does not reach the optimal range, the drilling rate increases with the increase in the vibration frequency. When the vibration frequency is greater than the optimal range, the rock-breaking efficiency decreases with the increase in the vibration frequency.

Sun [60] verified the above conclusion by loading granite with ultrasonic vibration tests of different frequencies. The test results are shown in Figure 2. The results show that when the vibration frequency is close to the natural frequency of the rock, the crushing effect of the rock is the best. At the same time, the test found that frequencies that were slightly lower than the natural frequency of the rock were more conducive to the crushing of the rock. This is because, with the continuous development of internal cracks in the rock, the degree of damage increases and the natural frequency decreases. Ultrasonic cavitation is also affected by the ultrasonic frequency, but the occurrence and cavitation effects are also affected by the ultrasonic amplitude, the physical properties of the liquid phase and the gas content of the liquid phase. Therefore, the ultrasonic frequency alone cannot determine the ultrasonic assistance in the rock-breaking effect [54–56,61,62].

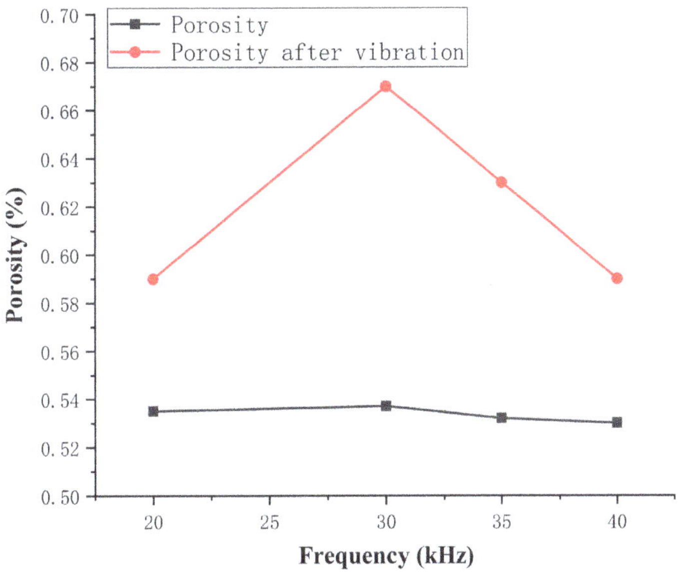

Figure 2. Variation curve of rock sample porosity caused by different frequencies.

3.2.2. Ultrasonic Power

Ultrasonic power represents the amount of energy that an ultrasonic wave can provide and is one of the key indicators in ultrasonic-assisted rock-breaking. The magnitude of ultrasonic power is determined by the frequency and amplitude of ultrasonic vibration. When the frequency of ultrasonic vibration is determined, the magnitude of ultrasonic power is characterized by the magnitude of amplitude, and the magnitude of ultrasonic power is positively correlated with the magnitude of amplitude.

As the ultrasonic power increases, the energy provided by the ultrasonic wave increases. When the vibration frequency is constant, the ultrasonic amplitude increases with the increase in power. The larger the amplitude, the greater the energy with which each vibration acts on the rock, which is more conducive to the crushing of the rock. At the same time, within a certain range, a high power will increase the effect of ultrasonic cavitation, but too high a power intensity will hinder the formation and collapse of cavitation bubbles [54]. Chemical bubbles often implode, limiting the energy stored in the bubbles. Tian [63] took the red sandstone specimen as the research object, carried out ultrasonic vibration crushing tests of the rock sample under the same frequency and different power conditions, and measured the time at which the rock sample was completely broken. The test results are shown in Figure 3, indicating that, with the increase in power, the fragmentation time of the rock sample is significantly reduced, and the rock-breaking effect is improved.

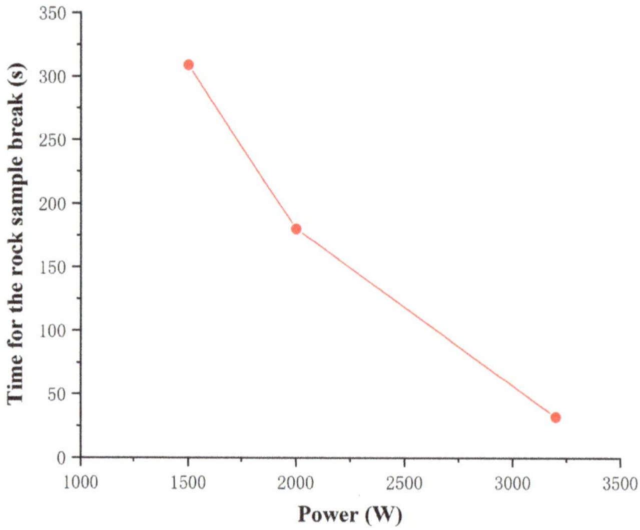

Figure 3. Variation curve of average rupture time of rock samples under different powers.

3.2.3. Static Load Pressure

The static load pressure is the prefabricated constant pressure above the rock. It is expressed by the WOB in actual drilling construction. As an important drilling process parameter, the WOB directly affects the rock-drilling efficiency. From the energy perspective, the static load pressure can temporarily close some micro-cracks and discontinuous positions in the rock, so that the rock medium is in a stable and continuous state, reducing the ultrasonic energy in the discontinuous position. Dissipation increases the utilization efficiency of ultrasonic energy and accelerates the destruction of rocks. With the destruction of the rock, the accumulation of cracks in the rock increases, and the influence of the static load pressure is no longer obvious, and the intrusion of the pressure head into the rock does not proceed smoothly as the pressure increases. However, when the pressure on the rock increases, and the tolerable limit is reached, the bit suddenly intrudes into the rock and the rock suddenly breaks [64]. Therefore, in the process of rock-breaking, it is not that the higher the static load pressure, the better the rock-breaking effect. There is a critical threshold for the static load pressure. When the pressure is greater than this value, the effect of the static load pressure on the rock-breaking efficiency is no longer significant [65–67]. Zhai [68] conducted ultrasonic vibration rock-crushing tests on cylindrical medium-grained granite samples under different axial static load pressures. The rock fragmentation effect is characterized by the change rate of porosity inside the rock after rock-fragmentation and the change rate of the compressive strength of the rock sample. The test results are shown in Figure 4, indicating that there is a threshold value for the axial static load pressure of the rock. When the static load pressure is near this threshold value, the rock-crushing efficiency is significantly enhanced.

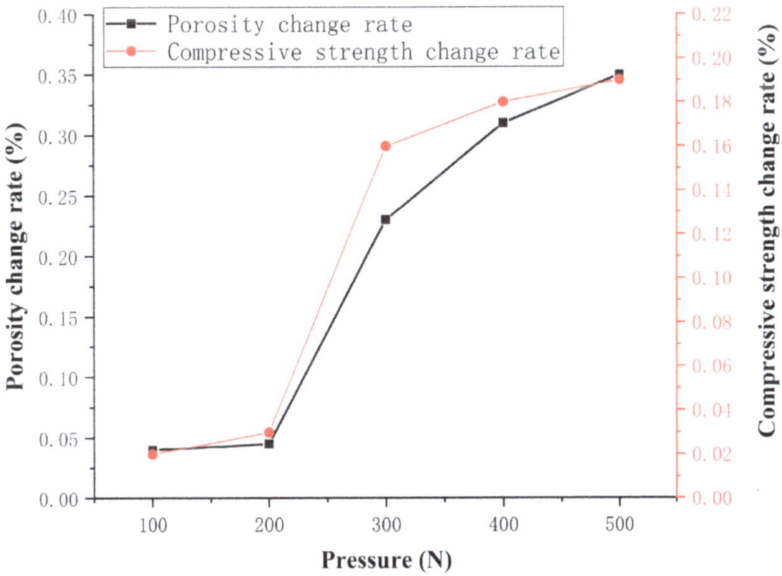

Figure 4. The relationship between static load pressure and rock porosity change rate and compressive strength change rate.

3.2.4. Vibration Time

When the ultrasonic parameters applied to the rock are fixed, the energy provided by the ultrasonic wave for rock-crushing is also fixed. With the increase in the action time and the continuous input of energy, the number of cracks and the extension length of the rock continue to increase, the compressive strength continues to decrease, and the crushing effect improves [69]. However, under the action of an ultrasonic wave, the expansion and strength changes in internal cracks in rocks are not stable and can be roughly divided into four processes: crack initiation, stable expansion, rapid expansion and fragmentation. Yuan et al. [70,71] conducted ultrasonic-vibration crushing tests on granite samples at different times. After the test, the porosity changes in the rock samples were measured by nuclear magnetic resonance technology. The results are shown in Figure 5. When the internal cracks are in the initiation state, the porosity slowly increases, and the rock strength decreases relatively slowly. With the increase in the action time, the internal cracks in the rock have a relatively stable growth stage and then rapidly expand. At this time, the internal porosity of the rock rapidly increases. The strength rapidly drops until the rock breaks apart. The damage that ultrasonic cavitation causes to other materials in the rock surface also exhibits development characteristics over time [72]. In the initial period of time, the surface and mechanical properties of the rock did not much change. After a period of time, micro-craters gradually appeared on the rock surface, and over time, cavities and cracks began to appear on the rock surface.

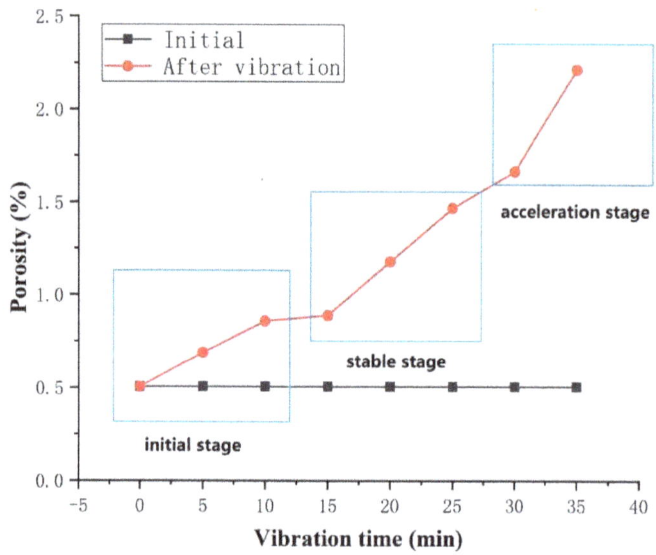

Figure 5. Variation curve of rock sample porosity caused by different vibration times.

4. Research Progress of Ultrasonic-Assisted Rock-Breaking Laboratory Test

At present, reports on the field application and testing of ultrasonic-assisted rock-crushing are extremely rare, and related research is still at the indoor research stage. The researchers tested a variety of lithological rock samples under different experimental conditions, and used some advanced methods to observe and characterize the macroscopic and microscopic changes in the rock samples to explore the change law. Table 2 summarizes the test conditions, test contents and main conclusions of the relevant laboratory test studies reported in the literature in recent years, and summarizes the main action forms of rock breakage in each test.

Table 2. Research progress of ultrasonic-assisted rock-breaking laboratory tests.

Frequency	Test Rock Properties	Test Content	The Conclusion	References
20 kHz	Granite	Under static load pressures of 100 N, 200 N, 300 N, 400 N and 500 N, the cylindrical rock samples were subjected to ultrasonic vibrations for 5, 10 and 15 min, respectively.	The compressive strength of the rock sample did not significantly change at the initial stage of the increase in the static load pressure. When the static load pressure increased from 200 N to 300 N, the compressive strength of the rock sample rapidly decreased. When the static load pressure increased to 400 N, the compressive strength continued to decline. However, the rate of descent slowed down.	[65]
20 kHz, 30 kHz, 35 kHz, 40 kHz	Granite	Under a static load pressure of 300 N, ultrasonic vibration is directly applied to the cylindrical rock sample through the ultrasonic transducer.	The porosity inside the rock is inversely proportional to the compressive strength; when the vibration frequency is 35 kHz, the rock-breaking effects are the best.	[73]
30 kHz	Granite	The ultrasonic vibration test was carried out on the cylindrical rock sample under a 300 N static load, and the infrared radiation temperature of the rock during the whole loading process was observed and measured using an infrared thermal imager.	The rock failure process can be divided into the following three stages: elastic deformation, micro-fracture and yield stage, macro-crack and failure stage; according to the axial temperature distribution of the sample, the rock sample can be divided into a fracture zone, plastic deformation zone and elastic deformation zone. The effective crack depth in the rock sample can reach 10 mm. Fatigue damage caused by ultrasonic vibrations and thermal damage caused by increases in temperature are the main factors used to destroy granite.	[74]

Table 2. *Cont.*

Frequency	Test Rock Properties	Test Content	The Conclusion	References
20 kHz	Sandstone and Shale	Ultrasonic cavitation tests on sandstone and shale samples in pure water and nanoparticle abrasives.	The roughness results show that the erosion effect of abrasive cavitation on shale and sandstone is stronger than that of pure water cavitation; the surface roughness of sandstone and shale samples is increased by 20% and 400%, respectively.	[57]
—	Sandstone, Mudstone and Shale	On the self-designed ultrasonic rock-breaking simulation test bench, the control variable method and the orthogonal test method were used to test the factors affecting the efficiency of ultrasonic high-frequency rotary rock-breaking.	Under normal temperature and pressure conditions in the laboratory, compared with the conventional rotary rock-breaking technology, the ultrasonic high-frequency rotary penetration drilling technology has a higher rock-breaking efficiency, with an average increase of 77.65%; the factors affecting the ultrasonic high-frequency rotary rock-breaking efficiency range from large to small. The order is: WOB, amplitude, bit diameter and rotation speed. WOB and amplitude have the most significant influence on the rock-breaking efficiency of ultrasonic high-frequency rotation, and the greater the amplitude, the higher the efficiency of ultrasonic high-frequency rotation at breaking through rock.	[75]
20 kHz	Sandstone, Shale and Granite	Ultrasonic cavitation tests were performed on cylindrical rock samples in a mixture of distilled water and silica water abrasives.	After the test, several micro-cracks and holes appeared on the surface of the rock sample; the average mass loss of each rock sample was 81.41% (sandstone), 557.38% (shale) and 188.16% (granite).	[76]
20 kHz	Red Sandstone	Under a static load pressure of 35 N, an ultrasonic vibration test is applied to cylindrical rock samples.	After 120 s of ultrasonic vibration, the compressive strength and elastic modulus of the rock samples were reduced by 55.3% and 26.9%, respectively; the natural frequency of the rock samples was reduced by 2.4% due to the changes in mass and elastic modulus caused by damage electron microscope scanning. When ultrasonic vibration occurs in the rock sample, the tensile stress between the rock crystals causes the original cracks in the rock sample to expand and penetrate, forming macroscopic fractures.	[77]
20 kHz, 30 kHz, 35 kHz, 40 kHz	Granite	Through the ultrasonic vibration crushing test of rock samples, the individual and coupled effects of vibration frequency, load and confining pressure on granite damage were analyzed.	The optimal vibration frequency of the sample is 30 kHz, which is close to its natural frequency. The axial static load pressure is the main factor affecting the local fracture degree of the rock sample under ultrasonic vibration. The reason for this is that, when the axial stress of the rock sample is high, the fatigue damage is more serious. The optimal static load pressure value in the test is about 300 N. The confining pressure determines the damage mechanism of the granite specimen under ultrasonic vibration. With the increase in confining pressure, under a confining pressure of about 7 MPa, the damage mechanism of granite changes from fatigue damage to fatigue and shear composite damage, which is further transformed into shear damage at about 9 MPa confining pressure.	[78]
30 kHz, 35 kHz, 40 kHz	Granite	The ultrasonic vibration crushing test was carried out on the rock sample under different parameter combinations, and the optimal coupling parameters of ultrasonic vibration were determined through the nuclear magnetic resonance test and uniaxial compressive strength test.	The optimal combination of parameters for vibration and crushing in the test is as follows: static load pressure of 400 N, amplitude of 100, and frequency of 35 kHz; within the effective resonance frequency range, the primary and secondary order affecting rock crushing is as follows: amplitude > static load pressure > frequency.	[79]

Table 2. *Cont.*

Frequency	Test Rock Properties	Test Content	The Conclusion	References
20 kHz, 30 kHz, 35 kHz, 40 kHz	Granite	Under a static load pressure of 200 N, ultrasonic vibration crushing tests were carried out at different temperatures.	Under the action of ultrasonic vibration, the number of rock cracks will increase with the increase in temperature; the accumulated thermal damage will cause a large number of internal cracks to expand and penetrate, resulting in macroscopic damage to the rock sample. The distribution of cracks in the rock sample has the characteristics of regional concentration, with the cracks' location concentrated near the upper end of the rock sample. When the temperature is lower than 150 °C, the increase in the porosity of the rock sample is not obvious; when the temperature of the rock sample reaches 150 °C, the increase rate significantly and steadily increases as the temperature continues to increase.	[80]

5. Problems and Challenges Faced by Ultrasonic-Assisted Rock-Breaking Technology in Oil- and Gas-Drilling Applications

Research on ultrasonic-assisted rock-breaking technology is at the initial stage, and most research is concentrated in the mining industry. There is little research on the oil- and gas-drilling industry that includes circulating medium and combined conventional rotary drilling under high-temperature and high-pressure conditions in the deep layer and ultrasonic layer, and research on the field application of ultrasonic-assisted rock-breaking is even rarer. Therefore, ultrasonic-assisted rock-breaking still faces many problems in oil- and gas-drilling applications.

5.1. Poor Conditions in Oil- and Gas-Drilling Downhole

Oil- and gas-drilling downhole has the characteristics of "large depth, harsh environment and high rock hardness", especially the deep and ultra-deep wells that are vigorously being developed at present; the downhole temperature can reach more than 300 °C, and the rock is under complex, high-stress conditions. At present, the physical properties and strength properties of the rock are different from those at the surface and at shallow depths, and the rock fragmentation conditions, forms and influencing factors are also different. In the current research on ultrasonic-assisted rock-breaking, the above factors are rarely considered in oil and gas drilling. Therefore, it is necessary to combine the above factors to further explore and improve the mechanism of ultrasonic-assisted rock-breaking.

5.2. The Influence of Circulating Medium

Oil and gas drilling requires the participation of the circulating medium, which can cool and lubricate the drill bit, carry cuttings and balance formation pressure. Except for jet-impact rock-breaking, most circulating media in conventional oil and gas drilling do not directly participate in rock-breaking. However, ultrasonic waves will cause ultrasonic cavitation in the liquid-phase circulating medium, which may damage the rock and affect the rock-breaking effect. Most of the existing theoretical studies do not consider the influence of the circulating medium, and experimental studies are mostly carried out directly in the atmosphere, without considering the participation of the circulating medium, and have not clarified the rock damage caused by the ultrasonic cavitation of the circulating medium in ultrasonic-assisted rock-breaking. The mechanism of action cannot provide a basis for the selection of circulating media in field applications.

5.3. Energy Transfer and Utilization of Ultrasonic Wave

The essence of ultrasonic-assisted rock-breaking is a way to provide additional rock-breaking energy for rock-breaking through ultrasonic wave, and its energy transmission and utilization efficiency have a great impact on the final rock-breaking effect. Ultrasound is essentially a mechanical wave. During its propagation, reflection, interference and other phenomena will occur in the well. The composition of the downhole medium encountered in the propagation is also complex, and a lot of energy will be absorbed or converted into heat energy during long-distance downhole propagation, resulting in high levels of loss. The energy that finally enters the rock medium for rock-breaking is very different from the power output rated by the ultrasonic generator. However, the existing research does not consider the propagation process of the ultrasonic wave, or the loss, and most of them directly use ultrasonic waves. The rated power of the generator calculates the energy provided by the ultrasonic wave, which is not consistent with the actual situation. Therefore, it is necessary to explore the propagation loss law of ultrasonic waves in the actual wellbore of oil and gas drilling, and to determine how to measure the energy of ultrasonic waves that are actually participating in rock-breaking, the actual utilization efficiency of ultrasonic energy, and how to efficiently use ultrasonic energy. The selection of ultrasonic equipment provides the basis for this.

In ultrasonic-assisted rock-breaking, the ultrasonic wave derives from an ultrasonic generator composed of piezoelectric ceramic sheets, etc., and is converted into high-frequency mechanical vibrations through the horn and transmitted through the medium to act on the rock. Sonic drilling rigs are used in mining, the sonic-vibration-generating device is placed at the wellhead, and the mechanical vibrations are transmitted to the bottom of the well through mediums such as a drill pipe to realize the vibration and crushing of the rock. The hole is small, and the rock is relatively hard. If the ultrasonic generator is placed at the wellhead, the high energy will be lost during transmission under long-distance, downhole, complex conditions. Therefore, the rock-breaking process of oil and gas drilling should be based on the characteristics of oil- and gas-drilling technology to reasonably determine the position of the ultrasonic generator.

5.4. Matching Drilling Tools and Products

Ultrasound has the characteristics of "high frequency". High-frequency percussion drilling is the method used in oil and gas drilling at present; however, because the principle of vibration is mostly pneumatic or hydraulic, this so-called "high frequency" mostly refers to the vibration of 10 Hz-level frequency, compared with an ultrasonic wave. In terms of vibration at a frequency of 10^3 Hz, there is a great difference between the two. Therefore, it is precisely because of the high-frequency characteristics of ultrasonic vibration that ultrasonic-assisted rock-breaking is facing many new challenges.

High-frequency vibration has higher requirements regarding the performance of drilling tools. High-frequency vibrations are equivalent to applying high-frequency periodic changing loads to the drilling tools, which makes the metal drilling tools extremely prone to fatigue, resulting in cracks and occurrences in the drilling tools. The fracture greatly reduces the service life of the drilling tool, and it is difficult for conventional drilling tools to meet these requirements. At the same time, the high-frequency impact of the drill bit on the formation causes the cutting teeth of the drill bit to grind more rapidly, and the vibration generated by the stick-slip effect is generated during the rotation of the drill bit and the high-frequency ultrasonic-vibration interaction law, as well as the coupling effect. At present, there are few studies on the characteristic impact on drilling tools, and the relevant influencing factors are not clear, which restricts the application of ultrasonic-assisted rock-breaking technology in oil- and gas-drilling sites.

High-frequency vibration will produce a high-frequency shearing effect on the circulating medium, and an ultrasonic wave will also lead to ultrasonic cavitation in the circulating medium. In addition, the drilling tool will generate a lot of heat under high-frequency vibrations, which will destroy the circulating medium. Its physical properties make the circulating medium function invalid. Therefore, in the rock-breaking of ultrasonic-assisted oil and gas drilling, higher requirements are put forward for the selection and process of the circulating medium, and new challenges are faced.

According to existing reports, the output power of high-power ultrasonic equipment can reach 10 kW, but the high-power output equipment of this power level cannot meet the auxiliary rock-breaking of hard rocks such as deep wells and ultra-deep wells. At the same time, there are many problems in the stable transmission of underground energy. At present, cables suitable for the complex underground environment have been successfully developed and used, which solves the problem of electric energy transmission. However, lowering the cable downhole, especially in deep wells and ultra-deep wells, and ensuring the continuous and stable transmission of electric energy, are urgent problems that must be solved in the field promotion and application of ultrasonic-assisted rock-breaking.

6. Application Prospect of Ultrasonic-Assisted Rock-Breaking Technology in Oil and Gas Drilling

Most of the existing research is focused on the shallow rock-breaking in the field of planetary drilling the mining field. There is little research on the problem of ultrasonic-assisted rock-breaking in oil and gas drilling, especially under more complex deep oil- and gas-drilling conditions. The influencing factors are also unclear. To introduce the advanced ultrasonic technology into the oil- and gas-drilling industry, speed up the drilling of deep and ultra-deep wells, and improve the efficiency of drilling operations, the following three perspectives are proposed.

6.1. Mechanism of Ultrasonic-Assisted Rock-Breaking with Characteristics of Oil and Gas Drilling

According to the characteristics of oil and gas drilling, the relevant factors that affect ultrasonic-assisted rock-breaking with the characteristics of oil and gas drilling are clarified, and factors that have not been considered in previous studies are taken as the key research objects. We focus on the influence of a high temperature, high stress and circulating medium on ultrasonic-assisted rock-breaking. In the future, with the development of test conditions and equipment, if conditions permit, we will carry out experimental research to explore rock crack initiation, and the length of crack propagation until stability. The change laws regarding rocks' physical parameters and strength during crushing, and the influence law of each factor is clarified. Combined with the classical theory, through an analysis of the research conclusions, the quantitative relationship between the rock change characteristics, and various influencing factors under the conditions of oil and gas drilling are established, and the rock fracture equation is established, with ultrasonic parameters as variables.

Through a summary of the test results and the theoretical derivation, the law of ultrasonic-assisted rock-breaking is further revealed, the mechanism of ultrasonic-assisted rock-breaking with the characteristics of oil and gas drilling is clarified, and the influence of rock type is considered. This lays the foundation for the application of ultrasonic-assisted rock-breaking in oil and gas drilling.

6.2. Energy Transfer and Utilization in Ultrasonic-Assisted Rock-Breaking Mechanical Wave

In research, the electrical characteristics and placement position of ultrasonic equipment should be fully considered, the theoretical relationship between the control parameters of ultrasonic generating equipment and the ultrasonic parameters that actually input into the well should be established, and the initial value of the actual energy provided by the ultrasonic generator should be clarified. Considering the characteristics of an ultrasonic wave, the research should use parameters such as the structure of the wellbore during construction, the physical parameters of the drilling fluid, and the characteristics of the formation rock to analyze the propagation path of the ultrasonic wave and the characteristics of the medium encountered by the propagation. The transmission distance and the dynamic changes in the transmission environment, etc., should be used to clarify the energy loss value of ultrasonic waves in borehole propagation. The geological parameters should be considered, including the characteristics of the downhole environment and the physical properties of the rock at the bottom of the well, based on the theories of rock mechanics, material mechanics, vibration mechanics, etc. This should be combined with the mechanism of ultrasonic-assisted rock-breaking and construction technology to determine the energy required for rock-breaking, namely, the ultrasonic actual utilization value.

This study aims to clarify the energy transfer path of ultrasonic-assisted rock-breaking and the main influencing factors affecting energy transfer and utilization, and provide a basis for the selection of ultrasonic generating equipment and the selection of construction parameters.

6.3. High-Power Ultrasonic Generator and Matching Products

Ultrasonic power generators have always been a concern in related fields. In recent years, China has carried out research in this field and achieved some breakthroughs. High-power ultrasonic generators and ultra-high-power ultrasonic generators have been successfully tested. The upper power limit of ultrasonic generators is constantly increasing. The author believes that the further development of technical conditions, especially following the breakthrough in the development of ultra-high-power generators, will vigorously promote the application of ultrasonic-assisted technology in oil and gas drilling. Increasing the research and development regarding supporting drilling tools, so that they have a high performance strength, can help to achieve the characteristics of ultrasonic high-frequency vibration, and lead to success in tests under the complex conditions of downhole oil and gas drilling. Using the premise of clarifying the failure characteristics of the circulating medium in ultrasonic-assisted rock-breaking, the functional characteristics and applicability of the ultrasonic-assisted rock-breaking drilling fluid were studied, and the drilling fluid system and suitable products were developed for ultrasonic-assisted rock-breaking. This provides a reliable hardware guarantee to accelerate the field application of ultrasonic-assisted oil and gas drilling and rock-breaking.

7. Conclusions

"How to carry out high-efficiency rock breaking" is an eternal topic of research and discussion in the oil- and gas-drilling industry. With the successive development of deep wells, ultra-deep wells and difficult-to-drill formations, the development of new rock-breaking methods is particularly urgent. Drawing on the successful experience in related fields, ultrasonic-assisted rock-breaking has great potential in the rock-breaking in oil- and gas-drilling industry as a new rock-breaking method with high efficiency and the possibility of environmental protection.

The development of an efficient rock-breaking method is a systematic project, ranging from mechanism exploration to indoor simulation tests, as well as field tests and the research and development of supporting equipment tools and products. At present, the existing research on ultrasonic-assisted rock-breaking is still at the initial stage, although the feasibility and efficiency of the technology have been verified, and the influencing factors and mechanisms of ultrasonic-assisted rock-breaking have been preliminarily clarified regarding the application of ultrasonic-assisted rock-breaking in oil and gas drilling.

In the future, the laws and mechanism of ultrasonic-assisted rock-breaking under the harsh conditions of deep oil and gas drilling and in a downhole environment should be improved. The energy-transfer path and energy utilization in ultrasonic-assisted rock-breaking should be clarified to improve the energy utilization efficiency in ultrasonic-assisted rock-breaking. The development of matching products should be increased, and combined with the development trend of ultrasonic technology, to form a complete and mature ultrasonic-assisted rock-breaking theory, as well as technology and equipment, and promote the popularization and application of this efficient rock-breaking method in oil and gas drilling. The efficient development of resources should also be increased.

Author Contributions: Conceptualization, J.F. and T.Y.; validation, J.F.; resources, Y.C.; writing—original draft preparation, J.F.; writing—review and editing, T.Y. and S.S. All authors have read and agreed to the published version of the manuscript.

Funding: This study was supported by the National Natural Science Foundation of China (grant No. 52174001).

Data Availability Statement: Data available on request from the authors.

Conflicts of Interest: The authors declare no conflict of interest.

References

1. He, H.; Fang, T.; Guo, X.; Yang, T.; Zheng, M.; Huang, F.; Gao, Y. Major achievements in oil and gas exploration of PetroChina during the 13th Five-Year Plan period and its development strategy for the 14th Five-Year Plan. *China Pet. Explor.* **2021**, *26*, 17–30.
2. Lei, Q.; Xu, Y.; Yang, Z.; Cai, B.; Wang, X.; Zhou, L.; Liu, H.; Xu, M.; Wang, L.; Li, S. Progress and Development Directions of Stimulation Techniques for Ultra-Deep Oil and Gas Reservoirs. *Pet. Explor. Dev.* **2021**, *48*, 221–231. [CrossRef]
3. Wang, H.; Huang, H.; Bi, W.; Ji, G.; Zhou, B.; Zhuo, L. Deep and Ultra-Deep Oil and Gas Well Drilling Technologies: Progress and Prospect. *Nat. Gas Ind. B* **2022**, *9*, 141–157. [CrossRef]
4. Wang, X.; Wang, X.; Wang, J.; Tian, Z. Feasibility Study and Prospects of Rock Fragmentation Using Ultrasonic Vibration Excitation. *Appl. Sci.* **2020**, *10*, 5868. [CrossRef]
5. Xu, F.; Chen, Q.; Zhu, H.; Wang, D.; Chen, J.; Liu, P.; Yao, G.; Zhang, K.; Huo, Z. Response Analysis of Shale Bedding Structure to Ultrasonic Characteristics and Its Application. *Pet. Explor. Dev.* **2019**, *46*, 82–92. [CrossRef]
6. Badescu, M.; Ressa, A.; Lee, H.J.; Bar-Cohen, Y.; Sherrit, S.; Zacny, K.; Paulsen, G.L.; Beegle, L.; Bao, X. *Auto-Gopher: A Wireline Deep Sampler Driven by Piezoelectric Percussive Actuator and EM Rotary Motor*; Lynch, J.P., Yun, C.-B., Wang, K.-W., Eds.; Jet Propulsion Laboratory, National Aeronautics and Space Administration: Pasadena, CA, USA, 2013; p. 86922S.
7. Fang, S.; Zhao, H.; Zhang, Q. The Application Status and Development Trends of Ultrasonic Machining Technology. *J. Mech. Eng.* **2017**, *53*, 22–32. [CrossRef]
8. Otumudia, E.; Hamidi, H.; Jadhawar, P.; Wu, K. The Utilization of Ultrasound for Improving Oil Recovery and Formation Damage Remediation in Petroleum Reservoirs: Review of Most Recent Researches. *Energies* **2022**, *15*, 4906. [CrossRef]
9. Thoe, T.; Aspinwall, D.; Wise, M. Review on Ultrasonic Machining. *Int. J. Mach. Tools Manuf.* **1998**, *38*, 239–255. [CrossRef]
10. Pei, Z.J.; Prabhakar, D.; Ferreira, P.M.; Haselkorn, M. A Mechanistic Approach To The Prediction Of Material Removal Rates In Rotary Ultrasonic Machining. *J. Eng. Ind.* **1993**, *64*, 142–151. [CrossRef]
11. Zhang, C.L.; Feng, P.F.; Wu, Z.J. Research on the Properties of Ultrasonic Vibration Amplitude and Actual Cutting Depth in Rotary Ultrasonic Machining. *Acta Armamentarii* **2013**, *34*, 883–888. [CrossRef]
12. Fang, G.F.; Wang, S.Y. Research on Multi-Frequency Ultrasonic Machining Holes in Ceramics. *Mach. Des. Manuf.* **2018**, *3*, 44–46.
13. Guo, Q.; Yan, J.; Yang, X. Review of the Gravity Field of Asteroids. *Prog. Geophys.* **2021**, *36*, 1–7.
14. Neumann, N.; Sattel, T. Set-Oriented Numerical Analysis of a Vibro-Impact Drilling System with Several Contact Interfaces. *J. Sound Vib.* **2007**, *308*, 831–844. [CrossRef]
15. Neumann, N.; Sattel, T.; Wallaschek, J. On Set-Oriented Numerical Methods for Global Analysis of Non-Smooth Mechanical Systems. *J. Vib. Control.* **2007**, *13*, 1393–1405. [CrossRef]
16. Khmelev, V.N.; Khmelev, S.S.; Khmelev, M.V.; Levin, S.V.; Kuzovnikov, Y.M. Development of Ultrasonic Specifically Drilling Technology and Improvement of Construction of Ultrasonic Machine Tools. In Proceedings of the IEEE International Conference & Seminar of Young Specialists on Micro/nanotechnologies & Electron Devices, Altai, Erlagol, Russia, 2–6 July 2012.
17. Potthast, C.; Twiefel, J.; Wallaschek, J. Modelling Approaches for an Ultrasonic Percussion Drill. *J. Sound Vib.* **2007**, *308*, 405–417. [CrossRef]
18. Zacny, K.; Paulsen, G.; Barcohen, Y.; Beegle, L.; Sherrit, S.; Badescu, M.; Mellerowicz, B.; Rzepiejewska, O.; Craft, J.; Sadick, S. Wireline Deep Drill for Exploration of Mars, Europa, and Enceladus. In Proceedings of the Aerospace Conference, Big Sky, MT, USA, 2–9 March 2013.
19. Zacny, K.; Mellerowicz, B.; Kim, D.; Paulsen, G.L.; Simonini, A.C. Auto-Gopher-II: A Wireline Rotary-Hammer Ultrasonic Drill That Operates Autonomously. In Proceedings of the Sensors and Smart Structures Technologies for Civil, Mechanical, and Aerospace Systems, Denver, CO, USA, 5–8 March 2018.
20. Wang, Y.; Quan, Q.; Yu, H.; Bai, D.; He, L.; Deng, Z. Rotary-Percussive Ultrasonic Drill: An Effective Subsurface Penetrating Tool for Minor Planet Exploration. *IEEE Access* **2018**, *6*, 37796–37806. [CrossRef]
21. Wang, Y.; Quan, Q.; Yu, H.; Bai, D.; Deng, Z. A Rotary-Percussive Ultrasonic Drill Driven by Single Piezoelectric Stack. *Beijing Hangkong Hangtian Daxue Xuebao/J. Beijing Univ. Aeronaut. Astronaut.* **2018**, *44*, 1850–1859. [CrossRef]
22. Pu, C.; Shi, D.; Zhao, S.; Xu, H.; Shen, H. Technology of Removing near Wellbore Inorganic Scale Damage by High Power Ultrasonic Treatment. *Pet. Explor. Dev.* **2011**, *38*, 243–248. [CrossRef]
23. Huang, X.; Liang, S. Mechanism and Experimental Study of Acoustic Oil Recovery. *Acta Pet. Sin.* **1993**, *14*, 110.
24. Du, C. *Application of Ultrasonic Plugging Removal Technology in Daqing Oilfield*; Northeast Petroleum University: Daqing, China, 2018.

25. Aarts, A.; Ooms, G.; Bil, K.J.; Bot, E. Enhancement of Liquid Flow Through a Porous Medium by Ultrasonic Radiation. In Proceedings of the European Petroleum Conference, London, UK, 20 October 1986.
26. Abramov, V.O.; Mullakaev, M.S.; Abramova, A.V.; Esipov, I.B.; Mason, T.J. Ultrasonic Technology for Enhanced Oil Recovery from Failing Oil Wells and the Equipment for Its Implemention. *Ultrason. Sonochem.* **2013**, *20*, 1289–1295. [CrossRef]
27. Abramov, V.O.; Abramova, A.V.; Bayazitov, V.M.; Altunina, L.K.; Gerasin, A.S.; Pashin, D.M.; Mason, T.J. Sonochemical Approaches to Enhanced Oil Recovery. *Ultrason. Sonochem.* **2015**, *25*, 76–81. [CrossRef] [PubMed]
28. Xiao, G.; Du, Z.; Li, G.; Shu, Z. High Frequency Vibration Recovery Enhancement Technology in the Heavy Oil Fields of China. *Soc. Pet. Eng.* **2004**. [CrossRef]
29. Yin, W.D.; Administration, S.P.; City, D.; Province, S.; Ping, W.; Dong, H. Development and Application of a Whole Set of High Power Ultrasonic Oil Recovery Equipment. *China Pet. Mach.* **2007**, *5*, 1–4 + 69.
30. Abramov, V.O.; Abramova, A.V.; Bayazitov, V.M.; Mullakaev, M.S.; Marnosov, A.V.; Ildiyakov, A.V. Acoustic and Sonochemical Methods for Altering the Viscosity of Oil during Recovery and Pipeline Transportation. *Ultrason. Sonochem.* **2017**, *35*, 389–396. [CrossRef] [PubMed]
31. Yang, Z.; Zhu, L.; Zhang, G.; Ni, C.; Lin, B. Review of Ultrasonic Vibration-Assisted Machining in Advanced Materials. *Int. J. Mach. Tools Manuf.* **2020**, *156*, 103594. [CrossRef]
32. Diehl, L.O.; Gatiboni, T.L.; Mello, P.A.; Muller, E.I.; Duarte, F.A.; Flores, E.M. Muller Ultrasound-Assisted Extraction of Rare-Earth Elements from Carbonatite Rocks. *Ultrason. Sonochem.* **2018**, *40*, 24–29. [CrossRef] [PubMed]
33. Yang, W. *Experimental Study on Physical and Mechanical Characteristics of Rock under Ultrasonic Excitation*; China University of mining and Technology: Xuzhou, China, 2017.
34. Li, S.; Li, W.; Yan, T.; Gao, H.; Bi, F.; Ma, H. A Study on the Rock Breaking Mechanism of Drill Bits under Combined Loads and Field Applications. *J. Vib. Shock* **2017**, *36*, 51–55.
35. Liu, J. Mechanic Behavior of Resonant Rubblization on Concrete Slab. *J. Chongqing Jiaotong Univ. (Nat. Sci.)* **2013**, *32*, 938.
36. Li, S.; Yan, T.; Li, W.; Bi, F. Simulation on Vibration Characteristics of Fractured Rock. *Rock Mech. Rock Eng.* **2016**, *49*, 4209. [CrossRef]
37. Huakai, W. *Analysis of Vehicle Road Coupling Vibration Characteristics of Resonant Crusher*; Wuhan University of Technology: Wuhan, China, 2014.
38. Wiercigroch, M.; Wojewoda, J.; Krivtsov, A.M. Dynamics of Ultrasonic Percussive Drilling of Hard Rocks. *J. Sound Vib.* **2005**, *280*, 739–757. [CrossRef]
39. Pavlovskaia, E.; Hendry, D.C.; Wiercigroch, M. Modelling of High Frequency Vibro-Impact Drilling. *Int. J. Mech. Sci.* **2015**, *91*, 110–119. [CrossRef]
40. Yang, W.; Lei, L.I.; Zhao, Y.X.; Xie, S.D.; Zhao, J. Resonates the Detritus the Preliminary Study. *Energy Technol. Manag.* **2007**, *4*, 7–9.
41. Wang, X. *Experimental Study on Rock Fracture Propagation Law under Ultrasonic Excitation*; China University of Mining and Technology: Xuzhou, China, 2019.
42. Zhao, D.; Zhang, S.; Wang, M. Microcrack Growth Properties of Granite under Ultrasonic High-Frequency Excitation. *Adv. Civ. Eng.* **2019**, *2019*, e3069029. [CrossRef]
43. Zhou, Y.; Zhao, D.; Li, B.; Wang, H.; Zhang, Z. Fatigue Damage Mechanism and Deformation Behaviour of Granite Under Ultrahigh-Frequency Cyclic Loading Conditions. *Rock Mech. Rock Eng.* **2021**, *54*, 4723–4739. [CrossRef]
44. Zhao, Y.; Zhang, C.; Zhang, Z.; Gao, K.; Zhao, D.; Sun, Z.; Lv, X.; Zhou, Y.; Zhai, G. Experimental and Simulation Study on Breaking Rock under Coupled Static Loading and Ultrasonic Vibration. *Shock Vib.* **2022**, *2022*, e5536358. [CrossRef]
45. Li, S. *Study on Rock Breaking Mechanism of Drill Bit under Resonant Excitation*; Northeast Petroleum University: Daqing, China, 2017.
46. Li, T.; ZHANG, D.; PAN, D.; LIN, M.; DENG, T.; YANG, D.; HAN, R.; YAO, J.; HE, Z. Failure Characteristics and Laws of Acoustic Emission for Phosphorite Under Triaxial Loading. *J. Wuhan Inst. Technol.* **2017**, *39*, 616–621.
47. Zhou, Z.L.; Xi-Bing, L.I.; Zhao, G.Y.; Liu-Qing, H.U. Three Dimensional Numerical Analysis of Perfect Loading Wave-Form of Rock With SHPB. *Min. Metall. Eng.* **2005**, *25*, 18–20. [CrossRef]
48. Zhang, J.; Liu, H. Constitutive Model of Jointed Rock Mass by Combining Macroscopic and Microcopic Composite Damage. *Coal Geol. Explor.* **2013**, *41*, 49–52.
49. Deng, Y.; Chen, M.; Jin, Y.; Yunhu, L.U.; Zou, D. Prediction Model and Numerical Simulation for Rock Fissure Length under Impact Load. *Pet. Drill. Tech.* **2016**, *44*, 41–46. [CrossRef]
50. Ni, H.; Wang, R.; Zhang, Y. Numerical Simulation Study on Rock Breaking Mechanism and Process under High Pressure Water Jet. *Appl. Math. Mech.-Engl. Ed.* **2005**, *26*, 1595–1604. [CrossRef]
51. Li, C.; Duan, L.; Tan, S.; Chikhotkin, V.; Fu, W. Damage Model and Numerical Experiment of High-Voltage Electro Pulse Boring in Granite. *Energies* **2019**, *12*, 727. [CrossRef]
52. Zhou, Y.; Liu, W. New Progress on PCDS Precise Pressure Management Drilling Technology. *Pet. Drill. Tech.* **2019**, *47*, 68–74.
53. Peng, K.; Tian, S.; Li, G.; Huang, Z.; Yang, R.; Guo, Z. Bubble Dynamics Characteristics and Influencing Factors on the Cavitation Collapse Intensity for Self-Resonating Cavitating Jets. *Pet. Explor. Dev.* **2018**, *45*, 343–350. [CrossRef]
54. Chen, H.; Li, J.; Chen, D.; Wang, J. Damages on Steel Surface at the Incubation Stage of the Vibration Cavitation Erosion in Water. *Wear* **2008**, *265*, 692–698. [CrossRef]
55. Neppiras, E.A. Acoustic Cavitation. *Phys. Rep.* **1980**, *61*, 159–251. [CrossRef]

56. Yoichiro, M. Influence of Homogeneous Condensation inside a Bubble on Cavitation Inception. *Trans. Jpn. Soc. Mech. Eng.* **1985**, *51*, 2036–2042. [CrossRef]
57. Sheng, M.; Zhang, X.; Zhang, R.; Zhang, Y.; Peng, C.; Tian, S. Enhanced Erosion on Sedimentary Rock by Adding Abrasive Nanoparticles in Ultrasonic Cavitation. *Appl. Nanosci.* **2020**, *10*, 1319–1330. [CrossRef]
58. Yang, S.; Zhang, N.; Feng, X.; Kan, J.; Pan, D.; Qian, D. Experimental Investigation of Sandstone under Cyclic Loading: Damage Assessment Using Ultrasonic Wave Velocities and Changes in Elastic Modulus. *Shock Vib.* **2018**, *2018*, 7845143. [CrossRef]
59. Xu, J.; Lu, X.; Zhang, J.; Wang, Z.; Bai, E. Research on Energy Properties of Rock Cyclical Impact Damage under Confining Pressure. *Chin. J. Rock Mech. Eng.* **2010**, *29*, 4159–4165.
60. Sun, Z. *Study on the Influence of Ultrasonic Vibration Frequency on the Fragmentation Law of Granite*; Jilin University: Changchun, China, 2017.
61. Luo, X.; Zhao, L.; Feng, C.; Yan, S.X.; Zhang, J. The development and collapse process of acoustic cavitation bubble. *J. Eng. Thermophys.* **2011**, *32*, 17–20.
62. Li, W.; Yan, T.; Li, S.; Zhang, X. Rock Fragmentation Mechanisms and an Experimental Study of Drilling Tools during High-Frequency Harmonic Vibration. *Pet. Sci.* **2013**, *10*, 205–211. [CrossRef]
63. Tian, Z. *Experimental Study on Influencing Factors of Rock Fragmentation under Ultrasonic Excitation*; China University of Mining and Technology: Xuzhou, China, 2018.
64. Guan, Z.; Hu, H.; Wang, B.; Sun, M.; Liu, Y.; Xu, Y. Experimental Study on Rock-Breaking Efficiency of PDC Bit Based on Mechanical Specific Energy and Sliding Frictional Coefficient. *J. China Univ. Pet. (Ed. Nat. Sci.)* **2019**, *43*, 92–100.
65. Yin, S.; Zhao, D.; Zhai, G. Investigation into the Characteristics of Rock Damage Caused by Ultrasonic Vibration. *Int. J. Rock Mech. Min. Sci.* **2016**, *84*, 159–164. [CrossRef]
66. Song, X.; Xu, Z.; Wang, M.; Li, G.; Shah, S.N.; Pang, Z. Experimental Study on the Wellbore-Cleaning Efficiency of Microhole-Horizontal-Well Drilling. *SPE J.* **2017**, *22*, 1189–1200. [CrossRef]
67. Yan, T.; Xu, R.; Sun, W.; Liu, W.; Hou, Z.; Yuan, Y.; Shao, Y. Similarity Evaluation of Stratum Anti-Drilling Ability and a New Method of Drill Bit Selection. *Pet. Explor. Dev.* **2021**, *48*, 450–459. [CrossRef]
68. Zhai, G. *Study on the Influence of Pressure on the Effect of Ultrasonic Vibration on Rock Breaking*; Jilin University: Changchun, China, 2016.
69. Wang, J.; Wang, X.; Chen, X.; Chen, L.; Yang, Z.; Chang, Z.; Zhang, L.; Niu, Z. Experimental Study on Failure Law and Mechanism of Red Sandstone under Ultrasonic Vibration Excitation. *Geofluids* **2022**, *2022*, 3078599. [CrossRef]
70. Peng, Y. *Study On The Failure Rate of Rock under Ultrasonic Vibration*; Jilin University: Changchun, China, 2017.
71. Zhao, D.; Yuan, P. Research on the Influence Rule of Ultrasonic Vibration Time on Granite Damage. *J. Min. Sci.* **2018**, *54*, 751–762. [CrossRef]
72. Prikhod'ko, V.M.; Aleksandrov, V.A.; Fatyukhin, D.S.; Petrova, L.G. Effect of Ultrasonic Cavitation on Nitrided Steel Surface Layer Condition. *Met. Sci. Heat Treat.* **2015**, *57*, 300–303. [CrossRef]
73. Liu, L.; Li, K.; Zhang, H. The Effects of Ultrasonic Vibration Frequency on the Rules of Rock Crushing. *J. Chang. Inst. Technol. (Nat. Sci. Ed.)* **2017**, *18*, 81–83 + 98.
74. Zhao, D.; Zhang, S.; Zhao, Y.; Wang, M. Experimental Study on Damage Characteristics of Granite under Ultrasonic Vibration Load Based on Infrared Thermography. *Environ. Earth Sci.* **2019**, *78*, 1–12. [CrossRef]
75. Lu, Z.; Zheng, J.; Jiang, Z.; Zhao, F. An Experimental Study on Rock Breaking Efficiency with Ultrasonic High-Frequency Rotary-Percussive Drilling Technology. *Pet. Drill. Tech.* **2021**, *49*, 20–25.
76. Peng, C.; Zhang, C.; Li, Q.; Zhang, S.; Su, Y.; Lin, H.; Fu, J. Erosion Characteristics and Failure Mechanism of Reservoir Rocks under the Synergistic Effect of Ultrasonic Cavitation and Micro-Abrasives. *Adv. Powder Technol.* **2021**, *32*, 4391–4407. [CrossRef]
77. Zhang, L.; Wang, X.; Wang, J.; Yang, Z. Mechanical Characteristics and Pore Evolution of Red Sandstone under Ultrasonic High-Frequency Vibration Excitation. *AIP Adv.* **2021**, *11*, 055202. [CrossRef]
78. Zhang, C.; Zhao, D.; Zhang, S.; Zhou, Y. Individual and Combined Influences of Main Loading Parameters on Granite Damage Development under Ultrasonic Vibration. *J. Mt. Sci.* **2021**, *18*, 3366–3379. [CrossRef]
79. Yuan, P. Study on Granite Crushing Law under Ultrasonic Multi-Parameter Vibration Based on Microscopic Damage. *J. Min. Sci.* **2021**, *57*, 1060–1074. [CrossRef]
80. Zhao, D.; Wu, J.; Li, Z. Simulation and Experimental Research on Ultrasonic Vibration High Temperature Rock. *J. Pet. Sci. Eng.* **2022**, *212*, 110255. [CrossRef]

energies

MDPI

Article

A Quantitative Evaluation Method of Anti-Sloughing Drilling Fluid Inhibition for Deep Mudstone

Kehao Bo [1,2], Yan Jin [1,2,*], Yunhu Lu [1,2], Hongtao Liu [3] and Jinzhi Zhu [3]

[1] State Key Laboratory of Petroleum Resources and Prospecting, China University of Petroleum, Beijing 102249, China; bkh1615@163.com (K.B.); luyh@cup.edu.cn (Y.L.)
[2] College of Petroleum Engineering, China University of Petroleum, Beijing 102249, China
[3] Tarim Oil Field Company, Petro China, Korla 841000, China; lhtcnpc@163.com (H.L.); zjzcnpc@163.com (J.Z.)
* Correspondence: jiny@cup.edu.cn

Abstract: Wellbore instability resulting from deep mudstone hydration severely restricts the development of oil and gas resources from deep reservoir in western China. Accurate evaluation of drilling fluid inhibition properties plays an important role in selecting drilling fluid that can control deep mudstone hydration and then sustain wellbore stability. The previous evaluations are conducted by qualitative analysis and cannot consider the influence of complex hydration conditions of deep mudstone (high temperature, high pressure and flushing action). The study proposes a quantitative method to evaluate drilling fluid's inhibition property for deep mudstone under natural drilling conditions. In this method, the cohesive strength of mudstone after hydration is adopted as the inhibition index of the tested drilling fluid. An experimental platform containing a newly designed HPHT (High pressure and high temperature) hydration experiment apparatus and mechanics characterization of mudstone after hydration based on scratch test is proposed to obtain the current inhibition index of tested drilling fluid under deep well drilling environments. Based on the mechanical–chemical wellbore stability model considering strength weakening characteristics of deep mudstone after hydration, a cross-correlation between drilling fluid density (collapse pressure) and required inhibition index (cohesive strength) for deep mudstone is provided as the quantitative evaluation criterion. Once the density of tested mud is known, one can confirm whether the inhibition property of tested mud is sufficient. In this study, the JDK mudstone of a K block in western China is selected as the application object of the proposed evaluation method. Firstly, the evaluation chart, which can demonstrate the required inhibition indexes of the tested fluids quantitatively with various densities for JDK mudstone, is constructed. Furthermore, the experimental evaluations of inhibition indexes of drilling fluids taken from two wells in K block are conducted under ambient and deep-well drilling conditions, respectively. In order to show the validity and advantage of the proposed method, a comparison between the laboratory evaluation results and field data is made. Results show that the laboratory evaluation results under deep-well drilling conditions are consistent with the field data. However, the evaluation under ambient conditions overestimates the inhibition property of the tested fluid and brings a risk of wellbore instability. The developed quantitative method can be a new way to evaluate and optimize the inhibition property of drilling fluid for deep mudstone.

Keywords: wellbore stability; deep mudstone; quantitative evaluation method; drilling fluid inhibition; HPHT

Citation: Bo, K.; Jin, Y.; Lu, Y.; Liu, H.; Zhu, J. A Quantitative Evaluation Method of Anti-Sloughing Drilling Fluid Inhibition for Deep Mudstone. *Energies* **2022**, *15*, 1226. https://doi.org/10.3390/en15031226

Academic Editor: Hossein Hamidi

Received: 12 January 2022
Accepted: 4 February 2022
Published: 8 February 2022

1. Introduction

The exploration and development of oil and gas resources from a deep reservoir in western China is showing great potential. However, wellbore instability resulting from deep mudstone hydration is still a troublesome problem to drilling engineers, which causes tremendous economic losses. The capability of drilling fluid to inhibit deep mudstone hydration is a key indicator of wellbore stability. To accurately evaluate drilling fluid inhibition property is a prerequisite for choosing the drilling fluid with good performance and then sustaining wellbore stability.

In recent years, many evaluation methods of drilling fluid inhibition property are proposed as summarized in Table 1. Based on the analysis of evaluation principles, these evaluation methods can be clarified into two types. (1) Methods to evaluate the physical and mechanical properties of solid medium after hydration (swelling, dispersion and strength): ① hot rolling dispersion test [1–3]; ② slake durability evaluation [2,4,5]; ③ swelling test [5–8]; ④ bulk hardness test [4,9]; ⑤ water adsorption test [10]; ⑥ mud ball immersing test [3,11]; ⑦ particle size distribution test [11,12]; ⑧ shale stability index test [8,13]; ⑨ cohesive strength test (Energy Sector Standard of the People's Republic of China: NB/T 10121-2018) [14]. (2) Methods to evaluate the chemical and rheological properties of liquid medium after hydration: ① rheological evaluation [15]; ② high-speed centrifugation [16,17]; ③ flocculation experiment [18,19]; ④ water separating ratio test [16]; ⑤ specific surface area test with methylene blue solution [20,21]; ⑥ capillary suction time test [22–24]; ⑦ accretion test [2,9]; ⑧ zeta potential test [8,25]; ⑨ X-ray diffraction analysis [15,25].

Although these common evaluation methods of drilling fluid inhibition performance have been widely used in the oil and gas industry, there are still many limitations to overcome, especially for deep mudstone. (1) The complex hydration conditions in a deep well cannot be provided (as shown in Table 1). However, numerous studies have demonstrated that well drilling conditions have a significant impact on the property of drilling fluid and the rock–fluid interaction [24,26–30]. As for deep mudstone in western China, a harsher and more hostile hydration environment will be encountered such as high temperature (~20 °C), high drilling differential pressure (~40 MPa) and flushing action of drilling fluid. Therefore, it is necessary to consider the effect of complex drilling conditions on the evaluation of drilling fluid inhibition property for deep mudstone. (2) Quantitative evaluation basis of drilling fluid inhibition property is not proposed. Previous qualitative/empirical evaluation methods are difficult to use to definitely identify the safe value of inhibition index. (3) The coupling of mechanical and chemical actions of drilling fluid on sustaining wellbore stability is not involved in the current estimation of drilling fluid inhibition property.

In this study, an experimental platform containing a newly designed HPHT hydration experiment and mechanical characterization of mudstone based on a scratch test is adopted to obtain the value of inhibition index of drilling fluid under deep-well drilling conditions. In addition, a cross-correlation between drilling fluid density (collapse pressure) and the matching inhibition index (cohesive strength) for deep mudstone, which is revealed by the mechanical–chemical wellbore stability model, is provided as the quantitative evaluation criterion. Based on the above works, a quantitative evaluation method for the inhibition property of drilling fluid for deep mudstone is proposed. In order to validate the proposed method, a comparison analysis of the laboratory evaluation results under different conditions (deep-well drilling condition and ambient condition) and field data of JDK mudstone (depth: about 5200–6800 m) of the K block in western China is conducted.

Table 1. Review of inhibition property evaluation methods.

Order	Method	Experimental Apparatus	Evaluation Condition		Tested Sample		Applicability Analysis	Reference
			Temperature	Pressure	Solid Medium	Liquid Medium		
1	Rheological evaluation	ZNN-D6 viscometer	Ambient temperature	Ambient pressure	Bentonite powder	Inhibitor solution	Qualitative evaluation	[15]
2	Hot rolling dispersion test	Roller oven	~160 °C	Ambient pressure	Shale cuttings	Inhibitor solution	Qualitative evaluation (dispersion tendency)	[1–3]
3	Slake durability evaluation	Slake durability test instrument	Ambient temperature	Ambient pressure	Shale cuttings	Inhibitor solution	Qualitative evaluation (dispersion test with more abrasive condition)	[2,4,5]
4	Swelling test	NP-02A shale expansion apparatus	~180 °C	~10 MPa	Shale/bentonite pellet	Inhibitor solution	Qualitative evaluation (swelling tendency)	[5–8]
5	Bulk hardness test	Bulk hardness tester	Hot rolling	Ambient pressure	Shale cuttings	Inhibitor solution	Qualitative evaluation (mechanical property)	[4,9]
6	High-speed centrifugation	Centrifuge	Ambient temperature	Ambient pressure	Bentonite powder	Inhibitor solution	Qualitative evaluation (swelling tendency); Poor performance to amine inhibitors	[16,17]
7	Flocculation experiment	-	Ambient temperature	Ambient pressure	Bentonite powder	Inhibitor solution	Qualitative evaluation (swelling tendency)	[18,19]
8	Water separating ratio test	Roller oven	Hot rolling	Ambient pressure	Bentonite powder	Inhibitor solution	Qualitative evaluation (swelling tendency)	[16]
9	Water adsorption test	-	Ambient temperature	Ambient pressure	Modified bentonite with inhibitors	Deionized water	Qualitative evaluation (swelling tendency)	[10]
10	Specific surface area test with methylene blue	731 spectrophotometer	Ambient temperature	Ambient pressure	Bentonite/cutting powder	Inhibitor solution	Qualitative evaluation (swelling tendency)	[20,21]

Table 1. *Cont.*

Order	Method	Experimental Apparatus	Evaluation Condition		Tested Sample		Applicability Analysis	Reference
			Temperature	Pressure	Solid Medium	Liquid Medium		
11	Mud ball immersing test	-	Ambient temperature	Ambient pressure	Bentonite	Inhibitor solution	Qualitative evaluation (dispersion tendency)	[3,11]
12	Capillary suction time test	Capillary suction time tester	Ambient temperature	Ambient pressure	Shale powder	Inhibitor solution	Qualitative evaluation; Simple operation; Poor repeatability.	[22–24]
13	Particle size distribution test	Laser particle size analyzer	Ambient temperature	Ambient pressure	Bentonite	Inhibitor solution	Qualitative evaluation (dispersion tendency)	[11,12]
14	Accretion test	Roller oven + steel bar	~160 °C	Ambient pressure	Shale cuttings	Inhibitor solution	Qualitative evaluation	[2,9]
15	Zeta potential test	Zeta sizer 3000 electric potential and granularity meter	Ambient temperature	Ambient pressure	Bentonite/shale powder	Inhibitor solution	Qualitative evaluation (dispersion tendency)	[8,25]
16	X-ray diffraction analysis	X-ray diffractometer	Ambient temperature	Ambient pressure	Bentonite/shale powder	Inhibitor solution	Qualitative evaluation (swelling tendency)	[15,25]
17	Shale stability index test	Penetrometer	~160 °C	Ambient pressure	Shale/bentonite pellet	Inhibitor solution	Qualitative evaluation (mechanical property)	[8,13]
18	Cohesive strength test	Triaxial compressive test system	Ambient temperature	Ambient pressure	Cylindrical shale/mudstone	Inhibitor solution	Qualitative evaluation (mechanical property); Complicated and destructive test	[14]

2. Methodology

The heart of the matter for the previous evaluation methods can be divided into two points. ① How to obtain inhibition property of drilling fluid under deep-well drilling conditions, and ② how to achieve the quantitative evaluation of drilling fluid inhibition property for deep mudstone. In order to overcome the two difficult problems above, the main work in this new proposed method are divided into two parts (as shown in Figure 1). ① The first part is to develop an experimental method of deep mudstone hydration characterization. In the first part, a newly designed hydration experimental apparatus is adopted to provide deep-well drilling conditions (high temperature, high pressure and flushing action of drilling fluid). Furthermore, the continuous scratch test is used to obtain the cohesive strength of mudstone after hydration, which can be regarded as the inhibition index of drilling fluid. ② The second part is to propose a method that can determine the quantitative correlation between drilling fluid density (collapse pressure) and required inhibition index (cohesive strength). Based on the mechanical–chemical wellbore stability model, wellbore collapse pressure, which is the critical minimum mud density for preventing wellbore collapse, is related to rock cohesive strength. In the second part, considering the weakening characteristics of the strength property of deep mudstone at different degrees of hydration, the correlation chart of drilling fluid density and matching inhibition index (cohesive strength) for deep mudstone is constructed. Once the density of tested drilling fluid is known, the required inhibition index for deep mudstone can be confirmed.

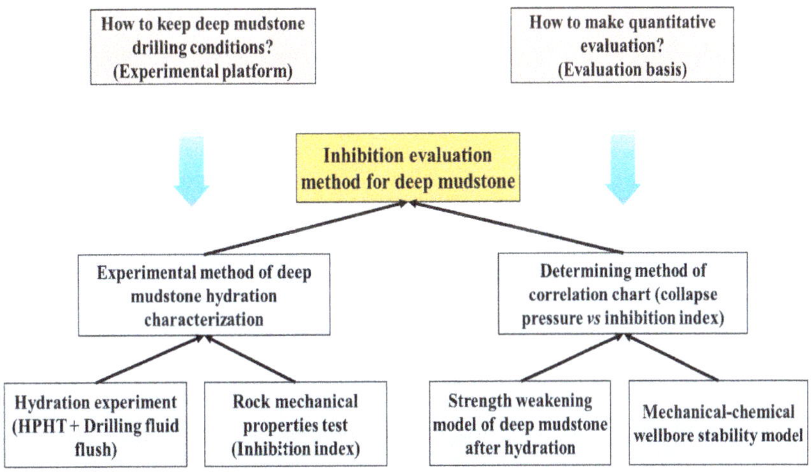

Figure 1. Framework of evaluation method of drilling fluid inhibition for deep mudstone.

2.1. Method of Hydration Characterization of Deep Mudstone

In our previous study [30], the new experimental platform containing a HPHT (high temperature and high pressure) hydration experiment and rock mechanical characterization based on scratch test is constructed to study the effect of hydration on rocks' mechanical properties. In this paper, because cohesive strength of mudstone after hydration can be regarded as the inhibition index of the drilling fluid, the experimental platform is proposed as a new experimental method to obtain inhibition performance of drilling fluid under deep-well drilling conditions.

(1) HPHT hydration experimental apparatus

As shown in Figure 2, the new designed HPHT experimental apparatus consists of hydration test section and core flow test section. A hydration test section, which can provide the hydration of several samples with different shapes under complex environments (maximum surrounding temperature of 200 °C, maximum surrounding pressure of 60 MPa and flushing action of fluid), is adopted in this study. The hydration test section is mainly made up of hydration kettle, electrical heating system, stirring system, pressure pump, monitoring system and data acquisition system.

(2) Estimation of rock mechanical properties based on continuous scratch test

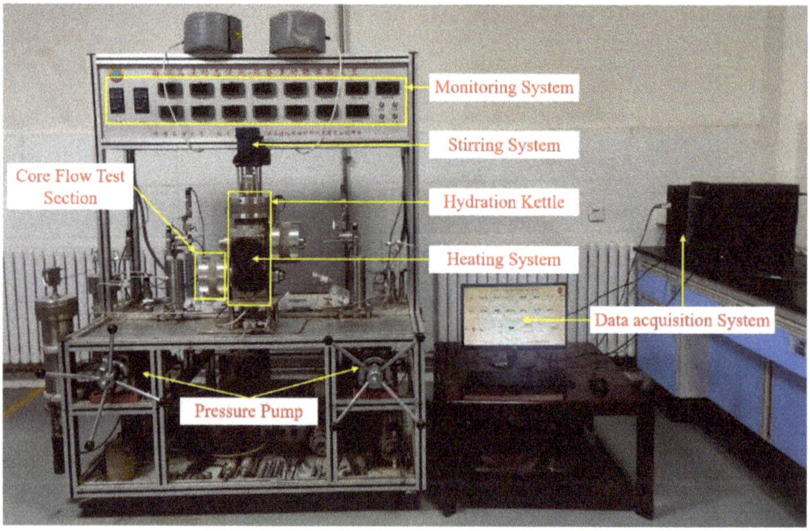

Figure 2. HPHT experimental apparatus for studying rock–fluid interaction (only use hydration experiment section in this study).

After immersed in tested fluid, the mudstone sample is used to evaluate the performance of tested fluid by strength test. Firstly, the uniaxial compressive strength (*UCS*) and internal friction angle are estimated by continuous scratch test with a sharp cutter and blunt cutter. The detailed principle and procedure are demonstrated by previous studies [30–33]. In addition, the Mohr–Coulomb criterion (as shown in Equation (1)), which reveals the relation between *UCS*, internal friction angle and cohesive strength is adopted to obtain the cohesive strength (inhibition index) in this study. The TerraTek continuous scratch test system used in this study is shown in Figure 3.

$$C = \frac{UCS \times (1 - \sin\varphi)}{2\cos\varphi} \tag{1}$$

where C is rock cohesive strength, MPa; φ is rock internal friction angle, °; *UCS* is rock uniaxial compressive strength, MPa.

2.2. Method of Determining Correlation between Drilling Fluid Density and Matching Inhibition Index

Preventing wellbore collapse relies not only on the chemical property of drilling fluid (inhibition performance) but also on mechanical support from drilling fluid density. Poor performance of drilling fluid in inhibiting mudstone hydration always requires adjustment of drilling fluid density, which provides wellbore collapse pressure. Based on the mechanical–chemical wellbore stability model, the relation between drilling fluid density (collapse pressure) and matching inhibition index (cohesive strength) can be constructed and then regarded as the quantitative criterion of estimating drilling fluid inhibition properties. The method of determining the quantitative relation consists of three parts as follows.

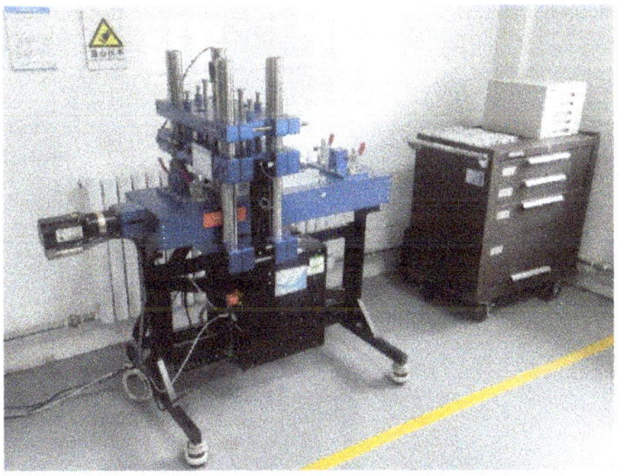

Figure 3. TerraTek continuous scratch test system.

(1) Framework of wellbore stability model

In this study, in order to ensure the popularization and application of the developed evaluation method, the framework of mechanical wellbore stability model, which is simple and widely used in drilling engineering, is selected to describe the relationship between collapse pressure (drilling fluid density) and cohesive strength (matching inhibition index).

The equations for stress distribution around wellbore are shown in Equation (2) [34,35].

$$\begin{cases} \sigma_r = & \frac{\sigma_{xx}+\sigma_{yy}}{2}\left(1-\frac{R_w{}^2}{r^2}\right)+\frac{\sigma_{xx}-\sigma_{yy}}{2}\left(1+3\frac{R_w{}^4}{r^4}-4\frac{R_w{}^2}{r^2}\right)\cos 2\theta \\ & +\tau_{xy}\left(1+3\frac{R_w{}^4}{r^4}-4\frac{R_w{}^2}{r^2}\right)\sin 2\theta+\frac{R_w{}^2}{r^2}p_m+\delta\left[\frac{a(1-2\nu)}{2(1-\nu)}\left(1-\frac{R_w{}^2}{r^2}\right)-\phi\right](p_m-p_p) \\ \sigma_\theta = & \frac{\sigma_{xx}+\sigma_{yy}}{2}\left(1+\frac{R_w{}^2}{r^2}\right)-\frac{\sigma_{xx}-\sigma_{yy}}{2}\left(1+3\frac{R_w{}^4}{r^4}\right)\cos 2\theta-\tau_{xy}\left(1+3\frac{R_w{}^4}{r^4}\right)\sin 2\theta-\frac{R_w{}^2}{r^2}p_m \\ & +\delta\left[\frac{a(1-2\nu)}{2(1-\nu)}\left(1+\frac{R_w{}^2}{r^2}\right)-\phi\right](p_m-p_p) \\ \sigma_z = & \sigma_{zz}-2\nu\left[(\sigma_{xx}-\sigma_{yy})\frac{R_w{}^2}{r^2}\cos 2\theta+2\tau_{xy}\frac{R_w{}^2}{r^2}\sin 2\theta\right]+\delta\left[\frac{a(1-2\nu)}{2(1-\nu)}-\phi\right](p_m-p_p) \\ \tau_{r\theta} = & \frac{\sigma_{yy}-\sigma_{xx}}{2}\left(1+2\frac{R_w{}^2}{r^2}-3\frac{R_w{}^4}{r^4}\right)\sin 2\theta+\tau_{xy}\left(1+2\frac{R_w{}^2}{r^2}-3\frac{R_w{}^4}{r^4}\right)\cos 2\theta \\ \tau_{\theta z} = & \tau_{yz}\left(1+\frac{R_w{}^2}{r^2}\right)\cos\theta-\tau_{xz}\left(1+\frac{R_w{}^2}{r^2}\right)\sin\theta \\ \tau_{rz} = & \tau_{xz}\left(1-\frac{R_w{}^2}{r^2}\right)\cos\theta+\tau_{yz}\left(1-\frac{R_w{}^2}{r^2}\right)\sin\theta \end{cases} \tag{2}$$

where σ_r is radial normal borehole stress, MPa; σ_θ is circumferential normal borehole stress, MPa; σ_z is axial normal borehole stress, MPa; $\tau_{r\theta}$, $\tau_{\theta z}$, τ_{rz} are three components of the shear stress, MPa; σ_{xx}, σ_{yy}, σ_{zz} are in situ normal stresses rotated to the borehole coordinate reference frame, MPa; τ_{xy}, τ_{yz}, τ_{xz} are in situ shear stresses rotated to the borehole coordinate reference frame, MPa; R_w is the radius of the borehole, m; r is radial distance away from wellbore, m; p_m is the downhole pressure, MPa; p_p is the formation pore pressure, MPa; θ is the point location angle, °; ν is Poisson's ratio of the formation, dimensionless; a is Biot effective stress coefficient, dimensionless; ϕ is formation porosity, %.

The stress distribution around wellbore also can be expressed by in situ stresses (σ_H, maximum horizontal stress; σ_h, minimum horizontal stress; σ_v, vertical stress) through a number of coordinate transformations. The detailed transformations can be found in previous studies [36,37].

The shear failure of rock is described by Mohr–Coulomb strength criterion (as shown in Equation (3)).

$$\sigma_1-\left[\sigma_3+2(C+\sigma_3\tan\varphi)\left(\sqrt{(\tan\varphi)^2+1}+\tan\varphi\right)\right]=0 \tag{3}$$

where σ_1, σ_3 are maximum principal stress and minimum principal stress, respectively, MPa; C is cohesive strength of rock, MPa; φ is internal friction angle of rock, °.

By coupling Equation (2) with Equation (3), we can obtain the minimum mud weight required to prevent wellbore collapse for mudstone with a given strength. However, it must be noted that chemical effect of hydration on deep mudstone strength properties and wellbore stability is not considered in the above model and needs to be updated.

(2) Deterioration model of mudstone mechanical properties after hydration

Based on the developed experimental apparatus, the empirical model describing the strength weakening characteristics of mudstone at various levels of hydration under well drilling conditions can be constructed. As shown in Equation (4), the strength change of mudstone after hydration is a function of its uptake water content $w(r,t)$, which is related to the radial distance from wellbore (r) and hydration time (t). The detailed descriptions about $w(r,t)$ in [38,39] are used in this study.

$$\begin{aligned} C_{(w(r,t))} &= C_i - A \times (w(r,t) - w_i) \\ \varphi_{(w(r,t))} &= \varphi_i - B \times (w(r,t) - w_i) \end{aligned} \tag{4}$$

where $C_{(w(r,t))}$, C_i are cohesive strength of mudstone after and before hydration, respectively, MPa; $\varphi_{(w(r,t))}$, φ_i are internal friction angle of mudstone after and before hydration, respectively, °; $w(r,t)$, w_i are uptake water content of mudstone after and before hydration, respectively, %; A, B are the empirical coefficients determined by experimental data.

Therefore, Equation (3) can be updated as follows

$$\sigma_1 - \left[\sigma_3 + 2\left(C_{(w(r,t))} + \sigma_3 \tan\varphi_{(w(r,t))}\right)\left(\sqrt{\left(\tan\varphi_{(w(r,t))}\right)^2 + 1} + \tan\varphi_{(w(r,t))}\right)\right] = 0 \tag{5}$$

(3) Quantitative relation between drilling fluid density and matching inhibition index

According to the above wellbore stability model considering strength-weakening characteristics of mudstone after hydration, the drilling fluid densities required to prevent wellbore collapse for mudstones with different hydration levels (different cohesive strengths) can be calculated. In addition, when a certain extent of failure regions around the wellbore can be allowed (acceptable maximum borehole enlargement rate is generally 15% in well-drilling engineering), the relationship of the required drilling fluid density, the matching inhibition index and the borehole enlargement rate (the ratio of the difference between borehole diameter and drilling bit diameter to the drilling bit diameter) can also be constructed.

2.3. Evaluation Procedure of Drilling Fluid Inhibition for Deep Mudstone

As shown in Figure 4, the detailed procedure of estimating drilling fluid inhibition property for deep mudstone is divided into two steps.

(1) Obtaining the cross-correlation chart of drilling fluid density and inhibition index for deep mudstone

On the basis of the method described in Section 2.2, the first stage is to develop the strength-weakening model of deep mudstone at various levels of hydration, and the next is to construct the correlation chart of the required drilling fluid density, matching inhibition index and borehole enlargement rate.

1. Establishing the strength deterioration model of deep mudstone after hydration

The developed HPHT experimental apparatus is utilized to conduct the hydration of deep mudstone for different times under deep-well drilling conditions. Furthermore, the mechanical properties of deep mudstone after hydration are estimated based on scratch test. At the same time, the uptake water content of mudstone is obtained by using a weighting method. Therefore, the empirical model describing the chemical effect of hydration on deep mudstone strength parameters is updated.

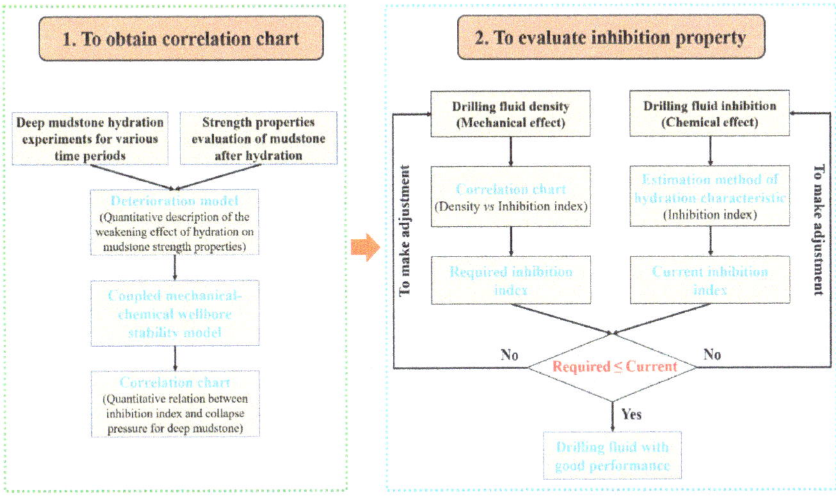

Figure 4. Detailed procedure of evaluation of the drilling fluid inhibition property for deep mudstone.

2. Constructing the quantitative correlation chart

According to the chemical–mechanical wellbore stability model for deep mudstone, the relationships between required drilling fluid density and matching inhibition index for different extent of wellbore collapse (borehole enlargement rate) can be constructed and are regarded as the quantitative evaluation criterion for drilling fluid inhibition property.

(2) Evaluating inhibition property of drilling fluid

1. Obtaining the required inhibition index of the tested drilling fluid based on the above correlation chart

According to the requirement of borehole enlargement rate in drilling engineering and the density of the tested drilling fluid, the required inhibition index for sustaining the wellbore stability in deep mudstone can be confirmed from the above correlation chart.

2. Estimating the current inhibition index of the tested drilling fluid for deep mudstone

The mudstone sample is immersed in the tested drilling fluid under the well drilling conditions for a constant time, which can provide fully hydration. Additionally, the cohesive strength of mudstone after hydration is then estimated as the current inhibition index of the tested drilling fluid. All of the above works is conducted on the developed experimental platform.

3. Judging the inhibition property of the tested drilling fluid

By comparing the above two kinds of inhibition indexes, it can be changed whether the tested drilling fluid have a good performance in terms of the inhibition property. As shown in Figure 4, if the current tested drilling fluid has a risk of causing wellbore instability, adjustments of drilling fluid density and inhibition property are both available choices.

3. Application and Analysis

The problem of downhole fall-blocks (as shown in Figure 5) and pipe-stuck accidents are frequent in the JDK mudstone formation of the K block (depth: about 5200–6800 m). Therefore, the application of the proposed evaluation method in this study is focused on the JDK mudstone in the K block.

Figure 5. Returned fall-blocks of JDK mudstone in the K block (western China).

3.1. Detailed Design and Preparation for Evaluation

According to the above evaluation procedure, the main work consists of three parts: experiment on the hydration characterization of the JDK mudstone, establishment of the evaluation chart for JDK mudstone in K block and inhibition property evaluations of drilling fluids taken from the engineering field. The detailed plan and preparation of the above work are described as follows.

(1) Hydration characterization of JDK mudstone

In order to more accurately describe the weakening characteristics of mechanical properties of JDK mudstone after hydration (strength deterioration model), two samples of JDK mudstone outcrops were prepared. The hydration times of mudstones were 0, 2, 4, 8, 12, 24 and 48 h, respectively. Based on the geological information of JDK mudstone formation in K block, the temperature and pressure of immersing mudstones in poly-potassium mud were 130 °C and 23 MPa, respectively. In addition, for the mechanical test, the sharp cutter with scratching depth of 0.16 mm and blunt cutter with scratching depth of 0.06–0.12 mm were used in this study.

(2) Establishment of the evaluation chart for JDK mudstone formation in the K block

The establishment of evaluation chart of drilling fluid inhibition property for JDK mudstone in K block is based on the calculation of the wellbore stability model considering the strength weakening characteristics of the JDK mudstone. Basic input parameters of the mechanical–chemical model are listed in Table 2. The mathematical calculations in this paper are conducted in MATLAB (The MathWorks, Inc., Natick, MA, USA).

(3) Case study: Inhibition evaluations of drilling fluids taken from engineering field

In this study, the inhibition performance of drilling fluids of well K 8-13 and well K 9-3 were estimated using the proposed method. In order to study the influence of the surrounding environments on evaluation results, the evaluation of the inhibition property of drilling fluids under both deep-well drilling conditions (130 °C and 23 MPa) and ambient conditions (50 °C and normal pressure) are conducted. In addition, the comparisons between laboratory evaluation results and field data are designed to confirm whether the developed method is feasible and valid.

Table 2. Input parameters for calculation.

Parameters	Unit	Value	Parameters	Unit	Value
Mudstone porosity	-	6%	Drilling fluid density	g/cm^3	Variable
Overburden stress	MPa	132	Azimuth angle of borehole	°	N150E
Maximum horizontal principle stress	MPa	180	Inclination angle of borehole	°	2
Minimum horizontal principle stress	MPa	108	Azimuth angle of minimum horizontal stress	°	N100E
Initial cohesive strength	MPa	13	Initial Poisson's ratio	-	0.23
Initial internal friction angle	°	31	Effective stress coefficient	-	0.8

3.2. Evaluation Chart of Drilling Fluid Inhibition Index for JDK Mudstone Formation

3.2.1. Hydration Characterization of Mudstone in JDK Formation

Two samples of JDK mudstones with different hydration levels under deep-well drilling conditions are shown in Figure 6. Because continuous scratch test is a kind of nondestructive method, the mechanical parameters of mudstone after hydration can be estimated using the same sample. Compared to the traditional compressive test (destructive test), it can avoid the interference of experimental sample and reveal the real hydration characteristics of deep mudstone.

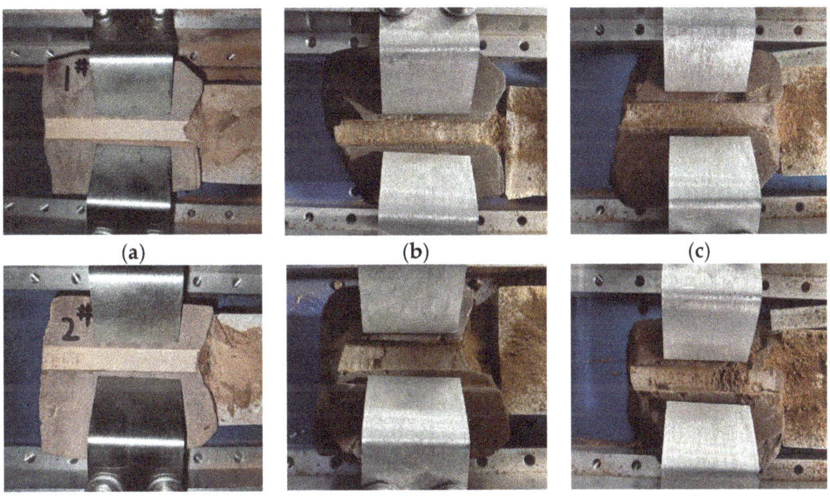

(a) (b) (c)

(d) (e) (f)

Figure 6. Photos of two samples of JDK mudstones after hydration. (**a**) 1#—After 0 h, (**b**) 1#—After 4 h, (**c**) 1#—After 24 h, (**d**) 2#—After 0 h, (**e**) 2#—After 4 h, (**f**) 2#—After 24 h.

The hydration characteristics of JDK mudstones under deep-well drilling conditions are presented in Figure 7. The uptake water contents of JDK mudstones after hydration increase with soaking time and reach the maximum value (0.045–0.06) when the soaking time is greater than 24 h. The changes in strength parameters of JDK mudstones after hydration have similar trends, which include a decrease in the soaking time until fully hydration, including uniaxial compressive strength (*UCS*), internal friction angle and cohesive strength. Therefore, the hydration time of 24 h is adopted in the following inhibition evaluation experiments of tested fluids.

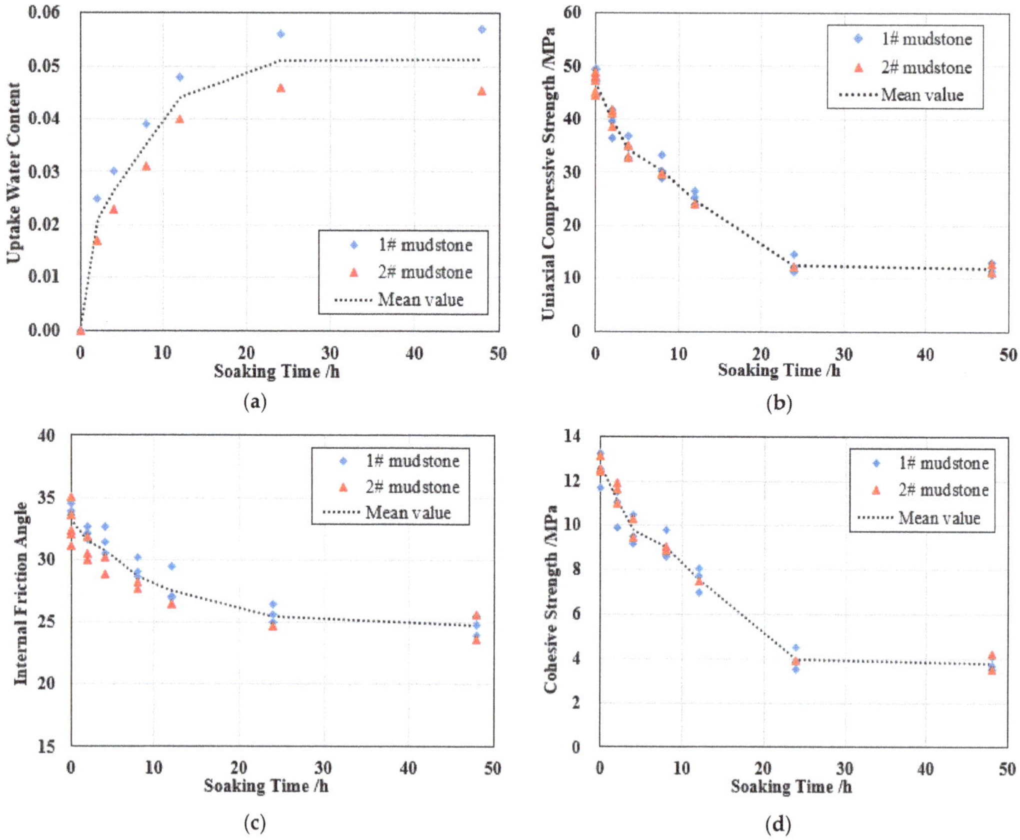

Figure 7. Hydration characteristics of JDK mudstones. (**a**) Uptake water content, (**b**) Uniaxial compressive strength, (**c**) Internal friction angle, (**d**) Cohesive strength.

Based on the above experimental results, the relationships between the shear strength properties and uptake water contents of JDK mudstones are described in Figure 8. The deterioration models of mechanical properties of JDK mudstones after hydration can be concluded as per Equation (6).

$$\begin{aligned} C_c &= C_i - 154.67 \times (w_c - w_i) \\ \varphi_c &= \varphi_i - 178.92 \times (w_c - w_i) \end{aligned} \tag{6}$$

where C_c, φ_c are current cohesive strength and internal friction angle of mudstone with a certain uptake water content, respectively, MPa; w_c is the current uptake water content of mudstone after hydration.

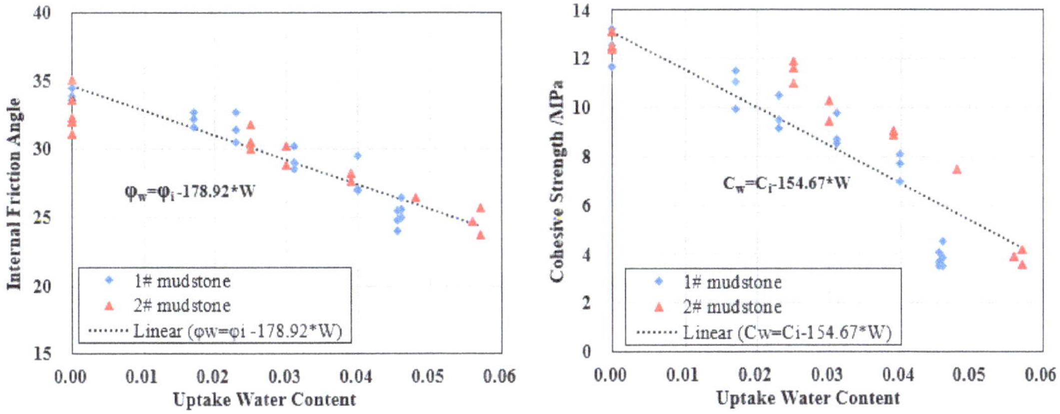

(**a**) (**b**)

Figure 8. The weakening equations of mechanical properties of JDK mudstones after hydration. (**a**) Internal friction angle, (**b**) Cohesive strength.

3.2.2. Quantitative Correlation Chart of Drilling Fluid Density and Required Inhibition Index for JDK Mudstone Formation

According to the wellbore stability model updated by the strength deterioration model of JDK mudstone after hydration (as described in Equation (6)) and basic parameters listed in Table 2, the wellbore collapse regions of JDK mudstones with different hydration levels (the poorer the inhibition performance of the drilling fluid is, the higher hydration level of mudstone is and the more serious the deterioration of the mudstone cohesive strength is) can be displayed visually. For instance, the borehole enlargement distributions of JDK mudstones at different hydration levels with a given drilling fluid density (1.80 g/cm^3) are shown in Figure 9. In the case of constant drilling fluid density, the wellbore collapse region (red area) enlarges gradually with the aggravation of JDK mudstone hydration caused by the poorer mud inhibition property. Based on the wellbore collapse contour map, the maximum borehole enlargement rates of JDK mudstones for different types of drilling fluid (i.e., drilling fluid density and inhibition property) can be directly revealed.

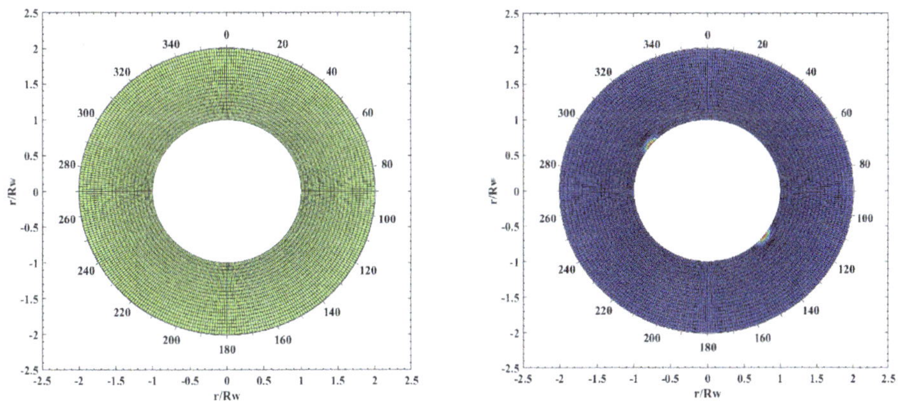

(**a**) (**b**)

Figure 9. *Cont.*

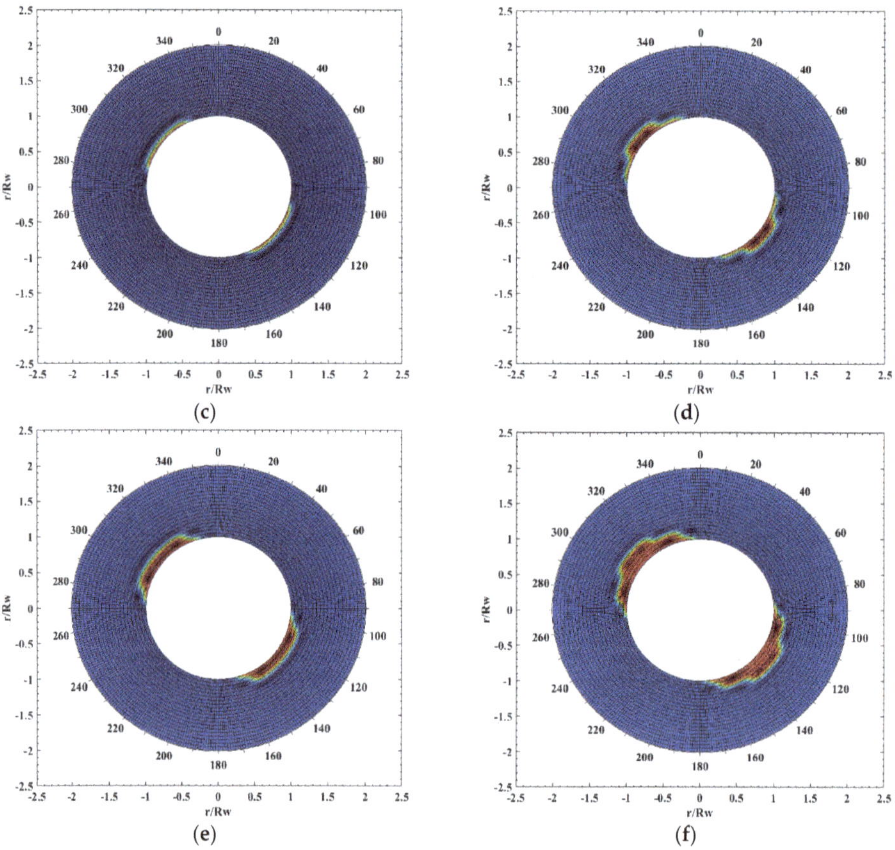

Figure 9. Wellbore collapse regions of JDK mudstones at different hydration levels with drilling fluid density of 1.80 g/cm^3 (**a–f**), the hydration level of mudstone continues to gradually increase; graph (**a**) represents the mudstone without hydration; red areas denote wellbore collapse region.

Figure 10 gives the quantitative cross-correlation of drilling fluid density and required inhibition index (cohesive strength of mudstone after immersed in tested drilling fluid) for sustaining different borehole enlargement rates in JDK mudstone. In this chart, only the data of five borehole enlargement rates, which are 0%, 5%, 10%, 15% and 20%, respectively, are presented. Based on the correlation chart, once the permissible borehole enlargement rates and tested drilling fluid densities are known, the required inhibition index of tested drilling fluid can be obtained.

3.3. Case 1: Evaluation of Drilling Fluid in Well K 8-13 for JDK Mudstone

The density of drilling fluid sample which is taken from well K 8-13 is 1.84 g/cm^3. As shown in Figure 10, the blue dashed line, which represents the mud density of well K 8-13, is marked and intersects with the solid lines with asterisk markers, which represent the data of different borehole enlargement rates. Because the permissible borehole enlargement rate in drilling engineering is 15%, the required inhibition index of the current drilling fluid sample for JDK mudstone is 4.76 MPa, as indicated by the blue solid line in Figure 10.

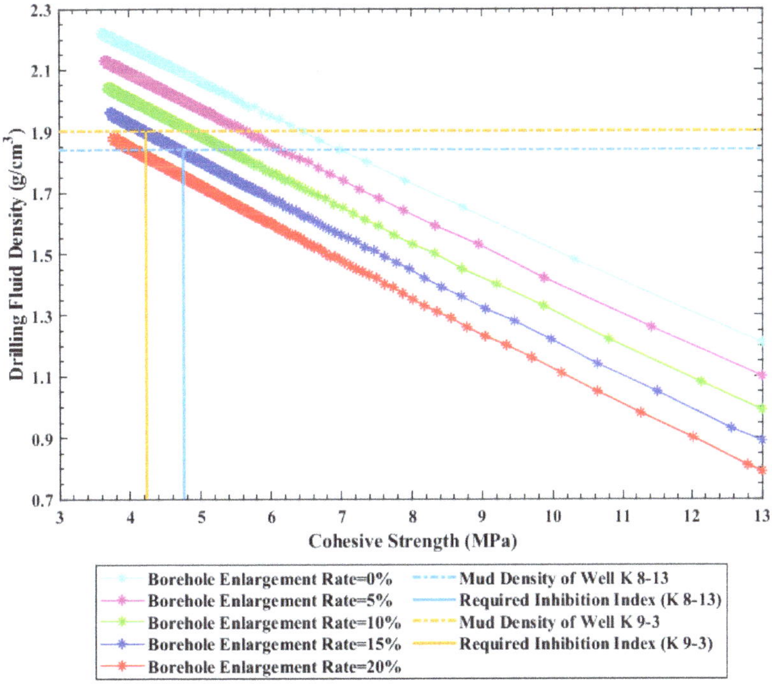

Figure 10. The cross-correlation chart of drilling fluid density and matching inhibition index for JDK mudstone.

3.3.1. Inhibition Evaluations under Deep-Well Drilling Conditions and Ambient Conditions

Figures 11 and 12 show the JDK mudstone samples used in inhibition-property evaluations of drilling fluid under real well drilling conditions and ambient conditions, respectively. From the qualitative comparison between the surface flatness of mudstones immersed in same drilling fluid under different conditions, it can be found that the damage to mudstone immersed in mud under deep-well drilling conditions is more serious. In other words, the inhibition property of drilling fluid under deep-well drilling conditions is worse.

Figure 13 illustrates the quantitative evaluation results regarding the inhibition property of drilling fluid under different conditions. The blue column denotes the initial cohesive strength of mudstone and the red column denotes cohesive strength of mudstone after hydration, which represents the inhibition index of the tested drilling fluid. The data of $1^{\#}$, $2^{\#}$ and $3^{\#}$ show that the inhibition property of the tested drilling fluid under deep-well drilling conditions cannot meet the required inhibition index as marked by the pink dashed line (i.e., the average value of the current inhibition index is about 3.88 MPa and less than 4.76 MPa). However, the results of $4^{\#}$ and $5^{\#}$ indicate that the current inhibition performance of the drilling fluid under ambient conditions is sufficient. Based on the contradictory results above, it can be inferred that considering the natural surrounding environments in inhibition evaluation of drilling fluid is very meaningful and necessary.

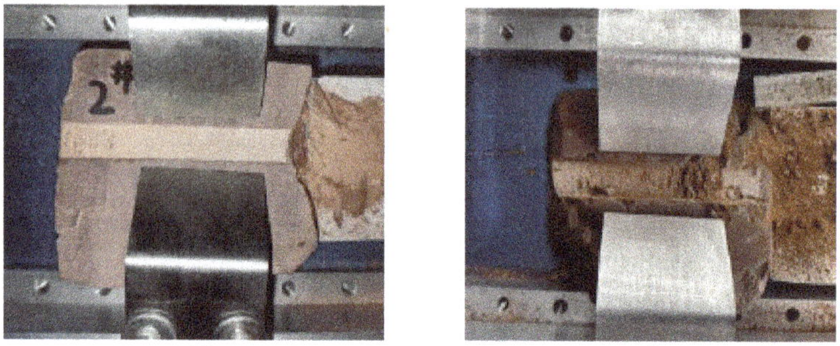

(**a**) (**b**)

Figure 11. JDK mudstone used in inhibition evaluation of drilling fluid in well K 8-13 under deep-well drilling conditions. (**a**) Before hydration, (**b**) After hydration.

(**a**) (**b**)

Figure 12. JDK mudstone used in inhibition evaluation of drilling fluid in well K 8-13 under ambient conditions. (**a**) Before hydration, (**b**) After hydration.

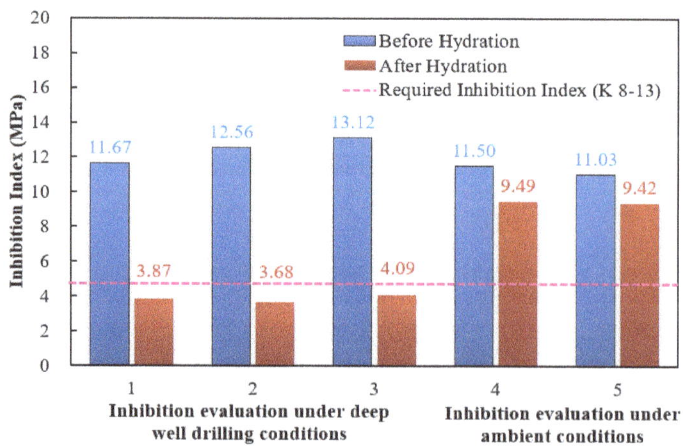

Figure 13. Inhibition evaluation results of drilling fluid in well K 8-13 under different hydration conditions (1#, 2# and 3# are the inhibition evaluation results of drilling fluid under deep-well drilling conditions; 4# and 5# are the inhibition evaluation results of drilling fluid under ambient conditions).

3.3.2. Field Validation

As shown in Figure 14, when the JDK mudstone formation is drilled by using lower mud density (1.84–1.90 g/cm^3), the complex accidents such as stuck pipe and caving are frequent and the borehole enlargement rate always stays at a high level. This indicates that the property of the current drilling fluid (e.g., mud density and inhibition index) cannot sustain wellbore stability of the JDK mudstone. In order to control the further wellbore collapse of the JDK mudstone and prevent more serious and complex accidents, the drilling engineers adopt the widely effective method of increasing mud density. When the mud density increases to 1.92 g/cm^3, the borehole enlargement rate of JDK mudstone is under 15%.

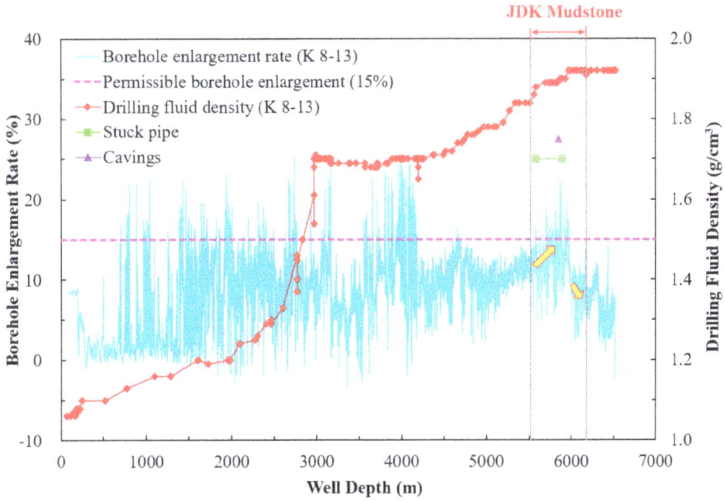

Figure 14. Drilling history of well K 8-13 (the orange area denotes the JDK mudstone formation).

In reality, this adopted solution in the engineering field is consistent with the guidance implied by the quantitative cross-correlation chart above. As described in Figure 4, except for increasing inhibition performance of the drilling fluid, adjustment of drilling fluid density is also an available option for controlling wellbore collapse. From Figure 10, it can be inferred that the density of drilling fluid needs to be increased to about 1.92–1.93 g/cm^3 if the current inhibition index of drilling fluid is not altered. Therefore, in some ways, the proposed inhibition evaluation method of drilling fluid in this paper is validated.

3.4. Case 2: Evaluation of Drilling Fluid in Well K 9-3 for JDK Mudstone

The density of the drilling fluid sample taken from well K 9-3 was 1.90 g/cm^3. As presented in Figure 10, the yellow dashed line means the mud density of well K 9-3 has an intersection point with the blue solid line marked by asterisks (the data of borehole enlargement rate = 15%). The value of abscissa of the intersection point is 4.23 MPa (i.e., the required inhibition index of drilling fluid sample) as marked by the yellow solid line in Figure 10.

3.4.1. Inhibition Evaluation under Deep-Well Drilling Conditions

Figure 15 is the JDK mudstone sample used in inhibition evaluation of drilling fluid in well K 9-3 under deep-well drilling conditions. The tested drilling fluid of well K 9-3 may have a good performance in the inhibition property from the point of view of integrity of the mudstone after hydration. Additionally, the quantitative result presented in Figure 16 validates this. The current inhibition index of the tested drilling fluid taken from well K 9-3 is about 5.86 MPa and exceeds the required inhibition index.

Figure 15. JDK mudstone used in inhibition evaluation of drilling fluid in well K 9-3 under deep-well drilling conditions. (**a**) Before hydration, (**b**) After hydration.

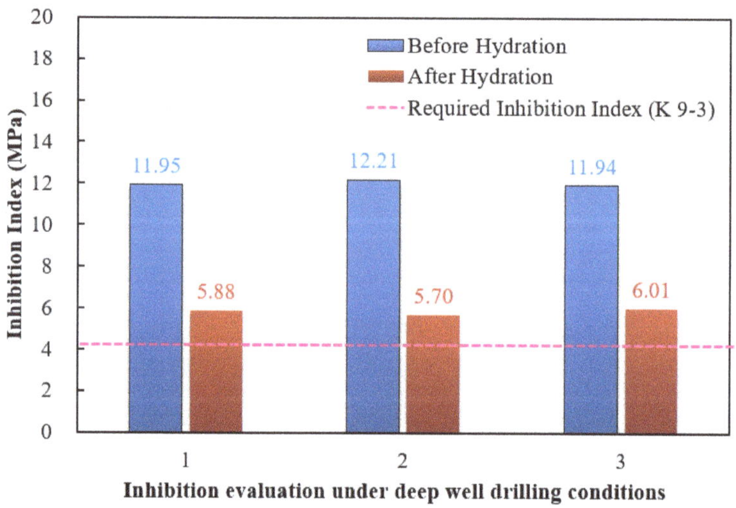

Figure 16. Inhibition evaluation result of drilling fluid in well K 9-3 under deep-well drilling conditions.

3.4.2. Field Validation

From Figure 17, it can be seen that the mud density used in drilling JDK mudstone formation is 1.88–1.90 g/cm^3 and the average borehole enlargement rate is less than 10%. Especially for the well drilling process with mud density of 1.90 g/cm^3, the borehole enlargement rate of JDK mudstone decreases gradually and stays within 5%. The above field data illustrate that the current inhibition index of the tested drilling fluid sample in well K 9-3 can meet the requirement of sustaining wellbore stability of JDK mudstone. This is consistent with our inhibition evaluation results provided by the proposed method.

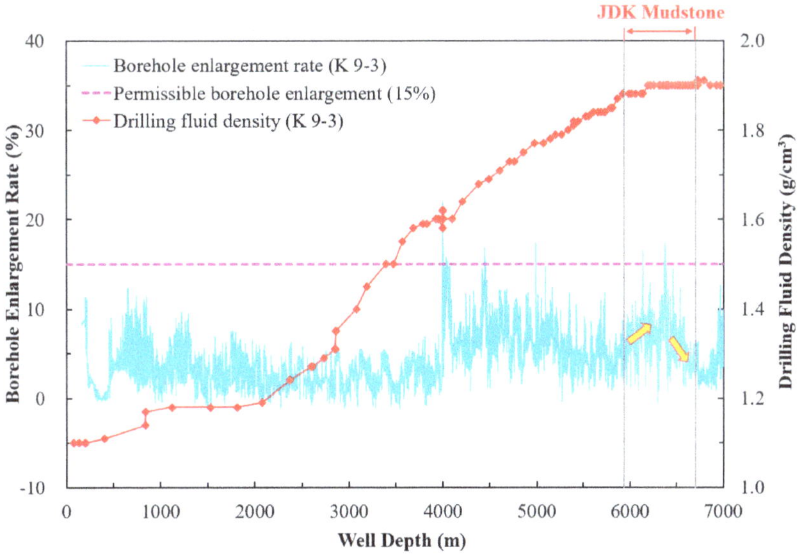

Figure 17. Drilling history of well K 9-3 (the orange area denotes the JDK mudstone formation).

4. Conclusions

(1) In order to achieve the quantitative evaluation of the inhibition property of drilling fluid under deep-well drilling conditions, a new evaluation method that adopts cohesive strength of mudstone after hydration as the inhibition index of drilling fluid is proposed in this study. In this method, a newly designed HPHT hydration experiment apparatus (maximum surrounding temperature of 200 °C, maximum surrounding pressure of 60 MPa and flushing action of fluid) is developed to provide the real hydration environments in the process of drilling deep well. Additionally, a method of estimating cohesive strength of mudstone after hydration based on a scratch test is established to obtain the inhibition index of the tested drilling fluid. In addition, a cross-correlation between drilling fluid density and the matching inhibition index for deep mudstone, which is revealed by the mechanical–chemical wellbore stability model considering hydration characteristic of deep mudstone, is constructed as the quantitative evaluation criterion.

(2) The comparison of inhibition evaluation results of drilling fluids under ambient conditions and natural well-drilling conditions illustrates that ignoring the natural surrounding environments in the process of inhibition evaluation of drilling fluid may cause overestimation of inhibition properties. Therefore, it is very necessary and meaningful to propose a new inhibition evaluation method of drilling fluid under real well-drilling conditions for wellbore stability in deep mudstone.

(3) The evaluation results of drilling fluids taken from the well K 8-13 and well K 9-3, which are provided by the proposed evaluation method, are in line with the field data. In some ways, the developed quantitative evaluation method of inhibition property of drilling fluid for deep mudstone is validated.

(4) The developed quantitative method can consider the coupling actions of mechanical and chemical actions of drilling fluid and provide strong support for field engineers to sustain wellbore stability through adjustments of drilling fluid density and inhibition properties.

(5) In order to achieve the widespread application of the proposed method in drilling engineering field, our research group is trying to make a miniaturized experimental evaluation platform and standardize the evaluation procedure.

Author Contributions: Conceptualization, Y.J., Y.L. and K.B.; methodology, K.B.; software, K.B.; validation, K.B., H.L. and J.Z.; writing—original draft preparation, K.B.; writing—review and editing, K.B., Y.J. and Y.L.; visualization, K.B.; supervision, Y.J. and Y.L. All authors have read and agreed to the published version of the manuscript.

Funding: This research was funded by National Natural Science Foundation of China (General Program: No. 52074314) and National Key Research and Development Program of China (2019YFA0708303-05).

Institutional Review Board Statement: Not applicable.

Informed Consent Statement: Not applicable.

Data Availability Statement: The data presented in this study are available on request from the corresponding author.

Conflicts of Interest: The authors declare no conflict of interest.

References

1. Aston, M.S.; Elliot, G.P. Water-based glycol drilling muds: Shale inhibition mechanisms. In Proceedings of the European Petroleum Conference, London, UK, 25–27 October 1994.
2. Patel, A.; Stamatakis, E.; Young, S.; Cliffe, S. Designing for the future—A review of the design, development and testing of a novel, inhibitive water-based drilling fluid. In Proceedings of the AADE 2002 Technical Conference, Houston, TX, USA, 2–3 April 2002.
3. Ghasemi, M.; Moslemizadeh, A.; Shahbazi, K.; Mohammadzadeh, O.; Zendehboudi, S.; Jafari, S. Primary evaluation of a natural surfactant for inhibiting clay swelling. *J. Pet. Sci. Eng.* **2019**, *178*, 878–891. [CrossRef]
4. Qiu, Z.; Zhong, H.; Huang, W.; Wan, J.; Xu, J. A new method for evaluation the properties of high-performance drilling fluids. *Drill. Fluid Completion Fluid* **2009**, *26*, 58–59.
5. Al-Arfaj, M.; Amanullah, M.; Sultan, A. Chemical and mechanical aspects of wellbore stability in shale formations: A literature review. In Proceedings of the Abu Dhabi International Petroleum Exhibition and Conference, Abu Dhabi, United Arab Emirates, 10–13 November 2014.
6. Stephens, M.; Gomez, S.; Churan, M. Laboratory methods to assess shale reactivity with drilling fluids. In Proceedings of the 2009 National Technical Conference and Exhibition, New Orleans, LA, USA, 12–14 April 2009.
7. Salles, F.; Bildstein, O.; Douillard, J.M.; Jullien, M.; Raynal, J.; Van, D.H. On the cation dependence of interlamellar and interparticular water and swelling in smectite clays. *Langmuir* **2010**, *26*, 5028–5037. [CrossRef]
8. Huang, W.A.; Li, X.; Qiu, Z.S.; Jia, J.H.; Wang, Y.; Li, X. Inhibiting the surface hydration of shale formation using preferred surfactant compound of polyamine and twelve alkyl two hydroxyethyl amine oxide for drilling. *J. Pet. Sci. Eng.* **2017**, *159*, 791–798. [CrossRef]
9. Patel, A.; Stamatakis, E.; Young, S.; Friedheim, J. Advances in inhibitive water-based drilling fluids—Can they replace oil-based muds? In Proceedings of the 2007 SPE International Symposium on Oilfield Chemistry, Houston, TX, USA, 28 February–2 March 2007.
10. Zhong, H.Y.; Qiu, Z.S.; Sun, D.; Zhang, D.M.; Huang, W.A. Inhibitive properties comparison of different polyetheramines in water-based drilling fluid. *J. Pet. Sci. Eng.* **2015**, *26*, 99–107. [CrossRef]
11. Chen, G.; Yan, J.; Li, L.; Zhang, J.; Gu, X.; Song, H. Preparation and performance of amine-tartaric salt as potential clay swelling inhibitor. *Appl. Clay Sci.* **2017**, *138*, 12–16. [CrossRef]
12. Hou, J.; Liu, Y.G.; Song, G.S.; Song, T.; Yan, J.; Zhao, X.Z. Synthesis and application of a new high temperature high performance salt resistant shale inhibitor. *Drill. Fluid Completion Fluid* **2016**, *33*, 22–27.
13. Dobson, J.W.; Harrison, J.C.; Hale, A.H.; Bernardi, L.A.; Kielty, J.M.; Albrecht, M.S.; Bruner, S.D. Laboratory development and field application of a novel water-based drill-in fluid for geopressured horizontal wells. In Proceedings of the SPE Technology Conference and Exhibition, Denver, CO, USA, 6–9 October 1998.
14. National Energy Administration. *Evaluating Method of Shale Inhibition for Drilling Fluids (Energy Sector Standard of the People's Republic of China): NB/T 10121-2018*; Petroleum Industry Press: Beijing, China, 2018.
15. Zhong, H.Y.; Qiu, Z.S.; Huang, W.A.; Cao, J. Shale inhibitive properties of polyether diamine in water-based drilling fluid. *J. Pet. Sci. Eng.* **2011**, *78*, 510–515. [CrossRef]
16. Li, H.K.; Tian, R.J.; Luo, J.S.; Zhang, X.L.; Liu, G.; Wang, N. New method for quantitative evaluation of amine inhibitor used in drilling fluid. *Drill. Fluid Completion Fluid* **2012**, *29*, 51–53.
17. Yin, Z.W.; Geng, D.S.; Wang, X.N. Evaluation method on inhibiting mud shale hydration and swelling in mud filtrate. *Drill. Fluid Completion Fluid* **2013**, *30*, 44–47.
18. Zhong, H.Y.; Sun, D.; Qiu, Z.S.; Huang, W.A.; Liu, Y.F. Evaluation and analysis on the inhibitive capacity of a new naphthenic amine. *Drill. Fluid Completion Fluid* **2014**, *31*, 1–5.
19. Shadizadeh, S.; Moslemizadeh, A.; Dezaki, A.S. A novel nonionic surfactant for inhibiting shale hydration. *Appl. Clay Sci.* **2015**, *118*, 74–86. [CrossRef]
20. Fan, Z.X.; Gao, J.P.; Guo, D.R. Comprehensive evaluation of mud shale hydration. *Pet. Drill. Tech.* **1999**, *27*, 26–27.

21. Luo, C.Z.; Zeng, F.D.; Xiang, X.J.; Xie, Q.P.; Lai, Y.L. Laboratory study on inhibitive evaluation method of drilling fluid and its treating agent. *Pet. Drill. Tech.* **2000**, *28*, 35–36.
22. Huang, L.J.; Luo, X.S. Study on the testing method of polymer inhibit evaluation by capillary suction time tester. *Drill. Fluid Completion Fluid* **1995**, *12*, 44–47.
23. Luo, Z.H.; Wang, L.X.; Yu, P.Z.; Chen, Z.X. Experimental study on the application of an ionic liquid as a shale inhibitor and inhibitive mechanism. *Appl. Clay Sci.* **2017**, *150*, 267–274. [CrossRef]
24. Gholami, R.; Elochukwu, H.; Fakhari, N.; Sarmadivaleh, M. A review on borehole instability in active shale formations: Interactions, mechanisms and inhibitors. *Earth-Sci. Rev.* **2018**, *177*, 2–13. [CrossRef]
25. Zhong, H.Y.; Huang, W.A.; Qiu, Z.S.; Wang, M.M.; Liu, Y.Q.; Jin, H.F.; Liu, C.Z. Experimental study of inhibition comparison between polyamine and formats. *Fault-Block Oil Gas Field* **2012**, *19*, 508–512.
26. Lu, Y.; Chen, M.; Jin, Y.; Teng, X.; Wu, W.; Liu, X. Experimental study of strength properties of deep mudstone under drilling fluid soaking. *Chin. J. Rock Mech. Eng.* **2012**, *31*, 1400–1405.
27. Mohamed, A.K.; Benaafi, M.; Elkatany, S.M.; Bageri, B.S. Effect of high-density water-based drilling fluid on the mechanical properties of the drilled formation in horizontal wells. In Proceedings of the 53rd US Rock Mechanics/Geomechanics Symposium, New York, NY, USA, 23–26 June 2019.
28. Gamal, H.; Elkatany, S.; Adebayo, A.; Bageri, B. Effect of exposure time on the compressive strength and formation damage of sandstone while drilling horizontal wells. *J. Pet. Sci. Eng.* **2020**, *195*, 107590. [CrossRef]
29. Taherdangkoo, R.; Tatomir, A.; Anighoro, T.; Sauter, M. Modeling fate and transport of hydraulic fracturing fluid in the presence of abandoned wells. *J. Contam. Hydrol.* **2019**, *221*, 58–68. [CrossRef] [PubMed]
30. Bo, K.H.; Jin, Y.; Lu, Y.H.; Wang, K.; Zhu, J.Z. An approach for studying effect of hydration under deep well drilling conditions on strength properties of mudstone. In Proceedings of the 55th US Rock Mechanics/Geomechanics Symposium, Houston, TX, USA, 20–23 June 2021.
31. Detournay, E.; Defourny, P. A phenomenological model for the drilling action of drag bits. *Int. J. Rock Mech. Min. Sci. Geomech. Abstr.* **1992**, *29*, 13–23. [CrossRef]
32. Richard, T.; Dagrain, F.; Poyol, E.; Detournay, E. Rock strength determination from scratch tests. *Eng. Geol.* **2012**, *147*, 91–100. [CrossRef]
33. Rostamsowlat, I.; Richard, T.; Evans, B. An experimental study of the effect of back rake angle in rock cutting. *Int. J. Rock Mech. Min. Sci.* **2018**, *107*, 224–232. [CrossRef]
34. Bradley, W.B. Failure of inclined boreholes. *J. Energy Resour. Technol.* **1979**, *101*, 232–239. [CrossRef]
35. Chen, M.; Jin, Y.; Zhang, G.Q. *Petroleum Related Rock Mechanics*; Science Press: Beijing, China, 2008; pp. 58–61.
36. Jin, Y.; Yuan, J.; Hou, B.; Chen, M.; Lu, Y.; Li, S.; Zou, Z. Analysis of the vertical borehole stability in anisotropic rock formations. *J. Pet. Explor. Prod. Technol.* **2012**, *2*, 197–207. [CrossRef]
37. Liu, M.; Jin, Y.; Lu, Y.; Chen, M.; Wen, X. Oil-based critical mud weight window analyses in HTHP fractured tight formation. *J. Pet. Sci. Eng.* **2015**, *135*, 750–764. [CrossRef]
38. Yew, C.H.; Chenevert, M.E.; Wang, C.L.; Osisanya, S.O. Wellbore stress distribution produced by moisture adsorption. *SPE Drill. Eng.* **1990**, *5*, 311–316. [CrossRef]
39. Huang, R.Z.; Chen, M.; Deng, J.G.; Wang, K.P.; Chen, Z.X. Study on shale stability of wellbore by mechanics coupling with chemistry method. *Drill. Fluid Completion Fluid* **1995**, *12*, 15–21.

energies

MDPI

Article

Eco-Friendly Drilling Fluid: Calcium Chloride-Based Natural Deep Eutectic Solvent (NADES) as an All-Rounder Additive

Muhammad Hammad Rasool [1,*], Maqsood Ahmad [1,*], Numair Ahmed Siddiqui [1,*] and Aisha Zahid Junejo [2]

1 Department of Petroleum Geosciences, Universiti Teknologi Petronas, Seri Iskander 31750, Malaysia
2 Department of Component Safety, Bundesanstalt für Materialforschung und Prüfung, 12205 Berlin, Germany; aisha-zahid.junejo@bam.de
* Correspondence: muhammad_19000949@utp.edu.my (M.H.R.); maqsood.ahmad@utp.edu.my (M.A.); numair.siddiqui@utp.edu.my (N.A.S.)

Abstract: Designing an effective drilling mud is a critical aspect of the drilling process. A well-designed drilling mud should not only provide efficient mud hydraulics but also fulfill three important functions: enhancing mud rheology, inhibiting hydrate formation in deepwater drilling, and suppressing shale swelling when drilling through shale formations. Achieving these functions often requires the use of various additives, but these additives are often expensive, non-biodegradable, and have significant environmental impacts. To address these concerns, researchers have explored the potential applications of ionic liquids and deep eutectic solvents in drilling mud design, which have shown promising results. However, an even more environmentally friendly alternative has emerged in the form of natural deep eutectic solvents (NADES). This research focuses on an in-house-prepared NADES based on calcium chloride and glycerine, with a ratio of 1:4, prepared at 60 °C, and utilizes it as a drilling mud additive following the API 13 B-1 standards and checks its candidacy as a rheology modifier, hydrates, and shale inhibitor. The findings of the study demonstrate that the NADES-based mud significantly improves the overall yield point to plastic viscosity ratio (YP/PV) of the mud, provides good gel strength, and inhibits hydrate formation by up to 80%. Additionally, it has shown an impressive 62.8% inhibition of shale swelling while allowing for 84.1% improved shale recovery. Moreover, the NADES-based mud exhibits a 28% and 25% reduction in mud filtrate and mud cake thickness, respectively, which is further supported by the results of XRD, zeta potential, and surface tension. Based on these positive outcomes, the calcium chloride–glycerine NADES-based mud is recommended as a versatile drilling mud additive suitable for various industrial applications. Furthermore, it presents a more environmentally friendly option compared to traditional additives, addressing concerns about cost, biodegradability, and environmental impact in the drilling process for an ultimate global impact.

Keywords: NADES; shale swelling; drilling fluid; mud rheology; hydrate inhibition

Citation: Rasool, M.H.; Ahmad, M.; Siddiqui, N.A.; Junejo, A.Z. Eco-Friendly Drilling Fluid: Calcium Chloride-Based Natural Deep Eutectic Solvent (NADES) as an All-Rounder Additive. *Energies* **2023**, *16*, 5533. https://doi.org/10.3390/en16145533

Academic Editor: Hossein Hamidi

Received: 6 June 2023
Revised: 19 June 2023
Accepted: 24 June 2023
Published: 21 July 2023

1. Introduction

Drilling mud, being a complex fluid system, often requires the incorporation of additives to enhance its performance and address specific challenges encountered during drilling operations [1]. These additives in general serve three major purposes: improving rheological properties, inhibiting hydrate formation during drilling, and suppressing shale swelling while drilling through shale rock [2]. However, it is important to consider the potential disadvantages associated with traditional additives, as well as explore alternative options that offer promising solutions. One of the major concerns with traditional drilling mud additives is their environmental impact [3]. Certain additives, such as synthetic polymers, can be detrimental to the environment. They may exhibit toxicity, lack biodegradability, or pose challenges in terms of treatment and disposal [4]. This raises concerns about potential contamination of soil and water resources, emphasizing the need for environmentally friendly alternatives. Another factor to consider is the cost of traditional additives. Some additives used in drilling mud formulations can be expensive, which can significantly impact drilling operations' overall cost [5]. This cost aspect necessitates exploring cost-effective alternatives that can maintain or improve drilling mud performance without straining the budget.

To address these concerns, researchers and industry professionals have been exploring the potential of innovative additives such as ionic liquids and deep eutectic solvents [6]. Ionic liquids are molten salts with unique properties, including low volatility, high thermal stability, and tunable chemical characteristics. These properties make them attractive candidates for various applications, including drilling mud formulations. Deep eutectic solvents, on the other hand, are a type of liquid mixture formed by combining a hydrogen bond donor and a hydrogen bond acceptor. These solvents exhibit characteristics such as low volatility, low toxicity, and biodegradability. They can be tailored to possess desired properties, making them a promising alternative to traditional drilling mud additives.

Various studies have demonstrated the efficacy of ionic liquids as shale inhibitors. In a study by Rahman et al. (2022), the use of tetramethylammonium chloride and 1-ethyl-3-methylimidazolium chloride in drilling mud resulted in a reduction of 23.40% and 15.66% in linear swelling, respectively [7]. Similarly, Rizwan et al. (2021) found that employing a trihexyltetradecyl phosphonium bis(2,4,4-trimethyl pentyl) phosphonate-based ionic liquid led to a 12.3% reduction in shale swelling compared to water [8,9]. Another study conducted by Huuang et al. (2020) revealed that the utilization of ionic liquids such as 1-hexyl-3-methylimidazolium bromide and 1,2-bis(3-hexylimidazolium1-yl) ethane bromide with Na-Bt pellets resulted in significant reductions of 86.43% and 94.17% in shale swelling, respectively [10]. Yang et al. (2019) employed 1-Vinyl-3-dodecylimidazolium bromide and 1-Vinyl-3-tetradecylimidazolium bromide, leading to reductions of 16.91% and 5.81% in shale swelling, respectively [11]. Ofei et al. (2017) utilized 1-butyl-3-methylimidazolium chloride (BMIM-Cl) in a water-based mud, observing up to a 50% decrease in mud cake thickness and a reduction in yield point/plastic viscosity (YP/PV) [12]. Furthermore, a study conducted by Yang et al. (2017) found that the use of 1-Vinyl-3-ethylimidazolium bromide resulted in a 31.62% reduction in shale swelling, accompanied by a shale recovery rate of 40.60%, as reported by Luo, Wang et al. (2017) [13]. In a study by Luo et al. (2017), the utilization of 1-octyl-3-methylimidazolium tetraflouroboreate resulted in an impressive 80% reduction in shale swelling [14].

In their study, Jia et al. (2019) investigated the use of deep eutectic solvents (DES) as shale inhibitors and observed significant inhibition of bentonite swelling. Different DES compositions, namely propanoic acid ChCl (1:1), 3-phenyl propanoic acid ChCl (1:2), and 3-mercapto propanoic acid + Itaconic acid + ChCl (1:1:2), demonstrated inhibition rates of 68%, 58%, and 58%, respectively [15]. Rasool et al. (2021) employed a glycerine/potassium carbonate DES (2:1) for swelling shale samples, achieving an impressive 87% inhibition of swelling [16,17]. Furthermore, Rasool et al. (2022) employed a combination of a potassium carbonate-based DES and poly (2-ethyl-2-Oxazoline) hydroxyl-terminated polymer as a double-action inhibitor in drilling mud, resulting in a remarkable 76% inhibition of swelling [18].

However, it is essential to acknowledge that even innovative additives such as ionic liquids and deep eutectic solvents have their own set of challenges [19]. Ionic liquids can sometimes have higher costs due to the complexity of their synthesis or purification processes. Additionally, the selection of the appropriate ionic liquid for a specific drilling mud application requires careful consideration of factors such as compatibility, stability, and environmental impact [20–22]. Similarly, the utilization of deep eutectic solvents as drilling mud additives may require further research and development and they are not green in a true sense as most of them utilize ammonium-based salts, which are not environmentally friendly [23]. The identification of suitable combinations of hydrogen bond donors and acceptors, as well as an in-depth understanding of their compatibility with other drilling mud components, is necessary. Additionally, a thorough evaluation of the performance and environmental impact of deep eutectic solvents is crucial before their widespread adoption.

In recent years, the potential of natural deep eutectic solvents (NADES) has emerged as an intriguing avenue for drilling mud additives and many other related applications in petroleum engineering [24]. NADES are derived from natural compounds, offering advantages such as biodegradability, lower production costs, and reduced environmental impact. The use of NADES as drilling mud additives holds promise for both enhanced performance and environmental sustainability. Utilizing natural deep eutectic solvents (NADES) as additives in drilling mud offers a multifaceted approach to enhance mud rheology, inhibit hydrate formation, and suppress shale swelling. NADES, derived from natural compounds, exert their effects through specific mechanisms.

For improving mud rheology, NADES act as rheology modifiers by adjusting viscosity, density, and gel strength [25,26]. The unique composition of NADES allows them to alter the flow properties of the drilling mud, enhancing its ability to carry and suspend solid particles, such as drill cuttings [6]. This optimization ensures the efficient removal of cuttings from the wellbore, reducing the risk of clogging and improving overall drilling performance. In terms of hydrate inhibition, NADES disrupt the conditions necessary for hydrate formation. The specific chemical properties of NADES, including their low freezing points and ability to interact with water molecules, hinder the formation of ice-like solids. By incorporating NADES into drilling mud, the risk of hydrate blockage is minimized, ensuring the uninterrupted flow of gas and preventing equipment damage [27]. NADES also exhibit shale inhibition properties by suppressing the swelling and disintegration of clay-rich shale formations. Through their chemical composition and interactions with shale particles, NADES reduce water absorption and minimize swelling. This stabilizes the shale formations and helps maintain wellbore integrity during drilling operations, mitigating the risks associated with wellbore instability and improving overall drilling efficiency [28].

By harnessing the unique mechanisms of NADES, drilling mud can benefit from improved mud rheology, enhanced hydrate inhibition, and effective suppression of shale swelling. The utilization of NADES as additives in drilling operations represents a promising approach for optimizing drilling performance while also emphasizing environmental sustainability through the use of natural compounds. Recently, an ascorbic acid-based NADES was utilized as a drilling mud additive for stabilization and had positive results [29]. NADES (natural deep eutectic solvents) offer a promising combination of biodegradability and cost-effectiveness, making them an attractive option in various applications. Their inherent biodegradability ensures that they can break down into environmentally benign components, minimizing potential ecological impacts. This characteristic aligns with sustainability goals and regulatory requirements. Additionally, NADES can be synthesized from readily available and economically viable components, making them a cost-effective alternative to traditional solvents [4,30]. Their sustainable nature and cost efficiency position NADES as a compelling choice for industries seeking environmentally friendly solutions without compromising on affordability.

The current research aims to explore the utilization of calcium chloride–glycerine-based natural deep eutectic solvents (NADES) as additives in mud. The motivation behind this study is to investigate the potential of these NADES in serving as effective rheology modifiers, hydrate inhibitors, and shale inhibitors. By incorporating these NADES into the mud, it is anticipated that their unique properties will contribute to enhancing the overall performance of the drilling mud. The rheological properties of the drilling mud can be optimized, ensuring efficient transport of drill cuttings and maintaining wellbore stability. Additionally, the potential of these NADES to inhibit hydrate formation will mitigate the risk of blockages and equipment damage, enabling a safe and uninterrupted flow of hydrocarbons. Furthermore, their ability to suppress shale swelling and improve shale recovery will also be evaluated. The current research will not only focus on the utilization of calcium chloride–glycerine-based natural deep eutectic solvents (NADES) as drilling mud additives but also involve comprehensive characterizations to provide a thorough understanding of their effects. Various characterization techniques, including X-ray diffraction (XRD), surface tension analysis, and zeta potential measurements, were used to elucidate the results and justify the underlying mechanism.

2. Materials and Methods

2.1. In-House Preparation of Calcium–Glycerine (CC:Gly) NADES Molar Ratios

The resulting mixture exhibited a transparent and uniform appearance, indicating the attainment of its eutectic composition. Experimental procedures involved the use of a METTLER Digital Balance for precise ingredient weighing and a Thermo Fisher Hot plate for controlled heating and stirring at a speed of 100 rpm. Additional characterization techniques, such as FTIR and NMR, will be conducted and the findings will be detailed in a subsequent publication.

Screening Criteria for Selecting HBD and HBA for NADES

The selection of hydrogen bond donor (HBD) and hydrogen bond acceptor (HBA) components for the natural deep eutectic solvents (NADES) screening criteria was based on prior research conducted by Rasool, Ahmad et al. (2022) [31]. This research demonstrated that components with a higher hydrogen bond count and polar functional groups exhibited greater effectiveness in shale inhibition.

2.2. Drilling Mud Preparation

The drilling mud was formulated in accordance with the API 13B-1 standard [32]. To prepare the mud, caustic soda (0.25 g) and sodium carbonate (0.25 g) were sourced from Sigma-Aldrich, Selangor, Malaysia. These ingredients, along with 22.5 g sodium bentonite (Na-Bt), were then combined with 350 mL of distilled water and thoroughly stirred in a multimixer for a duration of 40 min. Different concentrations of NADES (1%, 3%, and 5%) were added to the drilling fluid while mixing to optimize the hydrate inhibition and mud rheology results. The literature survey showed that various DES give optimum results at low concentrations, so we chose concentrations of 1–5% for this research work.

Drilling Mud Properties

Mud samples containing different concentrations of NADES were analyzed using a FANN Viscometer. Readings were recorded at speeds of 3 rpm, 6 rpm, 300 rpm, and 600 rpm, both before and after subjecting the samples to aging at temperatures of 100 °C and 150 °C. The aging process involved placing the mud samples in a rolling oven at a pressure of 1000 psia for a duration of 16 h. The obtained values were then used to calculate various parameters, including yield point (*YP*), plastic viscosity (*PV*), and 10 s and 10 min gel strength (G.S). The calculations were carried out according to API 13B-1 standards [32] and the corresponding Equations (1) and (2).

$$PV = \theta_{600} - \theta_{300} \tag{1}$$

$$YP = \theta_{300} - PV \tag{2}$$

The API 13B-1 guidelines [32] were followed to conduct an HPHT filtration test at a temperature of 150 °C and pressure of 400 psia. This test aimed to determine the amount of filtrate loss and the thickness of the mud cake in drilling mud samples. The testing equipment included a mud cell, which was pressurized using valve stems, and placed within a jacket that maintained a constant temperature through thermostatic control. The temperature was adjusted as required for the test. After pressurizing the cell, the filtrate volume and mud cake thickness were measured after a duration of 30 min. The obtained results are thoroughly analyzed in Section 3.

2.3. Shale Stabilization Studies

2.3.1. Bentonite Wafer Preparation

To investigate shale swelling, the research utilized bentonite wafers due to their composition containing the clay mineral smectite, which is responsible for shale swelling behavior. Acquiring authentic shale core samples is challenging since the coring process renders the shale unstable and often leads to the inclusion of other minerals such as sandstone and limestone. Conducting experiments on shale outcrops is not feasible as they typically lack smectite. In this study, Na-bentonite powder was compressed into pellets with a diameter of 2.54 cm and a weight of 11.5 g using a hydraulic press at a pressure of 1600 psi. The thickness of the pellets was measured prior to their placement in the linear swell meter (LSM) environment. Within the LSM, the pellets were submerged in drilling mud samples, including both the base sample and samples containing inhibitors. The LSM measured the change in thickness of the pellets at 60 s intervals for a duration of 24 h. The composition of sodium bentonite was determined using XRD, and Table 1 shows the percentage composition of bentonite.

Table 1. Percentage composition of sodium bentonite.

Mineral	Percentage
Quartz	29%
Hematite	4.7%
Carbonates	9%
Smectite	47%
Hatrurite	4%
Sodium Oxide	1.83%
Iron Silicates	2.17%

2.3.2. Linear Swell Meter

The Grace HPHT Linear Swell Meter (M4600) is a dedicated instrument utilized for measuring changes in sample thickness, enabling the evaluation of water-based drilling muds' effectiveness in preventing shale swelling. It comprises two main components: the wafer compactor and the linear swell meter (Model: M4600). Bentonite wafers were produced using the Grace core/wafer compactor, and the subsequent swelling tests were performed using the linear swell meter (LSM), which provides real-time data on the swelling phenomenon.

2.3.3. Shale Recovery Test

A shale sample was acquired from Niah, Miri District in Sarawak, Malaysia. To examine the clay composition and evaluate its suitability for dispersion and shale recovery experiments, the sample underwent X-ray diffraction (XRD) analysis. Among the clay minerals analyzed, illite accounts for 18% of the composition, making it a significant component. Kaolinite, on the other hand, makes up the largest portion, with a percentage of 31%. This suggests that the shale sample contains a substantial amount of kaolinite. Chlorite, another clay mineral, represents 22% of the composition, indicating its presence in notable proportions. Vermiculite, with a percentage of 10%, and mica, with 19%, also contribute to the overall clay mineral composition. These percentages provide insights into the relative abundance of each clay mineral in the shale sample and help in understanding its mineralogical characteristics.

The shale recovery test involved immersing shale cuttings in drilling mud and aging them for 24 h under 1000 psia at 150 °C before subjecting them to a series of sieves with different mesh sizes. The cuttings were then separated based on size, and the percentage recovery was calculated by calculating the non-dispersed cuttings that were able to pass through a particular sieve size. This test provides valuable information on the effectiveness of the drilling mud in controlling shale and its ability to remove cuttings during drilling operations.

2.4. Hydrate Inhibition Study

Nucleation, a highly intricate process involving numerous molecules, poses challenges for direct experimental observation. Although several methods exist for studying nucleation, the most dependable approach is the measurement of induction time, which combines nucleation theory with experimental observations. In this particular investigation, the inhibitory effects of different concentrations of deep eutectic solvents (DES) on hydrate formation were examined by adding them to the drilling mud.

To determine the induction time of various mud samples, micro-DSC (differential scanning calorimetry) was employed, which can be further studied in our previous work [27]. The samples were subjected to cooling within a temperature range of -20 °C to +20 °C, while maintaining a pressure of 114 bar and a cooling rate of 0.1 K/min. Methane gas was used to form SII-type hydrates. The resulting graph of heat flow versus time exhibited two distinct peaks for each sample. The smaller peak indicated the formation of hydrates, while the larger peak corresponded to ice formation. The onset time of hydrate formation was determined by analyzing the smaller peak. Interestingly, it was observed that at higher pressures (above 70 bar), hydrates formed prior to ice, while the reverse was true at lower pressures.

2.5. Characterization Techniques

2.5.1. Surface Tension

Surface tension is the term used to describe the tension present on the surface of a liquid, caused by cohesive forces. This tension influences the capillary action, which leads to the penetration of water cations into the micropores of shale. The degree of invasion is directly related to the surface tension. To measure the surface tension of different NADES-based mud samples with various concentrations (1%, 3%, and 5%), an interfacial tensiometer (IFT) was utilized.

2.5.2. D-Spacing

The d-spacing refers to the measurement of the distance between the layers of alumino-silicate within clay, plus the thickness of a single alumino-silicate layer. In order to investigate the incorporation of inhibitors into the layers of bentonite, an analysis using X-ray diffraction (XRD) was conducted. This analysis involved the examination of various samples, including dry sodium bentonite, base mud, and wet drilling mud samples. The wet drilling mud samples consisted of different concentrations (1%, 3%, and 5%) of CC:Gly-based NADES.

To obtain XRD peaks, a benchtop X-ray diffractometer (D2 phaser) was used. The diffractometer operated at a current of 40 mA and 45 kV, utilizing Cu-Kα radiation with a wavelength of 1.54059 Å. Once the XRD peaks were obtained for both the wet samples and dry Na-Bt, the d-spacing was calculated using Bragg's equation.

2.5.3. Zeta Potential Measurement

The zeta potential of mud samples containing various concentrations (1%, 3%, and 5%) of NADES was determined using a laser-based system known as the Malvern Zetasizer. The Malvern Zetasizer operates by subjecting suspended particles in the liquid to an electric field and measuring their velocity of motion. This measurement provides valuable information about the zeta potential of the particles. The obtained zeta potential data are analyzed and discussed in Section 3.

3. Results and Discussion

3.1. In-House Preparation of NADES

In this research, calcium chloride and glycerol were chosen as the components for the formation of NADES based on M.H. screening criteria. Calcium chloride, represented by the chemical formula $CaCl_2$, acts as a hydrogen bond acceptor in the NADES composition. It possesses two potential hydrogen bond acceptor sites, both located on chloride, allowing for hydrogen bonding interactions through the partial negative charge on the chlorine atom. Glycerol, also known as glycerine, is a widely used substance found in various products such as food, cosmetics, and pharmaceuticals. It functions as a hydrogen bond donor in the NADES formulation. Glycerol contains three potential hydrogen bond donor sites located on each of its hydroxyl (-OH) groups. These hydroxyl groups can engage in hydrogen bonding interactions through their respective lone pairs of electrons. The ability of glycerol to donate hydrogen bonds is significant in stabilizing proteins and enzymes in solution, as it forms hydrogen bonds with the amino acid residues present in proteins and enzymes [33].

A eutectic mixture was achieved at a molar ratio of 1:4 at a temperature of 60 °C while stirring the components at 100 rpm. The stirring process continued until a homogeneous, stable, and clear mixture was obtained. The composition of the CC:Gly NADES formulation is presented in Table 2 and Figure 1.

Table 2. Synthesis of CC:Gly NADES with different molar ratios at 60 °C.

Molar Ratio	Observation
1:1	Cloudy and turbid
1:2	Turbid with ppt
1:3	Precipitated
1:4	Transparent, no precipitate (eutectic mixture)
2:1	Cloudy
2:2	Intensely cloudy

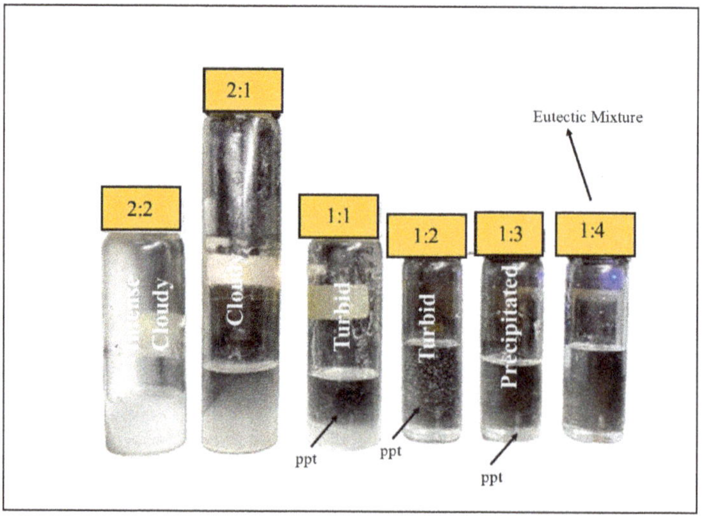

Figure 1. Preparation of calcium chloride–glycerine-based NADES with different molar ratios.

3.2. YP/PV of Mud Samples

The yield point (YP) is a parameter that measures the strength of the attractive forces between particles in a colloidal drilling mud slurry. Plastic viscosity (PV) quantifies the resistance offered by the solid particles and the liquid in the drilling mud. The ability of the mud to effectively transport cuttings to the surface depends on the ratio of YP to PV. Generally, increasing YP/PV values enhance the flow profile and cutting carrying capacity of the mud. However, excessively high values can lead to elevated annular frictional pressure losses and equivalent circulation density (ECD), potentially causing fractures in the formation. Existing literature suggests that YP/PV values within the range of 0.75–1 lbm/100 ft2/cp are optimal for efficient cutting transportation without undesired ECD effects.

The incorporation of deep eutectic solvents (NADES) into the mud helps improve the cutting carrying ability by reducing the YP/PV value closer to the ideal range. Similar to ionic liquids, NADES has the ability to modify the rheology of the mud by altering the structure of clay platelets. The YP/PV values of aged mud samples tend to approach the optimum range, as demonstrated in Figure 2. As temperature increases, the YP/PV ratio in the drilling mud decreases due to various factors. These factors include changes in the width of the electrical double layer around clay particles, reduced hydration, increased thermal energy of clay particles, and decreased viscosity of the colloidal solution. At higher temperatures, bentonite undergoes substantial dehydration, degradation, and mechanical shearing, causing the clay platelets to come closer together and resulting in increased attractive forces between them. This phenomenon affects the YP/PV ratio.

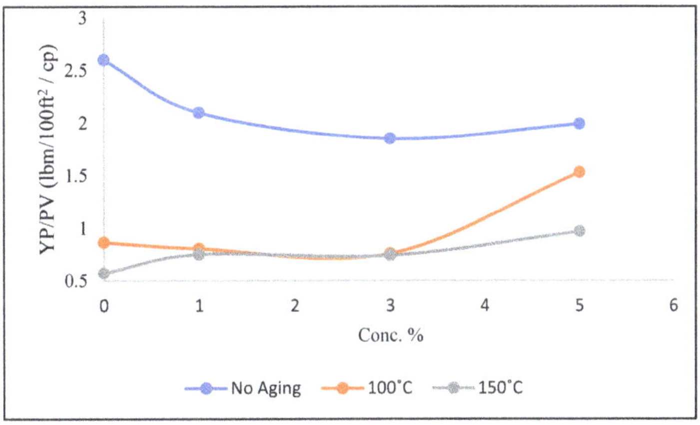

Figure 2. YP/PV of CC:Gly-based mud of aged and non-aged mud samples.

3.3. Filtration Properties of Mud

The performance of natural deep eutectic solvents (NADES) based on calcium chloride was remarkable, particularly under challenging conditions of high temperature and high pressure. The incorporation of NADES into the base drilling mud sample had dual benefits: it effectively reduced both the thickness of the mud cake and the volume of filtrate, as clearly depicted in Figures 3 and 4. Thick mud cakes have the potential to cause drill pipe sticking, which is why it is necessary to add substances that can make the mud thinner. NADES acts as a thinning agent by altering the wettability of clay platelets through interactive processes, thereby resulting in decreased mud cake thickness and filtrate volume. When 3% NADES was used without aging, the mud cake thickness was measured at 1.5 mm, and the filtrate loss was recorded as 18 mL, as demonstrated in Figures 3 and 4. Comparatively, the addition of 3% NADES led to a significant reduction of 25% in mud cake thickness and a 28% reduction in filtrate volume when compared to the base sample.

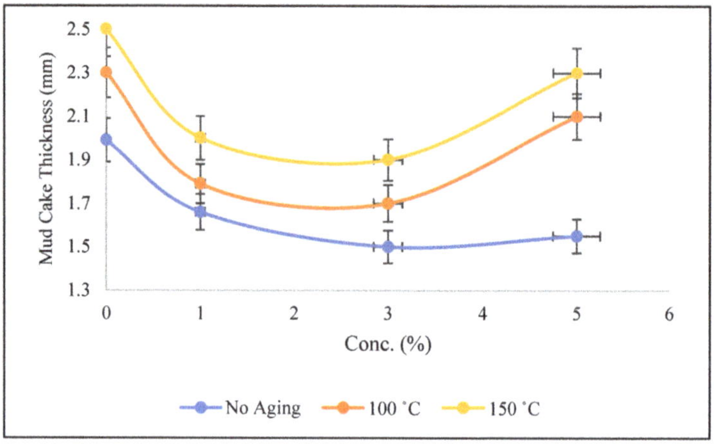

Figure 3. Mud cake thickness of CC:Gly-based mud for aged and non-aged mud samples.

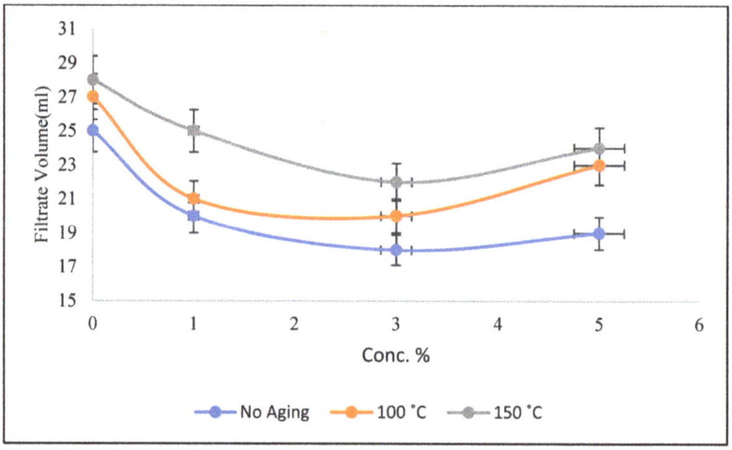

Figure 4. Filtrate volume of CC:Gly-based mudn for aged and non-aged mud samples.

It is important to note that as the aging temperature increases, the clay particles tend to flocculate and aggregate, leading to the formation of thicker mud cakes and increased filtrate loss, as illustrated in Figures 3 and 4. Error bars were drawn in Figures 3 and 4 to indicate 5% error margins.

3.4. Gel Strength

Gel strength refers to the mud's ability to retain suspended cuttings under static conditions. It is measured at three different time intervals: 10 s, 10 min, and 30 min. Although both gel strength and yield point are measured in the same units (lb/100 ft^2), yield point is a dynamic property, while gel strength can be considered a static property of the drilling mud. Gel strength is calculated by comparing the increase in gel strength from the previous reading at various time intervals. The larger the difference between readings, the higher the pump pressure needed to disrupt the gel and initiate circulation. In some cases, the mud may solidify, necessitating the addition of different chemicals to dilute it.

Examining Figure 5, it is apparent that the addition of NADES to the drilling mud did not significantly increase the gel strength. This observation holds true for both fresh and aged samples. All samples containing DES exhibited a hindrance in the increase in gel strength to a notable extent. However, the samples containing 3% NADES displayed the most substantial decline in the increment in gel strength.

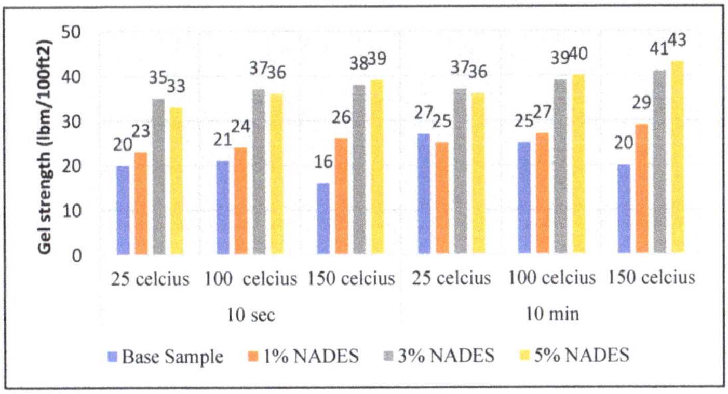

Figure 5. Gel strength of CC:Gly-based mud.

3.5. Induction Time of Hydrate Formation

The data presented in Table 3 and Figures 6 and 7 clearly indicate that the base sample effectively prevented hydrate formation for a duration of 4.185 h. Notably, when the mud contained NADES, the formation of hydrates was delayed even further, reaching a time of 6.523 h with the addition of 1% NADES. It is noteworthy that the inclusion of 3% NADES yielded the most favorable results (80% hydrate inhibition) in terms of inhibiting hydrate formation, while 5% NADES also gave close results. Further discussion of the results is given in Section 3.10.

Table 3. Hydrate and ICE onset times of NADES-based mud.

Conc.	Induction Time (h)	Peak Height (Hydrates) (mW)	Ice Onset Time (h)	Hydrate Onset Temperature (°C)
Base	4.185	7.948	5.439	−4.136
1%	6.523	7.284	6.214	−12.525
3%	7.573	12.715	6.361	−13.393
5%	7.213	80.97	6.42	−9.631

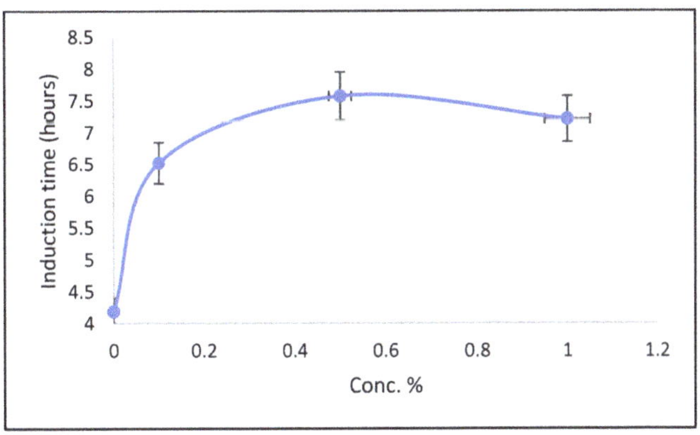

Figure 6. Hydrate induction time of CC:Gly-based mud.

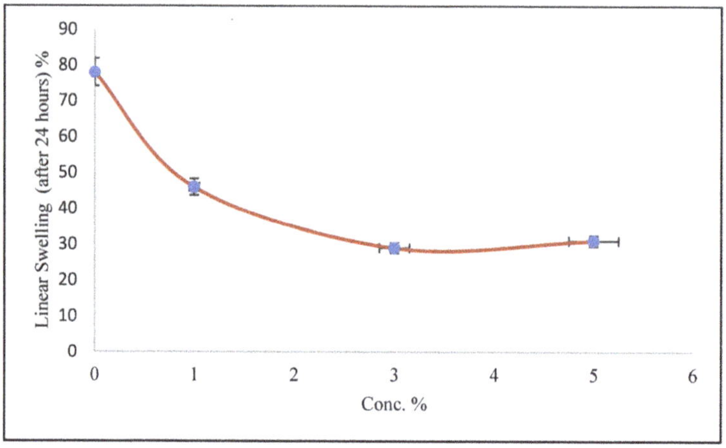

Figure 7. Linear swelling of CC:Gly-modified bentonite wafers.

3.6. Linear Swelling and Shale Recovery Tests (Dispersion Test)

The assessment of linear swelling was conducted to compare the effects of different concentrations of CC:Gly NADES-based mud samples with varying concentrations against the base drilling mud sample. Figure 7 visually represents the findings, demonstrating that the base sample without NADES exhibited a linear swelling of 78%. However, the addition of NADES significantly reduced the swelling to 29%, with the most favorable outcomes observed when 3% NADES was incorporated into the drilling mud. This concentration resulted in a mere 29% swelling or 62.8% inhibition.

Figure 8 presents the results of shale recovery, showing a 12% recovery rate for the base drilling mud and a remarkable 76% recovery rate when 3% NADES was employed. The improved shale recovery can be attributed to the stabilizing effect of NADES on the clay present in shale, which occurs through its interaction with the drilling mud. NADES forms bonds with the negatively charged clay particles, neutralizing the negative charge on the clay surface and thereby stabilizing the hydration process of shale. This capability is attributed to NADES' remarkable ability to form hydrogen bonds with clay particles. The experimental results for linear swelling find support in the analysis of d-spacing, which is further elaborated on in Section 3.7.

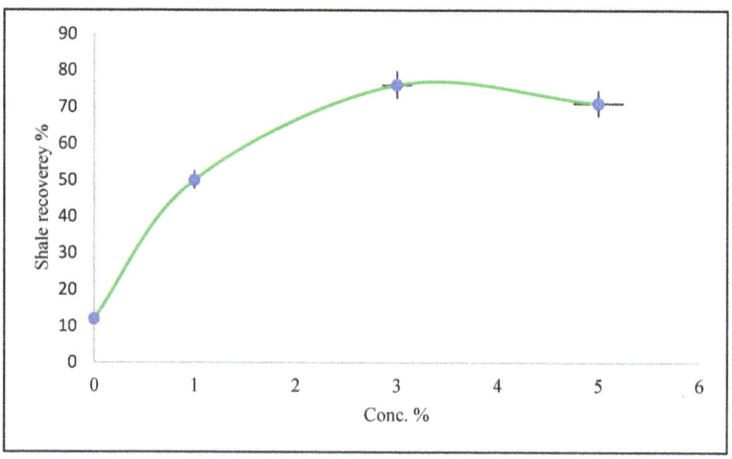

Figure 8. Shale recovery of CC:Gly-based mud.

3.7. D-Spacing of Inhibitor-Based Mud

D-spacing measures the combined length of the interlayer spacing of clay and one alumino-silicate layer within the clay. Figure 9 presents the d-spacing results for dry sodium bentonite (Na-Bt), hydrated mud samples, and NADES-based mud samples. The results help understand the intercalation of water and NADES into clay layers, with the d-spacing of dry Na-Bt increasing with the addition of water and decreasing with the addition of NADES, showing that NADES has a higher affinity for clay than water and expels water from the clay layers. The d-spacing of dry Na-Bt is 12.64 A°, increasing to 18.01 A° after hydration. The maximum elimination of water between the clay layers was observed with 3% and 5% NADES, resulting in a maximum decline in d-spacing of 14.23 A° and 14.99 A°, respectively, as shown in Figure 9.

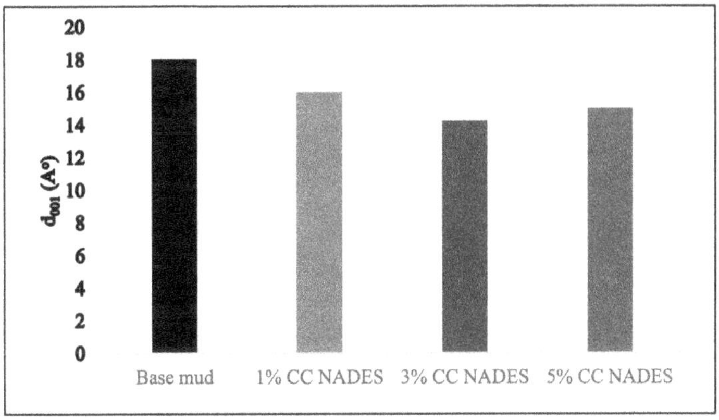

Figure 9. D-spacing of CC:Gly-based mud wafer.

3.8. Surface Tension

Surface tension, which refers to the cohesive forces between molecules at the surface of a liquid, plays a significant role in capillary pressure. Capillary pressure is the pressure exerted by a liquid as a result of the adhesion of its molecules to the container or air–liquid interface. The relationship between surface tension and capillary pressure is direct, meaning that higher surface tension results in higher capillary pressure. When it comes to shale inhibition, substances such as NADES can effectively reduce surface tension, preventing water cations from penetrating the layers of clay and thus mitigating shale swelling. The findings of surface tension testing revealed a significant decrease of up to 11.29% when 3% NADES was employed as shown in Figure 10. This decline can be attributed to the ability of NADES to form hydrogen bonds with clay, consequently altering its contact angle and behavior when exposed to water.

3.9. Zeta Potential

Zeta potential, which refers to the electrical potential at the interface between a solid and a liquid, plays a vital role in understanding the stability of suspensions and the behavior of colloidal systems [34]. In the context of clay swelling, zeta potential is closely related, as the negative charge present on clay particles attracts water molecules and promotes swelling [35]. However, the introduction of NADES can counterbalance this negative charge on the clay particles, leading to a reduction in swelling [9]. The results obtained from zeta potential testing demonstrated a significant decrease, with the most notable decline of 57.14% observed when using 5% NADES, which is closer to the Z.P of the 3% NADES-based sample as shown in Figure 11. This decline in zeta potential impacts the electrical double layer surrounding the clay particles, further influencing their behavior and stability.

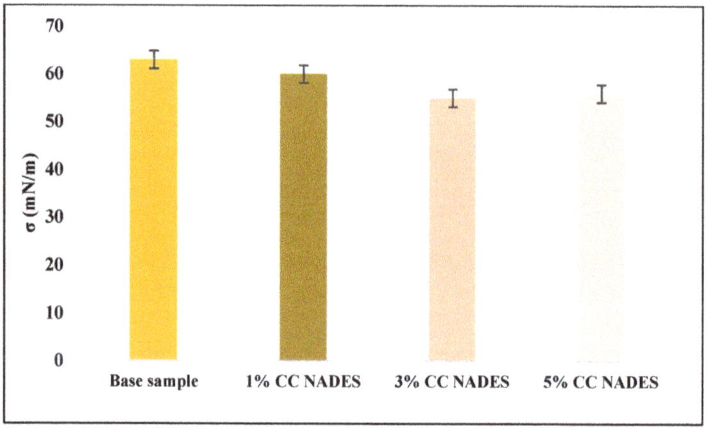

Figure 10. Surface tension of CC:Gly-based mud.

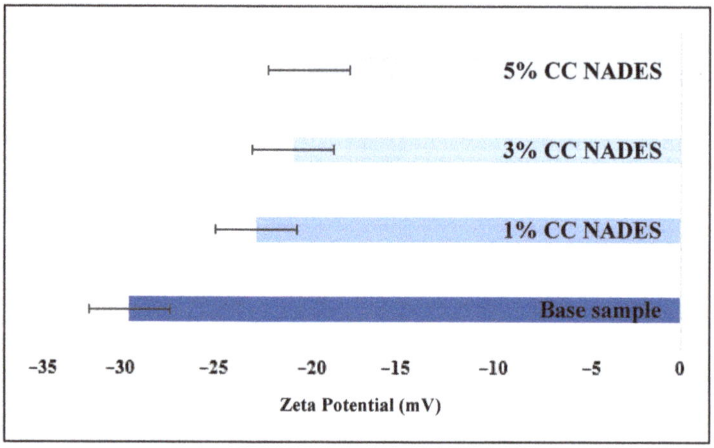

Figure 11. Zeta potential of CC:Gly-based mud.

In this research, calcium chloride and glycerol were chosen as the components for the formation of NADES based on M.H. screening criteria. Calcium chloride, represented by the chemical formula $CaCl_2$, acts as a hydrogen bond acceptor in the NADES composition. It possesses two potential hydrogen bond acceptor sites, with both located on chloride, allowing for hydrogen bonding interactions through the partial negative charge on the chlorine atom. Glycerol, also known as glycerine, is a widely used substance found in various products such as food, cosmetics, and pharmaceuticals. It functions as a hydrogen bond donor in the NADES formulation. Glycerol contains three potential hydrogen bond donor sites located on each of its hydroxyl (-OH) groups. These hydroxyl groups can engage in hydrogen bonding interactions through their respective lone pairs of electrons. The ability of glycerol to donate hydrogen bonds is significant in stabilizing proteins and enzymes in solution, as it forms hydrogen bonds with the amino acid residues present in proteins and enzymes [33].

3.10. Discussion on Underlying Mechanism

One of the key mechanisms through which NADES (natural deep eutectic solvents) can improve mud rheology is by influencing the face-to-edge orientation of bentonite particles.

Bentonite, a common clay mineral used in drilling mud, consists of thin, plate-like particles. These particles have a natural tendency to orient themselves with their faces parallel to each other, forming stacked layers. This face-to-face orientation leads to an increase in the mud viscosity and gel strength [36]. However, the addition of NADES to the drilling mud can modify this face-to-edge orientation of bentonite particles. NADES molecules have unique properties that can disrupt the face-to-face alignment of the bentonite particles. The NADES molecules penetrate the interlayer spaces of the bentonite, causing a reorientation of the particles from face to edge. This reorientation results in a more open and dispersed structure, reducing the viscosity and gel strength of the mud.

The mechanism behind NADES' ability to inhibit hydrate formation during drilling involves several factors, including induction time and nucleation inhibition. Induction time refers to the period required for the formation of the first hydrate crystals from the supersaturated drilling mud [37]. NADES can extend the induction time significantly, thereby delaying the onset of hydrate formation. This delay occurs due to the unique solvent properties of NADES, which interfere with the formation of hydrate crystal structures by making a hydrogen bond with it. NADES molecules have the ability to penetrate the water cages that form around gas molecules, disrupting the hydrogen bonding network necessary for hydrate formation. By occupying the spaces within the water cages, NADES hinder the efficient organization of water molecules into the stable hydrate lattice structure, which increases induction time and suppresses the growth of hydrates. Consequently, the formation of hydrate crystals is delayed, and the induction time is extended.

Furthermore, NADES can also inhibit nucleation, which is the initial step of hydrate formation when small clusters of hydrate molecules form and serve as the building blocks for larger crystals. NADES molecules act as inhibitors, preventing the clustering and growth of these hydrate nuclei. They achieve this by creating a physical barrier or steric hindrance that inhibits the aggregation of hydrate molecules and hinders the nucleation process. Moreover, the presence of NADES in the drilling mud alters the thermodynamic conditions required for hydrate formation. NADES can modify the water activity and salinity of the fluid, thereby shifting the hydrate formation conditions to less favorable regions of the phase diagram. This modification reduces the likelihood of hydrate formation and further contributes to hydrate inhibition [38].

NADES have the capability to prevent shale swelling by interacting with the clay minerals present in shale formations. Shale swelling occurs when clay minerals, such as smectite, absorb water and expand, leading to issues such as increased mud viscosity, compromised wellbore stability, and decreased hydrocarbon production [39]. The prevention of shale swelling by NADES is achieved through two main mechanisms. Firstly, NADES reduce the water absorption capacity of clay minerals by interacting with them through their hydrogen bond acceptor (HBA) component. This interaction forms hydrogen bonds with water molecules, thereby limiting the expansion of clay minerals and reducing overall shale swelling [40].

Additionally, NADES can prevent shale swelling by modifying the chemical and physical properties of clay mineral surfaces through their interaction with the hydrogen bond donor (HBD) component. This alteration of the surface charge of clay minerals affects their interaction with water and other fluids, thereby limiting their expansion and reducing shale swelling. The effectiveness of NADES or deep eutectic solvents (DES) in preventing shale swelling depends on various factors such as the specific HBA and HBD components used, the properties of the shale, and the particular application [41].

Moreover, NADES can enhance the filtration properties of drilling mud by reducing the presence of solid particles carried over by the mud. This is accomplished by the interaction of the HBA component with clay minerals, causing them to flocculate and settle out of the mud. It is important to note that while reducing surface tension through the use of drilling mud with low surface tension and high viscosity can improve shale inhibition, other factors such as clay mineral properties, drilling mud chemistry, and shale characteristics also play significant roles in shale inhibition. Therefore, it is crucial to conduct field testing and monitor the results to assess the effectiveness of the drilling mud in achieving shale inhibition may be divided by subheadings. This should provide a concise and precise description of the experimental results, their interpretation, as well as the experimental conclusions that can be drawn.

4. Conclusions

This study demonstrated that the NADES-based mud effectively improved mud rheology, inhibited hydrate formation, and suppressed shale swelling. The incorporation of NADES into the drilling mud led to significant enhancements in the mud's yield point to plastic viscosity ratio (YP/PV) and gel strength. This improvement in rheological properties contributes to better mud hydraulics and the ability to suspend cuttings in static conditions. Furthermore, the NADES formulation exhibited exceptional performance in inhibiting hydrate formation during drilling operations. By extending the induction time and impeding nucleation, the NADES significantly delayed the formation of hydrate crystals. This inhibitory effect can be attributed to the NADES' unique solvent properties, which disrupt the hydrogen bonding network necessary for hydrate formation. Moreover, the study revealed that the NADES-based mud effectively prevented shale swelling. By interacting with clay minerals present in shale formations, the NADES reduced water absorption and limited the expansion of clay minerals. This mitigated shale swelling, thereby ensuring improved wellbore stability, reduced mud viscosity, and enhanced hydrocarbon production.

In a nutshell, the results of the study indicate that the NADES-based mud offers significant improvements in the ratio of yield point to plastic viscosity (YP/PV) of the mud, as well as favorable gel strength and a remarkable 80% inhibition of hydrate formation. Moreover, it exhibits an impressive 62.8% reduction in shale swelling while enabling an 84.1% improvement in shale recovery. Additionally, the NADES-based mud demonstrates a notable decrease of 28% in mud filtrate and 25% in mud cake thickness, which is supported by the results obtained from XRD, zeta potential, and surface tension analyses. Based on these promising findings, the calcium chloride–glycerine NADES-based mud is recommended as a versatile additive for drilling mud, suitable for various industrial applications. Furthermore, this alternative presents a more environmentally friendly option compared to traditional additives, effectively addressing concerns related to cost, biodegradability, and environmental impact in the drilling process. Its potential for global impact cannot be overlooked, making it a valuable addition to the field of drilling operations.

Propelling Forward: It is recommended to conduct a detailed assessment for the comparison and compatibility of the combined effects of NADES and conventional additives on key drilling mud properties such as fluid loss control, shale inhibition, lubricity, rheological behavior, and thermal stability. This evaluation will help quantify the synergistic benefits and optimize the formulation for improved drilling performance.

Author Contributions: Conceptualization, M.H.R.; methodology, M.H.R. and M.A.; formal analysis, M.H.R.; investigation, M.H.R. and A.Z.J.; resources, N.A.S. and M.A.; writing—original draft preparation, M.H.R.; writing—review and editing, N.A.S. and M.A.; visualization, A.Z.J.; supervision, M.A. and N.A.S. All authors have read and agreed to the published version of the manuscript.

Funding: The APC of this research was funded by YUTP FRG 1/2021, grant no. 015LC0-363. The lab testing, shale sample procurement, and the postgraduate student involved in this research work were funded by YUTP, grant no. 015LC0-326.

Data Availability Statement: No Data is available for this research.

Conflicts of Interest: The authors declare no conflict of interest.

References

1. Agwu, O.E.; Akpabio, J.U.; Ekpenyong, M.E.; Inyang, U.G.; Asuquo, D.E.; Eyoh, I.J.; Adeoye, O.S. A critical review of drilling mud rheological models. *J. Pet. Sci. Eng.* **2021**, *203*, 108659. [CrossRef]
2. Zhang, J.R.; Xu, M.D.; Christidis, G.E.; Zhou, C.H. Clay minerals in drilling fluids: Functions and challenges. *Clay Miner.* **2020**, *55*, 1–11. [CrossRef]
3. Wastu, A.; Hamid, A.; Samsol, S. The effect of drilling mud on hole cleaning in oil and gas industry. *J. Phys. Conf. Ser.* **2019**, *1402*, 022054. [CrossRef]
4. Rana, A.; Murtaza, M.; Saleh, T.A.; Kamal, M.S.; Mahmoud, M. An efficient, cost-effective, and green natural extract in water-based drilling muds for clay swelling inhibition. *J. Pet. Sci. Eng.* **2022**, *214*, 110332. [CrossRef]
5. Hassan, A.; Murtaza, M.; Alade, O.; Tariq, Z.; Kamal, M.S.; Mahmoud, M. Chapter 3—Interactions of drilling and completion fluids during drilling and completion operations. In *Developments in Petroleum Science*; Hussein, I.A., Mahmoud, M., Eds.; Elsevier: Amsterdam, The Netherlands, 2023; Volume 78, pp. 41–74.
6. Zamir, A.; Elraies, K.A.; Rasool, M.H.; Ahmad, M.; Ayoub, M.; Abbas, M.A.; Ali, I. Influence of alkyl chain length in ionic liquid based drilling mud for rheology modification: A review. *J. Pet. Explor. Prod. Technol.* **2021**, *12*, 485–492. [CrossRef]
7. Rahman, M.T.; Negash, B.M.; Danso, D.K.; Idris, A.; Elryes, A.A.; Umar, I.A. Effects of imidazolium-and ammonium-based ionic liquids on clay swelling: Experimental and simulation approach. *J. Pet. Explor. Prod. Technol.* **2021**, *12*, 1841–1853. [CrossRef]
8. Khan, R.A.; Murtaza, M.; Ahmad, H.M.; Abdulraheem, A.; Kamal, M.S.; Mahmoud, M. Development of Novel Shale Swelling Inhibitors Using Hydrophobic Ionic Liquids and Gemini Surfactants for Water-Based Drilling Fluids. In *SPE Middle East Oil & Gas Show and Conference*; OnePetro: Richardson, TX, USA, 2021.
9. Khan, R.A.; Murtaza, M.; Abdulraheem, A.; Kamal, M.S.; Mahmoud, M. Imidazolium-Based Ionic Liquids as Clay Swelling Inhibitors: Mechanism, Performance Evaluation, and Effect of Different Anions. *ACS Omega* **2020**, *5*, 26682–26696. [CrossRef]
10. Huang, P.; Jia, H.; Han, Y.; Wang, Q.; Wei, X.; Luo, Q.; Dai, J.; Song, J.; Yan, H.; Liu, D. Designing novel high-performance shale inhibitors by optimizing the spacer length of imidazolium-based bola-form ionic liquids. *Energy Fuels* **2020**, *34*, 5838–5845. [CrossRef]
11. Yang, L.; Yang, X.; Wang, T.; Jiang, G.; Luckham, P.F.; Li, X.; Shi, H.; Luo, J. Effect of alkyl chain length on shale hydration inhibitive performance of vinylimidazolium-based ionic liquids. *Ind. Eng. Chem. Res.* **2019**, *58*, 8565–8577. [CrossRef]
12. Ofei, T.N.; Bavoh, C.B.; Rashidi, A.B. Insight into ionic liquid as potential drilling mud additive for high temperature wells. *J. Mol. Liq.* **2017**, *242*, 931–939. [CrossRef]
13. Yang, L.; Jiang, G.; Shi, Y.; Yang, X. Application of ionic liquid and polymeric ionic liquid as shale hydration inhibitors. *Energy Fuels* **2017**, *31*, 4308–4317. [CrossRef]
14. Luo, Z.; Wang, L.; Yu, P.; Chen, Z. Experimental study on the application of an ionic liquid as a shale inhibitor and inhibitive mechanism. *Appl. Clay Sci.* **2017**, *150*, 267–274. [CrossRef]
15. Jia, H.; Huang, P.; Wang, Q.; Han, Y.; Wang, S.; Zhang, F.; Pan, W.; Lv, K. Investigation of inhibition mechanism of three deep eutectic solvents as potential shale inhibitors in water-based drilling fluids. *Fuel* **2019**, *244*, 403–411. [CrossRef]
16. Rasool, M.H.; Zamir, A.; Elraies, K.A.; Ahmad, M.; Ayoub, M.; Abbas, M.A.; Ali, I. Rheological characterization of potassium carbonate deep eutectic solvent (DES) based drilling mud. *J. Pet. Explor. Prod. Technol.* **2021**, *12*, 1785–1795. [CrossRef]
17. Rasool, M.H.; Zamir, A.; Elraies, K.A.; Ahmad, M.; Ayoub, M.; Abbas, M.A. Potassium carbonate based deep eutectic solvent (DES) as a potential drilling fluid additive in deep water drilling applications. *Pet. Sci. Technol.* **2021**, *39*, 612–631. [CrossRef]
18. Rasool, M.H.; Ahmad, M.; Abbas, M.A. A Double Action PD (Polymer-Deep Eutectic Solvent) Based Shale Inhibitor in Drilling Mud. *J. Adv. Res. Fluid Mech. Therm. Sci.* **2022**, *99*, 149–157. [CrossRef]
19. Ghandi, K. A review of ionic liquids, their limits and applications. *Green Sustain. Chem.* **2014**, *27*, 2014. [CrossRef]
20. Shamsuri, A.A. Ionic liquids: Preparations and limitations. *Makara J. Sci.* **2011**, *14*, 101–106. [CrossRef]
21. Maciel, V.G.; Wales, D.J.; Seferin, M.; Ugaya, C.M.L.; Sans, V. State-of-the-art and limitations in the life cycle assessment of ionic liquids. *J. Clean. Prod.* **2019**, *217*, 844–858. [CrossRef]
22. Torregrosa-Crespo, J.; Marset, X.; Guillena, G.; Ramon, D.J.; Martínez-Espinosa, R.M. New guidelines for testing "Deep eutectic solvents" toxicity and their effects on the environment and living beings. *Sci. Total Environ.* **2020**, *704*, 135382. [CrossRef]
23. Elencovan, V.; Joseph, J.; Yahaya, N.; Samad, N.A.; Raoov, M.; Lim, V.; Zain, N.N.M. Exploring a novel deep eutectic solvents combined with vortex assisted dispersive liquid–liquid microextraction and its toxicity for organophosphorus pesticides analysis from honey and fruit samples. *Food Chem.* **2022**, *368*, 130835. [CrossRef]
24. Rasool, M.H.; Ahmad, M.; Ayoub, M. Selecting Geological Formations for CO_2 Storage: A Comparative Rating System. *Sustainability* **2023**, *15*, 6599. [CrossRef]
25. Al-Risheq, D.I.; Shaikh, S.M.; Nasser, M.S.; Almomani, F.; Hussein, I.A.; Hassan, M.K. Influence of combined natural deep eutectic solvent and polyacrylamide on the flocculation and rheological behaviors of bentonite dispersion. *Sep. Purif. Technol.* **2022**, *293*, 121109. [CrossRef]
26. Al-Risheq, D.I.; Nasser, M.; Qiblawey, H.; Hussein, I.A.; Benamor, A. Choline chloride based natural deep eutectic solvent for destabilization and separation of stable colloidal dispersions. *Sep. Purif. Technol.* **2021**, *255*, 117737. [CrossRef]
27. Rasool, M.H.; Zamir, A.; Elraies, K.A.; Ahmad, M.; Ayoub, M.; Abbas, M.A. A Deep Eutectic Solvent based novel drilling mud with modified rheology for hydrates inhibition in deep water drilling. *J. Pet. Sci. Eng.* **2022**, *211*, 110151. [CrossRef]

28. Rasool, M.H.; Zamir, A.; Elraies, K.A.; Ahmad, M.; Ayoub, M.; Abbas, M.A. Investigative review on cutting transportation ability of ionic liquid-based drilling mud. *J. Hunan Univ. Nat. Sci.* **2021**, *48*, 2.
29. Rasool, M.H.; Ahmad, M.; Ayoub, M.; Abbas, M.A. A Novel Ascorbic Acid Based Natural Deep Eutectic Solvent as a Drilling Mud Additive for Shale Stabilization. *Processes* **2023**, *11*, 1135. [CrossRef]
30. Murtaza, M.; Ahmad, H.M.; Zhou, X.; Al-Shehri, D.; Mahmoud, M.; Kamal, M.S. Okra mucilage as environment friendly and non-toxic shale swelling inhibitor in water based drilling fluids. *Fuel* **2022**, *320*, 123868. [CrossRef]
31. Rasool, M.H.; Ahmad, M.; Ayoub, M.; Zamir, A.; Abbas, M.A. A review of the usage of deep eutectic solvents as shale inhibitors in drilling mud. *J. Mol. Liq.* **2022**, *119673*, 2022.
32. *API RP 13B-1*; Field Testing Water-based Drilling Fluids. American Petroleum Institute (API): Washington, DC, USA, 2003.
33. Rasool, M.H.; Ahmad, M. Synthesis and physico-chemical characterization of novel Epsom salt based natural deep Eutectic solvent. *Chem. Data Collect.* **2023**, *44*, 101004. [CrossRef]
34. Clogston, J.D.; Patri, A.K. Zeta potential measurement. *Charact. Nanoparticles Intend. Drug Deliv.* **2011**, *697*, 63–70.
35. Ahmad, H.M.; Murtaza, M.; Gbadamosi, A.; Kamal, M.S.; Hussain, S.M.S.; Mahmoud, M.; Patil, S. Application of Novel Magnetic Surfactant-Based Drilling Fluids for Clay Swelling Inhibition. *Energy Fuels* **2023**, *37*, 8212–8223. [CrossRef]
36. İşçi, S.; Ünlü, C.H.; Atici, O.; Güngör, N. Rheology and structure of aqueous bentonite-polyvinyl alcohol dispersions. *Bull. Mater. Sci.* **2006**, *29*, 449–456. [CrossRef]
37. Farhadian, A.; Zhao, Y.; Naeiji, P.; Rahimi, A.; Berisha, A.; Zhang, L.; Rizi, Z.T.; Iravani, D.; Zhao, J. Simultaneous inhibition of natural gas hydrate formation and CO_2/H_2S corrosion for flow assurance inside the oil and gas pipelines. *Energy* **2023**, *269*, 126797. [CrossRef]
38. Lee, D.; Go, W.; Oh, J.; Lee, J.; Jo, I.; Kim, K.-S.; Seo, Y. Thermodynamic inhibition effects of an ionic liquid (choline chloride), a naturally derived substance (urea), and their mixture (deep eutectic solvent) on CH4 hydrates. *Chem. Eng. J.* **2020**, *399*, 125830. [CrossRef]
39. Ibrahim, M.A.; Saleh, T.A. Partially aminated acrylic acid grafted activated carbon as inexpensive shale hydration inhibitor. *Carbohydr. Res.* **2020**, *491*, 107960. [CrossRef]
40. Ding, L.; Shi, B.; Liu, Y.; Song, S.; Wang, W.; Wu, H.; Gong, J. Rheology of natural gas hydrate slurry: Effect of hydrate agglomeration and deposition. *Fuel* **2019**, *239*, 126–137. [CrossRef]
41. Rasool, M.H.; Ahmad, M. Understanding Shale Instability through the Lens of Clay Mineralogy and Zeta Potential. *Geol. Earth Mar. Sci.* **2023**, *5*, 1–10. [CrossRef]

energies

MDPI

Article

Optimization of Cement–Rubber Composites for Eco-Sustainable Well Completion: Rheological, Mechanical, Petrophysical, and Creep Properties

Abdennour C. Seibi [1], Fatick Nath [2,*], Adedapo B. Adeoye [3] and Kaustubh G. Sawant [4]

1 The Mechanical Engineering Program, Department of Engineering, Utah Valley University, Orem, UT 84058, USA; aseibi@uvu.edu
2 The Petroleum Engineering Program, School of Engineering, Texas A & M International University, Laredo, TX 78041, USA
3 The Dow Chemical Company, Freeport, TX 77541, USA; dapyday@gmail.com
4 United States Department of the Army, Houston, TX 77025, USA; kaustubhsawant23@gmail.com
* Correspondence: fatick.nath@tamiu.edu; Tel.: +1-361-720-2024

Abstract: To ensure well integrity, wellbore must be strongly cased using durable cement slurries with essential additives during downhole completion. The rubber materials that come from industrial waste are becoming extremely encouraged in the use as an additive in preparing cement slurries due to their growing environmental footprint. However, the proper design of cement slurry strongly depends on its rheological, mechanical, petrophysical, and creep properties, which can be altered by changing additives. This study aimed to examine the cement properties under alteration in different chemical admixtures to create efficient binding properties, and to estimate the optimum cement–rubber slurry composition for eco-sustainable completion. Three cement samples with different mesh sizes of the crumb rubber particles were prepared. This study examined the variation in rheological behaviors, elastic and failure characteristics, permeability, and creep behavior of the cement–rubber composites for petroleum well construction. The experimental study showed that the addition of 15% or more crumb rubber to the cement resulted in very thick slurries. Moreover, it was shown that the addition of crumb rubber with various particle sizes to the cement reduced the strength by more than 50%, especially for a higher amount of rubber added. It was also revealed that the addition of a superplasticizer resulted in an 11% increase in compressive strength. The results showed that cement–crumb-rubber composites with 12% by weight of cement (BWOC) represented the optimum composite, and considerably improved the properties of the cement slurry. Water-permeability tests indicated the addition of 12% BWOC with 200-mesh crumb rubber decreased the permeability by nearly 64% compared to the base cement. Creep tests at five different stress levels illustrated that the neat cement was brittle and did not experience strain recovery at all stress levels. Cement slurries with the largest rubber-particle size were elastic and demonstrated the highest amount of strain recovery. Finally, a relationship was established between the permeability, average strain, and mesh size of the rubber particles, which offered the strain recovery, satisfied the zonal isolation, and consequently reduced the microannulus problem to ensure the cement's integrity.

Keywords: portland cement; crumb-rubber particles; cement–rubber composites; rheology; thickening time; compressive strength; permeability; creep

Citation: Seibi, A.C.; Nath, F.; Adeoye, A.B.; Sawant, K.G. Optimization of Cement–Rubber Composites for Eco-Sustainable Well Completion: Rheological, Mechanical, Petrophysical, and Creep Properties. *Energies* **2022**, *15*, 2753. https://doi.org/10.3390/en15082753

Academic Editors: Mofazzal Hossain, Hisham Khaled Ben Mahmud and Md Motiur Rahman

Received: 22 February 2022
Accepted: 6 April 2022
Published: 8 April 2022

1. Introduction

Downhole casing cementing is one of the most essential parts of the drilling and completion process [1]. Cement slurries are essentially injected into the annulus between the wellbore and casing after successfully drilling a well [2]. The primary functions of cement are isolating the cross-flow-occurring zones in the formations, strongly supporting the casing weight, and upholding any applied load [3]. Other functions are actively preventing the casing from corrosion and protecting it, lost circulation, and unwanted downhole fluids from being produced [4], as well as controlling abnormal pore pressure. A petroleum well possibly will never achieve its highest potential of production without appropriate isolations of its trouble zones [5]. A schematic of petroleum well-cementing operations is provided [6] in Figure 1.

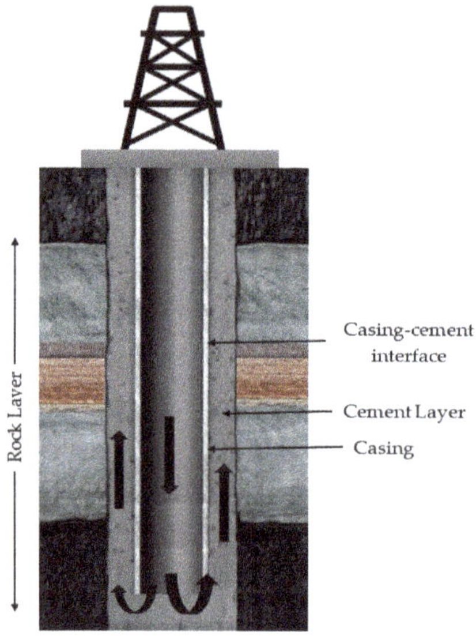

Figure 1. Representation of oil and gas well-cementing operations.

Portland cement has been used in downhole cementing applications for a long time, but this type of use has encountered numerous difficulties, and as a result, wellbore cement integrity is compromised with time [7]; The dominant problems due to cement failure include the development of a microannulus, radial cracking, chemical attacks, and loss of durability, thereby resulting in leakage and environmental damages [8]. Cement-sheath failure begins in the wellbore, mostly right after the cement sets. The process of casing cementing involves the pumping of cement slurry up to the desired depth. To ensure the wellbore integrity, high-quality cementing is crucial, which has triggered researchers to seek to design fit-for-purpose cement slurries [9].

Several different additives and materials are usually used to optimize the cement slurry design to enhance the different desirable properties in cement mixtures [10,11]. It is desirable that the selected cement composition for a particular downhole cementing should maintain the appropriate rheological, mechanical, petrophysical, and creep properties.

To optimize cement design and improve cement properties, researchers have been conducting experimental works using various materials; for example, weighting constituents, accelerators, retarders, plasticizers, dispersants, fluid-loss agents, and defoamers, [11–16], as well as mixing cement paste with various materials such as polypropylene fibers, laponite, nano clay, nanofibers, metakaolin, iron ore, biomass waste, granite sludge, tire waste, silicon carbide powder, olive waste, rice husk ash, and silica flour [17–29].

Cement pumping subjects the cement to phase changes, thereby affecting its rheological properties, while flowing leads to either shear thinning or thickening. The hydration superplasticizers are prolonged by adding other superplasticizers to the slurry, which also impacts the cement-hydration process, the formation structure, and the cement's characteristics [30]. Thickening time is one of the critical properties of cement slurries and is of the highest importance [31,32] among all properties, as it is largely affected by pressure and temperature. Based on real measurements, API recommended the use of thermal gradients of 0.9–1.9 °F per 100 ft by using the bottom-hole cementing temperature relationship [33]. Once the cement is pumped to the desired depth, it is usually left for more than 24 h to set and harden, thereby requiring an understanding of its postsetting mechanical behavior and its ability to resist cracks for well-integrity assurance [34]. The general field practices require that the cement should build to a minimum strength of 3.5 MPa within 24 h and 13.8 MPa during applications [35]; bulk cement shrinkage is the critical factor by which the formation of outer microannulus cracks [35,36] occurs.

Cement failure may also take place while executing the well-completion workflow, which comprises well stimulation, fluid injection, production, and workover operation. In addition, the higher temperature and pressure of the fluid developed inside the set casing cause the casing to expand circumferentially and longitudinally, thereby causing cement failure, and consequently, environmental hazards and high overburden cost [37–40].

Researchers have continuously increased their efforts to mitigate cement cracking by adding crumb rubber [41]. Although rubber particles have a progressive use in the construction industry, research has shown a substantial decrease in mechanical properties [42,43] and an increase in ductility [44]. Long et al. [44] showed that the compressive strength of cementitious materials was reduced by 4.5% for every 1% addition of crumb rubber by volume. In most cases, the added amount of crumb rubber to cement by weight was limited to 25%, and any further increase in rubber deteriorated the mechanical strength significantly [45–47]. This was mainly due to the hydrophobic nature and low Young's modulus of the rubber particles compared to the cement [48].

After completion of the operation, the cement settles down, but for continuous high pressure and temperature in the bottom-hole condition, the cement may become deformed over time. This time-dependent property is known as creep. Compared to other properties, ensuring creep recovery is critical to maintaining wellbore integrity, cement durability, and serviceability of the casing for long-term extended loading when the cement is subjected to constant bottom-hole pressure [49–54].

From an ecological perspective, rubber-material utilization in petroleum well-cementing is extremely encouraged, because inappropriate disposal of these wastes can lead to environmental and human health issues. This is why researchers have been attempting to use this industrial waste in any suitable applications in order to decrease the impacts of waste removal [55–60] and reduce the cost.

The objective of this study was to determine the optimum cement–rubber slurry composition that offered acceptable rheological, mechanical, petrophysical, and creep properties, and that enhanced the cement's properties in downhole cement applications when using crumb–rubber composites. Using different concentrations of crumb rubber, the current study conducted an extensive experimental investigation of the rheological (yield stress, plastic viscosity, thickening time, and segregation test), mechanical (compressive strength and failure pattern), petrophysical (permeability), scanning electron microscope (SEM) analysis (bonding of cement rubber), and creep properties (strain recovery) of cement–rubber slurries.

2. Materials and Methods

2.1. Materials Used and Elemental Analysis

Class H Portland cement and crumb rubber with various mesh sizes were used in this experiment. The chemical compositions of the as-received cement and crumb rubber are shown in Table 1. It can be seen that the 40-mesh rubber contained zinc (Zn), while the 200-mesh rubber exhibited the presence of sodium (Na) and aluminum (Al) (Figure 1). We believed that the occurrence of zinc was probably due to the vulcanization process of the rubber. According to the manufacturer, the rubber comprised cured natural rubber, styrene–butadiene (SBR), and 30% carbon black. The presence of sodium was not surprising, since in earlier days, SBR was called "Buna"; i.e., butadiene (Bu) and sodium (Na).

Table 1. Element analysis for neat cement, 40-mesh, and 200-mesh rubber particles.

Element	Element Concentration (wt %)		
	Neat Cement	w/40-Mesh Rubber	w/200-Mesh Rubber
C	10.695	73.99	81.982
O	19.671	12.146	7.025
Mg	1.097	0.00	0.00
Al	2.109	0.00	0.635
Si	14.327	10.41	5.67
S	1.401	1.816	2.463
K	0.228	0.00	0.00
Ca	47.044	0.469	0.752
Fe	3.427	0.00	0.00
Zn	0.00	1.168	0.00
Na	0.00	0.00	1.473
	Total		100 wt %

The apparent density of the rubber was approximately 1.02 g/cm^3. A lignosulfonate retarder (0.1–0.4% by weight) was used to lower the thickening time of the cement mixture by hindering hydration for easy pumping. Moreover, an ADVA 575 superplasticizer (0.03–0.2% by weight) was used as a dispersant and water-reducing agent to increase the bonding between the cement and rubber particles after solidification.

The particle-size distribution (PSD) of the cement and rubber particles with rated sizes of 40, 80, 140, and 200 mesh presented in Figure 2 indicated that the mean diameters were 328.6, 106.95, 64.51, and 59.11 µm, respectively. The reported mean values were equivalent to the spherical diameter of the cement and rubber particles measured on a volume basis. This paper intended to study the effects of the weight fraction of the rubber particles, rubber-particle sizes, retarder, and superplasticizer. Figure 3 shows the elemental concentrations of the samples investigated in this study.

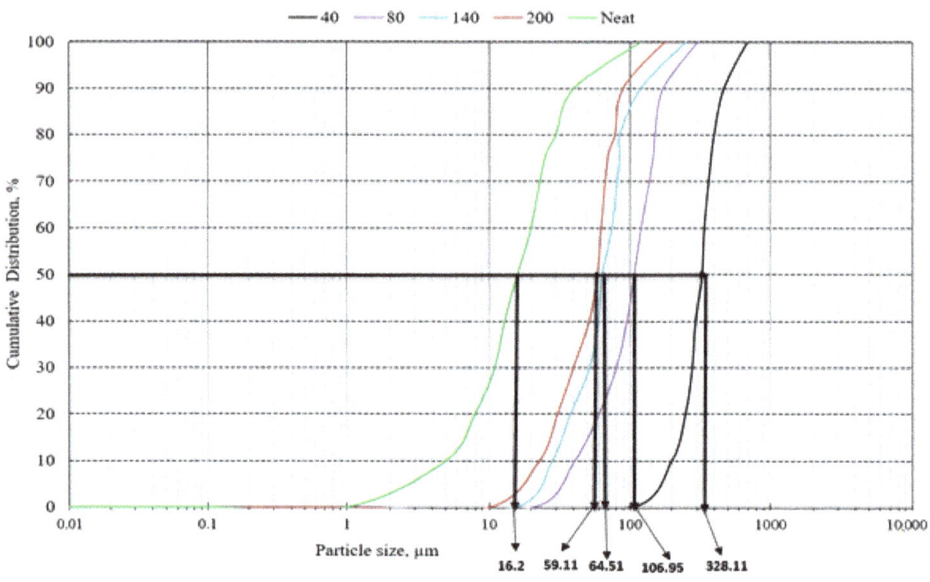

Figure 2. Particle-size distribution for 40-mesh and 200-mesh rubber and neat cement.

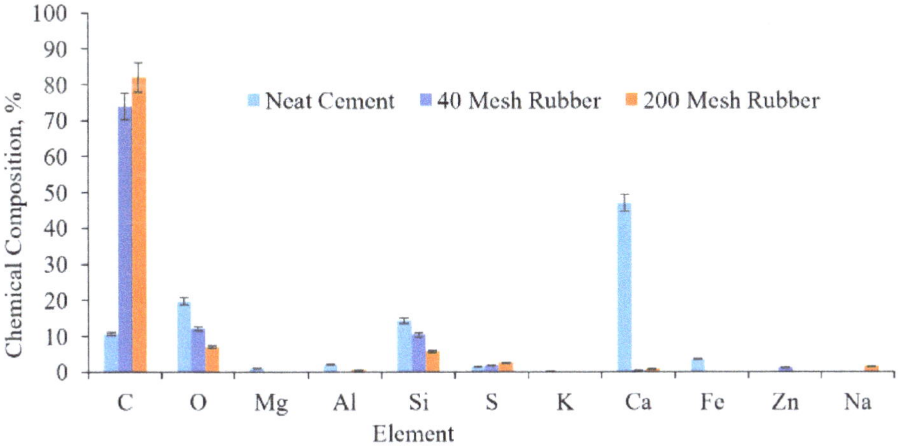

Figure 3. Elemental concentrations.

2.2. Sample Preparation

The effects of crumb rubber at various particle sizes in combination with a retarder and more superplastic on the rheology and compressive strength of the cement–crumb-rubber composites were evaluated. The thickening time of the cement–crumb-rubber slurry was measured using a consistometer set at a 7000 psi pressure and 200 °F temperature, reflecting the field conditions of high pressure–high temperature (HPHT) wells. Four mesh sizes of crumb rubber (40, 80, 140, and 200) were used in this study. Various cement–crumb-rubber slurries created by adding 2%, 12%, 15%, and 18% BWOC crumb-rubber particles were prepared, and their rheological parameters were determined using an OFITE 800 rotational viscometer.

The slurries were prepared by blending class H cement, rubber particles, water, and additives according to the API 10B-2 procedure [38]. All cement materials were dried, weighed, and mixed before mixing them into the liquid. Water and other liquid additives such as the retarder and superplasticizer were stirred together and measured in the mixer using a high-precision weighing balance. Neat cement and crumb-rubber particles were mixed and added to the liquid solution for slurry preparation. The density of the slurry was measured as suggested by API 10B-2 [38] and was found to be within the range of 1.8–2.0 g/cm^3.

After completion of the rheological tests, the slurry was poured inside 2 in (diameter) × 4 in (length) cylindrical molds, and Vaseline petroleum jelly was used to coat the molds internally so that the cement samples could be easily removed after curing. The molds filled with the slurries were then placed inside an oven set at 200 °F and cured for 24 h. The cured samples were removed from the molds and prepared for mechanical testing to measure the unconfined compressive strength (UCS) of the cement–crumb-rubber composites according to ASTM C 39-04a. Three samples for each material type were tested, from which an average value was obtained. A scanning electron microscope (SEM) was employed to examine the interfacial bonding between the cement and rubber particles. A flowchart of the statement and objectives of this study is shown in Figure 4.

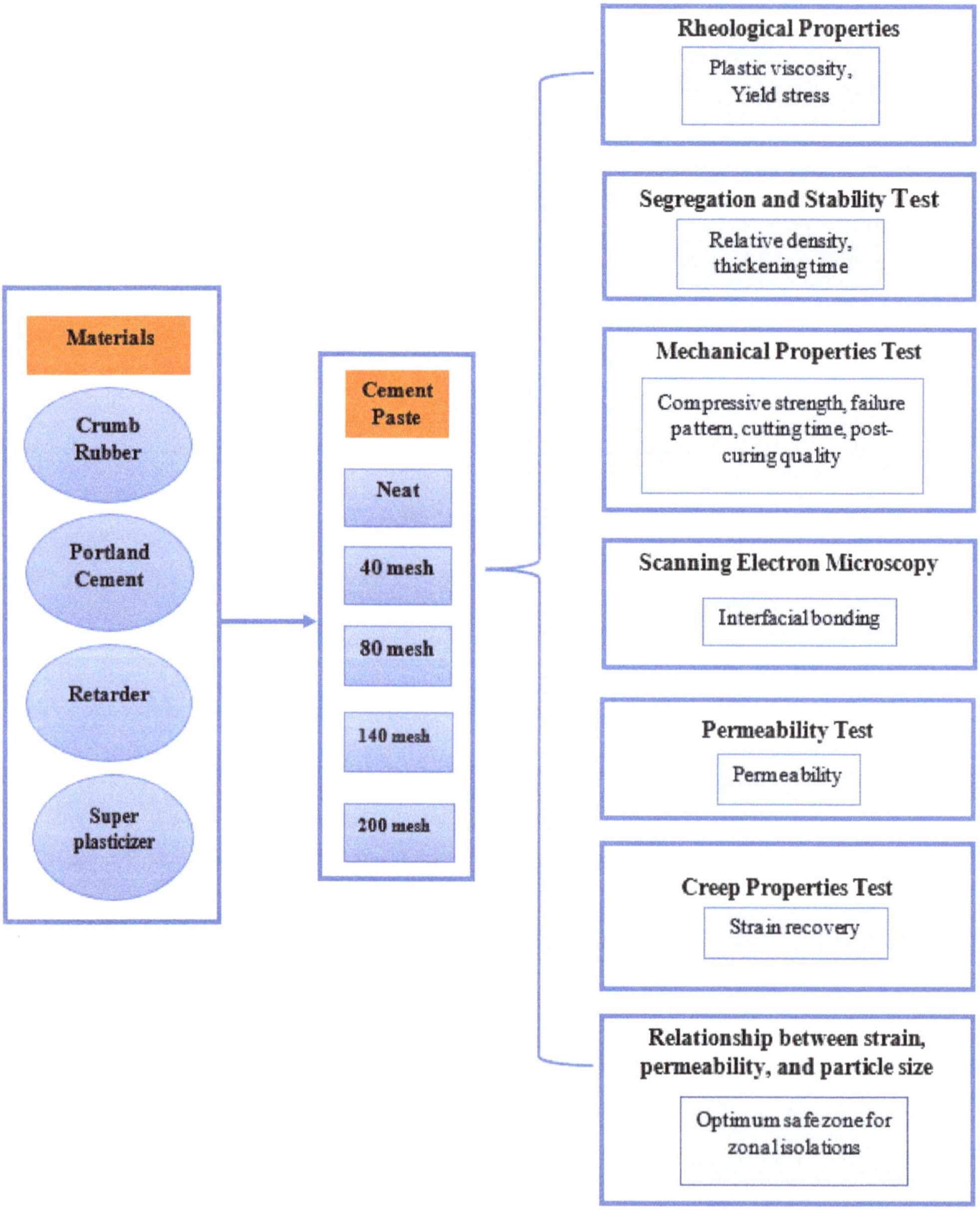

Figure 4. Flowchart of this study.

2.3. Testing Methods

2.3.1. Rheological Properties Measurement

In the initial phase of this research work, several tests were conducted using the finest and coarsest rubber-particle sizes (40- and 200-mesh sizes corresponding to particle sizes of 425 and 75 μm) to observe any significant differences in the thickening time and the rheological properties. Experiments were carried out to obtain a general understanding of the rheology of the slurries and to design subsequent cement slurries for quality purposes.

The preliminary results revealed that the plastic viscosity of the 200-mesh rubber was increased by a factor of two to three times that of the 40-mesh rubber, depending on the quantity of crumb rubber added to the cement.

It was also shown that the addition of 15% and 18% crumb-rubber particles to the neat cement combined with retarder and plasticizer resulted in very thick slurries, making it difficult for them to flow. This effect became more apparent for particles with mesh sizes higher than 80, resulting in inflow indexes higher than 1. In both cases, the slurry exhibited some shear thickening caused by the excessive addition of rubber particles, and thereby became a dilatant fluid in which the changes in viscosity and shear rate were proportional. To obtain an acceptable slurry viscosity for Portland cement, the recommended and optimum water percentage is 38% [61]. During the cement slurry preparation, it was observed that the slurries became very thick, and 38% water was not sufficient. Then, to reduce the dilatant effect, we decided to increase the amount of water to obtain the acceptable slurries. A total of four trials were conducted with rubber particles of sizes 140 and 200 mesh using the equations shown below, and several multiplying factors (0.2, 0.33, 0.32, and 0.42) were achieved after multiple trials:

$$\%\ \text{water in C2} = 38\% + (0.2 \times 15\%) = 41\% \tag{1}$$

$$\%\ \text{water in C3} = 38\% + (0.33 \times 18\%) = 44\% \tag{2}$$

$$\%\ \text{water in D2} = 38\% + (0.32 \times 15\%) = 43\% \tag{3}$$

$$\%\ \text{water in D3} = 38\% + (0.42 \times 18\%) = 46\% \tag{4}$$

where C2 and D2 were samples containing 15% rubber of 140 mesh and 200 mesh sizes, while C3 and D3 were samples containing 18% rubber of 140 mesh and 200 mesh sizes, respectively. The above proportions provided good guidance to prepare slurries with different particle sizes and weight percentages. The superplasticizer was added to enhance bonding between the cement and crumb-rubber particles, thereby providing an acceptable viscosity range. Several trials showed that the slurries containing 15% crumb-rubber particles BWOC or more continuously thickened. Therefore, it was decided to focus only on 2% and 12% cement–crumb-rubber slurries.

The rheological properties were evaluated using an OFITE 800 model viscometer at 200° F and atmospheric pressure. Cement slurries were subjected to a series of shear forces at rates of 3, 6, 100, 200, 300, and 600 rpm in the rotational viscometer. A total of three readings were carefully taken and recorded, and the arithmetic average was then calculated from the three readings. Then, the yield stress and plastic viscosity were calculated using these values.

2.3.2. Segregation and Stability Test

Using the API-suggested procedure 10B-2 [2], a segregation test was performed to compare the density variance between slurry and set cement, and in the entirely cured cement samples from top to bottom. The sample lengths of the cement–crumb-rubber composites were measured, marked at 0.75 inches from the top and bottom, and equally divided into four parts. Marked parts of the samples were carefully cut from top to bottom in that order using a saw (see Figure 5). The final dimensions of each cement sample was 2 inches in diameter and 0.625 inch in length.

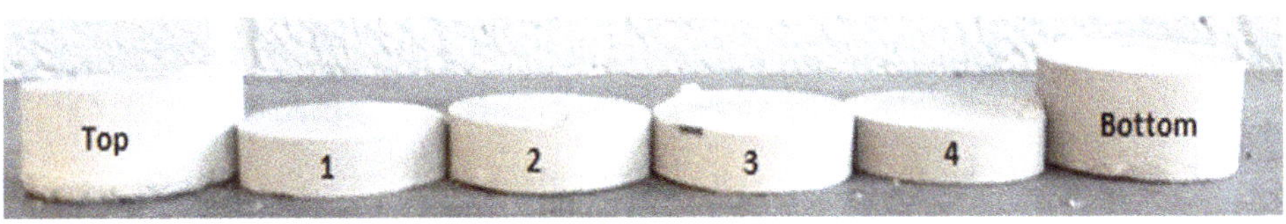

Figure 5. Cement sample cut into segments for stability test.

A beaker containing water and a high-precision digital balance were used for the density measurement. First, each segment was placed on the balance next to the beaker, and the mass was recorded. Then, dental floss was employed to suspend the sample segment in the beaker while ensuring that the segment did not touch the beaker. The wet-segment mass was recorded while ensuring that air bubbles did not adhere to the segment. This process was undertaken for all the sample segments, from which the relative density was estimated using Equation (5):

$$\rho_{rel} = \frac{m_{air}}{m_{water}} \tag{5}$$

where ρ_{rel} = relative density, m_{air} = mass (g) in air, and m_{water} = mass in water. The experiments showed that the prepared rubber–cement composite was stable (ρ_{rel} = 1.64–1.82) for all particle sizes, and that segregation rarely took place while curing. Therefore, we concluded that the rubber particles were evenly distributed throughout the sample volume.

2.3.3. Scanning Electron Microscopy Analysis

Interfacial bonding between the crumb rubber and cement sample was investigated using a scanning electron microscope (SEM). Four different samples were analyzed using a Hitachi SU3500 equivalent scanning electron microscope at the University of Louisiana at Lafayette's Microscopy Center. All samples were coated with gold to obtain the best image. The operating condition maintained in secondary electron (SE) mode with an accelerating voltage of 15 kV and magnification varied at 100X–200X with a fixed resolution of 500 μm.

2.3.4. Mechanical Properties Measurement

The compressive strengths and failure patterns were investigated in this study for the cement–crumb-rubber composites with varying meshes using an MTS brand, 100 kN (22 kips) universal testing machine at room temperature. The cement samples underwent curing for 24 h at 200 °F and 7000 psi. The compressive strength was estimated as an average from three observations for each cement sample with a 2 in diameter and a 0.625 in length.

2.3.5. Permeability Measurement

All of the permeability tests in this study were conducted at the Halliburton Cementing lab facility in Houston, TX, USA. The permeability tests were conducted using water flow and a cylindrical sample with a 2 in diameter × 0.625 in length. To hold the cylindrical sample in place, a rubber seal was used. On one side, high-pressure water was pumped into the sample until the water flowed through the cement sample and reached atmospheric pressure. Then, the permeability was estimated as a function of the flow rate of the water and the pressure drop, viscosity, and sample dimensions using Darcy's law,

$$q = \frac{A\,\Delta P}{\mu\,L} \tag{6}$$

where q = flow rate, A = cross-sectional area, μ = fluid viscosity, k = permeability, and $\Delta P/L$ = pressure drop across the length.

2.3.6. Strain-Recovery Tests

The time-dependent behavior of the cement slurries was determined by the creep test. This test was conducted by utilizing the MTS universal load frame and an extensometer. Before commencing the test, an extensometer was firmly attached to each cement sample to measure the strain and track the deformation behavior. During the test, the load frame and extensometer were synchronized to track the strain reading simultaneously. The test duration was one hour for each test. A constant load was applied for 30 min for strain development in the sample, and then kept unloaded for another 30 min for each sample to measure the amount of strain recovery. The strain evolution was continuously monitored during the recovery process.

3. Results and Discussion

3.1. Rheological Properties

The rheological results showed that the excessive addition of rubber particles (15% BWOC or higher) to the cement slurry resulted in very thick slurries that had much difficulty in flowing. The retarder's effects on the cement's rheological properties containing 2% and 12% crumb rubber by weight for the two mesh sizes of 40 and 200 were assessed. Figure 6 shows that the viscosity and yield stress decreased as the retarder was increased from 0.1% to 0.4% and the mesh increased. This indicated that the slurry was thinning with an increase in the retarder, which acted as a dispersant. Moreover, it can be seen that the addition of 2% rubber resulted in almost similar properties for the neat cement for the 200 mesh, but a higher plastic viscosity and a lower yield stress when 12% rubber particles was added.

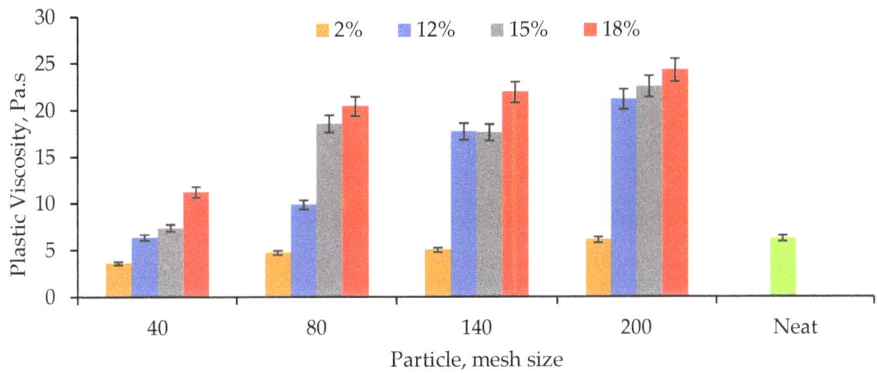

(a)

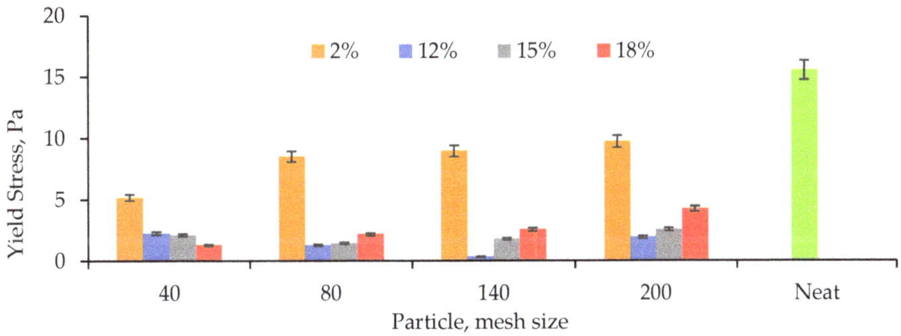

(b)

Figure 6. (**a**) Plastic viscosity and (**b**) yield stress for various particles sizes using three different weight percentages of rubber particles (for 0.3% retarder and various plasticizers (0.03–0.24%; plasticizers were varied and conducted several trials to obtain the optimum amount of plasticizer; e.g., 0.11% was found to be optimum for 40 mesh after trials from 0.03–0.11%; similarly, 0.16% was obtained for 80 mesh and 0.24% for both 140 and 200 mesh).

The effects of adding the superplasticizer to the cement, crumb rubber, and retarder mixture on the rheological properties of the composite slurries were also assessed. Figure 6a,b show the effect of the rubber-particle size on the plastic viscosity and yield stress for 2%, 12%, 15%, and 18% of rubber BWOC. The amount of retarder was adjusted according to the number of rubber particles present in the slurry. It was observed that the plastic viscosity increased, whereas the yield stress decreased, as the weight percentage and particle size increased. The plastic viscosity for the 2% and 12% rubber particles with mesh sizes less than 80 were not affected much as compared to the neat cement, which had a plastic viscosity of 6.2 Pa. However, the plastic viscosity became high for smaller particles sizes (higher mesh sizes) and higher rubber concentration (weight fractions). This increase in plastic viscosity indicated that the shear stress between the particles and cement increased. It became more dominant in the cement paste when the concentration of rubber particles was increased with higher-mesh-size particles. Moreover, the figure shows that the yield stress was a bit lower than that of the neat cement of 15.5 Pa. Similar rheological behaviors were also investigated by Long et al. [44].

3.2. Thickening Time

Thickening time is of great importance when cementing a long casing string. Shorter thickening times are hazardous in these situations, as the cement will harden before reaching the set depth. A consistometer was used to measure the thickening time for three slurries consisting of (i) 2% BWOC and retarder at 0.2% BWOC, (ii) 12% BWOC and retarder at 0.4% BWOC, and (iii) 12% containing 0.3% retarder and superplasticizer at varying concentrations. Figure 7 shows the thickening time for the cement slurries with different rubber sizes and concentrations. The figure also shows that the effect of the superplasticizer on the thickening time was significant with an increase in mesh size. For a given mass concentration of rubber particles in the cement, the rubber volume was inversely proportional to its density. Since the 200-mesh rubber had a lower density (0.43 g/cm^3) than that of the 40 mesh (0.5 g/cm^3), the volume occupied by finer particles was more than that of the coarser ones, resulting in a better distribution of rubber particles in the cement slurry. Thus, higher thickening times were produced for finer mesh sizes, accompanied by an increase in viscosity (time-independent).

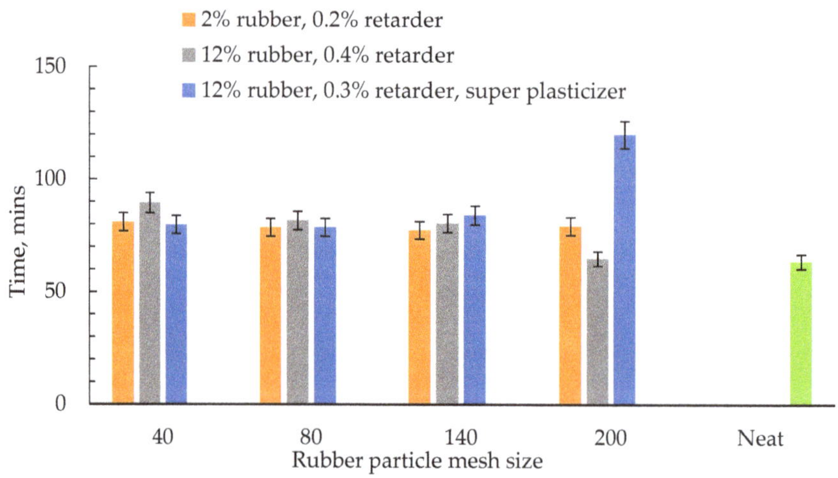

Figure 7. Thickening Time vs. Rubber Particle Size.

3.3. Cement–Crumb-Rubber Postcuring Quality

The quality of the cement slurries after curing was also studied. During the first phase of our investigation, it was found that most of the rubber particles migrated to the top of the slurry due to their hydrophobic nature while curing, thereby resulting in an uneven particle distribution within the samples and poor bonding between the cement and rubber (see Figure 8a) with no other additives. Thus, the samples could be chipped off with the tip of a finger. Based on these observations, we felt it was necessary to enhance the bonding strength between the cement and rubber particles by adding a retarder and superplasticizer. A transition from poor to excellent bonding is shown in Figure 8. This was achieved by adding the retarder to the cement and rubber particles, which moderately improved the bonding (see Figure 8b). The bonding was further enhanced (Figure 8c) by adding a superplasticizer, ADVA 575, which behaved as a water-reducing agent and dispersant in the slurry containing particles of the cement, rubber, and retarder. Proper care had to be taken while adding the superplasticizer, as an excess of this water-reducing agent in cement slurries may cause the slurry material to separate and become unstable.

Figure 8. Cement–crumb-rubber samples showing (**a**) poor, (**b**) moderate, and (**c**) excellent bonding strength between rubber particles and Portland cement.

3.4. Bonding between Cement and Crumb-Rubber Particles

SEM analyses of several cement pastes vividly illustrated the bonding between the cement and crumb-rubber particles (Figure 9). Figure 9a,b show an SEM picture of cement material for only rubber particles under 100X and 200X magnifications, respectively. The figure shows that the rubber particles were surrounded by a cement matrix with no bonding between the cement and rubber particles, resulting in poor-quality samples; whereas Figure 9c shows that the addition of a retarder lessened the space between the cement paste and rubber particles, leading to better bonding. The bonding was enhanced even further by adding a plasticizer with the retarder, as seen in Figure 9d in which the gap between the rubber particles and cement was almost closed, thereby leading to good bonding.

3.5. Mechanical Properties Measurement: Unconfined Compression Strength and Failure Pattern

This section summarizes the compressive strength of the cement–rubber composites for four different mesh sizes and containing 2% and 12% rubber particles. Figure 10 shows a reduction in the compressive strength for all cement composites. The figure shows that the addition of 12% rubber particles BWOC resulted in drops in strength of almost 55% and 40% for the 40- and 200-mesh sizes, respectively. This reduction was more pronounced for all meshes with 12% rubber particles. Nevertheless, this reduction was still acceptable, as the minimum required cement strength ranges between 3.4 and 17.0 MPa [38], thereby identifying the use of rubber particles at 12% BWOC as the optimum composition for field applications. It was found that the cement paste with 140-mesh rubber particles at 12% BWOC, 0.3% retarder, and 0.16% superplasticizer was the optimum concentration with favorable operating rheological properties (consistency index, K ~ 0.22 lb/(100 $ft^2 \times s^n$); flow behavior index, n ~ 0.7; yield stress ~ 6 Pa, and plastic viscosity ~ 15 Pa).

The addition of the superplasticizer was observed to augment the bonding between the cement and rubber particles. This effect is depicted in Figure 11; the compressive strength increased by almost 15% after adding a superplasticizer.

The effect of curing time was also studied, and revealed that the composite strength containing 12% crumb rubber and additives improved the compressive strength of the composite. Figure 12 shows an increase in the compressive strength for higher curing times and as the particle mesh size increased.

The fracture mode of the samples under compression was also investigated. It was observed that the neat cement exhibited a brittle type of failure (see Figure 13a); however, the cement slurries containing substantial amounts of crumb-rubber particles had high levels of ductility, which is a desirable quality in cementing operations. This behavior resulted in a reduction in crack development in the cement–rubber composite samples as compared to the neat cement (see Figure 13b,c). We also noticed that the composite samples bulged as a result of their ductile behavior, thereby delaying the development of cracks.

Figure 9. Cement sample containing: (**a**) only rubber particles at 100X; (**b**) only rubber particles at 200X; (**c**) rubber particles and retarder; (**d**) rubber particles, retarder, and plasticizer.

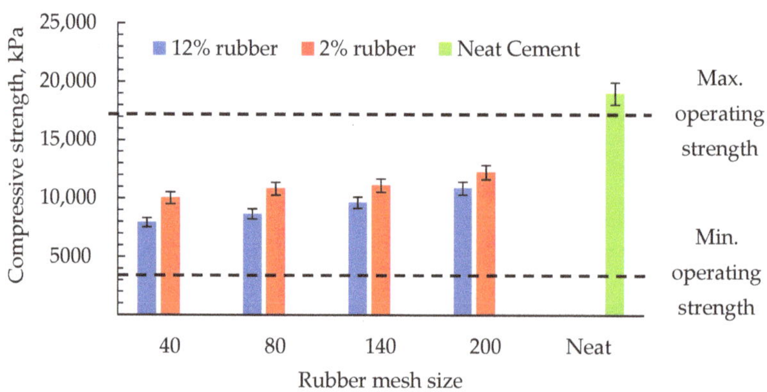

Figure 10. The effect of rubber-particle size on the compressive strength of cement slurries.

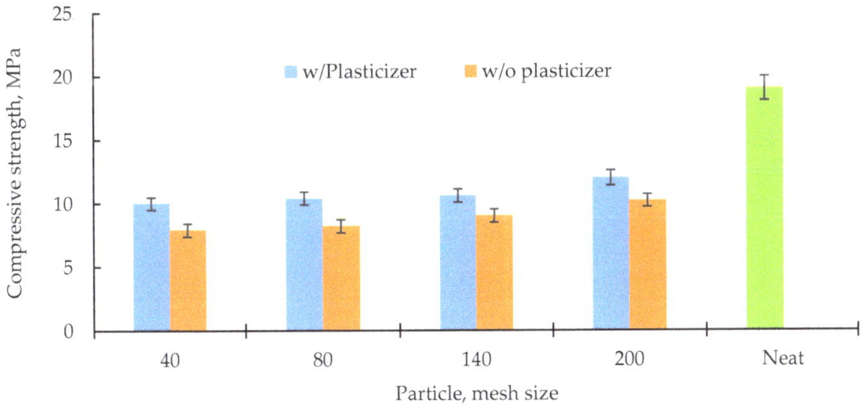

Figure 11. Effect of plasticizer on the compressive strength of cement slurries.

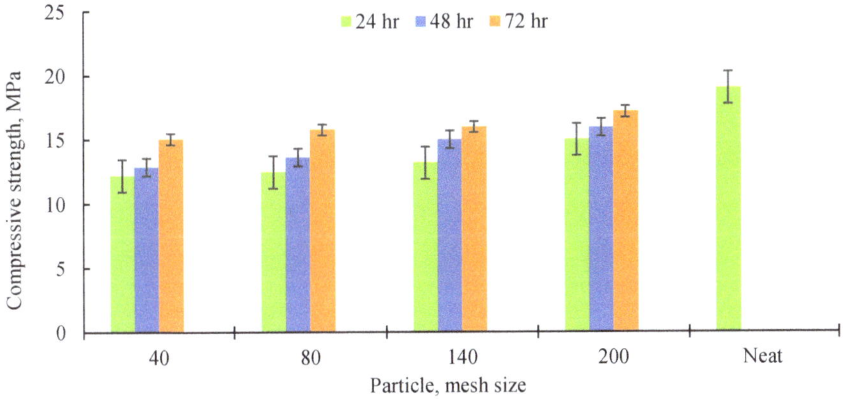

Figure 12. Effect of curing time on the compressive strength of cement slurries.

(**a**) (**b**) (**c**)

Figure 13. Comparison of fracture patterns (**a**) neat cement; (**b**) rubber–cement; (**c**) rubber–cement and additives.

3.6. Permeability Results

It is usually expected to have the lowest permeability present in the cement paste to ensure cement integrity and high-quality completion. The permeability results demonstrated that the cement–rubber slurry with the largest particle size (40 mesh) had the highest permeability (0.04102 md), whereas the smallest rubber-particle size (20 mesh) exhibited the lowest permeability, with a value of 0.00319 md. Thus, it was evident that the permeability was reduced with a reduction in the rubber-particle size compared to the neat cement's permeability, 0.0087 md. We observed that the addition of 12% BWOC with 200-mesh crumb rubber decreased the permeability by nearly 64% compared to the base cement. Figure 14 shows the permeability distribution of the cement paste for different sizes of rubber particles compared to that of the neat cement.

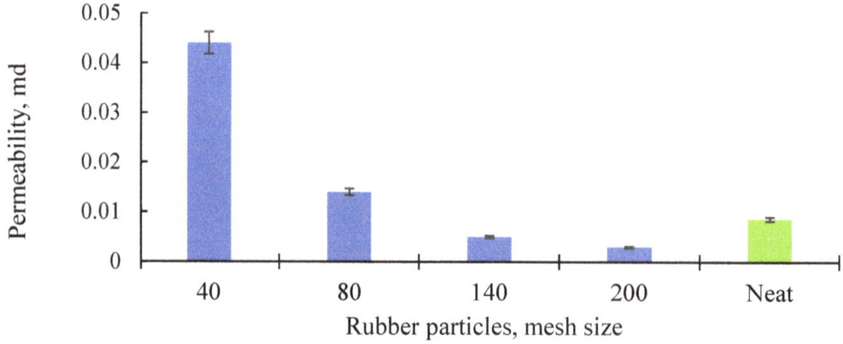

Figure 14. Permeability test plot.

The relationship between the permeability and rubber-grain size can be found in the following equation:

$$k = \left(2.098 \times 10^{-5}\right) d_r^{1.358} \quad (7)$$

where k is the permeability in md and d_r is the rubber grain size in μm. Figure 15 shows the relationship between the permeability and the rubber-grain size.

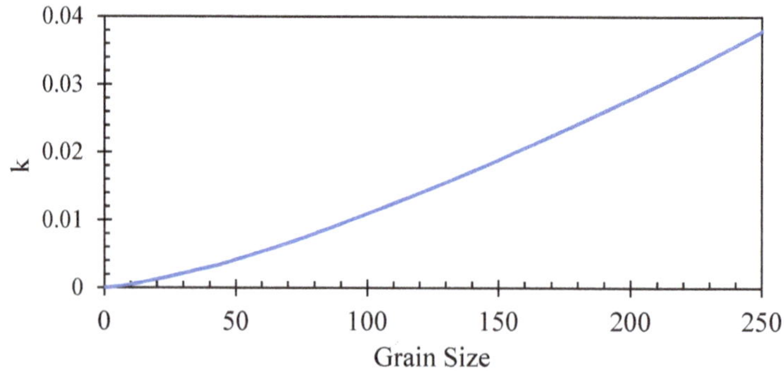

Figure 15. Relationship between permeability and rubber-grain size.

3.7. Creep Testing for Strain Recovery

Analyzing the deformation characteristics in cement materials is of great importance to sustainable well completion. Downhole cementing undergoes sustained and continuous in situ stress, which causes deformation in the cement materials; failure to ensure proper cement integrity to withstand such deformation in the long term has resulted in serious consequences in hydrocarbon production [54].

A creep test usually measures the strain developed in a material under a constant load with time. Creep is a time-dependent property of cement materials. When cement materials are loaded under compression, they exhibit a small amount of strain that is invisible and difficult to observe visually. That is why sensitive deformation-measuring devices such as extensometers are utilized to track the strain.

In this study, creep tests were conducted with cement slurries exposed to five different stress levels—1000, 1300, 1600, 1800, and 2000 pounds of load. The total test duration was 1 h, and the total time was split equally for loading and unloading to half an hour to quantify the mechanical response of the cement. Three samples were tested under the loading and unloading cycle, and the average strain was calculated from the three consecutive observations. Since the strain value was small, each strain value was multiplied by a factor of 106. Figure 16 shows the creep behaviors of the rubber–cement composites with different rubber meshes compared to that of the neat cement.

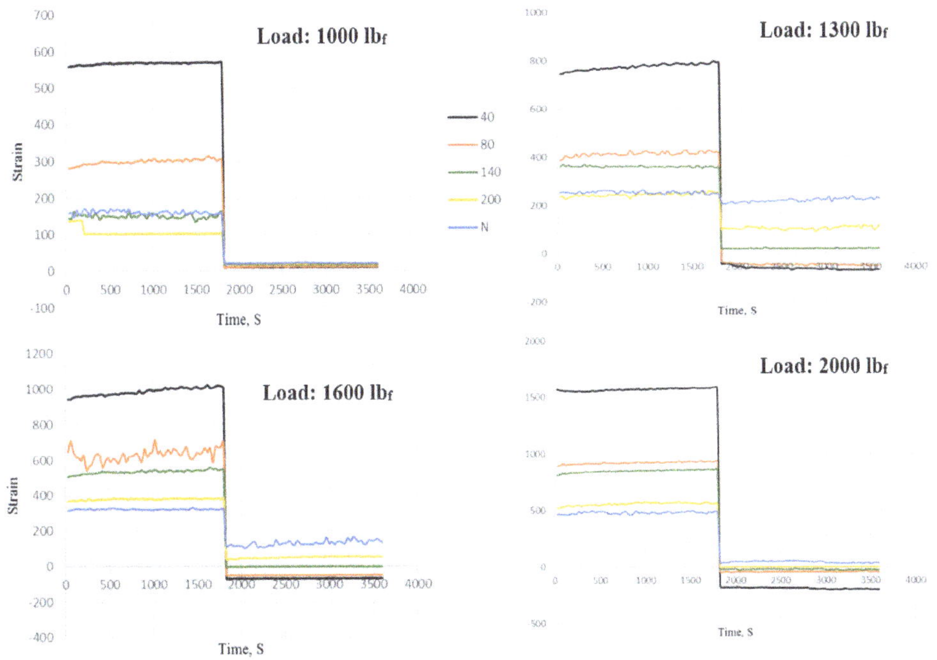

Figure 16. Strain plot during and after loading at 1000–2000 lbs.

Materials return to their original positions after unloading due to elasticity, and plasticity is responsible for not fully returning back to the original positions. The rubber-particle size affected the strain development in the specimens. The creep results illustrated the relationship between the rubber-particle size and the amount of the strain was proportional. After unloading, the cement with 40-mesh rubber showed a high amount of recovery by returning back to its original position, since it reached or crossed below the zero origins, as seen in Figure 16 (1300 lbf, 1600 lbf, and 2000 lbf).

The obtained results showed that the neat cement did not exhibit strain recovery at any of the applied loads, which substantiated the brittle nature of the neat cement. The cement slurries with the largest size of the crumb-rubber particles showed the highest strain recoveries, which was the signature of the elastic nature of the cement slurries, which might have had the capability to reduce the microcracking in the cement-casing bonding.

3.7.1. Strain Comparison for Different Mesh Sizes

To provide a better understanding of how the permeability and strain were related, an extensive comparison was made based on the obtained strain results in the experiment. Figure 17 shows the averaged strain recovery for all samples (40, 80, 140, and 200 mesh; and neat cement) after unloading at 1000 to 2000 lb_f. The graph was constructed using a total of 20 strain-recovery values at different applied stresses for each rubber-particle mesh size.

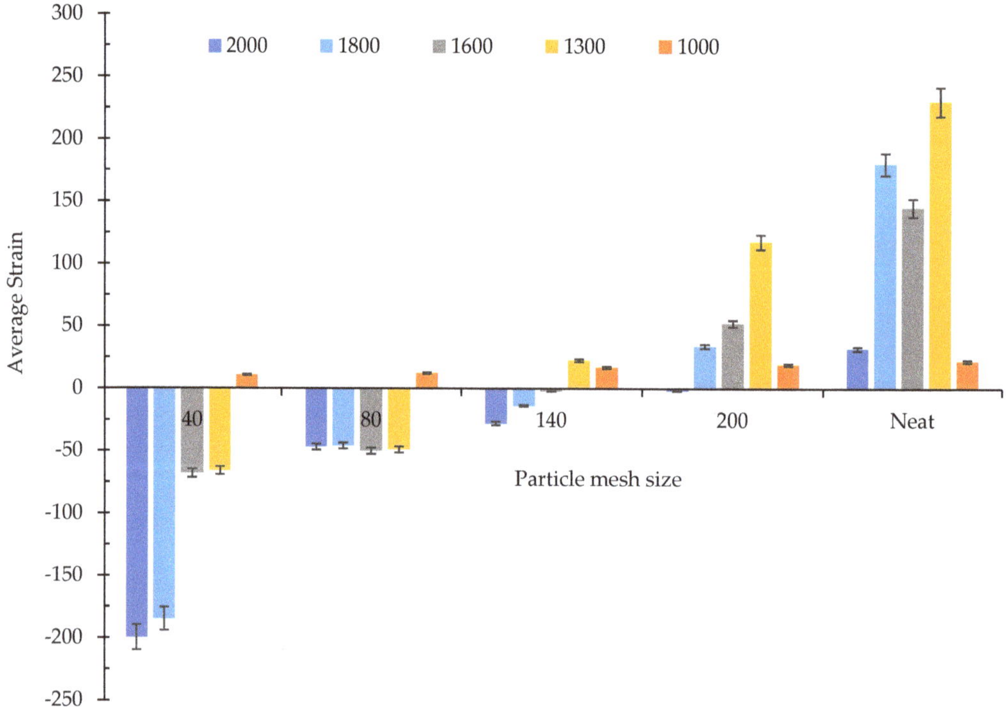

Figure 17. Averaged strain recovery for all samples (40, 80, 140, and 200 mesh; and neat cement) after unloading at 1000 to 2000 lbf.

To ensure strong wellbore integrity, the lowest permeability of the cement slurries was desirable. The current experiment showed that the cement–rubber composite with the smallest particle size (200 mesh) exhibited the lowest permeability, while the largest rubber-particle size (40 mesh) showed the highest permeability value. We believe that the finer the mesh, the better interlocking between particles. However, at all applied stresses, the 40-mesh-rubber–cement had the highest amount of strain recovery, and the 200-mesh formulation exhibited the lowest strain recovery.

3.7.2. Relationship between Permeability, Strain, and Mesh Size

With the help of the strain and permeability values, a relationship was developed between three parameters; namely, the permeability, average strain, and mesh size of rubber particles (Figure 18). As shown in the figure, the strain values for the 1000 lbf load were all positive, which was in contrast to what the scope of this research demanded. The majority of the strain values for the other loads fell well below the threshold value, which in this case was set at zero. The permeability values for the composite were expected to be lower than the values for the neat cement to satisfy zonal isolation. Concerning the mesh size of the rubber particles, the upper and lower limits were set to 140 mesh and 100 mesh. These mesh-size limits were determined based on the levels at which the optimum strain and permeability were obtained.

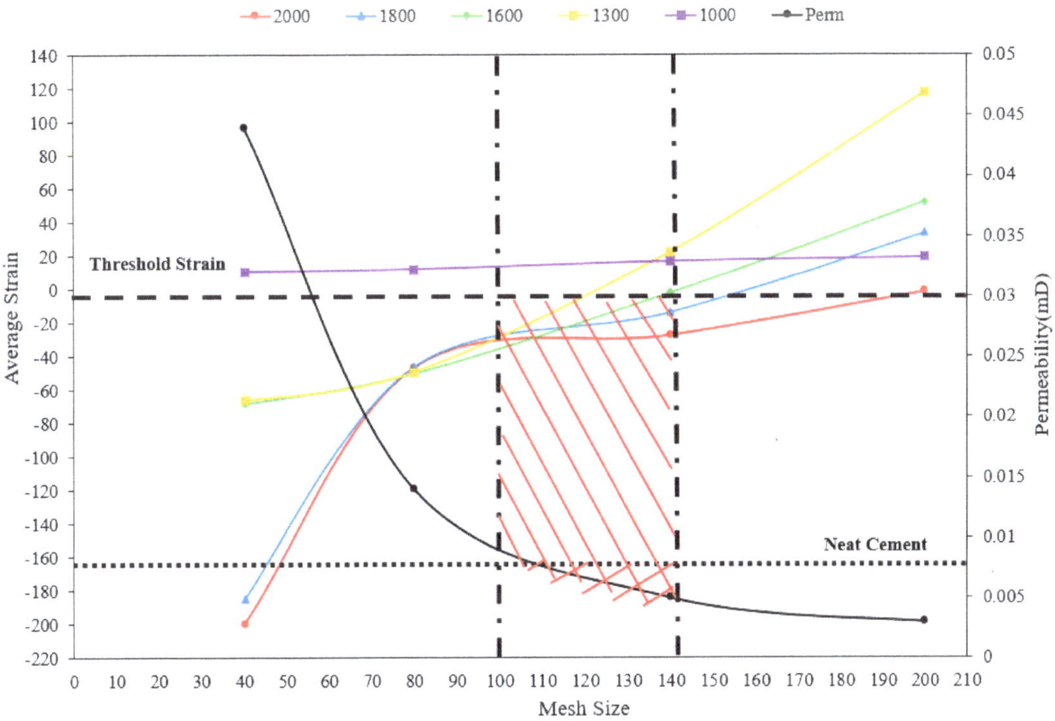

Figure 18. Relationship between permeability, average strain, and mesh size.

The inclined red line region in Figure 18 shows the optimum relationship between permeability, average strain, and mesh size. The optimum mesh operating region will result in better creep, and consequently, will help to lessen the problem of the microannulus.

4. Conclusions

Rubber particles of four different sizes (40, 80, 140, and 200) were used in this research. Cement slurries with 12%, 15%, and 18% BWOC of the crumb-rubber material were prepared in this study to assess the prospect of utilizing crumb-rubber composites to enhance the cement's rheology, mechanical properties, strain recovery, and permeability. The results demonstrated that the cement slurry, after mixing in the crumb-rubber material, had improved properties. The following conclusions were drawn from this study:

- The quantity and size of the rubber particles affected the rheology of the cement slurry, which showed a transition from thinning to thickening behavior; crack generation was minimized when crumb-rubber particles were added to the cement as compared to the neat cement. The addition of 15% and 18% crumb-rubber particles to the cement was not feasible for application in the field.
- The addition of rubber particles at 2% BWOC slightly affected the cement slurry's viscosity as compared to 12% rubber particles, and increased the viscosity of the cement slurry by 50% and 300% for 40- and 200-mesh sizes, respectively.
- The addition of the superplasticizer resulted in a 15% increase in the compressive strength as compared to that of the cement–crumb-rubber composite with the retarder, and improved the bonding between the cement and crumb-rubber particles. We also

found that the addition of rubber particles reduced the cement's strength, which mostly depended on the variation in the rubber-particle size and the quantity.

- Water-permeability tests showed that the addition of 12% BWOC with 200-mesh crumb rubber decreased the permeability by nearly 64% compared to the base cement.
- At all applied loads, the neat cement did not exhibit any strain recovery, which substantiated that the neat cement was brittle. The cement slurries with the largest rubber particles exhibited an elastic nature by demonstrating the highest strain recoveries, thereby offering the capability of reducing the microcracking of the cement-casing sheath.
- Based on the relationship between the permeability, strain, and mesh sizes at different loads, an optimum operating region was established that included cement pastes with 100–140-mesh rubber particles at 12% BWOC, retarder, and superplasticizer with favorable permeability and rheological properties. The plastic viscosity substantially increased beyond this concentration.

Author Contributions: Conceptualization, A.C.S. and F.N.; methodology, A.C.S. and F.N.; validation, A.C.S. and F.N.; formal analysis, F.N. and A.B.A.; investigation, A.C.S. and F.N.; resources, A.C.S. and F.N.; data curation, A.B.A., F.N., and K.G.S.; writing—original draft preparation, A.B.A., A.C.S., and F.N.; writing—review and editing, F.N.; supervision, A.C.S. and F.N.; project administration, A.C.S. and F.N.; funding acquisition, F.N. All authors have read and agreed to the published version of the manuscript.

Funding: This research received no external funding.

Institutional Review Board Statement: Not applicable.

Informed Consent Statement: Not applicable.

Data Availability Statement: The data presented in this study are available upon request from the corresponding author.

Acknowledgments: The authors appreciate the support of Halliburton, which provided the class H cement and its additives, as well as that of Lehigh Technologies Inc., which supplied various crumb-rubber particles used in this research work. The authors thank Texas A & M International University for supporting this work through a university research grant.

Conflicts of Interest: The authors declare no conflict of interest.

References

1. El-Gamal, S.M.A.; Hashem, F.S.; Amin, M.S. Influence of carbon nanotubes, nano-silica, and nanometakaolin on some morphological-mechanical properties of oil well cement pastes subjected to elevated water curing temperature and regular room air curing temperature. *Constr. Build. Mater.* **2017**, *146*, 531–546. [CrossRef]
2. Ahmed, R.M.; Takach, N.E.; Khan, U.M.; Taoutaou, S.; James, S.; Saasen, A.; Godøy, R. Rheology of foamed cement. *Cem. Concr. Res.* **2009**, *39*, 353–361. [CrossRef]
3. Rodrigues, E.C.; de Andrade Silva, F.; de Miranda, C.R.; de Sá Cavalcante, G.M.; de Souza Mendes, P.R. An appraisal of procedures to determine the flow curve of cement slurries. *J. Pet. Sci. Eng.* **2017**, *159*, 617–623. [CrossRef]
4. Mahmoud, A.A.; Elkatatny, S. Mitigating CO_2 Reaction with Hydrated Oil Well Cement under Geologic Carbon Sequestration Using Nanoclay Particles. *J. Nat. Gas Sci. Eng.* **2019**, *68*, 102902. [CrossRef]
5. Nelson, E.B. *Well Cementing*; Elsevier: New York, NY, USA, 1990.
6. Tao, C.; Kutchko, B.G.; Rosenbaum, E.; Wu, W.-T.; Massoudi, M. Steady Flow of a Cement Slurry. *Energies* **2019**, *12*, 2604. [CrossRef]
7. Kiran, R.; Teodoriu, C.; Dadmohammadi, Y.; Nygaard, R.; Wood, D.A.; Mokhtari, M.; Salehi, S. Identification and evaluation of well integrity and causes of failure of well integrity barriers (A review). *J. Nat. Gas Sci. Eng.* **2017**, *45*, 511–526. [CrossRef]
8. Nygaard, R.; Salehi, S.; Weideman, B.; Lavoie, R.G. Effect of Dynamic Loading on Wellbore Leakage for the Wabamun Area CO_2-Sequestration Project. *J. Can. Pet. Technol.* **2014**, *53*, 69–82. [CrossRef]
9. Nath, F.; Kimanzi, R.J.; Mokhtari, M.; Salehi, S. A novel method to investigate cement-casing bonding using digital image correlation. *J. Pet. Sci. Eng.* **2018**, *166*, 482–489. [CrossRef]
10. Kremieniewski, M.; Rzepka, M. Correlation of permeability and parameters describing the structure of hardened cement slurries used to seal boreholes in the area of the Pomeranian Basin. *Naft. Gaz.* **2015**, *10*, 737. [CrossRef]
11. Kremieniewski, M.; Rzepka, M. The causes and effects of gas flow in the cemented annular space of borehole and methods to prevent this phenomenon. *Naft. Gaz.* **2016**, *72*, 722. [CrossRef]
12. Taylor, H.F.W. *Cement Chemistry*, 2nd ed.; Thomas Telford Publishing: London, UK, 1997.

13. Ahmed, A.; Mahmoud, A.A.; Elkatatny, S.; Chen, W. The Effect of Weighting Materials on Oil-Well Cement Properties While Drilling Deep Wells. *Sustainability* **2019**, *11*, 6776. [CrossRef]
14. Kremieniewski, M. Recipe of Lightweight Slurry with High Early Strength of the Resultant Cement Sheath. *Energies* **2020**, *13*, 1583. [CrossRef]
15. Kremieniewski, M. Improvement of the early mechanical strength of cement sheath formed from lightweight cement slurry. *Naft. Gaz.* **2018**, *74*, 606. [CrossRef]
16. Eric, B.-B.; Joel, F.; Grace, O. Oil Well Cement Additives: A Review of the Common Types. *Oil Gas Res.* **2016**, *2*, 112. [CrossRef]
17. Calvert, D.G.; Smith, D.K. API oil well-cementing practices. *J. Pet. Technol.* **1990**, *42*, 1364–1373. [CrossRef]
18. Mahmoud, A.A.; Elkatatny, S. Improved durability of Saudi Class G oil-well cement sheath in CO2 rich environments using olive waste. *Constr. Build. Mater.* **2020**, *262*, 120623. [CrossRef]
19. Elkatatny, S. Development of a Homogenous Cement Slurry Using Synthetic Modified Phyllosilicate while Cementing HPHT Wells. *Sustainability* **2019**, *11*, 1923. [CrossRef]
20. Sun, X.; Wu, Q.; Zhang, J.; Qing, Y.; Wu, Y.; Lee, S. Rheology, curing temperature and mechanical performance of oil well cement: Combined effect of cellulose nanofibers and graphene nano-platelets. *Mater. Des.* **2017**, *114*, 92–101. [CrossRef]
21. Mahmoud, A.A.; Elkatatny, S.; Mahmoud, M. Improving Class G Cement Carbonation Resistance Using Nanoclay Particles for Geologic Carbon Sequestration Applications. In Proceedings of the 2018 Abu Dhabi International Petroleum Exhibition & Conference, Abu Dhabi, United Arab Emirates, 12–15 November 2018.
22. Mahmoud, A.A.; Elkatatny, S. Effect of the Temperature on the Strength of Nanoclay-Based Cement under Geologic Carbon Sequestration. In Proceedings of the 2019 AADE National Technical Conference and Exhibition, Denver, CO, USA, 9–10 April 2019.
23. Medina, G.; Sáez del Bosque, I.F.; Frías, M.; Sánchez de Rojas, M.I.; Medina, C. Energy performance and calorimetric behavior of cements bearing granite sludge. *Powder Technol.* **2019**, *356*, 517–527. [CrossRef]
24. Bu, Y.; Du, J.; Guo, S.; Liu, H.; Huang, C. Properties of oil well cement with a high dosage of metakaolin. *Constr. Build. Mater.* **2016**, *112*, 39–48. [CrossRef]
25. Soares, L.W.O.; Braga, R.M.; Freitas, J.C.O.; Ventura, R.A.; Pereira, D.S.S.; Melo, D.M.A. The effect of rice husk ash as pozzolan in addition to cement Portland class G for oil well cementing. *J. Pet. Sci. Eng.* **2015**, *131*, 80–85. [CrossRef]
26. Mahmoud, A.A.; Elkatatny, S. The Effect of Silica Content on the Changes in the Mechanical Properties of Class G Cement at High Temperature from Slurry to Set. In Proceedings of the 53rd US Rock Mechanics/Geomechanics Symposium, New York, NY, USA, 23–26 June 2019.
27. Jeon, I.K.; Qudoos, A.; Jakhrani, S.H.; Kim, H.G.; Ryou, J.-S. Investigation of sulfuric acid attack upon cement mortars containing silicon carbide powder. *Powder Technol.* **2020**, *359*, 181–189. [CrossRef]
28. Yao, G.; Wang, Q.; Wang, Z.; Wang, J.; Lyu, X. Activation of hydration properties of iron ore tailings and their application as supplementary cementitious materials in cement. *Powder Technol.* **2020**, *360*, 863–871. [CrossRef]
29. Anjos, M.A.S.; Martinelli, A.E.; Melo, D.M.A.; Renovato, T.; Souza, P.D.P.; Freitas, J.C. Hydration of oil well cement containing sugarcane biomass waste as a function of curing temperature and pressure. *J. Pet. Sci. Eng.* **2013**, *109*, 291–297. [CrossRef]
30. Nagrockiene, D.; Pundiene, I.; Kicaite, A. The Effect of Cement Type and Plasticizer Addition on Concrete Properties. *Constr. Mater.* **2013**, *45*, 324–331. [CrossRef]
31. Calvert, D.J.; Smith, D.K. API oil-well cementing practices. In Proceedings of the Offshore Technology Conference, Houston, TX, USA, 7 May 1990.
32. Bourgoyne, A.T.; Millheim, K.T.; Chenevert, M.E. *Applied Drilling Engineering*; Evers, J.F., Pye, D.S., Eds.; SPE Textbook Series: Richardson, TX, USA, 1991; Chapter 3; pp. 85–103.
33. Klieger, P. *Significance of Tests and Properties of Concrete and Concrete Materials*; Klieger, P., Lamond, J.F., Eds.; ASTM International: Washington, DC, USA, 1994; Issue 169 part.
34. Baumgarte, C.; Thiercelin, M.; Klaus, D. Case Studies of Expanding Cement to Prevent Microannular Formation. In Proceedings of the SPE Annual Technical Conference and Exhibition, Houston, TX, USA, 3–6 October 1999. [CrossRef]
35. Bois, A.P.; Garnier, A.; Rodot, F.; Saint-Marc, J. How to prevent loss of Zonal isolation through a comprehensive analysis of micro-annulus formation. SPE-124719-PA. *SPE Drill. Completion* **2011**, *26*, 13–31. [CrossRef]
36. Karakurt, C. Microstructure properties of waste tire rubber composites. *J. Mater. Cycles Waste Manag.* **2015**, *17*, 422–433. [CrossRef]
37. Abdul-Rahman, R.; Chong, A. Cementing multilateral well with latex cement. In Proceedings of the SPE/IADC 37623, SPE/IADC Drilling Conference, Amsterdam, The Netherlands, 4–6 March 1997. [CrossRef]
38. API. *API 10B-2 Recommended Practice for Testing Well Cements*, 2nd ed.; API: Washington, DC, USA, 2013.
39. Atrushi, D. Tensile and Compressive Creep of Early Age Concrete: Testing and Modeling. Ph.D. Thesis, Norwegian University of Science and Technology, Trondheim, Norway, March 2003.
40. Chou, L.; Lin, C. Improving rubber concrete by waste organic sulfur compounds. *Waste Manag. Resour.* **2009**, *28*, 29–35. [CrossRef]
41. Clark, P.E.; Sundaram, L.; Balakrishnan, M. Yield points in Oilfield cement Slurries. In Proceedings of the SPE Eastern Regional Meeting, Columbus, OH, USA, 31 October–2 November 1990. SPE 21279. [CrossRef]
42. Najim, K.B.; Hall, M.R. A review of the fresh/hardened properties and applications for plain- (PRC) and self-compacting rubberized concrete (SCRC). *Constr. Build. Mater.* **2010**, *24*, 2043–2051. [CrossRef]
43. Fiore, A.; Marano, G.C.; Marti, C. On the fresh/hardened properties of cement composites incorporating rubber particles from recycled tires. *Adv. Civ. Eng.* **2014**, *2014*, 876158. [CrossRef]

44. Long, G.; Fyes, D.; Khayat, K.; Yahia, A. Efficiency of waste tire rubber aggregate on the rheological properties and compressive strength of cementitious materials. *J. Sustain. Cem. -Based Materials.* **2014**, *3*, 201–211. [CrossRef]
45. Huang, B.; Shu, X.; Cao, J. A two-staged surface treatment to improve properties of rubber-modified cement composites. *Constr. Build. Mater.* **2012**, *7*, 270–274. [CrossRef]
46. Onana, D.D.; Goodwin, K.J.; McPherson, T.W. Elastomeric Composites for use in Well Cementing Operations. SPE-26572-MS. In Proceedings of the SPE Annual Technical Conference and Exhibition, Houston, TX, USA, 3–6 October 1993. [CrossRef]
47. Shafieyzadeh, M. Prediction of compressive strength of concretes containing silica fume and styrene-butadiene rubber (SBR) with the mechanical model. *Int. J. Concr. Struct. Mater.* **2013**, *7*, 295–301. [CrossRef]
48. Wu, H.; Lim, Y.M.; Li, V.C. Application of Recycled Tyre Cord in Concrete for Shrinkage Crack Control. *J. Mater. Sci. Lett.* **1996**, *15*, 1828–1831. [CrossRef]
49. Carrier, B.; Vandamme, M.; Pellenq, R.J.; Bornert, M.; Ferrage, E.; Hubert, F.; Van Damme, H. Effect of Water on Elastic and Creep Properties of Self-Standing Clay Films. *Langmuir* **2016**, *32*, 1370–1379. [CrossRef] [PubMed]
50. Al Awad, M.N.; Fattah, F. Utilization of Shredded Waste Car Tyres as a Fracture Seal Material (FSM) in Oil and Gas Drilling Operations. *J. Pet. Environ. Biotechnol.* **2017**, *8*, 21. [CrossRef]
51. Sprince, A.; Pakrastinsh, L.; Korjakins, A. Experimental Creep Tests on Concrete Made With Montmorillonite Nano Particles. In Proceedings of the 8th International Conference: 2012, Concrete in the Low Carbon Era, Dundee, UK, 9–11 July 2012.
52. Padevět, P.; Bittnar, P. Measuring the Creep and Material Properties of Cement Paste Specimens. *Wseas Trans. Appl. Theor. Mech.* **2010**, *5*, 81–92, ISSN: 1991-8747.
53. Reddy, D.H.; Ramaswamy, A. Experimental and numerical modeling of creep in different types of concrete. *Heliyon* **2018**, *4*, e00698. [CrossRef]
54. Kammouna, Z.; Briffaut, M.; Malecot, M. Experimental Study of the Creep Effect on the Mechanical Properties of Concrete. *Adv. Civ. Eng.* **2019**, *2019*, 5907923. [CrossRef]
55. Hanbing, L.; Wang, X.; Jiao, Y.; Sha, T. Experimental Investigation of the Mechanical and Durability Properties of Crumb Rubber Concrete. *Materials* **2016**, *9*, 172. [CrossRef]
56. Yong, Y.; Zhu, H. Influence of Rubber Size on Properties of Crumb Rubber Mortars. *Materials* **2016**, *9*, 527. [CrossRef]
57. Cheng, X.; Huang, S.; Guo, X.; Duan, W. Crumb waste tire rubber surface modification by plasma polymerization of ethanol and its application on oil-well cement. *Appl. Surf. Sci.* **2017**, *409*, 325–342. [CrossRef]
58. Gwon, S.; Jeong, Y.; Oh, J.E.; Shin, M. Sustainable sulfur composites with enhanced strength and lightweightness using waste rubber and fly ash. *Constr. Build. Mater.* **2017**, *135*, 650–664. [CrossRef]
59. Mohammed, Shahabas; Poornima, V Strength and durability study of sulfur concrete with replaced fine aggregate. *Mater. Today Proc.* **2018**, *5*, 23888–23897. [CrossRef]
60. Ahmed, A.; Mahmoud, A.A.; Elkatatny, S.; Gajbhiye, R. Improving Saudi Class G Oil-Well Cement Properties Using the Tire Waste Material. *ACS Omega* **2020**, *5*, 27685–27691. [CrossRef] [PubMed]
61. Brothers, L.; Onan, D.; Morgan, R. Well Cement Compositions Containing Rubber Particles and Methods of Cementing Subterranean Zones. U.S. Patent No. US 5779787A, 14 July 1998.

energies

Article

An Integrated Approach to Study Cement-to-Formation Bonding

Arpita P. Bathija * and Peter J. Boul

Houston Research Center, Aramco Americas, 16300 Park Row Blvd, Houston, TX 77084, USA
* Correspondence: arpita.bathija@aramcoamericas.com

Abstract: It is crucial to assess the bond strength of the cement–formation interface while developing novel cements for efficient zonal isolation. An integrated method is presented to investigate the failure mechanism in cement and formation rock under downhole reservoir temperature and pressure conditions using a triaxial experimental setup. The acoustic emission count, strain, and velocity data aid in inferring the fracture process that led to the failure of a specimen. Although most specimens investigated exhibit the three dominant events of compaction, multi-cracking, and sliding, there are variations in the basic structure of each specimen. Furthermore, the insight obtained about the internal structure of the specimen points to its strength and damage tolerance, both of which are vital requirements for bonding. This method can distinguish between a standard cement and modified cement very effectively and help in pairing the appropriate cement formulation for a formation rock.

Keywords: cement bond strength; cement shear bond; oil-well cementing

Citation: Bathija, A.P.; Boul, P.J. An Integrated Approach to Study Cement-to-Formation Bonding. *Energies* **2022**, *15*, 5949. https://doi.org/10.3390/en15165949

Academic Editor: Shu Tao

Received: 7 July 2022
Accepted: 14 August 2022
Published: 17 August 2022

1. Introduction

From the well construction phase to the well abandonment phase, poor cementation makes it difficult to achieve effective zonal isolation due to debonding at the interfaces of cements and rock formations. Stress fluctuations, thermal shocks, stimulation treatments, casing shrinkage/expansion during shut down or restart of production, chemical attack, wettability variations, and surface conditions of the casing and formation are all factors for cement bond failure [1–5].

The ability to isolate different geologic zones with varying pore pressures and fluid/gas chemistries, as well as the ability to prevent uncontrolled migration of pore fluids to shallower depths, is called zonal isolation. Cement shear bond strength measurements are used to evaluate a particularly important aspect of the cement's ability to provide effective zonal isolation [6]. Cement's bonding strength to casing and formations was studied extensively [2,7–10].

Testing the compressive strength of cubic cement samples on a loading frame where the vertical load is increased at a constant rate until the sample fails is a recommended practice [11]. Ultrasonic cement analyzers (UCA) employ a sonic nondestructive testing technique to correlate compressive strength to sonic compressional transit time. Strength correlations are specific to certain cement compositions. Although this common UCA approach is quick, simple, and API-recommended, new cement compositions may not fit the manufacturer's correlations, necessitating bespoke correlations.

Several push-out experiments were performed to determine shear bond strength, in which the cement is set with a cylindrical rock or casing sample within [12]. A load frame is used to apply a vertical load on the rock until it separates from the cement. Another approach is to create a composite core by vertically setting cement on top of a cylindrical rock [4]. The composite core is then loaded in the direction parallel to the bonding surface until failure. Earlier experimental methods for evaluating new cements were difficult to apply in a consistent and repeatable manner for all rock types. Separating the shear force to determine the shear bond strength is challenging and inconclusive because multiple forces act at the contact [13,14].

Our work will build upon that of researchers [14] wherein cement and shale samples were prepared with a 60° contacting angle for basic compressive strength testing in an unconfined and ambient laboratory condition. We are measuring the cement-to-formation shear bond strength while also monitoring velocities, acoustic emissions, and strain at downhole reservoir conditions. We have a quantitative analysis of the shear bond failure between the cement and the rock, as well as a value for the shear bond strength. Quantitative analysis gives us an insight into the strength and damage tolerance of the specimen.

2. Methodology

Two very different formations, the Berea and Torrey sandstones were used to demonstrate the method's applicability. Table 1 shows the mean values for porosity measured using Boyle's law helium porosimeter, permeability measured with flowing nitrogen, and grain densities. Berea [15] has higher permeability and porosity than Torrey, as shown in Table 1.

Table 1. Mean porosity, permeability, and grain density of formations.

Formation	Porosity (%)	Permeability (mD)		Grain Density (g/cc)
		k_air	Klinkenberg	
Berea	24.6	826	785	2.65
Torrey	17.7	3.33	2.74	2.73

Table 2 shows the results of the X-ray diffraction analysis. A PANalytical X'PertPro X-ray diffractometer was used to examine the mineral content of both the formations. Quartz is the main component in both sandstones. Torrey has 30% dolomite, whereas Berea [15] contains trace amounts of carbonate minerals.

Table 2. X-ray diffraction analysis results. Tr $\leq$ 0.5%.

Formation	Clay Minerals (wt.%)			Other Minerals (wt.%)				Carbonate Minerals (wt.%)		Totals (wt.%)		
	Illite-Smectite	Illite/Mica	Kaolinite	Quartz	K-Feldspar	Plagioclase	Pyrite	Calcite	Dolomite	Clay	Other	Carbonate
Berea	Tr	1	3	95	1	Tr	Tr	Tr	Tr	4	96	Tr
Torrey	Tr	3	9	54	5	Tr	Tr	Tr	29	12	59	29

API Class G cement [16,17] was used in the study as pure cement. Coring cylindrical formation specimens (2 inches in diameter by 4.5 inches in height) is the first step in the sample preparation process. A composite specimen is prepared by cutting the aforementioned formation specimen in half at a 60° angle from the horizontal direction [15]. Laboratory data, field results, and acoustic logs emphasize the change in cement bonding with roughness [8]. Hence, a preset cutting speed was used to control the roughness of the formation interface to achieve consistent results. Before setting the cement, the surface roughness was measured using an optical profiler (Nanovea). Berea was found to have a roughness of 230 $\pm$ 47 micrometers. It was verified that using the same cutting tool at a preset speed consistently maintained the surface roughness.

The method of preparing cement and formation samples with a 60° contacting angle is based on the theory of rock strength [18,19]. The theory estimates the angle of the bedding planes that lower the rock strength [20] (pp. 105–107). According to this theory, a slip on a bedding plane at a 60° angle would occur at a far lower stress level than that required to generate a new fault. It was verified experimentally [21] that foliation planes at this angle have the maximum effect, reducing the rock strength by 50%.

Cement was prepared according to the standard recommended procedure [11]. The prepared cement was poured into a cylindrical mold on top of the cut core specimen. The molds were cured for three days at 180 °F and 20.68 MPa. Another specimen (improved cement) was prepared by mixing API Class G cement with silica flour (35% by weight of cement). Improved cement (silica flour is known to improve cement by reducing shrinkage) was cured at 300 °F and 20.68 MPa. The cured specimen was trimmed to achieve a length-to-diameter ratio of at least two. Special attention was paid to maintaining equal amounts of cement and formation in the composite specimen. Figure 1a shows one of our composite specimens cured with API Class G cement and Berea cut at 60° to the horizontal plane, which is the weakest plane of failure. The ends of the specimen were paralleled for precise velocity acquisition. The specimen was wrapped in a copper sleeve for protection from confining fluid at 6.89 MPa. Vertical and horizontal strain gauges were attached to the specimen. Input wires were soldered to the strain gauges for monitoring deformation. Endcaps with acoustic transducers were attached on both ends of the specimen. The specimen was then placed in the triaxial experimental setup (Autolab 3000, NER), as shown in Figure 2, and all the wires properly connected to the feed-thrus for data acquisition. A constant confining pressure (P_c) of 10 megapascals (MPa) was applied at room temperature and reading calibrations were performed [22]. The cement curing (temperature and pressure) and data acquisition conditions (axial stress, confining pressure, and temperature) may be adjusted to match the reservoir conditions of interest. At a P_c of 10 MPa, a vertical compressive force was applied at a constant rate of 0.01 MPa per second until the specimen failed. Velocities, specimen deformation, and acoustic emission counts were measured at regular intervals. It should be noted that the measurements and data analyses were performed on multiple samples of each kind and the standard deviation in the data are also presented.

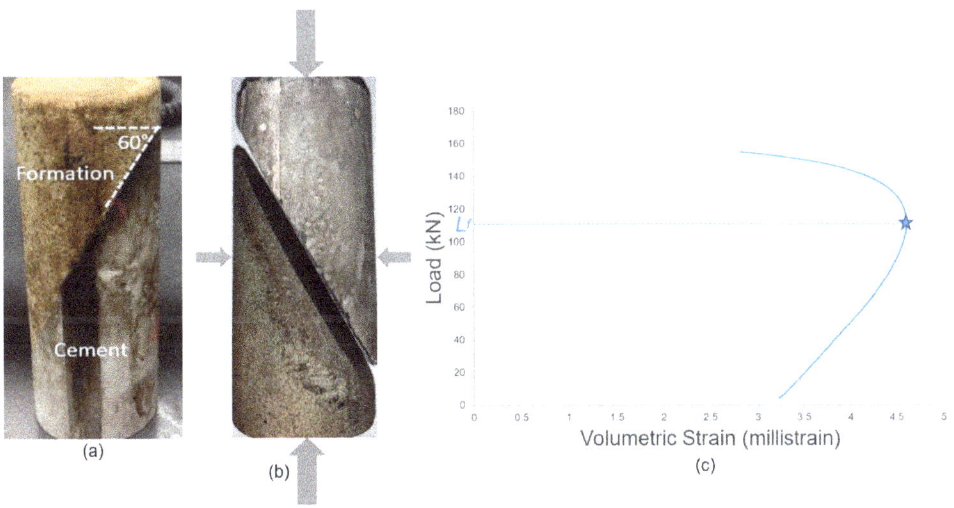

Figure 1. Failure in composite specimen. (**a**) Prior to setting the cement, the formation was cut at 60° to the horizontal in a composite specimen. (**b**) The composite specimen of cement (top) and Berea (bottom). Arrows indicate the direction of applied stresses. Vertical axial stress is applied, as well as confining stress all around the specimen. The size of the arrows indicates the relative magnitudes of the stresses that caused the shear failure. (**c**) Load versus volumetric strain data obtained from one of our specimens. Failure in a specimen is recognized when volumetric strain rolls over (at the star point).

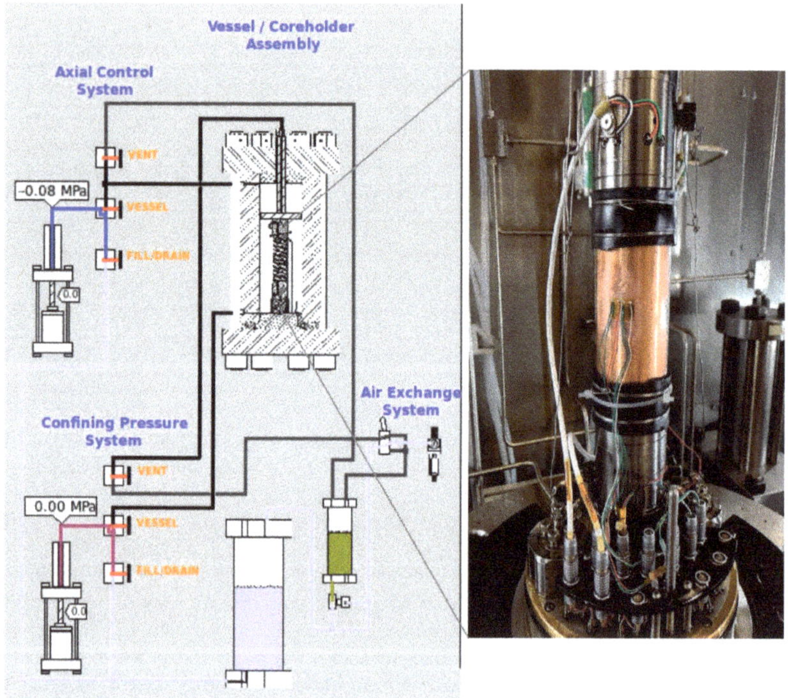

Figure 2. Triaxial experimental setup to pressurize the sample with confining and axial stress. An inset with a fully connected sample is shown.

The compressive strength (CS) at confined conditions (P_c) is the differential stress (axial stress P_c) at which the specimen fails. Figure 1b depicts one of our failed samples owing to interfacial shear slip. Figure 1c demonstrates that failure in a specimen is recognized when volumetric strain rolls over, giving us the load at failure (L_f). The shear bond strength may be obtained by dividing the shear force applied to the interface at failure by the interfacial area [14]. Figure 3 is reproduced with permission of the Cambridge University Press through PLSclear. We know the exact failure plane area with $\beta = 60°$ in the composite specimen (Figure 3a), where β is the angle between the maximum compressive stress (σ_1) and the normal to the fault. It is difficult to determine the failure plane for the whole cement or formation specimen, resulting in inaccurate shear strength measurements. Figure 3b [20] (p. 88) shows that the shear force can be calculated based on the Mohr–Coulomb failure analysis as:

$$\tau_f = \frac{L_f \, sin60° cos60°}{\pi r^2}$$

where, τ_f is the shear force at failure, r is the radius of the composite specimen, and L_f is the load at failure.

Acoustic emission method detects defects using sounds generated by the defects themselves. When a material is under stress, acoustic waves are generated by the transient release of strain energy caused by sudden internal displacements and spatial crack surface oscillations. Sensitive acoustic transducers are used to detect the waves. The various attributes of acoustic emission are counts, risetime, amplitude, threshold, and duration. These attributes can be employed alone or in combination to detect, localize, assess, or monitor failure.

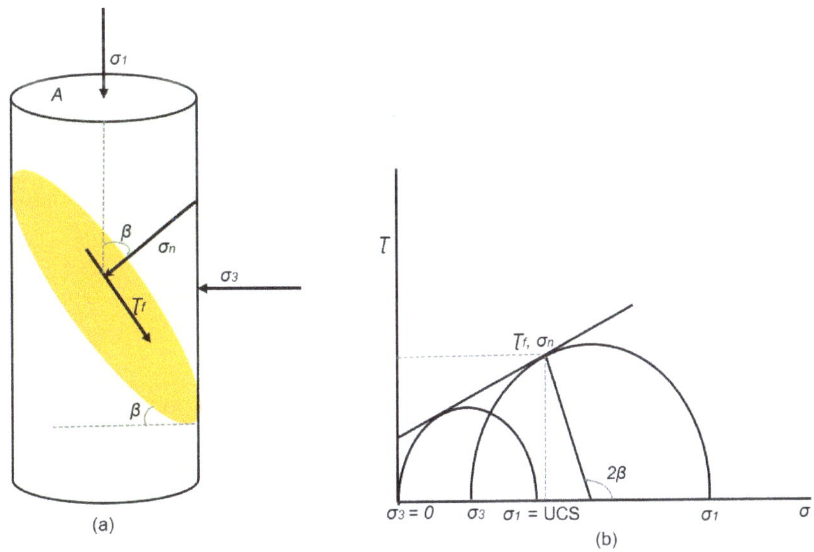

Figure 3. Stresses at failure. (**a**) Illustration of a cylindrical sample (top circle area A) with a bedding plane at an angle β under a confining pressure σ_3 and maximum compressive stress or axial stress σ_1. (**b**) The normal stress σ_n and shear stress τ_f on the plane correspond to the point where the Mohr circle hits the linearized Mohr failure envelope [20] (p. 88).

Only acoustic emission counts are acquired using one sensor mounted at the top of our specimen. It was discovered that there is a significant common trend in the variation in acoustic emission activity as a function of strain in each specimen. As a result, each specimen's failure is governed by similar fracture processes. Comparison of the acoustic emission count and variations in P- or S-wave velocities during deformation provides considerable insight into the mechanical processes at work [23].

A rock specimen's deformation history may be categorized into five distinct categories:

1. The specimen's initial compaction is caused by the preferential closure of cracks normal to the principal compression axis. This region is associated with low acoustic emission activity and an increase in P- and S-wave velocities.
2. A quasi-linear elastic region with constant elastic wave velocity.
3. Micro-crack initiation and extension, with an increasing acoustic emission count, when cracks develop predominantly parallel to the principal compression axis.
4. Macro-cracks form through coalescence of micro-cracks as the acoustic emission count saturates and the P- and S-wave velocities decrease sharply.
5. Shear faults are formed when macro-cracks combine. On the shear fault, frictional sliding occurs.

In most cases, it is difficult to distinguish all five regions since some are short-lived or absent in relation to nearby regions, or occur simultaneously, resulting in a combined region effect. The usage of velocity and acoustic emission activities as a function of strain was split into four stages in mortar [24]: crack closure (1), linear elastic (2), stable crack growth (3), and unstable crack growth stage (4). The third region of a rock specimen corresponds to the stable crack growth stage in mortar, whereas the fourth and fifth region correspond to the unstable crack growth stage.

A schematic diagram of distinct events/regions during the deformation of a rock/concrete specimen is shown in Figure 4. The green vertical arrows represent the direction of axial stress, which represents the principal compression axis. Region 2 is not depicted here since it lacks a visually distinct feature. Region 1 corresponds to a compaction event, whereas regions 3 and 4 correspond to a multi-cracking event, and region 5 corresponds to a sliding event. The green slanting arrows in region 5 represent the sliding motion.

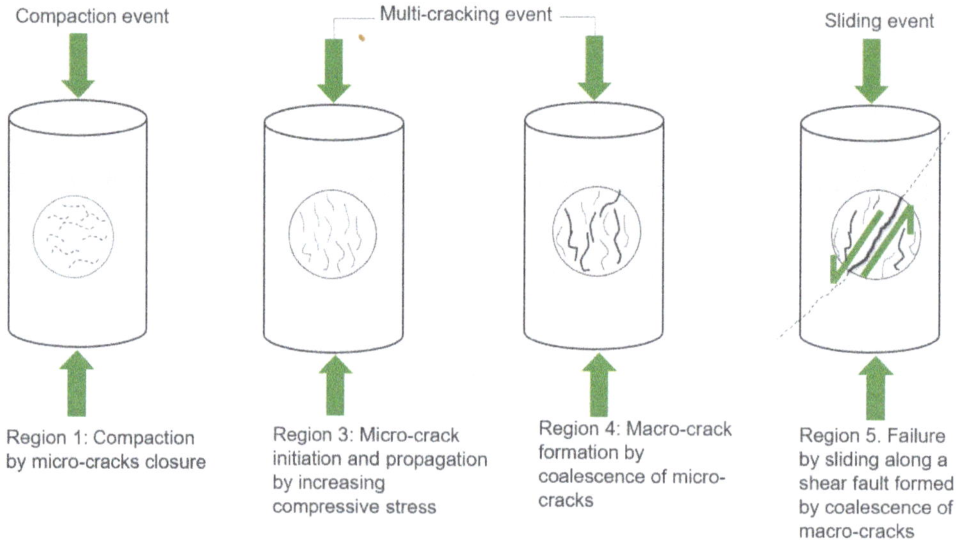

Figure 4. Schematic diagram of distinct events/regions during deformation of a rock/concrete specimen.

3. Results

The compressive strength of cement, formation, and composite cement–formation specimens, as well as the shear bond strength of the composite are shown in Table 3. Unconfined compressive strength (UCS) is the most widely used rock property. It is worth mentioning that Berea's UCS of 32.8 MPa [22] is lower than the CS of 58 MPa at P_c of 10 MPa (Table 3). Similarly, UCS of 50 MPa utilizing UCA [15] is lower than CS of 65.6 MPa at P_c of 10 Mpa with improved cement (Table 3).

Table 3. The mean compressive and shear bond strength of cement, formation, and composite cement–formation specimens at a confining pressure of 10 MPa.

Specimen	No. of Samples	Compressive Strength (MPa)	Shear Strength (MPa)
Berea	3	58 ± 2.65	-
Torrey	2	62	-
Cement	5	50 ± 3.24	-
Improved cement	2	65.6 ± 3.67	-
Cement–Berea	3	73.43 ± 3.83	31.8 ± 1.66
Cement–Torrey	3	54.3 ± 2.84	23.53 ± 1.23
Improved cement–Berea	4	71.35 ± 8.15	30.89 ± 3.53

Shear bond strength for cement–Castlegate sandstone of 4.5 Mpa [12] by push-out tests and 1.7 MPa [4] for cement–Berea sandstone were reported. Please note that [4] used Class H cement and their composite sample was made with cement and formation cores in a series arrangement. Shear bond strengths of 2.76–6.9 MPa were found between cement and shale [14]. Lower values in this study are due to the following factors: (a) unconfined conditions result in lower strengths [20] (p. 104); (b) multiple forces act at the interface, rendering shear bond studies in most tests inconclusive [14]; and (c) cement–shale shear bond strength is lower than cement–sandstone shear bond strength [4].

Initial compaction of the specimen is characterized by a short-lived acoustic emission activity and an increase in velocity. The multi-cracking stage is characterized by the formation and propagation of both micro-cracks and macro-cracks. As a result of the complex multi-scale structure, micro-cracking initiates and accumulates, forming a fracture zone surrounding the starting point of the eventual macro-crack [25]. During the macro-crack formation and propagation, the acoustic emission counts exceed the sensor's limit, leading to the acoustic emission peaks. The acoustic emission peaks may be sensor-dependent and/or arbitrarily set. As a result, it is critical to maintain the sensor and its setting throughout the experiment. As confirmed by the velocity decrease and acoustic emission counts, numerous cracking incidents have a detrimental effect on the material durability or damage tolerance. The cracking process then repeats itself, with the formation of a new micro-scale crack leading to a new macro-scale crack. With increasing stress, newly developed macro-cracks show larger strain amplitudes. The acoustic emission count remains constant in the post-cracking stage because matrix cracks are saturated and the accumulated damage hinders acoustic emission wave propagation.

3.1. Whole Specimen Analysis

Compaction, multi-cracking, and sliding are three distinct and dominant events in the Berea sandstone specimen (Figure 5a). Region 1 corresponds to a compaction event, whereas regions 3 and 4 correspond to a multi-cracking event, and region 5 corresponds to a sliding event. Similar to Berea, the Torrey sandstone specimen (Figure 5b) exhibits three dominant events: compaction, multi-cracking, and sliding. The acoustic emission peaks in event 1 show that Torrey has a larger number of macro-cracks of smaller strain amplitudes generated from pre-existing micro-cracks than Berea. The acoustic emission peaks in event 2 show that Torrey also has a higher number of macro-cracks generated from newly formed micro-cracks than Berea. The total number of acoustic emission peaks in Figure 5 for Berea is 5 and Torrey is 26. Higher porosity and permeability in Berea (Table 1) may result in more empty space for particles to slide and interconnect more efficiently than in Torrey. As a result, the stress distribution in Berea is better. Torrey has more disconnected/isolated macro-cracks before a few of them merge to form a failure plane for sliding. Please note that the compressive strength values of Torrey (62 MPa) and Berea (58 MPa) are quite similar (Table 3), therefore the strength values alone would not provide this insight into their internal structure.

The pure cement specimen (Figure 5c) contains fewer pre-existing micro-cracks than the sandstone specimen, resulting in short-lived compaction event 1. Macro-cracks generated from pre-existing micro-cracks (acoustic emission peaks in event 1) do not demonstrate a strain amplitude increase with stress. Macro-cracks generated from new micro-cracks increase in strain amplitude with stress, as shown in event 2 of the pure cement specimen. The total number of acoustic emission peaks in Figure 5c for pure cement is 12, which is more macro-cracks than Berea (5) but less than Torrey (26).

Improved cement with compressive strength of 65.6 MPA is higher than 50 MPa for pure cement (Table 3). In improved cement, the initial compaction event 1 is absent as there is no increase in velocity observed (Figure 5d). This is because there are few (closed) or no pre-existing micro-cracks. The total number of acoustic emission peaks in Figure 5d for improved cement is 37. As the structure is denser in improved cement than pure cement [15], a larger number of isolated macro-cracks occur before a few of them join to create a failure plane for sliding.

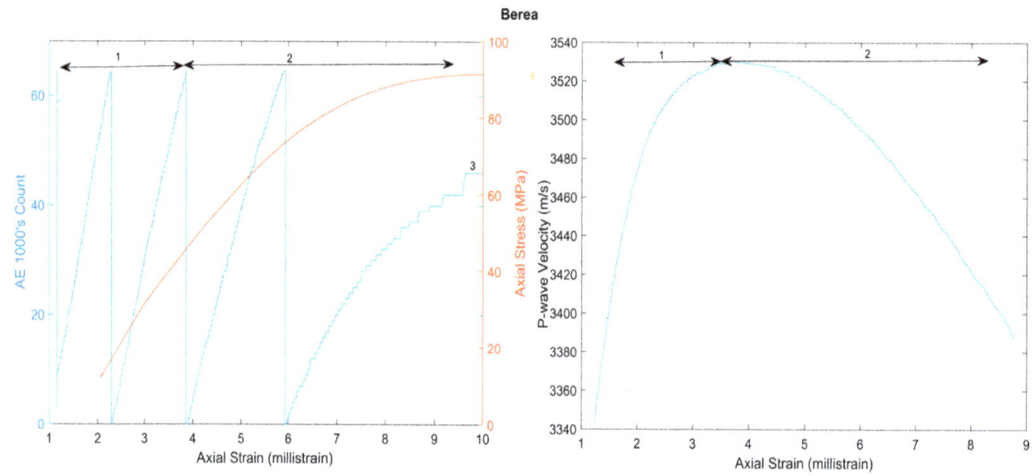

(a)

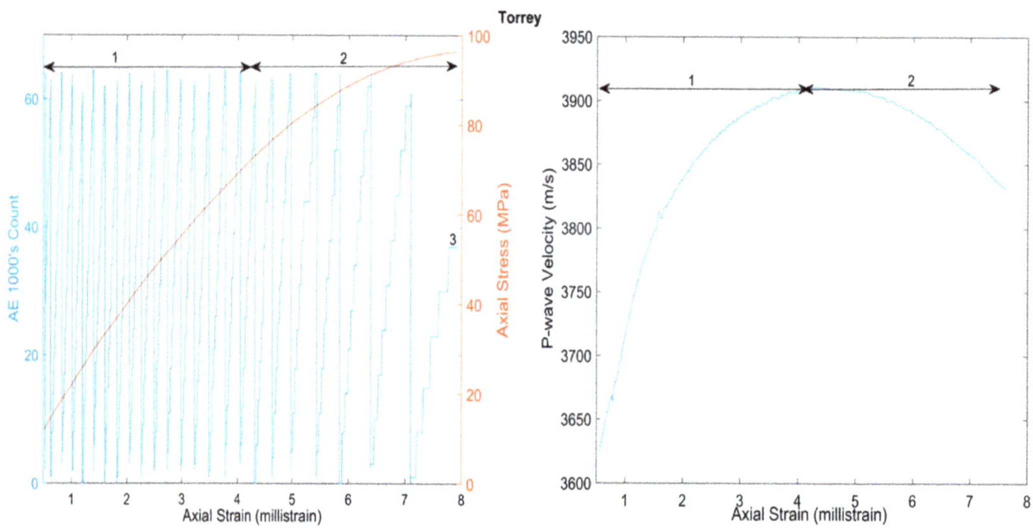

(b)

Figure 5. *Cont.*

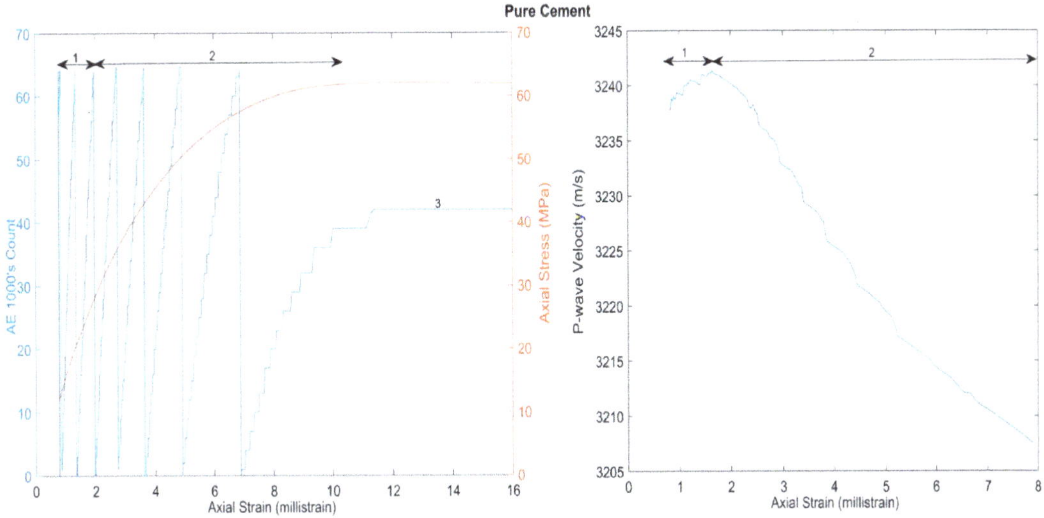

(**c**)

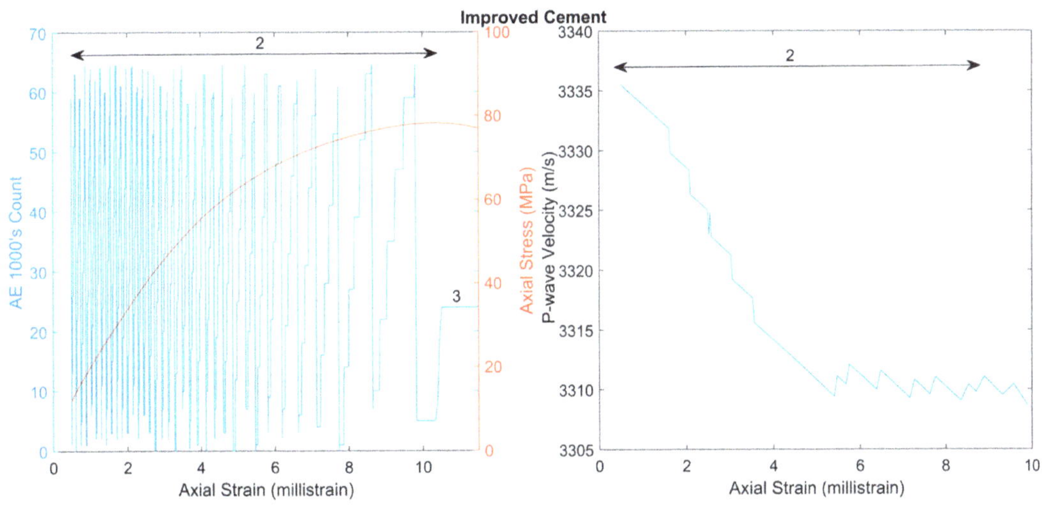

(**d**)

Figure 5. Failure analysis in a whole specimen. Acoustic emission (AE) counts in blue and axial stress in red are shown as a function of axial strain for (**a**) Berea (**b**) Torrey (**c**) pure cement and (**d**) improved cement in the left plots. The plots depict three distinct events: (1) compaction, (2) multi-cracking, and (3) sliding. The right plots show the P-wave velocities as a function of axial strain.

3.2. Composite Specimen Analysis

We found that the composite cement–sandstone specimen fails at the interface rather than in the cement because failure at the interface occurs at a lower stress level than that necessary to generate a new fault in the cement. Our composite specimens (Figure 6) reveal a combination of sandstone compaction (more pre-existing micro-cracks leading to macro-cracks from acoustic emission peaks in event 1) and cement multi-cracking (more macro-cracks from acoustic emission peaks in event 2) leading to shear failure. The cement–Berea (73.43 MPa) composite specimen has a higher compressive bond strength than any of its constituents, Berea (58 MPa) or cement (50 MPa). Initial compaction of the cement–Torrey composite specimen (Figure 6b) is characterized by long-lived acoustic emission activity and an increase in velocity, as Torrey has a larger number of pre-existing micro-cracks. Table 3 shows that cement–Berea (31.8 MPa) bonding is stronger than cement–Torrey (23.53 MPa) bonding. This strength difference may be attributed to Berea's higher porosity and permeability. Higher porosity might lead to better stress distribution and thus, attributing to the greater strength, as seen in the whole specimen case. Furthermore, higher permeability corresponds to a well-connected pore system, allowing more cement to fill the pore spaces, which leads to better bonding between the cement and formation [26].

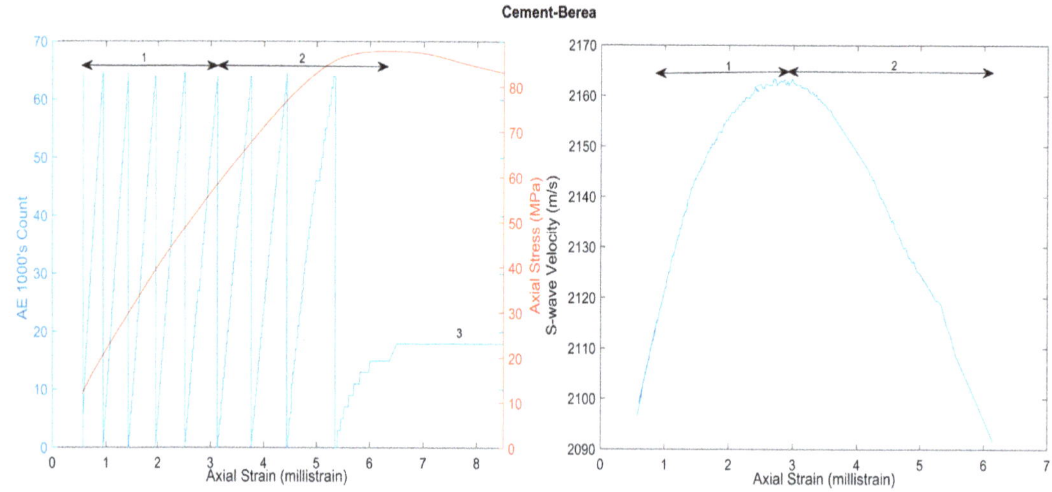

(a)

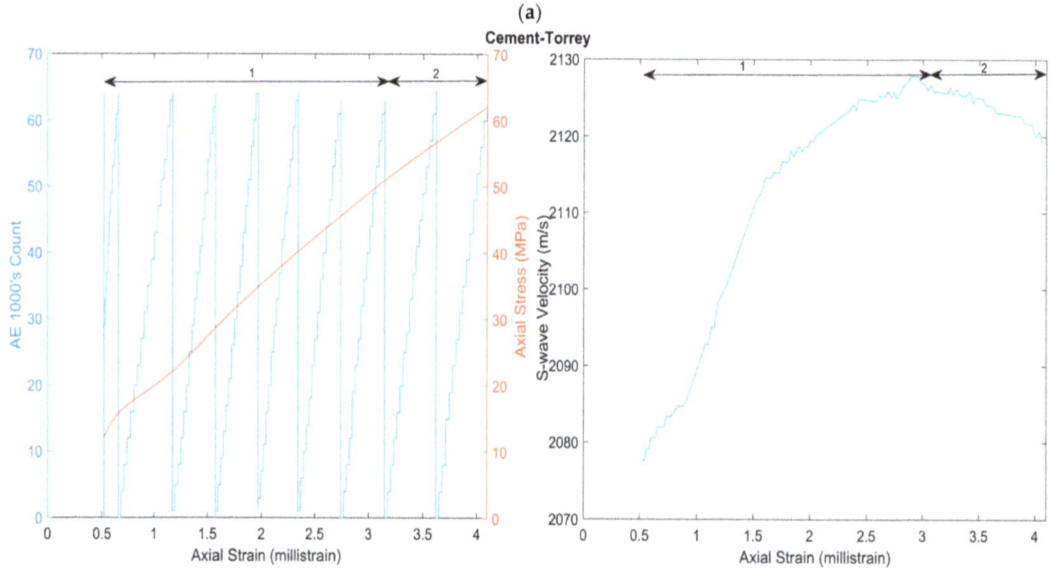

(b)

Figure 6. *Cont.*

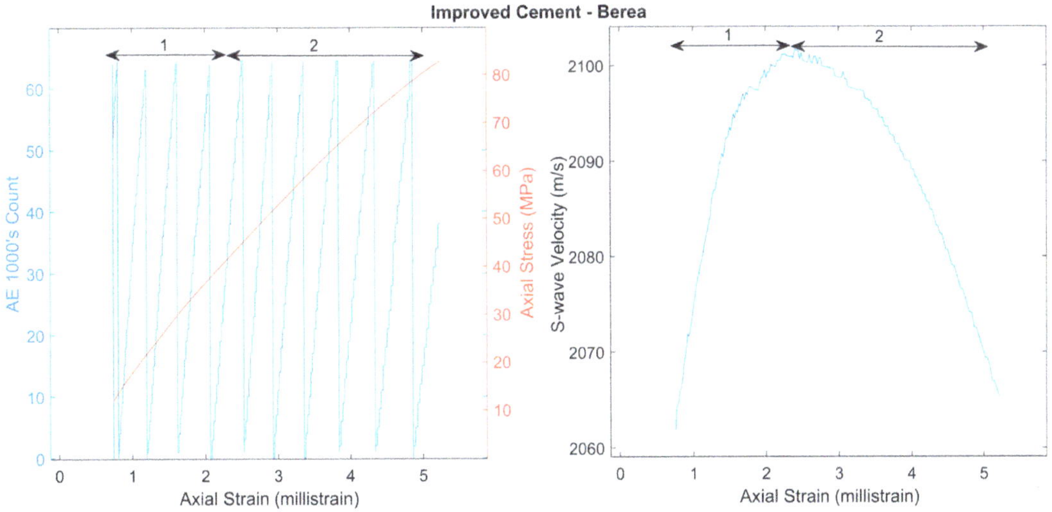

(c)

Figure 6. Failure analysis in a composite specimen. Acoustic emission (AE) counts in blue and axial stress in red are shown as a function of axial strain for (**a**) cement–Berea, (**b**) cement–Torrey, and (**c**) improved cement–Berea composite specimen in the left plots. The fracture mechanism is a combination of features observed in both cement and sandstone. The right plots show the S-wave velocities as a function of axial strain.

The shear bond strength of cement–Berea (31.8 MPa) and that of improved cement–Berea (30.89 MPa) are very close. The total number of acoustic emission peaks in Figure 6 for cement–Berea is 11 and that of improved cement–Berea is 12. Despite similar bond strengths and total number of macro-cracks, the composite specimen shows a slightly larger number of new isolated macro-cracks (acoustic emission peaks in event 2) in improved cement–Berea (7) than pure cement–Berea (5) in Figure 6. More isolated macro-cracks point to a lower damage tolerance in improved cement. Cement strength retrogression is an issue in some situations, such as deeper and thermal wells, where improved cement is still preferable [16].

Our method relates the material bonding to the number of macro-cracks generated by counting the acoustic emission peaks. When comparing materials, it is critical to use the same acoustic emission sensor and instrument setting to keep the acoustic emission counts that define the acoustic emission peaks consistent. This integrated approach is an excellent method to compare various specimens, however, the exact number of cracks generated may differ owing to the arbitrary definition of acoustic emission peaks. As a result, an imaging technique might be helpful in validating the findings for future research.

We attempted to keep our sample preparation protocol consistent, but unknown variables may still exist. One such factor is the variation due to change in curing and testing conditions. Future studies should identify and control more factors for enhanced accuracy.

4. Conclusions

Sample preparation procedures are essential for consistent results and accurate forecasts by reducing the number of variables. The fracture mechanism that causes the failure in cement formation and composite cement–formation specimens under downhole conditions is described using an acoustic emission count in conjunction with velocities and strain data. The ultimate failure in all specimens is caused by the development of fracture process zones from micro-cracks (acoustic emission count increase) progressing to macro-cracks (acoustic emission peaks), and the coalescence of macro-cracks leading to shear fault. Although the three dominant events of compaction, multi-cracking, and sliding may be present, there are some differences related to the basic structure of each specimen.

Because the compressive strength values of Torrey (62 MPa) and Berea (58 MPa) are so close, the strength values alone would not reveal anything about their internal structure: Torrey has more pre-existing micro-cracks than Berea, and Berea has more empty space for particle movement to slide and interconnect more effectively than Torrey.

A compaction event is short-lived (pure cement) to absent (improved cement) due to the decreasing number of pre-existing micro-cracks. A larger number of isolated macro-cracks are formed in improved cement, indicating a denser structure. The denser structure is also confirmed by a higher compressive strength in improved cement (65.6 MPa) compared to pure cement (50 MPa).

Pure cement binds better with Berea (31.8 MPa) than Torrey (23.53 MPa) based on the shear bond strength values of the composite specimen. Although the shear bond strengths of Berea with pure cement (31.8 MPa) and improved cement (30.89 MPa) are very close, pure cement might pair better with Berea than improved cement due to higher damage tolerance, except where cement retrogression issues exist. Thus, our approach offers additional information for pairing the appropriate cement formulation to a formation rock by assessing its strength and damage tolerance.

Author Contributions: Conceptualization, A.P.B. and P.J.B.; methodology, A.P.B.; data analysis, A.P.B.; investigation, A.P.B.; writing—original draft preparation, A.P.B.; writing—review and editing, P.J.B.; project administration, A.P.B. All authors have read and agreed to the published version of the manuscript.

Funding: This research received no external funding.

Acknowledgments: The authors thank Roland Martinez for sample preparation and acoustic measurements

Conflicts of Interest: The authors declare no conflict of interest.

References

1. Parcevaux, P.; Sault, P. Cement Shrinkage and Elasticity: A New Approach for a Good Zonal Isolation. In Proceedings of the SPE Annual Technical Conference and Exhibition, Houston, TX, USA, 16–19 September 1984. [CrossRef]
2. Carpenter, R.; Brady, J.; Blount, C.G. The Effects of Temperature and Cement Admixes on Bond Strength. *J. Pet. Technol.* **1992**, *44*, 936–941. [CrossRef]
3. Bois, A.-P.; Garnier, A.; Rodot, F.; Saint-Marc, J.; Aimard, N. How To Prevent Loss of Zonal Isolation Through a Comprehensive Analysis of Microannulus Formation. *SPE Drill. Complet.* **2011**, *26*, 13–31. [CrossRef]
4. Radonjic, M.; Oyibo, A. Experimental evaluation of wellbore cement-formation shear bond strength in presence of drilling fluid contamination. In Proceedings of the 5th International Conference on Porous Media and Their Applications in Science, Engineering and Industry, Kona, HI, USA, 22–27 June 2014.
5. De Andrade, J.; Sangesland, S.; Skorpa, R.; Todorovic, J.; Vrålstad, T. Experimental Laboratory Setup for Visualization and Quantification of Cement-Sheath Integrity. *SPE Drill. Complet.* **2016**, *31*, 317–326. [CrossRef]
6. Evans, G.W.; Carter, L.G. Bonding Studies of Cementing Compositions to Pipe and Formations. In *Drilling and Production Practice*; American Petroleum Institute: New York, NY, USA, 1962.
7. Peterson, B. Bond of cement compositions for cementing wells. In Proceedings of the 6th World Petroleum Congress, Frankfurt am Main, Germany, 19–26 June 1963.
8. Carter, L.; Evans, G. A Study of Cement-Pipe Bonding. *J. Pet. Technol.* **1964**, *16*, 157–160. [CrossRef]
9. Zhang, Z.; Scherer, G.W.; Prud'Homme, R.K. Adhesion and Bonding Between Steel Pipe and Cement/Spacer/Mud System. In Proceedings of the 9th International Conference on Fracture Mechanics of Concrete and Concrete Structures, Berkeley, CA, USA, 29 May–1 June 2016; pp. 22–25. [CrossRef]
10. Lamik, A.; Pittino, G.; Prohaska-Marchried, M.; Krishna, R.; Thonhauser, G.; Antretter, T. Evaluation of Cement-Casing & Cement-Rock Bond Integrity During Well Operations. In Proceedings of the SPE/IADC Middle East Drilling Technology Conference and Exhibition, Abu Dhabi, United Arab Emirates, 25–27 May 2021. [CrossRef]
11. API. *Recommended Practice for Testing Well Cements*; American Petroleum Institute: New York, NY, USA, 2005; pp. 1–146.
12. Opedal, N.; Todorovic, J.; Torsaeter, M.; Vrålstad, T.; Mushtaq, W. Experimental Study on the Cement-Formation Bonding. In Proceedings of the SPE International Symposium and Exhibition on Formation Damage Control, Lafayette, LA, USA, 26–28 February 2014. [CrossRef]

13. Ladva, H.K.J.; Craster, B.; Jones, T.G.J.; Goldsmith, G.; Scott, D. The Cement-to-Formation Interface in Zonal Isolation. In Proceedings of the IADC/SPE Asia Pacific Drilling Technology Conference and Exhibition, Kuala Lumpur, Malaysia, 13–15 September 2004. [CrossRef]
14. Liu, X.; Nair, S.D.; Cowan, M.; van Oort, E. A Novel Method to Evaluate Cement-Shale Bond Strength. In Proceedings of the SPE International Symposium on Oilfield Chemistry, The Woodlands, TX, USA, 13–15 April 2015. [CrossRef]
15. Bathija, A.P.; Martinez, R. An integrated approach to understand the failure mechanism in cement and formation. In *SEG Technical Program Expanded Abstracts*; Society of Exploration Geophysicists: Houston, TX, USA, 2020; pp. 2535–2539.
16. Murtaza, M.; Rahman, M.K.; Al-Majed, A.A.; Samad, A. Mechanical, Rheological and Microstructural Properties of Saudi Type G Cement Slurry with Silica Flour Used in Saudi Oil Field under HTHP Conditions. In Proceedings of the SPE Saudi Arabia Section Technical Symposium and Exhibition, Al-Khobar, Saudi Arabia, 19–22 May 2013. [CrossRef]
17. Reddy, B.R.; Zhang, J.; Ellis, M. Cement Strength Retrogression Issues in Offshore Deep Water Applications—Do We Know Enough for Safe Cementing? In Proceedings of the Offshore Technology Conference, Houston, TX, USA, 2–5 May 2016. [CrossRef]
18. Donath, F.A. Experimental Study of Shear Failure in Anisotropic Rocks. *GSA Bull.* **1961**, *72*, 985–989. [CrossRef]
19. Jaeger, J.C.; Cook, N.G.; Zimmerman, R. *Fundamentals of Rock Mechanics*; John Wiley & Sons: Hoboken, NJ, USA, 2009.
20. Zoback, M.D. *Reservoir Geomechanics*; Cambridge University Press: Cambridge, UK, 2010.
21. Vernik, L.; Lockner, D.; Zoback, M.D. Anisotropie strength of some typical metamorphic rocks from the KTB pilot hole, Germany. *Sci. Drill.* **1992**, *3*, 153–160.
22. Bathija, A.P.; Batzle, M.L.; Prasad, M. An experimental study of the dilation factor. *Geophysics* **2009**, *74*, E181–E191. [CrossRef]
23. Sammonds, P.; Ayling, M.; Meredith, P.; Murrell, S.; Jones, C. A laboratory investigation of acoustic emission and elastic wave velocity changes during rock failure under triaxial stresses. In Proceedings of the ISRM International Symposium, Pau, France, 30 August–2 September 1989.
24. Elaqra, H.; Godin, N.; Peix, G.; R'Mili, M.; Fantozzi, G. Damage evolution analysis in mortar, during compressive loading using acoustic emission and X-ray tomography: Effects of the sand/cement ratio. *Cem. Concr. Res.* **2007**, *37*, 703–713. [CrossRef]
25. Tsangouri, E.; Michels, L.; el Kadi, M.; Tysmans, T.; Aggelis, D.G. A fundamental investigation of textile reinforced cementitious composites tensile response by Acoustic Emission. *Cem. Concr. Res.* **2019**, *123*, 105776. [CrossRef]
26. Bathija, A.P.; Sengupta, M.; Eichmann, S.L. Cement to formation bond strength evaluation using image analysis. In Proceedings of the First International Meeting for Applied Geoscience & Energy, Denver, CO, USA, 26 September–1 October 2021.

energies

MDPI

Article

Numerical Simulation of Effective Hole Cleaning by Using an Innovative Elliptical Drillpipe in Horizontal Wellbore

Guoshuai Ju *, Tie Yan and Xiaofeng Sun

College of Petroleum Engineering, Northeast Petroleum University, Daqing 163318, China; yant@nepu.edu.cn (T.Y.); sunxf@nepu.edu.cn (X.S.)
* Correspondence: juguoshuai@nepu.edu.cn

Abstract: In the drilling of horizontal wells, the drill cuttings tend to settle down on the low side of the annulus due to gravity and form a stationary bed, which results in hole cleaning problems. In this paper, a novel type of drillpipe with an elliptical shape was proposed to alleviate inadequate hole cleaning during the drilling of horizontal wells. A three-dimensional computational fluid dynamic (CFD) Eulerian-Eulerian approach with the Realizable k-ε turbulence model was developed to predict the solid–liquid two-phase flow in the annular space. Numerical examples were given to investigate the influence of different parameters on cuttings' transport behavior, and the elliptical drillpipe was compared with the circular drillpipe. The annular cuttings concentration, annular pressure drop, and hole cleaning efficiency were evaluated. The numerical results clarify the potential of the elliptical drillpipe to enhance the hole cleaning efficiency without significantly increasing the annular pressure drop. Due to the swirl flow and secondary flow caused by the rotation of the curvature wall, the swaying phenomenon of drill cuttings' distribution along the rotation direction of drillpipe was observed and enhanced the cuttings transport ability. Using the elliptical drillpipe as a joint-type tool can improve hole cleaning performance. Under the optimum conditions applied in this study, the hole cleaning efficiency increased by nearly 18%.

Keywords: numerical study; horizontal well; hole cleaning; computational fluid dynamic (CFD); elliptical drill pipe

Citation: Ju, G.; Yan, T.; Sun, X. Numerical Simulation of Effective Hole Cleaning by Using an Innovative Elliptical Drillpipe in Horizontal Wellbore. *Energies* **2022**, *15*, 399. https://doi.org/10.3390/en15020399

Academic Editors: Mofazzal Hossain, Hisham Khaled Ben Mahmud and Md Motiur Rahman

Received: 29 November 2021
Accepted: 1 January 2022
Published: 6 January 2022

1. Introduction

With the rapid development of the world economy, energy demands are increasing. Oil and gas are still vital energy resources to ensure human survival and development over a sufficiently long time [1,2]. When compared with conventional drilling methods, the horizontal well has more advantages in improving well production and reducing costs. Horizontal well technology has broad application prospects in oil and gas reservoir exploration and development [3–7].

When drilling long horizontal wells, the drill cuttings tend to deposit on the low side of the annulus and form a stationary bed, causing hole-cleaning problems. Inadequate hole cleaning can cause downhole safety accidents, such as pipes becoming stuck, premature bit wear, slow rate of penetration (ROP), formation fracturing, excessive torque and drag on the drillstring, as well as difficulties in logging and cementing. Inadequate removal of drill cuttings is a primary challenge in horizontal well drilling and may increase the cost and risk of drilling while increasing the non-productive time (NPT) [3,8–14]. Therefore, it is essential to effectively alleviate inadequate hole cleaning during horizontal well drilling.

In the last decades, the cuttings transport phenomenon in the wellbore annulus has been extensively investigated. The results of previous studies show that many interdependent factors and the inherent complexity of the transport process pose challenges to efficient hole cleaning [15]. There are four groups of parameters that affect the cuttings' transport behavior, which are the cutting and fluid parameters (cutting density, cutting shape and size, fluid viscosity, fluid flow rate, etc.), the operational parameters (inclination, drillpipe rotation, geometry of the wellbore, and eccentricity), and the formation parameters (temperature, pressure, and porosity) [16]. These studies are of considerable significance in predicting cuttings' transport behavior. Among many variables that affect the ability of cutting transportation, it is considered that the annular fluid velocity and the rheological properties of drilling fluid are the most effective drilling parameter to prevent the development of the cuttings' bed [17,18].

However, drilling parameters, such as flow rate and rheological properties, are limited by hydraulic conditions and cannot be changed at will. Increasing the flow rate of drilling fluid will increase the energy consumption of the pumping system, increase the operational costs, and even cause other operational problems, such as wellbore erosion or lost circulation [19]. When drilling long horizontal wells, the annular dynamic pressure loss may reach the rock fracture limit. The further increase of the annular fluid velocity may cause this lost circulation problem. Besides, while the high costs of procurement and toxic waste management should be considered for promoting changes in the rheological properties of drilling fluid, its stability and cuttings lifting abilities still have significant limitations.

Although a detailed understanding of the cuttings' transport behavior in the annulus has been obtained, inadequate hole cleaning sometimes cannot be solved by actively changing drilling parameters [20]. In high-angle wells, rotating the drillpipe can enhance the turbulence intensity of drilling fluid [21]. The annular fluid forms a spiral axial flow with the drillpipe at the center, improving the cuttings' carrying efficiency of drilling fluid and improving hole cleaning efficiency. The proper modification of a conventional circular drillpipe may obtain a better hole cleaning performance. A more vigorous swirl motion can be induced under the drillpipe rotation [22].

Over the years, in many different engineering systems and processes, researchers have found that a vortex generator or rotating tube can drive the swirl or secondary flow. It can promote the fluid mixing between the flow core and the near-wall regions, strengthen the heat transfer performance, and improve the energy utilization efficiency [23–25]. One typical method is to design the tube as a twisted elliptical tube. Taking this as a reference, if the operational conditions permit, designing the drillpipe into an elliptical shape as a joint-type tool may better promote the cuttings' transport, under the action of pipe rotation. Applying such new drilling technologies can play a helpful role in the harsh conditions of extended-reach and deep-water drilling.

Thus, an improved understanding of the hole cleaning performance and cuttings transport behavior of the elliptical drillpipe would be beneficial to optimize and improve the design. However, the understanding of microscopic flow details caused by the elliptical drillpipe is limited in the experiment. With the development of computer technology, this understanding can be significantly improved using the computational fluid dynamics (CFD) technique, due to its advantages of revealing the internal flow mechanism of solid–liquid two-phase and the flow details [26,27].

In this study, the primary purpose was to enhance the hole cleaning efficiency of the elliptical drillpipe. A Eulerian-Eulerian approach was developed to numerically investigate the hydrodynamics of solid–liquid two-phase flow in the annulus, which was composed of an elliptical drillpipe, and compared the efficiency of hole cleaning with the conventional circular drillpipe. The model predictions were compared with the experimental results of annular cuttings' concentration, reported by Song et al., 2017 [13]. The validated simulation model was then further extended, and the effects of ROP, rotational speed, axis ratio, eccentricity, and pitch length ratio on the cuttings' transport behavior with an elliptical drillpipe were investigated, and thus paved an avenue toward using hole cleaning in the drilling of horizontal wells in petroleum engineering [28].

2. Mathematical Model

2.1. Governing Equations

The flow characteristics of solid–liquid two-phase flow were studied by using the three-dimensional transient CFD model as described below. The Eulerian-Eulerian approach, which uses a generalized form of Navier-Stokes equations, was adopted to describe the flow characteristics of each phase. The fluid and solid phases are treated mathematically as interpenetrating continua, and their volume fractions are assumed to be a continuous function of space and time. The sum of their volume fractions is unity. Equivalent conservation equations were used for each phase, and additional closure laws on the basis of the kinetic theory of granular flow (KTGF) were applied to describe solid–solid and solid–fluid interactions.

The continuity equation for the liquid phase (l) and solid phase (s) is expressed as:

$$\frac{\partial}{\partial t}(\alpha_l\rho_l) + \nabla\cdot(\alpha_l\rho_l\vec{u}_l) = 0 \tag{1}$$

$$\frac{\partial}{\partial t}(\alpha_s\rho_s) + \nabla\cdot(\alpha_s\rho_s\vec{u}_s) = 0 \tag{2}$$

where α, $\vec{u}$, and ρ stand for volume fraction, velocity vector, and density, respectively. Similarly, the momentum conservation equation for the liquid phase and solid phase are expressed as:

$$\frac{\partial}{\partial t}(\alpha_l\rho_l\vec{u}_l) + \nabla\cdot(\alpha_l\rho_l\vec{u}_l\vec{u}_l) = -\alpha_l\nabla p + \alpha_l\nabla\cdot\tau_l + \alpha_l\rho_l\vec{g} - \beta(\vec{u}_l - \vec{u}_s) \tag{3}$$

$$\frac{\partial}{\partial t}(\alpha_s\rho_s\vec{u}_s) + \nabla\cdot(\alpha_s\rho_s\vec{u}_s\vec{u}_s) = -\alpha_s\nabla p - \nabla p_s + \alpha_s\nabla\cdot\tau_s + \alpha_s\rho_s\vec{g} + \beta(\vec{u}_l - \vec{u}_s) \tag{4}$$

where $\vec{g}$ is the gravitational acceleration, $\overline{\overline{\tau}}$ is the stress tensor, p is the pressure, and p_s is the solid pressure. β is the fluid–particle exchange coefficient, which can be calculated by the Huilin-Gidaspow drag correction [29]. The Huilin-Gidaspow correction was verified to improve the Gidaspow correction by introducing a smoothing function, and better continuity in the transition from a low solid concentration to a high solid concentration is obtained. The fluid–particle exchange coefficient can be shown as:

$$\beta = \varphi\beta_{\mathrm{Ergun}} + (1-\varphi)\beta_{\mathrm{Wen\&Yu}} \tag{5}$$

$$\varphi = \frac{arctan[262.5(\alpha_s - 0.2)]}{\pi} + 0.5 \tag{6}$$

When $\alpha_l \le 0.8$, the fluid–particle exchange coefficient is calculated as:

$$\beta_{\mathrm{Ergun}} = 150\frac{\alpha_s(1-\alpha_l)\mu_l}{\alpha_l {d_s}^2} + 1.75\frac{\rho_l\alpha_s\left|\vec{u}_s - \vec{u}_l\right|}{d_s} \tag{7}$$

When $\alpha_l > 0.8$,

$$\beta_{\mathrm{Wen\&Yu}} = \frac{3}{4}C_D\frac{\alpha_s\alpha_l\rho_l\left|\vec{u}_s - \vec{u}_l\right|}{d_s}{\alpha_l}^{-2.65} \tag{8}$$

where the drag coefficient, C_D, is

$$C_D = \frac{24}{Re_s}\left(1 + 0.15{Re_s}^{0.687}\right),\ Re_s \le 1000 \tag{9}$$

$$C_D = 0.44,\ Re_s > 1000 \tag{10}$$

where the particle Reynolds number, Re_s, is given by:

$$Re_s = \frac{\rho_1 d_s \left| \vec{u}_s - \vec{u}_1 \right|}{\mu_1} \tag{11}$$

The Realizable *k*-ε turbulence model was employed to describe the phenomenon of turbulence. The modeled transport equations for turbulence kinetic energy (TKE), *k*, and its dissipation rate, ε, are expressed as follows:

$$\frac{\partial(\rho_1 k)}{\partial t} + \frac{\partial(\rho_1 k \vec{u_j})}{\partial x_j} = \frac{\partial}{\partial x_j}\left[(\mu_1 + \frac{\mu_t}{\sigma_k})\frac{\partial k}{\partial x_j}\right] + G_k - \rho_1 \varepsilon \tag{12}$$

$$\frac{\partial(\rho_1 \varepsilon)}{\partial t} + \frac{\partial(\rho_1 \varepsilon \vec{u_j})}{\partial x_j} = \frac{\partial}{\partial x_j}\left[(\mu_1 + \frac{\mu_t}{\sigma_\varepsilon})\frac{\partial \varepsilon}{\partial x_j}\right] + \rho_1 C_1 s \varepsilon - \rho_1 C_2 \frac{\varepsilon^2}{k + \sqrt{\frac{\mu_1}{\rho_1}\varepsilon}} \tag{13}$$

where σ_k and σ^ε are the turbulent Prandtl Numbers of the *k* equation and ε equation, respectively; G_k is the generation of TKE due to the mean velocity gradients; *S* is the modulus of the mean rate of strain tensor; the coefficients appeared in above equations are as following, $C_1 = \max\left[0.43, \frac{\eta}{\eta+5}\right]$, $\eta = S\frac{k}{\varepsilon}$; $C_2 = 1.9$, $\sigma_k = 1.0$ and $\sigma_\varepsilon = 1.2$.

2.2. Numerical Model

In this work, a three-dimensional simulation was conducted based on the previous experiment [13]. However, the pre-calculations showed that the calculation efficiency was not appropriate for the experimental analysis with a real scale in the present situation, so the annulus length adopted in this model was shortened to *L* = 2 m when compared with the experimental annulus length of *L* = 6 m. The main observation parameter of the simulation is the volume concentration of cuttings in a section of annulus length. It is a compromise between ensuring the accuracy of calculation and significantly reducing the computation cost. The annular geometry and the cross-section diagram of the circular drillpipe and elliptical drillpipe are presented in Figure 1. The detailed parameters of the annular space are listed in Table 1.

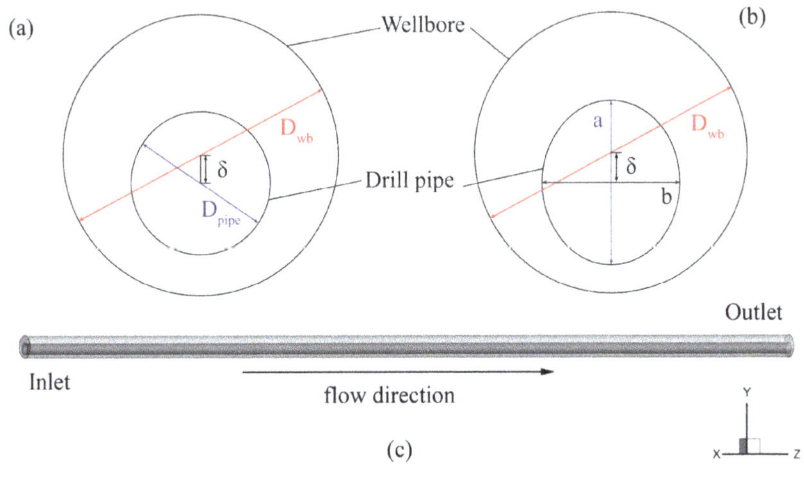

Figure 1. Physical model of the annulus. (**a**) cross section of the circular annulus; (**b**) cross section of the elliptical annulus; (**c**) detailed view of the annulus.

Table 1. Geometrical parameters of the annular space.

Geometrical Parameters	Value
Annulus length (L), mm	2000
Wellbore diameter (D_{wb}), mm	50
Drillpipe diameter (D_{pipe}), mm	24.6
Minor axis (b) (equal to D_{pipe}), mm	24.6
Major to minor axis ratio (a/b)	1.00; 1.05; 1.10; 1.15
Eccentricity (e)	0, 0.2, 0.4
Pitch length ratio (P/D_{pipe})	2.95, 5.91, 11.81
Inclination angle (θ), degrees	90

2.3. Boundary Conditions and Grid Independence Study

The boundary condition is the velocity inlet condition at the entrance of the annulus, and the turbulence intensity at the entrance is also given by:

$$I = \frac{u'}{\overline{u}} = 0.16 \times (Re)^{-0.125} \tag{14}$$

$$Re = \frac{\rho_l u D_h}{\mu} \tag{15}$$

The pressure outlet condition was adopted for the outlet, and the pressure was set equal to 1 atm. The no-slip condition was imposed on the walls of the wellbore and the drillpipe. The sliding mesh (SM) method was used to carefully deal with the interface between static and moving parts in the flow field. An interface was introduced to exchange all parameters in the liquid and solid governing equations between the static and moving zones. The governing equations of the model are discretized by the finite-volume method. The well-known phase coupled semi-implicit pressure-linked equation (SIMPLE) algorithm was adopted to couple the pressure and velocity fields. The equations of momentum, volume fraction, turbulent kinetic energy, and turbulent dissipation rate were discretized using the QUICK routine scheme, due to its better adaptation to hexahedral meshes. Time integration was done with a second-order implicit scheme. The numerical computation was considered converged when the scaled residuals of different variables are lower than 10^{-5}. The physical parameters of the liquid and solid phases in the simulation are shown in Table 2.

Table 2. The physical parameters of drill cuttings and drill fluid.

Physical Parameters	Value
Fluid density (ρ_l), kg/m^3	998.2
Fluid viscosity (μ_l), Pa.s	0.001
Cuttings density (ρ_s), kg/m^3	2650
Cuttings diameter (d_s), mm	1.2
ROP, (m/s)	0.00211; 0.00256; 0.00319;

As shown in Figure 2, the hexahedral structured mesh of the cylindrical annulus was adopted for the modeling. The standard wall function approach was used near the wall to ensure the accuracy of the CFD results. To guarantee the accuracy of the numerical results, four meshing schemes with different grid numbers (8.0×10^5, 1.6×10^6, 3.2×10^6, and 8.0×10^6 meshes) were used for grid independence and validation, and the verification results are shown in Figure 3. The maximum difference of annular pressure drop between 3.2×10^6 meshes, when the timestep was $\Delta t = 0.0002$ s, and 8.0×10^6 meshes, when the timestep is $\Delta t = 0.0001$ s, was less than 1%. Therefore, the elliptical annulus was considered to have a better grid independence with a grid of 3.2×10^6 meshes and $\Delta t = 0.0002$ s, and this mesh distribution and the timestep of $\Delta t = 0.0002$ s was utilized through the research.

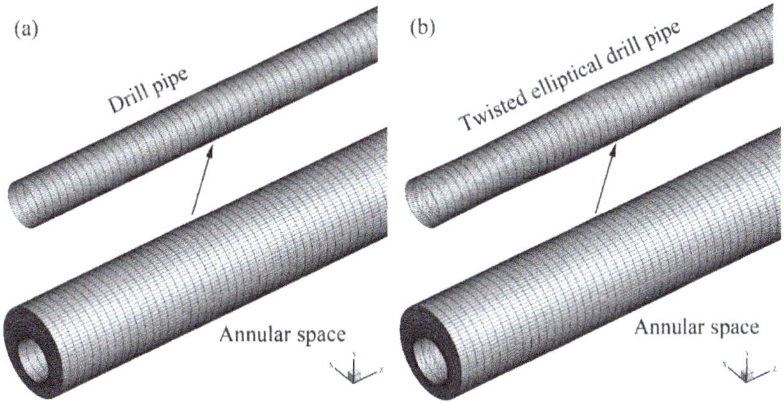

Figure 2. Detail view of computational mesh (grid number 3.2 × 10^6). (**a**) Circular drillpipe. (**b**) Twisted elliptical drillpipe.

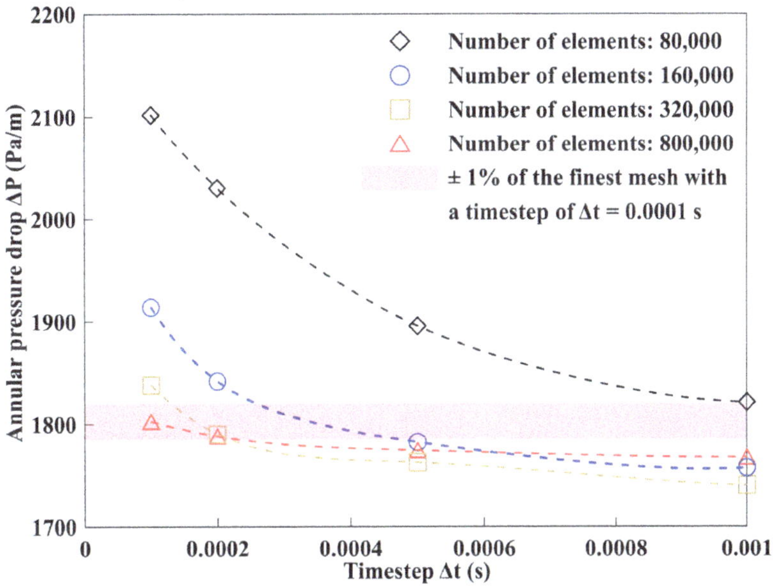

Figure 3. Grid independence test.

2.4. Verification of CFD-Based Simulation

To verify the reliability of the CFD model and the accuracy of this numerical method, the pre-mentioned CFD model used in this paper was verified. As seen in Figure 4, the simulations were obtained by comparing the annular cuttings concentration of a circular drillpipe between the experimental results [13]. Among the validation, the mean value of the relative error was 8.2%, which indicates that the present CFD-based simulation is reliable for the liquid–solid flow simulation in the annular space.

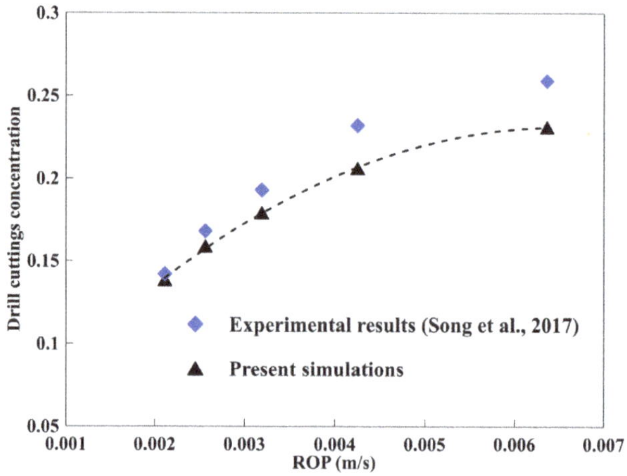

Figure 4. Comparison between numerical results and experimental results in a horizontal wellbore [13].

3. Results and Discussion

3.1. Effect of Drillpipe Rotation

The influences of varying pipe rotational speeds on annular cuttings' concentration and pressure drops are presented in Figure 5. As seen in Figure 5a, the annular cutting concentration decreases with the increase of rotational speed, due to the centrifugal force and shear force produced by rotating the drillpipe. This is the same as laboratory results of Ozbayoglu et al., 2012 [30]. At the rotational speeds of 50, 100, and 150 rpm, the cuttings' concentration in the annulus, composed of an elliptical drillpipe, had reduced by 7.7, 11.4, and 14.2% on average when compared with the circular drillpipe. This means that the higher the rotational speed, the better the hole cleaning performance of the elliptical drillpipe, but this improvement degree is no longer significant when the rotational speed is high. As seen in Figure 5b, the change of annular pressure drops almost keep the same trend of change as that of the cuttings' concentration; the change in the annular pressure drop is more sensitive to a high rotational speed. For example, at the conventional rotational speed of 50, 100, and 150 rpm, the pressure drops in the annulus that was composed of elliptical drillpipe were 5.9, 7.0, and 9.5% lower than those of the circular drillpipe. We infer that the combined effects of the decreased degree of hole cleaning improvement and the rise of the annular pressure drop yields this phenomenon. The velocity fluctuation and the collision between particles and particles/wall, caused by the increase of pipe rotational speed, are more substantial with the elliptical drillpipe, and the annular pressure drop increases.

In order to evaluate the hole cleaning efficiency of the elliptical drillpipe, compared with the circular drillpipe, the variation of the degree of hole cleaning improvement with different rotational speeds is shown in Figure 6. The hole cleaning ability of the elliptical drillpipe gradually improved as the rotational speed increased. The hole cleaning effect is more apparent when the value of ROP is high. For example, when the rotational speed was 50 rpm, in the ranges of ROP used in the simulation, the hole cleaning effect of the elliptical drillpipe with minimum ROP and maximum ROP increased by 6.2 and 10.2%, respectively. Although the ROP is an index to evaluate drill bit performance, poor hole cleaning in the annulus caused by the high value of ROP is also a problem that needs full attention. It can be observed that at the rotational speed of 50, 100, and 150 rpm, the hole cleaning effect of the elliptical drillpipe at ROP = 0.00636 m/s reached 10.2, 11.8, and 16.8%, respectively. It explains the fact that when the value of ROP is considerable, maintaining the elliptical drillpipe at a higher rotational speed will have a better hole cleaning performance.

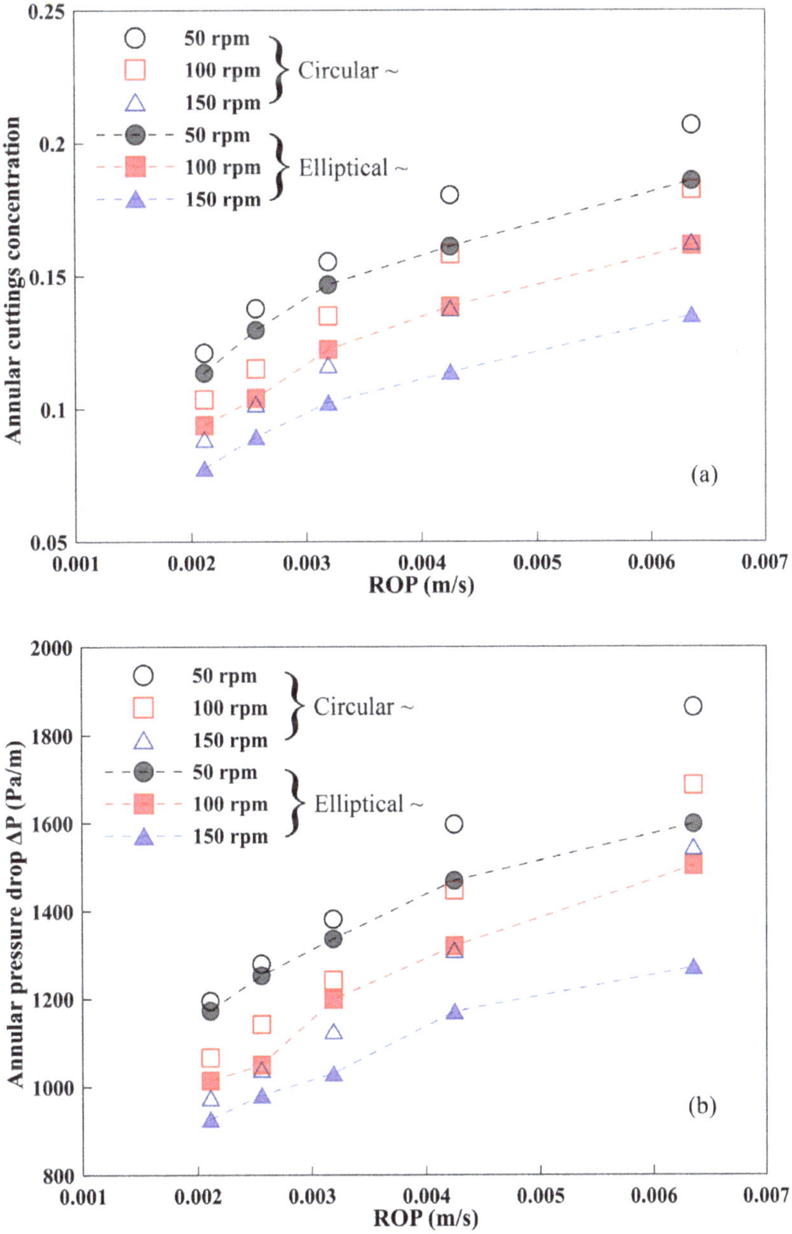

Figure 5. Annular cuttings' concentration and pressure drop with different rotational speeds. ($a/b = 1.1$ for elliptical drillpipe; e = 0.4). (**a**) Annular cuttings' concentration. (**b**) Annular pressure drop.

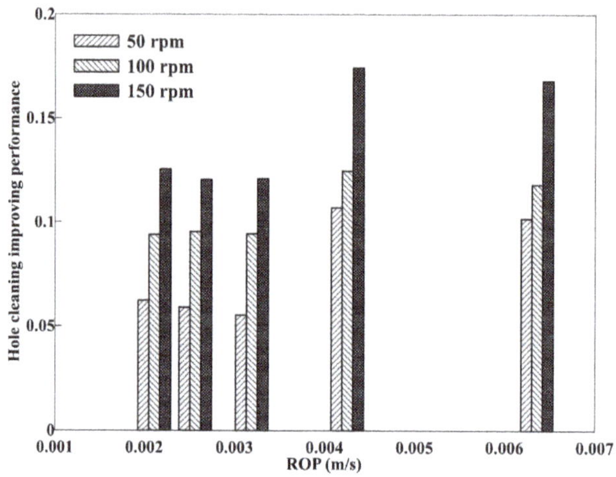

Figure 6. Hole cleaning efficiency of the elliptical drillpipe. (a/b = 1.1 for elliptical drillpipe; e = 0.4).

Figure 7 shows the cross-section of the cuttings' distribution in the annulus composed of the circular drillpipe and elliptical drillpipe under different rotational speeds. Similar to the experimental reports, the profile of the cuttings' bed presents the swaying phenomenon due to drillpipe rotation [14]. The swaying phenomenon is obvious due to higher drag forces applied to cuttings under the action of elliptical drillpipe rotation [31]. The cuttings bed distribution on the low side of the annulus is deflected towards the direction of drillpipe rotation. The cuttings bed on the high-side of the annulus reaches a dynamic equilibrium state under the combined action of axial high-speed fluid and the secondary flow caused by the elliptical drillpipe [22]. The high-side cuttings bed does not change obviously with the change of rotational speed, but the deflecting effect of the cuttings bed on the low-side is noticeable. Due to the friction and shear forces exerted by the rotating elliptical drillpipe, the tendency of cuttings particles to move in the direction of the rotation of the drillpipe is more significant, which is equivalent to reducing the number of cutting particles in the cuttings' bed and improving the hole cleaning efficiency.

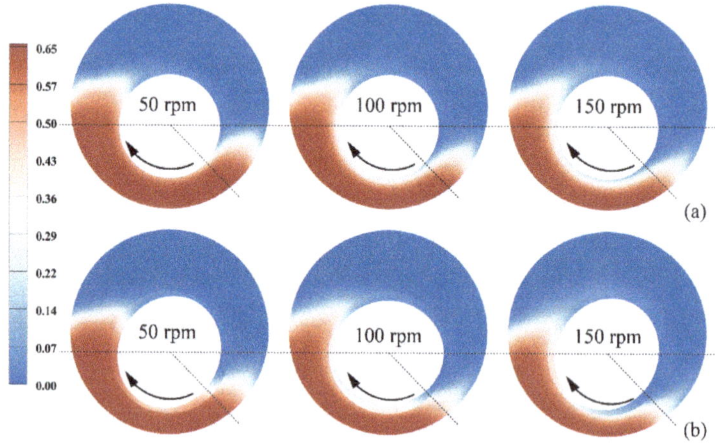

Figure 7. Contours of cuttings' concentration with different rotational speeds. (**a**) circular drillpipe (a/b = 1.0). (**b**) elliptical drillpipe (a/b = 1.1, non-twisted). (ROP = 0.00636 m/s, e = 0.4).

The shear force generated by rotating the drillpipe is equivalent to exerting tangential kinetic energy on the drilling fluid. As the carrier for cuttings, the change of tangential kinetic energy of drilling fluid will inevitably cause the tangential velocity of drill cuttings to change. Figure 8 depicts the tangential velocity distribution of drill cuttings along the radial direction. Rotating the drillpipe enables the cuttings to obtain a higher tangential kinetic energy from the drilling fluid, increasing the tangential velocity and improving transport efficiency. The cuttings' tangential velocity distribution decreases to the minimum at the wall and reaches the maximum at the core area inside the annular space. The tangential velocity of the cuttings increases as the rotational speed increases, and the change of tangential velocity caused by the elliptical drillpipe is more prominent. For example, when different rotational speeds were selected (50, 100, and 150 rpm), the maximum tangential velocity of cuttings under the action of the elliptical drillpipe increased by 412.5, 237.7, and 181.4%, respectively when compared with that of circular drillpipe. It indicates that the rotation of the elliptical drillpipe has a strong tangential disturbance effect on the drilling fluid, thus increasing the tangential velocity of the drill cuttings.

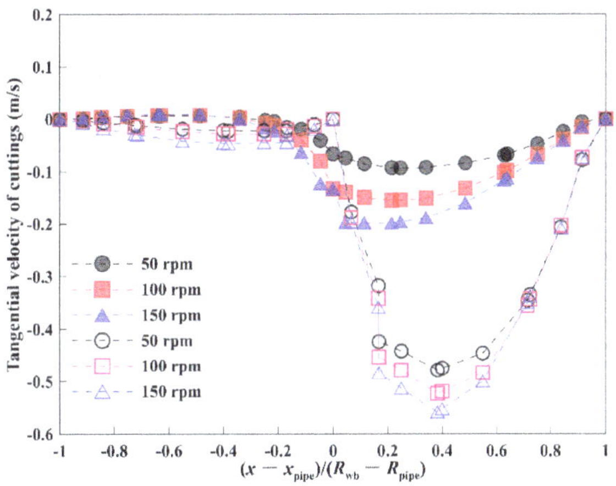

Figure 8. Tangential velocity of cuttings under different rotational speeds. (ROP = 0.00636 m/s, e = 0.4).

3.2. Effect of Major to Minor Axial Ratio a/b

The variation of the annular cuttings' concentration and annular pressure drop with the major to minor axial ratio is indicated in Figure 9. As shown in Figure 9a, the annular cuttings' concentration increases with the decrease of the axial ratio and the increase of ROP. At different ROPs, the cuttings concentration in the annulus composed of four types of elliptical drillpipe (a/b = 1.0, 1.05, 1.10, and 1.15) was 15.5, 15.0, 14.2, and 13.5% on average, respectively. When compared with the circular drillpipe, a degree of hole cleaning improvement in the elliptical drillpipe was obtained. When a/b = 1.05, 1.10, and 1.15, the cuttings removal efficiency was 3.2, 8.4, and 13.0%, respectively. The best hole cleaning effect was obtained when the axis ratio was large. As shown in Figure 9b, the annular pressure drop increases almost linearly with the decrease of the axis ratio and the increase of ROP. The height of the cuttings' bed had reduced, which increased the area of the annular flow channel, thereby reducing the annular pressure drop. Although the rise of axis ratio will also cause the increase in annular pressure drop, the effect of pressure drop reduction caused by the removal of the cuttings' bed is more significant. It is a good choice to increase the axial ratio when conditions permit, especially in areas with an obvious cuttings' deposition.

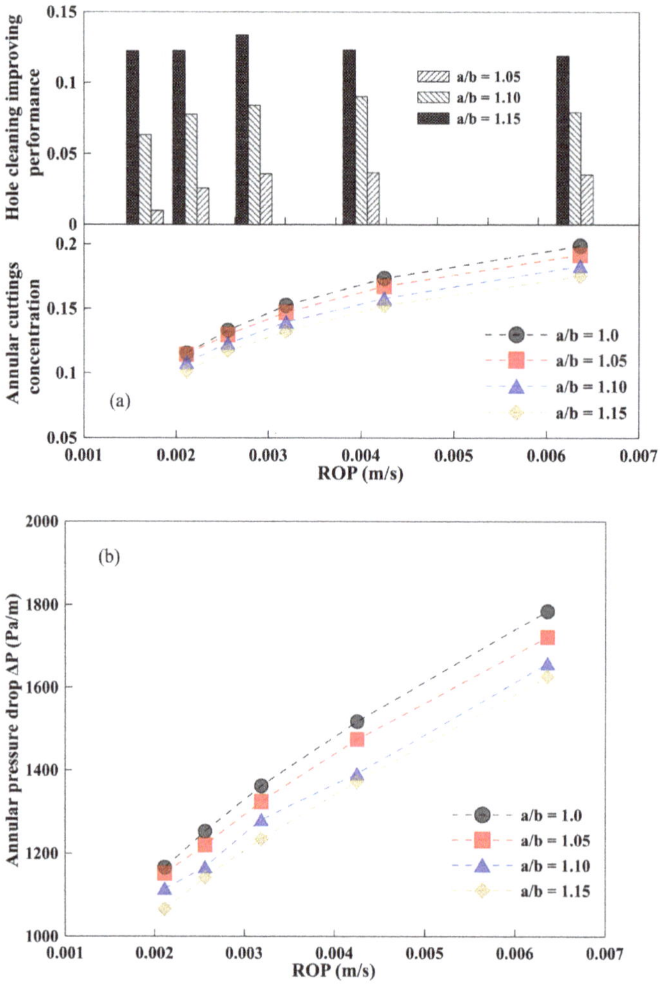

Figure 9. Annular cuttings' concentration and pressure drop with different axial ratios. (100 rpm; $e = 0.2$). (**a**) Annular cuttings' concentration. (**b**) Annular pressure drop.

The contours of the annular cuttings' concentration under different axial ratios are shown in Figure 10. It is observed that the cuttings concentration decreases slightly with the increase of the axis ratio, which indicates that the ability to promote cuttings' transportation was slightly enhanced. The interval of the flow channel at the low side of the annulus will periodically change with the pipe rotation. As the major axis of the elliptical drillpipe touched the cuttings' bed, pipe rotation continuously shook cuttings out of their bed into the moving stream above the bed, promoting the cuttings' bed erosion significantly [14]. The cuttings on the high side of the cuttings bed increased the transport ability due to the increase of the tangential momentum, which makes it possible to reach the minimum transport velocity (MTV) of cuttings' transportation. The cuttings' particles entered the main flow area of drilling fluid of the wide gap and left the annulus with the flow of drilling fluid.

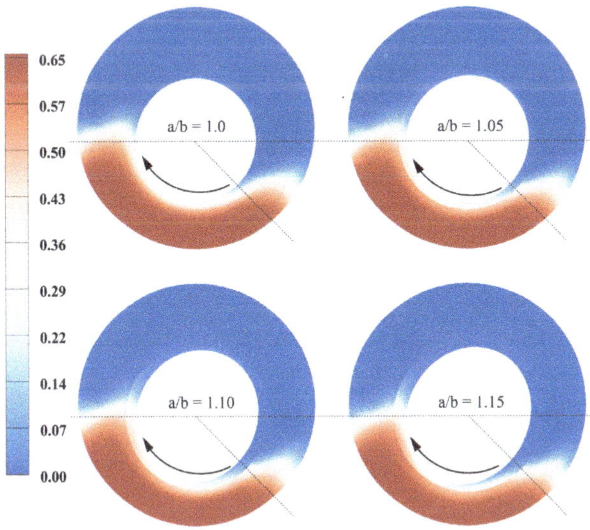

Figure 10. Contours of cuttings' concentration with different axial ratios. (100 rpm, ROP = 0.00425 m/s, $e = 0.2$).

Figure 11 represents the change of cuttings' tangential velocity with different axial ratios. The tangential velocity of cuttings at the pipe wall surface reaches its maximum, and then decays rapidly along the radial direction when the drillpipe is circular. In the case of an elliptical drillpipe, the tangential velocity distribution of the cuttings reached its peak inside of the annulus. This is due to the secondary flow phenomenon induced by the irregular shape of the ellipse, which increases the tangential kinetic energy intensity and the continuous range of the drilling fluid. When compared with the circular drillpipe with $a/b = 1.0$, the maximum tangential velocity of drill cuttings increases by 19.0% when the axial ratio is $a/b = 1.15$.

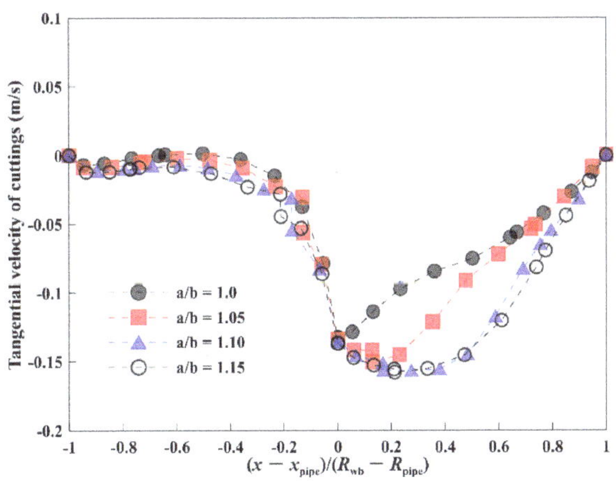

Figure 11. Tangential velocity of cuttings under different axis ratios. (ROP = 0.00425 m/s, $e = 0.2$).

3.3. Effect of Eccentricity

In the actual drilling process, the drillpipe eccentricity is a common issue due to gravity. Figure 12 illustrates the variations of annular cuttings' concentration in the elliptical drillpipe under different eccentricities. According to the numerical results, drillpipe eccentricity can increase the cuttings' accumulation [6]. As seen in Figure 12a, the annular cuttings concentration with the eccentricity of $e = 0$ and $e = 0.2$ is almost the same. In contrast, the annular cuttings concentration with the eccentricity of $e = 0.4$ is significantly reduced. It indicates that the hole cleaning performance of the elliptical drillpipe had gradually improved with the increase of eccentricity. The rise of eccentricity is more conducive to promoting the suspension of cuttings particles and improving transport efficiency. It can be seen from Figure 12b that the change in annular pressure drop is consistent with the change in the annular cuttings concentration. Unlike the circular drillpipe, the elliptical drillpipe will have a better hole cleaning effect when the eccentricity is large. So, the height of the bed, and the annular pressure drop are close at $e = 0$ and $e = 0.2$. A smaller annular pressure drop will be obtained when the eccentricity is further increased to $e = 0.4$, due to its strong ability to promote cuttings' transport.

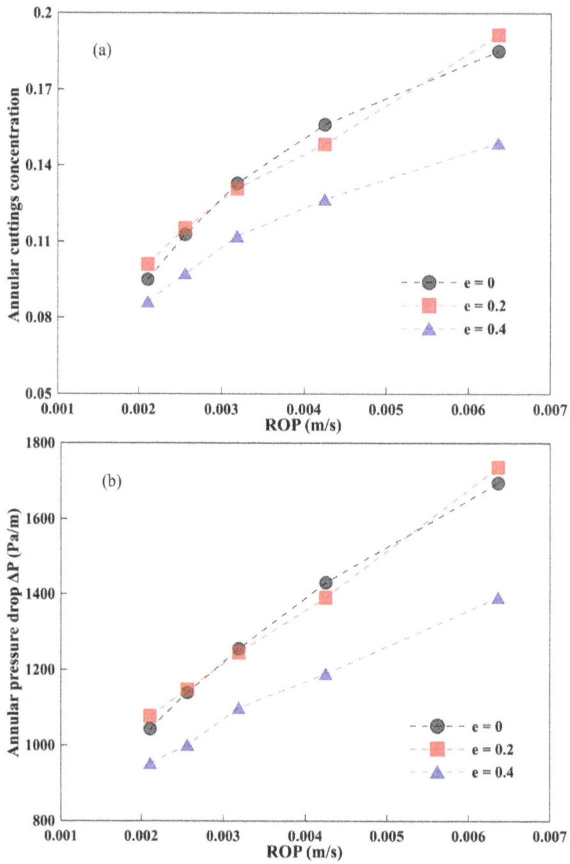

Figure 12. Annular cuttings' concentration and pressure drop with different eccentricities. ($a/b = 1.05$, 150 rpm). (**a**) Annular cuttings' concentration. (**b**) Annular pressure drop.

Figure 13 shows the distribution of annular cuttings' concentration under different eccentricities conditions. With the increase of the eccentricity, the height of the cuttings' bed decreases, but the circumferential distribution range of the cuttings bed becomes larger [6]. This performance is due to the cuttings' bed offset caused by the rotating elliptical drillpipe. When the eccentricity becomes larger, the offset effect becomes more and more apparent, and it will cause the particles at the low part of the bed to deflect towards the rotation direction. The particles at the high part of the bed will be more easily suspended in the drilling fluid and move out of the wellbore as the drilling fluid flows.

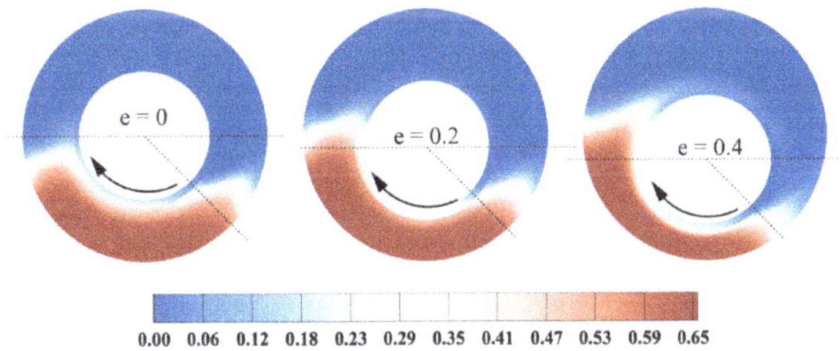

Figure 13. Contours of cuttings' concentration with different eccentricities. (150 rpm, a/b = 1.05, ROP = 0.00636 m/s).

Figure 14 shows the contour plots of the axial velocity distribution of drilling fluid in the annulus with different eccentricities conditions. It can be observed that the core area of axial velocity also deflects in consistency with the direction of the drillpipe rotation, and the larger the eccentricity is, the more obvious the effect of deflection is. With the increase of eccentricity, the effect of the elliptical drillpipe rotation on the distribution of the cuttings' bed, and the axial velocity of drilling fluid at wide gap is more obvious. The axial velocity of the drilling fluid in the core area decreased with the increase of the eccentricity, confirming that the area of the cuttings' bed decreased with the rise of eccentricity. It indicates that the reduction of the area of the cuttings' bed is equivalent to the increase of the fluid flow area in the annulus. At the same flow rate, the axial velocity of drilling fluid in the wide gap of the annulus decreases.

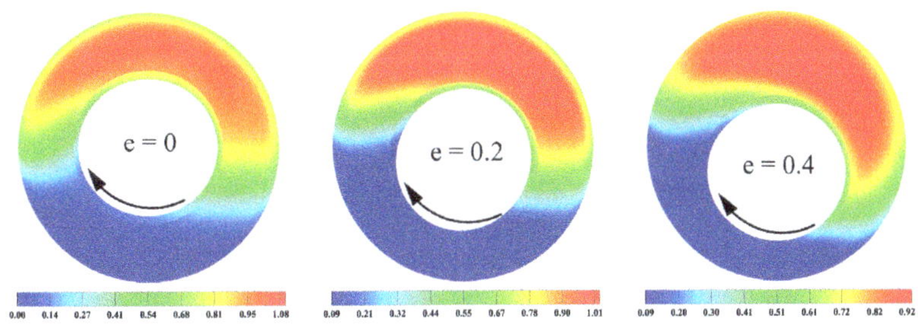

Figure 14. Contours of axial velocity of drilling fluid with different eccentricities. (150 rpm, a/b = 1.05, ROP = 0.00636 m/s).

3.4. Effect of Pitch Length Ratio (P/D_{pipe})

From some of the fields of industrial engineering, the application of a twisted elliptical tube may play a more active role [24,32]. The twisted elliptical tube can generate swirl flow or secondary flow, disturbing the boundary layer periodically and enhancing the heat transfer capacity. Therefore, the annular cuttings concentration and pressure drop of the elliptical drillpipe with different pitch length ratios are shown in Figure 15. The pitch length P is defined as the pipe length that undergoes 360° rotation; a smaller pitch length ratio means a more pronounced distortion of the pipe wall. As seen in Figure 15a, the hole cleaning ability of the elliptical drillpipe increased with the decrease in the pitch length ratio. The hole cleaning performance of the twist elliptical drillpipe was the strongest at pitch length ratio to pipe diameter (P/D) = 2.95. This can be attributed to the more vigorous fluid mixing caused by the intense twist degree of the pipe wall when the pitch length ratio is small. The decrease of pitch length ratio can promote cuttings' transport, but at the same time, will inevitably cause additional pressure loss. As seen in Figure 15b, the annular pressure drop under different ROPs did not change significantly with the increase of pitch length ratio. This can be explained by the fact that equilibrium is formed between the reduction of annular pressure drop caused by hole cleaning improvement, and the increase of pressure drop caused by twist geometry.

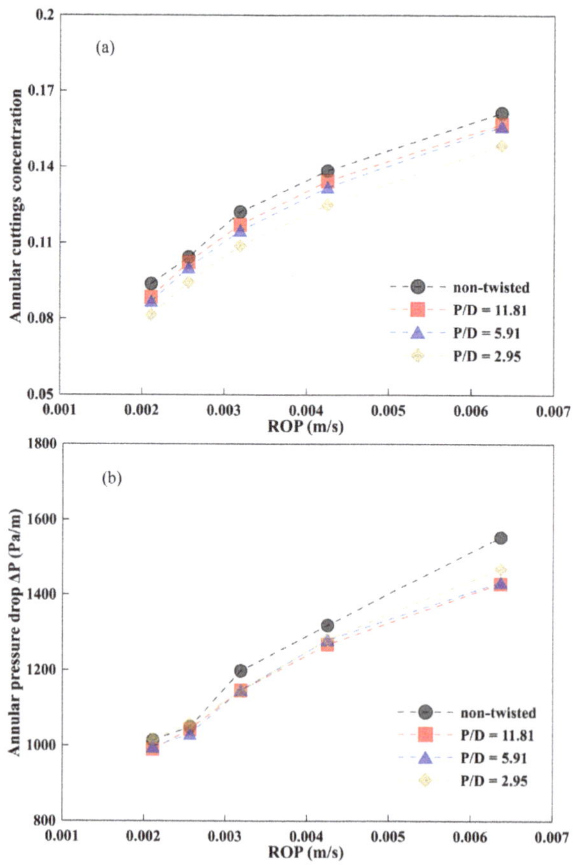

Figure 15. Annular cuttings concentration and pressure drop with different pitch length ratios. (a/b = 1.1, 100 rpm, e = 0.4). (**a**) Annular cuttings' concentration. (**b**) Annular pressure drop.

Figure 16 shows the turbulent kinetic energy (TKE) distribution under different pitch length ratios. The TKE can indicate the degree of turbulence in the flow caused by the rotation of the elliptical drillpipe. A higher value of TKE represents a minimal slip velocity and a higher cutting transport efficiency [21]. It can be regarded as an important parameter for cuttings particles that transport from the annulus. It can be observed that in the radial direction, the TKE obtained its maximum value inside of the annular space. With the increase of the pitch length ratio, the TKE value of the fluid in the annulus tends to increase. When P/D = 2.95, the maximum TKE value increased by 24.5% when compared with the circular drillpipe. The area with a higher value of the TKE represents the minimal slip velocity thereon, so the possibility of clustering of the cuttings in the form of the bed at the bottom could be avoided [33].

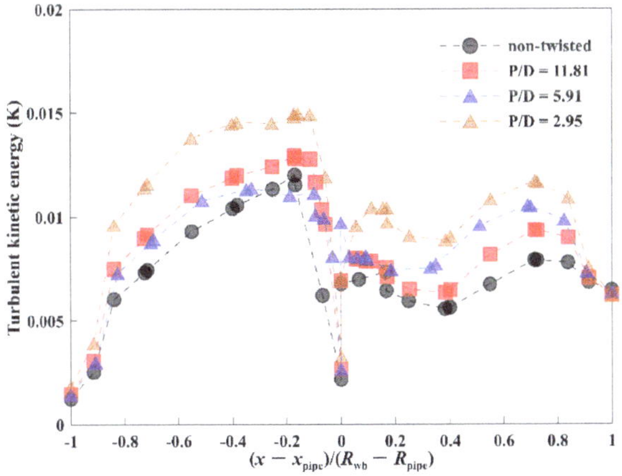

Figure 16. TKE distribution under different pitch length ratios. (a/b = 1.1, 100 rpm, e = 0.4, ROP = 0.00636 m/s).

In order to better understand the influence of the wall rotation of the elliptical drillpipe on the turbulence effect of the surrounding fluid. Figure 17 illustrates the contours of the TKE under different pitch length ratios. According to this Figure, by decreasing the pitch length ratio, TKE increases inside the annulus, resulting in a better hole cleaning effect [21]. The hole cleaning enhancement mechanisms using the elliptical drillpipe can be explained as (1) the swirl flow mixing between the core zone and the near-wall zone increases the kinetic energy of the drilling fluid; (2) the turbulent intensity is increased in the near rotating-wall zone. When the pitch length ratio is significant, the eddy current scale in the annulus is large, so the turbulence intensity is large, promoting the cuttings' transport and improving the hole cleaning efficiency.

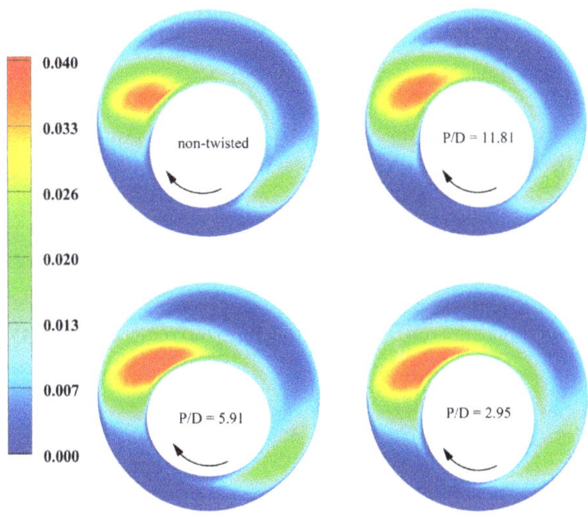

Figure 17. TKE distribution under different pitch length ratios. (a/b = 1.1, 100 rpm, e = 0.4, ROP = 0.00636 m/s).

4. Conclusions

In this study, the Eulerian-Eulerian approach was used to model and simulate the cuttings transport behavior in the annulus composed of the elliptical drillpipe. The results have been compared with the experimental data, and the accuracy and validity of the modeling and the computations were examined. Analysis of the solid–liquid flow was carried out in order to determine the effect of drillpipe rotation, ROP, eccentricity, axial ratio, and pitch length ratio on the pressure drop and cuttings' concentration. The hole cleaning ability of the elliptical drillpipe and the circular drillpipe was compared and evaluated.

The rotational speed played an essential role in improving the hole cleaning ability of the elliptical drillpipe. When the rotational speed becomes greater, the cuttings' transport ability can be effectively improved. At a higher ROP, high rotational speed will have a better hole cleaning performance. The increase of the major to minor axis ratio will also cause an increase of annular pressure drop and the decrease of annular cuttings' concentration. The larger the axial ratio of the elliptical drillpipe, the stronger the improving effect of hole cleaning. Due to the change of the relative position of the drillpipe and cuttings bed, unlike the circular drillpipe, the hole cleaning performance of the elliptical drillpipe gradually improved with the increase of eccentricity. The hole cleaning ability of the elliptical drillpipe increased with the decrease of the pitch length ratio. However, at the same time, it will inevitably cause additional pressure loss.

Therefore, the developed computational model has successfully captured the key features of cuttings' transport behavior in the annulus via an elliptical drillpipe. The model and results presented in this paper are helpful to extend the application of the CFD model for simulating the annular solid–liquid two-phase flow phenomenon with different kinds of devices used for hole cleaning. The present work makes a useful contribution to understanding the complex flow processes caused by the rotating elliptical drillpipe.

Author Contributions: Conceptualization, G.J.; methodology, T.Y.; validation, X.S.; writing—original draft preparation, G.J.; writing—review and editing, G.J. All authors have read and agreed to the published version of the manuscript.

Funding: This research was funded by the National Natural Science Foundation of China (grant number 51674087) and the National Science and Technology Major Project of the Ministry of Science and Technology of China (grant number 2017ZX05009-003).

Conflicts of Interest: The authors declare no conflict of interest.

References

1. Zhang, W.; Xie, P.; Li, Y.; Teng, L.; Zhu, J. Hydrodynamic characteristics and mass transfer performance of rotating packed bed for CO_2 removal by chemical absorption: A review. *J. Nat. Gas Sci. Eng.* **2020**, *79*, 103373. [CrossRef]
2. Zhang, J.; Li, Y.; Pan, Y.; Wang, X.; Yan, M.; Shi, X.; Zhou, X.; Li, H. Experiments and analysis on the influence of multiple closed cemented natural fractures on hydraulic fracture propagation in a tight sandstone reservoir. *Eng. Geol.* **2021**, *281*, 105981. [CrossRef]
3. Mahmoud, H.; Hamza, A.; Nasser, M.S.; Hussein, I.A.; Ahmed, R.; Karami, H. Hole cleaning and drilling fluid sweeps in horizontal and deviated wells: Comprehensive review. *J. Pet. Sci. Eng.* **2020**, *186*, 106748. [CrossRef]
4. Guan, Z.C.; Liu, Y.M.; Liu, Y.M.; Xu, Y.Q. Hole cleaning optimization of horizontal wells with the multi-dimensional ant colony algorithm. *J. Nat. Gas Sci. Eng.* **2016**, *28*, 347–355. [CrossRef]
5. Guo, J.; Zhang, L.; Zhu, Q. A quadruple-porosity model for transient production analysis of multiple-fractured horizontal wells in shale gas reservoirs. *Environ. Earth Sci.* **2015**, *73*, 5917–5931. [CrossRef]
6. Heydari, O.; Sahraei, E.; Skalle, P. Investigating the impact of drillpipe's rotation and eccentricity on cuttings transport phenomenon in various horizontal annuluses using computational fluid dynamics (CFD). *J. Pet. Sci. Eng.* **2017**, *156*, 801–813. [CrossRef]
7. Movahedi, H.; Vasheghani Farahani, M.; Jamshidi, S. Application of Hydrated Basil Seeds (HBS) as the herbal fiber on hole cleaning and filtration control. *J. Pet. Sci. Eng.* **2017**, *152*, 212–228. [CrossRef]
8. Van Puymbroeck, L.; El Bachiri, K.; Williams, H. New generation drill pipe to increase drilling performance in ERD wells. In Proceedings of the SPE/IADC Middle East Drilling Technology Conference & Exhibition, Dubai, United Arab Emirates, 7–9 October 2013; pp. 7–9.
9. Duan, M.; Miska, S.Z.; Yu, M.; Takach, N.E.; Zettner, R.M.A.C.M. Critical Conditions for Effective Sand-Sized Solids Transport in Horizontal and High-Angle Wells. *SPE Drill. Complet.* **2009**, *24*, 229–238. [CrossRef]
10. Han, S.M.; Woo, N.S.; Hwang, Y.K. Solid-liquid mixture flow through a slim hole annulus with rotating inner cylinder. *J. Mech. Sci. Technol.* **2009**, *23*, 569–577. [CrossRef]

11. Ulker, E.; Sorgun, M. Comparison of computational intelligence models for cuttings transport in horizontal and deviated wells. *J. Pet. Sci. Eng.* **2016**, *146*, 832–837. [CrossRef]
12. Epelle, E.I.; Gerogiorgis, D.I. CFD modelling and simulation of drill cuttings transport efficiency in annular bends: Effect of particle sphericity. *J. Pet. Sci. Eng.* **2018**, *170*, 992–1004. [CrossRef]
13. Song, X.; Xu, Z.; Wang, M.; Li, G.; Shah, S.N.; Pang, Z. Experimental Study on the Wellbore-Cleaning Efficiency of Microhole-Horizontal-Well Drilling. *SPE J.* **2017**, *22*, 1189–1200. [CrossRef]
14. Tomren, P.H. Experiment study of cuttings transport in Directional Wells. *SPE Drill. Eng.* **1986**, *1*, 43–56. [CrossRef]
15. Epelle, E.I.; Gerogiorgis, D.I. A review of technological advances and open challenges for oil and gas drilling systems engineering. *AIChE J.* **2020**, *66*. [CrossRef]
16. Yan, T.; Wang, K.; Sun, X.; Luan, S.; Shao, S. State-of-the-art cuttings transport with aerated liquid and foam in complex structure wells. *Renew. Sustain. Energy Rev.* **2014**, *37*, 560–568. [CrossRef]
17. Boyou, N.V.; Ismail, I.; Wan Sulaiman, W.R.; Sharifi Haddad, A.; Husein, N.; Hui, H.T.; Nadaraja, K. Experimental investigation of hole cleaning in directional drilling by using nano-enhanced water-based drilling fluids. *J. Pet. Sci. Eng.* **2019**, *176*, 220–231. [CrossRef]
18. Song, X.Z.; Li, G.S.; Huang, Z.W.; Wang, H.Z.; Tian, S.C.; Shi, H.Z. Experimental study on horizontal wellbore cleanout by rotating jets. *J. Pet. Sci. Eng.* **2010**, *75*, 71–76. [CrossRef]
19. Allahvirdizadeh, P.; Kuru, E.; Parlaktuna, M. Experimental investigation of solids transport in horizontal concentric annuli using water and drag reducing polymer-based fluids. *J. Nat. Gas Sci. Eng.* **2016**, *35*, 1070–1078. [CrossRef]
20. Busahmin, B.; Saeid, N.H.; Alusta, G.; Zahran, E.S.M.M. Review on hole cleaning for horizontal wells. *ARPN J. Eng. Appl. Sci.* **2017**, *12*, 4697–4708. [CrossRef]
21. Keshavarz Moraveji, M.; Sabah, M.; Shahryari, A.; Ghaffarkhah, A. Investigation of drill pipe rotation effect on cutting transport with aerated mud using CFD approach. *Adv. Powder Technol.* **2017**, *28*, 1141–1153. [CrossRef]
22. Yan, T.; Qu, J.; Sun, X.; Chen, Y.; Hu, Q.; Li, W.; Zhang, H. Numerical investigation on horizontal wellbore hole cleaning with a four-lobed drill pipe using CFD-DEM method. *Powder Technol.* **2020**, *375*, 249–261. [CrossRef]
23. Kunlabud, S.; Chuwattanakul, V.; Kongkaitpaiboon, V.; Promthaisong, P.; Eiamsa-ard, S. Heat transfer in turbulent tube flow inserted with loose-fit multi-channel twisted tapes as swirl generators. *Theor. Appl. Mech. Lett.* **2017**, *7*, 372–378. [CrossRef]
24. Tang, X.; Dai, X.; Zhu, D. International Journal of Heat and Mass Transfer Experimental and numerical investigation of convective heat transfer and fluid flow in twisted spiral tube. *Int. J. Heat Mass Transf.* **2015**, *90*, 523–541. [CrossRef]
25. Zhang, C.; Wang, D.; Ren, K.; Han, Y.; Zhu, Y.; Peng, X.; Deng, J.; Zhang, X. A comparative review of self-rotating and stationary twisted tape inserts in heat exchanger. *Renew. Sustain. Energy Rev.* **2016**, *53*, 433–449. [CrossRef]
26. GhasemiKafrudi, E.; Hashemabadi, S.H. Numerical study on cuttings transport in vertical wells with eccentric drillpipe. *J. Pet. Sci. Eng.* **2016**, *140*, 85–96. [CrossRef]
27. Mohammadzadeh, K.; Hashemabadi, S.H.; Akbari, S. CFD simulation of viscosity modifier effect on cutting transport by oil based drilling fluid in wellbore. *J. Nat. Gas Sci. Eng.* **2016**, *29*, 355–364. [CrossRef]
28. Liang, C.; Jiang, H.; Lin, S.; Li, H.; Wang, B. Intelligent Generation of Evolutionary Series in a Time-Variant Physical System via Series Pattern Recognition. *Adv. Intell. Syst.* **2021**, *3*, 2000172. [CrossRef]
29. Huilin, L.; Gidaspow, D.; Bouillard, J.; Wentie, L. Hydrodynamic simulation of gas-solid flow in a riser using kinetic theory of granular flow. *Chem. Eng. J.* **2003**, *95*, 1–13. [CrossRef]
30. Ozbayoglu, E.M.; Osgouei, R.E.; Ozbayoglu, A.M.; Yuksel, H.E. Hole-cleaning performance of gasified drilling fluids in horizontal well sections. *SPE J.* **2012**, *17*, 912–923. [CrossRef]
31. Amanna, B.; Khorsand Movaghar, M.R. Cuttings transport behavior in directional drilling using computational fluid dynamics (CFD). *J. Nat. Gas Sci. Eng.* **2016**, *34*, 670–679. [CrossRef]
32. Jafari, M.; Dabiri, S.; Farhadi, M.; Sedighi, K. Effects of a three-lobe swirl generator on the thermal and flow fields in a heat exchanging tube: An experimental and numerical approach. *Energy Convers. Manag.* **2017**, *148*, 1358–1371. [CrossRef]
33. Dewangan, S.K.; Sinha, S.L. Exploring the hole cleaning parameters of horizontal wellbore using two-phase Eulerian CFD approach. *J. Comput. Multiph. Flows* **2016**, *8*, 15–39. [CrossRef]

minerals

Article

Study on the Shale Hydration Inhibition Performance of Triethylammonium Acetate

Yuanzhi Qu [1], Ren Wang [1,2,*], Shifeng Gao [1], Hongjun Huang [1], Zhilei Zhang [1], Han Ren [1], Yuehui Yuan [1], Qibing Wang [1,2], Xiangyun Wang [3] and Weichao Du [3,*]

1 CNPC Engineering Technology R & D Company Limited, Beijing 102206, China; quyz_dri@cnpc.com.cn (Y.Q.); gaosfdr@cnpc.com.cn (S.G.); huanghj_dri@cnpc.com.cn (H.H.); zhangzhidr@cnpc.com.cn (Z.Z.); renhandr@cnpc.com.cn (H.R.); yuanyhdr@cnpc.com.cn (Y.Y.); wangqbdr@cnpc.com.cn (Q.W.)
2 Department of Petroleum Engineering, China University of Petroleum (East China), Qingdao 266580, China
3 Shaanxi Province Key Laboratory of Environmental Pollution Control and Reservoir Protection Technology of Oilfields, School of Chemistry and Chemical Engineering, Xi'an Shiyou University, Xi'an 710065, China; 18298818107@163.com
* Correspondence: wangrdr@cnpc.com.cn (R.W.); duweichao@xsyu.edu.cn (W.D.)

Abstract: Shale inhibitor is an additive for drilling fluids that can be used to inhibit shale hydration expansion and dispersion, and prevent wellbore collapse. Small molecular quaternary ammonium salt can enter the interlayer of clay crystal, and enables an excellent shale inhibition performance. In this paper, a novel ionic shale inhibitor, triethylammonium acetate (TEYA), was obtained by solvent-free synthesis by using acetic acid and triethylamine as raw materials. The final product was identified as the target product by Fourier transform infrared spectroscopy (FT-IR). The inhibition performance of TEYA was studied by the mud ball immersion test, linear expansion test, rolling recovery test and particle size distribution test. The results demonstrated that the shale inhibitor shows a good shale hydration inhibition performance. The inhibition mechanism was studied by FT-IR and X-ray diffraction (XRD), respectively; the results showed that triethylammonium acetate TEYA could enter the crystal layer of clay and inhibit it through physical adsorption.

Keywords: drilling fluids; hydration swelling; shale inhibitor; shale gas

Citation: Qu, Y.; Wang, R.; Gao, S.; Huang, H.; Zhang, Z.; Ren, H.; Yuan, Y.; Wang, Q.; Wang, X.; Du, W. Study on the Shale Hydration Inhibition Performance of Triethylammonium Acetate. *Minerals* **2022**, *12*, 620. https://doi.org/10.3390/min12050620

Academic Editors: Mofazzal Hossain, Hisham Khaled Ben Mahmud and Md Motiur Rahman

Received: 13 April 2022
Accepted: 6 May 2022
Published: 13 May 2022

1. Introduction

With the development of oil and gas exploration to deep wells, ultra-deep wells, offshore wells and wells in complex sections [1,2], a series of drilling problems caused by the hydration expansion and dispersion of shale, such as wellbore instability, bit mud bag and borehole purification, would occur frequently, which increases the drilling difficulty and non-operation time of the oil industry [3–6]. Drilling fluid, commonly known as the blood of drilling engineering, plays the role of suspending drilling cuttings, cleaning and cooling the bit, balancing formation pressure, reservoir protection, and maintaining wellbore stability. According to different mobile phases, drilling fluids can be divided into water-based drilling fluids, oil-based drilling fluids and synthetic-based drilling fluids. Water-based drilling fluids have the advantages of less components, a simple preparation process, environmental protection and low cost, which make them widely used all over the world. The performance of drilling fluid plays a decisive role in the stability of wellbore; therefore, special requirements are put forward for the inhibition of the drilling fluid system [7–10].

Adding shale inhibitor to drilling fluid can prevent wellbore collapse to a certain extent. There are two main action mechanisms of inhibitors [11–13]. One is that inhibitors are wrapped on the surface of shale to separate it from water. Another way is that the inhibitor enters the clay crystal layer, the water between the clay crystal layers is replaced, and the inhibitor is adsorbed inside the crystal layer, so as to play an excellent inhibition effect [14–17]. The high molecular weight inhibitors will play the role of wrapping inhibition, which have a certain impact on the performance of drilling fluids, so its application is greatly limited. However, intercalation inhibitors gradually show a good development prospect in shale inhibitors because of their unique molecular structure advantages [18–20].

The commonly used shale inhibitors mainly include inorganic salts, asphalts, polymers and humic acids, but these shale inhibitors have some shortcomings; for example, asphalt is not conducive to environmental protection, silicate will make the rheology of the system difficult to control, glycosides cannot solve the drilling problem of active shale, etc. [21–26]. In recent years, amine-based drilling fluid has become the focus of petroleum researchers all over the world because of its environmental protection and low cost. Its essence is that amine (ammonium) shale inhibitor is added to drilling fluid. Ammonium-based shale inhibitors are usually prepared by an acid–base reaction between small molecular weight secondary tertiary amine and matrix acid. They have low molecular weight and low cost. They are used in drilling fluid and oil production, but their economy and temperature resistance still need to be further improved.

Based on the above analysis, a new ionic shale inhibitor triethylammonium acetate was developed with acetic acid and triethylamine as raw materials. Its structure was determined by infrared spectroscopy, its inhibition performance was comprehensively studied, and its inhibition mechanism was analyzed.

2. Experimental

2.1. Materials

Triethylamine, acetic acid, acetone, and absolute ethanol were purchased from Chengdu Kelong Chemical Reagent Factory (Chengdu, China) (the above reagents were analytically pure), and sodium bentonite (Na-MMT) was purchased from Xinjiang Xiazijie Bentonite Co., Ltd. (Tacheng, China).

2.2. Preparation of TEYA

Triethylamine (0.25 mol) was added into a three-neck flask, then acetic acid (0.2 mol) was slowly added with stirring, and then heated at the temperature required until the reaction was completed. After the reaction, the reaction solution was recrystallized with a mixture of ethyl acetate and anhydrous ethanol (volume ratio 8:2). TEYA was left to stand for 24 h to obtain an acicular white solid. The solid was vacuum dried at 45 °C for 8 h and TEYA was obtained. The chemical formula and the synthetic route of TEYA is shown in Scheme 1.

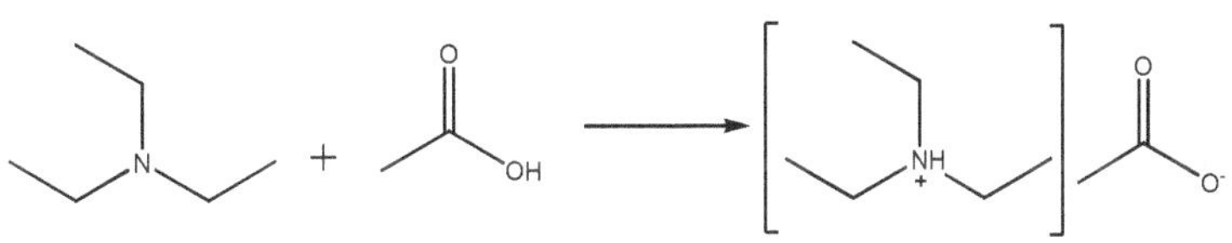

Scheme 1. The synthetic route of triethylammonium acetate (TEYA).

2.3. Inhibition Performance Study

2.3.1. Mud Ball Immersing Tests

Bentonite (50 g) was added to 25 g distilled water and stirred evenly, which was used to make a 30 g mud ball. Then, the mud ball was immersed in inhibitor solutions, and the change in the mud ball size was observed after a period of time.

2.3.2. Linear Expansion Tests

A total of 10 g Na-MMT dried at 105 °C was compressed at 10 MPa for 5 min by using a pressing machine, and the swelling height of Na-MMT was tested using a shale swelling instrument.

2.3.3. Rolling Recovery Tests

Mesh shale samples (50 g) were put into the high-temperature aging tanks filled with 350 mL solutions. The hot-rolling tests were carried out at 100 °C for 16 h. When these experiments were completed, the shale samples were carefully rinsed and filtered by a 40-mesh sieve, then dried to constant weight at 105 °C, and weighed. We calculated the hot rolling recovery rate by the equation as below:

$$R = \frac{M}{50} \times 100\%$$

where M is the recycling mass, g, and R is the rolling recovery rate, %.

2.3.4. Particle Distribution Tests

Different types of inhibitors were added to 350 mL 4% Na-MMT based slurry and fully stirred. The effect of inhibitor on the particle size of bentonite was tested after 24 h. Particle size analysis was measured with a laser diffraction technique (HORIBA, Kyoto, Japan). The operating temperature was 25 °C, the circulation speed was 2000 r/min, the ultrasonic system was 40 Hz, 70 W, and the stirring speed was 100–475 rpm.

2.4. Inhibition Mechanism Study

2.4.1. Fourier Transform Infrared Spectroscopy (FT-IR) Analysis

The Na-MMT treated with 1.2% triethylammonium acetate (TEYA) was centrifuged and filtered. During the filtration process, the Na-MMT was carefully rinsed by the mixed solution of absolute ethanol and ethyl acetate. The obtained Na-MMT was vacuum dried at 50 °C for 24 h and then stored in a closed manner. The prepared sample powder was mixed with KBr and pressed into pellets for FT-IR analysis. The FT-IR spectra in the wavelength range of 4000–500 cm^{-1} were recorded by a WQF-520 Fourier transform infrared spectrometer (Thermo Electron Co., Waltham, MA, USA).

2.4.2. X-ray Diffraction (XRD) Analysis

Na MMT (1 g) was added into 100 mL 1.2% TEYA solutions. The solutions were stirred for 2 h and then stirred at high speed for 30 min. After it was centrifuged, the bottom precipitation was washed, filtered and dried it at 50 °C for XRD tests. X-ray diffraction tests were performed on an X Pert PRO MPD diffractometer. Scans were taken from a 2θ angle from 3° to 20°, step size 0.1, and scan time per step of 10 s.

3. Results and Discussion

3.1. FT-IR Analysis

The results of TEYA are shown in Figure 1.

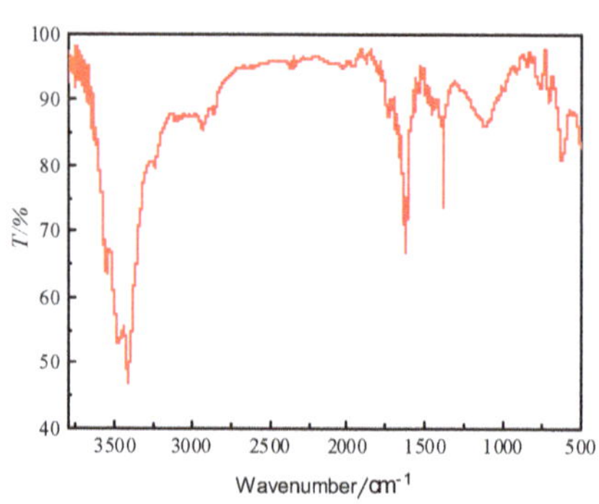

Figure 1. The Fourier transform infrared spectroscopy (FT-IR) spectrum of TEYA.

It can be seen from Figure 1 that a strong and wide absorption peak appeared near 3485 cm^{-1}, which was the stretching vibration absorption peak of -COOH; 2850 cm^{-1} was the stretching vibration absorption peak of C-H bond in $-CH_2-$ and $-CH_3$; and the peak at 1630 cm^{-1} was the stretching vibration of C-N. According to the above analysis, the synthetic product was the target product [27–30].

3.2. Mud Ball Immersing Tests

The hydration swelling tests of the inhibitor on the mud ball was studied, and the results are shown in Figure 2.

Figure 2. The status of mud balls immersed in different fluids.

It can be seen from Figure 2 that the final shape of the two mud balls was very different after immersing in different solutions for a period of time. The mud ball in water presented as expanded and broken; however, the mud ball in inhibitor solutions was relatively intact, indicating that the shale inhibitor shows an obvious inhibitory effect on the hydration of shale. This experiment directly proved that shale inhibitor TEYA could effectively prevent the hydration expansion of shale [31].

3.3. Linear Expansion Tests

The linear swelling effect of 1.2% inhibitor on shale was studied, and the experimental results are shown in Table 1.

Table 1. Linear swelling tests of shale.

Experiment Number	Test Solution	Height/mm	Swelling/mm	Swelling Rate/%
1	TEYA	11.04	2.93	26.53985507
4	KCl	10.96	3.45	31.47810219
5	Water	11.84	6.15	51.94256757

The main reason for swelling of Na-MMT is that water molecules enter into the crystal layer of minerals and the cation exchange of bentonite minerals, which increases the crystal layer spacing of bentonite. It can be seen from Table 1 that the prepared inhibitor showed the most obvious inhibitory effect on shale swelling, followed by potassium chloride (KCl). The above experimental results showed that KCl also had a certain inhibitory effect; the reason was that the diameter of K^+ was equivalent to the diameter of the hexagonal oxygen ring on the bottom of the clay silica tetrahedron, and the hydration energy of K^+ was small, which was easily inserted into the clay crystal layer. Therefore, Ca^{2+} and Na^+ with hydration radius and hydration energy greater than K^+ can be replaced. K^+ entering the hexagonal oxygen ring can be firmly connected with the clay wafer, and it was not easy to separate and inhibit the hydration expansion of the shale. TEYA can play an inhibition role through the synergistic effect of crystal layer substitution and concentration difference, so as to play a nice shale hydration inhibition effect [32–34].

3.4. Rolling Recovery Tests

The rolling recovery rates of shale in different solutions were compared to evaluate the inhibition performance of TEYA, as shown in Figure 3.

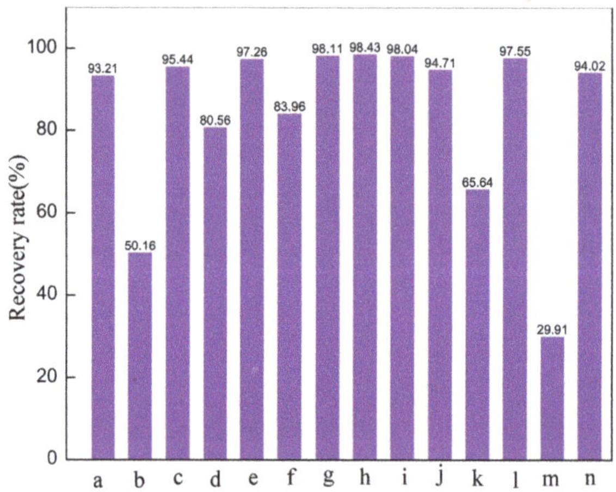

Figure 3. Effect of different inhibitors on the rolling recovery rate of shale (a, 2 %TEYA; b, water; c, 3% polydimethylamine; d, 5% emulsified asphalt; e, 5# white oil; f, 3% small cation; g, 2# oil-based drilling fluids; h, 3# oil-based drilling fluids; i, 5% KCl; j, 5%KCl+ 5% emulsified asphalt; k, 1% $CaCl_2$; l, 5% K_2SiO_3; m, 5% $FeCl_3$).

As can be seen from Figure 3, the rolling recovery of 5% $FeCl_3$ was lower than 40% and distilled water was only about 50%; the others were higher than 60%. The highest rolling recoveries were 5# white oil, oil-based drilling fluids, 5% potassium sorbate, 5% KCl + 5% emulsified asphalt, 5% K_2SiO_3, 5% KCl, 1% CaO and TEYA (93.21%). It can be seen that TEYA demonstrated a good inhibition performance.

3.5. Particle Distribution Tests

The median diameter and average diameter of the water group were compared with those with added TEYA to study the adsorption of TEYA on clay. The results are shown in Figure 4.

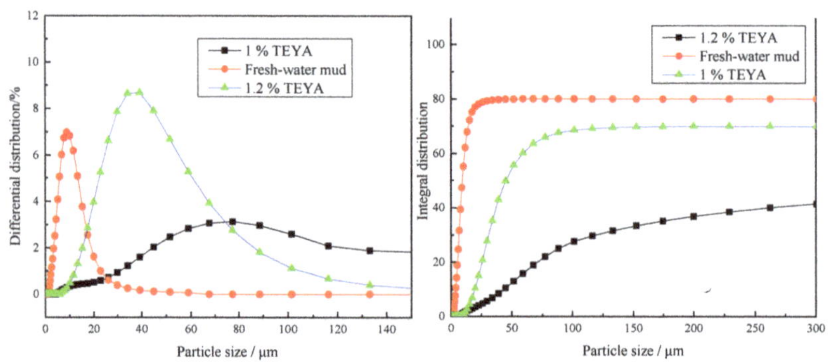

Figure 4. The differential distribution of bentonite treated by several anti-swelling agents: differential distribution/% (**left**); integral distribution (**right**).

According to Figure 4, the pitch diameter of water-based slurry was 8.05068 μm and the average diameter was 21.80842 μm. When the inhibitor was added, the median diameter was raised to 56.19773 μm and the average diameter was raised to 60.04097 μm. It can be seen that the particle diameter obviously increased, which played a certain role in inhibiting the hydration and dispersion of shale.

3.6. Inhibition Mechanism Study

The interaction between Na-MMT and clay was studied by FT-IR. Figure 5 shows the infrared spectra of Na MMT, TEYA and TEYA/Na-MMT.

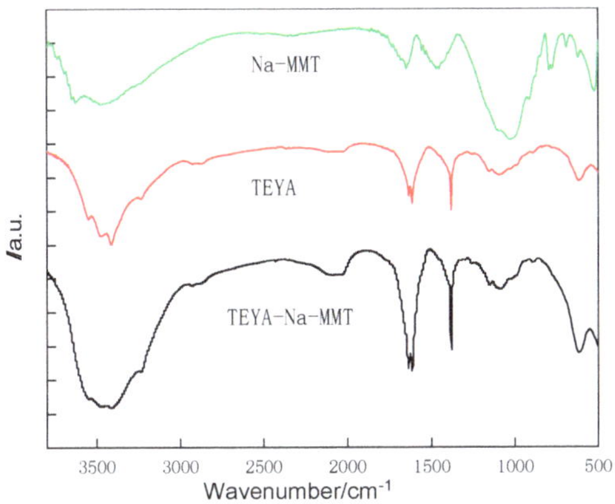

Figure 5. FT-IR spectra of Na-MMT, TEYA and Na-MMT/TEYA.

For the curve of Na-MMT, 3451 cm^{-1}, 1020 cm^{-1} and 1645 cm^{-1} correspond to the absorption peaks of -OH stretching vibration, Si-O stretching vibration and variable angle vibration of interlayer water molecules in bentonite, respectively [35]. The infrared spectra of TEYA/Na-MMT contained all the characteristic absorption peaks of the TEYA and Na-MMT, and no new peak appeared, which indicated that TEYA and Na-MMT were effectively combined together. However, the position of the absorption peak of Na-MMT was changed slightly. The variable angle vibration absorption peak of water molecules in bentonite had a red shift from 1645 cm^{-1} in low frequency to 1613 cm^{-1} in high frequency, and the peak width had narrowed and the peak strength was weakened, which indicated that the interlayer water of Na-MMT had been greatly reduced. Since TEYA contains quaternary ammonium salt, it is beneficial for TEYA to squeeze out the water molecules between the crystalline layers of bentonite, and therefore presents a nice inhibitory performance [35,36].

The XRD of dry Na-MMT treated with 1.2 wt% TEYA is shown in Figure 6. From Figure 6 we can see that the crystal layer spacing of dry Na-MMT is 1.28 nm [36], and the crystal layer spacing of TEYA/Na MMT is changed to 1.47 nm, which indicated that TEYA had successfully entered into the clay crystal layer and expanded the crystal layer spacing. Due to the occurrence of lattice substitution, the clay surface was negatively charged. The positively charged quaternary ammonium group on TEYA can ensure its strong adsorption on the clay. The molecular weight of TEYA was small and can enter the Na-MMT crystal layer to pull the crystal layer spacing.

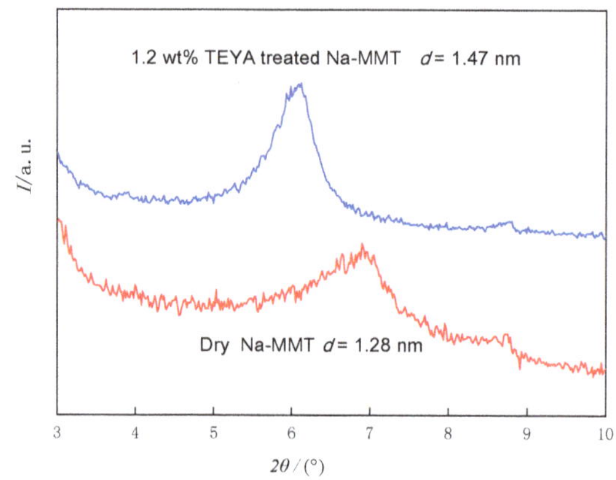

Figure 6. X-ray diffraction (XRD) of dry Na-MMT and TEYA/Na-MMT.

The inhibition mechanism is shown in Figure 7. In the water environment, water molecules increased the spacing of Na-MMT layers, and TEYA entered into the clay crystal layer to replace water molecules, so as to effectively prevent the hydration and expansion of clay [36].

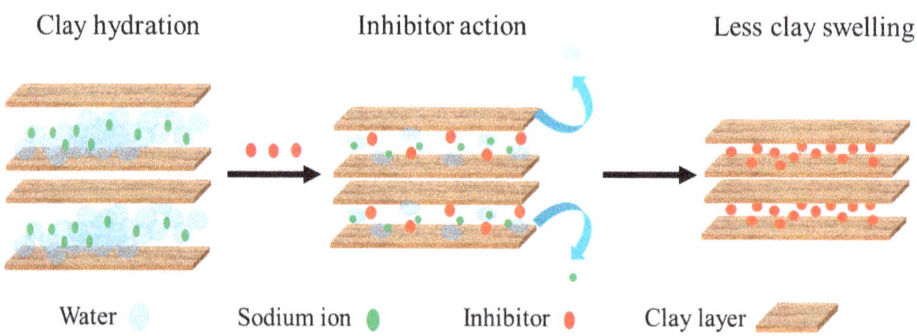

Figure 7. Inhibition mechanism of TEYA.

4. Conclusions

In this paper, a new ionic shale inhibitor TEYA was obtained by solvent-free synthesis with acetic acid and triethylamine as raw materials. The final product was identified as the target product by FT-IR. The mud ball immersion test shows that the mud ball in the inhibitor solution was relatively intact after 24 h. The linear expansion experiment showed that the linear expansion rate of clay in the inhibitor solution is only 26.54% after 16 h, the rolling recovery rate of shale in the inhibitor solution was as high as 93.21%, and the average particle size of clay in the inhibitor solution was 60.04097 μm. The results show that the shale inhibitor had a certain shale inhibition effect.

Author Contributions: Conceptualization, Y.Q.; methodology, Y.Y.; software, Y.Y.; validation, H.R.; formal analysis, R.W. and Q.W.; investigation, R.W. and S.G.; resources, X.W.; data curation, Z.Z.; writing—original draft preparation, H.H., Z.Z. and W.D.; writing—review and editing, Y.Q., Q.W., X.W. and W.D.; supervision, S.G and H.H.; project administration, Q.W. and W.D.; funding acquisition, R.W. and H.R. All authors have read and agreed to the published version of the manuscript.

Funding: This research was funded by CNPC's Major Science and Technology Projects (2021DJ4407), National Natural Science Foundation of China (41902323, 52174016).

Acknowledgments: The authors would like to thank the CNPC's Major Science and Technology Projects (2021DJ4407), National Natural Science Foundation of China (41902323,52174016) for financial support.

Conflicts of Interest: The authors declare no conflict of interest.

References

1. Lai, N.J.; Tang, L.; Jia, N.; Qiao, D.Y.; Chen, J.L.; Wang, Y.; Zhao, X.B. Feasibility study of applying modified nano-SiO_2 hyperbranched copolymers for enhanced oil recovery in low-mid permeability reservoirs. *Polymers* **2019**, *11*, 1483. [CrossRef] [PubMed]
2. Hong, S.H.; Jo Jin, H.; Choi, M.J.; Jang, H.W.; Kim Ju, Y.; Hwang, W.R.; Kim, S.Y. Influence of MoS_2 nanosheet size on performance of drilling mud. *Polymers* **2019**, *11*, 321. [CrossRef] [PubMed]
3. Wilson, M.J.; Wilson, L. Clay mineralogy and shale instability: An alternative conceptual analysis. *Clay Miner.* **2014**, *49*, 127–145. [CrossRef]
4. Anderson, R.L.; Ratcliffe, I.; Greenwell, H.C.; Williams, P.A.; Cliffe, S.; Coveney, P.V. Clay swelling-A challenge in the oilfield. *Earth-Sci. Rev.* **2010**, *98*, 201–216. [CrossRef]
5. Gholami, R.; Elochukwu, H.; Fakhari, N.; Sarmadivaleh, M. A review on borehole instability in active shale formations: Interactions, mechanisms and inhibitors. *Earth-Sci. Rev.* **2018**, *177*, 2–13. [CrossRef]
6. Du, W.; Wang, X.; Chen, G.; Zhang, J.; Slaný, M. Synthesis, property and mechanism analysis of a novel polyhydroxy organic amine shale hydration inhibitor. *Minerals* **2020**, *10*, 128. [CrossRef]
7. Rana, A.; Arfaj, M.K.; Saleh, T.A. Advanced developments in shale inhibitors for oil production with low environmental footprints: A review. *Fuel* **2019**, *247*, 237–249. [CrossRef]
8. Ahmed, H.M.; Kamal, M.S.; Al-Harthi, M. Polymeric and low molecular weight shale inhibitors: A review. *Fuel* **2019**, *251*, 187–217. [CrossRef]
9. Gou, S.H.; Yin, T.; Liu, K.; Guo, Q.P. Water-soluble complexes of an acrylamide copolymer and ionic liquids for inhibiting shale hydration. *New J. Chem.* **2015**, *39*, 2155–2161. [CrossRef]
10. Liu, X.J.; Liu, K.; Gou, S.H.; Liang, L.X.; Luo, C.; Guo, Q.P. Water-soluble acrylamide sulfonate copolymer for inhibiting shale hydration. *Ind. Eng. Chem. Res.* **2014**, *53*, 2903–2910. [CrossRef]
11. Jia, H.; Huang, P.; Wang, Q.X.; Han, Y.G.; Wang, S.Y.; Zhang, F.; Pan, W.; Lv, K.H. Investigation of inhibition mechanism of three deep eutectic solvents as potential shale inhibitors in water-based drilling fluids. *Fuel* **2019**, *244*, 403–411. [CrossRef]
12. Zhao, X.; Qiu, Z.S.; Zhang, Y.J.; Zhong, H.Y.; Huang, W.A.; Tang, Z.C. Zwitterionic polymer P(AM-DMC-AMPS) as a low-molecular-weight encapsulator in deepwater drilling fluid. *Appl. Sci.* **2017**, *7*, 594. [CrossRef]
13. Ma, X.; Zhu, Z.; Shi, W.; Hu, Y. Synthesis and application of a novel betaine-type copolymer as fluid loss additive for water-based drilling fluid. *Colloid Polym. Sci.* **2017**, *295*, 53–66. [CrossRef]
14. Zhao, Y.-F.; Zhang, P.-B.; Sun, J.; Liu, C.-J.; Yi, Z.; Zhu, L.-P.; Xu, Y.-Y. Versatile antifouling polyethersulfone filtration membranes modified via surface grafting of zwitterionic polymers from a reactive amphiphilic copolymer additive. *J. Colloid Interface Sci.* **2015**, *448*, 380–388. [CrossRef] [PubMed]
15. Morimoto, N.; Wakamura, M.; Muramatsu, K.; Toita, S.; Nakayama, M.; Shoji, W.; Suzuki, M.; Winnik, F.M. Membrane Translocation and Organelle-Selective Delivery Steered by Polymeric Zwitterionic Nanospheres. *Biomacromolecules* **2016**, *17*, 1523–1535. [CrossRef]
16. Vu-Bac, N.; Lahmer, T.; Zhuang, X.; Nguyen-Thoi, T.; Rabczuk, T. A software framework for probabilistic sensitivity analysis for computationally expensive models. *Adv. Eng. Softw.* **2016**, *100*, 19–31. [CrossRef]
17. Vu-Bac, N.; Rafiee, R.; Zhuang, X.; Lahmer, T.; Rabczuk, T. Uncertainty quantification for multiscale modeling of polymer nanocomposites with correlated parameters. *Compos. Part B Eng.* **2015**, *68*, 446–464. [CrossRef]
18. Vu-Bac, N.; Silani, M.; Lahmer, T.; Zhuang, X.; Rabczuk, T. A unified framework for stochastic predictions of mechanical properties of polymeric nanocomposites. *Comp. Mater. Sci.* **2015**, *96*, 520–535. [CrossRef]
19. Vu-Bac, N.; Lahmer, T.; Zhang, Y.; Zhuang, X.; Rabczuk, T. Stochastic predictions of interfacial characteristic of polymeric nanocomposites (PNCs). *Compos. Part B Eng.* **2014**, *59*, 80–95. [CrossRef]
20. Pu, X.L.; Du, W.C.; Sun, J.S.; Luo, X.; Zhang, H.D. Synthesis and application of a novel polyhydroxy amine clay anti-swelling agent. *Petrochem. Technol.* **2016**, *45*, 595–600.
21. Silva, F.; Siopa, F.; Figueiredo, B.; Gonçalves, A.; Pereira, J.L.; Gonçalves, F.; Coutinho, J.; Afonso, C.; Ventura, S. Sustainable design for environment-friendly monoand dicationic cholinium-based ionic liquids. *Ecotox. Environ. Safe* **2014**, *108*, 302–310. [CrossRef] [PubMed]
22. Slaný, M.; Jankovič, L'.; Madejová, J. Structural characterization of organo-montmorillonites prepared from a series of primary alkylamines salts: Mid-IR and near-IR study. *Appl. Clay Sci.* **2019**, *176*, 11–20. [CrossRef]
23. Uranta, K.G.; Rezaei-Gomari, S.; Russell Paul-Hamad, F. Studying the effectiveness of polyacrylamide (PAM) application in hydrocarbon reservoirs at different operational conditions. *Energies* **2018**, *11*, 2201. [CrossRef]
24. Wang, Z.H.; Qiu, Y.L.; Guo, P.; Du, J.F.; Liu, H.; Hu, Y.S.; Zeng, F.H. Experimental Investigation of the Damage Mechanisms of Drilling Mud in Fractured Tight gas Reservoir. *J. Energy Resour. Technol.* **2019**, *141*, 092907. [CrossRef]

25. Caglar, B.; Topcu, C.; Coldur, F.; Sarp, G.; Caglar, S.; Tabak, A.; Sahin, E. Structural, thermal, morphological and surface charge properties of dodecyl trimethylammonium-smectite composites. *J. Mol. Struct.* **2016**, *1105*, 70–79. [CrossRef]
26. Madejová, J.; Komadel, P. Baseline studies of the Clay Minerals Society source clays: Infrared methods. *Clays Clay Miner.* **2001**, *49*, 410–432. [CrossRef]
27. Madejová, J.; Sekeráková, L'.; Bizovská, V.; Slaný, M.; Jankovič, L'. Near-infrared spectroscopy as an effective tool for monitoring the conformation of alkylammonium surfactants in montmorillonite interlayers. *Vib. Spectrosc.* **2016**, *84*, 44–52. [CrossRef]
28. Júnior, L.P.C.; Silva, D.B.R.; Aguiar, M.F.; Melo, C.P.; Alves, K.G.B. Preparation and characterization of polypyrrole/organophilic montmorillonite nanofibers obtained by electrospinning. *J. Mol. Liq.* **2019**, *275*, 452–462. [CrossRef]
29. Adewole, J.K.; Najimu-Musa, O. A Study on the Effects of Date Pit-Based Additive on the Performance of Water-Based Drilling Fluid. *J. Energy Resour. Technol.* **2019**, *140*, 052903. [CrossRef]
30. Zhao, X.; Qiu, Z.S.; Wang, M.L.; Huang, W.A.; Zhang, S.F. Performance Evaluation of a Highly Inhibitive Water-Based Drilling Fluid for Ultralow Temperature Wells. *J. Energy Resour. Technol.* **2017**, *140*, 012906. [CrossRef]
31. Du, W.C.; Pu, X.L.; Sun, J.S.; Luo, X.; Zhang, Y.N.; Li, L. Synthesis and evaluation of a novel monomeric amine as sodium montmorillonite swelling inhibitor. *Adsorpt. Sci. Technol.* **2018**, *36*, 655–668. [CrossRef]
32. Pérez, A.; Montes, M.; Molina, R.; Moreno, S. Modified clays as catalysts for the catalytic oxidation of ethanol. *Appl. Clay Sci.* **2014**, *95*, 18–24. [CrossRef]
33. Li, M.C.; Wu, Q.L.; Song, K.L.; French, A.D.; Mei, C.T.; Lei, T.Z. pH-responsive water-based drilling fluids containing bentonite and chitin nanocrystals. *ACS Sustain. Chem. Eng.* **2018**, *6*, 33783–33795. [CrossRef]
34. Li, M.C.; Ren, S.X.; Zhang, X.Q.; Dong, L.L.; Lei, T.Z.; Lee, S.Y.; Wu, Q.L. Surface-chemistry-tuned cellulose nanocrystals in a bentonite suspension for water-based drilling fluids. *ACS Appl. Nano Mater.* **2018**, *1*, 7039–7051. [CrossRef]
35. Suter, J.L.; Coveney, P.V.; Anderson, R.L.; Greenwell, H.C.; Cliffe, S. Rule based design of clay-swelling inhibitor. *Energy Environ. Sci.* **2011**, *4*, 4572–4586. [CrossRef]
36. Du, W.; Slaný, M.; Wang, X.; Chen, G.; Zhang, J. The inhibition property and mechanism of a novel low molecular weight zwitterionic copolymer for improving wellbore stability. *Polymers* **2020**, *12*, 708. [CrossRef]

energies

MDPI

Article

A Novel Prediction Model of the Drag Coefficient of Shale Cuttings in Herschel–Bulkley Fluid

Xiaofeng Sun [1], Minghao Sun [1,*] and Zijian Li [2]

1 Key Laboratory of Enhanced Oil Recovery, Ministry of Education, College of Petroleum Engineering, Northeast Petroleum University, Daqing 163318, China; sunxf@nepu.edu.cn
2 Faculty of Engineering and Applied Science, Memorial University of Newfoundland, St. John's, NF A1B3X5, Canada; zl5082@mun.ca
* Correspondence: sunminghao@stu.nepu.edu.cn

Abstract: In the drilling industry, it is of great significance to accurately predict the drag coefficient and settling velocity of drill cuttings falling in the non-Newtonian drilling fluid. However, the irregular shape of drill cuttings and the non-Newtonian rheological properties of drilling fluid (e.g., shear-thinning and yield stress behavior) make it challenging to predict the settling velocity. In this study, the velocity of particle settlement was studied by a visual device and high-speed camera system. Experimental data of the free settlement of 224 irregular drilling cuttings and 105 spherical particles in the Herschel–Bulkley fluid were obtained. A mechanical model dependent on the force balance of settlement particles was adopted to conduct a detailed statistical analysis of the experimental results, and a prediction model of the drag coefficient of spherical particles in the Herschel–Bulkley fluid was established. A two-dimensional shape description parameter is introduced to establish a model for predicting the drag coefficient of irregular-shaped cuttings in a Herschel–Bulkley fluid. The model has high prediction accuracy for the settling velocity of irregular shale cuttings in Herschel–Bulkley fluid. The average relative error is 7.14%, verifying the model's accuracy.

Keywords: Herschel–Bulkley fluid; hole cleaning; cuttings; settling velocity; drag coefficient

Citation: Sun, X.; Sun, M.; Li, Z. A Novel Prediction Model of the Drag Coefficient of Shale Cuttings in Herschel–Bulkley Fluid. *Energies* **2022**, *15*, 4496. https://doi.org/10.3390/en15124496

Academic Editor: Hossein Hamidi

Received: 23 May 2022
Accepted: 13 June 2022
Published: 20 June 2022

1. Introduction

In the process of oil and gas drilling, using mud (non-Newtonian fluid) to remove solids (cuttings) of different shapes in the wellbore is an essential step [1–4]. However, due to operational limitations (e.g., bit change, single operation) or emergencies, the drilling operation must be stopped briefly, and the circulation of drilling fluid in the annulus will also be interrupted briefly. The cuttings initially transported to the wellhead through drilling fluid circulation begin to settle and deposit at the bottom of the well, causing hole cleaning problems [5,6]. The settling velocity can be used to estimate the formation depth of cuttings. Through cuttings analysis, important formation information such as lithology, porosity, permeability, and pore pressure can be obtained. The settling velocity can also be used as an input parameter for drilling optimization and design. For example, given the condition that the upward velocity of drilling fluid in the annulus needs to be greater than the settling velocity of cuttings, it can guide the mud pump displacement required for the drilling fluid to effectively carry cuttings. The settling velocity of cuttings can usually be calculated according to the prediction model of the particle settling drag coefficient. The in-depth study of the settlement law of cuttings in the mud drilling fluid of horizontal wells and accurate calculation of the settlement drag coefficient is the basic work needed to understand the hole cleaning problem of horizontal wells, which can provide a basis for optimizing the rheological properties and circulation speed of drilling fluid [7,8].

Revealing the law of cuttings settlement can start from the study of spherical particle settlement. Since Stokes (Stokes, 1851, $C_D = 24/Re_s$), researchers have carried out many studies on using the C_D-Re_s relationship to accurately predict the drag coefficient. The drag coefficient (C_D) is defined as the ratio of viscous drag to kinetic energy acting on particles [6,7], and its expression is:

$$C_D = \frac{4(\rho_p - \rho_l) d_e g}{3\rho_l V_t^2} \tag{1}$$

where ρ_p is the density of particles, kg/m^3; ρ_l is the density of fluid, kg/m^3; g is the acceleration of gravity, m/s^2; d_e is particle diameter, m; and V_t is the settling velocity, m/s. If the particle is non-spherical, then d_e is the equivalent diameter of the particle, and its expression is $d_e = \sqrt[3]{6m_p/\pi\rho_p}$, where m_p is the mass of the particle, kg. The particle Reynolds number (Re_s) is defined as:

$$Re_s = \frac{\rho_p V_t d_e}{\mu} \tag{2}$$

where μ is the Newton viscosity of the fluid, Pa·s.

There is much research on particles' drag coefficient in Newtonian and non-Newtonian fluids [8–30]. For example, Shah [21] proposed a method using a drag coefficient to analyze particle settling velocity in the power-law non-Newtonian fluid. This method takes into account the influence of fluid flow index n and establishes the functional relationship between particle Reynolds number Re_s and $C_D{}^{2-n}$. In data processing, Shah fitted $C_D{}^{2-n}$ and Re_s to a straight line on the coordinate diagram. At a low Reynolds number, the particle settling velocity is greatly affected by the fluid flow index, which decreases with the increase in the Reynolds number. Especially when n = 1, the drag coefficient decreases to the Newton curve of C_D and Re_s. The method has been tested with the data of previous studies, and the results show that the method has high prediction accuracy.

The settling velocity of particles in a liquid is affected by fluid properties. The non-Newtonian characteristics of fluid have a great influence on the settlement rate, especially the non-Newtonian fluid with yield stress (τ_0). Dedegil [31] established a prediction model of the sedimentation drag coefficient of particles in Herschel–Bulkley fluid, considering viscous resistance, gravity and buoyancy. The yield stress is included in the influencing conditions of the settling velocity of the model, which is verified by the published clay suspension data.

Certain yield stress is a unique feature of Herschel–Bulkley fluid. When the vertical force acting on it reaches equilibrium, a single particle can be completely suspended. If other parameters remain unchanged, the net downward force increases with the increase in particle size. The fluid used in the petroleum industry not only has yield stress but also has time-dependent and viscoelastic rheological behavior [29]. Through the study of particle settlement in a viscoelastic fluid, it can be found that fluid viscoelasticity will significantly affect the settling velocity in the transient region and has no effect on the settling velocity in the laminar region. The experimental phenomenon reveals that the drag coefficient has also changed substantially due to the changes in particle Reynolds number (Re_s), particle shape, and viscoelastic fluid parameters. Since the Herschel–Bulkley fluid has certain yield stress, and the settlement law of particles in a non-Newtonian fluid is also different from that in a Newtonian fluid, and to ensure that the settlement results conform to the distribution form of the standard Newton drag coefficient curve, it is necessary to modify the definition of the Reynolds number and apparent viscosity. The modified particle Reynolds number Re_s in the Herschel–Bulkley fluid is expressed as [32];

$$Re_s = \frac{d_e^n \rho_l V_t^{2-n}}{\tau_0 \left(\frac{d_e}{V_t}\right)^n + K} \tag{3}$$

where τ_0 is the yield stress, Pa; n is the fluid behavior index, dimensionless; and K is the Herschel–Bulkley fluid consistency index, Pa·s^n. The fluid behavior index is the indicator of shear behavior, and the consistency coefficient is the indicator of fluid viscosity, both of which are used to define the rheological parameters of Herschel–Bulkley fluid. The expression is as follows:

$$\tau = \tau_0 + K\dot{\gamma}^n \quad (4)$$

where τ is the shear stress, Pa, and $\dot{\gamma}$ is the shear rate, 1/s.

The shape characteristics of the particles are one of the factors affecting the settlement behavior of irregular particles in the fluid. Drill cuttings are irregularly shaped non-spherical particles. It can be seen from the public literature that researchers have established many correlations to predict the drag coefficient of irregular particles, indicating that the shape factor has an impact on the drag coefficient. However, few studies have developed drag coefficient models for irregular particles in Herschel–Bulkley fluids.

Among the parameters describing the shape characteristics of particles in the drag coefficient prediction model, sphericity is a very important one [33,34]. However, it is not easy to measure the surface area of irregular particles, especially for highly irregular particles such as shale cuttings, and it is also difficult to operate in the field. Therefore, other particle shape characteristic parameters, such as circularity c [35,36], can be applied. Circularity $\boldsymbol{c}$ is the ratio of the circumference of the maximum projection surface of a particle to the rim of its equivalent circle. Since it is a two-dimensional shape parameter sensitive to the irregularity of the particle contour, it is relatively easy to measure [37]. Its definition is as follows:

$$c = \frac{\pi d_A}{P} \quad (5)$$

where d_A is the diameter of the equivalent circle of the maximum projection surface of the particle, m; P is the entire projection perimeter of the particle, m.

This paper studies the deposition characteristics of shale cuttings in Herschel–Bulkley fluid and verifies the correctness of the experimental process by using particle deposition experiments in Newtonian fluid. By introducing the description parameters representing the cutting morphology, and based on the regression analysis of experimental data, a prediction model for the settlement drag coefficient of spherical particles and shale cuttings in Herschel–Bulkley fluid is proposed. The iterative method was used to calculate the particle settling velocity, which was compared with the particle settling velocity measured in the experiment. The results show that the model has good prediction accuracy for drag coefficient and settling velocity of the Herschel–Bulkley fluid of spherical particles and shale cuttings. These results have important guiding significance for clean hydraulic parameters of shale oil wells in the petroleum industry.

2. Experimental Study

2.1. Experimental Materials

Although high-density materials are not encountered in most practical operations, we hope to broaden the scope and applicability of the experiment by adding some high-density materials to the investigation. In this study, the precise spherical particles of glass, steel, and zirconium and the irregular shape of shale cuttings obtained at the operation site were used. The physical parameters of the particles involved in the experiment are listed in Table 1.

Table 1. Main physical parameters of particles involved in experiments.

Material	Particle Equivalent Diameter ds (mm)	Density (kg/m^3)
Shale cuttings	2.1–5.7	2073
Glass	1, 2, 3, 4, 5	2510
Zirconium oxide	1, 2, 3, 4, 5	6080
Steel	1, 2, 3, 4, 5	7940

2.2. Experimental Fluid

Carbopol polymer (Lubrizol, Wycliffe, OH, USA) and propylene glycol (Fuchen, Tianjin, China) were used to prepare the test fluids. Carbopol aqueous solution was used as the base fluid of Herschel–Bulkley fluid (Carbopol aqueous solution with yield stress and shear dilution conforms to the rheological model of Herschel–Bulkley fluid), and propylene glycol aqueous solution was used as the base fluid of the Newtonian fluid. The mass concentration range of carbopol aqueous solution is 0.075–0.2 wt%, and the mass concentration range of propylene glycol aqueous solution is 80–100 wt%. In the preparation of Herschel–Bulkley fluid, Carbopol polymer was first mixed with deionized water. Then, the mixture was neutralized with triethanolamine (scientific chemical, Chengdu, China) to form a transparent gel. All pH values were maintained between 6.5 and 7.0, and the viscosity depends on the degree of neutralization. Therefore, two similar polymer concentrations may eventually have different rheological properties. Newtonian fluid is a mixture of propylene glycol and water.

An Anton Paar MCR 92 advanced rheometer (Figure 1) was used to test the rheological properties of the experimental fluid at the experimental temperature. According to the rheological model of Herschel–Bulkley fluid and Newton fluid, the rheological parameters were fitted by data analysis software. The temperature of the experimental liquid was controlled by the temperature control system of the rheometer and was consistent with the temperature in the experiment. The rheological properties of the Carbopol aqueous solution and glycol aqueous solution were tested, and the results are shown in Table 2.

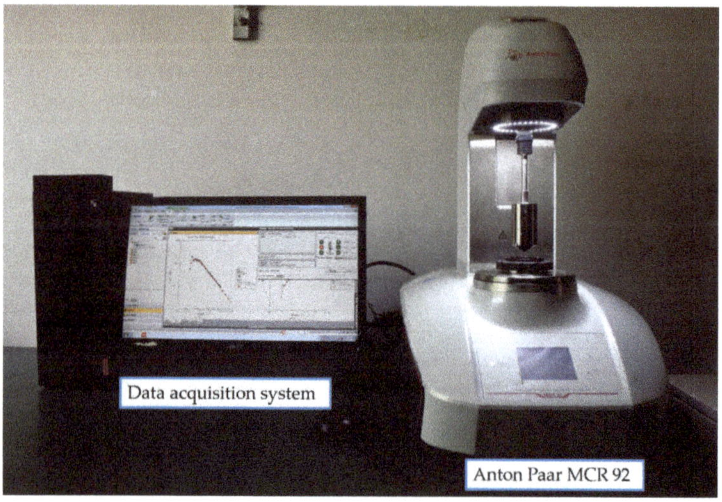

Figure 1. Advanced rheometer (Anto Paar MCR 92).

Table 2. Rheological test results of Carbopol aqueous solution and glycol aqueous solution for experiments.

Testing Liquid	Concentration (wt%)	Temperature (°C)	Density (kg/m^3)	Rheological Parameter		
				τ_0 (Pa)	K (Pa·s^n)	n
Carbopol	0.2	17	1000	5.6575	3.45189	0.59372
	0.175	17.4	1000	2.4828	1.99123	0.64663
	0.15	17.4	1000	1.2732	1.00723	0.70503
	0.125	16.6	1000	0.71474	0.56533	0.74698
	0.1	17.6	1000	0.30369	0.24412	0.8189

Table 2. *Cont.*

Testing Liquid	Concentration (wt%)	Temperature (°C)	Density (kg /m^3)	Rheological Parameter		
				τ_0 (Pa)	K (Pa·s^n)	n
	0.0875	17.4	1000	0.23224	0.18042	0.84364
	0.075	17.5	1000	0.13798	0.10341	0.88723
glycerol	100	18.6	1260	0	1.4313	0
	95	19.0	1246	0	0.5825	0
	90	18.6	1231	0	0.2265	0
	80	19.0	1205	0	0.0656	0

2.3. Characterization of Shale Cuttings Shape

The shale cuttings used in the experiment are shown in Figure 2.

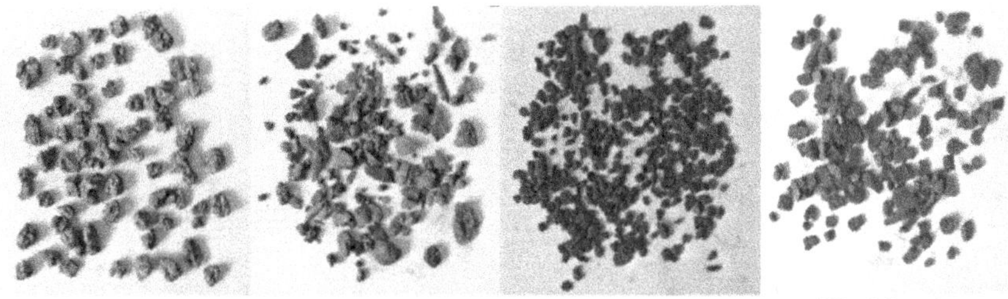

Figure 2. Schematic diagram of shale cuttings morphology in different well sections.

The circularity of the target particle was measured by using the 'analysis particle' function of Image J [38]. The circularity in the Image J user guide is defined as:

$$c = 4\pi \times \frac{A_P}{P^2} \quad (6)$$

where A_P is the surface area of the largest projection surface of particles, m^2.

For spherical particles, $c = 1$, and for any other shape particles, $c < 1$ [39]. We selected some shale cutting image conversion examples, as shown in Figure 3.

The distribution of circularity and equivalent diameter parameters of 224 groups of effective irregular shale cuttings used in the experiment are shown in Figure 4. The equivalent diameter of the shale cuttings used in the experiment is 2.5–5.9 mm, with a median of 4.2 mm, and the circularity is 0.36–0.88, with a median of 0.62. The circularity of most shale cuttings is concentrated between 0.6 and 0.85.

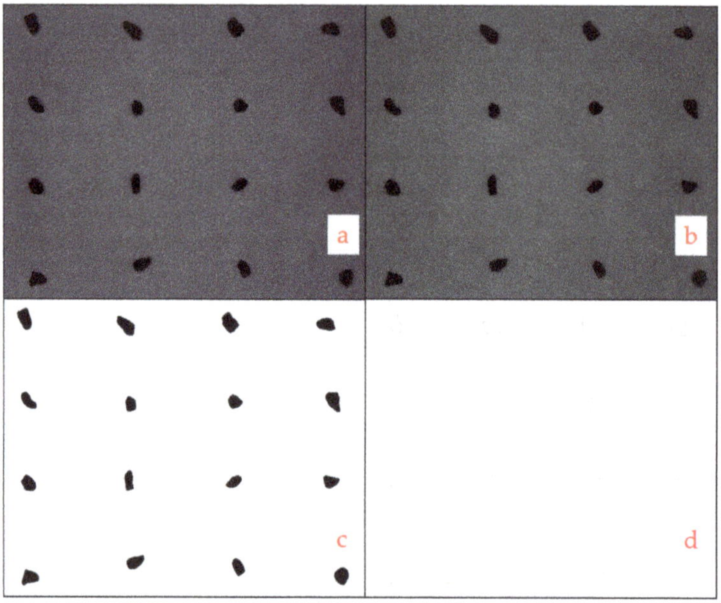

Figure 3. Examples of image conversion of shale cuttings: (**a**) Original RGB image; (**b**) Gray image; (**c**) Image threshold; (**d**) Drawing contour.

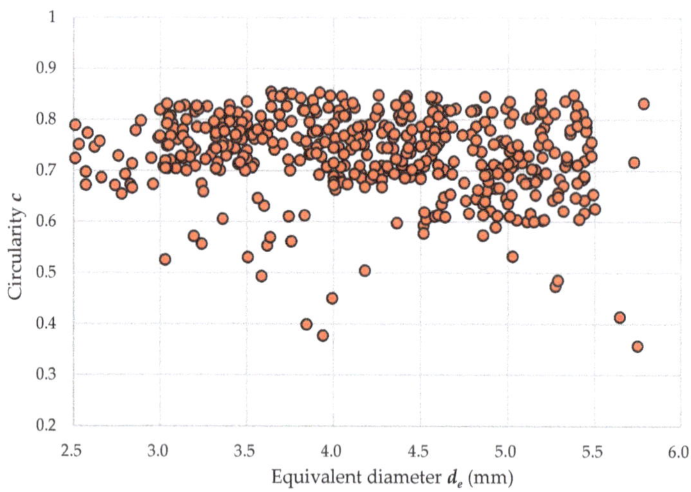

Figure 4. Relationship between circularity of shale cuttings and equivalent diameter distribution.

2.4. Experimental Apparatus and Process

As shown in Figure 5, the settlement experimental device is an organic glass tube with 1500 mm height and 100 mm inner diameter, using a Re-vealer 2F04C high-speed camera to shoot the shale cuttings' deposition trajectory at 100 frames/second. We connected the high-speed camera to the computer for data storage and analysis. After the feasibility calculation at the beginning of the experiment, the liquid level height above 700 mm in the observation area of this experiment can make the particles reach a stable settlement state. The effect of the end effect on settling velocity is also considered. Therefore, the image acquisition area is set below 700 mm from the top of the liquid inside the tube and above 300 mm from the bottom of the liquid inside the tube.

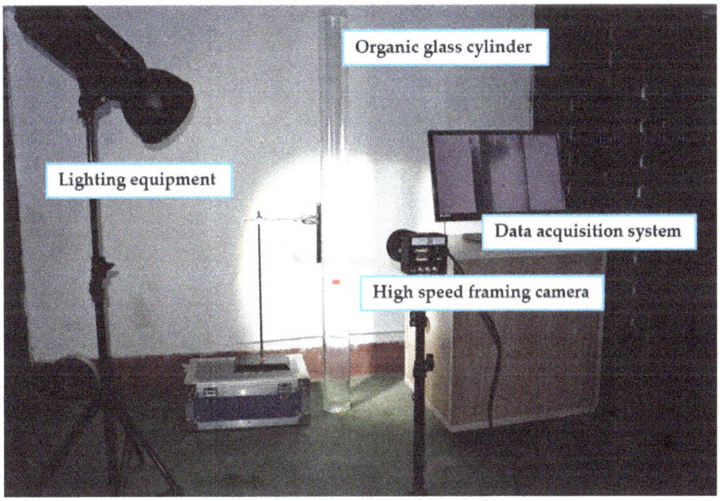

Figure 5. Schematic diagram of particle settlement experimental device.

The configured practical solution was poured into the settling device and stood for 12 h until the bubble was discharged. Before each test, the particles were immersed in a small amount of test fluid to ensure that the particles and the test fluid could come fully into contact during the experiment. During the settling experiment, the particles should be put under the main liquid level of the cylinder container. The temperature should be kept constant throughout the investigation. At the same time, to reduce the interference of uncertainty in the experiment, each spherical particle should be repeated at least three times in the same density fluid. In the subsequent data processing, the experimental data with a relative error greater than 5% are eliminated, and the remaining data are used to fit the drag coefficient relationship.

In addition, the settlement direction of particles in the projection area can be determined and calculated by high-speed camera images and image processing software. Detailed processing and examples are shown in Figure 6.

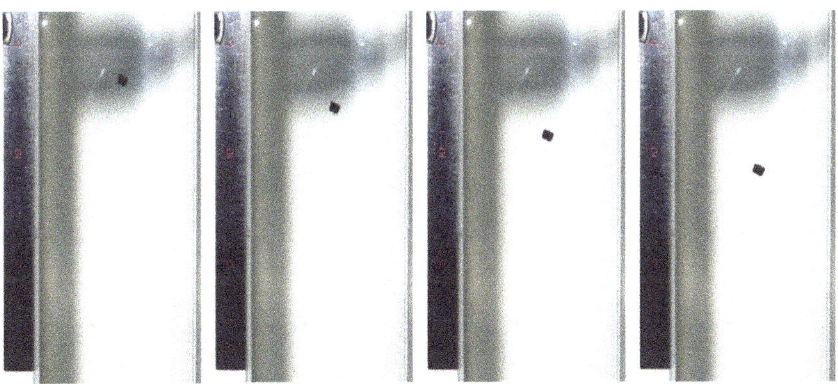

Figure 6. Four consecutive pictures captured during the settlement of shale cuttings.

3. Results and Discussion

3.1. Settlement of Spherical Particles in a Fluid

Firstly, the settlement experiment of spherical particles was carried out, and the relationship between C_D (Equation (1)) and Re_s (Equation (3)) was established. The experimental settlement data of 165 groups (105 groups in Herschel–Bulkley fluid and 60 groups in Newtonian fluid) of spherical particles were analyzed, and the relationship between C_D and Re_s was plotted in the form of logarithmic coordinates (Figure 7). It can be seen from the figure that under the condition of a low Reynolds number, the settlement drag coefficient C_D distribution of spherical particles in different fluids conforms to the prediction trend of the Stokes equation ($C_D = 24/Re_s$). The curve of the settling drag coefficient of particles in a Newtonian fluid with a Reynolds number is in good agreement with the Stokes equation, but there is an obvious error in predicting the settling drag coefficient of spherical particles in Herschel–Bulkley fluid by the Stokes equation. For example, when Re_s is less than 0.1, the average relative error between the test results of spherical particles in the Newtonian fluid and the results calculated by the Stokes equation is only 2.17%. The average relative error in Herschel–Bulkley fluid is as high as 19.96%. The results show that the rheological properties of non-Newtonian fluids have an important influence on the fluid–particle interaction. Therefore, using the correlation of Newtonian fluid to calculate the drag coefficient of spherical settlement in the non-Newtonian fluid will produce many errors.

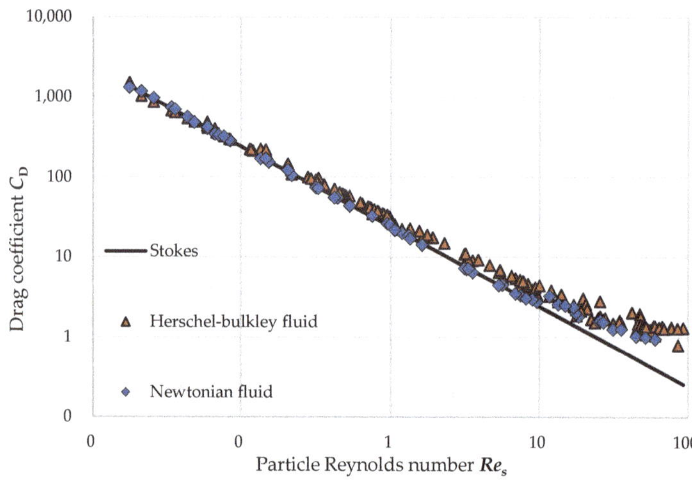

Figure 7. Variation in C_D-Re_s relationship of 165 groups of spherical particles obtained from sedimentation experiment.

According to the experimental data of spherical particle settlement, through the parameter fitting of different prediction models of settlement drag coefficient in Table 3, it is found that the five-parameter model (Equation (7)) proposed by Clift (1970) has the best fitting degree, and its expression is:

$$C_D = \frac{24}{Re_s}\left(1 + A{Re_s}^{B}\right) + \frac{C}{1 + D/{Re_s}^{E}} \tag{7}$$

where A, B, C, D, and E are correlation coefficients. Since the total drag F_D of particles under any flow state can be regarded as the sum of the components of drag F_D in laminar and turbulent flows, the drag coefficient C_D can be improved by adding a complex turbulent term to the extended Stokes equation. From Equation (7), it can be seen that the first item on the right side of the equation shows that under laminar flow conditions, the drag coefficient decreases with the increase in particle Reynolds number. The second item on the right side of the formula shows that under turbulent conditions, the drag coefficient decreases gradually with the increase in particle Reynolds number.

Table 3. Representative C_D-Re_s correlation.

Author	Equation	Number
Clift [11]	$C_D = \frac{24}{Re_s}(1+0.15{Re_s}^{0.687}) + \frac{0.42}{1+\frac{42500}{{Re_s}^{1.16}}}(Re_s < 10^5)$	(8)
Shah [28]	$\begin{cases} \sqrt{{C_D}^{2-n}{Re_s}^2} = A(Re_s)^B \\ A = 6.9148(n^2) - 24.838(n) + 22.642 \\ B = -0.5067(n^2) + 1.3234(n) - 0.1744 \end{cases}$	(9)
Elgaddafi [29]	$\begin{cases} C_D = 24/Re_s(Re_s < 1) \\ C_D = 27.41/{Re_s}^{0.85} \end{cases}$	(10)
Yang and Zeng [30]	$C_D = \frac{24}{Re_s}(2-\phi)^{1.29} + \frac{24}{Re_s}(1+0.15{Re_s}^{0.687})\phi^{-{Re_s}^{0.134}} + \frac{0.42}{1+\frac{42500}{{Re_s}^{1.16}}}e^{1.43(1-\phi)}$	(11)
Khan [31]	$C_D = (2.25{Re_s}^{-0.31} + 0.36{Re_s}^{0.06})^{3.45}(1 < Re_s < 1000)$	(12)
Machač [33]	$\begin{cases} C_D = \frac{24}{Re_s}X(n)(Re_s < 1) \\ C_D = \frac{24}{Re_s}X(n) + \frac{10.5n-3.5}{{Re_s}^{0.32n+0.13}}(1 < Re_s < 1000) \\ X(n) = 3^{(3n-3)/2}\frac{33n^5-64n^4-11n^3+97n^2+16n}{4n^2(n+1)(n+2)(2n+1)} \end{cases}$	(13)
Okesanya [9]	$C_D = \frac{24}{Re_s}(1+0.12Re_s)^{0.35} + 0.398[1-\exp(-0.01{Re_s}^{0.9282})]$	(14)

The experimental settlement data of 105 groups of spherical particles were fitted and regressed, and the relationship between the settlement drag coefficient of spherical particles in Herschel–Bulkley fluid was obtained.

$$C_D = \frac{24}{Re_s}(1 + 1.773{Re_s}^{0.551}) - \frac{61.377}{1 + \frac{1.889}{{Re_s}^{0.484}}}(0.01 < Re_s < 96) \quad (15)$$

By comparing Equation (15) with the representative relationship of the settlement drag coefficient in the Herschel–Bulkley fluid (Equations (12)–(14)) in the literature, the prediction of the proposed correlation of spherical settlement drag coefficient in Herschel–Bulkley fluid is evaluated by using three statistical parameters: the average relative error MRE, the maximum average relative error Max MRE and the logarithmic root mean square error RMSLE [40,41]. The results are shown in Table 4. The expressions of MRE and RMSLE are:

$$MRE = \left(\sum_{i=1}^{N} \frac{|C_{D,c,i} - C_{D,m,i}|}{C_{D,m,i}}\right)/N \times 100\% \quad (16)$$

$$RMSLE = \sqrt{\frac{\sum_{i=1}^{N}(\ln C_{D,c,i} - \ln C_{D,m,i})^2}{N}} \times 100\% \quad (17)$$

where N is the total number; $C_{D,c}$ is the predicted drag coefficient, dimensionless; and $C_{D,m}$ is the experimentally measured drag coefficient, dimensionless.

Table 4. Error Statistics of Drag Coefficient of Spherical Particles Settlement in Herschel–Bulkley Fluid.

Re_s	Author	Error		
		MRE	RMS	Max MRE
$0.01 < Re_s < 96$	Khan and Richadson (1987)	16.19%	0.21	45.6%
	Machač. (1995)	18.42%	0.26	32.73%
	Okesanya (2020)	12.25%	0.16	34.8%
	This chapter (Equation (8))	6.85%	0.07	16.5%

In this study, the sedimentation velocity V_{ts} of the sphere was obtained by the experimental method. According to the data obtained in the experiment, the relationship between the settlement drag coefficient C_D and the particle Reynolds number Res was fitted by the fitting software. According to the proposed C_D-Re_s correlation, the iterative trial-and-error method was used to calculate the settling drag coefficient C_D and settling velocity V_t of particles in the fluid [35]. The iteration process is shown in Figure 8.

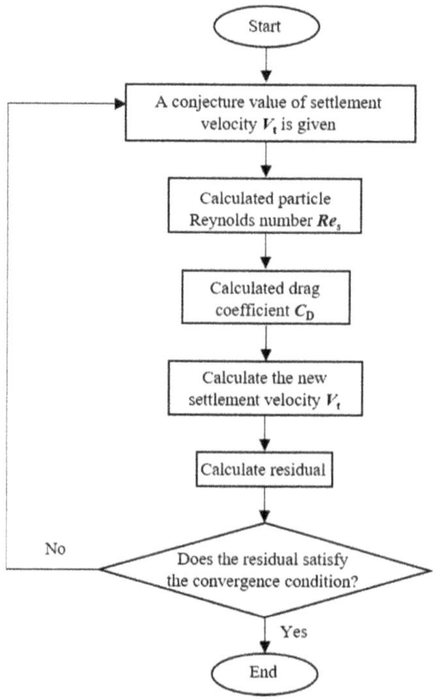

Figure 8. Iterative trial and error flow chart.

According to the proposed prediction model of spherical particle drag coefficient (15), the drag coefficient C_D and settling velocity V_t of settling particles are calculated by the trial-and-error process shown in Figure 8, as shown in Figures 9 and 10. It can be seen from Figure 9 and Table 4 that for the Herschel–Bulkley fluid, the predicted values of Equations (12)–(14) are close to the measured values in the experiment, and the MRE is about 15.62%. The MRE of the model (Equation (15)) is 4.34%. Compared with other models, the three quantitative evaluation parameters are reduced to some extent, and the prediction accuracy of the experimental results is higher. In addition, as shown in Figure 10, the MRE between the predicted and measured sedimentation velocity of spherical particles is only 6.85%. Therefore, the correlation proposed in this paper can better predict the settling drag coefficient C_D and settling velocity V_t of spherical particles in the Herschel–Bulkley fluid.

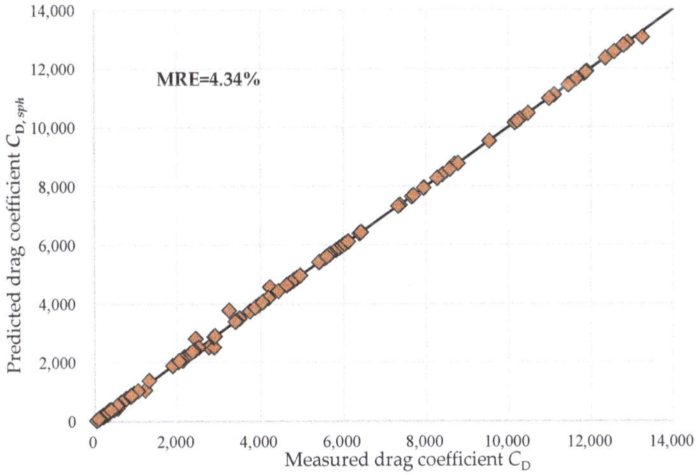

Figure 9. Comparison between the predicted value of this model and the experimentally measured value of the settling drag coefficient of spherical particles.

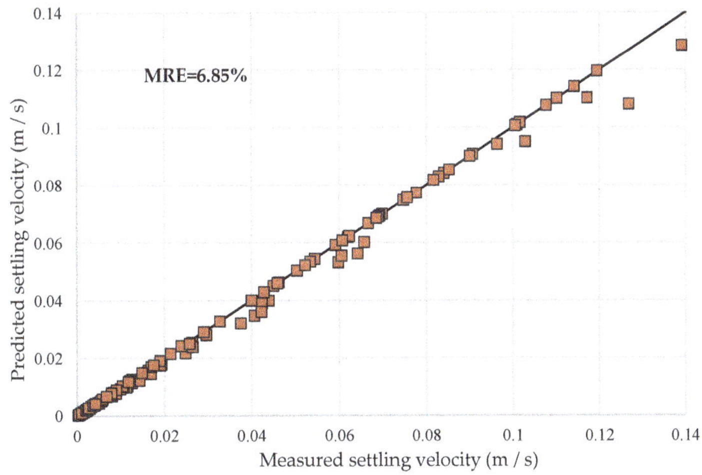

Figure 10. Comparison between the predicted value and the experimentally measured value of settling velocity of spherical particles.

3.2. Establishment and Analysis of Shale Cutting Settling Drag Coefficient Model

Based on the above-proposed prediction model of spherical particle drag coefficient (Equation (15)), the settling drag coefficient C_D relationship suitable for shale cuttings is established by introducing particle circularity *c*. At any given Reynolds number, the drag force on the sphere is less than that on shale cuttings [32,37]. The irregular surface of shale cuttings will increase drag and greater flow separation, resulting in lower settling velocity than spherical particles [42]. Some scholars pointed out that particle property changes lead to two different settling modes [43]: When $Re_s > 100$, cutting sedimentation is swinging, while when $Re_s < 100$, cutting sedimentation is stable. Circularity *c* is more suitable for establishing the prediction model without considering the secondary motion in particle falling. Therefore, in the case of a particle with low Reynolds number with $Re_s < 100$, it is feasible to select the prediction model of shale cuttings' settling drag coefficient in Herschel–Bulkley fluid by introducing circularity. Through experimental observation (Figure 11), under the same conditions, the ratio of shale cutting drag coefficient C_D to spherical particle drag coefficient $C_{D,sph}$, is slightly greater than 1. At a high Reynolds number, the percentage will be more significant due to the influence of shape.

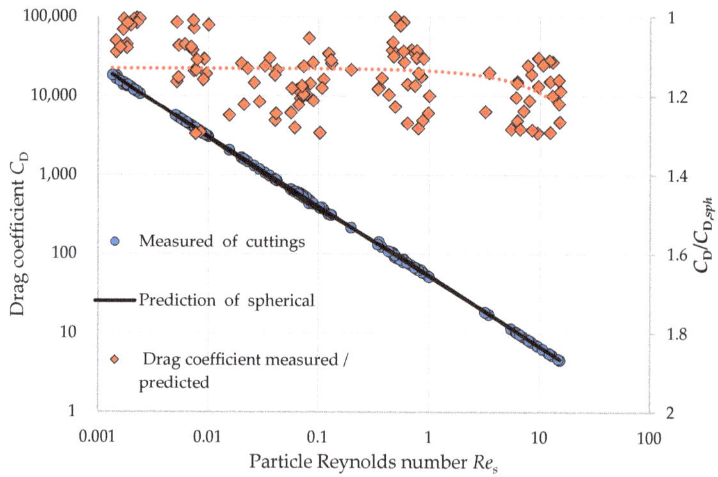

Figure 11. C_D-Re_s relationship of 224 shale cutting data.

By analyzing the difference between the drag coefficient C_D obtained by the shale cutting experiment and the predicted drag coefficient $C_{D,sph}$ of spherical particle sedimentation under the same conditions, the exponential function of circularity function f (c) based on natural number e is established, and the functional relationship of $C_D/C_{D,sph}$ is also established. The expression is as follows:

$$C_D = C_{D,sph}\exp[f(c)] \tag{18}$$

In special cases, such as $c = 1$, the drag coefficient of shale cuttings is equal to that of spherical particles under the same parameters. Combined with 224 sets of experimental data, f (c) is determined by Equation (19).

$$f(c) = \alpha Re_s^{\beta}(1-c)^{\eta} \tag{19}$$

where α, β, and η are empirical coefficients, which can be obtained by nonlinear fitting.

The expression of the drag coefficient C_D of shale cuttings in Herschel–Bulkley fluid is as follows:

$$C_D = C_{D,sph}exp\left[0.45{Re_s}^{0.36}(1-c)^{0.95}\right](0.08 < Re_s < 15) \tag{20}$$

Figure 12 shows the relationship between the predicted value of shale cuttings drag coefficient and the measured drag coefficient C_D and the measured value calculated by Equation (20). Table 5 lists the evaluation results of the model, and the MRE of the predicted drag coefficient is 6.39%.

According to the proposed model (Equation (20)), the settling velocity of shale cuttings in Herschel–Bulkley fluid is calculated and compared with the experimentally measured value (Figure 13). The comparison shows that the average relative error of the predicted value of settling velocity V_t of shale cuttings in Herschel–Bulkley fluid is 7.14%. Figure 13 shows that although the prediction results of the model have a certain dispersion, the data are uniformly distributed on the coordinate axis line, so the model can better predict the deposition rate of shale cuttings in the Herschel–Bulkley fluid.

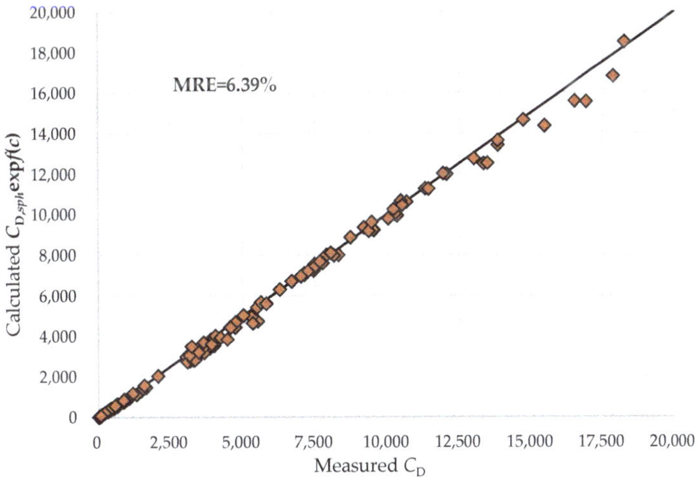

Figure 12. Comparison of prediction results of drag coefficient C_D and $C_{D,sph}$ exp$f(c)$ measured by shale cutting settling experiment.

Table 5. Error statistics of shale cuttings drag coefficient.

Fluid	Coefficient CD Relationship	Error		
		MRE	RMS	Max Error
Herschel–Buckley fluid	0.08 < coefficient CD relationship < 15	6.39%	0.069	16.74%

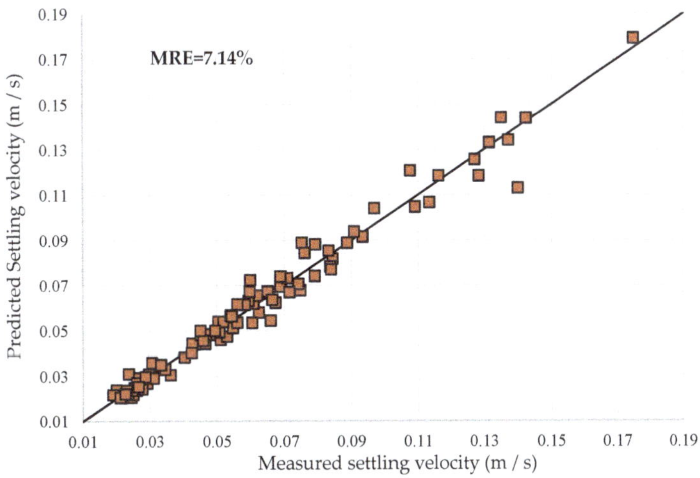

Figure 13. Comparison results between experimental measured values and model predicted values of shale cuttings settlement velocity.

4. Discussion on the Limitations of the Model

The limitation of the prediction model of shale cuttings' drag coefficient proposed in this study is that the range of particle Reynolds number is relatively small (0.08~15). With increasing the Reynolds number of the particles, either by increasing the particle size or density or by reducing the viscosity of the fluid, particles collide frequently with the column wall during its settlement, even if the inner diameter of the resin glass column we studied is 100 mm. Therefore, these experimental data are invalid because the velocity after particle collision is very different from the original velocity.

We plan to conduct further experiments after increasing the wellbore diameter. With the discovery of relevant fields, the model will continue to evolve, and with more additional data, the rationality and application scope of our results will gradually improve.

In summary, for the test of spherical and shale cuttings, the prediction model proposed in this paper is effective in the range of particle Reynolds numbers and roundness. However, the use of these prediction models outside these ranges may lead to inaccurate results. Therefore, the correlation needs to be confirmed by additional experimental data, such as large particle Reynolds numbers, which will be pursued in our future work.

5. Conclusions

In this paper, the effective experimental data of 105 groups of spherical particles and 224 groups of shale cuttings in Herschel–Bulkley fluid are obtained through laboratory experiments. Through the regression analysis of the experimental data, the following conclusions are drawn:

(1) For spherical particles, a model for predicting the settling drag coefficient of spherical particles in Herschel–Bulkley fluid is established by modifying the correlation proposed by Clift (1970) and considering the rheological properties of the fluid.

(2) For non-spherical shale cuttings, based on the spherical particle prediction model, using image analysis technology and introducing two-dimensional shape description parameters, a resistance coefficient model for predicting the settlement of shale cuttings in Herschel–Bulkley fluid is established.

(3) According to the prediction model, the settlement rate is calculated by an iterative method. The average relative errors of spherical particles and shale cuttings are 6.85% and 7.14%, respectively. The results show that the model has good prediction ability. This work can provide a valuable reference for those interested in cuttings' settling velocity, cuttings' migration, and hole cleaning in drilling engineering.

Author Contributions: Conceptualization, X.S.; Validation, Z.L.; Writing—original draft, M.S. All authors have read and agreed to the published version of the manuscript.

Funding: This research was funded by National Natural Science Foundation of China (Grant No. 52174002) and (Grant No. 52004064).

Conflicts of Interest: The authors declare no conflict of interest.

Nomenclature

C_D	drag coefficient, dimensionless
Re_s	particle Reynolds number
ρ_p	density of particles (kg/m^3)
ρ_l	density of fluid (kg/m^3)
g	acceleration of gravity (m/s^2)
d_e	particle diameter (m)
V_t	settling velocity (m/s)
m_p	mass of the particle (kg)
μ	Newton viscosity of the fluid (Pa·s)
τ_0	yield stress (Pa)
n	fluid behavior index, dimensionless
K	Herschel–Bulkley fluid consistency index ($Pa{\cdot}s^n$)
τ	shear stress (Pa)
$\dot{\gamma}$	shear rate (1/s)
d_A	diameter of the equivalent circle of the maximum projection surface of the particle (m)

P	entire projection perimeter of the particle (m)
c	circularity
A_P	The surface area of the largest projection surface of particles (m^2)
A, B, C, D, and E	correlation coefficients Equation (7)
N	total number
$C_{D,c}$	predicted drag coefficient, dimensionless
$C_{D,m}$	experimentally measured drag coefficient, dimensionless
$C_{D,sph}$	predicted drag coefficient of spherical particle settlings, dimensionless
α, β, and η	empirical coefficient Equation (19)

References

1. Duan, M.; Miska, S.; Yu, M.; Takach, N.; Ahmed, R.; Zettner, C. Transport of Small Cuttings in Extended-Reach Drilling. *SPE Drill. Complet.* **2008**, *23*, 258–265. [CrossRef]
2. Sorgun, M. Simple Correlations and Analysis of Cuttings Transport with Newtonian and Non-Newtonian Fluids in Horizontal and Deviated Wells. *J. Energy Resour. Technol.* **2013**, *135*, 032903. [CrossRef]
3. Khormali, A.; Petrakov, D.G.; Farmanzade, A.R. Prediction and Inhibition of Inorganic Salt Formation under Static and Dynamic Conditions-Effect of Pressure, Temperature, and Mixing Ratio. *Int. J. Technol.* **2016**, *7*, 943–951. [CrossRef]
4. Wells, H.; Larsen, T.I.; Corp, U.; Pilehvari, A.A.; Texas, A.; Azar, J.J.; Tulsa, U. Development of a new cuttings-transport model for high-angle wellbores including horizontal wells. *SPE Drill. Complet.* **1997**, *12*, 129–135.
5. Moreira, B.A.; de Oliveira Arouca, F.; Damasceno, J.J.R. Analysis of Suspension Sedimentation in Fluids with Rheological Shear-Thinning Properties and Thixotropic Effects. *Powder Technol.* **2017**, *308*, 290–297. [CrossRef]
6. Khormali, A.; Sharifov, A.R.; Torba, D.I. Experimental and Modeling Analysis of Asphaltene Precipitation in the near Wellbore Region of Oil Wells. *Pet. Sci. Technol.* **2018**, *36*, 1030–1036. [CrossRef]
7. Sun, X.; Zhang, K.; Chen, Y.; Li, W.; Qu, J. Study on the Settling Velocity of Drilling Cuttings in the Power Law Fluid. *Powder Technol.* **2020**, *362*, 278–287. [CrossRef]
8. Khormali, A.; Moghadasi, R.; Kazemzadeh, Y.; Struchkov, I. Development of a New Chemical Solvent Package for Increasing the Asphaltene Removal Performance under Static and Dynamic Conditions. *J. Pet. Sci. Eng.* **2021**, *206*, 109066. [CrossRef]
9. Okesanya, T.; Kuru, E.; Sun, Y. A New Generalized Model for Predicting the Drag Coefficient and the Settling Velocity of Rigid Spheres in Viscoplastic Fluids. *SPE J.* **2020**, *25*, 3217–3235. [CrossRef]
10. Rushd, S.; Hassan, I.; Sultan, R.A.; Kelessidis, V.C.; Rahman, A.; Hasan, H.S.; Hasan, A. Terminal Settling Velocity of a Single Sphere in Drilling Fluid. *Part. Sci. Technol.* **2019**, *37*, 939–948. [CrossRef]
11. Clift, R.; Gauvin, W.H. Motion of Entrained Particles in Gas Streams. *Can. J. Chem. Eng.* **1971**, *49*, 439–448. [CrossRef]
12. Turton, R.; Levenspiel, O. A Short Note on the Drag Correlation for Spheres. *Powder Technol.* **1986**, *47*, 83–86. [CrossRef]
13. Haider, A.; Levenspiel, O. Drag Coefficient and Terminal Velocity of Spherical and Nonspherical Particles. *Ind. Eng. Chem.* **1967**, *59*, 55–58. [CrossRef]
14. Chhabra, R.P.; Peri, S.S. Simple Method for the Estimation of Free-Fall Velocity of Spherical Particles in Power Law Liquids. *Powder Technol.* **1991**, *67*, 287–290. [CrossRef]
15. Saha, G.; Purohit, N.K.; Mitra, A.K. Spherical Particle Terminal Settling Velocity and Drag in Bingham Liquids. *Int. J. Miner. Process.* **1992**, *36*, 273–281. [CrossRef]
16. Chien, S.F. Settling Velocity of Irregularly Shaped Particles. *Proc.—SPE Annu. Tech. Conf. Exhib.* **1994**, *4*, 281–289. [CrossRef]
17. Hill, K.B.; Shook, C.A. Pipeline Transport of Coarse Particles by Water and by Fluids with Yield Stresses. *Part. Sci. Technol.* **1998**, *16*, 163–183. [CrossRef]
18. Brown, P.P.; Lawler, D.F. Sphere Drag and Settling Velocity Revisited. *J. Environ. Eng.* **2003**, *129*, 222–231. [CrossRef]
19. Wilson, K.C.; Horsley, R.R.; Kealy, T.; Reizes, J.A.; Horsley, M. Direct Prediction of Fall Velocities in Non-Newtonian Materials. *Int. J. Miner. Process.* **2003**, *71*, 17–30. [CrossRef]
20. Kelessidis, V.C.; Mpandelis, G. Measurements and Prediction of Terminal Velocity of Solid Spheres Falling through Stagnant Pseudoplastic Liquids. *Powder Technol.* **2004**, *147*, 117–125. [CrossRef]
21. Shah, S.N.; El Fadili, Y.; Chhabra, R.P. New Model for Single Spherical Particle Settling Velocity in Power Law (Visco-Inelastic) Fluids. *Int. J. Multiph. Flow* **2007**, *33*, 51–66. [CrossRef]
22. Cheng, N.S. Comparison of Formulas for Drag Coefficient and Settling Velocity of Spherical Particles. *Powder Technol.* **2009**, *189*, 395–398. [CrossRef]
23. Vlasak, P.; Chara, Z. Conveying of Solid Particles in Newtonian and Non-Newtonian Carriers. *Part. Sci. Technol.* **2009**, *27*, 428–443. [CrossRef]
24. Gumulya, M.M.; Horsley, R.R.; Wilson, K.C.; Pareek, V. A New Fluid Model for Particles Settling in a Viscoplastic Fluid. *Chem. Eng. Sci.* **2011**, *66*, 729–739. [CrossRef]
25. Mikhailov, M.D.; Freire, A.P.S. The Drag Coefficient of a Sphere: An Approximation Using Shanks Transform. *Powder Technol.* **2013**, *237*, 432–435. [CrossRef]
26. Pani, G.K.; Rath, P.; Barik, R.; Senapati, P.K. The Effect of Selective Additives on the Rheological Behavior of Power Plant Ash Slurry. *Part. Sci. Technol.* **2015**, *33*, 418–422. [CrossRef]

27. Arabi, A.S.; Sanders, R.S. Particle Terminal Settling Velocities in Non-Newtonian Viscoplastic Fluids. *Can. J. Chem. Eng.* **2016**, *94*, 1092–1101. [CrossRef]
28. Arnipally, S.K.; Kuru, E. Settling Velocity of Particles in Viscoelastic Fluids: A Comparison of the Shear-Viscosity and Elasticity Effects. *SPE J.* **2018**, *23*, 1689–1705. [CrossRef]
29. Elgaddafi, R.; Ahmed, R.; Growcock, F. Settling Behavior of Particles in Fiber-Containing Herschel Bulkley Fluid. *Powder Technol.* **2016**, *301*, 782–793. [CrossRef]
30. Yang, F.; Zeng, Y.H.; Huai, W.X. A New Model for Settling Velocity of Non-Spherical Particles. *Environ. Sci. Pollut. Res.* **2021**, *28*, 61636–61646. [CrossRef]
31. Dedegil, M.Y. Drag Coefficient and Settling Velocity of Particles in Non-Newtonian Suspensions. *J. Fluids Eng. Trans. ASME* **1987**, *109*, 319–323. [CrossRef]
32. Machač, I.; Ulbrichová, I.; Elson, T.P.; Cheesman, D.J. Fall of Spherical Particles through Non-Newtonian Suspensions. *Chem. Eng. Sci.* **1995**, *50*, 3323–3327. [CrossRef]
33. Khan, A.R.; Richardson, J.F. The Resistance to Motion of a Solid Sphere in a Fluid. *Chem. Eng. Commun.* **1987**, *62*, 135–150. [CrossRef]
34. Yan, T.; Qu, J.Y.; Sun, X.F.; Li, W.; Chen, Y.; Hu, Q.B. A Novel Predictive Model of Drag Coefficient and Settling Velocity of Drill Cuttings in Non-Newtonian Drilling Fluids. *Pet. Sci.* 2021. [CrossRef]
35. Dioguardi, F.; Mele, D.; Dellino, P. A New One-Equation Model of Fluid Drag for Irregularly Shaped Particles Valid Over a Wide Range of Reynolds Number. *J. Geophys. Res. Solid Earth* **2018**, *123*, 144–156. [CrossRef]
36. Tran-Cong, S.; Gay, M.; Michaelides, E.E. Drag Coefficients of Irregularly Shaped Particles. *Powder Technol.* **2004**, *139*, 21–32. [CrossRef]
37. Dioguardi, F.; Mele, D. A New Shape Dependent Drag Correlation Formula for Non-Spherical Rough Particles. *Experiments and Results. Powder Technol.* **2015**, *277*, 222–230. [CrossRef]
38. Carlier, T.; Ferrer, L.; Berruchon, J.B.; Cuissard, R.; Martineau, A.; Loonis, P.; Couturier, O. Quality Controls for Gamma Cameras and PET Cameras: Development of a Free Open-Source ImageJ Program. *Phys. Med. Imaging* **2005**, *5745*, 1237–1246. [CrossRef]
39. Breakey, D.E.S.; Vaezi, G.F.; Masliyah, J.H.; Sanders, R.S. Side-View-Only Determination of Drag Coefficient and Settling Velocity for Non-Spherical Particles. *Powder Technol.* **2018**, *339*, 182–191. [CrossRef]
40. Barati, R.; Neyshabouri, S.A.A.S.; Ahmadi, G. Development of Empirical Models with High Accuracy for Estimation of Drag Coefficient of Flow around a Smooth Sphere: An Evolutionary Approach. *Powder Technol.* **2014**, *257*, 11–19. [CrossRef]
41. Goossens, W.R.A. A New Explicit Equation for the Terminal Velocity of a Settling Sphere. *Powder Technol.* **2020**, *362*, 54–56. [CrossRef]
42. Shahi, S.; Kuru, E. An Experimental Investigation of Settling Velocity of Natural Sands in Water Using Particle Image Shadowgraph. *Powder Technol.* **2015**, *281*, 184–192. [CrossRef]
43. Fang, G. An Experimental Study of Free Settling of Cuttings in Newtonian and Non-Newtonian Drillings Fluids: Drag Coefficient and Settling Velocity. *SPE J.* **1992**, *4*, 1–13.

energies

MDPI

Article

Reservoir Pore Characteristics Based on Depositional Microfacies Control in the Neogene Guantao Formation, Bohai Bay Basin, China

Zhao Wang, Hongming Tang *, Jun Yang and Lu Huang

School of Geoscience and Technology, Southwest Petroleum University, Chengdu 610500, China; wangzhao_528@163.com (Z.W.); yj060197@163.com (J.Y.); hl20001971@163.com (L.H.)
* Correspondence: swpithm@vip.163.com; Tel.: +86-028-8303-7128

Abstract: The Neogene Guantao Formation Reservoir in the PLOilfield is a unconsolidated sandstone with high porosity and high permeability. The reservoir diagenesis is weak and dominated by compaction diagenesis. At present, insufficient research into the relationship between the pore characteristics and sedimentary microfacies of shallow delta sandstone reservoirs restricts the prediction of favorable reservoir distribution. This article takes the unconsolidated sandstone reservoir of the Guantao Formation as the research object and analyzes the potential coupling relationship between pore characteristics and sedimentary microfacies. In this study, seven typical sedimentary microfacies were identified and the microscopic characteristics of different sedimentary microfacies reservoirs are described. The results show that the pore structure of various sedimentary microfacies is comprehensively influenced by the sedimentary rock texture, siliceous minerals, and clay mineral distribution. Characterized by more abundant hard quartz and feldspar minerals, positive skewness, and lower pore sorting coefficients, the arenaceous microfacies possess larger pore throat radius and lower pore fractal dimensions than argillaceous microfacies. Finally, due to the difficulty of coring in offshore oilfields and the lack of data on mercury injection experiments, empirical formulas for the porosity, permeability, and pore throat radius of conventional core tests were established, and the pore radius that corresponded to 35% mercury saturation (r_{35}) was used as the characteristic pore structure parameter. The grey correlation method was used to analyze the influence of sedimentary structure factors and mineral content differences on r_{35} and to determine the main controlling factors.

Keywords: sandstone reservoir; pore throat characteristics; fractal dimension; depositional microfacies; r_{35}

Citation: Wang, Z.; Tang, H.; Yang, J.; Huang, L. Reservoir Pore Characteristics Based on Depositional Microfacies Control in the Neogene Guantao Formation, Bohai Bay Basin, China. *Energies* **2022**, *15*, 2870. https://doi.org/10.3390/en15082870

Academic Editors: Mofazzal Hossain, Hisham Khaled Ben Mahmud and Md Motiur Rahman

Received: 1 March 2022
Accepted: 12 April 2022
Published: 14 April 2022

1. Introduction

According to the oilfield development experience, microscopic pore structure controls the reservoir capacity and seepage characteristics and ultimately exerts a profound influence on oilfield recovery factors [1,2]. Pore structure is a comprehensive reflection of pore throat size, shape, and connectivity [3]. As a result of the non-homogeneity of pore structure, significant problems have arisen during the development process such as the high injection pressure, rapid rise in water content, and high start-up pressure [4].

The processes of deposition and diagenesis occurring together can potentially cause the complex processes of diverse pore types, complex pore structures, and non-homogeneity in the reservoirs to occur simultaneously [5,6]. Sedimentation has an obvious effect on clastic rock mineral's composition, structure, sorting, rounding, and miscellaneous base content, which, in turn, exerts a profound influence on its porosity and permeability [7–9]. The increase in burial depth will make the microscopic pore structure more complex [10,11]. Dissolution can increase the secondary porosity of the reservoir and has positive effects on reservoir quality whilst cementation and compaction have negative effects [12].

The PL Oilfield is one of the largest offshore oil fields discovered in China to date [13] and its main oil-bearing zone, the Neogene Guantao Formation, is a highly porous and highly permeable unconsolidated sandstone characterized by weak diagenesis and is dominated mainly by compaction, and the reservoir pore structure is controlled by deposition. The change in depositional environment has a pivotal and direct role in the control of clastic reservoir properties [7]. The long oil-bearing zone of the PL Oilfield, the longitudinal evolution of the depositional mechanism, the lateral changes in the sedimentary microfacies, and hydrodynamic factors have led to large differences in reservoir pore structure both within and between layers, resulting in many problems in the development process such as quickly decreasing fluid production and quickly decreasing water absorption capacity. In recent years, extensive research has been undertaken in relation to the target layer of this study, the majority of which have focused on the sedimentary evolution, oil gas reservoir-forming conditions, and provenance systems, however, the coupling between sedimentary microfacies and pore structure was found to be relatively insignificant.

Reservoir rock pore structure parameters are crucial indicators for reservoir evaluation, and how to objectively determine these parameters is a problem that many petrologists have long been trying to solve. At present, the commonly used research methods include μ-CT [14–16], high-pressure mercury injection (HPMI) [17–20], scanning electron microscopy (SEM), nuclear magnetic resonance (NMR) [21,22], logging methods, etc. These methods, which characterize the pore structure, have their advantages, but HPMI is the most appropriate experimental method based on the reservoir characteristics in this study area [23].

The difficulties of coring in offshore fields make it impossible to test the reservoir sections of all wells with mercury injection data. In recent decades, geophysicists have established a series of empirical equations to determine easily accessible physical property parameters and pore throat radius with the objective to replace mercury intrusion experiments with other easier and safer methods.

Winland (1972) [24] chose a pore throat radius at 35% mercury saturation to divide the cutoff value for net production from clastic reservoirs because the best correlation is obtained at this point.

$$\log r_{35} = 0.732 + 0.588 \log K_{air} - 0.864 \log \varnothing_{core} \tag{1}$$

where r_{35} is the pore throat radius at 35% mercury saturation; $\varnothing_{core}$ is the core porosity; and K_{air} is the core permeability measured using air flow.

The R35 has been applied in many oil fields after being proposed by Winland (1972) [24]. However, studies in different oil fields have shown that the correlation between porosity, permeability, and pore throat radius is inconsistent with Winland's conclusion. Pittman (1992) [18] gave empirical equations to calculate pore throat radius for mercury injection saturations from 10% to 75%, and identified the best correlation at the mercury injection saturation as 25%.

$$\log K_{air} = -1.221 + 1.415 \varnothing_{core} + 1.512 \log r_{25} \tag{2}$$

Spearing et al. (2001) [25] considered this value to be 45% using core analysis. Nabawy (2009) [26] suggested that it is more reliable to establish the equation when mercury saturation is at 30%.

$$\log r_{30} = 0.529 \log K_{air} + 2.758 \log \varnothing_{core} - 4.918 \tag{3}$$

In summary, this paper combines the sedimentary microfacies and their mineral compositions, obtained by core description, XRD mineral analysis, and logging interpretation, with the pore structure characteristics obtained by HPMI, casting thin sections and SEM, focusing on the following aspects: (1) studying sedimentary systems and identifying sedimentary microfacies; (2) finding the difference in the pore structures of different sedimentary microfacies including the pore throat type, pore size distribution, and fractal dimension representing the pore structure complexity; (3) establishing pore throat radius matching relationships for porosity and permeability and then achieving a quantitative characterization of the pore throat radius in non-cored well areas; and (4) conducting a grey correlation analysis based on the similarity between the geometric shapes of the sequence curves to judge whether the relationship is close and the calculation amount is small, and if there is no requirement for the number of samples, applying the grey correlation theory to quantitatively investigate the major control elements affecting the pore structure. This paper studied the coupling relationship between the shallow-water braided river delta sedimentary system and the pore structure to fill the gaps in this type of sedimentary system and proposed comprehensive pore structure parameters that are suitable for unconsolidated sandstone reservoirs, providing a reference for the characterization of pore structure in this kind of oilfield, which is lacking a core.

2. Geological Background

The Bohai Bay Basin is a typical Cenozoic continental petroliferous basin in eastern China, with a total area of 20×10^4 km^2 (Figure 1A) [27,28]. The PL structure is a faulted anticline structure, which developed at the base of the Bonan Uplift (Figure 1C). Two strike-slip faults parallel to the structural strike developed in the PL Oilfield. The strike-slip faults and their near-northeast-trending derivative normal faults cut the PL anticline into fault blocks of horsts and grabens. The high part of the central structure between the two strike-slip faults is the main production area of the PL Oilfield. Several previous studies have used heavy minerals, paleomorphology, and the ZTR index in the PL Oilfield to conclude that the provenance of the Guantao Formation was mainly the Jiaoliao Uplift and Liaodong Uplift. The sedimentary environment of the Neogene Guantao reservoir is shallow braided river delta [29].

The reservoir rocks of the PL Oilfield are the Neogene Guantao (Ng) and Lower Member of Minghuazhen (Nm) formations of sandstone with a shallow burial depth (from −745 m to −1540 m). This oil field is divided into 13 oil-bearing groups and 47 small layers (Figure 1B). The oil-bearing interval of the lower Minghuazhen formation is divided into five oil groups (L00~L40 oil groups), the upper Guantao formation is divided into three oil groups (L50~L70 oil groups), and the lower Guantao formation is divided into five oil groups (L80~L120 oil groups) (Figure 1B). Oil and gas distribution is controlled by the distribution characteristics of sand bodies and can be classified as a typical lithologic-structural reservoir (Figure 2).

The results of these rock core experiments indicate that the average permeability is 1321 mD and average porosity is 27%. The reservoir lithology is medium-fine-grained feldspar sandstone with loose particle cementation and poor sorting. The grain size is mainly distributed between 1 mm and 0.05 mm and the roundness ranged from angular to suborbicular. The content of silt and mud in the reservoir is 5.9~42.5%, with an average of 25.2%. The content of small particles is relatively high, and the reservoir is prone to particle migration.

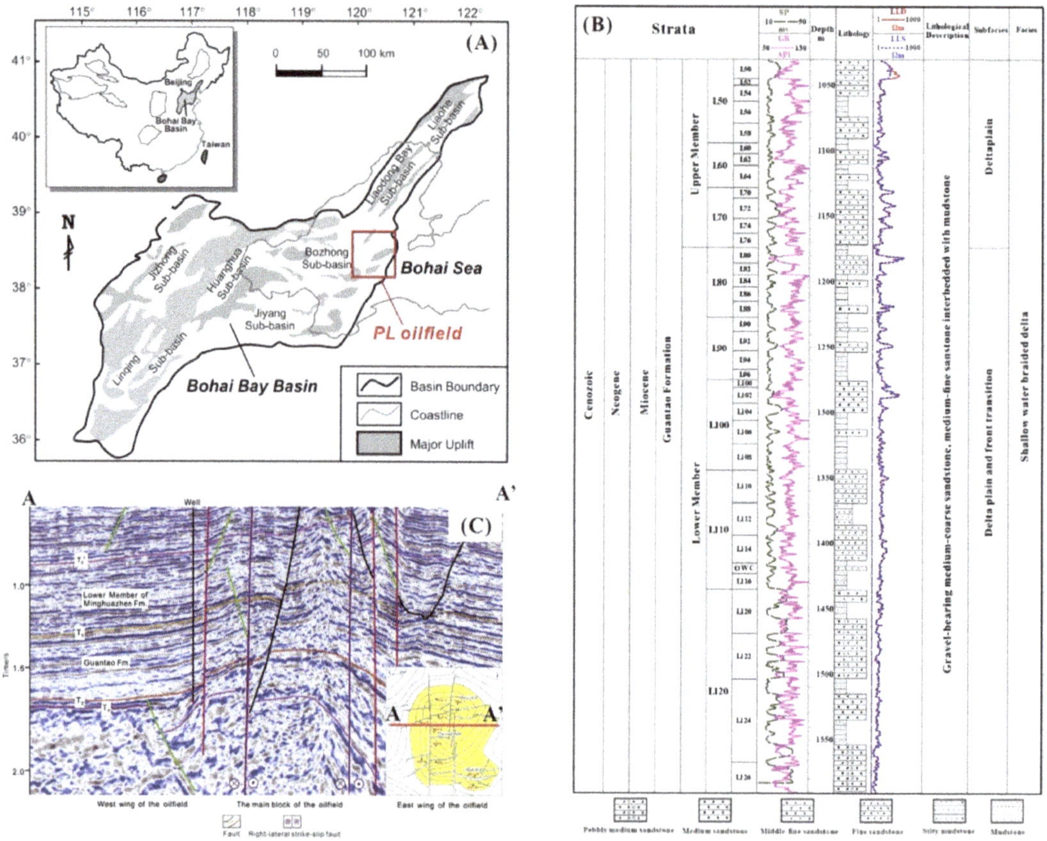

Figure 1. (**A**) The location of the PL Oilfield [30]. (**B**) The geologic column of the Guantao Formation in well PL19−3−b. (**C**) PL: The E−W seismic profile of the PL oilfield is steep in the west and slow in the east [29].

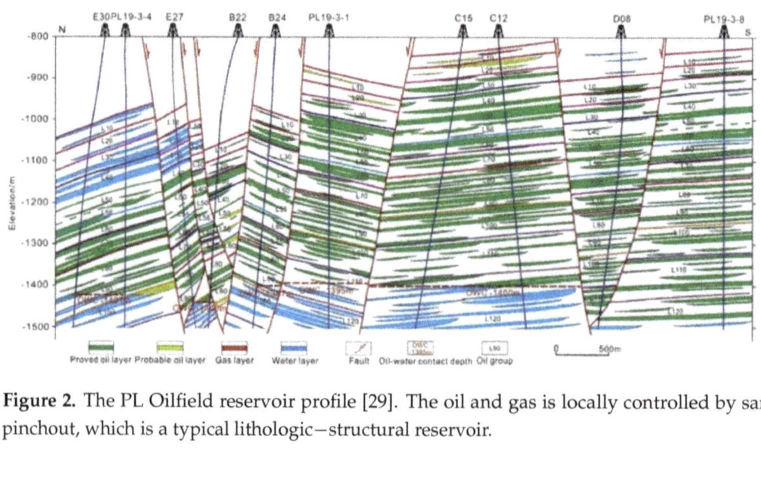

Figure 2. The PL Oilfield reservoir profile [29]. The oil and gas is locally controlled by sandstone pinchout, which is a typical lithologic−structural reservoir.

3. Materials and Methods

Core data were obtained from four sealed coring wells and the sedimentary units were identified and described by the naked eye, hand microscopes, and binocular microscopes, and the sedimentary structure. Sediment color and grain size characteristics were fully examined. Combined with logging curves (AC, GR, SP, etc.), the logging model of sedimentary microfacies was established to analyze the vertical and planar distribution characteristics of sedimentary microfacies (Table 1). The X-ray diffraction method and the mineral clay quantitative analysis method were used to study the mineral composition of 81 samples from seven wells. The casting thin section method analyzed the minerals, diagenesis, and pore throats of 188 samples from 14 wells. A total of 129 samples from 13 wells were selected for SEM analysis. Characterizing the particle size, sorting, and skewness of sample particles using a laser particle size analyzer (LPSA) showed the depositional environment and the lithological features of the sediments from 234 samples of 18 wells.

Table 1. Sedimentary characteristics and logging response.

Sedimentary Facies	Sedimentary Subfacies	Sedimentary Microfacies	Lithology	Sedimentary Structure	Log Response
Shallow braided river delta	Delta plain (L50–L70)	Braided Channel (BC)	Gray, off-white medium-fine-grained feldspar sandstone and lithic feldspar sandstone	Trough cross-bedding, plate cross-bedding, and erosional basal surface	Dentate small cylindrical or bell shaped, medium-high amplitude, GR and SP curves characteristics
		Distributary Bar/Channel Bar(DB)	Gravel-bearing coarse sandstone to medium-fine sandstone	Plate cross bedding, trough cross bedding and parallel bedding	Homogeneity, box-shaped, high amplitude, curves characteristics.
		Flood Plain (FP)	Reddish-brown, khaki, variegated and gray-green silty mudstones and mudstones	Horizontal bedding, with massive structure	The GR curve is high and dentate, and the SP curve has the characteristics of low amplitude and smooth logging.
	Delta front and Delta plain interact (L80–L120)	Distributary Channel (DC)	Gravelly sandstone, medium sandstone, fine sandstone	Slotted cross bedding, parallel bedding, and small cross bedding	Positive rhythm, box, or bell shaped, medium to high amplitude, dentate GR and SP curves characteristics
		Interdistributary Bay (IB)	Gray green, light gray mudstone, and silty mudstone	Horizontal bedding, with massive structure	The GR curve is high and dentate, and the SP curve has the characteristics of low amplitude and smooth logging.
		Mouth Bar (MB)	Gray to gray-black fine sandstone	Low energy bedding development	The GR curve is funnel-shaped, with obvious anti-rhythm
		Sheet Sand (SS)	Gray to gray-black fine sandstone and siltstone	Parallel bedding	The logging curves are in the shape of tines and fingers

High-pressure mercury reservoirs (HPMI) were tested with an Auto Pore IV9500 mercury porosimeter and the measurement conditions were as follows: the measurement of the pore diameter was from 0.003 to 950 μm and had a maximum pressure of 6000 psi. A total of 41 samples were completed, and the mercury saturation of the samples under different capillary pressures was obtained. The pore throat radius can be obtained by using the Washbrun equation as follows [31]:

$$r = \frac{2\alpha \cos\theta}{P} \tag{4}$$

where P is the capillary pressure (MPa); r is the pore radius (μm); α is the surface tension (mN/m); and θ is the contact angle (°) [31–33].

Fractal dimension is a parameter that characterizes the degree of self-similarity of objects with fractal characteristics, which can be applied to reveal the heterogeneity of the pore throat structure [34]. The fractal dimension can be calculated using Equation (5) [35,36]. If the pore structure in the rock sample satisfies the fractal characteristics, the connection between mercury saturation and capillary pressure should satisfy a power function, which is a straight line under double logarithmic coordinates. The slope of the straight line represents the fractal dimension of the rock sample [37].

$$\lg(1 - S_{Hg}) = (D_{f-}3)\lg P_c + \left(3 - D_f\right)\lg P_{min} \tag{5}$$

where Pc is the capillary pressure corresponding to the pore throat radius (r) (MPa); Pmin is the capillary pressure to the maximum pore throat radius (r_{max}) (MPa); S_{Hg} is the accumulative mercury saturation (%), and its slope can be used to calculate the fractal dimension using Equation (6) [38]:

$$D_f = S + 3 \tag{6}$$

The fractal dimension D_f is the slope S of $\lg(1 - S_{Hg}) - \lg(Pc)$ plots. The range of fractal dimension is 2~3. The closer the D_f is to 3, the greater the tortuosity of the pore structure.

4. Results

4.1. Sedimentary Facies

4.1.1. Sedimentary Facies Signs

1. Mudstone color

The authigenic color of depositional rocks can directly determine the climatic conditions, the water medium conditions, and the quality of source rocks. In particular, the color of mudstone plays a pivotal role in distinguishing the sedimentary environment. According to the observation of the naked eye, hand mirrors, and binocular microscopes, the mudstone color of the Guantao Formation reservoir in the PL Oilfield is mainly gray (Figure 3A), light gray (Figure 3B), gray-green (Figure 3C), brown-red variegated (Figure 3D), and red (Figure 3E), indicating that the depositional environment changes from shallow water exposed to the environment. On the plain, the adjacent well sections change rapidly, indicating that the depositional environment also changes rapidly. According to the change in color characteristics, it can be seen that the depositional environment of the PL Oilfield undergoes a transition from a weak reduction to an oxidation environment, from the bottom to the top.

2. Rock type

By analyzing the identification results of thin rock sections from seven wells, the sandstone composition data of different interest intervals were drawn into the sandstone composition triangle (Figure 4). The triangle chart shows that the sandstone composition of the lower Guantao Formation and upper Guantao Formation is mainly Lithic arkose sandstone and feldspathic litharenite, indicating that the rock composition maturity is medium. From the lower Guantao Formation to the upper Guantao Formation, the types of rocks are gradually concentrated and the range of lithology changes is gradually reduced, indicating that the provenance and depositional environment tend to be stable. In addition, the low-maturity rock types gradually decreased, indicating that the provenance tended to concentrate on the distant provenance.

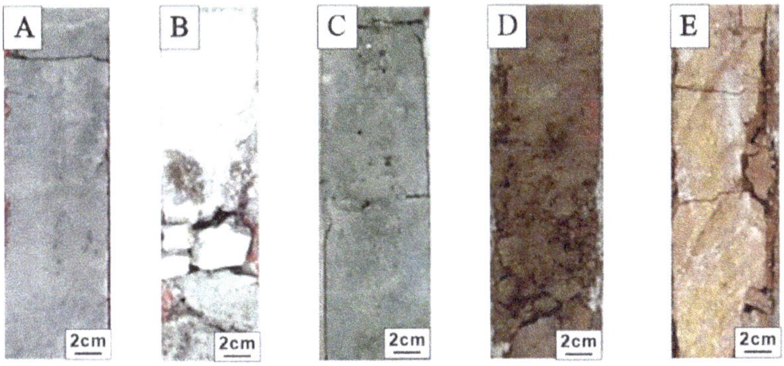

Figure 3. Mudstone color of well PL−19−H. (**A**) Gray mudstone, L124, 1713.4 m. (**B**) Light gray mudstone, L100, 1475.3 m. (**C**) Gray green mudstone, L50, 1180.5 m. (**D**) Brown red mudstone, L70, 1304.6 m. (**E**) Red mudstone, L120, 1574.8 m.

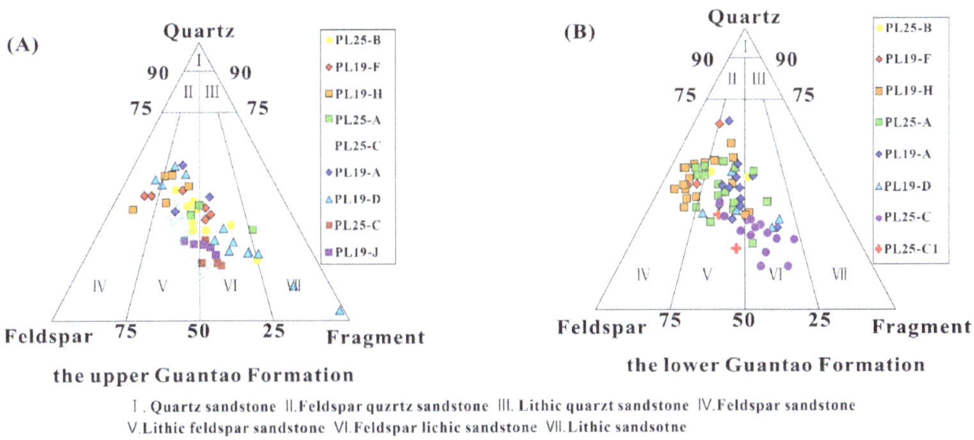

Figure 4. Sandstone composition triangle. (**A**) The triangle chart of sandstone classification in the upper Guantao Formation. (**B**) The triangle chart of sandstone classification in the lower Guantao Formation.

3. Grain size characteristics

The grain size distribution of sediments is mainly affected by hydrodynamic changes. Therefore, grain size analysis reflects the original sedimentary conditions and the strength of the hydrodynamic conditions [5]. The grain size probability curve shows that the reservoir is mainly composed of three sections, reflecting rolling, jumping, and suspension transport modes, of which jumping is the key component (Figure 5A–D). As shown in the C–M figures (Figure 5E,F), the tractive flow deposition features are prominent, indicating fluvial deposition characteristics. In summary, the lithology of the main oil-bearing intervals in the study area gradually becomes thinner from the bottom to top, the sorting becomes better, and the maturity increases, which reflects the depositional process of hydrodynamic conditions from strong to weak, and the depositional environment tends to be more stable.

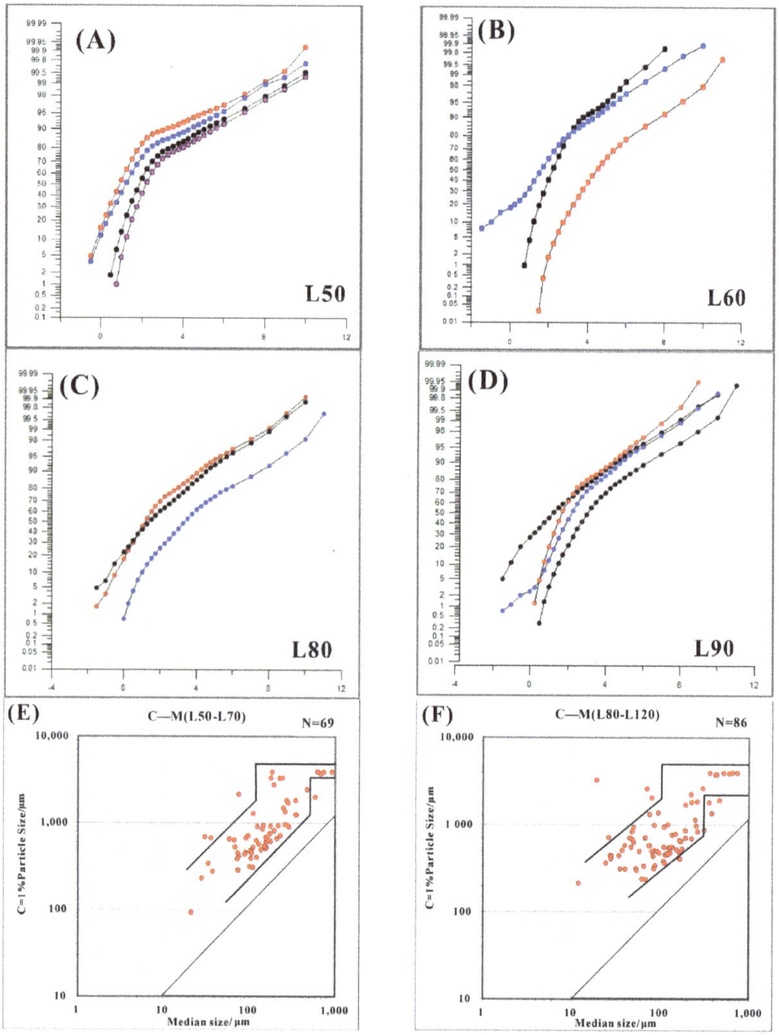

Figure 5. The probability cumulative grain size curves and C−M diagram. (**A**) The probability cumulative grain size curves of the PL19−3−A well. (**B**) The probability cumulative grain size curves of the PL19−3−B well. (**C**) The probability cumulative grain size curves of the PL25−C well. (**D**) The probability cumulative grain size curves of the PL25−C well. (**E**) C−M diagram of the upper Guantao Formation (L50−L70, N = 69). (**F**) C−M diagram of the lower Guantao Formation (L80−L120, N = 86).

4. Sedimentary structure and texture

The sedimentary structure reflects the deposition rate, hydrodynamic strength, and depositional medium so the sedimentary structure can be used to determine the environment when the rock was formed, and then the further divide in the depositional environment. On the basis of core observation, there are sedimentary structures such as granular sequence bedding (Figure 6A), parallel bedding (Figure 6B), plate cross bedding (Figure 6C), trough cross bedding (Figure 6D), corrugated cross bedding (Figure 6E), and a scoured surface (Figure 6F), which demonstrate that the hydrodynamic strength changes greatly.

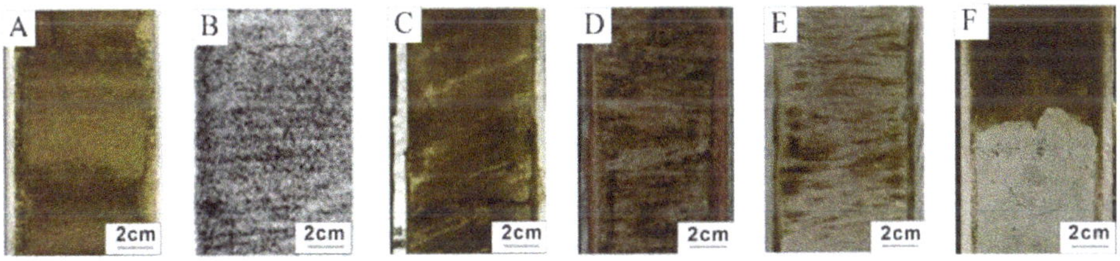

Figure 6. Typical sedimentary structure. (**A**) Graded bedding, PL19−3−H, 1059.2~1059.9 m. (**B**) Parallel bedding, PL25−6−C, 1450.5 m. (**C**) Tabular cross-bedding, PL19−3−F, 1190.5 m. (**D**) Crescent type cross bedding, PL19−3−F, 1181.3 m. (**E**) Sand cross bedding, PL19-3-F, 1166.2 m. (**F**) Erosional basal surface, PL19−3−D, 1364 m.

5. Paleontological features

Paleontological analysis confirmed the presence of Bohaidina, Para-bohaidina, Filisphaeridium, Sentusidinium, Granodiscus, and Leio-spaeridia (Figure 7), representing shallow lacustrine facies in the green mudstone of the Guantao Formation, and the fossils of lacustrine assemblage such as Pseudocandona and Sinocytheridea impressa indicate that the oilfield changed from a fluvial environment to a shallow lacustrine environment [29].

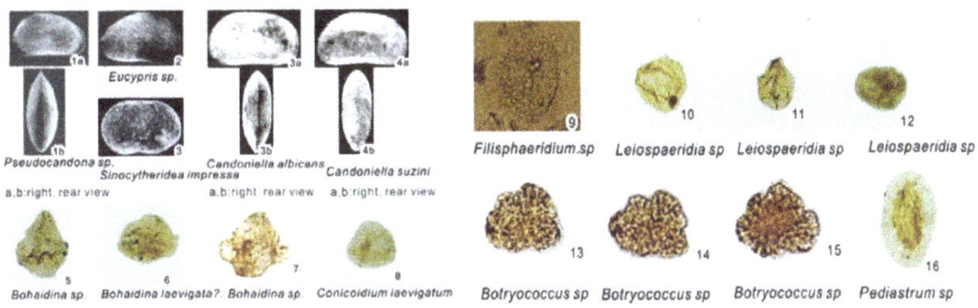

Figure 7. Neogene shallow lacustrine paleontological assemblages in the PL Oilfield [29].

4.1.2. Sedimentary Microfacies

By analyzing the core data, well logging facies and previously completed research, we established the opinion that the Guantao Formation is a shallow-water braided delta deposit. The following seven key depositional microfacies were identified: distributary bars (DBs), braided channels (BCs), mouth bars (MBs), flood plains (FPs), distributary channels (DCs), sheet sands (SSs), and interdistributary bays (IBs) (Table 1).

1. Braided channels (BCs)

BC is a common type of microfacies in delta plain deposits, deposited in the area between DB. BC sand bodies are mainly gray, off-white medium-fine-grained feldspar sandstone (Figure 8A). BCs mainly developed through cross-bedding, followed by plate cross-bedding (Figure 8B). This description indicates that the channel scour was strong and frequent during the sedimentary period. The BC sedimentary sequence has classic normal sequence characteristics. The GR curve is in the form of a serrated bell-shaped, small cylinder, or bell–cylinder combination, and the sand body has a wide distribution range (Figure 9A).

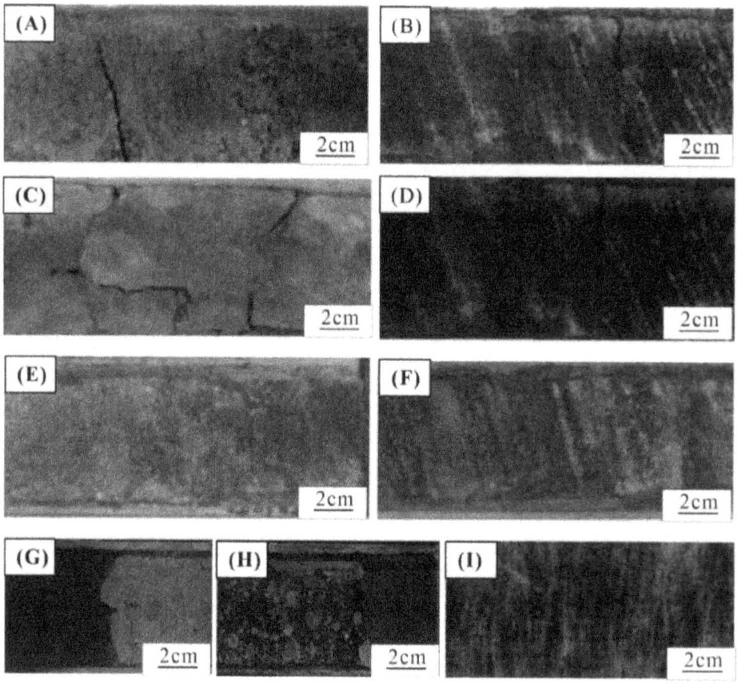

Figure 8. (**A**) PL19−3−F, 1127.8 m, BC, Gray medium sandstone. (**B**) PL19−3−F, 1190.4 m, BC, Cross−bedding. (**C**) PL19−3−E, 1190.1 m, DB, Gray fine sandstone. (**D**) PL19−3−F, 1190.43 m, DB, Tabular cross-bedding. (**E**) PL19−3−H, 1157.1 m, DC, Brown pebbly sandstone. (**F**) PL19−3−H, 1158.1 m, DC, Cross−bedding. (**G**) PL19-3-D, 1369.0 m, DC, Erosional basal surface. (**H**) PL19−3−H, 1421.1 m, DC, Erosional basal surface. (**I**) PL19−3−F, 1135.2 m, MB, Gray−black fine sandstone. DC = distributary channel; IB = interdistributary bay; MB = mouth bar; FP = flood plain; SS = sheet sand; BC = braided channel; DB = distributary bar.

2. Distributary bars (DBs)

The DB is located between BCs. Vertically, the grain size has the characteristics of normal sequence or compound rhythm, the bedding scale gradually decreases and the lithology gradually becomes finer, following gravel-bearing coarse sandstone (Figure 8C). The plate cross-bedding (Figure 8D) is the most common in the DBs and indicates that the Guantao Formation had a strong hydrodynamic force and strong river scouring during the deposition period. The DBs generally have the characteristics of box-shaped SP and GR logging curves with large thickness and good plain continuity (Figure 9B).

3. Flood plains (FPs)

FPs mainly refer to the floodplain between the rivers. These are formed by the river overflowing from the river channel during the flood period. Therefore, their sedimentation is the same as the river's sedimentation source, but more granular. They are mainly composed of suspended sediment in rivers and some silty sandstones. After the river sediments are accumulated, the river water recedes, and the sediments will be exposed. FPs are formed by the river channels bursting in different periods, so they have the characteristics of multiple layers. The characteristics of different layers may be quite similar or may be different (Figure 9B).

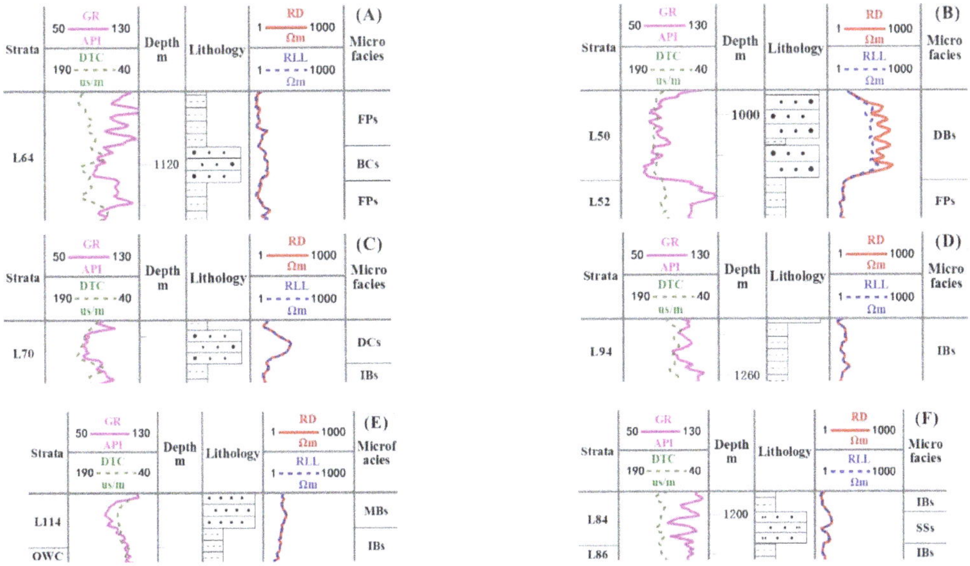

Figure 9. (**A–F**) Log response of typical sedimentary microfacies in wells PL19−3−A, PL19−3−B, and PL19−3−D. DC = distributary channel; BC = braided channel; FP = flood plain; DB = distributary bar; IB = interdistributary bay; MB = mouth bar; SS = sheet sand.

4. Distributary channels (DCs)

The lithology of the distributary channel sand body is mainly brown pebbly sandstone (Figure 8E), medium sandstone, and fine sandstone. Cross bedding (Figure 8F) and wavy bedding are common in sedimentary structures with an eroded basement surface containing mud and gravel at the bottom (Figure 8G,H). The GR curves are in the form of mid-to-high amplitude, and are slightly dentate bell-shaped or box-shaped. There is a positive rhythm in the vertical direction and the grain size becomes finer in the upward direction, reflecting the lateral movement of the channel (Figure 9C).

5. Interdistributary bays (IBs)

The IBs are low-lying bay areas between distributary channels. These are the separation zones or receding zones between different distributary channels, similar to the FPs. The IBs are mainly deposited with gray-green and light gray mudstones, intercalated with thin layers of siltstone. They sometimes develop small-scale sand beddings. SP and GR curves are linear and slightly serrated with low amplitude (Figure 9D).

6. Mouth bars (MBs)

Due to the rapid advancement of shallow water deltas and the gentle slope in the basement, the estuary dams and remote sand dams deposited in the early stage are easily washed and thinned by the later diversion channels or even, in some circumstances, disappear altogether. The mouth bar and far sand bar deposited in the early stage are easily scoured and thinned by the later diversion channels or even, in some circumstances, disappear altogether. Therefore, the MB sand body in the PL Oilfield is less developed, and the thickness is generally very thin. The lithological gray-black fine sandstone (Figure 8I) and siltstone are developed in the main. There is an obvious anti-rhythm in the sedimentary sequence, low-energy bedding is developed, and the thickness is thin. The GR curve is funnel-shaped, with obvious anti-rhythm features (Figure 9E).

7. Sheet sands (SSs)

The SSs are mainly thin layers of gray or gray-black silt sandstones, which are often interbedded with argillaceous and vertical sediments. Parallel bedding, wavy bedding, and convolution bedding are more common, and the GR curves have features such as tines and finger features (Figure 9F).

4.1.3. Sand Body Characteristics and Evolution Analysis

The thickness of the target layer does not change much and the overall thickness is characterized by being thin in the west and thick in the east. In the vertical direction, multiple sets of shallow braided river delta sedimentary sand bodies are superimposed, BC sand bodies, and DB bodies in the upper Guantao Formation are relatively developed, and a high number of BC sand bodies in the lower Guantao Formation are developed. Reservoir sand bodies have a variety of superimposed styles, mainly of independent, multilateral, and cut superimposed types.

1. Lower Guantao Formation

The L100 oil group in the lower Guantao Formation is in the low capacity space background, forming shallow-water braided river delta upper plain deposits. When the base level rises, sediments accumulate; when the base level drops, erosion and scouring occur. During the deposition period of the L80–L90 oil group in the lower Guantao Formation, the capacity space was close to the sediment supply and the shallow braided river delta deposits were in a transitional environment, where water and land frequently interacted. This period was dominated by the transitional environmental deposits of the delta plain and the delta front. The sedimentary structure is dominated by small-scale cross-bedding and corrugated cross-bedding, reflecting weak hydrodynamic energy. Braided channel sedimentary structures have major cross-bedding, indicating frequent hydrodynamic changes (Figure 10).

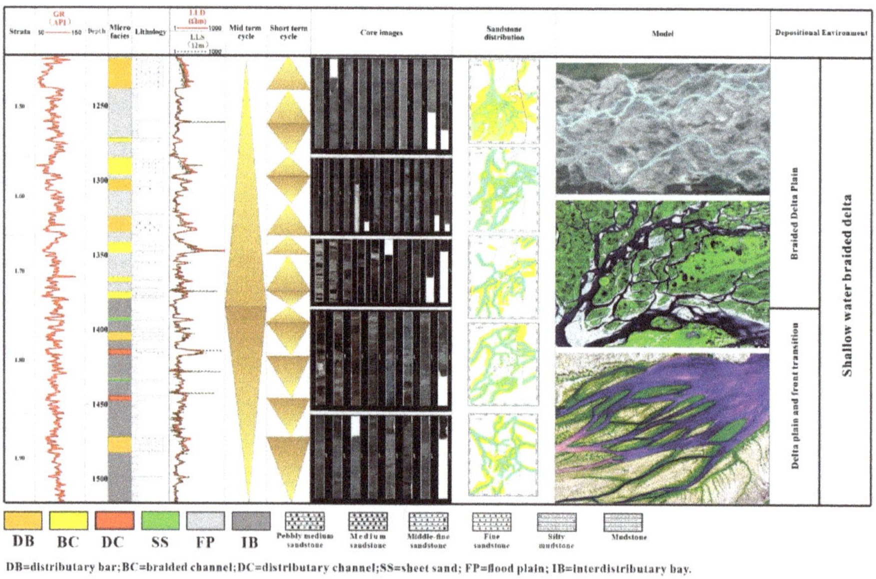

Figure 10. Geologic column of the core section of well PL25−6−A. This vertical geologic column includes logging facies, sedimentary microfacies, sedimentary facies spatial distribution, sedimentary rhythm, lithology information, and a similar modern delta sedimentary map, which can intuitively show the vertical changes in geological features.

2. Upper Guantao Formation

The lower L60 and L70 oil groups in the upper Guantao Formation formed shallow braided river delta deposits dominated by narrow channels, against the background of high-capacity space. The deposition period from the upper part of the L60 oil group to the L50 oil group is the shallow braided river delta upper plain deposited under the background of low-capacity space, mainly including distributary bars, braided channels, flood plains, and other microfacies. The sorting and roundness of the rock are both medium–poor grade, the proportion of unstable cuttings is high, and the structure and composition maturity are generally low. From the bottom and upward, the sand bodies have the characteristics of strips, thin strips, and continuous flakes. The sand layer is mainly composed of multi-stage sand bodies that overlap with each other, and there are obvious scouring surfaces between different channels. The later channel has obvious cutting, erosion, and scouring damage in comparison to the previous channel (Figure 10).

4.2. Pore Characteristics

4.2.1. Pore Types

By means of casting thin sections and SEM, the reservoir space of the Guantao Formation was found to mainly comprise primary intergranular pores, followed by intergranular and intragranular dissolved pores (Figure 11A,D,E,F). Most of the pores are irregular and parts of the pores are filled with clay minerals (Figure 11E,F). In this study area, the primary intergranular pores account for more than 90% of the total pore volume. The throat type is point throat (Figure 11C), followed by necking and laminated throat (Figure 11B). The throat radius is generally between 0.01 mm and 0.03 mm.

4.2.2. Pore Size Distribution

The pore size distribution can be obtained by using the change in mercury in the unit increment of the logarithmic value of the pore radius (dV/dlogD) to measure the rate of change in the pore volume with the pore diameter [39,40] (Figure 12B).

The study area is controlled by shallow water delta deposits with strong reservoir heterogeneity and extremely uneven pore throat distribution. The mercury intrusion experiments show that the length and dip angle of the mercury intrusion curves of different types reservoirs have obvious differences. The better the reservoir quality, the longer the horizontal steps and the weaker the heterogeneity of the pore structure (Figure 12A). The pore size distribution curve shows that the pore throat radius distribution is heterogeneous with single peaks, double peaks, and even triple peaks (Figure 12B). The smaller the peak in the pore throat radius, the worse the quality of the reservoir.

4.3. Mineral Composition

XRD shows that the mineral content is as shown in Table 2. The Guantao Formation is relatively enriched in quartz and feldspar minerals. The content of siliceous minerals (quartz, feldspar) is 74.59%, among which the content of quartz is 42.27%, orthoclase content is 12.88%, and plagioclase content is 16.44%.

Table 2. Mineral composition of the Guantao Formation.

Parameter	Absolute Mineral Content (%)						Clay Mineral Relative Content (%)				N
	Quartz	Orthoclase	Plagioclase	Clay	Calcite	Dolomite	Illite/smectite	Kaolinite	Illite	Chlorite	
range	17.31~81.14	4.35~35.16	3.05~42.41	2.18~59.22	0~5.42	0~4.32	3.21~90.15	1.02~78.25	2.21~58.87	1.15~23.21	81
average	45.27	12.88	16.44	23.14	0.8	0.99	48.23	31.78	11.43	8.56	

The content of clay minerals in samples from different sedimentary environments is also significantly different. The content of clay minerals varies between 2.18% and 59.22%, with an average of 23.14%. The types and contents of clay minerals are shown in Table 2.

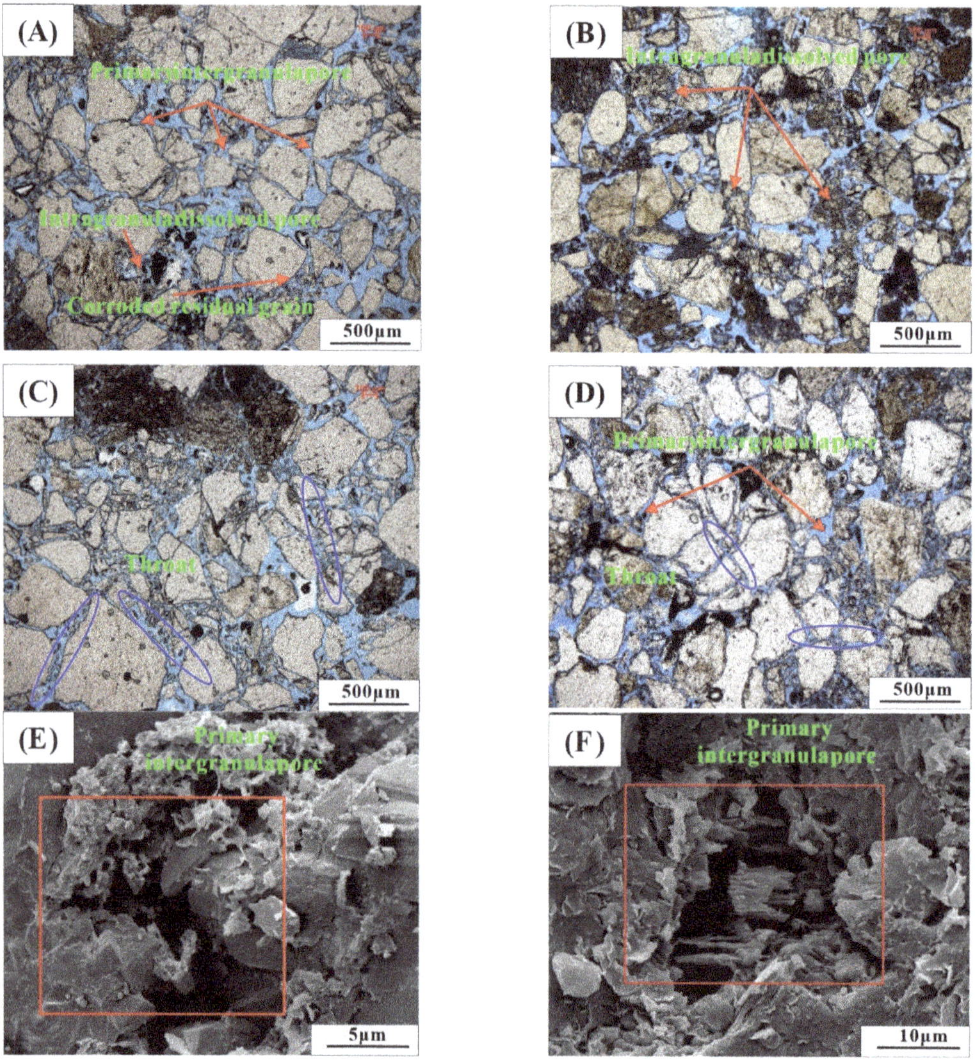

Figure 11. (**A**) The primary intergranular pores are the main ones, with a small amount of dissolution pores, and the face ratio is 25%, PL19−3−A well, L50, 1224.00 m. (**B**) There are two main types of neck-shaped and point-shaped throats, PL19−3−A well, L70, 1319.50 m. (**C**) The point type throat PL19−9−A well, L82, 1552.73 m. (**D**) The primary intergranular pores, PL19−9−A well, L82, 1553.08 m; (**E**) The primary intergranular pores, PL19−9−A well, L82, 1553.08 m. (**F**) The primary intergranular pores, PL19−9−A well, L50, 1379.61 m.

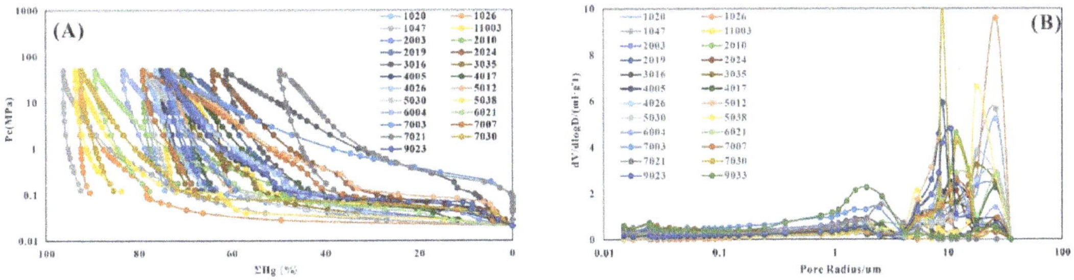

Figure 12. (**A**) Capillary pressure curves. (**B**) Pore throat radius distribution curve. The mercury intrusion curve is comprehensively controlled by the pore throat size, sorting, and skewness, and is a comprehensive reflection of the pore throat structure characteristics. (**A**) is the original mercury intrusion curve, and (**B**) is the pore size distribution curve calculated from (**A**).

5. Discussion

5.1. Pore Structure of Different Sedimentary Microfacies

5.1.1. Pore Throat Characteristics of Different Sedimentary Microfacies

As shown in Table 3, the samples of DB, BC, and DC microfacies had the largest mercury saturation (>75%), the largest pore throat radius (26.23~34.50 μm), and the lowest displacement pressure (0.02~0.03 MPa). The sorting ranged from 3.23 to 3.92, and the pore throat skewness ranged from 0.64 to 1.42. These three types of sedimentary microfacies are the best reservoir types in the PL Oilfield. The pore throat characteristics within different sedimentary microfacies showed significant differences.

Table 3. Pore throat characteristics of the Guantao Formation.

Microfacies	Por/%	Perm/mD	Vp/cm^3/100 g	$\bar{x}/\phi$	Skp	Sp	Dr
DBs	33.14	2256.15	17.68	6.61	1.42	3.23	0.49
BCs	31.77	1937.55	17.06	8.80	0.47	3.92	0.45
DCs	30.30	1826.25	16.01	8.35	0.64	3.43	0.43
SSs	20.19	115.89	12.51	10.33	−0.10	2.65	0.28
MBs	20.90	121.26	13.42	9.73	0.28	2.85	0.29
FPs	13.12	64.83	10.87	11.09	−0.70	3.15	0.22
IBs	12.13	58.65	11.42	11.77	−0.60	2.82	0.21
Microfacies	**P50/MPa**	**R50/μm**	**Pd/MPa**	**Smax/%**	**Rmax/μm**	**We/%**	**R35/μm**
DBs	0.07	18.81	0.02	89.15	34.50	3.84	20.76
BCs	0.20	4.60	0.03	74.51	29.88	10.70	12.88
DCs	0.24	6.33	0.03	82.76	26.23	11.38	12.03
SSs	2.65	0.28	0.12	69.76	6.36	19.78	0.79
MBs	3.12	1.22	0.14	67.72	5.53	23.14	1.52
FPs	9.71	0.08	0.38	61.12	1.96	28.48	0.61
IBs	7.20	0.10	0.30	59.15	2.60	29.58	0.68

Por = porosity; Perm = permeability; Vp = total pore volume; $\bar{x}$ = pore throat even size; Skp = skewness; Sp = pore throat sorting coefficient; Dr = pore throat variation coefficient; P50 = median pressure; R50 = median pore throat radius; Pd = displacement pressure; Smax = maximum mercury saturation; Rmax = maximum pore throat radius; We = mercury withdrawal efficiency; R35 = the pore throat radius that corresponds to 35% mercury saturation.

SS and MB microfacies samples had low mercury saturation (67.72% to 69.76%), a small pore throat radius, high displacement pressure (0.12 MPa to 0.14 MPa), and the maximum pore throat radius between 5.53 μm and 6.36 μm. Skewness ranged between 0.10 and 0.28. Shallow water deltas have fast deltaic progradation rates and a very slow slope of basement. The mouth bar and far sand bar deposited in the early stage are easily eroded by distributary channels in the later stage and eventually disappear. Therefore, the mouth bar sand in this study area is less developed and is very thin.

IB and FP had the smallest pore size, the lowest mercury saturation, and the highest displacement pressure. The sorting coefficient was 2.82~3.15 and the skewness was relatively fine. Compared with other microfacies, these two types of argillaceous facies reservoirs had a lower porosity, lower permeability, and poor storage capacity.

The pore characteristics within different depositional microfacies showed that parameters such as the mean pore throat, coefficient of variation, skewness, sorting, fractal dimension, or permeability had some correlations and their major characteristics were as follows: (1) The mean value of pore throat (ϕ) was negatively correlated with the variation coefficient; (2) the pore throat skewness of good reservoirs (DBs, BCs, DCs) were all positive, and poor reservoirs were mainly negative; (3) pore throat radius was positively correlated with reservoir physical properties; and (4) the sorting of the study area was complicated. It is not simply that the lower the sorting, the better the quality of the reservoir. This is because the pore throat distribution range of the argillaceous microfacies is smaller and the distribution probability is more concentrated. However, the high-quality microfacies have a wider distribution of pore throats, which will cause the sorting coefficient of some arenaceous microfacies to be larger than that of argillaceous microfacies. Therefore, the sorting coefficient should be compared within the same sedimentary microfacies type.

5.1.2. Pore Size Distribution of Different Sedimentary Microfacies

The DBs in the PSD curves are mainly single-peak characteristics (Figure 13A). The PSD range corresponding to the main peak is 10~25 μm and the pore volume contribution rate in this range is about 40%. DBs have large pore throat and are relatively homogeneous.

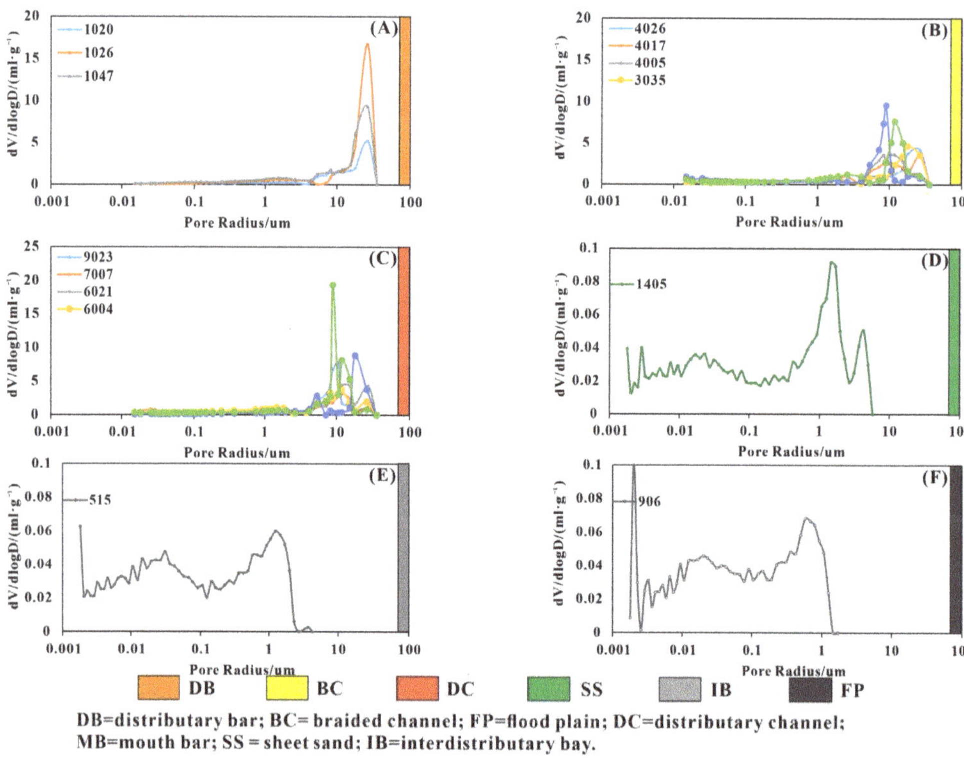

Figure 13. Pore size distribution of different sedimentary microfacies reservoirs. (**A**) Pore size distribution of DBs. (**B**) Pore size distribution of BCs. (**C**) Pore size distribution of DC.s (**D**) Pore size distribution of SSs. (**E**) Pore size distribution of FPs. (**F**) Pore size distribution of IBs.

The PSD curves of BCs and DCs had similar multi-peak characteristics (Figure 13B,C). The pore throat radius of the main peak ranged from 8 μm to 20 μm and the pore volume contribution rate was 35%. The pore throat radius of the secondary peak ranged from 5 μm to 8 μm and the pore volume contribution rate was 10%. The BCs and DCs were more heterogeneous compared with the DBs.

SSs (Figure 13D), FPs (Figure 13E), and IBs (Figure 13F) had poorly developed microfacies pores. The main peak of SS pores was less than 5 μm, and the main peaks of FP and IB pores were less than 2 μm. They all showed small-scale and multi-peak characteristics.

The pore size distribution of the DB, BC, and DC microfacies were significantly larger than those of other microfacies. Therefore, the DB, BC, and DC sandstone reservoirs are the main exploration target of the oil field.

5.1.3. Fractal Dimension of Porous Structure

As shown in Figure 14, the correlations (R^2) of all plots were more than 0.9, indicating that they are consistent with the fractal theory [41]. In addition, the sedimentary microfacies control the fractal dimension of the pore structure. The average Df of DB is 2.70 (Figure 14A,B), which is much lower than other microfacies. Both BCs and DCs had lower fractal dimensions, 2.81 (Figure 14C,D) and 2.79 (Figure 14E,F), respectively. However, the average fractal dimension of FPs was 2.93 (Figure 14G) and IBs was 2.94 (Figure 14H), reflecting that the pore structure of argillaceous microfacies has strong heterogeneity.

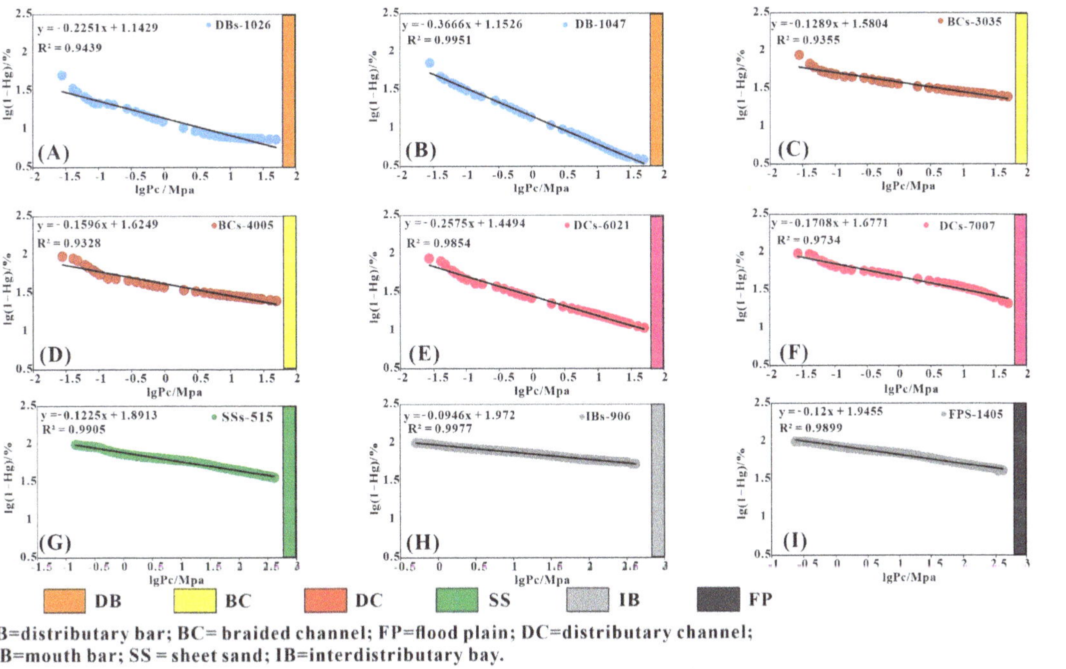

Figure 14. Representative sample plots of lg(Pc) vs. lg(1 − Hg). The correlation coefficients (R^2) of all the graphs were greater than 0.9, indicating that the samples conform to fractal characteristics. As the quality of the reservoir deteriorates, the fractal dimension gradually increases. (**A**) Fractal dimension of DB. (**B**) Fractal dimension of DB. (**C**) Fractal dimension of BC. (**D**) Fractal dimension of BC. (**E**) Fractal dimension of DC. (**F**) Fractal dimension of DC. (**G**) Fractal dimension of SS. (**H**) Fractal dimension of IB. (**I**) Fractal dimension of FP.

The fractal dimension is usually correlated with reservoir property [42]. In the PL Oilfield, the larger the fractal dimension, the smaller the pore volume (Figure 15A,B). Arenaceous microfacies are better quality reservoirs because they have a lower fractal dimension. The microfacies between IBs and FPs have the highest fractal dimension, suggesting that they have strong pore structure heterogeneity and small pore volume, which is not a high-quality storage space for oil and gas.

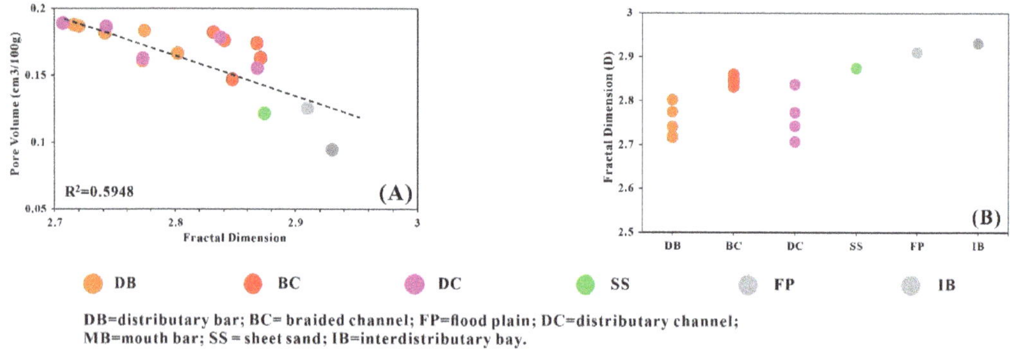

Figure 15. (**A**) Fractal dimension and pore volume scatter plots of different sedimentary microfacies. The pore volume decreases as the fractal dimension increases. (**B**) Fractal dimension of different sedimentary microfacies. With the improvement in reservoir quality, the fractal dimension gradually decreases.

5.2. Sedimentary Control of Pore Throat Characteristics

5.2.1. Sedimentary Rock Texture

Sedimentation and diagenesis are the main controlling factors affecting the pore structure characteristics of clastic rocks [42]. These two processes are controlled by sedimentary rock texture and clastic composition [43–45]. The texture refers to grain size, sorting, shape, roundness, and accumulation [7].

The texture and the physical properties of the reservoir have the following characteristics: (1) A coarse-grain size is beneficial to the preservation of intergranular pores so the physical properties of arenaceous sedimentary reservoirs are better than that of fine argillaceous reservoirs (Table 4); (2) the physical properties of the reservoir are negatively correlated with sorting, indicating that when the grain size is similar, the better-sorted reservoirs have better physical properties; and (3) arenaceous microfacies deposited in high-energy environments have greater skewness and poor reservoirs have less skewness, even resulting in negative skewness.

Particles that are well-sorted and coarse-grained will form better pore spaces during the burial process, and poorly sorted, fine-grained sandstone will fill the larger pores during the deposition process, greatly reducing the porosity. Accumulation controls porosity, and the different accumulation methods of particles affect the change in intergranular pores. Therefore, a reservoir pore structure is influenced by various aspects of sedimentary rock texture including grain size, sorting, and accumulation. As can be seen from the cast thin section, compared with argillaceous microfacies, DB, DC, and BC microfacies have a coarser grain size, and the grain contact is mainly point contact, which has better storage space.

Table 4. Petrological characteristics and sedimentary environment of shallow braided river delta sedimentary microfacies.

Microfacies	Distributary Bar (DB)	Braided Channel (BC)	Distributary Channel (DC)	Sheet Sand (SS)	Flood Plain (FP)	Interdistributary Bay (IB)
Sedimentary hydrodynamics	High energy	High energy	High–moderate energy	Moderate–low energy	Low energy	Low energy
Flow manner	Bidirectional flow	Bidirectional flow	Unidirectional flow	Unidirectional flow	Unidirectional flow	Unidirectional flow
Water depth	Shallow	Shallow	Shallow	Medium	Deep	Deep
Provenance distance	Proximal	Proximal	Proximal	Medium	Distant	Distant
Compositional maturity	Medium	Medium	Medium	Low	Low	Low
Median grain size/μm	220	180	110	60	25	20
Sorting	1.62	2.30	1.50	2.89	3.09	3.07
Skewness	0.62	0.	0.54	0.21	0.18	0.15
Particle contacts	Point	Point	Point	Line	Line-surface	Line-surface
Porosity/%	34.15%	32.05	32.01	25.80	20.24	19.15
Permeability/mD	2885.18	1504.17	1259.28	142.10	35.21	40.12

5.2.2. Minerals

1. Quartz and feldspar minerals

As shown in Table 4, the rigid mineral grains (quartz, orthoclase, and plagioclase) of DB, BC, and DC microfacies were significantly higher than other microfacies (Figure 16A). Previous studies have shown that rigid particles are conducive to the preservation of intergranular pores because rigid particles have stronger resistance to compaction [46]. Conversely, the higher the content of clay and plastic debris, the stronger the compressibility, leading to a greater reduction in porosity. In this study, the greater the maturity of the mineral composition, the higher the quartz content, and the greater the pore volume (Figure 17A,B). On the other hand, the diagenetic dissolution of feldspar improves the physical properties and seepage capacity of the reservoir. Atmospheric leaching occurs more in sedimentary microfacies with high physical properties. In the channel microfacies, there are a few intergranular and intra-granular dissolution pores, while the intra-granular dissolution pores of argillaceous microfacies are basically not developed.

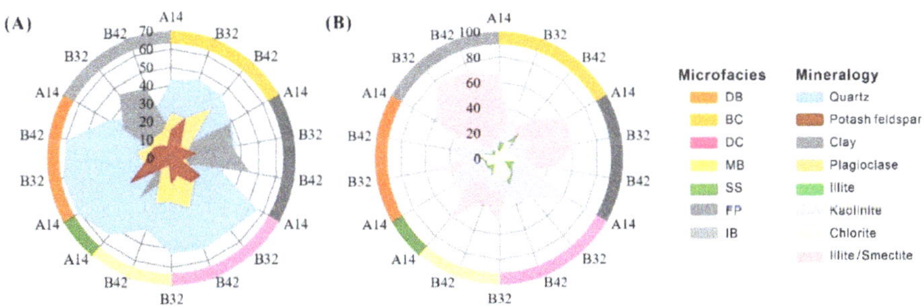

Figure 16. Star diagram shows the mineral distribution characteristics of different sedimentary microfacies. (**A**) X-ray diffraction method. (**B**) The quantitative mineral clay analysis method.

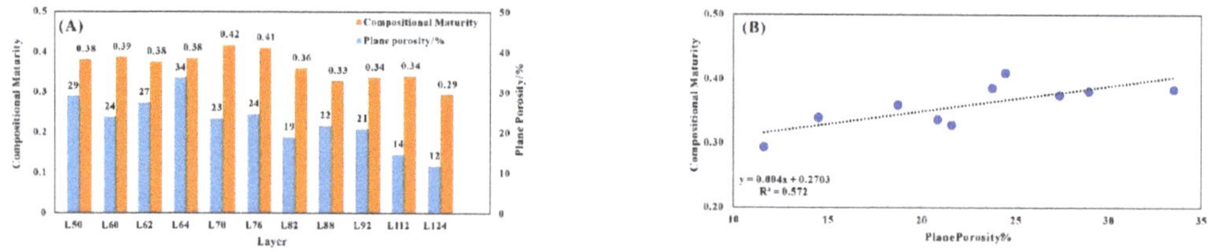

Figure 17. (**A**) Compositional maturity and plane porosity distribution. (**B**) Correlation between compositional maturity and plane porosity.

2. Clay minerals

Compared with clean sandstone, even a small amount of detrital clay will cause a lot of pore loss. Clay minerals are more susceptible to diagenetic compaction [43,44]. Compared with DB, BC, and DC microfacies, IB and FP microfacies contain higher clay mineral content (Table 5). The study area is an unconsolidated sandstone reservoir with relatively weak diagenesis, which only experiences compaction and cementation. The compaction and cementation are weak, and the cement content is low, mainly clay mineral cementation. According to the experimental results, the Guantao Formation reservoir is rich in authentically generated clay minerals including kaolinite, illite, illite/smectite, and chlorite. The occurrence of clay minerals has obvious depositional microfacies' zonation, which has a great influence on diagenesis, pore characteristics, and reservoir quality. The content of clay minerals (4%, 10%, and 6%) in the microfacies of DB, BC, and DC is far less than that of the IB and FP microfacies (35% and 38%). In particular, compared with other clay minerals, the microfacies of DB, BC, and DC developed relatively more kaolinite (Figure 16B).

Table 5. Mineral content of different sedimentary microfacies.

Microfacies	Siliceous Minerals (%)			Carbonate Minerals (%)		Clay Minerals (%)			
	Quartz	Orthoclase	Plagioclase	Calcite	Dolomite	Illite	Kaolinite	Chlorite	Illite/Smectite
BC	46.38	13.38	26.38	0.75	0.88	17.13	19.75	9.13	54
FP	36.89	9.94	11.82	0.83	0.67	5.39	25.17	7	62.44
DC	54.95	17.75	19.16	0.75	1.6	15.65	51.75	10.7	21.9
MB	45	9.67	21.67	0	1.5	10	22.67	15	52.33
SS	55	6	9	0	0	11	21	6	62
DB	64.14	17.43	16	0.29	1.14	26.29	37	14.43	22.29
IB	37.25	10.21	13.77	1.08	0.7	6.42	24.17	5.33	64.08

Illite fills the pores in filamentous and bridged form (Figure 18A,B), which damages and reduces the pore volume of the reservoir. Illite is mostly generated in deep water environments because when the water is deep, the mud content is high and the clay grains increase, which will generate a high amount of infiltration smectite. During the diagenesis process, the smectite will be transformed into illite, and finally formed illite. Therefore, in high-energy microfacies such as DBs and BCs, the relative content of illite is relatively small and the fine-grained depositional material in the shallow water environment usually contains more authigenic illite. The distribution of chlorite and illite has obvious sedimentary facies zoning and its content varies significantly among different sedimentary microfacies, affecting the heterogeneity of the pore structure. IBs and FPs have higher levels of chlorite (Figure 18C,D) and illite (Figure 18E), which deteriorate the quality of the reservoir.

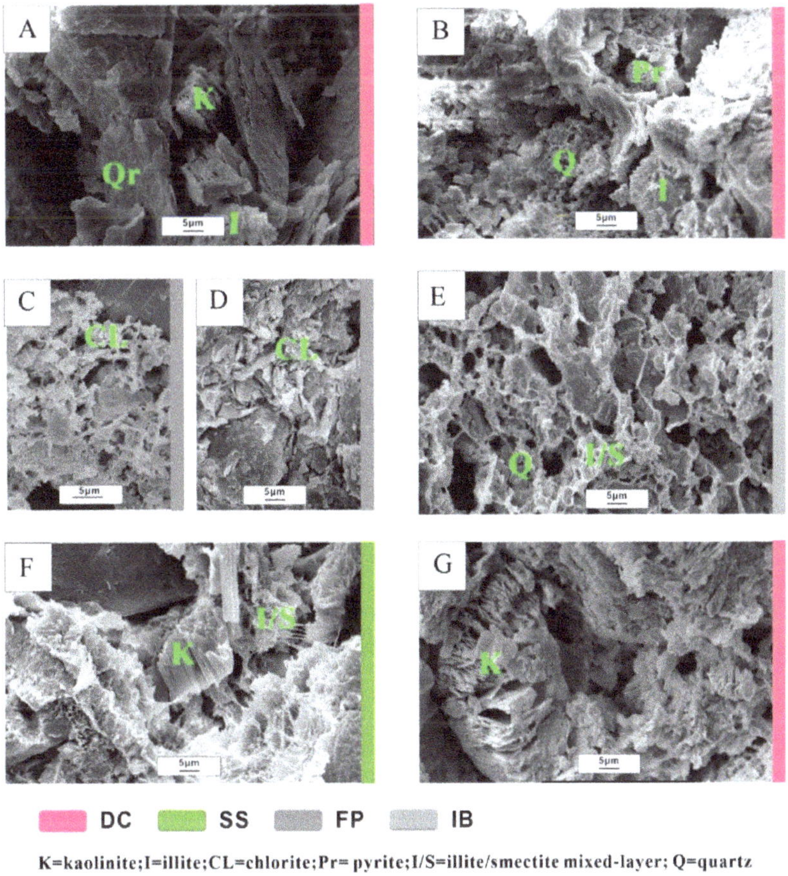

Figure 18. SEM images of representative clay mineral samples from the Guantao Formation. (**A**) PL25-C1, L80. The dissolution of feldspar produces intragranular pores and transforms into silky and scale kaolinite. (**B**) PL25-C, L80. Octahedral pyrite, silky illite filling intergranular pores. (**C**,**D**) PL19-B, L50. (**E**) PL25-3C1, L60. The filamentous illite/smectite mixed-layer was attached to the particle surface and filled the intergranular pores. (**F**) PL25-C, L100. A high number of silk-flocculent illite/smectite mixed-layer filled the intergranular pores, and some potassium feldspar particles were dissolved and transformed into book-shaped kaolinite. (**G**) PL25-C1, L80.Worm-shape kaolinite fills the intergranular pore throats, and the kaolinite transforms into silky illite.

The illite/smectite layer is the transition product of smectite to illite, which is mainly generated after the early diagenetic stage A, and filled in the pores in the honeycomb shape analyzed by SEM (Figure 18E,F). As shown in Table 5, the content of mixed-layer illite/smectite in the Guantao reservoir was the highest, followed by kaolinite, and the differences among different sedimentary microfacies were also great.

Authoritic kaolinite usually fills pores in a book and wormlike form (Figure 18F,G), dividing large pores into smaller ones, blocking rough channels into thin throats and reducing reservoir properties. Kaolinite takes shape when feldspar is corroded by an acidic medium. Therefore, the permeability of sedimentary rocks will directly affect the formation of kaolinite. For these reasons, compared with other clay minerals, kaolinite has a higher content in terms of arenaceous microfacies with a stronger hydrodynamic force, relatively better sorting, less mud content, good permeability, and coarser grain size. In contrast, it is less developed in the argillaceous microfacies with fine grain size, high mud content, and poor permeability.

5.3. Pore Throat Characteristics Comprehensive Evaluation

It is apparent that the reservoir quality of DB, BC, and DC microfacies are better than other microfacies, mainly for the following reasons: (1) with a better depositional structure, the grain size and sorting of arenaceous microfacies are significantly better than that of argillaceous microfacies; (2) it has a higher content of siliceous minerals with good compression resistance; and (3) it contains relatively higher content of kaolinite content and lower illite and chlorite content. As shown in Figure 19, the clastic particles in the DB, DC, and BC microfacies were mainly in point contact and there was little mud filling in the primary intergranular pores. The lower fractal dimension, coarser grain size, and high porosity of BC, DB, and DC microfacies indicate that good reservoirs have better reservoir heterogeneity and better storage space. This shows that the difference in pore structure of various sedimentary microfacies is comprehensively affected by the sedimentary rock texture (grain size, sorting, accumulation, etc.) and the differential distribution of siliceous minerals and clay minerals.

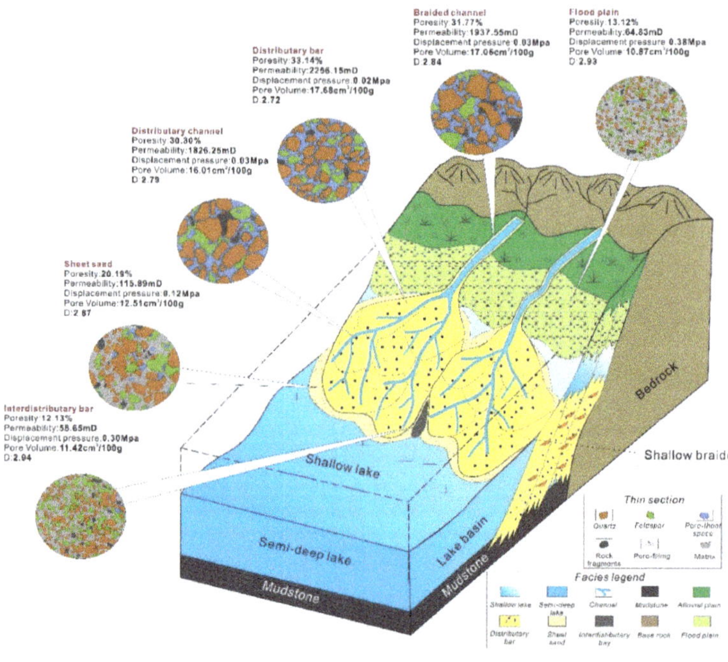

Figure 19. A summary diagram that displays pore characteristics controlled by microfacies in a deltaic depositional environment.

Previous investigations have conducted some work on the major controlling factors of microscopic pore structures [47]. However, a variety of the research results are limited to qualitative analysis and a description of the influencing factors in reservoir pore throat development, with few containing a quantitative evaluation. In the selection of influencing factors, sorting, grain size, and skewness have been taken as the representatives of primary depositional structure, while quartz, feldspar, and clay mineral content have been taken as the representatives of mineral distribution differences. This study was based on these quantitative parameters.

5.3.1. Feature Pore Characteristics Parameters Selection

Based on the mercury intrusion test, using Pittman's method, the empirical equations of porosity, permeability, and pore throat radius in the 10–75% mercury saturation interval were established (Table 6). The equation results show that the porosity, permeability, and pore throat radius had the best correlation at 35% mercury saturation, which was similar to the experimental results on high porosity and high permeability sandstone by Nabawy (2009) [26].

Table 6. The empirical equation of porosity, permeability, and pore throat radius at 10~75% mercury saturation.

Mercury Saturation (%)	Fitting Formula (φ,%; K, mD)	Correlation Coefficient (R^2)
10	lgR10 = 0.992 + 0.533lgK − 0.928lgφ (N = 41)	0.72
15	lgR15 = 0.977 + 0.570lgK − 1.048lgφ (N = 41)	0.85
20	lgR20 = 0.875 + 0.587lgK − 1.053lgφ (N = 41)	0.90
25	lgR25 = 0.677 + 0.632lgK − 1.050lgφ (N = 41)	0.91
30	lgR30 = 0.490 + 0.698lgK − 1.098lgφ (N = 41)	0.91
35	lgR35 = 0.321 + 0.791lgK − 1.227lgφ (N = 41)	0.92
40	lgR40 = 0.581 + 0.919lgK − 1.725lgφ (N = 41)	0.86
45	lgR45 = 0.110 + 1.018lgK − 1.684lgφ (N = 41)	0.80
50	lgR50 = −0.044 + 0.991lgK − 1.606lgφ (N = 40)	0.77
55	lgR55 = 1.146 + 1.117lgK − 2.779lgφ (N = 40)	0.70
60	lgR60 = 2.429 + 1.233lgK − 4.025lgφ (N = 40)	0.60
65	lgR65 = 2.914 + 1.155lgK − 4.335lgφ (N = 37)	0.45
70	lgR70 = 4.061 + 1.166lgK − 5.339lgφ (N = 37)	0.34
75	lgR75 = 0.222 + 0.794lgK − 2.156lgφ (N = 32)	0.18

Where R is the pore throat radius of different mercury saturations; φ is the core porosity; and K is the core permeability measured using air flow.

Moreover, the correlation between R35 and other characteristic parameters obtained from mercury injection experiments showed that R35 has a good correlation with other parameters (Figure 20), especially with permeability, r50, and skewness coefficients greater than 0.8. Therefore, r35 can be used as the feature pore characteristics parameter.

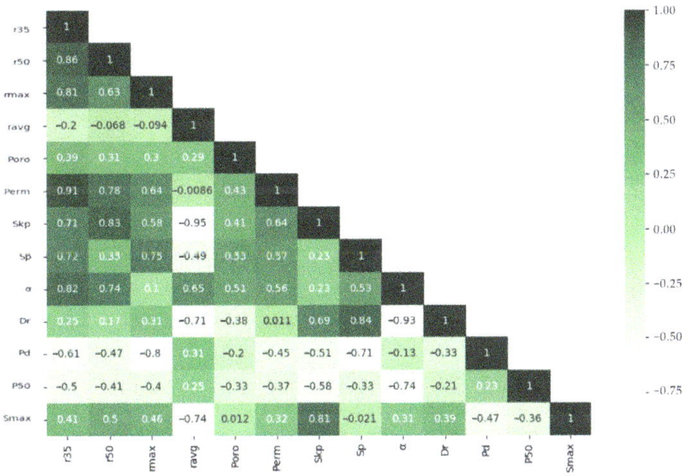

Figure 20. The correlation between R35 and other pore characteristic parameters.

5.3.2. Major Controlling Factors of Pore Characteristics

The grey system theory was founded by Professor Julong Deng (1989) [48] and has been widely used in many fields such as economics, ecology, military, medicine, and education. Gray relational analysis is not only an important part of gray theory, but also the cornerstone of gray system analysis, prediction, and decision-making. The gray relational analysis basic idea is to judge whether the connection is close according to the similarity of the geometric shape of the sequence curve from things or factors. The closer the curves, the greater the correlation between the corresponding sequences, and vice versa. The gray correlation analysis method makes up for the shortcomings caused by the mathematical statistics method in the system analysis. It is equally applicable to the number of sample sizes and irregular samples, and the calculation is small, and thus very easy to use. In gray correlation analysis, the quantitative results will not be inconsistent with the qualitative results.

1. Determine the analysis sequence

(1) The original sequence

The data sequence that can reflect the behavior characteristics of the system is called the original sequence. R35 is a comprehensive function of the pore structure. Therefore, this study defines R35 as the original sequence.

(2) The reference sequence

The data sequence comprising factors that affect the system behavior is called the reference sequence, which selects median grain size, sorting, skewness, quartz, orthoclase, potash feldspar, plagioclase, mix-layer illite/smectite, chlorite, illite, and kaolinite as a reference sequence of pore evaluation characteristics.

$$X_0 = (x_0(1), x_0(2), \cdots x_0(n)) \tag{7}$$

$$X_i = (x_i(1), x_i(2), \cdots x_i(n)) \tag{8}$$

where X_0 is the original sequence; X_i is the reference sequence; n is the sequence length; and $i = 1, 2, \cdots$, m is the number of reference sequence.

2. Data pre-processing

The data of the original sequence and the reference sequence should be preprocessed for two purposes. One is to de-dimensionalize, so that the data of different dimensions can also be compared. The other is to narrow the range of variables and improve calculation efficiency. In this study, the maximum value method was used to process the data without dimension.

Generally speaking, for the median grain size, skewness, orthoclase, potash feldspar, plagioclase, and kaolinite, the larger the value, the better the reservoir quality. For these parameters, the sequence can be normalized as follows:

$$X' = \frac{x}{x_{max}} \tag{9}$$

For sorting, mix-layer illite/smectite, orthoclase, potash feldspar, plagioclase, and kaolinite, the larger the value, the worse the reservoir quality. For these parameters, the sequence should be normalized as follows:

$$X' = \frac{x_{max} - x}{x_{max}} \tag{10}$$

3. Gray correlation

In gray correlation analysis, the degree of correlation between two systems or two sequences is defined as the degree of gray correlation. The gray relation coefficient $\varepsilon_{oi}(j)$

for the jth performance characteristics in the ith experiment can be expressed as Equation (11) after data pre-processing.

$$\varepsilon_{oi}(j) = \frac{\Delta min + \rho \Delta max}{\Delta_{oi}(j) + \rho \Delta max} \tag{11}$$

where $\varepsilon_{oi}(j)$ is the gray relational coefficient. $\Delta_{oi}(j) = |X_0(j) - X_i(j)|$ is the absolute difference between the original sequence and the reference sequence. Δmax and Δmin are the maximum and minimum absolute difference, respectively. $\rho = 0.5$ is generally used.

4. Gray correlation degree

The gray correlation degree is defined as:

$$r_{oi}(j) = \frac{1}{n} \sum_{j=1}^{n} \varepsilon_{oi}(j) \tag{12}$$

where $r_{oi}(j)$ is the gray relational grade and n is the number of parameters.

The gray relational grade is between [0, 1], and the closer it is to 1, the higher the correlation degree with the original sequence. Finally, the parameters are sorted according to the gray correlation degree, from large to small, where the order is as follows: mean grain size > median grain size > potassium feldspar > plagioclase > kaolinite > sorting> mix-layer illite/smectite > skewness > quartz > chlorite > illite.

The gray relational grade of each parameter and R35 shows that particle size has the greatest influence on the R35 among the depositional structure factors (Table 7). The sedimentary environment determines the grain size of sedimentary rocks. Grain size is the most significant structural feature of clastic rock particles. Grain size is the most significant sedimentary rock textural characteristic of clastic rock grains, which can point to their lithology and ultimately affect the physical properties of the reservoir. The pore structure of the reservoir is most closely related to grain size, which indicates that the primary sedimentary environment is the most important factor controlling the pore structure.

Table 7. Gray relational results table.

Gray Relational Grade	Original Sequence	Reference Sequence N = 70										
	R35 µm	Mz µm	Md µm	Pot %	Pla %	Kao %	Sp	I/S %	Skp	Qtz %	Chl %	Ill %
	1	0.831	0.822	0.66	0.656	0.654	0.653	0.638	0.637	0.545	0.504	0.492

R35 = the pore throat radius of 35% mercury saturation; Mz = mean grain size; Md = middle grain size; Pot. = potash feldspar; Pla. = plagioclase; Kao. = kaolinite; S p = pore–throat sorting coefficient; I/S = mixed-layer illite/smectite; Skp = skewness; Qtz. = quartz; Chl. = chlorite; Ill. = illite.

The feldspar content has the greatest influence of all the mineral content factors, followed by kaolinite and mix-layer illite/smectite. The high content of feldspar minerals can offer the reservoir stronger compaction resistance; on the other hand, it can make the reservoir develop more secondary dissolution pores. Kaolinite and mix-layer illite/smectite are the two clay minerals with the highest content in the Guantao Formation reservoirs. Kaolinite is more developed in high-energy and good permeability reservoirs and mix-layer illite/smectite is relatively high in argillaceous microfacies. Therefore, these two clay minerals have the most significant influence on pore characteristics.

Unconsolidated sandstone reservoirs are distributed in various oil fields throughout the world, and typical oil fields are shown in Table 8. This kind of oil reservoir usually has the characteristics of a shallow burial depth (buried depth < 1800 m), high porosity and permeability, coarse lithology, and loose cementation. Unconsolidated sandstone reservoirs are mainly distributed in Tertiary strata, and some are also distributed in Carboniferous strata such as the heavy oil fields in Alberta, Canada. The research methods and conclusions of this paper are not only applicable to the study area, but also have reference significance for the research work of all unconsolidated sandstone reservoirs.

Table 8. Characteristics of some unconsolidated sandstone reservoirs.

Country	Region	Oil Field	Burial Depth (m)	Porosity (%)	Permeability (mD)	Sedimentary Facies Type
China	Bohai Bay Basin	Gudao Oilfield	1120~1350	33	1639	Fluvial facies
		PL 19-3 Oilfield	910~1501	26	1351	Shallow braided river delta
		Qinhuangdao 32-6 Oilfield	950	35	3000	Fluvial facies
	Nanxiang Basin	Jinglou Oilfield	100~830	31.7	2642	Fan delta
		Gucheng Oilfield	150~1047	31.3	2250	River delta
	Junggar Basin	No.9 District of Karamay Oilfeld	215~350	31.7	3000	Braided river
Sudan	Muglide Basin	Fula Oilfield	1200~1500	29	2041	Fluvial facies
Canada	Eastern Alberta	Frog Lake Oilfield	424~600	30	1000–2000	Fluvial facies
United States	Gulf of Mexico, California	Kern River Oilfield	220~420	29.3	3140	Braided alluvial
		Wilmington Oilfield	63~2240	30.1	4260	Fluvial facies

6. Conclusions

This work contributes to the research on how the shallow-water braided river delta sedimentary system controls the reservoir pore structure, and also establishes empirical formulas for the pore throat radius, porosity, and permeability of high-porosity and high-permeability sandstone reservoirs. The conclusions of this study are as follows:

1. A total of seven typical sedimentary microfacies were identified in the study area; distributary bars (DBs), braided channels (BCs), distributary channels (DCs), mouth bars (MBs), sheet sands (SSs), interdistributary bays (IBs), and flood plains (FPs).
2. The pore structure of various sedimentary microfacies was found to be comprehensively affected by the textural properties of clastic sediments (grain size, sorting, accumulation, etc.) and the differential distribution of quartz, feldspar, and clay minerals.
3. The pore throat radius with 35% mercury saturation was selected as a comprehensive parameter of reservoir microscopic pore structure. R35 not only has the best correlation with porosity and permeability, but also has a close relationship with other mercury injection parameters.
4. Considering the depositional rock texture and mineral fraction, characteristic parameters (grain size, sorting, skewness, quartz, feldspar, kaolinite, chlorite, illite, illite/smectite) were selected to analyze the dominant controlling factor of the pore throat characteristics. Among them, grain size was selected as the most important control factor, indicating that the sedimentary environment is the main controlling factor for pore throat characteristics. It has been established that feldspar, kaolinite, and mixed-layer illite/smectite are the three types of minerals that have the greatest impact on pore characteristics.

Author Contributions: Z.W.: Methodology, Investigation, Data curation, Writing-original draft, Writing—review & editing, Visualization. H.T.: Conceptualization, Methodology, Validation, Formal analysis, Investigation, Resources, Data curation, Writing—review & editing, Supervision, Project administration, Funding acquisition. J.Y.: Visualization, Software, Investigation Experiment, Formal analysis. L.H.: Visualization, Software, Investigation Experiment. All authors have read and agreed to the published version of the manuscript.

Funding: This research received no external funding.

Institutional Review Board Statement: The study did not involve humans.

Informed Consent Statement: The study did not involve humans.

Data Availability Statement: The study did not report any data.

Acknowledgments: This work was supported by the General Project of Chongqing Natural Science Foundation (cstc2021jcyj-msxmX0897).

Conflicts of Interest: The authors declare no conflict of interest. The funders had no role in the design of the study; in the collection, analyses, or interpretation of data; in the writing of the manuscript, or in the decision to publish the results.

References

1. Lai, J.; Wang, G.; Cao, J.; Xiao, C.; Wang, S.; Pang, X.; Dai, Q.; He, Z.; Fan, X.; Yang, L.; et al. Investigation of pore structure and petrophysical property in tight sandstones. *Mar. Pet. Geol.* **2018**, *91*, 179–189. [CrossRef]
2. Liu, C.; Yin, C.; Lu, J.; Sun, L.; Wang, Y.; Hu, B.; Li, J. Pore structure and physical properties of sandy conglomerate reservoirs in the Xujiaweizi depression, northern Songliao Basin, China. *J. Pet. Sci. Eng.* **2020**, *192*, 107217. [CrossRef]
3. Wang, M.; Tang, H.; Tang, H.; Liu, S.; Zhang, L.; Zeng, M.; Cheng, Y. Impact of Differential Densification on the Pore Structure of Tight Gas Sandstone: Evidence from the Permian Shihezi and Shanxi Formations, Eastern Sulige Gas Field, Ordos Basin, China. *Geofluids* **2019**, *2019*, 4754601. [CrossRef]
4. Sakhaee-Pour, A.; Bryant, S.L. Effect of pore structure on the producibility of tight-gas sandstones. *AAPG Bull.* **2014**, *98*, 663–694. [CrossRef]
5. Ajdukiewicz, J.M.; Lander, R.H. Sandstone reservoir quality prediction: The state of the art. *AAPG Bull.* **2010**, *94*, 1083–1091. [CrossRef]
6. Wang, M.; Tang, H.; Zhao, F.; Liu, S.; Yang, Y.; Zhang, L.; Liao, J.; Lu, H. Controlling factor analysis and prediction of the quality of tight sandstone reservoirs: A case study of the He8 Member in the eastern Sulige Gas Field, Ordos Basin, China. *J. Nat. Gas Sci. Eng.* **2017**, *46*, 680–698. [CrossRef]
7. Lai, J.; Wang, G.; Ran, Y.; Zhou, Z.; Cui, Y. Impact of diagenesis on the reservoir quality of tight oil sandstones: The case of Upper Triassic Yanchang Formation Chang 7 oil layers in Ordos Basin, China. *J. Pet. Sci. Eng.* **2016**, *145*, 54–65. [CrossRef]
8. Morad, S.; Al-Ramadan, K.; Ketzer, J.M.; De Ros, L.F. The impact of diagenesis on the heterogeneity of sandstone reservoirs: A review of the role of depositional facies and sequence stratigraphy. *AAPG Bull.* **2010**, *94*, 1267–1309. [CrossRef]
9. Ozkan, A.; Cumella, S.P.; Milliken, K.L.; Laubach, S.E. Prediction of lithofacies and reservoir quality using well logs, Late Cretaceous Williams Fork Formation, Mamm Creek field, Piceance Basin, Colorado. *AAPG Bull.* **2011**, *95*, 1699–1723. [CrossRef]
10. Nabawy, B.S.; Géraud, Y. Impacts of pore- and petro-fabrics, mineral composition and diagenetic history on the bulk thermal conductivity of sandstones. *J. Afr. Earth Sci.* **2016**, *115*, 48–62. [CrossRef]
11. Sadhukhan, S.; Dutta, T.; Tarafdar, S. Simulation of diagenesis and permeability variation in two-dimensional rock structure. *Geophys. J. Int.* **2007**, *169*, 1366–1375. [CrossRef]
12. Lai, J.; Wang, G.; Wang, Z.; Chen, J.; Pang, X.; Wang, S.; Zhou, Z.; He, Z.; Qin, Z.; Fan, X. A review on pore structure characterization in tight sandstones. *Earth-Sci. Rev.* **2018**, *177*, 436–457. [CrossRef]
13. Tian, J.; Hao, F.; Zhou, X.; Zou, H.; Lan, L. Charging of the Penglai 9-1 oil field, Bohai Bay basin, China: Functions of the delta on accumulating petroleum. *Mar. Pet. Geol.* **2014**, *57*, 603–618. [CrossRef]
14. Blunt, M.J.; Bijeljic, B.; Dong, H.; Gharbi, O.; Iglauer, S.; Mostaghimi, P.; Paluszny, A.; Pentland, C. Pore-scale imaging and modelling. *Adv. Water Resour.* **2013**, *51*, 197–216. [CrossRef]
15. Dong, H.; Blunt, M.J. Pore-network extraction from micro-computerized-tomography images. *Phys. Rev. E Stat. Nonlinear Soft Matter Phys.* **2009**, *80*, 036307. [CrossRef]
16. Zeng, J.; Feng, X.; Feng, S.; Zhang, Y.; Qiao, J.; Yang, Z. Influence of Tight Sandstone Micro-Nano Pore-Throat Structures on Petroleum Accumulation: Evidence from Experimental Simulation Combining X-ray Tomography. *J. Nanosci. Nanotechnol.* **2017**, *17*, 6459–6469. [CrossRef]
17. Amaefule, J.O.; Altunbay, M.; Tiab, D.; Kersey, D.G.; Keelan, D.K. Enhanced Reservoir Description: Using Core and Log Data to Identify Hydraulic (Flow) Units and Predict Permeability in Uncored Intervals/Wells. In Proceedings of the SPE Annual Technical Conference and Exhibition, Huston, TX, USA, 3–6 October 1993.
18. Pittman, E.D. Relationship of Porosity and Permeability to Various Parameters Derived from Mercury Injection-Capillary Pressure Curves for Sandstone1. *AAPG Bull.* **1992**, *76*, 191–198. [CrossRef]
19. Weger, R.J.; Eberli, G.P.; Baechle, G.T.; Massaferro, J.L.; Sun, Y.-F. Quantification of pore structure and its effect on sonic velocity and permeability in carbonates. *AAPG Bull.* **2009**, *93*, 1297–1317. [CrossRef]
20. Zhang, K.; Pang, X.; Zhao, Z.; Shao, X.; Zhang, X.; Li, W.; Wang, K. Pore structure and fractal analysis of Lower Carboniferous carbonate reservoirs in the Marsel area, Chu-Sarysu basin. *Mar. Pet. Geol.* **2018**, *93*, 451–467. [CrossRef]

21. Pape, H.; Clauser, C. Improved Interpretation of Nuclear Magnetic Resonance T1 and T2 Distributions for Permeability Prediction: Simulation of Diffusion Coupling for a Fractal Cluster of Pores. *Pure Appl. Geophys.* **2009**, *166*, 949–968. [CrossRef]
22. Zhao, P.; Wang, Z.; Sun, Z.; Cai, J.; Wang, L. Investigation on the pore structure and multifractal characteristics of tight oil reservoirs using NMR measurements: Permian Lucaogou Formation in Jimusaer Sag, Junggar Basin. *Mar. Pet. Geol.* **2017**, *86*, 1067–1081. [CrossRef]
23. Li, P.; Zheng, M.; Bi, H.; Wu, S.; Wang, X. Pore throat structure and fractal characteristics of tight oil sandstone: A case study in the Ordos Basin, China. *J. Pet. Sci. Eng.* **2017**, *149*, 665–674. [CrossRef]
24. Winland, H.D. Oil accumulation in response to pore size changes. In *Amoco Production Research Report*; Field, W., Ed.; Scientific Research Publishing Inc.: Wuhan, China, 1972.
25. Spearing, M.; Allen, T.; McAulay, G. Review of the Winland R35 method for net pay definition and its application in low permeability sands. In Proceedings of the 2001 International Symposium of the Society of Core Analysts, Edinburgh, UK, 17–19 September 2001.
26. Nabawy, B.S.; Géraud, Y.; Rochette, P.; Bur, N. Pore-throat characterization in highly porous and permeable sandstones. *AAPG Bull.* **2009**, *93*, 719–739. [CrossRef]
27. Allen, M.B.; Macdonald, D.I.M.; Xun, Z.; Vincent, S.J.; Brouet-Menzies, C. Early Cenozoic two-phase extension and late Cenozoic thermal subsidence and inversion of the Bohai Basin, northern China. *Mar. Pet. Geol.* **1997**, *14*, 951–972. [CrossRef]
28. Liang, J.; Wang, H.; Bai, Y.; Ji, X.; Duo, X. Cenozoic tectonic evolution of the Bohai Bay Basin and its coupling relationship with Pacific Plate subduction. *J. Asian Earth Sci.* **2016**, *127*, 257–266. [CrossRef]
29. Xue, Y.a.; Deng, Y.; Wang, D.; Yang, H.; Lv, D.; Kang, K. Hydrocarbon accumulation conditions and key exploration and development technologies for PL 19–3 oilfield. *Pet. Res.* **2019**, *4*, 29–51. [CrossRef]
30. Hao, F.; Zou, H.; Gong, Z.; Deng, Y. Petroleum migration and accumulation in the Bozhong sub-basin, Bohai Bay basin, China: Significance of preferential petroleum migration pathways (PPMP) for the formation of large oilfields in lacustrine fault basins. *Mar. Pet. Geol.* **2007**, *24*, 1–13. [CrossRef]
31. Washburn, E.W. The Dynamics of Capillary Flow. *Phys. Rev.* **1921**, *17*, 273–283. [CrossRef]
32. Ziarani, A.S.; Aguilera, R. Pore-throat radius and tortuosity estimation from formation resistivity data for tight-gas sandstone reservoirs. *J. Appl. Geophys.* **2012**, *83*, 65–73. [CrossRef]
33. Schmitt, M.; Fernandes, C.P.; da Cunha Neto, J.A.B.; Wolf, F.G.; dos Santos, V.S.S. Characterization of pore systems in seal rocks using Nitrogen Gas Adsorption combined with Mercury Injection Capillary Pressure techniques. *Mar. Pet. Geol.* **2013**, *39*, 138–149. [CrossRef]
34. Song, Z.; Liu, G.; Yang, W.; Zou, H.; Sun, M.; Wang, X. Multi-fractal distribution analysis for pore structure characterization of tight sandstone—A case study of the Upper Paleozoic tight formations in the Longdong District, Ordos Basin. *Mar. Pet. Geol.* **2018**, *92*, 842–854. [CrossRef]
35. Giri, A.; Tarafdar, S.; Gouze, P.; Dutta, T. Fractal pore structure of sedimentary rocks: Simulation in 2-d using a relaxed bidisperse ballistic deposition model. *J. Appl. Geophys.* **2012**, *87*, 40–45. [CrossRef]
36. Li, K. Analytical derivation of Brooks–Corey type capillary pressure models using fractal geometry and evaluation of rock heterogeneity. *J. Pet. Sci. Eng.* **2010**, *73*, 20–26. [CrossRef]
37. Lu, H.; Tang, H.; Wang, M.; Li, X.; Zhang, L.; Wang, Q.; Zhao, Y.; Zhao, F.; Liao, J.; Mahmoud Aly, A. Pore Structure Characteristics and Permeability Prediction Model in a Cretaceous Carbonate Reservoir, North Persian Gulf Basin. *Geofluids* **2021**, *2021*, 8876679. [CrossRef]
38. Lai, J.; Wang, G. Fractal analysis of tight gas sandstones using high-pressure mercury intrusion techniques. *J. Nat. Gas Sci. Eng.* **2015**, *24*, 185–196. [CrossRef]
39. Clarkson, C.R.; Solano, N.; Bustin, R.M.; Bustin, A.M.M.; Chalmers, G.R.L.; He, L.; Melnichenko, Y.B.; Radliński, A.P.; Blach, T.P. Pore structure characterization of North American shale gas reservoirs using USANS/SANS, gas adsorption, and mercury intrusion. *Fuel* **2013**, *103*, 606–616. [CrossRef]
40. Clarkson, C.R.R.; Wood, J.M.M.; Burgis, S.E.E.; Aquino, S.D.D.; Freeman, M. Nanopore-Structure Analysis and Permeability Predictions for a Tight Gas Siltstone Reservoir by Use of Low-Pressure Adsorption and Mercury-Intrusion Techniques. *SPE Reserv. Eval. Eng.* **2012**, *15*, 648–661. [CrossRef]
41. Kuila, U.; Prasad, M. Specific surface area and pore-size distribution in clays and shales. *Geophys. Prospect.* **2013**, *61*, 341–362. [CrossRef]
42. Hao, L.; Tang, J.; Wang, Q.; Tao, H.; Ma, X.; Ma, D.; Ji, H. Fractal characteristics of tight sandstone reservoirs: A case from the Upper Triassic Yanchang Formation, Ordos Basin, China. *J. Pet. Sci. Eng.* **2017**, *158*, 243–252. [CrossRef]
43. Ramm, M. Reservoir quality and its relationship to facies and provenance in Middle to Upper Jurassic sequences, northeastern North Sea. *Clay Miner.* **2000**, *35*, 77–94. [CrossRef]
44. Bjørlykke, K. Relationships between depositional environments, burial history and rock properties. Some principal aspects of diagenetic process in sedimentary basins. *Sediment. Geol.* **2014**, *301*, 1–14. [CrossRef]
45. Xi, K.; Cao, Y.; Jahren, J.; Zhu, R.; Bjørlykke, K.; Haile, B.G.; Zheng, L.; Hellevang, H. Diagenesis and reservoir quality of the Lower Cretaceous Quantou Formation tight sandstones in the southern Songliao Basin, China. *Sediment. Geol.* **2015**, *330*, 90–107. [CrossRef]

46. Bloch, S.; Lander, R.H.; Bonnell, L. Anomalously High Porosity and Permeability in Deeply Buried Sandstone Reservoirs: Origin and Predictability. *AAPG Bull.* **2002**, *86*, 301–328. [CrossRef]
47. Yang, W.; Wang, Q.; Wang, Y.; Jiang, Z.; Song, Y.; Li, Y.; Liu, D.; Zuo, R.; Gu, X.; Zhang, F. Pore characteristic responses to categories of depositional microfacies of delta-lacustrine tight reservoirs in the Upper Triassic Yanchang Formation, Ordos Basin, NW China. *Mar. Pet. Geol.* **2020**, *118*, 104423. [CrossRef]
48. Deng, J.L. Introduction to grey system theory. *J. Grey Syst.* **1989**, *1*, 1–24.

energies

MDPI

Article

Reservoir Characteristics of Tight Sandstone and Sweet Spot Prediction of Dibei Gas Field in Eastern Kuqa Depression, Northwest China

Guangjie Zhao [1,2], Xianqing Li [1,2,*], Mancang Liu [3], Caiyuan Dong [3], Daye Chen [4] and Jizhen Zhang [5,6]

1 State Key Laboratory of Coal Resources and Safe Mining, China University of Mining and Technology (Beijing), Beijing 100083, China; zhaoguangjie23@126.com
2 College of Geoscience and Surveying Engineering, China University of Mining and Technology (Beijing), Beijing 100083, China
3 Research Institute of Petroleum Exploration and Development, PetroChina, Beijing 100083, China; mcliu1970@163.com (M.L.); d1404320223@126.com (C.D.)
4 China Coal Research Institute, Beijing 100013, China; chendaye@mail.ccri.ccteg.cn
5 Key Laboratory of Exploration Technologies for Oil and Gas Resources, Ministry of Education, Yangtze University, Wuhan 430100, China; zhangjz1991@126.com
6 College of Resources and Environment, Yangtze University, Wuhan 430100, China
* Correspondence: lixq@cumtb.edu.cn

Abstract: Great progress has been made in the exploration of tight sandstone gas resources in Kuqa depression. Great progress has been made in Dibei structural belt, which proves the previously unproven favorable area for tight sandstone gas. The physical properties, controlling factors, and characteristics of tight sandstone from the Ahe (J_1a) Formation in the Dibei gas reservoir are analyzed. The results show that the tight sandstone of the J_1a Formation is mainly feldspar lithic sandstone, with low porosity (average 9.1%) and low permeability (average 0.09 mD). Compaction (average compaction rate 61.9%) reduces porosity more than cementation (average cementation rate 14.3%). Secondary dissolution pores (average thin section porosity is 3.4%) dominate. The homogenization temperature has two peaks; the first peak is 85–110 °C, and the other peak is 115–140 °C, indicating that oil and gas experienced two filling stages at 12 Ma and 4.5 Ma, respectively. Eodiagenesis, A substage of mesogenetic diagenesis, and B substage of mesogenetic diagenesis happened in the area. Tight sandstone is developed in the B substage of mesogenetic diagenesis. The main controlling factors of diagenesis are: strong dissolution and structural pore increase; oil and gas charging and overpressure. The reservoir forming mode of the Dibei gas reservoir is: crude oil filling in the Late Neogene (12 Ma); reservoir densification in the late deposition of Kangcun Formation (7 Ma), mature natural gas filling in the early deposition of Kuqa Formation (4.5 Ma), and gas reservoir formed after transformation and adjustment in the deposition period of Quaternary (2 Ma). According to this model, it is predicted that the favorable area of the Dibei gas reservoir is in the southeast of the Yinan 2 well. This study provides guidance for the study of tight sandstone gas in other areas of the Kuqa Depression.

Keywords: diagenetic evolution; sweet spot; tight sandstone reservoir; controlling factors; Dibei gas field; Kuqa depression

Citation: Zhao, G.; Li, X.; Liu, M.; Dong, C.; Chen, D.; Zhang, J. Reservoir Characteristics of Tight Sandstone and Sweet Spot Prediction of Dibei Gas Field in Eastern Kuqa Depression, Northwest China. *Energies* **2022**, *15*, 3135. https://doi.org/10.3390/en15093135

Academic Editors: Mofazzal Hossain, Hisham Khaled Ben Mahmud and Md Motiur Rahman

Received: 17 March 2022
Accepted: 21 April 2022
Published: 25 April 2022

1. Introduction

With the advancement of oil and gas exploration, unconventional oil and gas have become an attractive subject [1–4]. Tight gas is an unconventional natural gas resource with huge reserves in many countries around the world, such as China, the United States, Canada, Russia, Europe, and Saudi Arabia in the Middle East [4–9]. Tight sandstone reservoirs were discovered in China's Tarim, Sichuan, Ordos Basin, Songliao, and Bohai Bay Basin, and great achievement has been made [10–14]. The favorable exploration area of tight oil in these basins is 16 × 10^4 km^2, and the oil geological resources are about (160–200) × 10^8 t [15].

The Dibei tight gas field is large and is discovered deep in the Jurassic in the Kuqa depression [16]. By contrast, the Dina, Dabei, Keshen, and other large tight gas fields are located in the Cretaceous and Eocene layers [17–19]. The reserves report provided by the PetroChina Tarim Oilfield (PCTO) predicts that the reserve of the Dibei gas field is 1.64 × 10^{11} m^3 [16]. Moreover, compared with other gas fields in the depression, the distribution of gas and water in the area is irregular [16]. Therefore, studying the sweet spot prediction and accumulation mode of the Dibei gas field is beneficial to provide a basis for the deep Jurassic tight gas exploration in the basin. There are a lot of natural gas and a little liquid hydrocarbon in the Dibei gas reservoir. The Jurassic Ahe Formation (J_1a) has a commercial-grade gas layer, and the overlying Lower Jurassic Yangxia Formation (J_1y) has a poor gas layer [20].

The sandstone reservoir of the J_1a Formation is a typical tight reservoir with poor porosity and permeability and high pore pressure [21]. In order to explore the formation mechanism of relatively high-quality reservoirs, many studies have been carried out [16,22–24]. Diagenesis is a key geological process that affects reservoir quality [25,26]. However, the development of good-quality tight sandstones in different oil and gas basins is affected by different diagenesis [27–29]. Therefore, the influence of diagenesis needs to be further studied. The tight sandstone of the J_1a Formation has strong heterogeneity [22], which provides a useful example for researching the changes of tight sandstone diagenesis. In addition, studying the process of diagenesis provides a basis for the development of good-quality tight sandstones.

The research objectives are as follows: (a) to study in detail the physical characteristics (composition, structural pore system, physical properties, and diagenetic minerals) of the J_1a tight sandstone; (b) to reveal the types, characteristics, and diagenetic stages of diagenesis; (c) to reveal the controlling factors of tight sandstone reservoir; (d) to establish the accumulation mode of tight sandstone gas reservoirs and point out favorable sweet spots. The result provides a scientific basis for other tight sandstone reservoirs in the Kuqa depression and other areas.

2. Geological Setting

Kuqa depression is in the northern Tarim Basin, which is adjacent to the Tianshan fold belt in the north [30] (Figure 1). It is a Mesozoic–Cenozoic foreland basin, which developed in the late Hercynian, and the superposition of multiple tectonic movements occurred, such as the extension depression stage during Jurassic–Paleogene and the development stage of intracontinental foreland thrust during Neogene–Quaternary [31]. It includes eight secondary structural units, including four structural belts: Northern monoclinal, Kelasu, Yiqikelike, Qiulitage, and Southern slope belt as well as three sags: Baicheng, Yangxia, and Wushi Sag [32,33].

There are sedimentary strata in Kuqa depression: Triassic, Jurassic, Cretaceous, Paleogene, and Quaternary sediments [34]. Source rocks in the area mainly include lacustrine mudstones in the Triassic and the coal seams, carbonaceous mudstone, and lacustrine mudstone in the Jurassic [35] (Figure 2). The favorable reservoirs comprise J_1a and J_1y Formation in Jurassic, Cretaceous Bashijiqike (K_1bs), Suweiyi (E_3s), and bottom conglomerate of Kumugeliemu (E_{1-2} km) in Paleogene, Jidike (N_1j) Formation in Neogene. The cap rocks of Jurassic and Cretaceous strata in Mesozoic are mainly mudstone, which is widely distributed throughout the depression. The Neozoic cap rocks are mainly in the E_{1-2} km and the N_1j strata, which mainly include the gypsum and salt layers.

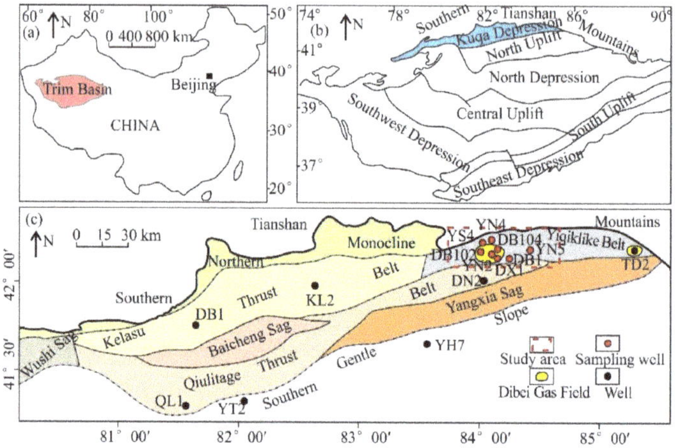

Figure 1. Tectonic units and location of Dibei gas field in Kuqa Depression. (**a**) Location of Tarim Basin; (**b**) location of Kuqa Depression; (**c**) sample wells and structural outline of Dibei gas field in Kuqa Depression.

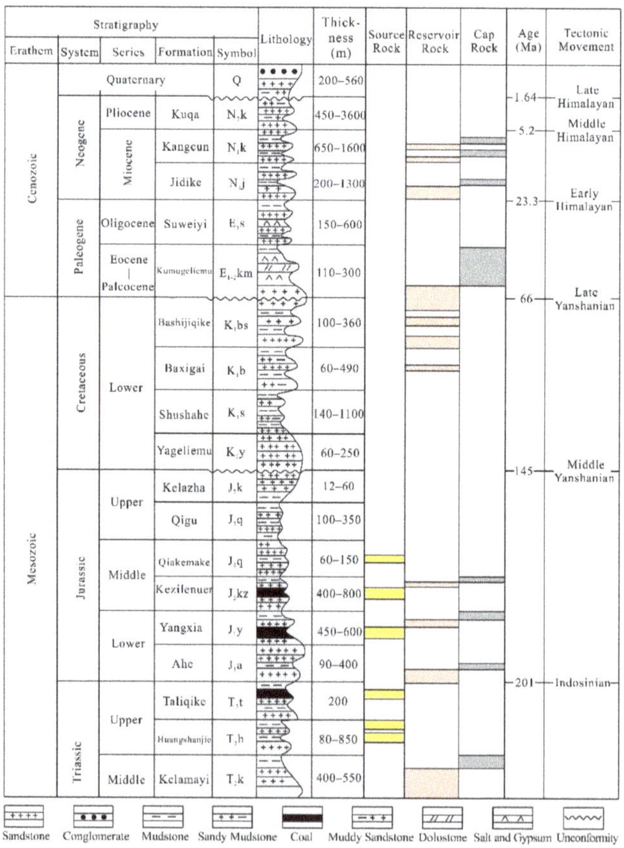

Figure 2. Mesozoic–Cenozoic stratigraphic system in the Dibei tectonic belt, Kuqa Depression (modified from Xiongqi Pang, 2019 [16] and Xiaowen Guo, 2016 [17]).

The Dibei tight gas field is in the middle of the Yiqiklike thrust belt. Oil, gas, and water all exist at the top of the anticline in this tight gas field, and there is a lot of natural gas in the tight reservoirs on the slope [16,36]. According to the oil and gas exploration results, commercial gas was discovered in the YN2, DX1, DB102, and DB104 wells along the slope, while the YN4 and YS4 far away from the slope have no commercial value. Studies on source rocks indicated that the oil and gas in the J_1a reservoir mainly came from the Jurassic source rocks, and the Middle-Upper Triassic strata also provided some natural gas [37,38].

3. Samples and Methods

In order to study the porosity and permeability and diagenesis characteristics of the reservoir, core samples were collected from PCTO. These samples are relatively uniformly distributed from 8 typical wells, with depths ranging from 4000 to 5100 m.

The CMS-300 automatic porosity and permeability measuring instrument was utilized to determine the porosity and permeability under a confining pressure of about 30 MPa. Scanning electron microscopy (SEM) was used to detect the types of clay minerals in the reservoir to determine the reservoir space type and condition of the clay minerals in the reservoir. QUANTA 200 SEM was used to detect representative samples. A total of 105 samples from 8 wells were analyzed for particle size, diagenetic characteristics, and porosity. The thin section is dyed with blue epoxy resin to mark the reservoir space, and the porosity of the thin section is calculated by the point method.

The DSG600 transparent reflection polarization fluorescence microscope was used to analyze the fluid inclusions of 31 samples from 8 wells in the Dibei gas reservoir to determine the hydrocarbon charging time. Petrographic and micro temperature methods were used to analyze the shape, size, distribution, color, and fluorescence of inclusions. The fluid inclusion was heated from room temperature to 0.5 °C/min until the phase boundary disappeared; the temperature was recorded in the state and maintained at a stationary temperature for 2 min. Then, the temperature of fluid inclusions dropped. When there were some other bubbles, the process was repeated to achieve the same uniform temperature.

By integrating data such as formation thickness, lithology, absolute age, erosion thickness, measured vitrinite reflectance (Ro), and borehole temperature, the burial and temperature history was reconstructed using BasinMod 1D software. In order to reconstruct the evolution of tight sandstone reservoirs and point out the influence of diagenesis, the model from Ruifei Wang, 2011 [39], was used to estimate the influence of different diagenesis on the reservoir. The model links reservoir porosity evolution to diagenetic stages, thereby quantifying the contributions of different diagenesis.

4. Results

4.1. Lithofacies Characteristics

According to the classification scheme of Folk [40] and the results of thin section analysis, the sandstone from the J_1a Formation in the Dibei structural belt is dominated by lithic sandstone and feldspar lithic sandstone (Figure 3). In the lithic sandstone, the relative content of rock fragments is 27.3%~86.4%, with an average of 81.2%; the content of feldspar is 5.65%~62.7%, with an average of 39.4%; the content of quartz is 1.2%~73.6%, with an average of 52.7%. The sorting of sandstone is medium-good, the roundness is mainly sub-angular-sub-round, the compositional maturity of the rock is low, and the structural maturity is medium. The sorting of sandstone is medium-good, the rounding is mainly sub-angular and sub-circular, the compositional maturity of the rock is low, and the structural maturity is medium.

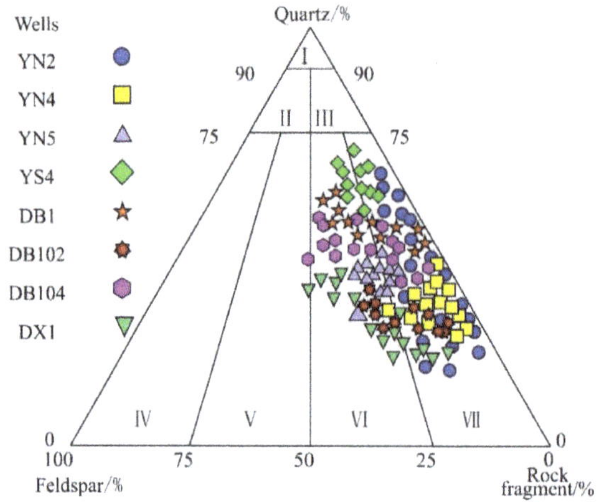

Figure 3. Figure indicating the composition of J_1a tight sandstone from Dibei gas field, eastern Kuqa Depression (number of samples = 105). Note: I—quartz sandstone; II—feldspar quartz sandstone; III—lithic quartz sandstone; IV—arkose sandstone; V—lithic arkose; VI—feldspar lithic sandstone; VII—lithic sandstone.

4.2. Reservoir's Porosity and Permeability

The relationship between porosity and permeability of 93 sandstone core samples from J_1a sandstone is indicated in Figure 4. The porosity values of the J_1a Formation range from 1% to 12%. The porosity values of most samples are between 2% and 10%, with an average of 9.1%. The range of permeability values is between 0.01 and 86.8 md, with 85% of the values varying from 0.01 to 1 mD (average 0.09 mD), which has the typical characteristics of tight sandstone gas.

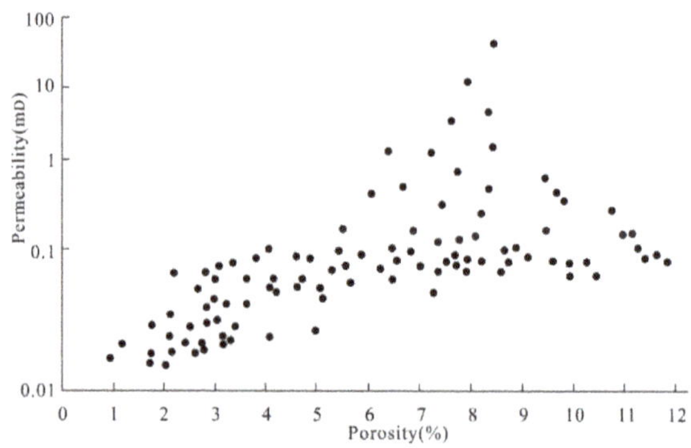

Figure 4. Diagram of the relationship between permeability and porosity.

4.3. Pore Systems

Based on the thin section observation and SEM analysis, there exist three types of storage space, namely primary intergranular pores (Figure 5a), secondary dissolution pores (Figure 5b–d), and fractures (Figure 5e,f). The thin section porosity (by point counting) of the tight reservoir of the J_1a Formation ranges from a trace level (less than 1%) to 9.9% (average 4.4%). The range of primary intergranular porosity is between a trace level and 5.6% (average 1.8%). The skeleton particles are dissolved to form secondary pores, which range from trace to 7.8% (average 3.4%). The J_1a Formation is buried deeply (more than 4000 m), so the original pores are severely damaged; therefore, primary intergranular pores are not the major storage space in the reservoir. The pores formed by dissolution, namely intergranular dissolution pores (Figure 5b) and intragranular dissolution pores (Figure 5c,d), are the second type of storage space. Because of the high ratio of feldspar and rock cuttings, lots of dissolution pores are generated, providing good storage space for oil and gas. The third type is a fracture (Figure 5e,f), which can be observed in the samples.

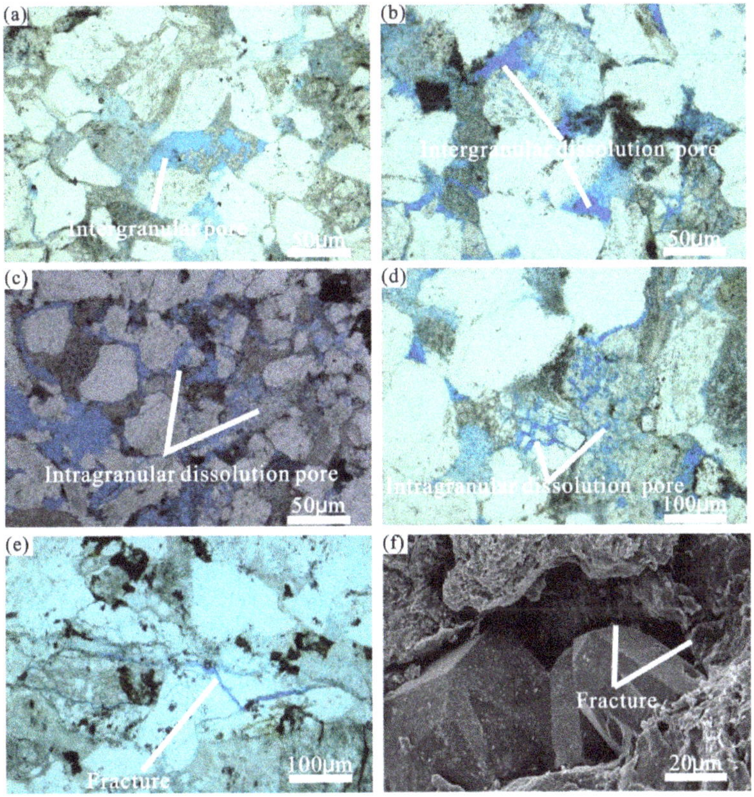

Figure 5. Space type of J_1a sandstone from Dibei area. (**a**) YN5, 4931.82 m, intergranular pores; (**b**) YN2, 4845.8 mm, intergranular dissolution pores; (**c**,**d**) YN4, 4189.6 m, intragranular dissolution pores; (**e**,**f**) DB102, 5056.5 m, fractures.

4.4. Diagenesis Types

4.4.1. Compaction

The main factor causing the decrease in porosity of clastic rocks is compaction [7,41]. The degree of compaction can be determined by the contact relationship of the particles. Compaction significantly reduces the porosity of tight reservoirs in the J_1a Formation. Firstly, the cuttings deform during the compaction process, resulting in point contact, line contact, and suture contact, which fills the intergranular pore and blocks pore throats (Figure 6a,b). Secondly, in the middle diagenetic stage, the acid generated from organic matter dissolved the feldspar debris and led to the weakening of the rock framework, which further compacted the framework particles [42]. Finally, the J_1a Formation is currently buried at a depth of about 4000–5000 m, which results in highly effective stress in the overburden and strong compaction through dissolution.

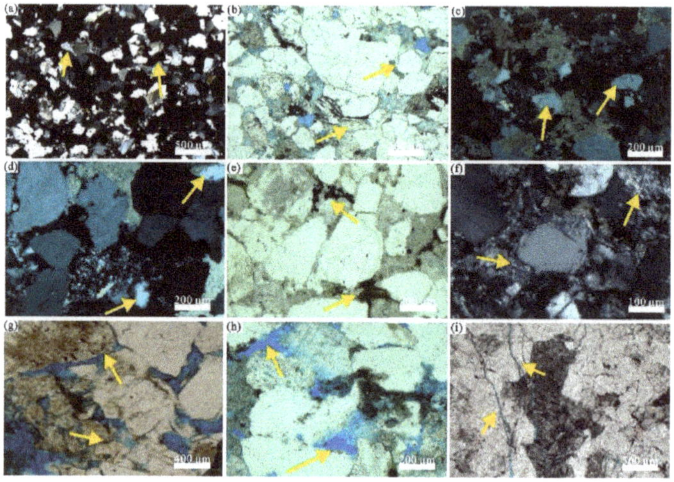

Figure 6. Diagenetic features of the J_1a sandstone reservoir. (**a**) PPL, grain line contact and stylolite contact (Well YN5, 4922.09 m); (**b**) PPL, grain line contact (Well YN4, 4121.58 m); (**c**) XPL, calcite cement (Well YN4, 4127.69 m); (**d**) PPL, calcite cement (Well YN2, 4841.63 m); (**e**) XPL, quartz overgrowth (Well DX 1, 4854.32 m); (**f**) PPL, quartz overgrowth (Well DB 102, 5004.39 m); (**g**) PPL, feldspar overgrowth (Well DX 1, 4858.74 m); (**h**) PPL, feldspar dissolution (Well YN2, 4839.26 m); (**i**) PPL, structural fractures (Well YN5, 4839.86 m). PPL—plane-polarized light; XPL—cross-polarized light.

The relationship between total intergranular volume and cement content shows that compaction reduces porosity more than cementation (Figure 7), with three exceptions. The compaction rate is 27.9%~76.8% (average 61.9%), and the cementation rate is 4.3%~62.9% (average 14.3%). According to the porosity evolution model of [39], the porosity loss during compaction is calculated. The average reduction rate of compacted porosity is 62.3%, and the average reduction rate of cemented porosity is 15.2%, which basically corresponds to the result in Figure 7. Comprehensive analysis shows that compaction is the main diagenesis causing porosity damage.

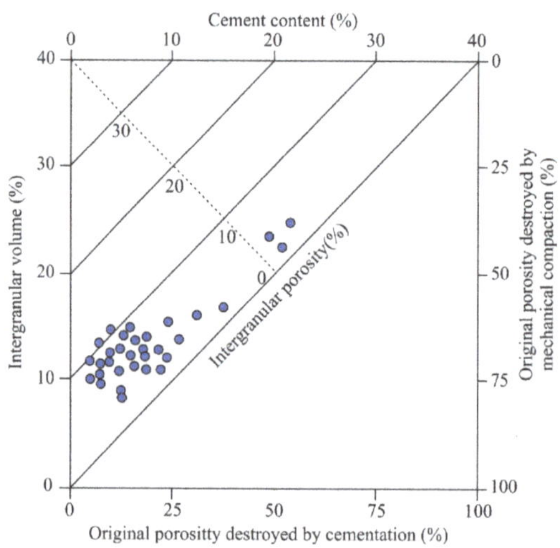

Figure 7. The relationship between the intergranular volume and the cementation volume of the J_1a tight reservoir (modified from Longlong Liu, 2020 [43]).

4.4.2. Cementation

Cementation is another major factor affecting the porosity of the reservoir. Calcite is a widely distributed cement in the J_1a unit, and its content is uneven (0–31%, average of 18.6%). Occasionally, calcite blocks of cement were distributed around the detrital grains (Figures 6c and 8a), but calcite cement usually filled the intergranular pores as microcrystalline (Figure 6d), indicating a high degree of carbonate cementation. Although cement can limit compaction, some of the original porosity damaged by compaction is preserved, but in general, the destructive effect of carbonate cement on the reservoir is more apparent.

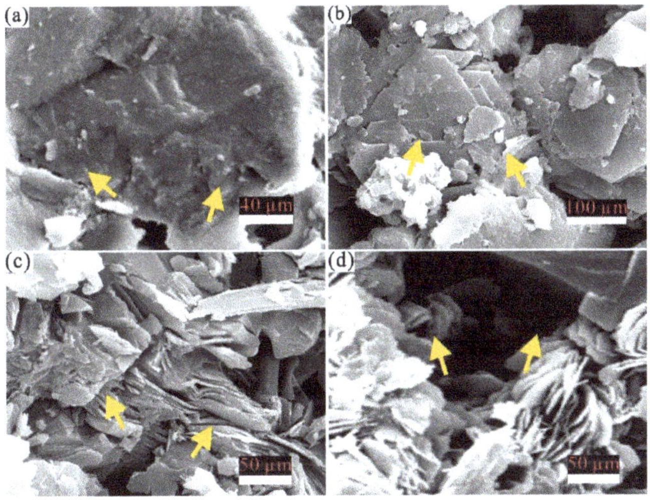

Figure 8. Diagenetic characteristics of the Ahe reservoir shown under the SEM. (**a**) Calcite cement (Well YN2, 4841.63 m); (**b**) Kaolinite cement (Well DX1, 4855.45 m); (**c**) Illite cement (Well YN5, 4930.58 m); (**d**) Chlorite cement (Well DX1, 4856.85 m).

Silica cementation is a key diagenetic process that causes the deterioration of sandstone reservoir quality. Silica cement is overgrown around single-crystal quartz grains. Quartz overgrowth is distributed on the surface of the quartz particles (Figure 6e,f). The volume content of quartz overgrowth in sandstone ranges from 51% to 79% (average of 67.9%). Feldspar overgrowth was occasionally detected (Figure 6g), which provided material for the corrosion process.

The total content of clay minerals determined by XRD was from 3.9% to 14.5%, with an average of 12.3%. There are three kinds of clay minerals in the sample, and their contents are: kaolinite (10.9%~27.8%, average of 38.6%), illite (7.9%~25.2%, average of 36.3%), and chlorite (8.5%~26.2%, average of 24.6%). Kaolinite is an important clay mineral in the J_1a Formation. The morphology of kaolinite is usually booklet form, and it occupies intragranular and intergranular pores. The kaolinite clay minerals are closely associated with dissolved feldspar crystals, indicating that the kaolinite minerals are formed by the dissolution of unstable feldspar. Moreover, the chemical reaction of feldspar with CO_2 resulted in the formation of kaolinite on the feldspar surface. Kaolinite is usually associated with illite and mixed illite/montmorillonite layers. During the middle diagenetic stage, a large amount of unstable feldspar was dissolved by the meteoric water, which may cause the feldspar to transform into kaolinite. It may be the reason for the low feldspar content [44]. Illite is another important clay mineral found in the J_1a Formation. Illite usually exists in a flake form. At times, illite seems to grow at the expense of kaolinite. However, illite is not fully developed (Figure 8c), probably because the kaolinite that formed illite was not buried deeply. Chlorite is another minor clay mineral in this sandstone. The morphology of chlorite clay is dominated by pore-filling crystals. Chlorite cladding is at the edge (Figure 8d), which is thought to form during the early diagenetic stage. The intergranular pores are mainly filled with chlorite, accompanied by minerals such as kaolinite and illite.

4.4.3. Dissolution

Dissolution is a critical factor in forming good-quality reservoirs [45,46]. According to the above analysis, compaction and cementation severely damaged the primary pores. However, the dissolution of feldspar and clastic rocks in the later stage formed secondary pores, which are the main pores of the J_1a tight sandstone. The thin section observation results indicate the secondary pores of the J_1a Formation are widely distributed, indicating that dissolution is major diagenesis. Feldspar usually dissolves partially along cleavage planes and fractures, resulting in the formation of numerous secondary intragranular pores (Figure 6h). In the regions with strong dissolution, the secondary intergranular and intragranular pores are widely developed, and the porosity is high. Therefore, the dissolution of feldspar is a decisive factor in the development of secondary pores [47].

It is worth mentioning that the thrust structure in the Dibei area is strong and lateral pressure is obvious, which causes the pores to be reduced [22]. However, structural fractures also make the fractures more developed (Figure 6i).

4.5. Burial History and Fluid Inclusion Characteristics

The burial-thermal evolution history was reconstructed to clarify the diagenetic process and pore evolution, as shown in Figure 9. The burial curve is taken from well DB102. In the Paleogene, the J_1a group began to deposit at a faster rate, and when the J_1a was deposited, the group was uplifted and eroded; Then, the J_1a layer continued to be buried rapidly. At about 4.5 Ma ago, the deposition rate increased significantly; in the early Quaternary, the J_1a Formation was buried at a depth of more than 4000 m.

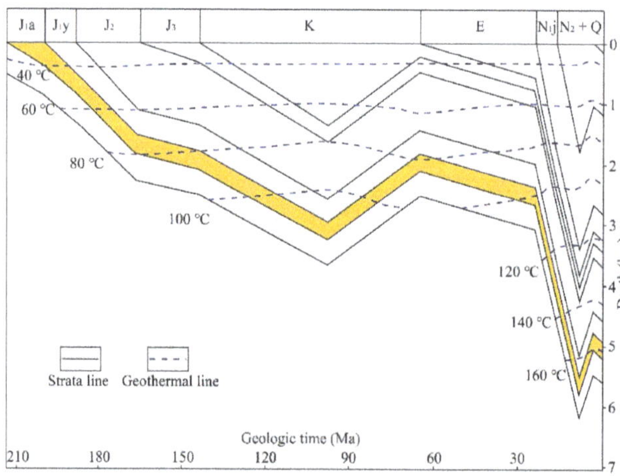

Figure 9. Burial, thermal history, and geothermal line of well DB102.

The fluid inclusion characteristics of the samples are shown in Figure 10. The aqueous inclusions in the same period as the hydrocarbon inclusions exist in wells DB102, YN2, YN4, and YN5. The homogenization temperature of the inclusion has two peaks, indicating the two periods of oil and gas charging. The temperature range of the first peak is 85~110 °C (12 Ma), and the second is 115~140 °C (4.5 Ma) (Figure 10). Under ultraviolet light, the first group of fluid inclusions showed yellow fluorescence (Figure 11a), indicating that the charged hydrocarbons have low maturity. The second group shows blue fluorescence (Figure 11b), suggesting that the charged hydrocarbon was of high maturity.

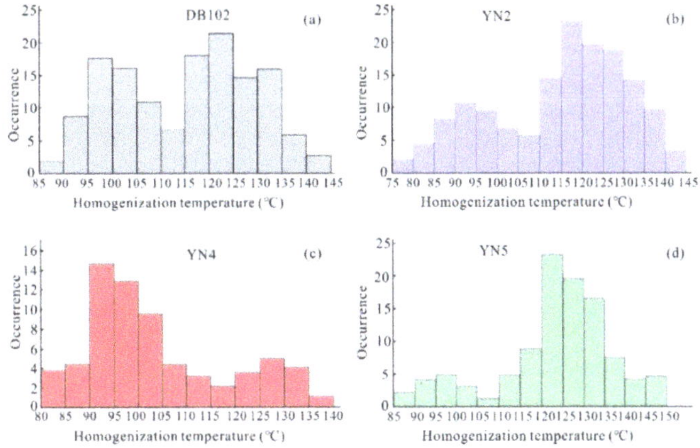

Figure 10. Homogenization temperatures of aqueous inclusions coeval with hydrocarbon inclusions in the Dibei area. (**a–d**) Homogenization temperatures for DB102 well, YN2 well, YN4 well, and YN5 well, respectively.

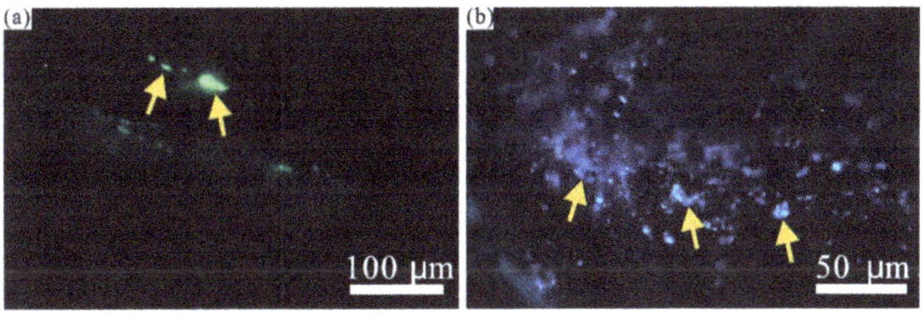

Figure 11. Micrographs of hydrocarbon inclusions in the J_1a Formation under UV light in the Well DB102. (**a**) Yellow fluorescence; (**b**) blue-white fluorescence.

5. Discussion

5.1. Diagenetic Evolution

According to the structural relationship [16,33] and diagenetic characteristics, three basic diagenetic stages in the Dibei area have been determined. Based on the standard "Diagenetic Stages of Clastic Rocks" (SY/T5477-2003) from China, the J_1a Formation has experienced three stages: Eodiagenesis, A and B substage of mesogenetic diagenesis. During Eodiagenesis, the paleo-temperature range was from paleo-normal temperature to 85 °C. The pore types were mainly primary pores with a little secondary pore. Ro was usually less than 0.5%; during A substage of mesogenetic diagenesis, the paleotemperature ranged from 85 °C to 14 °C, and the Ro ranges from 0.5% to 1.3%. The organic acid yield was high, and the pore types were mainly secondary pores; during the B substage of mesogenetic diagenesis, the paleotemperature range is 140–175 °C, the Ro range is 1.3–2.0%, and the rocks are densified with fractures. The diagenetic stages and pore evolution are shown in Figure 12.

Eodiagenesis: The first stage: the sandstone of the J_1a Formation has undergone long-term shallow burial and late rapid deep burial. It is the Eodiagenesis stage dominated by compaction, with the development of siliceous cement and kaolinite; A substage of mesogenetic diagenesis: this stage is the Paleocene–late Pliocene, low-maturity organic matter produces organic acid, and secondary pores are formed in soluble minerals such as dissolved feldspar and rock fragments. The third stage is the B sub-stage of Mesozoic diagenesis from Late Neogene to Quaternary, with high maturity of organic matter [22]. The content of organic acid decreases, the carbonate cementation increases in the late stage, and the reservoir quality continues to deteriorate [16]. Under the strong tectonic compression, the compaction is further strengthened, and the porosity is further reduced [22].

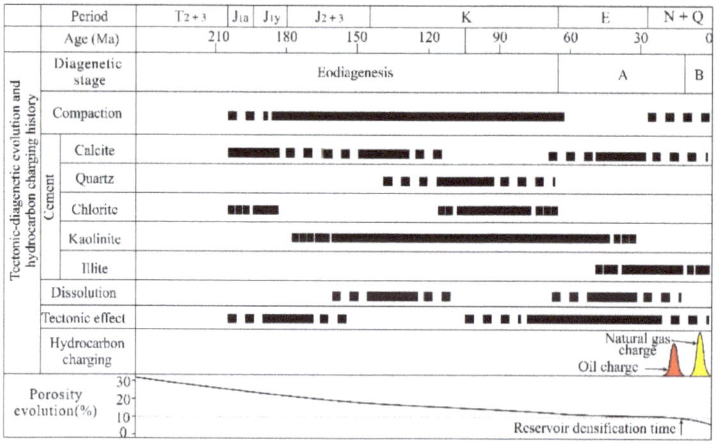

Figure 12. Diagenesis and porosity evolution of the J_1a Formation. A is A substage of mesogenetic diagenesis; B is B substage of mesogenetic diagenesis.

5.2. Controlling Factors of Tight Sandstone

5.2.1. Diagenesis

Destructive Diagenesis

According to Figure 8, the compaction rate is 27.9%~76.8% (average 61.9%). Therefore, compaction is considered to be the main diagenetic process causing porosity deterioration.

Cementation affects reservoir quality [48–50]. However, the cementation is not obvious based on the result calculated by [39] (the average cementation rate is 15.2%). In addition, the compaction rate of samples with a high cementation rate is low, indicating that early strong cementation inhibited mechanical compaction (Figure 7). The reason is that carbonate cement is formed by the formation of water with high ion content, and these types of cement fill the space between particles. With the increase in burial, the types of clay minerals change greatly, affecting the reservoir quality [51]. Moreover, because of the enrichment of alkaline and K^+ by dissolution, kaolinite is transformed into illite, further reducing the physical properties of the rock, but the depth of illite is not low enough, so illite cement has not been well developed.

Constructive Diagenesis

Although the compaction degree of sandstone is strong, good-quality reservoirs still exist in some areas. The research shows that the secondary dissolution porosity is the main storage space (average thin section porosity is 3.1%), so good-quality tight sandstone reservoirs are mainly affected by dissolution. To evaluate the dissolution effect, the diagram between sheet porosity and feldspar content is drawn. Feldspar is conducive to the occurrence of dissolution, resulting in the generation of secondary dissolution pores (Figure 13). Based on the model from Ruifei Wang, 2011 [39], the dissolution of feldspar and rock fragments increases the porosity by about 10%. Therefore, for high-quality tight sandstone, dissolution is the most important diagenetic process. In addition, the enrichment of kaolinite is also an important part of a high-quality reservoir. The enrichment of kaolinite is also evidence of the strong dissolution of feldspar because the dissolution of feldspar is associated with kaolinite [52].

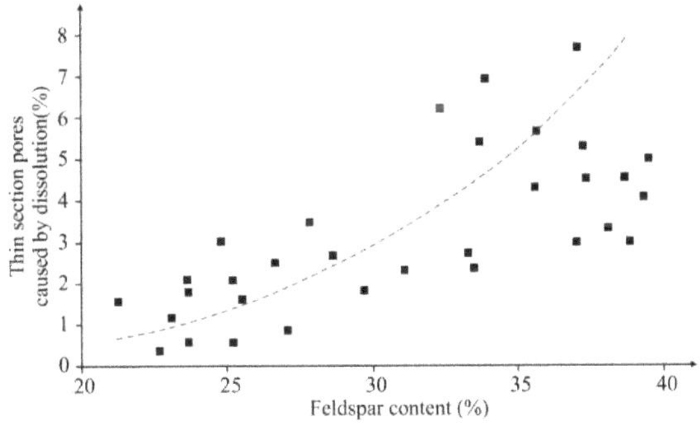

Figure 13. Secondary dissolution pores versus feldspar content.

Tectonism Has Both Destructive and Constructive Effects

Tectonism mostly carries out the secondary transformation of tight reservoirs. This transformation has both advantages and disadvantages for reservoir physical properties, and the overall advantages outweigh the disadvantages. Specifically, in the study area, on the one hand, structural compression leads to further compaction of the reservoir, resulting in the decrease in porosity [53], which has an adverse impact on the physical properties of the reservoir. On the other hand, tectonism will cause the formation of structural fractures in tight reservoirs [54]. Some structural fractures will be further dissolved and expanded under the action of organic acid to form structural dissolution fractures and become a good reservoir space. Some reticular and extended fractures will communicate with isolated residual pores, which greatly improves the permeability of reservoir. Statistics of some core data and cast thin section analysis data [16] show that under the same conditions, the average porosity and permeability values of the reservoir with fracture development are obviously higher than those of the reservoir without fracture development (Table 1). Fractures can effectively improve the porosity and permeability of tight reservoirs. The fracture development area formed by tectonic movement is often a favorable position to form tight sandstone reservoirs. This is consistent with the research results of the literature [2,12].

Table 1. Porosity and permeability of the J_1a sandstone in Dibei area (part of the data is from Hailiang Kang [24]).

Wells	Measured Porosity/%						Measured Permeability/mD					
	Samples with Developed Fractures			Samples without Developed Fractures			Samples with Developed Fractures			Samples without Developed Fractures		
	Min	Max	Ave	Min	Max	Ave	Min	Max	Ave	Min	Max	Ave
YN4	1.19	13.18	9.21	0.89	12.97	7.39	0.038	3160	97.6	0.006	63	1.936
YS4	3.26	12.35	8.73	1.63	16.21	8.36	0.09	926	81.2	0.009	89	1.942
YN2	2.63	13.96	5.62	0.28	15.86	5.48	0.349	392	36.4	0.013	52	1.09
YN5	2.36	9.18	5.65	0.69	12.63	5.91	0.98	2659	258.8	0.021	476	5.27
Ave		8.23			6.95			128.37			2.54	

Note: Min = Minimum; Max = Maximum; Ave = Average.

5.2.2. Oil and Gas Charge and Overpressure

The water in the reservoir can be displaced by charged oil and gas, thereby altering the environment and inhibiting late cementation [55]. Inclusions contain information on the hydrocarbon generation and accumulation, so fluid inclusion is a feasible index to determine the oil and gas charging time [56,57]. The homogenization temperature of fluid inclusions shows that hydrocarbon charging can be divided into two stages at 12 Ma and 4.5 Ma. Late cementation did not form during hydrocarbon charging (Figure 13); therefore, hydrocarbon charging inhibits the formation of late carbonate types of cement, which can also serve as evidence for the lack of late carbonate types of cement. In addition, the overpressure preserves the primary pores well and enables favorable conditions for dissolution [58,59]. The overpressure in the Dibei structural belt originates from hydrocarbon generation [60]. According to the above analysis, the Upper Triassic and Middle-Lower Jurassic source rocks are dominant in the Kuqa depression. With the evolution of source rocks, overpressure originating from hydrocarbon generation is transferred to nearby reservoirs, resulting in abnormal reservoir pressure [61]. The source rocks of the Triassic and Jurassic reached the hydrocarbon generation peak and formed overpressure at 23–12 Ma. In this stage, the sandstone reservoir was not densified (porosity greater than 10%) (Figure 9); the pressure coefficient of the Dibei structural belt generally exceeds 1.7, and the pressure coefficient in some areas exceeds 1.8 [61]. Therefore, the occurrence of overpressure is conducive to preserving the primary pores.

5.3. *Sweet Spot Prediction*

The geological characteristics of the Dibei gas reservoir are as follows: the faults formed by the tectonic movement in Yanshanian and Himalayan periods connect the source rocks and reservoirs, as well as the source rocks and caprocks, forming an oil and gas transportation system [54]. Because it is close to the center of Yangxia sag and the distance between hydrocarbon and reservoir is close, natural gas has obvious vertical, lateral, and near-source migration paths [16]; there exist two periods of hydrocarbon charging, namely, early oil charging and late gas charging.

A uniform gas water interface does not exist, and the gas reservoir formed at the high part of the structure is damaged by water and becomes a water layer (Figure 14); reservoir densification is earlier than natural gas charging. The reservoir forming mode of the Dibei gas reservoir is: crude oil filling in the Late Neogene (12 Ma); in the late sedimentary period of Kangcun Formation, the reservoir gradually densified (7 Ma); mature natural gas filling in the early deposition of Kuqa Formation (4.5 Ma), and gas reservoir formed after transformation and adjustment in the Quaternary period (2 Ma). The result basically corresponds to the previous conclusion [22].

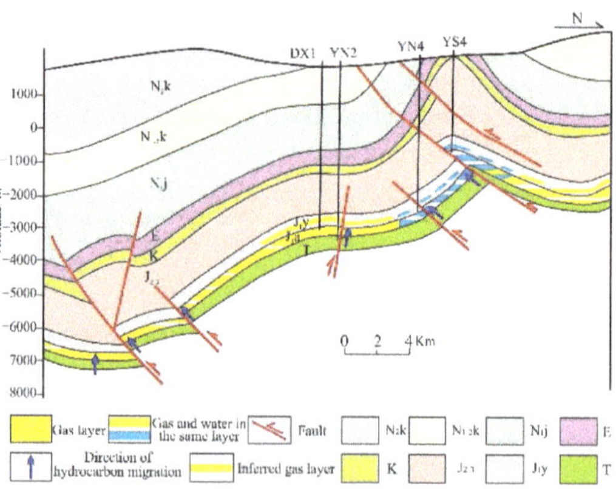

Figure 14. Comprehensive reservoir formation pattern in Dibei area (modified from Xiongqi Pang, 2019 [16] and Song Guo, 2018 [22]).

Based on the previous studies, the natural gas of the J_1a Formation mainly comes from the mixture of Jurassic and Triassic, and mainly from Jurassic [37,38]. Moreover, source rocks can provide a sufficient source for oil and gas accumulation [22].

Fractures can not only serve as good seepage channels but also improve the reservoir space of tight sandstone reservoirs [62,63]. The reservoir of the J_1a Formation in Dibei section of the Kuqa depression is tighter than most tight reservoirs (porosity < 14%, permeability < 0.1 mD) that produce gas in America [16], but the permeability of section with many fractures is generally high. Through the quantitative relationship between the gas production of drilled wells and the degree of fracture development, it can be found that there is a positive relationship between the degree of fracture development and tight gas production [64]; that is, the higher the degree of fracture development, the higher the tight gas production.

Fracture development areas are mainly distributed at the edge of the Dibei structural belt (Figure 15), while relatively few are in the middle [24,65]. Different from conventional reservoir forming conditions, tight sandstone gas reservoirs are less affected by faults and caprocks [16,22]. Diagenesis is widely developed in this area, which is consistent with the research result of previous scholars [66]. The results of this study and previous studies indicate that dissolution is well developed in the Dibei tectonic belt [24]. Coupling the effective gas intensity of source rock, the fracture development area of the J_1a Formation reservoir, and the favorable area of diagenesis, it is predicted that the sweet spot of tight gas of J_1a Formation in the Dibei area is in the southeast of YN 2 well (Figure 15).

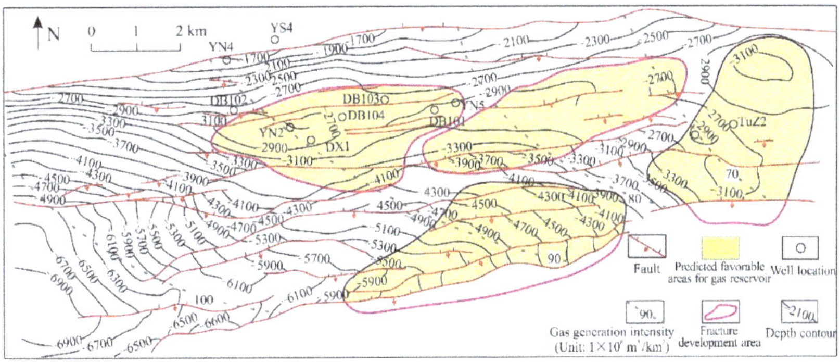

Figure 15. Sweet spots area for J_1a tight sandstone in the Dibei reservoir (the source rock data are from Caineng Zou, 2011 [37]; fracture data are modified from Hailiang Kang, 2016 [24]).

6. Conclusions

This study aimed to reveal the accumulation model of the Dibei gas reservoir in the Kuqa depression and predict the sweet spot area. By comprehensively using petrology, SEM, thermal burial history simulation, and fluid inclusion analysis, the reservoir physical properties, diagenesis, and accumulation mode of tight sandstone gas reservoirs were studied. This study provides a reference for the exploration and development of other areas in the Kuqa depression. The main conclusions are as follows: The tight sandstone of J_1a Formation in the Dibei gas field of Kuqa depression is mainly lithic arkose and feldspathic lithic sandstone, with moderate composition and moderate structural maturity. Secondary dissolution pores are a major part of the pore system. Four important diagenesis processes are determined. Compaction is the main diagenesis that causes porosity reduction, and the contact of suture particles is linear and evenly distributed. The types of cement mainly include carbonate, quartz, and clay cement. The main mineral for dissolution in the J_1a tight sandstone is feldspar. The tectonic movement has constructive and destructive effects on diagenetic evolution. Controlling factors to form good-quality tight sandstone include dissolution and overpressure, and dissolution is dominant. Based on the analysis of diagenetic evolution and reservoir controlling factors, the accumulation mode of the Dibei tight gas reservoir is proposed, and the sweet spot prediction is made according to the accumulation mode. The gas generation intensity of Jurassic source rocks, the fracture development area of the J_1a formation reservoir, and dissolution are the main factors controlling the distribution of sweet spots. It is pointed out that the sweet spot is mainly on the southeast edge of the Dibei gas reservoir. This model and sweet spot prediction method can be used for the exploration and deployment of other gas reservoirs in the Kuqa depression.

Author Contributions: Conceptualization, G.Z. and X.L.; methodology, M.L.; software, J.Z.; validation, M.L.; resources, C.D.; data curation, J.Z.; writing—original draft preparation, G.Z.; writing—review and editing, D.C.; funding acquisition, X.L. All authors have read and agreed to the published version of the manuscript.

Funding: This research was funded by the National Science and Technology Major Project of China (No. 2016ZX05007–003), Science and Technology Projects of PetroChina (No. T11083), the National Natural Science Foundation of China (No. U1810201), and the Fundamental Research Funds for the Central Universities (Nos. 2020YJSMT02, 2021YJSMT09).

Institutional Review Board Statement: Not applicable.

Informed Consent Statement: Not applicable.

Data Availability Statement: Not applicable.

Acknowledgments: We are indebted to Zhongyao Xiao (Tarim Oilfield Company, PetroChina) for his insightful suggestions and help in the manuscript.

Conflicts of Interest: The authors declare no conflict of interest.

References

1. Dai, J.; Ni, Y.; Qin, S.; Huang, S.; Peng, W.; Han, W. Geochemical characteristics of ultra-deep natural gas in the Sichuan Basin, SW China. *Pet. Explor. Dev.* **2018**, *45*, 619–628. [CrossRef]
2. Shanley, K.W.; Cluff, R.M.; Robinson, J.W.; Lantana, K.W.S. Factors controlling prolific gas production from low-permeability sandstone reservoirs: Implications for resource assessment, prospect development, and risk analysis. *AAPG Bull.* **2004**, *8*, 1083–1121. [CrossRef]
3. Law, B.E.; Curtis, J.B. Introduction to unconventional petroleum systems. *AAPG Bull.* **2002**, *86*, 1851–1852.
4. Alqubalee, A.; Abdullatif, A.; Babalola, L. Characteristics of Paleozoic tight gas sandstone reservoir: Integration of lithofacies, paleoenvironments, and spectral gamma-ray analyses, Rub' al Khali Basin, Saudi Arabia. *Arab. J. Geosci.* **2019**, *12*, 182742801. [CrossRef]
5. Zou, C.; Zhu, R.; Liu, K.; Ling, K.; Su, L.; Bai, B. Tight gas sandstone reservoirs in China: Characteristics and recognition criteria-ScienceDirect. *J. Pet. Sci. Eng.* **2012**, *88–89*, 82–91. [CrossRef]
6. Wüstefeld, P.; Hilse, U.; Lüders, V. Kilometer-scale fault-related thermal anomalies in tight gas sandstones. *Mar. Pet. Geol.* **2017**, *86*, 288–303. [CrossRef]
7. Stroker, T.M.; Harris, N.B.; Elliott, W. Diagenesis of a tight gas sand reservoir: Upper Cretaceous Mesaverde Group, Piceance Basin, Colorado. *Mar. Pet. Geol.* **2013**, *88–89*, 48–68. [CrossRef]
8. Chalmers, G.; Bustin, R.M. Geological evaluation of Halfway–Doig–Montney hybrid gas shale–tight gas reservoir, northeastern British Columbia. *Mar. Pet. Geol.* **2012**, *38*, 53–72. [CrossRef]
9. U.S. Energy Information Administration. Gas resources in Russia's tight sedimentary basins anil their future commercial development. *Gas Ind. Rus.* **2012**, *4*, 34–38.
10. Guo, X.B.; Huang, Z.L.; Zhao, L.B.; Han, W.; Ding, C.; Sun, X.W.; Yan, R.T.; Zhang, T.H.; Yang, X.J.; Wang, R.M. Pore structure and multi-fractal analysis of tight sandstone using MIP, NMR and NMRC methods: A case study from the Kuqa depression, China. *J. Pet. Sci. Eng.* **2019**, *178*, 544–558. [CrossRef]
11. Huyan, Y.Y.; Pang, X.Q.; Jiang, F.J.; Li, L.L.; Zheng, D.Y.; Shao, X.H. Coupling relationship between tight sandstone reservoir and gas charging: An example from lower Permian Taiyuan Formation in Kangning field, northeastern Ordos Basin, China. *Mar. Pet. Geol.* **2019**, *105*, 238–250. [CrossRef]
12. Wang, E.; Wang, Z.; Pang, X.; Zhang, Z.; Wang, Z.M.; Wu, Z.; Liang, Y.; Feng, Y.; Zhang, Z. Key factors controlling hydrocarbon enrichment in a deep petroleum system in a terrestrial rift basin—A case study of the uppermost member of the upper Paleogene Shahejie Formation, Nanpu Sag, Bohai Bay Basin, NE China. *Mar. Pet. Geol.* **2019**, *107*, 572–590. [CrossRef]
13. Zheng, D.Y.; Pang, X.Q.; Ma, X.H.; Li, C.R.; Zheng, T.Y.; Zhou, L.M. Hydrocarbon generation and expulsion characteristics of the source rocks in the third member of the Upper Triassic Xujiahe Formation and its effect on conventional and unconventional hydrocarbon resource potential in the Sichuan Basin. *Mar. Pet. Geol.* **2019**, *109*, 175–192. [CrossRef]
14. Xu, Z.; Jiang, S.; Liu, L.; Wu, K.; Li, R.; Liu, Z. Natural gas accumulation processes of tight sandstone reservoirs in deep formations of Songliao Basin, NE China. *J. Nat. Gas Sci. Eng.* **2020**, *83*, 103610. [CrossRef]
15. Chen, Q.; Deng, Y.; Wei, J.; Ma, G.; Long, L.; Xiao, W.; Li, W.; Zhang, L. Types, distribution and play targets of Lower Cretaceous tight oil in Jiuquan Basin, NW China. *Pet. Explor. Dev.* **2018**, *45*, 212–222. [CrossRef]

16. Pang, X.Q.; Peng, J.W.; Jiang, Z.X.; Yang, H.J.; Wang, P.W.; Jiang, F.J.; Wang, K. Hydrocarbon accumulation processes and mechanisms in Lower Jurassic tight sandstone reservoirs in the Kuqa subbasin, Tarim Basin, northwest China: A case study of the Dibei tight gas field. *Am. Assoc. Pet. Geol. Bull.* **2019**, *103*, 769–796. [CrossRef]
17. Guo, X.; Liu, K.; Jia, C.; Song, Y. Hydrocarbon accumulation processes in the Dabei tight-gas reservoirs, Kuqa Subbasin, Tarim Basin, northwest China. *Am. Assoc. Pet. Geol. Bull.* **2016**, *10*, 1501–1521. [CrossRef]
18. Lai, J.; Fan, X.; Pang, X.; Zhang, X.; Xiao, C.; Zhao, X.; Han, C.; Wang, G.; Qin, Z. Correlating diagenetic facies with well logs (conventional and image) in sandstones: The Eocene–Oligocene Suweiyi Formation in Dina 2 Gasfield, Kuqa depression of China. *J. Pet. Sci. Eng.* **2019**, *174*, 617–636. [CrossRef]
19. Sun, S.; Hou, G.T.; Zheng, C.F. Prediction of tensile fractures in KS2 trap, Kuqa depression, NW China. *Mar. Pet. Geol.* **2019**, *101*, 108–116. [CrossRef]
20. Lu, H.; Lu, X.; Fan, J.; Wang, X.; Fu, X.; Wei, H.; Zhang, B. The controlling effects of fractures on gas accumulation and production in tight sandstone: -a case of jurassic Dibei gas reservoir in the east Kuqa foreland basin. *J. Nat. Gas Geosci.* **2015**, *26*, 1047–1056.
21. Shi, C.; Li, Y.; Yuan, W.; Jiang, J.; Xie, Y.; Zhang, R.; Zhou, S. Characteristics on reservoir architecture and quality of tight sandstone reservoirs: Taking Jurassic Ahe formation in Dibei area of Kuga foreland basin as an example. *J. China Univ. Min. Technol.* **2021**, *50*, 877–892.
22. Guo, S.; Lyu, X.X.; Zhang, Y. Relationship between tight sandstone reservoir formation and hydrocarbon charging: A case study of a Jurassic reservoir in the eastern Kuqa depression, Tarim Basin, NW China. *J. Nat. Gas Sci. Eng.* **2018**, *52*, 304–316. [CrossRef]
23. Ju, W.; Wang, K.; Hou, G.; Sun, W.; Yu, X. Prediction of natural fractures in the Lower Jurassic Ahe Formation of the Dibei Gasfield, Kuqa depression, Tarim Basin, NW China. *Geosci. J.* **2018**, *22*, 241–252. [CrossRef]
24. Kang, H.; Lin, C.; Li, H.; Wang, K. Reservoir characteristics and favorable zone prediction of tight sandstone gas of the Ahe Formation in Yinan area Kuqa depression. *Pet. Geol. Exp.* **2016**, *38*, 162–169.
25. Ajdukiewicz, J.M.; Nicholson, P.H.; Esch, W.L. Prediction of deep reservoir quality using early diagenetic process models in the jurassic norphlet formation, Gulf of Mexico. *AAPG Bull.* **2010**, *94*, 1189–1227. [CrossRef]
26. Lai, J.; Wang, G.; Chai, Y.; Ran, Y.; Zhang, X. Depositional and diagenetic controls on pore structure of tight gas sandstone reservoirs: Evidence from lower Cretaceous Bashijiqike formation in Kelasu thrust belts, Kuqa Depression in Tarim Basin of West China. *Resour. Geol.* **2015**, *65*, 55–75. [CrossRef]
27. Okunuwadje, S.E.; Bowden, S.A.; Macdonald, D.I.M. Diagenesis and reservoir quality in high-resolution sandstone sequences: An example from the Middle Jurassic Ravenscar sandstones, Yorkshire CoastUK. *Mar. Pet. Geol.* **2020**, *118*, 104426. [CrossRef]
28. Rafaela, M.; Michael, S.; Turrero, M.J. Diagenetic processes influencing porosity in sandstones from the Triassic Buntsandstein of the Iberian Range, Spain. *Sediment. Geol.* **1996**, *105*, 203–219.
29. Su, N.; Song, F.; Qiu, L.; Zhang, W. Diagenetic evolution and densification mechanism of the Upper Paleozoic tight sandstones in the Ordos Basin, Northern China. *J. Asian Earth Sci.* **2021**, *205*, 104613. [CrossRef]
30. Du, Z.L.; Wang, F.Y.; Zhang, S.C.; Zhang, B.M.; Liang, D.G. Gas generation history of Mesozoic hydrocarbon kitchen in Kuqa depression, Tarim basin. *Geochimica* **2006**, *35*, 419–431.
31. Zhao, W.; Zhang, S.; Wang, F.; Cramer, B.; Chen, J.; Sun, Y.; Zhang, B.; Zhao, M. Gas systems in the Kuche Depression of the Tarim Basin: Source rock distributions, generation kinetics and gas accumulation history. *Org. Geochem.* **2005**, *36*, 1583–1601. [CrossRef]
32. Liang, D.G.; Zhang, S.C.; Chen, J.P.; Wang, F.Y.; Wang, P.R. Organic geochemistry of oil and gas in the Kuqa depression, Tarim Basin, NW China. *Org. Geochem.* **2003**, *34*, 873–888. [CrossRef]
33. Shen, Y.Q.; Lü, X.X.; Guo, S.; Song, X.; Zhao, J. Effective evaluation of gas migration in deep and ultra-deep tight sandstone reservoirs of Keshen structural belt, Kuqa depression. *J. Nat. Gas Sci. Eng.* **2017**, *46*, 119–131. [CrossRef]
34. Jia, C.Z. *Characteristics of Mesozoic and Cenozoic Structures and Petroleum Occurrence in the Tarim Basin*; Petroleum Industry Press: Beijing, China, 2004.
35. Jia, C.Z.; Li, Q.M. Petroleum geology of Kela-2, the most productive gas field in China. *Mar. Pet. Geol.* **2008**, *25*, 335–343.
36. Pang, X.Q.; Jiang, Z.X.; Jiang, F.J.; Huang, H.D.; Chen, D.X. Formation mechanisms, distribution models, and prediction of superimposed, continuous hydrocarbon reservoirs. *Acta Pet. Sin.* **2014**, *35*, 795–828.
37. Zou, C.N.; Jia, J.H.; Tao, S.Z.; Tao, X.W. Analysis of Reservoir Forming Conditions and Prediction of Continuous Tight Gas Reservoirs for the Deep Jurassic in the Eastern Kuqa Depression, Tarim Basin. *Acta Geol. Sin.* **2011**, *85*, 1173–1186. [CrossRef]
38. Zhao, M.J.; Zhang, B.M. Source rock conditions for the formation of large gas field in Kuqa foreland depression. *Chin. J. Geol.* **2002**, *37*, 35–44.
39. Wang, R.; Shen, P.; Zhao, L. Diagenesis of deep sandstone reservoirs and a quantitative model of porosity evolution: Taking the third member of Shahejie Formation in the Wendong Oilfield, Dongpu Sag, as an example. *Pet. Explor. Dev.* **2011**, *38*, 552–559. [CrossRef]
40. Folk, R.L. *Petrology of Sedimentary Rocks*; Hemphill Publishing: Austin, TX, USA, 1980.
41. Rezaee, R.; Saeedi, A.; Clennell, B. Tight gas sands permeability estimation from mercury injection capillary pressure and nuclear magnetic resonance data. *J. Pet. Sci. Eng.* **2012**, *88–89*, 92–99. [CrossRef]
42. Bjørlykke, K.; Høeg, K. Effects of burial diagenesis on stresses, compaction and fluid flow in sedimentary basins. *Mar. Pet. Geol.* **1997**, *14*, 267–276. [CrossRef]
43. Liu, L.; Li, Y.; Dong, H.; Sun, Z. Diagenesis and reservoir quality of Paleocene tight sandstones, Lishui Sag, East China Sea Shelf Basin. *J. Pet. Sci. Eng.* **2020**, *195*, 107615. [CrossRef]

44. Bjørlykke, K.; Jahren, J. *Sandstones and Sandstone Reservoirs. Petroleum Geoscience: From Sedimentary Environments to Rock Physics*, 2nd ed.; Springer: Berlin/Heidelberg, Germany, 2015.
45. Dou, W.; Liu, L.; Wu, K.; Xu, Z.; Feng, X. Origin and significance of secondary porosity: A case study of upper Triassic tight sandstones of Yanchang Formation in Ordos basin, China. *J. Pet. Sci. Eng.* **2017**, *149*, 485–496. [CrossRef]
46. Zhang, Y.; Tian, J.; Zhang, X.; Li, J.; Liang, Q.; Zheng, X. Diagenesis Evolution and Pore Types in Tight Sandstone of Shanxi Formation Reservoir in Hangjinqi Area, Ordos Basin, Northern China. *Energies* **2022**, *15*, 470. [CrossRef]
47. Higgs, K.E.; Zwingmann, H.; Reyes, A.G.; Funnell, R.H. Diagenesis, porosity evolution, and petroleum emplacement in tight gas reservoirs, Taranaki Basin, New Zealand. *J. Sediment. Res.* **2007**, *77*, 1003–1025. [CrossRef]
48. Oluwadebi, A.G.; Taylor, K.G.; Dowey, P.J. Diagenetic controls on the reservoir quality of the tight gas Collyhurst Sandstone Formation, Lower Permian, East Irish Sea Basin, United Kingdom. *Sediment. Geol.* **2018**, *371*, 55–74. [CrossRef]
49. Busch, B.; Becker, I.; Koehrer, B.; Adelmann, D.; Hilgers, C. Porosity evolution of two Upper Carboniferous tight-gas-fluvial sandstone reservoirs: Impact of fractures and total cement volumes on reservoir quality. *Mar. Pet. Geol.* **2019**, *100*, 376–390. [CrossRef]
50. Wu, H.; Zhao, J.; Wu, W.; Li, J.; Huang, Y.; Chen, M. Formation and diagenetic characteristics of tight sandstones in closed to semi-closed systems: Typical example from the Permian Sulige gas field. *J. Pet. Sci. Eng.* **2021**, *199*, 108248. [CrossRef]
51. Luo, L.; Meng, W.; Gluyas, J.; Tan, X.; Gao, X.; Feng, M.; Kong, X.; Shao, H. Diagenetic characteristics, evolution, controlling factors of diagenetic system and their impacts on reservoir quality in tight deltaic sandstones: Typical example from the Xujiahe Formation in Western Sichuan Foreland Basin, SW China. *Mar. Pet. Geol.* **2019**, *103*, 231–254. [CrossRef]
52. Sun, D.; Liu, X.; Li, W.; Lu, S.; He, T.; Zhu, P.; Zhao, H. Quantitative evaluation the physical properties evolution of sandstone reservoirs constrained by burial and thermal evolution reconstruction: A case study from the Lower Cretaceous Baxigai Formation of the western Yingmaili Area in the Tabei Uplift, Tarim. *J. Pet. Sci. Eng.* **2022**, *208*, 109460. [CrossRef]
53. Zhang, N.; Liu, L.; Su, T.; Dai, Q.; Zhao, Y. Reservoir characteristics and main controlling factors of the lower Jurassic tight sandstone in eastern Kuga Depression. *Acta Sedimentol. Sin.* **2015**, *33*, 160–169.
54. Liu, G.; Zeng, L.; Zhu, R.; Gong, L.; Ostadhassan, M.; Mao, Z. Effective fractures and their contribution to the reservoirs in deep tight sandstones in the Kuqa Depression, Tarim Basin, China. *Mar. Pet. Geol.* **2020**, *124*, 104824. [CrossRef]
55. Yuan, D.; Zhang, Z.; Liu, H. Effect of hydrocarbon charging on cementation of late carbonate minerals. *J. Oil Gas Technol.* **2005**, *27*, 298–300.
56. Tobin, R.C.; McClain, T.; Lieber, R.B.; Ozkan, A.; Banfield, L.A.; Marchand, A.M.E.; McRae, L.E. Reservoir quality modeling of tight-gas sands in Wamsutter field: Integration of diagenesis, petroleum systems, and production data. *Am. Assoc. Pet. Geol. Bull.* **2010**, *94*, 1229–1266. [CrossRef]
57. Xu, Z.; Liu, L.; Wang, T.; Gao, X.; Dou, W.; Xiao, F.; Zhang, N.; Song, X.; Ji, H.; Xu, Z.; et al. Application of fluid inclusions to the charging process of the lacustrine tight oil reservoir in the Triassic Yanchang Formation in the Ordos Basin, China. *J. Pet. Sci. Eng.* **2017**, *149*, 40–55. [CrossRef]
58. Duan, W.; Li, C.F.; Luo, C.; Chen, X.G.; Bao, X. Effect of formation overpressure on the reservoir diagenesis and its petroleum geological significance for the DF11 block of the Yinggehai Basin, the South China Sea. *Mar. Pet. Geol.* **2018**, *97*, 49–65. [CrossRef]
59. Yang, T.; Cao, Y.; Friis, H.; Liu, K.; Wang, Y.; Zhou, L.; Yuan, G.; Xi, K.; Zhang, S. Diagenesis and reservoir quality of lacustrine deep-water gravity-flow sandstones in the Eocene Shahejie Formation in the Dongying sag, Jiyang depression, eastern China. *Am. Assoc. Pet. Geol. Bull.* **2020**, *104*, 1045–1073. [CrossRef]
60. Wang, X.; Wei, H.; Shi, W.; Wang, Y. Formation pressure characteristics and hydrocarbon accumulation in eastern Kuqa Depression. *Geol. Sci. Technol. Inf.* **2016**, *35*, 68–73.
61. Ju, Y.; Sun, X.; Liu, L.; Xie, Y.; Wei, H. Characteristics of Jurassic Tight Sandstone Gas Reservoir in Dibei Area of Kuqa Depression, Tarim Basin. *Xinjiang Pet. Geol.* **2014**, *35*, 264–267.
62. Yue, D.; Wu, S.; Xu, Z.; Xiong, L.; Chen, D.; Ji, Y.; Zhou, Y. Reservoir quality, natural fractures, and gas productivity of upper Triassic Xujiahe tight gas sandstones in western Sichuan Basin, China. *Mar. Pet. Geol.* **2018**, *89*, 370–386. [CrossRef]
63. Tokan-Lawal, A.; Prodanovic, M.; Eichhubl, P. Investigating flow properties of partially cemented fractures in Travis Peak Formation using image-based pore-scale modeling. *J. Geophys. Res. Solid Earth.* **2015**, *120*, 5453–5466. [CrossRef]
64. Lu, X.; Zhao, M.; Liu, K.; Zhuo, Q.; Fan, J.; Yu, Z.; Gong, Y. Forming condition and mechanism of highly effective deep tight sandstone gas reservoir in Kuqa foreland basin. *Acta Pet. Sin.* **2018**, *39*, 365–378.
65. Li, J.; Wang, C.; Li, J.; Ma, W.; Zhang, H.; Lu, Y.; Li, D.; Liu, M. Source and exploration direction of tight oil and gas in the Dibei section of northern Kuqa Depression. *China Pet. Explor.* **2019**, *24*, 485–497.
66. Wei, G.; Zhang, R.; Zhi, F.; Wang, K.; Yu, C.; Dong, C. Formation conditions and exploration directions of Mesozoic structural lithologic stratigraphic reservoirs in the eastern Kuqa Depression. *Acta Pet. Sin.* **2021**, *42*, 1113–1125.

energies
MDPI

Article

Parameter Effect Analysis of Non-Darcy Flow and a Method for Choosing a Fluid Flow Equation in Fractured Karstic Carbonate Reservoirs

Yueying Wang *, Jun Yao and Zhaoqin Huang

School of Petroleum Engineering, China University of Petroleum (East China), Qingdao 266580, China; rcogfr_upc@126.com (J.Y.); huangzhqin@upc.edu.cn (Z.H.)
* Correspondence: 20070095@upc.edu.cn

Abstract: Fractured karstic carbonate reservoirs have obvious multi-scale characteristics and severe heterogeneity due to the development of abundant karst caves and fractures with different scales. Darcy and non-Darcy flows coexist due to this property. Therefore, selecting the appropriate flow equations for different regions in the numerical simulation of fluid flows, particularly two-phase and multiphase flows, is a critical topic. This paper compares and analyses the displacement distance differences of waterfront travel using the Darcy, Forchheimer and Barree–Conway equations, as well as analyzes the influence of the Forchheimer constant, fluid viscosity, flow rate and absolute permeability on inertia action based on the Buckley–Leverett theory. The results show that the Forchheimer number/Reynolds number of water/oil two-phase flow is not a constant value and varies with water saturation, making it difficult to determine whether the inertial action should be considered solely based on these values; the influence of inertial action can be measured well by comparing the difference between the displacement distances of the waterflood front, and the quantitative standard is given for the selection of the flow equation of different regions by calculating the allowable error of the displacement distance of the waterflood front. The magnitude of the inertial effect is affected by the physical properties of the fluid and reservoir medium and the fluid velocity. The smaller the difference in the viscosity of the oil/water fluid, the smaller the inertial effect is. This technique was used a preliminary attempt to analyze the fractured karstic carbonate reservoirs at Tarim, and the results confirmed the validity of the method described in this article.

Keywords: fractured karstic carbonate reservoir; non-Darcy flow; inertial effect; Forchheimer number; Reynolds number; distance deviation of waterflood front

Citation: Wang, Y.; Yao, J.; Huang, Z. Parameter Effect Analysis of Non-Darcy Flow and a Method for Choosing a Fluid Flow Equation in Fractured Karstic Carbonate Reservoirs. *Energies* **2022**, *15*, 3623. https://doi.org/10.3390/en15103623

Academic Editors: Mofazzal Hossain, Md Motiur Rahman, Hisham Khaled Ben Mahmud and Vamegh Rasouli

Received: 26 March 2022
Accepted: 11 May 2022
Published: 15 May 2022

1. Introduction

Carbonate reservoirs are an important type of reservoir that contain 60% of the world's total oil and gas reserves [1]; however, they differ from conventional reservoirs. Their storage space includes primary porosity, fractures, and karst caves; hence, their porosity scale ranges from millimeter to meter levels. Therefore, two types of flow regimes are present in fractured karst carbonate reservoirs: Darcy flow and non-Darcy flow. One group of researchers used reduced-order models and machine learning to forecast the production of flow by building a high-fidelity DFN model in an unconventional reservoir with fractures [2]. Furthermore, another study proposed the Darcy/non-Darcy coupled model for fluid flow where Darcy and non-Darcy flow coexist [3] (Yao, 2018). Thus, the choice of flow equations according to the fluid flow regimes is an essential part of reservoir simulations. At present, researchers depend mainly on qualitative analysis or the Reynolds number to distinguish the flowing regimes and flow equations. For example, non-Darcy flow usually occurs in near-wellbore regions, large-scale fractures, large karst caves and unconsolidated areas, and at Re = 10–1000. These expressions are qualitative or vague rather than quantitative. There is no specific definition or criteria for predicting non-Darcy flow in fractured karstic carbonate reservoirs; moreover, we do not know the accurate range of the near-wellbore region for non-Darcy flow at a given production rate, or under what conditions non-Darcy flow occurs for the given fractured karstic regions.

Researchers have set up several different fluid flow equations to describe Darcy and non-Darcy flows, including the Darcy, Forchheimer and Barree–Conway equations. The Darcy equations [4] (1856) are widely accepted to describe the behavior of linear fluids, while Forchheimer [5] (1901) proposed the Forchheimer equation to describe non-Darcy flow. Researchers [6–10] (Carman, 1974; Fand, 1987; Montillet, 2004) have found that the Forchheimer equation is unsuitable for high-velocity flow. Barree and Conway [11,12] (2004) proposed the Barree–Conway equation, which is a more general equation for non-linear flow. The Barree–Conway equation has a similar expression as the Darcy equation if the pressure gradient is expressed in terms of the apparent permeability. The Barree–Conway equation describes the entire range of flow velocities versus pressure gradient from Darcy to non-Darcy flow regimes. Since fluids change from Darcy flow to non-Darcy flow as the velocity increases, these flow regimes are located in different regions of the same reservoirs. Thus, how to distinguish these two types of flow regimes has become a significant issue.

Until now, two criteria have been widely used to identify the transition to non-Darcy flow from Darcy flow, namely the Reynolds number and the Forchheimer number. Chilton [13] (1931) obtained and refined the Reynolds number equation through fluid flow experiments with packed particles based on the turbulent flow equation according to the previous belief that non-Darcy flow was similar to turbulent flow. Owing to the difficulty of determining the particle diameter, Green [14] (1951) used the permeability (k) and non-Darcy coefficient (β) instead of the particle diameter. Ma [15] (1993) redefined the Reynolds number (Re) and gave a new criterion, the Forchheimer number (Fo). Not only are the definitions of the Reynolds number and the Forchheimer number inconsistent, but the critical values obtained in experiments are different for non-Darcy flow in porous media [15–19] recommended the Forchheimer number as a criterion for identifying non-Darcy flow due to its advantages of clear definition and wide applicability; the Forchheimer number is the ratio of the liquid–solid interaction pressure gradient to that of viscous resistance. The Reynolds number and Forchheimer number are both used to distinguish between Darcy flow and non-Darcy flow. However, there are still a few problems and obstacles. Furthermore, the critical values become more complex for multiphase flow.

It follows that the Darcy and non-Darcy flows are different flow regimes; however, they usually coexist when fluids flow in the same fractured karstic carbonate reservoir. For multiphase flow in the same fractured karstic reservoir, the non-Darcy equations are more complex than the Darcy equations. The Forchheimer coefficient in the Forchheimer equation is no longer a constant and instead varies with the parameters, which are related to the rock's structure and the fluids' properties, the denominator of the Barree–Conway equation becomes very complicated because the expression of apparent permeability is extremely complex. Thus, it not only requires considerable computation to implement the numerical simulation only using the Forchheimer or Barree–Conway equation for the whole reservoir, but also necessitates a greater effort to achieve stability in the numerical simulation. Therefore, the coupled model is a good choice [3] (Yao, 2018). Furthermore, most commercial software is developed based on the Darcy equations, and consequently there will be deviation in some cases with Darcy and non-Darcy flow coexistence. Thus, the following two problems arise: first, distinguishing Darcy flow and non-Darcy flow for choosing the flow equations, and second, evaluating and correcting the deviation in the numerical simulation using the Darcy equation to describe non-Darcy flow. Therefore, it is necessary to establish a method for evaluating the inertial effect and choosing the flow equations.

In this paper, the evolution of the non-Darcy flow mechanism and equations are reviewed. Then, analytical solutions are used to analyze the flow characteristics, compare the difference between Darcy flow and non-Darcy flow and analyze the effect of the parameters. Finally, a method is obtained for choosing the flow equations and estimating the displacement distance deviation of a waterflood front using the Darcy equation to describe non-Darcy flow.

2. Method of Comparing Darcy Flow and Non-Darcy Flows

2.1. Non-Darcy Flow Mechanism and Equations

Through physical experiments, researchers have found that after velocity increases to a certain value, the pressure gradient is no longer proportional to the flow velocity (Figure 1). The physical understanding of non-Darcy flow remains unclear, and there are two main opinions on it. One is that the non-Darcy effect is owing to turbulence, and the other is that inertial force causes the non-Darcy effect. Most researchers advocate for the second opinion, and an equation has been proposed to describe the additional pressure drop observed in non-Darcy flow at a high flow velocity. Bear [20] (1972) systematically gave three reasons to exclude turbulence as the cause for the non-Darcy effect. First, the linear term in the equation does not exist in turbulent flow; second, the transition from laminar to turbulent flow is rather sharp; and third, the critical Reynolds number (R_e) at which the transition starts is several orders of magnitude higher than that at which the non-Darcy effect begins. Moreover, it was concluded that the non-Darcy effect occurs because microscopic inertial effects alter the velocity and pressure fields.

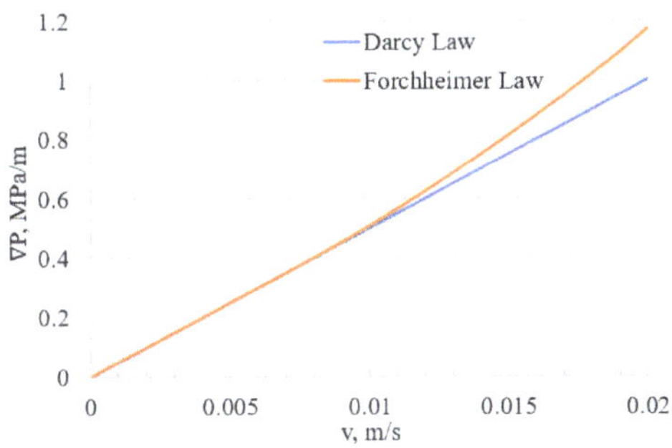

Figure 1. Relationship between pressure gradient and velocity.

The Forchheimer equation and Barree–Conway equation are the two main equations to describe non-Darcy flow. Although the equations are different from the Darcy equation in form, they have something in common. To some extent, the Forchheimer equation evolved from the Darcy equation according to the non-Darcy flow mechanism by adding an extra term; the Barree–Conway equation has a similar expression to the Darcy equation. The Forchheimer equation is widely accepted as follows:

$$-\nabla\phi = \frac{\mu v}{k} + \beta\rho v|v| \tag{1a}$$

$$F_0 = \frac{k\beta\rho v}{\mu} \tag{1b}$$

in which ϕ is the potential; μ is the viscosity; v is the superficial velocity; k is the absolute permeability; β is the Forchheimer coefficient; ρ is the fluid density; and F_0 is the Forchheimer number. On the right-hand side of the equation, the first term is the viscous term and the second term is the inertial term. When the viscous flow is at high rates, the second term becomes significant. The inertial term causes the differences in the flow behaviors and characteristics between Darcy flow and non-Darcy flow, and the value of β is substantial for the inertial effect. Usually, β is predicted using empirical formulas, and there are many different equations concerning the calculation of β. In this paper, we select one of them, which is based on an experimental study of single-phase flow.

$$\beta = \frac{C_\beta}{k^{5/4}\phi^{3/4}} \tag{2}$$

The Forchheimer equation has a limited range of applicability for high-rate flow conditions; thus, the Barree–Conway equation [11,12] (Barree and Conway, 2004) was presented to describe the relationship between rate and potential gradient for the flow of the entire range of the Reynolds number:

$$-\nabla\phi = \frac{\mu v}{k_{app}} \tag{3a}$$

$$k_{app} = k\left(k_{mr} + \frac{1-k_{mr}}{(1+R_e)}\right) \tag{3b}$$

$$k_{mr} = \frac{k_{min}}{k} \tag{3c}$$

$$R_e = \frac{\rho v}{\mu\tau} \tag{3d}$$

where $k_{\rm app}$ is apparent permeability; $R_{\rm e}$ is the Reynolds number; τ is the inverse of the characteristic length, which is related to the mean particle size of sand; k_{min} is the minimum permeability; and k_{mr} is the ratio of minimum permeability to Darcy permeability. T and k_{mr} are the key parameters in the equation, and τ is related to the permeability k and the mean size of the particles [11,12] (Barree, Conway, 2004). Usually, τ is a given value according to the particle size (Figure 2). The Reynolds number of Barree–Conway equation depends on the flow velocity and the inverse of the characteristic length τ. Theoretically, when $k_{\rm mr}$ is equal to 1, the Barree–Conway equation is reduced to the Darcy equation with constant permeability k; when k_{mr} is equal to zero, it converges to the Forchheimer equation. When the value of k_{mr} is set to zero, the Barree–Conway equation reduces to the form of the Forchheimer equation and we can obtain the relationship between these two equations and the relationship between τ and β for the single-phase.

$$\tau = \frac{1}{k\beta} \tag{4}$$

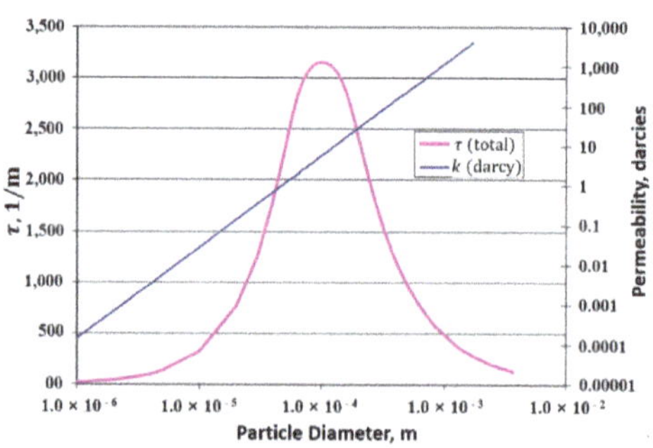

Figure 2. τ and permeability related to mean particle size (Barree, 2004) [11].

The main difference between the Forchheimer equation and the Barree-Conway equation is the different values of their parameters.

Darcy and non-Darcy flow are different but sometimes coexist underground, especially in complex fractured karstic carbonate media. The Forchheimer or Barree-Conway equations are used in the numerical simulation regardless of the magnitude of the non-Darcy effect, which will take large amounts of internal storage and calculating time. Similarly, the Darcy equation is not valid for high velocities because the inertial effect is ignored. The criteria for distinguishing the flow regimes of multiphase flow are more complex than those for single-phase flow. Establishing a method for evaluating the difference between them is necessary.

2.2. Analytical Solution of Darcy and Non-Darcy Equations

In an isothermal system, the mass-balance equations are as follows:

$$\frac{\partial}{\partial t}(\phi S_\alpha \rho_\alpha) = -\nabla \cdot (\rho_\alpha v_\alpha) + q_\alpha \ ; \ \alpha = w, o \tag{5}$$

where ϕ is the porosity of the medium, $(-)$; S_α is the saturation of fluid α, $(-)$; ρ_α is the density of fluid α $(\mathrm{kg/m^3})$; v_α is the volumetric flow rate of fluid α, $(\mathrm{m/s})$; q_α is the mass source/sink term for fluid α $(\mathrm{m^3/s})$. The Forchheimer equation and the Barree–Conway equation can be extended for multiphase flow:

$$-\nabla P_\alpha = \frac{\mu_\alpha}{kk_{r\alpha}} v_\alpha + \beta_\alpha \rho_\alpha v_\alpha |v_\alpha|; \ \alpha = w, o \tag{6a}$$

$$\beta_\alpha = \frac{C_\beta}{(kk_{r\alpha})^{5/4}[\phi(S_\alpha - S_{\alpha r})]^{3/4}}; \ \alpha = w, o \tag{6b}$$

$$-\nabla P_\alpha = \frac{\mu_\alpha v_\alpha}{kk_{r\alpha}\left[k_{mr} + \frac{(1-k_{mr})\mu_\alpha \tau}{\mu_\alpha \tau + \rho_\alpha |v_\alpha|}\right]}; \ \alpha = w, o \tag{6c}$$

$$k_{mr} = \frac{k_{min}}{k} \tag{6d}$$

Two additional equations are required:

$$\sum S_\alpha = 1, \ \alpha = w, \ o, \tag{7a}$$

$$P_w = P_o - P_c \tag{7b}$$

where P_c is the water-oil capillary pressure (Pa).

In an analytical solution of multiphase flow, the saturation profile can be obtained by using the method of characteristics [21,22] (Wu 1991; Wu 2001). The Buckley–Leverett flow conditions are assumed as follows: the multiphase fluids and porous media are incompressible; the capillary pressure gradient and gravity are ignored; the model is one-dimensional; and the fluid flows along the x-coordinate of a semi-infinite linear flow system. Based on these assumptions, acceptable approximations can be obtained. For one-dimensional flow, v_α can be expressed using the Forchheimer equation and Barree–Conway equation as follows:

$$v_\alpha = \frac{1}{2k\beta_\alpha \rho_\alpha}\left[-\frac{\mu_\alpha}{k_{r\alpha}} + \sqrt{\left(\frac{\mu_\alpha}{k_{r\alpha}}\right)^2 - 4k\beta_\alpha \rho_\alpha \frac{\partial P_\alpha}{\partial x}}\right]; \ \alpha = w, o \tag{8a}$$

$$v_\alpha = -\frac{1}{2\mu_\alpha \rho_\alpha}\left(\mu_\alpha^2 S_\alpha \tau + k_d k_{r\alpha} k_{mr} \rho_\alpha \frac{\partial P_\alpha}{\partial x}\right) + \frac{1}{2\mu_\alpha \rho_\alpha}\sqrt{\left(\mu_\alpha^2 S_\alpha \tau + k_d k_{r\alpha} k_{mr} \rho_\alpha \frac{\partial P_\alpha}{\partial x}\right)^2 - 4\rho_\alpha k_d k_{r\alpha} k_{mr} \mu_\alpha^2 S_\alpha \tau \frac{\partial P_\alpha}{\partial x}}; \ \alpha = w, o \tag{8b}$$

where $\frac{\partial P_\alpha}{\partial x}$ is along the x-coordinate. Then, the mass balance equation can be changed as follows:

$$-\frac{\partial v_\alpha}{\partial x} = \phi \frac{\partial S_\alpha}{\partial t}; \ \alpha = w, o \tag{9}$$

The fractional flow function for the wetting phase may be expressed in the following form:

$$f_w = \frac{v_w}{v_w + v_o} \tag{10}$$

$$f_o + f_w = 1 \tag{11}$$

Under the condition of the same pressure gradient, the inertial effect will affect the flow velocity of the fluid; hence, the displacement distances of waterflood fronts are different for the same volume of water injected. For the same fractured karstic media, the displacement distance calculated using the Darcy equation ignoring inertia is different from that using the non-Darcy equation; thus, there will be a displacement distance deviation of the waterflood front for non-Darcy flow when using the Darcy equation to simulate the fluid flow. In this paper, we introduce the distance deviation of the waterflood front to evaluate the difference between Darcy flow and non-Darcy flow:

$$Dev = \frac{L_D - L_{nD}}{L_D} \times 100\% \tag{12}$$

where Dev is the distance deviation of the waterflood front (%); L_D is the distance of the waterflood front based on the Darcy equation (m); and L_{nD} is the distance of the waterflood front based on the non-Darcy equation (m).

Assuming that the physical flow model is a one-dimensional linear porous medium, and at first uniformly saturated with a nonwetting fluid (S_o = 0.8) and a wetting fluid (S_{rw} = 0.2), the volumetric injection rate is constant with the wetting fluid and imposed at the inlet (x = 0) starting from t = 0. The relative permeability can be estimated using the Brooks–Corey model [23] (Brooks and Corey, 1964). In this paper, all the fluids and formation parameters of the physical flow model are summarized in Table 1.

Table 1. Parameters for physical flow model.

Parameters	Value	Unit
Porosity, ϕ	0.30	
Absolute permeability, k	9.869×10^{-13}	m^2
Cross section area, A	1.0	m^2
Length of model, L	5.0	m
Injection rate, q_t	5.0×10^{-4}	m^3/s
Water viscosity, μ_w	1.0×10^{-3}	Pa·s
Oil viscosity, μ_o	5.0×10^{-3}	Pa·s
Residual water saturation, S_{wr}	0.2	
Residual oil saturation, S_{or}	0.2	
Maximum relative permeability, $k_{rw,max}$	0.8	
Maximum relative permeability, $k_{ro,max}$	0.8	
Power index of relative permeability, n_w	2	
Power index of relative permeability, n_o	2	
Density of water, ρ_w	1000	kg/m^3
Density of oil, ρ_o	800	kg/m^3
Forchheimer flow constant, C_β	3.2×10^{-7}	$m^{3/2}$
Minimum permeability fraction, k_{mr}	0.01	
Inverse of characteristic length, τ	2.1×10^3	m^{-1}

Figure 3 shows the difference in the displacement distance of waterfront travel, water fractional flow, and flow velocities among the Darcy, Forchheimer, and Barree–Conway equations (τ = 2100, Figure 2) after the same injection volume of water. The solid lines represent the results of three equations and the dotted lines represent the results of modelling the Darcy and Forchheimer equations using the Barree–Conway equation by changing the values of k_{mr} and τ. In the water saturation profiles (Figure 3a), the result of the Barree–Conway equation is located between those of the Darcy and Forchheimer equations using the parameters in Table 1. The Barree–Conway equation will fit well with the result of the Darcy equation when k_{mr} is set to 1. However, when using the Barree–Conway equation to fit the result of the Forchheimer equation, k_{mr} is set to zero, and τ must be set to 700. In any case, the Barree–Conway equation fits the results of the Darcy and Forchheimer equations well in the water saturation profile. Comparing the flow velocities and the fractional flow (Figure 3b,c), when k_{mr} is set to 1, the results using the Barree–Conway equations coincide with those of the Darcy equation; however, the values using the Barree–Conway equation are inconsistent with those of Forchheimer equation when k_{mr} is set to zero and τ is set to 700. The reason for this difference is that the parameter values are different; τ is constantly dependent on the particle size and is equal to 700 in the Barree–Conway equation, but β_α varies with the parameters of the fluids and rock in the Forchheimer equation for two-phase flow. The variation of the Forchheimer/Reynolds number with the water saturation is shown in Figure 3d. Although changing the τ value of the Barree–Conway equation yields the same result in the water saturation profile as the Forchheimer equation, the Reynolds number differs from the Forchheimer number. The Reynolds numbers are the same when k_{mr} is equal to 0.01 and k_{mr} is equal to 0.0, because τ is equal to 2100.

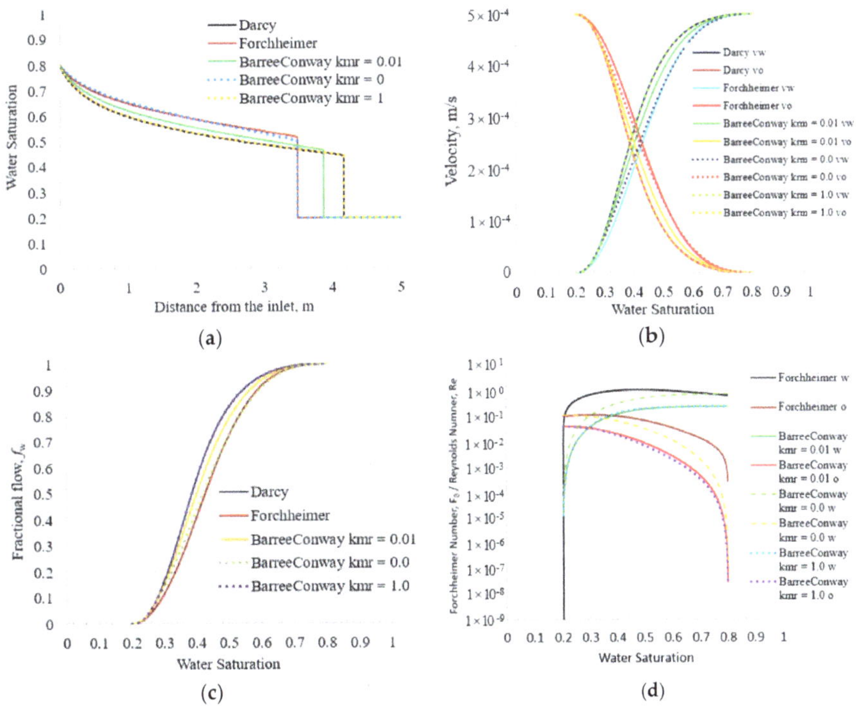

Figure 3. Comparison of Darcy, Forchheimer and Barree–Conway equations. (**a**) Water saturation profiles. (**b**) Velocity. (**c**) Water fractional flow curves. (**d**) Forchheimer number of Forchheimer equation and Reynolds number of Barree–Conway equation.

In summary, non-Darcy flow is more complex than Darcy flow for multiphase flow, and the Forchheimer and Reynolds numbers are no longer constant and vary with the water saturation. The results of the three equations depend on the values of the parameters. Even if the water saturation of the Forchheimer equation and the Barree–Conway equation are the same, the values of flow velocity, Forchheimer number, and Reynolds number are different. The Forchheimer number has a definite physical meaning and represents the ratio of the viscous term to the inertial term. The water saturation profile of displacement can describe the deviation of non-Darcy flow from Darcy flow with respect to the distance of waterfront travel and water saturation.

3. Comparison of Darcy and Non-Darcy Equations Based on Analytical Solution

3.1. Comparison of Darcy and Forchheimer Equations

The inertial term in the Forchheimer equation contains the following parameters: flow velocity, Forchheimer constant C_β, porosity, absolute permeability, relative permeability and fluid saturation. The difference between the Forchheimer equation and the Darcy equation will now be compared and analyzed under the influence of the different parameters for two-phase flow. During the comparisons of the key parameters, the other parameters are given the values shown in Table 1.

3.1.1. Forchheimer Constant C_β

Here, C_β is an important parameter for β_α and is related to the structure of the rock. Usually, the value of C_β is obtained through a physical experiment on rock cores. According to experimental tests, the value of C_β varies from 10^{-9} to 10^{-7} $m^{3/2}$. In this section, we will analyze the non-Darcy effect of rocks with a different structure using several different values of C_β from 3.2×10^{-9} to 3.2×10^{-7} $m^{3/2}$.

Figure 4 shows that the value of β_w changes with C_β and S_w for the same porous media. The β_w value changes from a very large value at the minimum water saturation to a minimum value at the maximum water saturation; conversely, the β_o value changes from a minimum value at low water saturation to a very large value at the maximum water saturation. Thus, the value of β_α is no longer constant and varies with the water saturation and C_β for multiphase fluids. The inertial force effect is largest when the oil begins to be displaced by the water and the inertial effect is especially substantial for very large or infinite at the beginning or end of the water saturation.

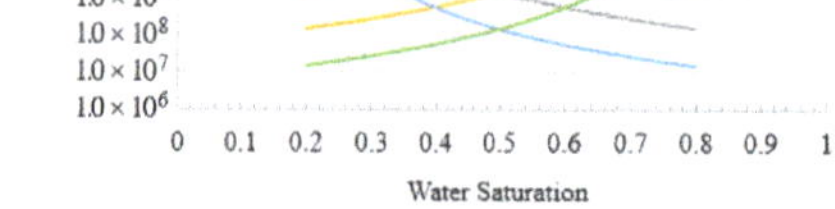

Figure 4. Forchheimer coefficient curves.

Figure 5 shows a comparison of the results of the Darcy equation and Forchheimer equation after the same volume of water is injected. The results of water saturation (Figure 5a) are different for the two-phase flow. The larger value of C_β displays a better sweeping efficiency and shorter front travel distance. When the value of C_β decreases, the line in the water saturation profile tends to be close to that of Darcy flow. When C_β is reduced to 3.2×10^{-9} $m^{3/2}$, the displacement saturation distributions of the Forchheimer equation are close to that of Darcy equation, which means the inertial force effect can be ignored. Figure 5b) displays the values of the Forchheimer number F_0 with the different water saturation and C_β values. Here, F_0 is also no longer constant and varies with the water saturation S_w and C_β. When C_β is equal to 3.2×10^{-7} $m^{3/2}$, the F_0 values of the water phase are larger than the others; the maximum is up to more than 1.00, and most values are between 0.10 and 1.10. The F_0 values of the oil phase are between 0.01 and 0.11, though they are smaller than those of the water phase. This means that the inertial force effect is obvious and cannot be ignored. When C_β equals 3.2×10^{-9} $m^{3/2}$, the F_0 values of the water phase are 0.013–0.006, and those of the oil phase are much smaller than 0.001. According to Zeng's [24] (2006) results, the fluid flow belongs to Darcy flow, and the inertial force effect is negligible. The velocities of the fluids and the fractional flow curves all vary with the value of C_β (Figure 5e,d). When the value of C_β equals 3.2×10^{-9} $m^{3/2}$, the velocities and fractional flow curves nearly coincide with those of Darcy flow. The distance deviation of the waterflood front is also shown with a different Forchheimer constant C_β (Figure 5e). When the value of C_β is less than 5.0×10^{-8}, the distance deviation of the waterflood front is less than 5.0%.

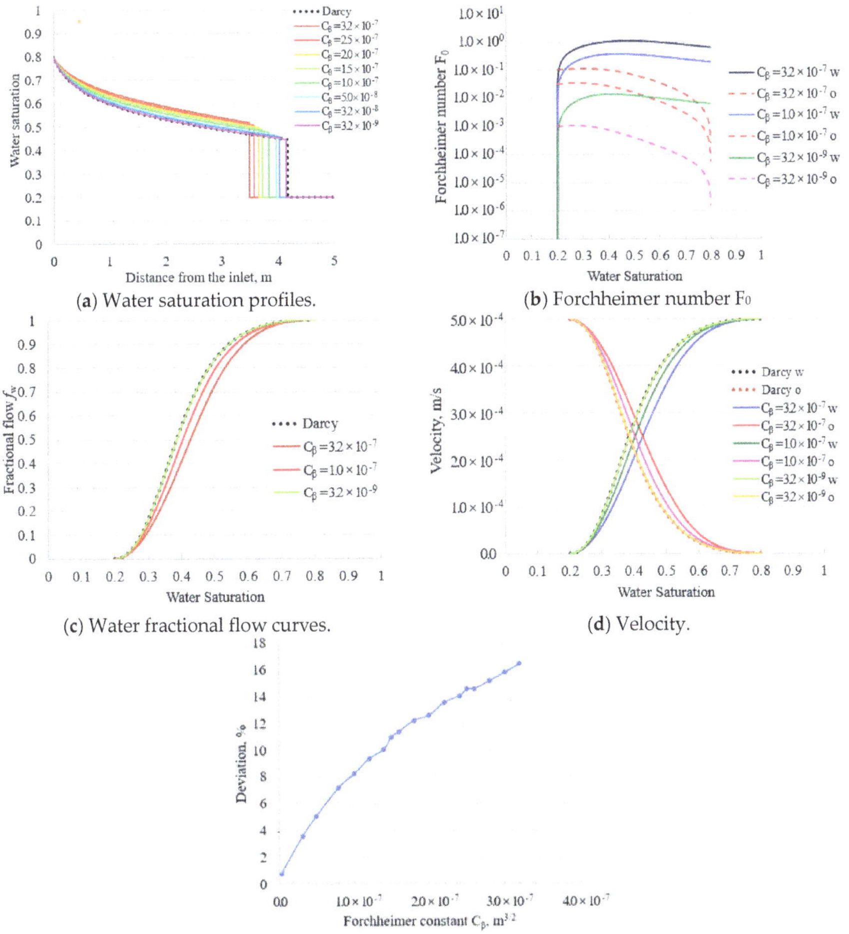

(**a**) Water saturation profiles. (**b**) Forchheimer number Fo (**c**) Water fractional flow curves. (**d**) Velocity. (**e**) Distance deviation of waterfront travel distance.

Figure 5. Comparison of Darcy equation versus Forchheimer equation with different C_β values.

In summary, distinguishing between Darcy and non-Darcy flows and choosing the flow equation based on the value of the Forchheimer number is difficult, because the Forchheimer coefficient C_β and Forchheimer number F_0 of multiphase flows are not constant and vary with the water saturation. The β_w (or β_o) cannot be replaced by a constant value, because the values of β_w (or β_o) are very large or infinite at the beginning (or end) of the water saturation; however, the area scope is very small, and the inertial effect cannot be ignored. The displacement distance deviation of the waterflood front can quantify the difference between Darcy and non-Darcy flows well and help us evaluate the effect of non-Darcy flows in fractured karst reservoirs, choose the appropriate model and correct errors in the numerical simulation.

3.1.2. Fluid Viscosity

In this section, we will analyze the effect of fluid viscosity on the inertia, assuming that the constant viscosity of oil is 0.005 Pa·s and water viscosity takes the following values: 0.0005, 0.001, 0.002, 0.003, 0.004 and 0.005 Pa·s.

Figure 6 shows the water saturation profiles of the Darcy and Forchheimer flows with different ratios of fluid viscosity. When the viscosity difference between oil and water is the largest—namely, the water viscosity is 0.0005 Pa·s and oil viscosity is 0.005 Pa·s—the influence of the inertial effect is obvious in the displacement distance of waterfront travel and sweeping efficiency regardless of the values of C_β. As the water viscosity increases, the displacement distance of waterfront travel decreases. When the water viscosity is equal to that of the oil, the inertial effect tends to weaken because the displacement is similar to piston displacement. In general, the smaller the viscosity difference between the oil and water, the better the oil displacement efficiency.

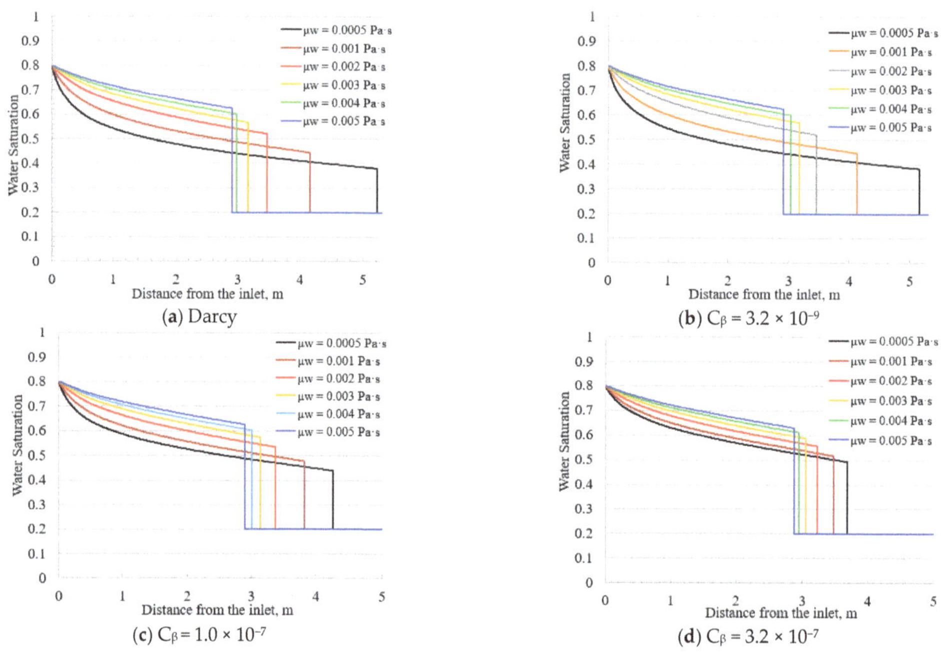

Figure 6. Water saturation profiles of Darcy versus Forchheimer flow with different fluid viscosities.

For the same C_β value (Figure 7), the change in water viscosity affects the Forchheimer number. The water Forchheimer number decreases as the water viscosity increases, whereas the oil Forchheimer number increases. As C_β decreases from 3.2×10^{-7} to 3.2×10^{-9}, the Forchheimer numbers of water and oil decrease as well. When C_β is equal to 3.2×10^{-7}, most of the water Forchheimer numbers are greater than 0.1 and those of oil are greater than 0.01. The effect of inertial force cannot be ignored. According to the results in Figure 7e, the distance deviations of the waterflood front are less than 0.05 when the water viscosity is greater than or equal to 0.03. Therefore, the deviation is more credible for evaluating the effect of inertial force compared to the Forchheimer number. The distance deviations of the waterflood front with the different C_β and water viscosities are shown in Figure 7e. When the oil viscosity is equal to 0.005 Pa·s and the water viscosity is not greater than 0.003 Pa·s, the distance deviation of the waterflood front is not greater than 0.05, regardless of the value of C_β.

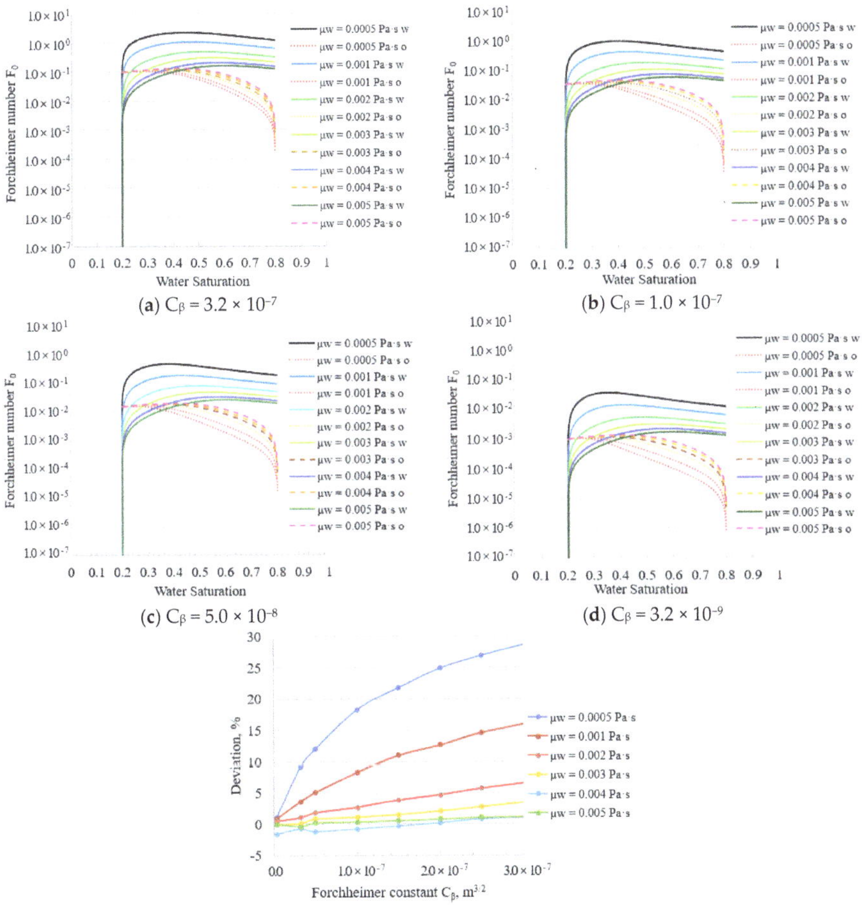

(**a**) $C_\beta = 3.2 \times 10^{-7}$ (**b**) $C_\beta = 1.0 \times 10^{-7}$ (**c**) $C_\beta = 5.0 \times 10^{-8}$ (**d**) $C_\beta = 3.2 \times 10^{-9}$ (**e**) Distance deviation of waterflood front.

Figure 7. Forchheimer number F_0 of Forchheimer flow with different fluid viscosities.

In summary, the inertial effect decreases as the oil–water viscosity ratio decreases. When the viscosities of the fluids are equal (i.e., the water viscosity increases from 0.0005 Pa·s to 0.005 Pa·s), the non-Darcy effect is minimal and can even be ignored. Determining whether to consider the inertial force only according to the value of the Forchheimer number is difficult, and the distance deviation of the waterflood front is more credible to evaluate the inertial effect and to correct the numerical simulation error.

3.1.3. Flow Velocity

When fluids flow in the subsurface, the flow velocity cannot be constant and keeps changing. For instance, the flow velocity is very high near the borehole; as the waterfront advances along the radial direction in the porous media, the flow velocity will gradually decrease. In addition, the velocity may be influenced by the porosity and connectivity. In this section, we will analyze the effect of velocity on non-Darcy behaviors after the same volume of fluids is injected. Here, we assume that the velocity varies from 5.0×10^{-4} m/s to 5.0×10^{-6} m/s.

Figure 8 shows the comparison of different flow velocities when C_β equals 3.2×10^{-7} m$^{3/2}$. When the flow velocity is 5.0×10^{-4} m/s, the sweeping efficiency is the highest and the displacement distance is the shortest. Then, as the velocity decreases, the displacement distance gets longer and becomes close to that of Darcy flow. The values of the Forchheimer number all decrease with a decrease in velocity. When C_β equals 3.2×10^{-7} m$^{3/2}$, most values of the water Forchheimer number are more than 0.01 and those of oil are more than 0.001 when the velocity is larger than 5.0×10^{-5} m/s. In these cases, the distance deviation of the waterflood front is larger than 0.03 (Figure 8c). When the velocity is no more than 5.0×10^{-5} m/s, the distance deviation of the waterflood front is less than 0.05, regardless of the C_β value. This means that the inertial force is very small, and the characteristic of the Forchheimer flow is close to that of Darcy flow.

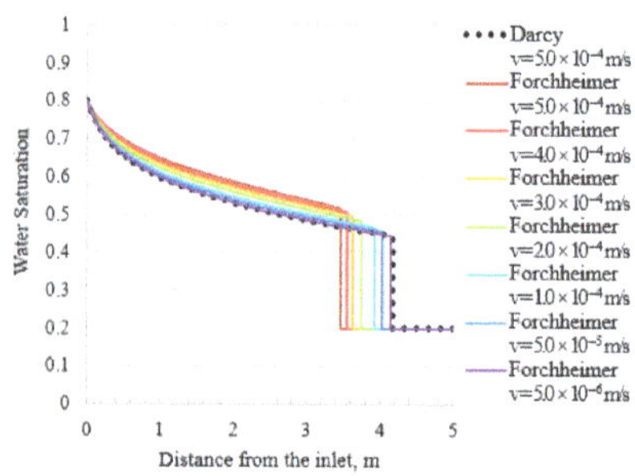

(**a**) Water saturation profiles.

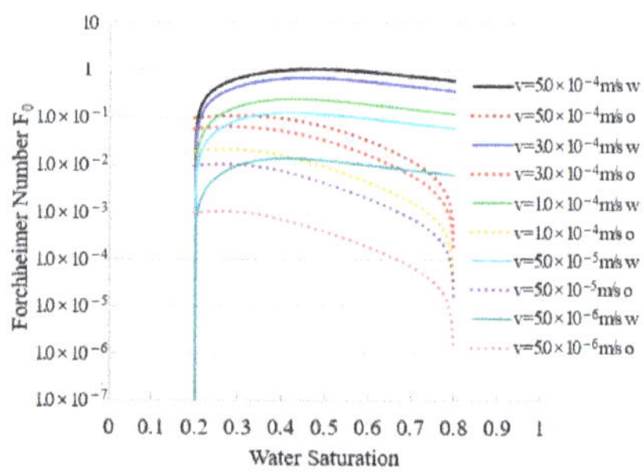

(**b**) Forchheimer number F_0.

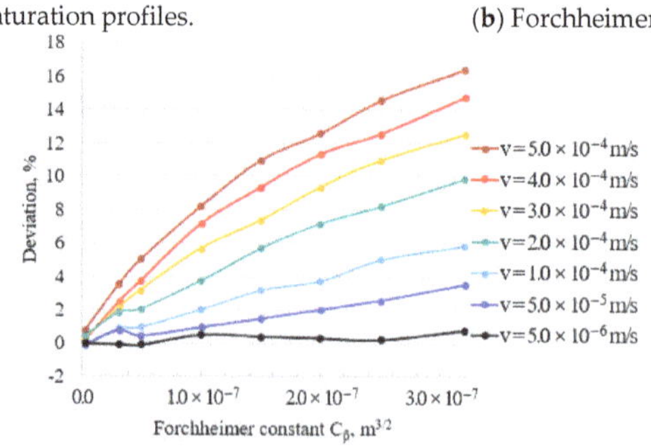

(**c**) Distance deviation of waterflood front with different rates and different C_β values.

Figure 8. Comparison of Darcy flow versus Forchheimer flow at different flow rates.

In summary, velocity is the key parameter for the characteristics of fluid flow. When the velocity is larger than a certain value, the non-Darcy effect cannot be ignored, and this threshold value of the velocity is critical. It is much easier to obtain the threshold value of velocity by comparing the distance deviation of the waterflood front than to do so according to the values of the Forchheimer number.

3.1.4. Absolute Permeability

Usually, most reservoirs are heterogeneous and their absolute permeability is not constant. Absolute permeability directly influences the Forchheimer coefficient for non-Darcy flow. In this section, a comparison of the Forchheimer equation with different absolute permeability values will be conducted to evaluate its distance deviations of the waterflood front from Darcy flow. Here, C_β is equal to 1.0×10^{-7} $m^{3/2}$, and the values of absolute permeability are 9.869×10^{-12}, 9.869×10^{-13} and 9.869×10^{-14} m^2.

Figure 9 shows a comparison of the results of the Darcy and Forchheimer equations with varying absolute permeabilities. According to the water saturation profile (Figure 9a), the displacement distance of waterfront travel is shortest for the lowest absolute permeabilities because of the high flow resistance for the same injection volume. Here, F_0 increases as the absolute permeability decreases (Figure 9b). This means that the lower absolute permeability will enhance the non-Darcy effect and lead to better displacement efficiency. When the absolute permeability is 9.869×10^{-12} m^2, the distance deviation of the waterflood front between the Forchheimer and Darcy equations is 10.5%. As the absolute permeability decreases, the distance deviation of the waterflood front increases, reaching 20.1% when the absolute permeability is 4.935×10^{-14} m^2.

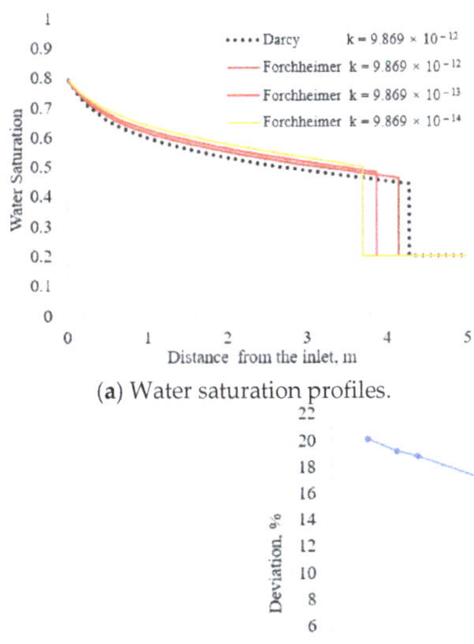

(a) Water saturation profiles.

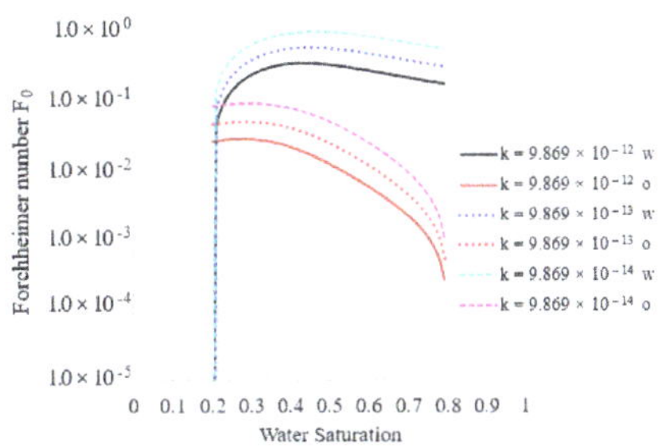

(b) Forchheimer number F_0.

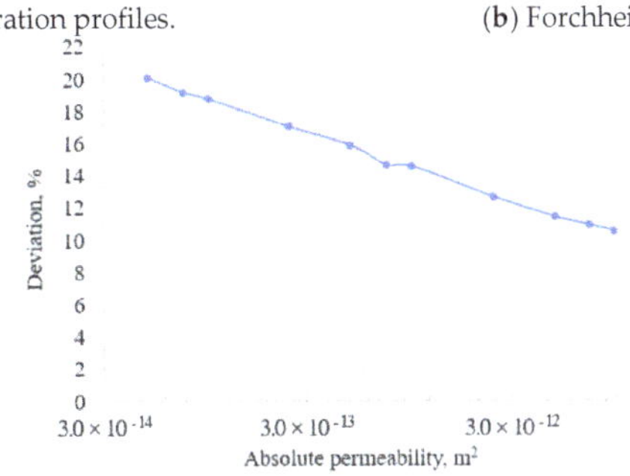

(c) Distance deviation of waterflood front with various absolute permeabilities.

Figure 9. Comparison of Darcy vs. Forchheimer equations with various absolute permeabilities.

In summary, more attention should be paid to the lower absolute permeability region, because lower absolute permeability results in high flow resistance, and the inertial effect increases. Thus, when the fluid flows from high absolute permeability to lower absolute permeability, the flow regime may change from Darcy flow to non-Darcy flow.

3.1.5. Effect of Parameters

As previously stated, there are many factors related to fluids and rock that can affect the inertial force, and the distance deviation of the waterflood front can describe the magnitude of the inertial effect well. In this section, we will set the water viscosity to 0.001 mPa·s to analyze the inertial effect with different Forchheimer constants, absolute permeability and flow velocity. The distribution characteristic and laws of the distance deviation of the waterflood front for waterfront travel under different conditions are shown in Figure 10. When the flow rate is less than 5.0×10^{-5} m/s, the distance deviation of the waterflood front is less than 6.0% and the non-Darcy effect is insignificant. When the Forchheimer constant is less than 3.2×10^{-8} m$^{3/2}$, the distance deviation of the waterflood front is within 5.0% and the non-Darcy effect can be ignored. In the other cases, the distance deviation of the waterflood front (Figure 10) can assist in determining whether to consider the non-Darcy effect and choosing the suitable flow equation; it can also aid in evaluating the numerical simulation error for a heterogeneous reservoir.

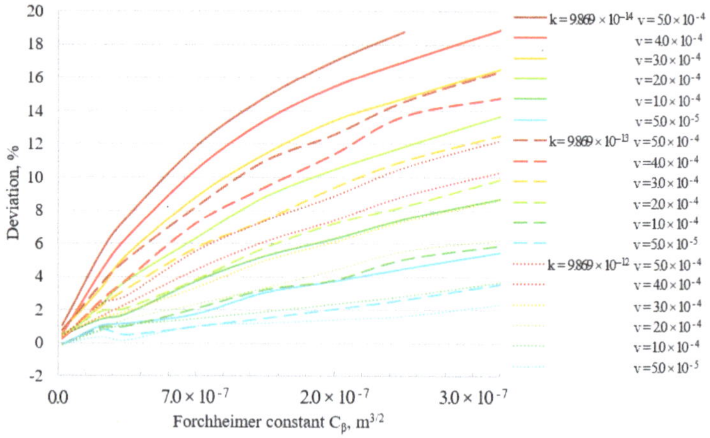

Figure 10. Distance deviation of waterflood front with various parameters.

3.2. Comparison of Darcy and Barree–Conway Equations

3.2.1. Fluid Viscosity

Here, we assume that the constant viscosity of oil is 0.005 Pa·s and water viscosity takes the following values: 0.0005, 0.001, 0.002, 0.003, 0.004 and 0.005 Pa·s.

Figure 11 shows the water saturation profiles of the Darcy and Barree–Conway equations for different oil–water viscosity ratios. Similarly, as the oil–water viscosity decreases, the influence of the non-Darcy effect weakens; when the water viscosity is greater than 0.002 Pa·s, the distance deviation of the waterflood front is less than 4% (Figure 11d). When the oil–water viscosity ratio decreases, the Reynolds number of the water decreases, whereas that of the oil increases (Figure 11c). We cannot evaluate the non-Darcy effect only according to the Forchheimer numbers of the water and oil phases. The distance deviations of the waterflood front with different flow rates are shown in Figure 11d. Even when the flow rate is very high ($v = 5.0 \times 10^{-4}$ m/s) and the water viscosity is no less than 0.002 Pa·s, the distance deviation of the waterflood front is less than 4.0%, indicating that the non-Darcy effect is insignificant. When the oil–water viscosity ratio is extremely high, the non-Darcy effect increases as the flow rate increases.

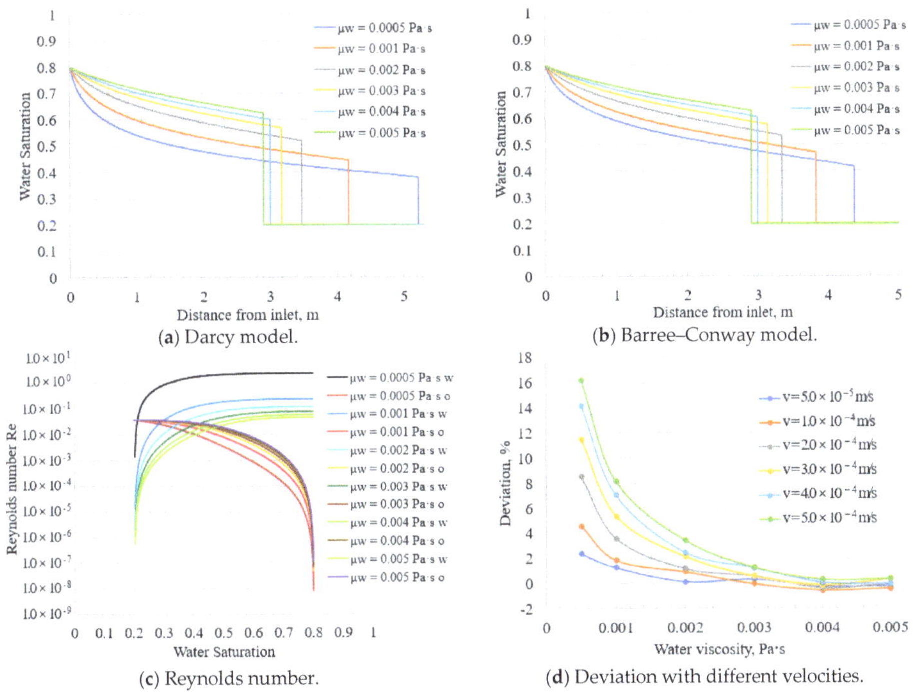

Figure 11. Comparison of Darcy model versus Barree–Conway model for different water viscosities.

3.2.2. Flow Velocity

In this section, we will analyze the effect of velocity on the non-Darcy behavior for the same porous media, assuming that the water viscosity is 0.001 Pa·s and the velocity varies from 5.0×10^{-4} m/s to 5.0×10^{-5} m/s.

Figure 12 shows the comparison of different flow velocities. When the flow rate is 5.0×10^{-4} m/s, the sweeping efficiency is the highest and the displacement distance is the shortest for the same porous media. Thereafter, as the rate decreases, the displacement distance increases and comes close but does not overlap with that of Darcy flow; the distance deviation of the waterflood front still is greater than 1% even when the flow rate is 0.5×10^{-4} m/s (Figure 12c), and the values of Reynolds number all decrease as the flow rate decreases. When the flow rate is less than 2.0×10^{-4} m/s, the distance deviation of the waterflood front is less than 4% and the Reynolds number is still within several orders of magnitude. In summary, velocity is still the critical parameter in the Barree–Conway model. When the velocity is less than a certain value, the distance deviation of the waterflood front is significantly small, and the non-Darcy effect can be ignored; however, the Reynolds number is not a constant value and will be within several orders of magnitude.

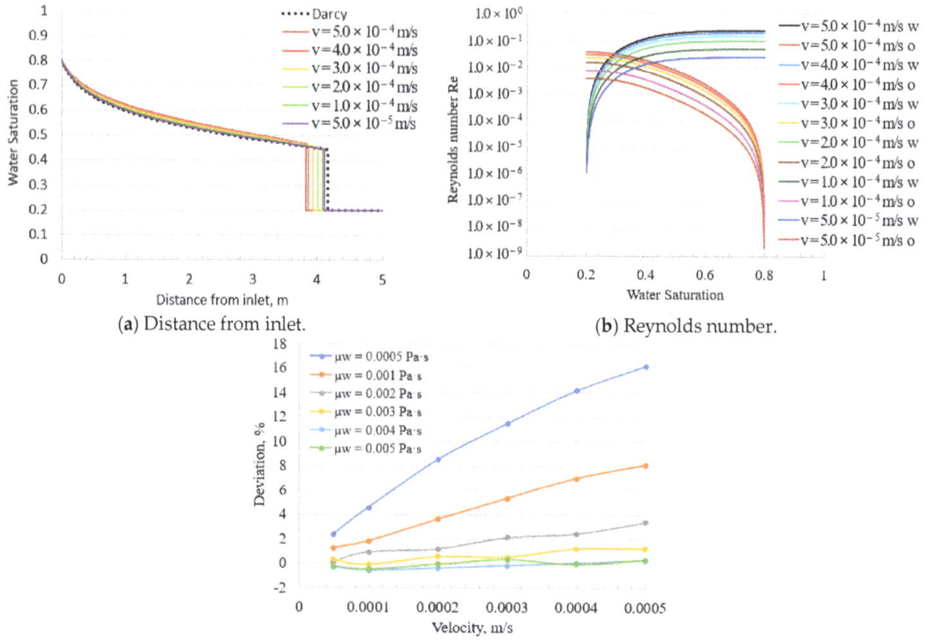

(**a**) Distance from inlet. (**b**) Reynolds number. (**c**) Deviation with different water viscosities.

Figure 12. Comparison of Darcy model vs. Barree–Conway model for different flow velocities.

3.2.3. Absolute Permeability

The results of the Barree–Conway model with different absolute permeabilities will be compared in this section to evaluate the distance deviations of the waterflood front with respect to Darcy flow. Here, the absolute permeability value ranges from 9.869×10^{-12} to 7.402×10^{-14} m^2 and the value of τ is obtained using the relationship in Figure 2 (Table 2).

Table 2. Parameter of Barree–Conway model (Figure 2).

	Absolute Permeability k, m^2	Inverse of Characteristic Length τ, m^{-1}
1	9.869×10^{-12}	3100
2	7.402×10^{-12}	3100
3	4.935×10^{-12}	3150
4	2.467×10^{-12}	3000
5	9.869×10^{-13}	2100
6	7.402×10^{-13}	1900
7	4.935×10^{-13}	1500
8	2.467×10^{-13}	1100
9	9.869×10^{-14}	650
10	7.402×10^{-14}	550

Figure 13 shows the comparison results of the Darcy and Barree–Conway models with various absolute permeabilities when the flow velocity is 5.0×10^{-4} m/s. According to the water saturation profile (Figure 13a), the displacement distance of waterfront travel is shortest for the lowest absolute permeabilities with the same injection volume, and the distance deviation of the waterflood front decreases as the absolute permeability increases. When the absolute permeability is greater than 2.467 × 10^{-12} m^2, the distance deviation of the waterflood front remains around 6% because there is little difference in the value of another key parameter, τ (Table 2). The Reynolds number decreases as the absolute permeability increases (Figure 13b). The distance deviation of the waterflood front decreases as the flow rate decreases (Figure 13c). For the same flow rate, the absolute permeability increases from 2.467 × 10^{-12} m^2 to 9.869 × 10^{-12} m^2, while the distance deviation of the waterflood front is kept nearly constant because the value of τ is almost the same.

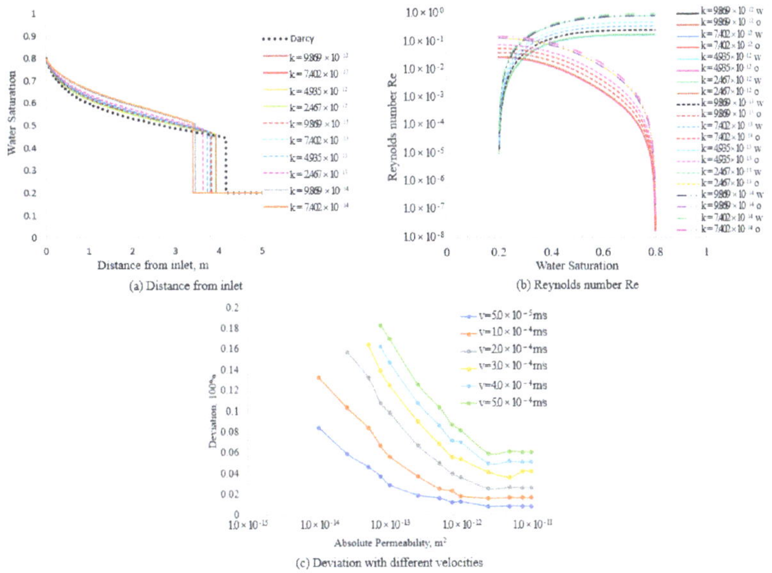

Figure 13. Comparison of Darcy model versus Barree–Conway model for different flow velocities.

3.2.4. Effect of Parameters

The parameters relating to fluids and rock influence the non-Darcy effect, and the distance deviation of the waterflood front can directly describe the magnitude of the non-Darcy effect. In this section, we will set the oil viscosity to 0.005 Pa·s and analyze the non-Darcy effect with different water viscosities, absolute permeabilities, and flow rates (Figure 14). When the water viscosity is greater than or equal to 0.004 Pa·s, the distance deviation of the waterflood front is less than 3%; when the water viscosity is greater than or equal to 0.003 Pa·s, the distance deviation of the waterflood front is less than 5% and the non-Darcy effect still is not evident; when the water viscosity is equal to 0.002 Pa·s, the non-Darcy effect cannot be ignored in cases with a high flow rate in the low permeability area; when the viscosity ratio of oil–water is extremely large (μ_o = 0.005 Pa·s and μ_w = 0.0005 Pa·s), the distance deviation of the waterflood front is evident and the inertial effect cannot be ignored, especially for a high flow rate and low absolute permeability. The distribution of the distance deviation of the waterflood front with the different parameters is shown in Figure 14.

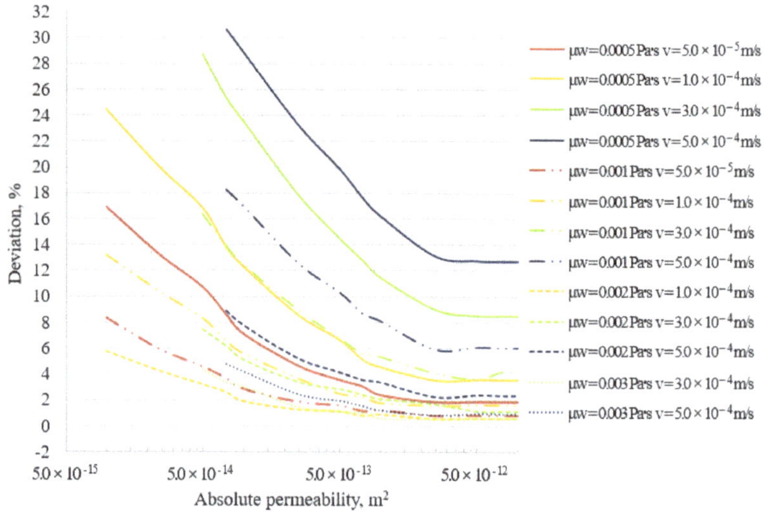

Figure 14. Distance deviation of Darcy equation versus Barree–Conway equation.

4. Applications of Tarim Fractured Karstic Carbonate Reservoir

We used this method to analyze the distance deviation of the waterflood front at the fractured karstic region in the Tarim reservoir. We chose one well area developing fractures and karstic caves, which were verified from the logging curves, seismic profiles and obtained cores (Figure 15). There were obvious signs of karstic caves and fractures in the seismic profiles and logging curves, and it was hard to obtain a core from the target strata. According to the geophysical method, the average porosity of underground reservoirs is widely distributed: about 2.0% in the low porosity area and up to 50% in the dissolution pores and karstic cave areas. The average permeability varies from 4.93×10^{-13} m^2 to 9.869×10^{-11} m^2. According to previous exploration and developments, the production well is located in the fracture cave regions, the porosity is about 45% and the initial effective permeability is 4.93×10^{-13} m^2. The oil viscosity is 0.05 Pa·s and the density is 860 kg/m^3. The water viscosity is 0.0005 Pa·s, and the density is 1000.0 kg/m^3. The daily production is about 300.0–500.0 m^3, which is clearly a high flow rate; thus, it is a high yield well.

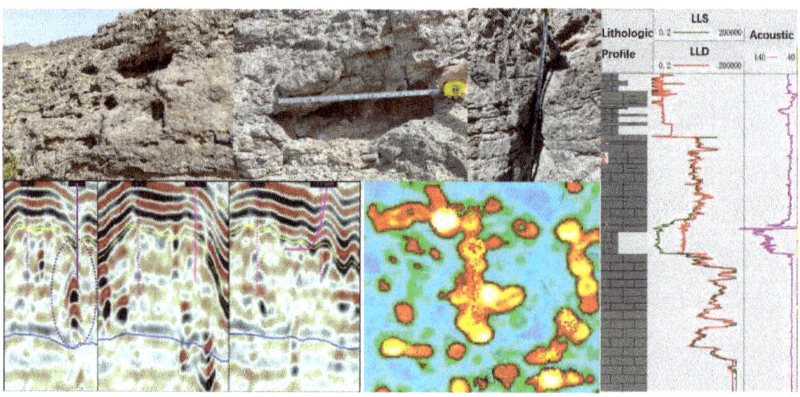

Figure 15. Geology and geophysics response of fractured karstic carbonate rocks.

The Eclipse software based on Darcy flow was used to simulate the fluid flows. The geological model was established according to seismic data and logging data, and the fracture and karstic cave development area was finely carved. The karstic cave area was set with high porosity and high permeability according to the seismic data and logging data, and the model was discretized by a 20 m × 20 m × 20 m grid. The well's conditions were set according of the well performance data, then a numerical simulation using the Eclipse software was carried out from 2013 to 2017. Usually, the next step is to modify the geological model for historical fitting according to the production dynamics. We analyzed the production curves of this production well (Figure 16) before modifying the geological model. The red dotted line is the actual production curve, and the blue line is the result of the numerical simulation using the Eclipse software. Figure 16 shows the difference between them on the production well, especially in the green dotted area. There was an obvious deviation of the water breakthrough time between them on the production well, and the time of water appearance in the numerical simulation was obviously earlier than that in the actual production. The water began to occur around the year 2015 in the actual production curves; however, the time of water appearance was obviously earlier in the numerical simulation using the Darcy equation. The flow velocity calculated by the numerical simulation was greater than the actual flow velocity because the inertial force was ignored; as a result, the water appeared earlier in the numerical simulation than in the actual production. Usually, we do not take the inertial effect into account and directly modify the geological model or phase permeability curve to achieve the historical fitting.

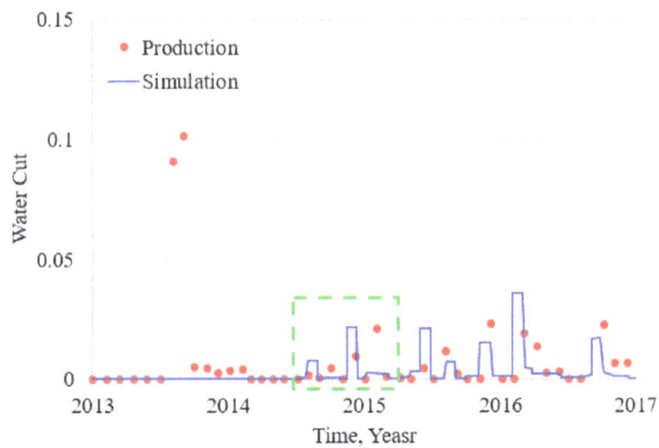

Figure 16. Results of production and numerical simulation.

We selected a region within 500.0 m of this production well and used the method described in this paper to analyze the distance deviation of the waterflood front. Since the main storage space was made up of fractures, karst cracks and dissolution pores, the flow rate was set to 5.0×10^{-5} m/s. Since there was no non-Darcy flow equation parameter determined by the physical experiment, we used the mild values (the large parameter value indicates that the inertial effect is strong, while the small value indicates that the inertial effect can be ignored). We assumed that the Forchheimer constant C_β was equal to 3.2×10^{-8} $m^{3/2}$ and τ was equal to 1500 m^{-1} in the Barree–Conway equation (Barree and Conway, 2004); these two values describe non-Darcy flow in which the inertial effect is not very strong. The waterflood displacement front and water saturation distribution curves were obtained by the three equations after injecting the same volume multiple times (as shown in Figure 17).

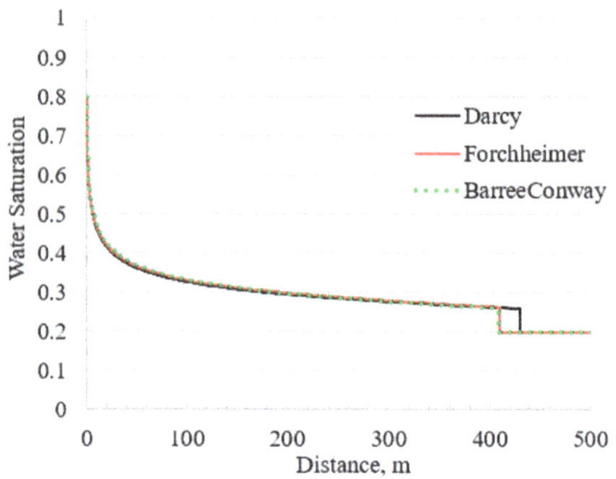

Figure 17. Displacement distance of waterfront travel.

Figure 17 shows that the waterflood frontal movement of Darcy flow was fast and the displacement distance of waterfront travel was the longest once the inertial effect was ignored, and the results of non-Darcy flow considering inertia were obviously different from those of Darcy flow. The distance deviation of the waterflood front was 4.69% using the Forchheimer equation and 5.04% using the Barree–Conway equation. The inertial effect should be one of the important factors, although there are many factors affecting the time of water emergence. With the reservoir developing, the distance deviation of the waterflood front between the simulation and the actual production ignoring the inertial effect will increase, thus further affecting the modification of the geological model, historical fitting and yield prediction. In reservoir development research, it is generally accepted that geological parameters are modified to achieve consistency in the dynamic production curve, then the prediction of future development dynamics is performed. Here, the geological model and the actual reservoir were equivalent models based on different flow states. In other words, a geological model based on Darcy flow is equivalent to an actual reservoir based on non-Darcy flow, which will lead to further errors in subsequent oil and gas field development and the prediction of remaining oil.

5. Conclusions

Several conclusions can be drawn from the comparison and analysis carried out in this paper:

1. It is not very realistic that the non-Darcy flow equation is used to simulate fluid flow in fractured karstic carbonate reservoirs. It is the inertial term in Forchheimer model and the apparent permeability in the Barree–Conway model that make these equations more complex than the Darcy equation. Furthermore, it will dramatically increase the difficulty and expense of numerical simulation.
2. The Forchheimer number and Reynolds number do not quantify the magnitude of the inertial effect. In addition, the crucial values of the Forchheimer number and Reynolds number are not constant and vary with the parameters of the rock and fluid properties. It is very difficult to determine whether to consider the non-Darcy effect only according to the Forchheimer/Reynolds number.
3. The distance deviation of the waterflood front can better describe the inertial effect of multiphase flow. In numerical simulation, we are concerned with the displacement range and the sweeping efficiency. According to the distance deviation of the waterflood front between the Darcy and non-Darcy equations, we will not only choose the flow equation within the numerical error allowed, but also obtain the evaluation of the numerical error for a complex heterogeneous reservoir.

Author Contributions: Conceptualization, J.Y. and Y.W.; methodology, Y.W.; software, Y.W.; validation, Y.W. and Z.H.; formal analysis, Y.W. and Z.H.; investigation, Z.H.; resources, Y.W. and Z.H.; data curation, Y.W.; writing—original draft preparation, Y.W.; writing—review and editing, Y.W.; visualization, Y.W.; supervision, Y.W. and Z.H. project administration, Y.W.; funding acquisition, Y.W. All authors have read and agreed to the published version of the manuscript.

Funding: This work was supported by the China Scholarship Council (Grant CSC201906455038), National Natural Science Foundation of China (No. 52074336), National Major Science and Technology Projects of China (No. 2015ZX05014-003) and Major Science and Technology Projects of China National Petroleum Corporation (No. ZD2019-183-008).

Institutional Review Board Statement: Not applicable.

Informed Consent Statement: Informed consent was obtained from all subjects involved in the study.

Data Availability Statement: The data that support the findings of this study are available from the corresponding author, [Y.W.], upon reasonable request.

Conflicts of Interest: The authors declare no conflict of interest.

References

1. Akbar, M.; Vissapragada, B.; Alghamdi, A.H. A snapshot of carbonate reservoir evaluation. *Oilfield Rev.* **2000**, *12*, 20–21.
2. Srinivasan, S.; O'Malley, D.; Mudunuru, M.K. A machine learning framework for rapid forecasting and history matching in unconventional reservoirs. *Sci. Rep.* **2021**, *11*, 21730. [CrossRef] [PubMed]
3. Yao, J.; Sun, H.; Li, A. Modern System of multiphase flow in porous media and its development trend. *Sci. China Press* **2018**, *63*, 425–451. [CrossRef]
4. Darcy, H. *Les Fontaines Publiques de la Ville de Dijon*; Technical Report; Dalmont: Paris, France, 1856.
5. Forchheimer, P. Wasserbewegung durch boden. *Z. Ver Deutsh. Ing.* **1901**, *45*, 1782–1788.
6. Whitaker, S. Advances in theory of fluid motion in porous media. *Ind. Eng. Chem.* **1969**, *61*, 14–28. [CrossRef]
7. Jones, S.C. A Rapid Unsteady-State Klinkenberg Permeameter. *Soc. Pet. Eng. J.* **1972**, *12*, 383–397. [CrossRef]
8. Geertsma, J. Estimating the Coefficient of Inertial Resistance in Fluid Flow through Porous Media. *Soc. Pet. Eng. J.* **1974**, *14*, 445–450. [CrossRef]
9. Firoozabadi, A.; Katz, D.L. An analysis of high-velocity gas flow through porous media. *J. Petrol. Technol* **1979**, *31*, 211–216. [CrossRef]
10. Ruth, D.; Ma, H. On the derivation of the Forchheimer equation by means of the averaging theorem. *Transp. Porous Media* **1992**, *7*, 255–264. [CrossRef]
11. Barree, R.D.; Conway, W. Beyond Beta Factors: A Complete Model for Darcy, Forchheimer, and Trans-Forchheimer Flow in Porous Media. In Proceedings of the SPE Annual Technical Conference and Exhibition (SPE 2004), Houston, TX, USA, 26–29 September 2004; pp. 1–8.
12. Barree, R.D.; Conway, M.W. Multiphase non-darcy flow in proppant packs. In Proceedings of the SPE Annual Technical Conference and Exhibition (SPE 2007), Anaheim, CA, USA, 11–14 November 2007; pp. 1–10.
13. Chilton, T.H.; Colburn, A.P. Pressure drop in packed tubes. *Ind Engng. Chem.* **1931**, *23*, 913–919. [CrossRef]
14. Green, L., Jr.; Duwez, P. Fluid flow through porous metals. *J. Appl. Mech.* **1951**, *18*, 39–45. [CrossRef]
15. Ma, H.; Ruth, D.W. The microscopic analysis of high Forchheimer number flow in porous media. *Transp. Porous Media* **1993**, *13*, 139–160. [CrossRef]
16. Ergun, S. Fluid flow through packed columns. *Chem. Engng. Prog.* **1952**, *48*, 89–94.
17. Evans, R.D.; Hudson, C.S.; Greenlee, J.E. The effect of an immobile liquid saturation on the non-darcy flow coefficient in porous media. *SPE Prod. Eng.* **1987**, *2*, 331–338. [CrossRef]
18. Scheidegger, A.E. *The Physics of Flow through Porous Media*, 3rd ed.; University of Toronto Press, Scholarly Publishing: Toronto, ON, Canada, 1974; pp. 31–150.
19. Blick, E.F.; Civan, F. Porous media momentum equation for highly accelerated flow. *SPE Reserv. Engng.* **1988**, *3*, 1048–1052. [CrossRef]
20. Bear, J. *Hydraulics of Groundwater*; McGraw-Hill: New York, NY, USA, 1979; pp. 66–82.
21. Wu, Y. Non-Darcy displacement of immiscible fluids in porous media. *Water Resour. Res.* **2001**, *37*, 2943–2950. [CrossRef]
22. Bitao, L.; Jennifer, L.; Wu, Y.-S. Non-darcy porous media flow according to the barree and conway model: Laboratory and numerical modeling studies. In Proceeding of the SPE Rocky Mountain Petroleum Technology Conference (SPE 2009), Denver, CO, USA, 14–16 April 2009; pp. 1–10.
23. Brooks, R.H.; Corey, A.T. Hydraulic Properties of Porous Media. In *Hydrology Papers*; Colorado State University: Fort Collins, CO, USA, 1964.
24. Zhengwen, Z.; Reid, G. A criterion for Non-Darcy Flow in Porous Media. *Transp. Porous Media* **2006**, *63*, 57–69.

energies

MDPI

Review

Advances in Cryogenic Fracturing of Coalbed Methane Reservoirs with LN_2

Sotirios Nik. Longinos [1,*], Lei Wang [1,2] and Randy Hazlett [1]

[1] Department of Petroleum Engineering, Nazarbayev University, Nur-Sultan 010000, Kazakhstan
[2] College of Energy, Chengdu University of Technology, Chengdu 610059, China
* Correspondence: sotirios.longinos@nu.edu.kz

Abstract: Coalbed methane (CBM) is a significant unconventional natural gas resource existing in matrix pores and fractures of coal seams and is a cleaner energy resource compared to coal and crude oil. To produce CBM, stimulation operations are required, given that the coal permeability is generally too low. Hydraulic fracturing is the most widely used technology for reservoir stimulation; however, there are a few challenging issues associated with it, e.g., huge water consumption. In the past decade, the use of liquid nitrogen (LN_2) as a fracturing fluid has been intensively studied for stimulating CBM reservoirs, achieving considerable progress in understanding fracturing mechanisms and optimizing fracturing techniques. This paper presents a thorough review of experimental design and observations, modeling procedures and results, field applications, and published patents. Existing studies are divided into five different groups for discussion and comparison, including immersion tests, injection tests, jet drilling tests, numerical modeling, and field applications. Based on the comprehensive evaluation of the outcomes, it is obvious that cryogenic fracturing using LN_2 is a promising eco-friendly fracturing technique that can effectively enhance coal rock permeability to increase the production of CBM.

Keywords: cryogenic fracturing; coalbed methane; immersion tests; injection experiments; numerical modeling

Citation: Longinos, S.N.; Wang, L.; Hazlett, R. Advances in Cryogenic Fracturing of Coalbed Methane Reservoirs with LN_2. *Energies* **2022**, *15*, 9464. https://doi.org/10.3390/en15249464

Academic Editor: Shu Tao

Received: 28 September 2022
Accepted: 10 November 2022
Published: 14 December 2022

1. Introduction

In 2020, the total amount of the world's energy consumption was 556.63 exajoules, 83.1% of which were provided by hydrocarbons [1,2], and the status will not vary crucially in coming years [3,4]. Coalbed methane, as an unconventional resource of natural gas [5], has seen its production increase to 1.4×10^{14} m^3, which accounts for approximately 20% of global natural gas resources [6]. While coal can be extensively fractured, the production of coalbed methane can be assisted tremendously by the creation of fractures emanating from and leading to wells that serve as the gathering system. Water is the primary power fluid for induced fracturing. The initial test in hydraulic fracturing took place in 1947, and the first industrial trial occurred in 1949 [7]. One problem with extracting CBM with conventional hydraulic fracturing is the consumption of large quantities of water, which brings about critical concern in arid regions. In addition, the accidental breakthrough of fluid during hydraulic fracturing operations can generate contamination to underground aquifers and the surrounding area, with many adverse consequences for the environment. Other negative consequences of the use of conventional hydraulic fracturing are serious formation damage and productivity loss, along with waste disposal [8–10]. Consequently, waterless fracturing methods would be a welcome solution from an ecological perspective. Oil-based and CO_2 energized oil fracturing, propellant and explosive fracturing, gas fracturing, gelled alcohol and liquefied petroleum gas fracturing, liquid/supercritical CO_2 fracturing, and LN_2 fracturing are possible alternatives [11–13].

Advantages in the use of LN_2 for fracturing include the elimination of water consumption without the need for chemical additives and the avoidance of formation damage.

Hence, LN_2 could be considered a candidate waterless fracturing fluid to stimulate CBM reservoirs [14–17]. Previous studies showed that the use of LN_2 augments the propagation of micro-cracks in specimens, resulting in the impairment of mechanical properties and in the amelioration of rock permeability. Furthermore, the porosity of wet coal specimens augmented with the increase of water saturation since LN_2 cooling leads to the enlargement of connections in water-filled cracks of coal [18–21].

Interest in LN_2 fracturing operations has been increasing among researchers over the last ten years. Until now, there have been no literature reviews explaining and comparing the outcomes of LN_2 fracturing in CBM reservoirs. This study systematically reviews laboratory methods for evaluating and illustrating the efficacy of LN_2 cryogenic fracturing for CBM reservoirs, modeling studies, and field pilots. Furthermore, we conducted critical discussions and proposed future research directions.

2. Laboratory Experimental Studies

Different modes have been developed to test the efficacy of the cryogenic treatment of rocks, and various analytical techniques have been adopted to characterize cryogenic fractures. Two modes of laboratory experimentation frequently used to test cryogenic fracturing processes are immersion tests [22] and injection tests [12]. In these tests, analytical and imaging methods were typically used, such as nuclear magnetic resonance (NMR), X-ray computer tomography scan (X-ray CT scan), scanning electron microscope (SEM), cryo-scanning electron microscope (cryo-SEM), and stereomicroscopic imaging. In this section, three cryogenic treatment modes are introduced, and the results obtained by fracture characterization techniques are illustrated.

2.1. Immersion Tests and Evaluation

In an immersion test, coal rock samples of relatively high temperature (from 50 °C to 100 °C) are partially or completely immersed into LN_2 to examine the fracturing capability of a LN_2 thermal shock. Immersion tests can be repeated several times on the same coal rock sample. That is, rock samples are first cooled down to the boiling point of LN_2, then warmed back to the original temperature, constituting one freezing and thawing cycle. Multiple cycles can be performed.

NMR is a technique extensively used for the characterization of pore size distributions and pore volumes in reservoirs [23]. NMR can test a greater range of pore sizes compared to other processes, such as optical microscopy (OM)-SEM-transmission electron microscopy (TEM) (qualitative analysis), small angle X-ray scattering (SAXS), small angle neutron scattering (SANS)-microcomputer tomography (μCT), and N_2 adsorption-CO_2 adsorption (quantitative analysis). It is a quick, developed, non-damaging, and handy process for examining the pore structure and range of pore magnitude of coal specimens. Hence, the alteration of pore structure in coal specimens before and after freezing–thawing cycles can be examined [24].

Cai et al. (2014) [25] conducted immersion tests for coal and sandstone specimens and characterized them using NMR. The specimens were first dried at 60 °C (coal) and 80 °C (sandstone) for 48 h and then were saturated for 48 h. Pore size distributions were then tested using NMR. Additional drying processes to obtain partial water saturations were carried out, with saturation determined by weight measurements. Finally, the specimens were submerged to LN_2 for different time periods. The outcomes showed that, in all coal samples, T_2 amplitudes augmented after LN_2 treatment. More specifically, the pore structure of specimens indicated enlargement after the use of LN_2, and the bigger pores corresponding to $T_2 > 33$ ms indicated increasing alteration with water saturation, with an exception for $S_w = 100\%$). However, sandstone specimens showed different behaviors with respect to water saturation. There was a downward trend in T_2 distribution after immersion tests in sandstones at low saturation, while specimens with water saturation from 30% to 70% showed an increasing augmentation in T_2 amplitude. Nonetheless, for T_2

greater than 33 ms, there was a reduction in amplitude. On the other hand, samples with a water saturation of 100% experienced a significant increase.

The NMR relaxation time is sensitive to the surface-to-volume ratio of a sample, which, in turn, can be related to a characteristic pore size and shape factor as

$$\frac{1}{T_2} = \rho_2 \left(\frac{S}{V}\right)_{pore} = F_S \frac{\rho_2}{r_c} \tag{1}$$

where T_2 is transverse relaxation time (ms), ρ_2 is a factor expressing the relaxation surface strength (μm/ms), $(S/V)_{pore}$ is the ratio of pore surface to volume (cm^{-1}), F_s is the shape factor, and r_c is the pore size. Furthermore, NMR outcomes indicated that after LN_2 treatment, coal specimens had more significant alterations compared to sandstone specimens, while there was amelioration in the pore space connected with free water in all water-containing coal specimens [25]. Figure 1 shows both the measured T_2 distributions of fully water-saturated specimens before and after freezing.

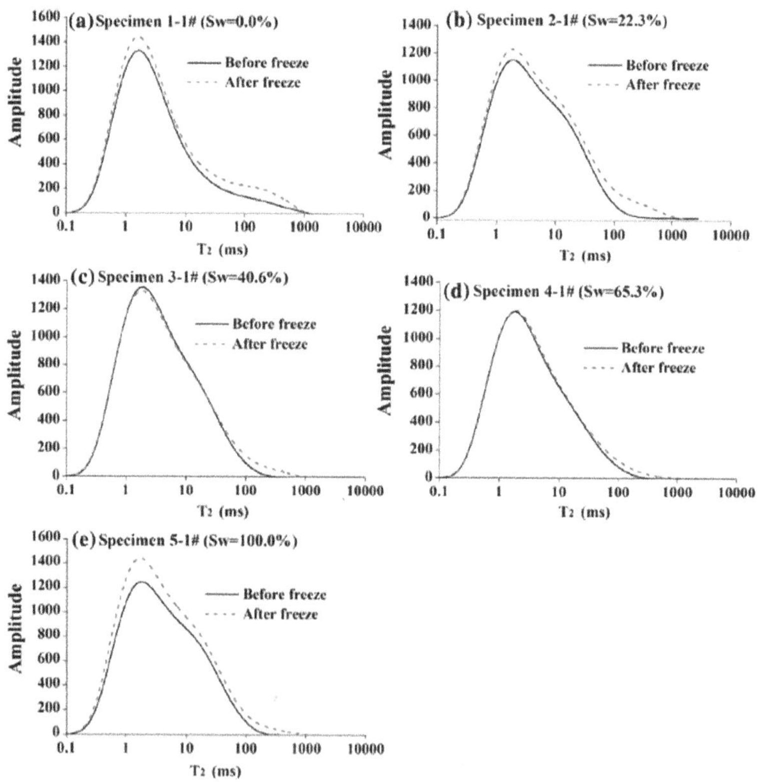

Figure 1. T_2 distributions for coal samples before and after freezing with LN_2.

Cai et al. (2015) [18] examined uniaxial compression before and after LN_2 treatment. Their outcomes showed that thermal stress caused a greater reduction in stress–strain curves compared to specimens without thermal shock, indicating the augmentation of micro-fracturing in specimens with LN_2. The compressive strength of specimens with the use of LN_2 was reduced from 16.18% to 33.74% compared with intact specimens. Moreover, the LN_2 treated specimens possessed more distinct brittle characteristics, and the largest reduction in axial peak was 20.61%. Furthermore, they examined the permeability of the samples before and after exposure to LN_2, indicating improvements of 48.89–93.55%. There were more open fractures after the use of LN_2 compared to intact specimens. Furthermore, LN_2 shrank the coal matrix to generate thermal stress; hence, there was cracking of inter-grain cementation and creation of local stress concentration at the tip of pre-existing cracks. With the augmentation of cracks inside the specimen, pore structure connectivity was greater, which translated into an augmentation in permeability [18]. Cai et al. (2016) [26] also compared coal and shale samples after immersion into LN_2. Their outcomes showed that the permeability of coal and shale samples augmented significantly after LN_2 cooling, confirming that LN_2 could promote the expansion of micro-cracks and induce rock fracturing [26].

Zhai et al. (2016) [14] examined coal samples from the Shengli coal mine in Inner Mongolia (China) using the processes of freezing and freezing–thawing cycles. They first dried the coal samples in the oven at 60 °C. Then, the samples were submerged in a water saturation device for 12 h operating under a vacuum to eliminate air in the system. The specimens were then treated by LN_2 freezing-thawing. Later, the specimens were resaturated before magnetic resonance image (MRI) analysis was carried out. Finally, the coal samples were centrifuged for 90 min to expel water within the connective macropores. According to T_2 curves, the widths and amplitudes for water-saturated specimens show a positive correlation with LN_2 freezing time. Moreover, the number of pores and freely connected pores increased with both the freezing time and the number of freezing–thawing cycles. On the other hand, the centrifuged specimens indicated a negative relationship between T_2 amplitude and freezing time, while the widths showed no changes, confirming that the number of closed pores reduced as freezing time increased [14].

Examining the porosity evolution and quantification after the use of LN_2, the outcomes suggested that the $T_{2cutoff}$ value and residual porosity reduce as the total porosity and effective porosity augment. The indication is that the number of fractures in coal increases after the use of LN_2 in the freezing–thawing process, and the augmented pore space gave the opportunity for CBM to desorb from the coal. The relationships between residual, total, and effective porosity are defined through the inclusion of free and bound fluid indices.

$$\varphi N_B = \varphi_N \times \frac{BVI}{BVI + FFI} \tag{2}$$

$$\varphi_{NF} = \varphi_N \times \frac{FFI}{BVI + FFI} \tag{3}$$

where φ_{NB}, φ_{NF} and φ_N are residual porosity, effective porosity, and total porosity, respectively. FFI and BVI are the free fluid index and the bound fluid index, respectively.

As quantified by the outcomes, the rate of change for both total ($\Delta\varphi_N\%$) and effective porosity ($\Delta\varphi_{NF}\%$) benefit from LN_2 freezing time and freezing–thawing cycles, while the residual porosity ($\Delta\varphi_{NB}\%$) is influenced negatively by these elements.

$$\Delta\varphi = \varphi_{post} - \varphi_{pre} \tag{4}$$

$$\Delta\varphi\% - \frac{\Delta\varphi}{\varphi_{pre}} \tag{5}$$

where $\Delta\varphi$ represents the porosity increment given by φ_{pre} and φ_{post} before and after the coals were then frozen and thawed, respectively. $\Delta\varphi\%$ is the rate of change in porosity under different freezing–thawing conditions. Furthermore, the incremental rates of the three porosities (total, residual, and effective) are linearly related to freezing time. On the other hand, in the freezing–thawing process, the rates showed nonlinear behavior. Specifically, as the number of freezing–thawing cycles increased, the alteration range of the porosity increment rate increased.

Zhai et al. (2016) [14] examined and described alterations in the number of pores versus freezing–thawing cycles. Dt expresses the rate of alteration in terms of the number of pores per unit volume of a coal specimen in T_2 spectrum area.

$$D_t = \frac{\Delta S}{S_{pre}} = \frac{S_{pre} - S_{post}}{S_{pre}} \tag{6}$$

where S is the region under T_2, and S_{pre} and S_{post} are the relevant regions in T_2 spectra of specimen before and after freezing–thawing cycles. Concerning pore growth and freezing time, cyclic freezing–thawing produces enhanced alteration compared to a single freezing–thawing treatment. Freezing–thawing experiments assisted in the creation of large seepage pores [14].

Moreover, Zhai et al. (2016) [14] described the influence of freezing–thawing cycles in axial, hoop, and volumetric strains (ε_v) in the coal specimen.

$$\varepsilon_v = \varepsilon_a + 2\varepsilon_r \tag{7}$$

where ε_a and ε_r represent the axial and hoop strains, respectively.

According to volumetric strain, freezing–thawing cycles were separated into four different smaller periods: freezing shrinkage period I (frost power related to the phase change of water into ice in the cracks), frost heave period II (expansion of frost power due to vaporization of LN_2), second freezing shrinkage period III (solid-state cryoturbation in the specimen at very low temperatures), and final frost heave period IV (state of in situ stress, σ). Hence, the coal specimen's morphology can be distorted through the freezing–thawing cycle method. Specimens broke when water and ice created swelling stress greater than the tensile strength of coal [14]. Figure 2 shows the failure caused after freezing–thawing, along with stress analysis for the fractures in coal.

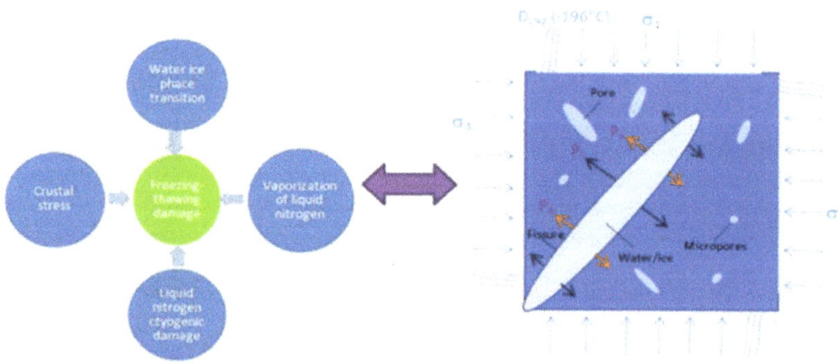

Figure 2. Schematic showing freezing–thawing damage caused by LN_2 and stress analysis for the fractures in coal [14].

Qin et al. (2016) [19] examined two sets of coal samples from the Shengli coal mine in Inner Mongolia (China) using the processes of freezing and freezing–thawing cycles at different moisture concentrations. Their outcomes indicated that the absolute values of axial and hoop strains diminished as the amount of freezing time, freezing–thawing cycles, and concentration of moisture increased. At the same time, the coal porosities were augmented by 17.5% and 68.1%, and Poisson's ratios by 7.14% and 28.6%, respectively. Due to the decrease in the coal's mechanical strength, the elastic strain stage was shortened, and the peak yield point and plastic deformation were accelerated. There was a negative relationship between uniaxial compressive strengths and elastic moduli with freezing time and freezing–thawing cycles, while the uniaxial compressive strengths and elastic moduli showed positive characteristics with increasing moisture content. Furthermore, the uniaxial test outcomes were compatible with the acoustic emission outcomes [19]. Figure 3 shows the variations in the coal's elastic modulus and uniaxial compressive strength with the length of the freezing time and the number of freezing–thawing cycles.

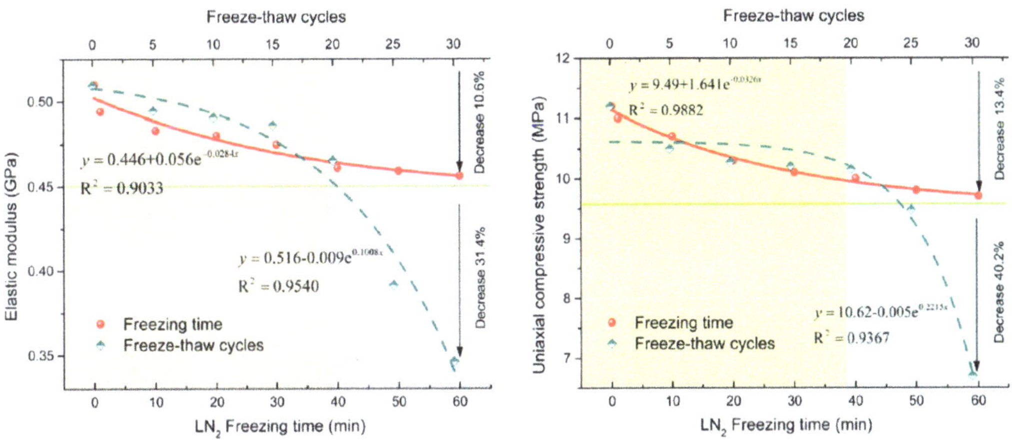

Figure 3. Changes in the elastic modulus and compressive strength after freezing–thawing cycles.

In addition, Qin et al. (2017a) [15] examined coal samples from Shengli (lignite), the Jungar (bituminous) coal mine in Inner Mongolia, and the Xinbei coalfield (anthracite) in Gansu, China. The first group of coal specimens was dried and then saturated with water. The samples were frozen and subjected to the freezing–thawing cycle process. The second group of experiments with different moistures (controlled by drying oven) were frozen for 90 min. Coal specimens of Shengli lignite were analyzed with different freezing times, different freezing–thawing cycles, and different moisture levels. The cyclic exposure samples were then put for vacuum saturation for 12 h. Finally, they were put in a centrifuge at 200 psi for 2 h. The results indicated that freezing time had a small influence on the permeability and porosity of the specimen. On the other hand, the number of freezing–thawing cycles imposed a great influence on pore structure. As the number of freezing–thawing cycles increased, porosity, and permeability were augmented, which is favorable for CBM extraction. As for samples containing moisture, the outcomes suggested that the greater the moisture concentration in coal specimens, the larger the increment of permeability. It should be mentioned that the influence of a high moisture concentration has limits. Comparing LN_2 freezing time effects and on lignite, anthracite, and bitumen, both the growth rate and total number of pores followed a downward trend of effectiveness: lignite > anthracite > bitumen. In contrast, the growth rate for seepage pores indicated a different behavior. The highest augmentation took place in anthracite, followed by bitumen, and then lignite. The influence of coal rank on the growth rates of porosity is also worth noting. The growth rates by rank, listed in downward trend, are lignite (low-rank), anthracite (middle-rank), and bituminous coal (high-rank). After the specimens were frozen and thawed using LN_2, the growth rates of their effective porosity and total porosity were all positive, but the growth rates of their residual porosity were negative. This means that after the coal samples were frozen and thawed, the closed pores were opened, and the open pores were augmented in magnitude and connectivity. This indicates improved reservoir conditions for extracting CBM [15].

They also went further to examine permeability after freezing–thawing cycles with the use of porosity parameters and the T_2 spectrum as a surrogate for pore size. They used three different models: the Schlumberger Doll Research (SDR) model, the Timur–Coates model, and the Producible Porosity model. The last two models indicated high error values; hence, only the outcomes of the SDR model were analyzed. Coals of disparate ranks show different increments of permeability augmentation due to different initial coal porosity. The downward trending permeability enhancement created by LN_2 treatments, in order, is lignite, anthracite, and bituminous coal. Qin et al. (2017a) [15] also used SEM analysis, apart from NMR. SEM images showed that with LN_2 freezing of 10 min, the largest crack width was 13.3 μm, while for samples following 30 freezing–thawing cycles, the largest crack width was 60 μm, suggesting that the length and width of cracks augmented with freezing time or freezing–thawing cycles. LN_2 treatments also improved the fracture density per volume of specimens [19]. Furthermore, it was observed that the frost heaving influence of the water–ice phase transition first creates single primary cracks. Then, many other secondary cracks expand along the primary cracks. As the freezing–thawing cycles continue, the amplitude and complexity of the cracks slowly augment [15].

Qin et al. (2017b) [27] also analyzed the freezing–thawing damage variable, D,

$$D = 1 - \frac{E_n}{E_0} \tag{8}$$

where E_o and E_n are elastic moduli of the original and the frozen–thawed coal specimens.

Their outcomes through uniaxial stress analysis indicated that augmenting the freezing time brought the value of D up to about 0.12. On the other hand, moisture is negatively correlated with D, such that the smaller the moisture, the greater the damage done in the elastic deformation phase. Nevertheless, D values connected to moisture are curbed by a specimen's highest water saturation. Furthermore, fractal dimensions obtained using the box counting process on fractures rose from 1.06 to 1.23 for specimens undergoing 5 and 25 freezing–thawing cycles, respectively. The ultrasonic wave velocities indicated a negative connection with coal permeability, and cyclic freezing–thawing is efficient in augmenting coal permeability [27].

Li et al. (2018) [28] investigated coal specimens from the Yima Group, Henan Province. After freezing–thawing cycles, there was an observation of augmentation in permeability, pore diameter, and porosity in coal specimens. This augmentation became bigger in samples with water saturation greater than 70%. There was also an observation that thermal degradation dry specimens were due to cold shock, while in water-saturated specimens, apart from thermal degradation, water phase change is also responsible. They indicated that permeability enhancement pauses when the reduction of temperature extends to the maximum, mentioning that the progress of temperature reduction is connected to pore diameter, heat transfer properties, and freezing period [28].

Jin et al. (2019) [29] examined anthracite specimens of high metamorphism from Zhaogu No.1. Coal mine in Jiaozuo city and coking coal specimens of middle metamorphism from No.8 coal mine in Pingdingshan city and found that elastic modulus, compressive strength, and brittleness index are lower in specimens with LN_2 treatment compared to others without LN_2 treatment. For specimens with anthracite, the maximum strain of LN_2 cryogenic coal specimen augmented from 3.65% (non-cryogenic value) to 6.52%. For specimens with coking coal, the maximum strain of LN_2 cryogenic coal specimens increased from 4.75% (non-cryogenic value) to 5.69%. The average elastic modulus of LN_2 cryogenic anthracite was 1.75 GPa, which is lower than the average elastic modulus (2.48 GPa) of the non-cryogenic samples, with a reduction rate of 41.5%. The average elastic modulus of the coking coal samples diminished from 3.57 GPa to 1.75 GPa, with a reduction rate of 50.84% [29].

Chen et al. (2019) [30] examined bituminous coal samples from the Xutuan coal mine of Anhui province in China using the methods of freezing time and freezing–thawing cycles. Their outcomes indicated that a single LN_2 treatment under a stable confining pressure augmented permeability exponentially with the augmentation of gas injection pressure. On the other hand, freezing–thawing cycles initially caused reduction and then augmentation with increasing gas injection pressure. Furthermore, the permeability of coal specimens augments exponentially with the number of cycles, but the enhancement rate decreases slowly. Generally, LN_2 was successful in promoting the formation of macroscopic fractures in coal and in increasing sample permeability. The level of damage to the coal structure under LN_2 circulation was much larger than that related to a single method, and the level of damage augmented with the number of cycles [30].

Chu et al. (2020) [24] examined anthracite coal samples from a Baijiao coal mine in China. Their tests started by drying samples for 24 h at 80 °C and then placing them in a water saturation device for 12 h with −0.1 MPa to achieve full saturation. They measured T_2 spectrum distribution and porosity through NMR. Finally, coal specimens were placed in a centrifuge machine at 200 psi pressure for 1.5 h to reach irreducible saturation. NMR examination was used again to estimate T_2 spectrum distribution and porosity. After these initial measurements, they compared freezing time and freezing thawing cycle methods at 60, 120, 180, 240, 300, and 360 min and 2, 4, 6, 8, 10, and 12 cycles (each freezing–thawing cycle comprised of 30 min of freezing and 30 min of thawing at room temperature), respectively. Outcomes indicated that with increasing length and number of freezing–thawing cycles, area, and amplitude of mesopores (100 nm to 1 μm) increased, inferring that considerable pore extension took place in micropores (<100 nm), which could turn into mesopores. The use of LN_2 can advance the link between pores, expedite the creation of fractures, significantly increase the ratio of mesopores to macropores (>1 μm), and notably augment the shape of seepage pores. Moreover, there is a tendency for stable enhancement with augmentation in freezing time and freezing–thawing cycles by the relative area change rate. The relative change area of the adsorption pores indicates an inclination for stable enhancement and then a steady reduction during LN_2 treatment, while the relative area change of the seepage pores shows an inclination for quick enhancement. Total porosity, effective porosity, and residual porosity were positively impacted by the use of LN_2 treatment time. There was also the observation that micropores constantly extend and then link to create more mesopores, macropores, and fractures driven by expansion and frost heaving pressure. Permeability measurements showed a significant augmentation effect after 4 freezing–thawing cycles; hence, more LN_2 freezing–thawing cycles can enhance the permeability by generating a larger number of mesopores and macropores, as well as enhancing the connectivity of pores. Their SEM analysis showed that the pore structure of coal specimens was constantly impaired and defaced, while fractures were created on the surfaces of the specimens. Increasing time and cycles, the cracks extended with an obvious increase in depth and width. New cracks created close to the initial ones can form a crack network by linking them to each other [24].

Du et al. (2020) [31] compared coal samples from the Yulin coal mine in Shanxi, China, and sandstone samples from Sichuan, China. The samples were dried for 12 h at 60 °C, weighted, and evaluated for P-wave velocity (V_p). The samples were submerged into LN_2 for 60 min and then analyzed. The results indicated that the micropores slowly expanded and linked with each other to create a network, augmenting the permeability in all coal samples. The use of LN_2 created both vertical and parallel fractures on coal surfaces; however, no obvious fractures took place on sandstone surfaces. This outcome showed that thermal stress compromised the integrity of the coal specimens by creating fractures. Hence, LN_2 is more effective in coal fracturing than sandstone fracturing [31], which was confirmed by Cai et al. (2014) in their NMR analysis [25]. Du et al. (2020) [31] also used an ultrasonic technique to compare coal and sandstone samples after the use of LN_2. Differentiation in wave velocities relates to the creation of new crevices or extending pre-existing crevices in rock specimens [32]. Broadly, waves of sound spread more quickly in solids than in liquids or gaseous. Hence, when V_p diminishes, more crevices may be created inside the rock specimen. After the use of LN_2, reduction in V_p was 24.7–38.1% with an average value of 30.63%, showing that the use of LN_2 had a beneficial effect in fracturing coal specimens. Furthermore, in one coal specimen, there was a great reduction in V_p, suggesting that large crevices may be created after the use of LN_2. On the contrary, V_p reduction for sandstone samples was only 0–0.06%, suggesting that LN_2 treatments of sandstone specimens created very little damage, which was also not visible [31].

Yan et al. (2020a) [10] examined coal samples from the Fenxi mining region of Shanxi, China. They followed an experimental process in which the coal sample was (a) heated to 60 °C in a drying oven for 24 h, (b) cooled to room temperature for 6 h, (c) examined by μCT scan, (d) immersed in LN_2 tank for approximately 1 h, and (e) equilibrated to room temperature for 30 min. Steps a to c were also repeated for the treated samples. The results indicated that LN_2 substantially influences pore distribution characteristics and strengthens their connectivity. According to stereoscopic diagrams, there was an augmentation of coal porosity by 183% after the use of LN_2. The surface porosity augmentation rate after thermal shocking is inversely related to the initial surface porosity of coal, suggesting that LN_2 fracturing is able to meet expectations for stimulating CBM reservoirs with low porosity and permeability. SEM analysis indicated that there were many deep and shallow scratch layers that showed progressive patterns. Moreover, several new cracks propagated, and sections of the pre-existing cracks extended and connected each other. Some embedded attachments are scoured, generating a few micro-holes or adventitious fractures on the surface. This fact affects the coal sample strength and could significantly ameliorate CBM reservoir recovery [10].

Su et al. (2020) [33] investigated coal specimens from a Yulin coal mine in Shanxi Province, China. Their outcomes showed that the tensile strength of coal samples under loading vertical and parallel to the coal bedding direction was reduced by 14.68% and 23.23%, respectively. Moreover, more complex fractures were observed following Brazilian splitting tests, while openings and the density of paths were also greater after LN_2 treatment. Lastly, SEM analysis suggested that amelioration in the permeability of a coal seam was due to multiple micro-cracks that were formed from matrix, original pores, and micro-cracks of coal after LN_2 treatment [33].

Akhondzadeh et al. (2020) [34] examined bituminous coal specimens from Morgantown, USA. The coal specimens were vacuum dried at 60 °C for 4 h in an oven. Then, the specimens were submerged in LN_2 for 60 min. The opening of the fractures before the use of LN_2 was not more than 10 μm, while after the treatment, an augmentation in the fracture aperture to 13 μm was noticed. Furthermore, it was observed that the created fractures were interconnected to the original cleats; hence, there was an augmentation in the apparent permeability of the specimen. According to the 3D analysis, the outcomes indicated that LN_2 not only created new fractures, but also interconnected the original isolated pores and fractures with the initial cleat network. SEM analysis performed in the pre-treatment and post-treatment states indicated that there were initially no important fractures in the SEM images of the specimen base and sides. After the use of LN_2, the largest crack opening was 9 μm. Furthermore, the fractures appeared on both sides and bases, forming a new fracture network inside the specimen. They also examined the mechanical properties of fracture rocks through nanoindentation. The indentation moduli, directly related to Young's modulus, of the specimen measured with 10 mN and 35 mN indentation forces were reduced by 5.1% and 14.4%, respectively. Such an important decrease corresponds to the existence of notably extended new fractures inside coal [34].

Zhang et al. (2020) [35] investigated the 3D microstructure and permeability changes in bituminous coal before and after LN_2 immersion. The outcomes indicated that accurate division of fractures was acquired, but that inordinate division could be avoided. Furthermore, according to the box-counting method, the 2D fractal dimensions of pore fractures in both directions augmented considerably after LN_2 treatment. The magnitude and quantity of fractures and crevices were augmented after LN_2 treatment. The summed volume of fractures augmented from 306.48 mm^3 to 967.55 mm^3, and the porosity augmented from 0.38% to 1.22%. Lastly, permeability values augmented exponentially with increasing inlet pressure under constant confining stress [35].

Zhang et al. (2020) [36] examined anthracite and bituminous coals subjected to both LN_2 cold shock and heating-cooling shock. Their specimens were from the Jiaozuo mining region of Henan province in China. Their results indicated a 2.11–3.89-fold improvement in permeability related to a sudden temperature increase from 180 °C to 360 °C. Their 3D

CT scan analysis indicated that thermal shocks increased the extension and linkage among fractures and effectively ameliorated the permeability of coal specimens. The results also indicated the formation of bifurcated and wing cracks on the surfaces of the coal samples during thermal shocks. The number of new cracks created by the heating–cooling method was greater than the single LN_2 cold shock. The crack propagation was also more complex in the heating-cooling method, indicating that the heating-cooling method achieved better outcomes in damaging and breaking the specimen than the cooling method alone [36].

Yan et al. (2020b) [13] examined coal specimens from the Zhongxing mine in Shanxi Province, China. They used two different samples with the same diameter of 25 mm and different heights of 42 mm and 50 mm for examining LN_2 quenching, along with various initial fracture magnitudes and saturations. The process was similar with LN_2 freezing–thawing cycle process with dry and water saturated specimens. Outcomes indicated that fracture thickness, fracture volume, fracture connectivity, and porosity augment with LN_2 freezing–thawing cycle process, notably, with the first cycle as the most significant. With the use of LN_2, large initial porosity was found to bear a smaller growth rate of surface porosity than small initial porosity, indicating greater prospects for the significant enhancement of rock specimens with low permeability. Furthermore, after the use of LN_2, there was continuous and gradual augmentation in the volume of fractures for the saturated coal specimens [13].

Lin et al. (2020) [37] examined coal specimens from the Hengyi Coal Mine in Shaanxi Province, China. They observed that fractures on the surface of coal specimens propagated after the use of LN_2. The authors reported expansion of fracture width with linear augmentation 3.3 times more with water-saturated specimens as compared to dried specimens. The damage factors expressed by fracture width, longitudinal wave velocity, and porosity were augmented exponentially. With the freezing–thawing cycle process, the longitudinal wave velocity in coal specimens was moderately reduced, and there was a positive exponential relationship between the decrease in wave velocity in coal specimens and the moisture concentration. Moreover, the freezing–thawing cycle process with LN_2 showed that micropores and small pores in coal mass were moderately changed to mesopores and macropores, in addition to the creation of new micropores and small pores [37].

Qin et al. (2020) [38] studied the pore size distribution of coal samples frozen by LN_2 with mercury porosimetery by combining the low-temperature nitrogen adsorption of lignite coal specimens from Shengli Coal Field, Inner Mongolia Autonomous Region, China. After freezing specimens for 180 min with LN_2, the number of small pores and macropores in the coal specimens increased. After 30 freezing thawing cycles, the quantity of small pores was reduced, but the number of macropores was augmented. Moreover, there is a beneficial exponential correlation with freezing time for both total pore volume and cumulative pore volume, but the augmentation gradually diminishes as freezing times augment. Nevertheless, the cumulative pore volume enhancement follows a quadratic function with the number of freezing–thawing cycles, and the increase slowly augments as the number of freezing–thawing cycles accrues [38].

Li et al. (2020) [39] studied anthracite coal samples collected from the Jiu Lishan Mine in Jiaozuo, China. According to their interpreted NMR results, there was augmentation in the number of small pores and micropores inside the coal specimens after LN_2 treatment. The water-saturated coal specimens had less thermal stress than the dried ones. The damage to water-saturated coal specimens was due to phase change, while in dry coal specimens, mechanical integrity was compromised due to thermal stress between particles. Generally, the temperature alteration of saturated coal is slower than that of dry coal, and the total to reaching equilibrium after thermal shock has a larger duration [39].

Liu et al. (2020) [40] collected samples from different regions. The specimens in the study were taken from the fat coal of the Pingmei No. 5 Coal Mine, the lean coal of the Daping Coal Mine, and the anthracite coal of the Guhanshan Coal Mine. The cryogenic nitrogen adsorption curves of the fat coal, lean coal, and anthracite before and after LN_2 cold soaking followed a lying "S" type. When $p/p_0 < 0.1$, there was a transition from monolayer to multi-molecular layer adsorption, and the curve is convex. When $0.8 < p/p_0 < 1.0$, the adsorption curve is concave. Moreover, samples showed an enlargement in pores, mesopore volume and pore-specific surface area, total pore volume, and total pore-specific surface area for every coal specimen. Lastly, the thermal stress created during LN_2 treatment was larger than the tensile strength of coal. LN_2 treatment plays the role of augmenting and expanding the pore and advances the growth of the coal pore structure [40].

Akhondzadeh et al. (2021) [41] examined coal specimens using a multiscale tomographic study. Their outcomes indicated that in micro (μ-CT) and macro (medical CT) scales, LN_2 fracturing in anthracite specimens was poor. On the other hand, in bituminous and sub-bituminous specimens, porosity evolution augmented 14% and 119%, respectively. In macroscale bituminous and sub-bituminous specimens, there was substantial augmentation in their porosity. Connectivity assessment through skeletonization and Euler number analysis showed that sub-bituminous connectivity increased by 20-fold for small pores (<3 μm), while it doubled in bituminous [41].

Sun et al. (2021) [42] examined bituminous coal samples from the Datong coal mine in Shanxi province of China using the freezing–thawing cycle method. The process started by putting coal specimens in a vacuum saturator for 12 h to extract air to saturate with water. NMR analysis then took place. Later, the specimens were placed in a centrifuge at 8000 rpm for 1 h for the removal of free water, followed by further NMR analysis. Centrifugal specimens were dried for 12 h; then, ultrasonic emission and stereomicroscope image analyses were performed. The specimens were also saturated and then immersed in LN_2 for 10 min, followed by water bath (40 °C) immersion for 10 min. The outcomes indicated that the width and range of the T_2 curve in coal specimens were augmented. With a higher number of cold shock periods, total, and effective porosity, together with free fluid proportion, are augmented linearly, while both $T_{2cutoff}$ and residual porosity are reduced linearly. Furthermore, the contribution of micropores plays a leading role in the augmentation of effective porosity. The augmentation of open holes due to exposure to LN_2 was more advantageous for seepage and gas diffusion. They also analyzed fracture evolution by stereoscopy, observing that with increasing cold shock times, the surface porosity and the number of pores augmented, and the pore distribution movement tended to become more random. They concluded that short fractures in specimens slowly emerged, expanded, and extended to long fractures. As the secondary branch fractures expanded, ternary branch fractures formed and became visible. In the weak structural areas of the specimen, weak planes developed due to the high fracture density. A fracture network was created, and the specimen body was damaged. When the local weak area was totally damaged, a large amount of pulverized coal was generated. Furthermore, they also conducted ultrasonic wave analysis, indicating that, following the use of LN_2, there was a reduction in the amplitude of the first peak. The primary frequency of the ultrasonic wave diminished while the signal attenuation was augmented rapidly. The P-wave and S-wave velocities diminished gradually with the number of thermal socks. The structure of the coal specimen became layered instead of loose. Furthermore, the anisotropy augmented with the expansion of primary fractures but lessened with the creation of a primary-secondary fracture network. The dynamic mechanical parameters were reduced as the number of cold shocks increased, while the most crucial reduction took place in the dynamic bulk modulus [42].

Shao et al. (2021) [43] examined bituminous coal samples from the Tazigou coal mine in Liaoning, China. First, the samples were polished and put in a drying oven for 48 h at 45 °C. Then, half of the specimens were saturated with water. Furthermore, half the specimens from both groups were wrapped to investigate the impact of LN_2 on the physical and mechanical properties. Apart from cylindrical samples, there were also Brazilian discs tested at the same conditions. All the samples were submerged in LN_2 for 36 h and then allowed to warm to room temperature before starting the other measurements. SEM analysis of the saturated samples after the use of LN_2 indicated similar failure and crevice characteristics, with more significant damage than dry coal.

Nevertheless, interconnected fracture networks were created on the surfaces of the dry samples. Relatively short micro-fractures with low connectivity and narrow openings were created in the interior of dry coal specimens, notably without much penetration of LN_2. On the other hand, big failure regions and large fractures with connections were created in the interiors of the water-saturated specimens. P-wave velocity was also examined in all 4 different groups (water saturated wrapped and unwrapped, and dry wrapped and unwrapped). LN_2 reduced the P-wave velocity in all four groups. The impact order from high to low was: saturated wrapped specimen > saturated unwrapped specimen > dry unwrapped specimen > dry wrapped specimen. A higher cryogenic fracturing efficiency was observed in water-saturated specimens than in dry ones. Concerning gas permeability, there was significant augmentation in all four groups [43].

Lin et al. (2021) [44] examined the use of LN_2 in coal specimens from thermodynamic perspectives. The experiments started by drying 10 g of coal particles at 353 °K for 6 h in an oven. Then, the coal specimen was placed in a sealed freezing–thawing tank to return to room temperature. LN_2 was added to the system, and the tank was put in an oven for 2 h until the specimens returned to room temperature. After the freezing–thawing process, the oven was turned off, and coal specimens remained in a sealed container for 6 h until returning to room conditions (i.e., the completion of 1 cycle). After the appropriate number of freezing–thawing cycles, the specimen was again put in an oven and dried at 353 °K for 6 h to totally evaporate the free water created in the coal specimen during the freezing–thawing process. Then, the frozen–thawed specimen was weighed and placed into a high-pressure adsorption chamber, where the specimen was degassed by a vacuum pump, and helium was injected to estimate the fixed volume. By setting the specific experimental temperature, the maximum experimental pressure was 5 MPa, the pressure gradient was 1 MPa, and the system was filled by CH_4. When the experimental pressure equaled the set pressure and adsorption reached equilibrium, the experimental process terminated. The above steps were implemented to complete CH_4 adsorption experiments for coal specimens with different freezing–thawing cycles at different temperatures. Outcomes showed that the freezing–thawing process can obviously augment the CH_4 adsorption quantity of a coal specimen ($V_{\text{unfrozen}} < V_{\text{freezing 5 cycles}} < V_{\text{freezing 10 cycles}}$). The adsorption enthalpy H^{θ}, adsorption entropy S^{θ} and adsorption Gibbs free energy G^{θ} all increased as the freezing–thawing process continued. With the use of LN_2, the isosteric heat adsorption augments, because the LN_2 freezing–thawing process can greatly expand the adsorption sites of coal specimens. Furthermore, the heterogeneity of the coal specimen matrix augmented with the freezing–thawing process of LN_2. As the number of freezing–thawing cycles increased, the quantity of larger pores was augmented. The shedding of particles on the coal specimen surface and the extension of microchannels shows propitious signs for CBM adsorption [44].

Liu et al. (2021) [45] examined the effect of different LN_2 soaking times on the temperature distribution of coal specimens and explored the temperature evolution mechanism of anthracite-lignite-bituminous coals. Their experimental process was described as follows: the dried coal specimens were placed in an incubator at the same distance from a LN_2 injection hole. LN_2 was injected into the incubator through the hole until the coal specimen temperature stabilized. There were measurements taken to characterize coal specimens, and the process was repeated with alternative LN_2 soaking times. The outcomes indicated that with increasing LN_2 soaking times, the temperature of all three coal specimens exhibited three phases, including a reduction in their acceleration initially, continuation with small decline, and then stabilization at low temperature. The order to reach phase III (anthracite > bituminite > lignite) was opposite compared to the temperature-changing rate. Furthermore, thermal stress created in coal specimens helped in the extension and expansion of crevices and accelerated heat transfer. The temperature increase curve of the three coal specimens was augmented with time as a logarithmic function and indicated a unimodal distribution [45].

Qin et al. (2021) [46] investigated the effect of LN_2 on coals of different ranks (anthracite, lignite, bituminite) and gas adsorption. After 3 h immersion of samples in LN_2, outcomes indicated that the greatest amounts of adsorbed gas took place in lignite and bituminite, while the anthracite specimen showed a reduction. Furthermore, the contact angle for the anthracite specimen also decreased. The gas adsorption capacities of the specimens indicated a positive linear relation with the analogies of aliphatic hydrocarbon but a negative linear relation with the analogy of substituted benzene functional groups and oxygen-comprising functional groups. As the contact angles became greater, the coal gas adsorption capacities initially decreased but then augmented [46].

Sun et al. (2021) [47] examined coal specimens from Qinshui coal in China. They used the immersion method with SEM and NMR analyses. The initial NMR porosity of deep coal was smaller than that of shallow coal from the same coalfield and coal seam. There was an improvement in connectivity between mesopores and macropores in both deep and shallow specimens after the freezing–thawing cycle process. Furthermore, the quantity of micropores in the total pore structure volume diminishes, and the quantity of mesopores and macropores in the total pore structure volume accrues [47].

Yang et al. (2021) [48] developed a non-destructive geophysical technique using seismic measurements to probe fluid flow through coal and ascertain the effectiveness of cryogenic fracturing. Their results showed that the ultrasonic velocity of dry and saturated coal samples reduced overall with freezing time because of induced thermal and frost failure. For a gas-filled sample, both normal and shear fracture stiffness were reduced with freezing time, as more cracks formed in the coal bulk. On the other hand, for water-filled samples, ice formation due to cryogenic treatment is guided to the bottom of coal bulk [48].

Qin et al. (2022) [49] selected bituminous coal from the Hengyi Coal Mine in Shaanxi, China. They studied the influence of various LN_2 freezing variables on mechanical properties. Their outcomes showed that the mechanical strength of samples can be augmented for a period up to 100 min in LN_2 treatment where tensile stress elastic modulus and uniaxial compressive elastic modulus were augmented by 30.1% and 54.6%, respectively. However, as freezing times augment, there will be a reduction in the mechanical properties of coal. The freeze enhancement factor, I, augmented with increased freezing time, while the damage factor, D, augmented exponentially as the number of freezing–thawing cycles increased [49].

Yuan et al. (2022) [50] investigated specimens from Huainan Coalfield and Qinshui Coalfield by examining adsorption and desorption from middle- and high-rank coal specimens. The curve shape outcomes showed that coal specimens were typically mesoporous. Comparing middle-rank and high-rank coal specimens in terms of pore volume and average pore size, the results showed that in middle-rank coals, values were higher compared to high-rank coals [50].

Akhondzadeh et al. (2022) [51] investigated the LN_2 freezing–thawing process in coal fracturing, focusing on the influence of different freezing cycles examined in μ-Computed Tomography (μ-CT) images. The freezing–thawing process showed promising efficiency in cleat network evolution after three freezing–thawing cycles. The maximum opening of the initial fractures was 15 μm, while several new fractures had a maximum opening of 10 μm. According to connectivity analysis, the number of pores augmented 50%, while the number of interconnected pores almost doubled. SEM analysis indicated that there was creation of a fully penetrating fracture, which accrued in aperture magnitude and extended the cleat network of the specimen, especially in latter freezing–thawing cycles [51].

Hou et al. (2022) [11] examined coal specimens from the Yulin mining area of Shanxi Province. Their results indicated that coal after LN_2 treatment, in comparison with untreated specimens, increased compressive strength by 52.85%, in P-wave velocity 10.97%, in fracture toughness 495.07%, and in elastic modulus 38.76%. On the other hand, tensile stress decreased by 26.17%. Their outcomes were different in the freezing–thawing experiments. There was an important diminution for P-wave velocity of 26.14%, for elastic modulus 23.20%, for fracture toughness 58.33%, for compressive strength 21.20%, and for tensile strength 42.99%. Micro-level outcomes showed that the failure mode, failure path, crack propagation direction, and fracture surface morphology of specimens in mechanical experiments became more complex, and the freezing–thawing specimens had the best failure effect [11].

Li et al. (2022) [52] selected coal specimens from the Qianjin coal mine in the Bijie coalfield, Guizhou, China. SEM outcomes indicated that as moisture content increased, the magnitude and width of pores augmented. NMR outcomes indicated that the proportion of micropores and transition pores from FT LN_2 treatment specimens was smaller than the untreated samples. Total pore volume, porosity, and average pore diameter were larger after LN_2 treatment. Moreover, permeability was increased after LN_2 treatment; the higher the moisture content, the higher the permeability of the coal samples after LN_2 treatment [52].

2.2. Injection Tests and Fracture Evaluation

LN_2 has been injected into the borehole drilled in a rock specimen to investigate its fracturing efficiency on the borehole wall. In most cases, confining stresses are exerted to mimic well stimulation scenarios under in situ stress conditions.

Cai et al. (2018) [53] examined coal specimens from Ordos in Inner Mongolia, China, using the method of LN_2 injection. A coal sample with a hole and injection tube was sealed with cement around it to make a cubic rock test specimen. Their acoustic emission (AE) outcomes indicated that as the specimens came into touch with LN_2, there was a continuous period of damage and cracking. The average AE ring-down count rate and cumulative ring-down counts during LN_2 injection were, at most, 14.45 and 12.08 times larger than before injection. Furthermore, the coal specimens had smaller tensile strengths than the specimens in their original state. In the cooling state, the specimens showed an average reduction of 17.39%, while in the cool-treated state, there was an augmentation of 31.43%. Reduction of wave velocity indicated that LN_2 can assist and expand the micro-fissures in the interior side of the specimen. In the cooling state, the average reduction in wave velocity was 14.46%, while in the cool-treated state, there was an augmentation of 18.68%. Additionally, during the LN_2 cooling process, AE signals were mainly produced from coal cracking [53].

After immersion tests with NMR and processing of SEM images, Qin et al. (2018) [54] continued to analyze the effect of LN_2 in CBM reservoirs with AE for both freezing and freezing–thawing injection methods. For the freezing injection method, the AE signals indicated that failure in the specimens took place near the injection tube and did not propagate. On the other hand, when they used the freezing–thawing injection, there was a plastic deformation region formed after the primary cracks were created and interconnected. Figure 4 shows the AE energy signals that were monitored during the single and cyclic LN_2 injections and the relationship between the AE energy and injection time.

Furthermore, after the use of LN_2 with a freezing time of 10,000 s and cyclic injection for 9 cycles for the same period, the lowest temperatures on the surface were -12 °C for the freezing method and -107 °C for cyclic injection method. In a single injection, the temperatures inside and outside the specimen ceased to diminish when they attained a specific value. Moreover, there was no effective fracture network in the tested specimens, and the heat transfer was through the solid material [54]. Mesoscopic analysis showed that the lower the density of the specimen, the higher the original water content, with larger associated frost heaving forces from water expansion during LN_2 injection. A greater temperature gradient forced more water to move, which resulted in more frost heaving due to water migration and ice accumulation. Furthermore, there was a connection between the shrinkage deformation of particles and the physical properties of the particles and a relationship between compression deformation and confining pressure [54].

Yin et al. (2018) [55] examined coal specimens from the Pingdingshan coal field of Henan Province, China, before and after the use of LN_2. Outcomes indicated that the micro-crevice threshold of coal specimens was far less than the frost heaving stress. Two main changes that took place after the use of LN_2 were the nonuniform shrinkage deformation of coal specimens and the creation of pores, the majority of which were on the micropore and macropore scale. Concerning permeability, in low permeable and heterogeneous coal specimens, there was augmentation due to thermal cracking and sporadic opening of seepage pores due to phase change of free water in pores or micro crevices.

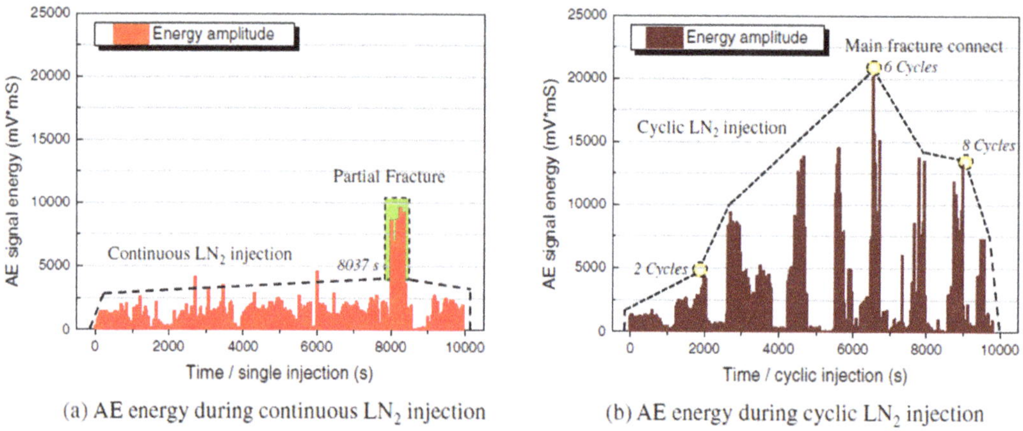

Figure 4. Changes in acoustic emission energy with time during the injection of LN_2.

Cong et al. (2021) [56] examined variations in the temperature field and strain field of coal around the drill hole after the injection of LN_2 into the drill hole. Upon heating from −30 °C to 30 °C, the expansion capacity augmented slowly until it leveled off. The average linear thermal expansion coefficient of coal in this temperature range was $\alpha = 36.06 \times 10^{-6}$/°C. Moreover, the low-temperature zone of coal around the drill hole expanded in a semicircular shape at a decelerated rate after LN_2 was injected into the drill hole. The strain and action range of frost contraction (long term) were larger than those of frost heave (short term) [56].

Yang et al. (2021) [57] examined synthetic coal cubes made of water, cement, sand, and coal particles with a ratio of 3:2:1:4. A 14-mm diameter, 3-mm thick stainless steel casing was then attached to the borehole with epoxy-resin adhesive. The casing extended 120 mm into the borehole, with an open-hole section of 60 mm. High pressure N_2 pressurized the LN_2 tank and pushed the LN_2 into boreholes under high pressure to perform the fracturing treatment. An insulation pipe was also applied between the high pressure LN_2 container and the borehole. The high pressure LN_2 tank can withstand a maximum pressure of 35 MPa and a minimum temperature of −196 °C. Moreover, there were also valves and fittings along LN_2 transportation lines that could withstand cryogenic conditions. In this research work, there was a comparison of LN_2 and water injection. Their outcomes indicated that the breakdown pressure for the LN_2-fractured specimen under higher confining stress may be lower than that for the water-fractured specimen subjected to a lower confining stress state. Furthermore, if the horizontal-stress-difference ratio is reduced from 1 to 0, the breakdown pressure of a specimen fractured by water will augment more considerably than that fractured by LN_2. The highly permeable regions were larger for LN_2-fractured specimens with multiple fractures starting from and adjacent to the borehole, while smaller stimulated areas were obtained in water-fractured samples, representing fractures with limited conductivity. The fracture planes of water-fractured samples were not as smooth as LN_2-fractured samples, and some of the water-fractured samples were not fully separated along the major fractures, showing that the fracture extension length was longer for LN_2 fracturing than water-based hydraulic fracturing. Hence, LN_2 fracturing tends to create more conductive and longer fractures than water fracturing. Concerning conductivity, the fracture conductivity created by LN_2 fracturing can be 8 to 280 times as large as that by water fracturing. Their analysis with cryo-SEM enabled the imaging of specimens at a temperature of 190 °C and SEM analysis took place at room temperature. Comparing coal and sandstone samples, pre-existing pores in coal started creating fractures after the use of LN_2 with the maximum fracture width of 1.3 μm. On the other hand, there were no long fractures but only short cracks generated in sandstone specimens [57].

Du et al. (2021) [58] investigated coal specimens from the Zhang Minggou coal mine in Shaanxi, China. Their results showed that LN_2 cooling can cause significant thermal damage inside the coal, along with a reduction in P-wave velocity (approximately 15.8%). At the macroscopic level, the damage is stated as distortions in coal's mechanical properties, comprising tensile strength, peak strain, and absorbed energy. Lastly, the process of macrocrack propagation is closely related to the bedding angle [58]. Furthermore, the creation of cryogenic fractures or fissures on the surfaces of the coal specimens due to LN_2 were associated with the removal of some coal particles from the specimen's surface, which could act as self-propping agents to impede fracture closure in the thawing stage.

2.3. Jet Drilling

Jet drilling utilizes high-velocity fluids to break down rock samples of high strength. Water is the most used jet fluid. Recently, LN_2 has been tested as a jet fluid, seeing that thermal shock could boost its mechanical impact against rock specimens. Yang et al. (2019) [59] used a new method for CBM stimulation called Abrasive LN_2 Radial Jet Drilling (ALN-RJD), which drilled multiple radiating laterals instead of a single vertical wellbore to create more expanses in an eco-friendlier way. Their outcomes indicated that the coal-breakage threshold pressure of ALN-RJD was 87% and 89% of those of LN_2 jet and water jet, respectively. Furthermore, the pre-mixed abrasive LN_2 jets indicated better outcomes compared to post-mixed abrasive LN_2 jets in the field of coal breakage. ALN-RJD, due to its ability to create conductive pathways for gas transport in complex lateral-fracture networks, can augment cumulative gas production by 10% compared to the conventional RJD process. Thus, the greatest Net Present Value (NPV) could be achieved [59].

3. Modeling Studies

There have also been a few studies trying to simulate cryogenic fracturing processes. Cai et al. (2016) [26] simulated a LN_2 jet in abrading coal specimens using a computational fluid dynamics method to examine the effect of a LN_2 jet on particle removal along with cracking effects of LN_2 on coal specimens compared to a water jet. Simulation results indicated that under the same nozzle pressure difference, the LN_2 jet showed better conduct in abrasive particle removal and pressure boosting in the perforation cavity in coal than did the water jet. Furthermore, the impact influence caused by LN_2 was similar to that generated by water [26].

By integrating the Mogi–Coulomb failure criterion into TOUGH2-EGS (finite difference scheme), which couples thermal-hydraulic-mechanical processes, Yao et al. (2017) [60] successfully reproduced the bulb-shape fracture morphology generated by LN_2 treatment of boreholes in rock cubes and quantitatively matched the permeability enhancements.

Cai et al. (2018) also carried out temperature distribution simulations in rock samples. Their results indicated that even though the sample was subjected to injection by LN_2 for 850 s, the cement part around the coal remained at a relatively high temperature [53].

Along with immersion experimental studies, Jin et al. (2019) [29] examined LN_2 treatment in coal specimens using a strength statistical damage (SSD) constitutive model. Compared with the Mazars model, the SSD modeling reflects the stress–strain relationship of LN_2 cryogenic coal, which is consistent with experimental outcomes. Furthermore, the substantial causes for brittleness-ductility were cohesive force reduction and friction force augmentation [29].

Apart from the experimental study of abrasive jet drilling, Yang et al. (2019) [59] simulated a stimulated vertical CBM well located in the Horseshoe Canyon coals of Western

Canada. They modeled a rectangular reservoir with a dimension of 500 m × 500 m × 8.99 m, which was discretized into 100 × 100 × 1 grid blocks. The simulation time was 750 days. The RJD well indicated an amelioration in gas production rate of 65.8% at the end of the simulation period compared with the horizontal well. Furthermore, the cumulative gas production of the ALN-RJD well was 10.7% higher than that of the RJD well, showing that thermal crevices induced by the LN_2 jet played a significant role in production improvement [59].

The simulation results of Du et al. (2020) [31] demonstrated that the area with low temperature slowly expanded inside samples with time and created a tensile area in the samples. Moreover, in coal specimens, the heat transfer rate is lower than in sandstone specimens due to the higher thermal diffusivity of sandstone. The outcomes from a maximum normal stress calculation indicated that the maximum tensile stress created by momentary contact between coal and LN_2 was 3.41 MPa, while for sandstone samples, it was 4.87 MPa. This was probably because of the higher elastic modulus and thermal diffusivity values for the sandstone specimens. Finally, the simulation outcomes showed that the maximum failure region occurred 3.3 mm from the outer surface for coal specimens, while the failure region for sandstone specimens was almost zero, which confirmed the SEM and P-wave outcomes reviewed above in this work [31].

Cong et al. (2021a) [56] used a hierarchical model to calculate the thermal stress around a borehole. The absolute values of thermal stresses at the measuring points augment initially quickly and then increase steadily. Furthermore, the outcomes showed that thermal stress and thermal stress gradient are higher close to the drilled hole. The thermal stress was over 1.3 MPa and below 0.2 MPa. For the strain analysis results, the borehole underwent frost heave under the action of LN_2 cold shock (LNCS) on both sides. As the LNCS augmented, the frost heave area became the frost contraction area and kept expanding outward [60]. Moreover, Cong et al. (2021b) [61] used FLUENT software to show that the coal specimen side wall temperature quickly decreases and then levels out. Moreover, apart from the effect of a gas layer between the LN_2 interface and the coal specimen side wall, the low-temperature coal specimen side wall with high thermal resistance plays an important role in heat transfer [61].

Ma et al. (2021) [62] modeled the effective freezing radius during the LN_2 injection process. In a sample with a moisture content of 12%, the thermal conductivity of the coal sample was enhanced linearly with the reduction of temperature. The nearer to the freezing hole, the greater the temperature gradient of coal, but the temperature gradient reduced quickly at a certain distance. With an increment in the freezing time, the freezing front became larger outwards, but the rate of freezing front advancement was progressively reduced [62].

Wen et al. (2022) [63] developed a 3D, unsteady-state, fluid flow, and heat transfer model for LN_2 fracturing in a CBM reservoir, which considered the phase transition of nitrogen, thermophysical properties variation of coal, and the heat transfer between nitrogen and the formation. The outcomes indicated that thermal stress had a minor effect on the propagation of the primary fracture near the tip due to the small temperature difference. Moreover, the fracture aperture, injection velocity, reservoir temperature, injection fluid temperature, fracture propagation pressure, and coal cleat porosity affected the effectiveness of LN_2 fracturing in a coal seam [63].

4. Field Applications

Field applications of cryogenic fracturing are scarce and insufficient. Four CBM wells (A, B, C, and D) and one tight sandstone well (E) were stimulated with the use of LN_2 [64]. Well B was originally completed with nitrogen foam fracturing and had an average gas production of 85 Mcf/D in a period of 24 weeks before re-stimulation with cryogenic LN_2. After the use of LN_2 the initial production was 1600 Mcf/D at a 42% N_2 cut. The rate had then undergone a high reduction, and the well was produced sporadically. The rates for the next 16 weeks produced an average of 70 Mcf/D, while the rate after 12 months of the use of LN_2 returned to approximately 85 Mcf/D. Well D had three attempts for stimulation, with the first two finished very early (first attempt with nitrogen foam and second with borate crosslinked gel). In the third attempt at stimulation, borate crosslinked gel was again used for better proppant transferring ability. Its gas production had an average of 42 Mcf/D in the period of 24 months before the use of LN_2. After LN_2 stimulation, the initial production rate was 400 Mcf/D while its average production rate was approximately 60 Mcf/D, indicating an amelioration of 43% [65].

It should also be mentioned that there are a few patents claiming case studies of cryogenic fracturing. The earliest patent referred to the recovery of methane gas from subterranean coal seams with the use of LN_2 to augment the permeability of the portion of a coal seam penetrated by a wellbore [65]. Zhang et al. (2016) [66] used LN_2 as a fracturing fluid and indicated that it is appropriate for the primary fracturing of non-fractured CBM wells and wells with existing natural fractures, as well as the re-fracturing of low productivity wells. Shen et al. (2017, 2018) [67,68] claimed that the LN_2 freezing–thawing method can effectively ameliorate a low-permeability coalbed; hence, the recovery of CBM can be improved. Zhai et al. (2018) [69] claimed a freezing–thawing procedure aimed at improving coal seam permeability by injecting LN_2 in a high-pressure pulsing manner. They indicated that its use is easy, practicable, and economically affordable. Furthermore, they claimed that their implementation could ameliorate both single-pore extraction and extraction concentrations of CBM, in addition to prolonging the attenuation time of gas concentration. Xu et al. (2020) [70] physically simulated LN_2 injection freezing–drying to augment the yield of CBM. They injected the LN_2 to fast freeze water in the coal briquette into ice to open gaps. Then, ice was heated to be directly sublimated into water vapor by combining vacuuming, which resulted in augmenting the yield of CBM.

5. Conclusions

During this transition time toward cleaner forms of energy and the push for eco-friendly solutions to conserve and protect fresh water, cryogenic fracturing with LN_2 offers distinct advantages over traditional water-based stimulation techniques in CBM exploitation.

Based on a thorough review of the use of LN_2 in stimulating CBM reservoirs, cryogenic fracturing can be an effective fracturing method for creating cracks and fracture networks in coal seams. Hundreds of experimental tests on coal demonstrated that existing fractures can be expanded and that new fractures can be initiated by LN_2, bringing about multifold permeability enhancement. Numerical modeling furthered our understanding of the cryogenic fracturing process by visualizing the distributions of temperature, stress, and the fractured region in a complementary manner. The limited number of field pilots carried out in coal seams 25 years ago can be considered successful, right after, despite the 15-year gap in research and development that followed. More recently, cryogenic fracturing has gathered momentum in terms of published research and case scenarios in patents.

Future studies recommended include fracturing mechanism revelation, simulation, large-scale experiments, and field pilot tests of LN_2 stimulation. In laboratory experiments, some cryogenic fractures may be too small for observation; hence, more advanced microscopy and fluorescence processes are imperative for the thorough characterization of thermal shock fractures. In addition, both freezing and thawing contribute to coal cracking; thus, optimization of their relative timings in a cycle deserves more effort. Moreover, more injection studies are recommended since almost 80% of research activities refer to immersion tests.

Additionally, simulation studies of cryogenic fracturing are limited. Thermal-hydraulic-mechanical processes should be coupled with fracturing criteria in simulators and scaled to design field pilot tests. In addition, thermal gradients, fracture formation and distribution, and the magnitude of permeability augmentation with cryogenic stimulation differ with specimen size. Hence, large-scale laboratory experiments using 1 cubic meter or more of coal or concrete/coal could be substantial for LN_2 delivery processes and advantageous audit tools for execution at the field scale. Cryogenic fractures partially close as temperature returns to ambient, indicating that proppants are necessary to hold fractures open to maintain conductivity during or after a cryogenic stimulation treatment, both as experimental and as modeling studies. Apart from thermal effects, LN_2 has different compressibility and can evaporate; hence, future studies must also examine how these physical effects may contribute to the result, since evaporation can significantly increase the volume and create longer fractures. Another future focus area should be LN_2 viscosity and the ability to transport proppants.

Author Contributions: Conceptualization, S.N.L.; methodology, S.N.L.; software, S.N.L.; validation S.N.L. and R.H.; formal analysis, S.N.L.; investigation, S.N.L.; resources, S.N.L.; data curation, S.N.L.; writing—original draft preparation, S.N.L.; writing—review and editing, S.N.L.; visualization, S.N.L.; supervision, L.W. and R.H.; project administration, R.H.; funding acquisition, R.H. All authors have read and agreed to the published version of the manuscript.

Funding: This research was funded by NAZARBAYEV UNIVERSITY, grant number 021220CRP2022" and "The APC was funded by NAZARBAYEV UNIVERSITY.

Acknowledgments: The authors are thankful to Nazarbayev University, Kazakhstan, for financial support in the form of Collaborative Research Program grant: 021220CRP2022.

Conflicts of Interest: The authors declare no conflict of interest.

References

1. Longinos, S.N.; Parlaktuna, M. Examination of behavior of lysine on methane (95%)–propane (5%) hydrate formation by the use of different impellers. *J. Pet. Explor. Prod.* **2021**, *11*, 1823–1831. [CrossRef]
2. British Petroleum. *Statistical Review of World Energy*, 7th ed.; British Petroleum: London, UK, 2021.
3. Longinos, S.N.; Parlaktuna, M. Are the amino acids inhibitors or promoters on methane (95%)–propane (5%) hydrate formation? *React. Kinet. Mech. Catal.* **2021**, *132*, 795–809. [CrossRef]
4. Merey, Ş.; Longinos, S. The gas hydrate potential of the Eastern Mediterranean basin. *Bull. Miner. Res. Explor.* **2019**, *160*, 1–10. [CrossRef]
5. Tian, F.C.; Liang, Y.T.; Wang, D.M.; Jin, K. Effects of caprock sealing capacities on coalbed methane preservation: Experimental investigation and case study. *J. Cent. South Univ.* **2019**, *26*, 925–937. [CrossRef]
6. Crow, D.J.; Balcombe, P.; Brandon, N.; Hawkes, A.D. Assessing the impact of future greenhouse gas emissions from natural gas production. *Sci. Total Environ.* **2019**, *668*, 1242–1258. [CrossRef]
7. Liew, M.S.; Danyaro, K.U.; Zawawi, N.A.W.A. A Comprehensive Guide to Different Fracturing Technologies: A Review. *Energies* **2020**, *13*, 3326. [CrossRef]
8. Jackson, R.E.; Gorody, A.W.; Mayer, B.; Roy, J.W.; Ryan, M.C.; Van Stempvoort, D.R. Groundwater protection and unconventional gas extraction: The critical need for field-based hydrogeological research. *Groundwater* **2013**, *51*, 488–510. [CrossRef]
9. Boudet, H.; Clarke, C.; Bugden, D.; Maibach, E.; Roser-Renouf, C.; Leiserowitz, A. "Fracking" controversy and communication: Using national survey data to understand public perceptions of hydraulic fracturing. *Energy Policy* **2014**, *65*, 57–67. [CrossRef]
10. Yan, H.; Tian, L.P.; Feng, R.M.; Mitri, H.; Chen, J.Z.; Zhang, B. Fracture evolution in coalbed methane reservoirs subjected to liquid nitrogen thermal shocking. *J. Cent. South Univ.* **2020**, *27*, 1846–1860. [CrossRef]
11. Hou, P.; Su, S.; Gao, F.; Liang, X.; Wang, S.; Gao, Y.; Cai, C. Influence of Liquid Nitrogen Cooling State on Mechanical Properties and Fracture Characteristics of Coal. *Rock Mech. Rock Eng.* **2022**, *40*, 1–20. [CrossRef]
12. Wang, L.; Yao, B.; Cha, M.; Alqahtani, N.B.; Patterson, T.W.; Kneafsey, T.J.; Miskimins, J.L.; Yin, X.; Wu, Y.S. Waterless fracturing technologies for unconventional reservoirs-opportunities for liquid nitrogen. *J. Nat. Gas Sci. Eng.* **2016**, *35*, 160–174. [CrossRef]
13. Yan, H.; Tian, L.; Feng, R.; Mitri, H.; Chen, J.; He, K.; Zhang, Y.; Yang, S.; Xu, Z. Liquid nitrogen waterless fracking for the environmental protection of arid areas during unconventional resource extraction. *Sci. Total Environ.* **2020**, *721*, 137719. [CrossRef] [PubMed]
14. Zhai, C.; Qin, L.; Liu, S.; Xu, J.; Tang, Z.; Wu, S. Pore structure in coal: Pore evolution after cryogenic freezing with cyclic liquid nitrogen injection and its implication on coalbed methane extraction. *Energy Fuels* **2016**, *30*, 6009–6020. [CrossRef]
15. Qin, L.; Zhai, C.; Liu, S.; Xu, J.; Yu, G.; Sun, Y. Changes in the petrophysical properties of coal subjected to liquid nitrogen freeze-thaw–A nuclear magnetic resonance investigation. *Fuel* **2017**, *194*, 102–114. [CrossRef]
16. Ding, Z.; Jia, J.; Feng, R. Effect of the vertical stress on CO_2 flow behavior and permeability variation in coalbed methane reservoirs. *Energy Sci. Eng.* **2019**, *7*, 1937–1947. [CrossRef]
17. Memon, K.R.; Mahesar, A.A.; Ali, M.; Tunio, A.H.; Mohanty, U.S.; Akhondzadeh, H.; Awan, F.U.R.; Iglauer, S.; Keshavarz, A. Influence of cryogenic liquid nitrogen on petro-physical characteristics of mancos shale: An experimental investigation. *Energy Fuels* **2020**, *34*, 2160–2168. [CrossRef]
18. Cai, C.; Li, G.; Huang, Z.; Tian, S.; Shen, Z.; Fu, X. Experiment of coal damage due to super-cooling with liquid nitrogen. *J. Nat. Gas Sci. Eng.* **2015**, *22*, 42–48. [CrossRef]

19. Qin, L.; Zhai, C.; Liu, S.; Xu, J.; Tang, Z.; Yu, G. Failure mechanism of coal after cryogenic freezing with cyclic liquid nitrogen and its influences on coalbed methane exploitation. *Energy Fuels* **2016**, *30*, 8567–8578. [CrossRef]
20. Zhao, D.; Wang, Q.; Li, D.; Feng, Z. Experimental study on infiltration and freeze–thaw damage of water-bearing coal samples with cryogenic liquid nitrogen. *J. Nat. Gas Sci. Eng.* **2018**, *60*, 24–31. [CrossRef]
21. Qin, L.; Zhai, C.; Xu, J.; Liu, S.; Zhong, C.; Yu, G. Evolution of the pore structure in coal subjected to freeze− thaw using liquid nitrogen to enhance coalbed methane extraction. *J. Pet. Sci. Eng.* **2019**, *175*, 129–139. [CrossRef]
22. Cha, M.; Yin, X.; Kneafsey, T.; Johanson, B.; Alqahtani, N.; Miskimins, J.; Patterson, T.; Wu, Y.S. Cryogenic fracturing for reservoir stimulation–Laboratory studies. *J. Pet. Sci. Eng.* **2014**, *124*, 436–450. [CrossRef]
23. AlGhamdi, T.M.; Arns, C.H.; Eyvazzadeh, R.Y. Correlations between NMR-relaxation response and relative permeability from tomographic reservoir-rock images. *SPE Reserv. Eval. Eng.* **2013**, *16*, 369–377. [CrossRef]
24. Chu, Y.; Sun, H.; Zhang, D.; Yu, G. Nuclear magnetic resonance study of the influence of the liquid nitrogen freeze-thaw process on the pore structure of anthracite coal. *Energy Sci. Eng.* **2020**, *8*, 1681–1692. [CrossRef]
25. Cai, C.; Li, G.; Huang, Z.; Shen, Z.; Tian, S. Rock pore structure damage due to freeze during liquid nitrogen fracturing. *Arab. J. Sci. Eng.* **2014**, *39*, 9249–9257. [CrossRef]
26. Cai, C.; Huang, Z.; Li, G.; Gao, F.; Wei, J.; Li, R. Feasibility of reservoir fracturing stimulation with liquid nitrogen jet. *J. Pet. Sci. Eng.* **2016**, *144*, 59–65. [CrossRef]
27. Qin, L.; Zhai, C.; Liu, S.; Xu, J. Factors controlling the mechanical properties degradation and permeability of coal subjected to liquid nitrogen freeze-thaw. *Sci. Rep.* **2017**, *7*, 3675. [CrossRef]
28. Li, B.; Zhang, L.; Wei, J.; Ren, Y. Pore damage properties and permeability change of coal caused by freeze-thaw action of liquid nitrogen. *Adv. Civ. Eng.* **2018**, *2018*, 5076391. [CrossRef]
29. Jin, X.; Gao, J.; Su, C.; Liu, J. Influence of liquid nitrogen cryotherapy on mechanic properties of coal and constitutive model study. *Energy Sources Part A Recovery Util. Environ. Eff.* **2019**, *41*, 2364–2376. [CrossRef]
30. Chen, S.; Zhang, L.; Zhang, C.; Huang, M. Experimental study on the seepage characteristics of bituminous coal under the conditions of liquid nitrogen fracturing. *Energy Sci. Eng.* **2019**, *7*, 2138–2154. [CrossRef]
31. Du, M.; Gao, F.; Cai, C.; Su, S.; Wang, Z. Study on the surface crack propagation mechanism of coal and sandstone subjected to cryogenic cooling with liquid nitrogen. *J. Nat. Gas Sci. Eng.* **2020**, *81*, 103436. [CrossRef]
32. Li, H.; Dong, Z.; Ouyang, Z.; Liu, B.; Yuan, W.; Yin, H. Experimental investigation on the deformability, ultrasonic wave propagation, and acoustic emission of rock salt under triaxial compression. *Appl. Sci.* **2019**, *9*, 635. [CrossRef]
33. Su, S.J.; Gao, F.; Cai, C.Z.; Du, M.L.; Wang, Z.K. Effect of low temperature damage on tensile strength of coal under the liquid nitrogen freezing. *Therm. Sci.* **2020**, *24*, 3979–3986. [CrossRef]
34. Akhondzadeh, H.; Keshavarz, A.; Al-Yaseri, A.Z.; Ali, M.; Awan, F.U.R.; Wang, X.; Yang, Y.F.; Iglauer, S.; Lebedev, M. Pore-scale analysis of coal cleat network evolution through liquid nitrogen treatment: A micro-computed tomography investigation. *Int. J. Coal Geol.* **2020**, *219*, 103370. [CrossRef]
35. Zhang, L.; Chen, S.; Zhang, C.; Fang, X.; Li, S. The characterization of bituminous coal microstructure and permeability by liquid nitrogen fracturing based on μCT technology. *Fuel* **2020**, *262*, 116635. [CrossRef]
36. Zhang, H.; Wang, D.; Yu, C.; Wei, J.; Liu, S.; Fu, J. Microcrack evolution and permeability enhancement due to thermal shocks in coal. *PLoS ONE* **2020**, *15*, e0232182. [CrossRef] [PubMed]
37. Lin, H.; Li, J.; Yan, M.; Li, S.; Qin, L.; Zhang, Y. Damage caused by freeze-thaw treatment with liquid nitrogen on pore and fracture structures in a water-bearing coal mass. *Energy Sci. Eng.* **2020**, *8*, 1667–1680. [CrossRef]
38. Qin, L.; Li, S.; Zhai, C.; Lin, H.; Zhao, P.; Shi, Y.; Bai, Y. Changes in the pore structure of lignite after repeated cycles of liquid nitrogen freezing as determined by nitrogen adsorption and mercury intrusion. *Fuel* **2020**, *267*, 117214. [CrossRef]
39. Li, B.; Ren, Y.; Lv, X. The evolution of thermal conductivity and pore structure for coal under liquid nitrogen soaking. *Adv. Civ. Eng.* **2020**, *2020*, 2748092. [CrossRef]
40. Liu, S.; Li, X.; Wang, D.; Zhang, D. Investigations on the mechanism of the microstructural evolution of different coal ranks under liquid nitrogen cold soaking. In *Energy Sources, Part A: Recovery, Utilization, and Environmental Effects*; Taylor & Francis: Oxfordshire, UK, 2020; pp. 1–17.
41. Akhondzadeh, H.; Keshavarz, A.; Awan, F.U.R.; Ali, M.; Al-Yaseri, A.; Liu, C.; Yang, Y.; Iglauer, S.; Gurevich, B.; Lebedev, M. Liquid nitrogen fracturing efficiency as a function of coal rank: A multi-scale tomographic study. *J. Nat. Gas Sci. Eng.* **2021**, *95*, 104177. [CrossRef]
42. Sun, Y.; Zhai, C.; Xu, J.; Cong, Y.; Zheng, Y. Experimental study on pore structure evolution of coal in macroscopic, mesoscopic, and microscopic scales during liquid nitrogen cyclic cold-shock fracturing. *Fuel* **2021**, *291*, 120150. [CrossRef]
43. Shao, Z.; Ye, S.; Tao, S.; Feng, X.; Wang, Y. Experimental study of the effect of liquid nitrogen penetration on damage and fracture characteristics in dry and saturated coals. *J. Pet. Sci. Eng.* **2021**, *201*, 108374. [CrossRef]
44. Lin, H.F.; Long, H.; Yan, M.; Li, S.G.; Shu, C.M.; Bai, Y. Methane adsorption thermodynamics of coal sample subjected to liquid nitrogen freezing–thawing process. *J. Nat. Gas Sci. Eng.* **2021**, *90*, 103896. [CrossRef]
45. Liu, S.; Li, X.; Wang, D.; Zhang, D. Experimental study on temperature response of different ranks of coal to liquid nitrogen soaking. *Nat. Resour. Res.* **2021**, *30*, 1467–1480. [CrossRef]
46. Qin, L.; Wang, P.; Li, S.; Lin, H.; Wang, R.; Wang, P.; Ma, C. Gas adsorption capacity changes in coals of different ranks after liquid nitrogen freezing. *Fuel* **2021**, *292*, 120404. [CrossRef]

47. Sun, Y.; Zhao, Y.; Li, Y.; Danesh, N.N.; Zhang, Z. Characteristics Evolution of Multiscale Structures in Deep Coal under Liquid Nitrogen Freeze-Thaw Cycles. *Geofluids* **2021**, *2021*. [CrossRef]
48. Yang, Y.; Liu, S.; Chang, X. Fracture stiffness evaluation with waterless cryogenic treatment and its implication in fluid flowability of treated coals. *Int. J. Rock Mech. Min. Sci.* **2021**, *142*, 104727. [CrossRef]
49. Qin, L.; Ma, C.; Li, S.; Lin, H.; Wang, P.; Long, H.; Yan, D. Mechanical damage mechanism of frozen coal subjected to liquid nitrogen freezing. *Fuel* **2022**, *309*, 122124. [CrossRef]
50. Yuan, Y.; Cai, F.; Yang, L. Pore structure characteristics and fractal structure evaluation of medium-and high-rank coal. *Energy Explor. Exploit.* **2022**, *40*, 328–342. [CrossRef]
51. Akhondzadeh, H.; Keshavarz, A.; Awan, F.U.R.; Zamani, A.; Iglauer, S.; Lebedev, M. Coal cleat network evolution through liquid nitrogen freeze-thaw cycling. *Fuel* **2022**, *314*, 123069. [CrossRef]
52. Li, C.; Nie, B.; Feng, Z.; Wang, Q.; Yao, H.; Cheng, C. Experimental Study of the Influence of Moisture Content on the Pore Structure and Permeability of Anthracite Treated by Liquid Nitrogen Freeze–Thaw. *ACS Omega* **2022**, *7*, 7777–7790. [CrossRef]
53. Cai, C.; Gao, F.; Yang, Y. The effect of liquid nitrogen cooling on coal cracking and mechanical properties. *Energy Explor. Exploit.* **2018**, *36*, 1609–1628. [CrossRef]
54. Qin, L.; Zhai, C.; Liu, S.; Xu, J. Mechanical behavior and fracture spatial propagation of coal injected with liquid nitrogen under triaxial stress applied for coalbed methane recovery. *Eng. Geol.* **2018**, *233*, 1–10. [CrossRef]
55. Yin, G.; Shang, D.; Li, M.; Huang, J.; Gong, T.; Song, Z.; Deng, B.; Liu, C.; Xie, Z. Permeability evolution and mesoscopic cracking behaviors of liquid nitrogen cryogenic freeze fracturing in low permeable and heterogeneous coal. *Powder Technol.* **2018**, *325*, 234–246. [CrossRef]
56. Cong, Y.; Zhai, C.; Sun, Y.; Xu, J.; Tang, W.; Zheng, Y. Visualized study on the mechanism of temperature effect on coal during liquid nitrogen cold shock. *Appl. Therm. Eng.* **2021**, *194*, 116988. [CrossRef]
57. Yang, R.; Hong, C.; Huang, Z.; Wen, H.; Li, X.; Huang, P.; Liu, W.; Chen, J. Liquid nitrogen fracturing in boreholes under true triaxial stresses: Laboratory investigation on fractures initiation and morphology. *SPE J.* **2021**, *26*, 135–154. [CrossRef]
58. Du, M.; Gao, F.; Cai, C.; Su, S.; Wang, Z. Experimental Study on the Damage and Cracking Characteristics of Bedded Coal Subjected to Liquid Nitrogen Cooling. *Rock Mech. Rock Eng.* **2021**, *54*, 5731–5744. [CrossRef]
59. Yang, R.; Hong, C.; Huang, Z.; Song, X.; Zhang, S.; Wen, H. Coal breakage using abrasive liquid nitrogen jet and its implications for coalbed methane recovery. *Appl. Energy* **2019**, *253*, 113485. [CrossRef]
60. Yao, B.; Wang, L.; Patterson, T.; Kneafsey, T.J.; Yin, X.; Wu, Y. Experimental study and modeling of cryogenic fracturing treatment of synthetic rock samples using liquid nitrogen under tri-axial stresses. In Proceedings of the SPE Unconventional Resources Conference, Calgary, AB, Canada, 15–16 February 2017.
61. Cong, Y.; Zhai, C.; Sun, Y.; Xu, J.; Tang, W.; Zheng, Y.; Zhu, X.; Yang, L. Study on transient boiling heat transfer of coal with different bedding angles quenched by liquid nitrogen. *Case Stud. Therm. Eng.* **2021**, *28*, 101463. [CrossRef]
62. Ma, H.; Zhu, C.; Zhao, P.; Wang, B. Freezing method for rock cross-cut coal uncovering: Aging characteristic of effective freezing distance on injecting liquid nitrogen into coal seam. *Adv. Civ. Eng.* **2021**, *2021*. [CrossRef]
63. Wen, H.; Yang, R.; Huang, Z.; Hu, X.; Hong, C.; Song, G. Flow and heat transfer of nitrogen during liquid nitrogen fracturing in coalbed methane reservoirs. *J. Pet. Sci. Eng.* **2022**, *209*, 109900. [CrossRef]
64. McDaniel, B.W.; Grundmann, S.R.; Kendrick, W.D.; Wilson, D.R.; Jordan, S.W. Field applications of cryogenic nitrogen as a hydraulic fracturing fluid. In Proceedings of the SPE Annual Technical Conference and Exhibition, San Antonio, TX, USA, 5–8 October 1997.
65. Wilson, D.R.; Siebert, R.M.; Lively, P. Conoco Inc. Cryogenic Coal Bed Gas Well Stimulation Method. U.S. Patent 5,464,061, 7 November 1995.
66. Zhang, L.; Ren, S.; Fan, Z.; Luo, J.; Che, H.; Yang, Y. Method of Low-Temperature Gas-Assisted Coalbed Methane Fracturing Technology. Chinese Patent CN103726819A, 24 June 2016.
67. Shen, C.; Zhang, L.; Wang, D.; Guo, J.; Gao, X.; Zhang, H.; Hou, J.; Jiang, L. Liquid Nitrogen Freeze-Thawing Cracking Coal-Bed Permeability-Increasing Coal-Bed Gas Mining Method. Chinese Patent CN104963660A, 24 October 2017.
68. Shen, C.; Zhang, L.; Wang, D.; Shu, L.; Guo, J.; Gao, X.; Zhang, H.; Qi, Q. Low Permeability Coal Bed Liquid Nitrogen Freeze Thawing Cracking Permeability Increasing Method. Chinese Patent CN104963674A, 2 March 2018.
69. Zhai, C.; Qin, L.; Xu, J.; Yu, G.; Sun, Y.; Dong, R.; Wu, S. A Kind of Method of the Anti-Reflection Coal Body Extraction Coal Bed Gas of Cycle Pulse Formula Low Temperature Freeze Thawing. Chinese Patent CN106337672B, 16 November 2018.
70. Xu, H.; Xin, F.; Wang, Q.; Wei, H.; Liu, D. Test Device and Test Method for Simulating Liquid Nitrogen Injection Freeze-Drying for Increasing Coal Bed Gas Yield. Chinese Patent CN103726819A, 27 September 2020.

resources MDPI

Article

A 30-Year Probability Map for Oil Spill Trajectories in the Barents Sea to Assess Potential Environmental and Socio-Economic Threats

Victor Pavlov [1,*], Victor Cesar Martins de Aguiar [2], Lars Robert Hole [3] and Eva Pongrácz [1]

1 Water, Energy and Environmental Engineering Research Unit, University of Oulu, P.O. Box 4300, FI-90014 Oulu, Finland; eva.pongracz@oulu.fi
2 Department of Physics and Technology, UiT The Arctic University of Norway, N-9037 Tromso, Norway; victor.d.aguiar@uit.no
3 Oceanography and Marine Meteorology, Norwegian Meteorological Institute, P.O. Box 7800, N-5020 Bergen, Norway; lrh@met.no
* Correspondence: victor.pavlov@oulu.fi

Abstract: Increasing exploration and exploitation activity in the Arctic Ocean has intensified maritime traffic in the Barents Sea. Due to the sparse population and insufficient oil spill response infrastructure on the extensive Barents Sea shoreline, it is necessary to address the possibility of offshore accidents and study hazards to the local environment and its resources. Simulations of surface oil spills were conducted in south-east of the Barents Sea to identify oil pollution trajectories. The objective of this research was to focus on one geographical location, which lies along popular maritime routes and also borders with sensitive ecological marine and terrestrial areas. As a sample of traditional heavy bunker oil, IFO-180LS (2014) was selected for the study of oil spills and used for the 30-year simulations. The second oil case was medium oil type: Volve (2006)—to give a broader picture for oil spill accident scenarios. Simulations for four annual seasons were run with the open source OpenDrift modelling tool using oceanographic and atmospheric data from the period of 1988–2018. The modelling produced a 30-year probability map, which was overlapped with environmental data of the area to discuss likely impacts to local marine ecosystems, applicable oil spill response tools and favourable shipping seasons. Based on available data regarding the environmental and socio-economic baselines of the studied region, we recommend to address potential threats to marine resources and local communities in more detail in a separate study.

Keywords: Arctic marine resources; oil drift; OpenDrift; OpenOil; oil spill simulations

Citation: Pavlov, V.; Aguiar, V.C.M.d.; Hole, L.R.; Pongrácz, E. A 30-Year Probability Map for Oil Spill Trajectories in the Barents Sea to Assess Potential Environmental and Socio-Economic Threats. *Resources* **2022**, *11*, 1. https://doi.org/10.3390/resources11010001

Academic Editors: Mofazzal Hossain, Hisham Khaled Ben Mahmud and Md Motiur Rahman

Received: 30 November 2021
Accepted: 22 December 2021
Published: 24 December 2021

Publisher's Note: MDPI stays neutral with regard to jurisdictional claims in published maps and institutional affiliations.

1. Introduction

Oil spill response research commenced after the Torrey Canyon spill in 1967, the first major supertanker disaster [1]. Even though the yearly number and severity of spills have since reduced, oil spills are still a recognised problem worldwide [2]. Lately, a lot of attention focuses on the Arctic [3], which has thus far been spared of major incidents. There is a variety of technological oil spill response applicable solutions, but they all have certain operational limitations due to regional environmental challenges [4]. As a result of rising oil exploration and industrial activities in the circumpolar seas (e.g., Norwegian Sea, Barents Sea, Kara Sea), the risk of oil spill accidents is much higher than before [5–8]. The situation is also aggravated by the Arctic ice cap melting and the resulting increase in maritime traffic in the area [9,10]. In light of the projected increase of shipping activities and the geographical scarcity of response centres, here, we address the following research questions: how likely is it that an oil spill would reach the sensitive areas; what are seasonal and spatial variations, and how applicable can oil spill response tools be for the local conditions? The objective of this research is to address potential threats to marine resources and local communities from oil spill accidents. More specifically, this paper aims at defining potential hazards to the local environment based on the simulated accidents of two fuel oil types. Our approach takes into account oil mass weathering during spreading, emulsification, vertical mixing, natural dispersion by wave breaking and evaporation as well as being based on high resolution ocean models. Oil weathering and drift over 30 years from continuous oil spills at one given location along popular maritime routes are calculated.

The focus of this study is on the Barents Sea of the Arctic Ocean. It accommodates popular maritime traffic lines, which then join into the Northern Sea Route. Despite its recognised high economic and biological value and availability of various environmental data sources, published results are scarce regarding simulating and analysis of offshore oil spill emergencies in the Barents Sea [11,12].

There are a number of tools to simulate oil spill accidents globally (e.g., GNOME, OSCAR, Seatrack Web, Oilmap, MEDSLIK, as well as 12 other models) and predict their behaviour and fate in water [13]. Many of them are commercial access only and have some additional functions. The probability map for this study was generated in OpenDrift due to its broad functionality, open access and ease of use.

Similar studies were conducted in the Eastern Mediterranean Sea. The approach was based on another open-source code Lagrangian oil spill model—MEDSLIK. Shoreline susceptibility analysis was conducted using Environmental Susceptibility Index (ESI) which was defined in Adler and Inbar [14–17]. Our work also includes reference to a sensitivity study of the Barents Sea region but it is based on offshore biological distribution of local species instead of bathymetric and geomorphological data and ESI index allocation for the shoreline. Both OpenDrift and MEDSLIK represent comprehensive operational oil spill models, which are most widely-used by researchers due to their input data requirements and characteristics [13].

2. Resources of the Barents Sea

2.1. Environment

The Barents Sea, being part of the Arctic Ocean, is positioned between 81° N and 66° N latitudes and 16° E and 68° E longitudes, and has the mean depth of 350 m, with the total area of 1,405,000 km^2. The sea is situated above the Polar Circle, where periods of darkness may last annually up to 4 months. The southern border of the sea is drawn by the Eurasian continent, as illustrated in Figure 1. All seas of the Arctic Ocean are young in geological terms and have the same origin. In times of the ice age, there was a large land area covered with ice instead of the ocean. After the ice age era, all lowlands were submerged in water, whereas highlands remained, turning into islands and peninsulas [18,19]. A large part of the Arctic Ocean, especially its periphery, is shallow—about 200 m in depth. There are many islands in the sea area, the coastline is sophisticated with fjords and bays as well as being mostly rocky and steep [20–22].

The Atlantic Ocean current coming from the west through the Norwegian Sea has an influence on the sea in terms of salinity, temperature and wind. The water is warmer, compared, for example, to the neighbouring on the east of the Kara Sea, which has low temperatures due to its central location in the Arctic Ocean. The water temperature of the Barents Sea ranges from 0 °C in winter to 9 °C in summer. The stratification in the water column is almost absent; instead, there is a characteristic feature of intensive water mass mixing in the sea. The salt content is about 35 g of salt per water kilogram [23]. Due to warm water masses brought from the Atlantic, the south-east of the sea never becomes frozen, and oil spill response (OSR) operations differentiate essentially for the given conditions. The Barents Sea, for example, in comparison with the Kara Sea, is biologically rich and productive: also due to active interaction with the Atlantic Ocean. During spring bloom, March to June, there may be primary production of phytoplankton of up to 400 mg m^{-3} day^{-1}. During summer, the upper 30 m of water column are abundant with it. Hence, there is also a large presence of zooplankton—up to 500 mg m^{-3} [24]. Even though the sea is relatively pure historically from chemical contamination, its ecosystems are less metabolically active due to local northern climatic conditions and function rather slowly compared to ones in tropical latitudes [12].

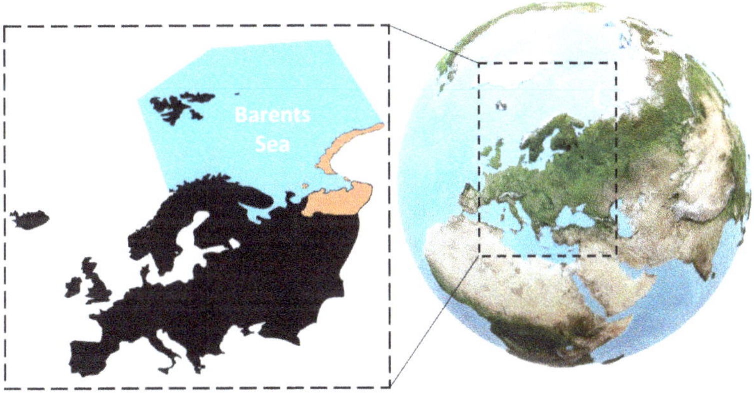

Figure 1. The Barents Sea: its location in relation to continental Europe and the study area in orange.

The polar marine climate of the Barents Sea favours long winters, short cold summers, strong winds in spring and autumn, and high relative humidity in the air. Overall, the weather is rapidly varying. Sudden changes in winds and precipitations are possible. Since the location of the sea is far north from the equator, the amount of incident solar heat is very low. The climate can be described as harsh and is characterised by extreme conditions of the Arctic [19]. In the perspective of potential oil spill response, this sea provides unfavourable prerequisites for natural oil slick photo- and biodegradation [25]. Due to its unstable water surface because of generated waves, the OSR operations may also be hindered.

2.1.1. Protected Areas

The Arctic is unique in terms of ecosystem functioning and vulnerability. Many land areas around the Arctic Ocean are ecologically sensitive and are on a list of protected areas. Only along the Russian Arctic coastline, there are about 35 nature reserves, which have the total area of about 39,000 km^2 [26]. The Barents Sea is not an exception from this either. In the north, on Svalbard archipelago, there are flora and fauna protective regulations with an emphasis on polar bear population protection. All touristic and domestic activities are limited in the area in order to protect its sensitive natural heritage [27,28]. In the west, Kola Peninsula has four of both terrestrial and nearshore protected areas along with several others inland. In the southern part of the sea, according to the World Database on Protected Areas, there are also several reserves [29]. On the mainland, there are five protected areas in proximity to the southern coast of the sea as seen in green in Figure 2: "Kolguevsky", "Nenetsky", "Haipudyrsky", "Yugorsky" and "Vaygach". As for fauna in these territories, there are bird habitats across the region but also various mammals. Among rare species, the representatives are such birds as cormorant, swan, sea eagle, sea ducks, and such mammals as grey lemming, wolverine and lynx to mention a few [24]. Many of them are on the list of the Russian Red Data Book and International Union for Conservation of Nature Red List of endangered species [28]. All these places have a special ecological emphasis on these local natural territories and demand a high level of OSR preparedness around them.

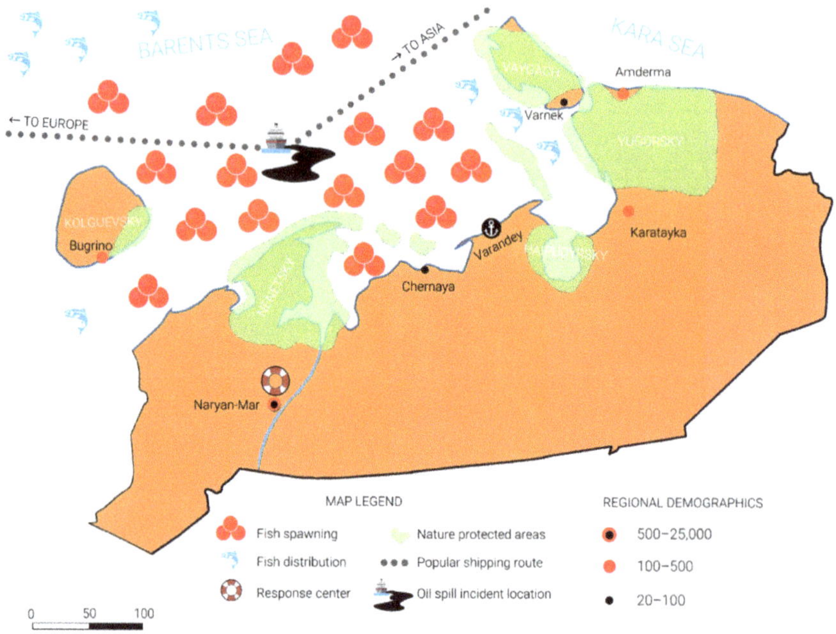

Figure 2. Nenets autonomous region in the hypothetical oil spill emergency.

2.1.2. Marine Resources

Regarding the abundance of marine ecosystems and fish species of the sea, such representatives as Greenland shark; Arctic skate; glacier lanternfish; deepwater redfish; sculpin (such species as Atlantic hooker, shorthorn, moustache, bigeye, polar, Arctic staghorn, twohorn, fourhorn, ribbed); lumpsucker (Atlantic spiny and leatherfin species); sea tadpole; eelpout (with representative species such as pale, polar, Arctic, threespot and longear eelpout); snakeblenny; Greenland halibut; Atlantic capelin; Atlantic and Polar cod; hamecon; Arctic alligatorfish; Atlantic poacher; lumpfish; stout and slender eelblenny, and daubed shanny—belong to the Barents Sea habitat. As an example, both polar cod and lumpfish have the same spawning and distribution areas in the sea, as shown in Figure 2 [30]. The only difference is in the spawning season: for polar cod, it is December to March, whereas for lumpfish—March to May [31,32]. For some fish species, the Barents Sea serves as a prevalent feeding area before spawning. Atlantic cod, being one of them, May to January, is often found here by commercial and local fishing vessels. In 2019, the cod stock was registered at about 1.5 million tonnes—to show a level of scale of its population [33]. Beluga, walrus, polar bear and white whale, for instance, being representatives of marine mammals, also have their habitats in the sea [24,34].

2.2. Economic Activities

From an economic perspective, the Arctic is an actively diversified region. Examples of activities present in the region may be fisheries and the oil and gas industry. Recent economic developments also include aquaculture, tourism, mining activities and sea transportation of oil products. By 2050, according to Corbett et al. [35], at least up to 5% increase of total shipping activities along the Arctic passages is expected. The main driving forces for it are global resource demand, growing population and availability of new maritime transport routes. In 2012, there were 1347 vessels registered in the Arctic region via the Automatic Identification System ship tracking. One third of them was carrying heavy fuel oil. The others were bulk carriers, fishing vessels, passenger cruises, general cargo carriers, oil tankers and chemical tankers. Among them, larger ships operating in the area have to comply with the Polar Code and have certain ice class for crossing the Arctic Seas [36]. As regards to the recent data, 1628 vessels entered the Arctic area in 2019 [37]. This is about 20% higher than in 2012.

2.2.1. Marine Resources

Historically, the Barents Sea has been a reliable supplier of such benthic species as cod, haddock, halibut, and also other fish species such as polar cod, herring, mackerel and capelin [30,38]. This is not surprising since, as mentioned above, the sea is very biologically productive. In 2016, for instance, Norway reported record statistics of whole seafood export—2.6 million tonnes. The main importer is the European Union, buying about 70% of Norwegian seafood. The total value of 2017 selling was 9.8 billion euros. In case of the Barents Sea, there always have been certain quotas for the annual fish catch, binding both Norway and Russia, which mutually share the sea. This is a good practice supporting the idea of sustainable fisheries management, which has led to the local fish stock being one of the healthiest in the world. For 2017, in particular, the quota for cod catch in the Barents Sea was set at about 900,000 t, and for 2018—775,000 t [39]. It has approximately been the same for the last several years [40–43]. For 2020, the fish catch limit was recommended at about 690,000 t [33]. In the case of Russia, and its Murmansk sea port in the Kola Bay, being the largest above the Polar Circle, and the only ice-free port in the country—it also relies extensively on fisheries originating from the Barents Sea. The fish plays an important role in food supply in the region and beyond its borders: e.g., St. Petersburg area [25,44,45].

2.2.2. Transportation

Apart from fisheries, the Barents Sea is also considered as a very important transport route: not only for oil transportation, but also other goods. Regarding oil carriage, about 15 million tonnes were carried along the sea only within the Russian export scale during 2004–2009. The largest volume of oil transported was registered in 2010. The oil was mainly coming from an oil field developed in the Pechora Sea—the southeastern part of the Barents Sea, and counted for 15 million tonnes of crude oil [44]. In 2013, there was a start of "Prirazlomnaya" oil deposit development also in the Pechora Sea, and the first oil tanker shipping took place with its arrival to Rotterdam port in the EU. The annual deposit capacity to be reached by 2021 is 6.5 million tonnes of oil. As the size of specially built tankers "Mikhail Ulyanov" and "Kirill Lavrov" is 70,000 tonnes, then the calculated yearly number of round trips from the EU to the deposit would be 93 [46]. As a part of the Northern Sea Route, the Kara Sea has the key importance. Due to large developments in the mining industry and oil and gas sector, there is also ongoing export through the Barents Sea of gas condensate and multiple products of "Norilsk Nickel", which mainly represent non-ferrous metals: palladium, nickel, platinum, cobalt, copper and others [44,47]. In general, there is a number of terminals located in the Barents Sea, which make it busy in terms of maritime traffic [48]. This fact can be visually observed with a live traffic density map on a ship tracking website [49].

2.2.3. Oil and Gas Industry

The third important economic activity of the Barents Sea after fisheries and transport is energy resource deposits. Together with the Kara Sea, the sea contains about 70% of hydrocarbon resources of the continental shelf [11]. Norwegian surveys claim that the Barents Sea only holds about 2.8 billion m^3 of oil equivalent [50]. Consequently, it can be foreseen that the Barents Sea is being explored and will be explored in the upcoming years. In 2017, Norway, for instance, announced that 93 blocks were available for oil companies for license rounds [51]. Previously, oil and gas were explored strictly by the shore and within shallow waters—up to 70 m. Nowadays, fields of up to 700 m are considered for resource development [52].

2.3. Local Communities

One of the representatives of inland local communities is people of the Nenets autonomous region of the Russian Federation. It is the most sparsely populated region of the country, with a population of approximately 44,000 people. It is also the least populated area in geographical Europe. The population density is 0.25 people per square kilometre. This is about eight times less than in the most sparsely populated region of the least populated country in the EU—Lapland of Finland, and about five times less than the least populated state of the USA—Alaska [53,54]. As Figure 2 shows, there are several settlements of different population sizes by the seacoast of Nenets: Bugrino, populated with 417 inhabitants, Chernaya—with 22, Karatayka—with 544, Varnek—with 101 and Amderma—with 577. The capital of the region with 25,000 people—Naryan-Mar, is marked to demonstrate remoteness of the settlements from the regional centre, which also has the closest OSR centre and required equipment for operations at sea [55,56].

2.3.1. Indigenous Peoples

As overall regional statistics, the largest share of population in Nenets is demographically Russian—63%. The second and third largest are Nenets and Komi peoples, respectively 18% and 9%. Although Komi is only 9% of the population and it is twice less than Nenets, their population is about 350,000 people and they are widespread in the whole country around its various regions. Therefore, they are not considered in the section as indigenous for this particular study area. Nenets people, in turn, are known for their traditional reindeer herding, hunting and fishing in the region [57]. Reindeer herding is the dominant activity [58]. There are about 1000 people involved in it, managing more than 135,000 reindeers over 12 million ha. All reindeers are divided between 18 reindeer cooperative societies and are distributed across the region. Several of them are present around Chernaya, Karatayka, Varnek and Amderma [59]. In winter, reindeers are usually taken inland, and in all the other seasons, to the sea coastal areas. This takes place everywhere along the regional coast [60]. Indigenous hunting is mostly done for hoof and fur animals. Animal husbandry and cell breeding are also common these days. Fish catch is mostly traditional rather than commercial and related mostly to species of cisco and salmon. Salmon is represented by ten species, with its migration area in the study site. Among other popular fish species are navaga, Polar cod and Pacific herring, which feed on local phyto- and zooplankton. The spawning for navaga is from December to January, and for Polar cod—December to March. Polar cod is also an important link in the trophic chain, being a food for other fish, mammals and birds [24]. As an example of proper establishment of fishing, Bugrino has a fish cooperative society [61]. In total, there are 22 traditional cooperative societies in Nenets autonomous region. The remaining four deal specifically with fishing. Annually, October to March, up to 500 local fishermen are winter fishing at sea. Besides the previously mentioned species, they also catch Arctic flounder and Pacific rainbow smelt. In addition to indigenous traditional activities, there are also certain places of cultural heritage present in the region, where spiritual traditions are practiced [55,62]. Many of sacred places are found on the island, where Varnek is located—Vaygach island [63].

2.3.2. Regional Development

The region is part of the Arctic and has severe northern climatic conditions. This inhibits its economic development, primarily due to infrastructure insufficiencies. However, there are a number of local industries which are established in the area. One of the most developed economic sectors is fossil fuels extraction: both oil and gas, and mineral resources [57]. The share of both is the largest, when comparing with other sectors of the economy. Due to the presence of the onshore and offshore oil and gas industry, an oil terminal was opened in an abandoned regional settlement—Varandey. Nowadays, there are about 200 people living there. Apart from a couple of regular families, it is mostly inhabited by industrial workers, operating for oil and gas industry purposes locally in the region [64]. In Nenets, Varandey is the only seaport, except for Amderma and Naryan-Mar down the river. Varandey terminal is marked in Figure 2, where oil is usually transported from to Murmansk, then further on across Scandinavia and towards the EU centre [55]. Besides the oil and gas industry, there is also a noticeable operation of such sectors such as tourism and traditional activities. As mentioned, it is primarily reindeer herding and fishing. The region is seen as a cultural touristic attraction because of expressed indigenous traditions and lifestyle. Similar cultural sights might be seen in Finnish Lapland. Since the territory is remote and pristine with several nature reserves, ecotourism is also developing in Nenets: e.g., in "Kolguevsky", "Nenetsky" and "Vaygach" protected areas. All these were mentioned in the development strategy of the region by 2030 and on the official Nenets website [60,65].

3. Location, Methods and Oil Characteristics

3.1. Location

The purpose of the work is to overlap several layers of data (environmental set of factors with oil spill-related probability map, proximity of response infrastructure as well as technical drawbacks of response technologies) into one discussion of hypothetical risks of oil spill accident in the Barents Sea to the environment. The selected geographical location of the oil spill, having the coordinates at 69.49° N 53.42° E, is near Novaya Zemlya and in close proximity to the Kara Strait. The choice of that particular study spot is explained by several reasons. Firstly, it lies along one of the most popular maritime passages in the local area, which makes it lively in terms of traffic activities [65]. This has been confirmed, for example, by Arctic shipping status reports and online marine traffic density maps—showing the scale of several hundred ships a year in the area [49,66,67]. Secondly, the selected regional spot is also characterised by local oil and gas activities. There are several offshore oil deposits northeast from Chernaya settlement. Despite the presence of the oil and gas industry in Nenets with its oil terminal in Varandey, West-East and East-West transportation routes are mostly transit without stops on the mainland [60]. Thirdly, it has a large economic activity besides the oil and gas industry. Several sources confirm the minor presence of commercial fisheries in the study area: mostly from September to December on an annual basis [30,60,68]. Among the examples may be navaga, Polar cod and Pacific herring [24]. Fourthly, there is also rich natural heritage and large biological diversity in the local environment including, for example, harp seals, polar bears, walruses and belugas. Out of all marine fish species of the Arctic Ocean, the Barents Sea hosts approximately one fourth of its overall stock [38,69].

The location of hypothetical accidents was intentionally chosen in the remote site. It is placed far from the regional search and rescue OSR centres, which are located at Murmansk, Archangelsk and Vorkuta. The closest OSR centre, Naryan-Mar, is 250 km away from the accident site (Figure 2). The distance from the spill to the nearest nature reserve in the south—"Nenetsky zapovednik"—is 70 km.

3.2. Methods

Oil drift simulations were performed with OpenDrift (Release v. 1.2.0), an open-source Lagrangian framework developed in Python by the Norwegian Meteorological Institute [70]. It is a programme used to simulate movement of objects in ocean and atmosphere. The current work required only oceanographic application. OpenDrift contains sub-modules that simulate the trajectory of different floating objects in the ocean, ranging from search and rescue to oil drift. Among possible examples of modelled trajectories may be such objects as microplastics, drifting vessels, boats, bodies and many others. For oil spill drift trajectory modelling, there is a specialised module—OpenOil. It has been used in this study to give realistic results: as, for example, with Deepwater Horizon oil drift simulations [71]. OpenOil simulates oil spill transformation, and also tracks oil changes in time and space. The slick pathway is visualised as a trajectory of individual dots in movement. Based on the generated visual picture, it is possible to make relevant general conclusions about mitigation tools and clean up strategies as well as environmental or economic risks in the polluted areas. OpenDrift works with input oceanographic and atmospheric data: in our case, we used ERA Interim reanalysis and SVIM. OpenOil is configured with such physical processes as natural dispersion, evaporation, emulsification and biodegradation as weathering processes, using Euler interpolation scheme and a certain time step. The oil spills are simulated using 3D module, with vertical mixing due to oceanic waves and currents, resurfacing of oil because of buoyancy—thus taking into consideration oil physical properties. All these help to study a range of oils rather than concentrating only on one particular type.

Our experiments were designed based on the World Meteorological Organization's suggestion that a climate's normal periods are usually 30 years. Hence, the duration of each study set was also three decade long. Due to its availability, past weather data was used: from 1988 to 2018. Four seasons of the year were studied to observe seasonal differences and produce probability maps for each. Seasonal month settings were as commonly used: winter—December to February, and so on. The source point for all simulations stayed the same. The experiments were run with a 1-hour time step. Oil spill was released daily on a continuous basis, 100 oil particles per day during the selected study period (30 years). Two oil types were tested to generate the probability maps presented in Figure 3. Here, oil particles are considered stranded once they touch/are in nearest proximity to the coastal black line. In the produced maps, the remaining oil mass particles will reduce mass due to evaporation, and physical influence of sea such as waving activity causing dispersion of oil and other processes. Oil particles with remaining mass by the end of simulation will be weeks to months old. Thus, the particles in Figure 3 mainly illustrate the most likely oil transport from the source point marked with a red star. In real life, such continuous oil spills are unrealistic. Neither well blow-out, tanker accident nor pipe leakage can cause spill for such a stretched period of time. However, these experiments were made to offer statistically relevant data for the Barents Sea: footprint of oil contamination associated with the potential spill. They provided the necessary scenarios of oil drifting trajectories under local dynamic oceanographic conditions to demonstrate the most probable pathways of oil transport at sea surface and to envision coastal locations at risk of oil contamination.

In addition to the modelling tool, two publicly available online tools were applied for the study region: circumpolar oil spill response viability analysis tool and the Arctic risk map [72,73]. Analytical and synthetic methods were also used in the research: theoretical analysis and synthesis of various literary sources. These included local economic practices, strategies of regional development, its implementation plans, national legislation and regulations, publicly available materials from government websites, multiple academic journal articles and other relevant publications.

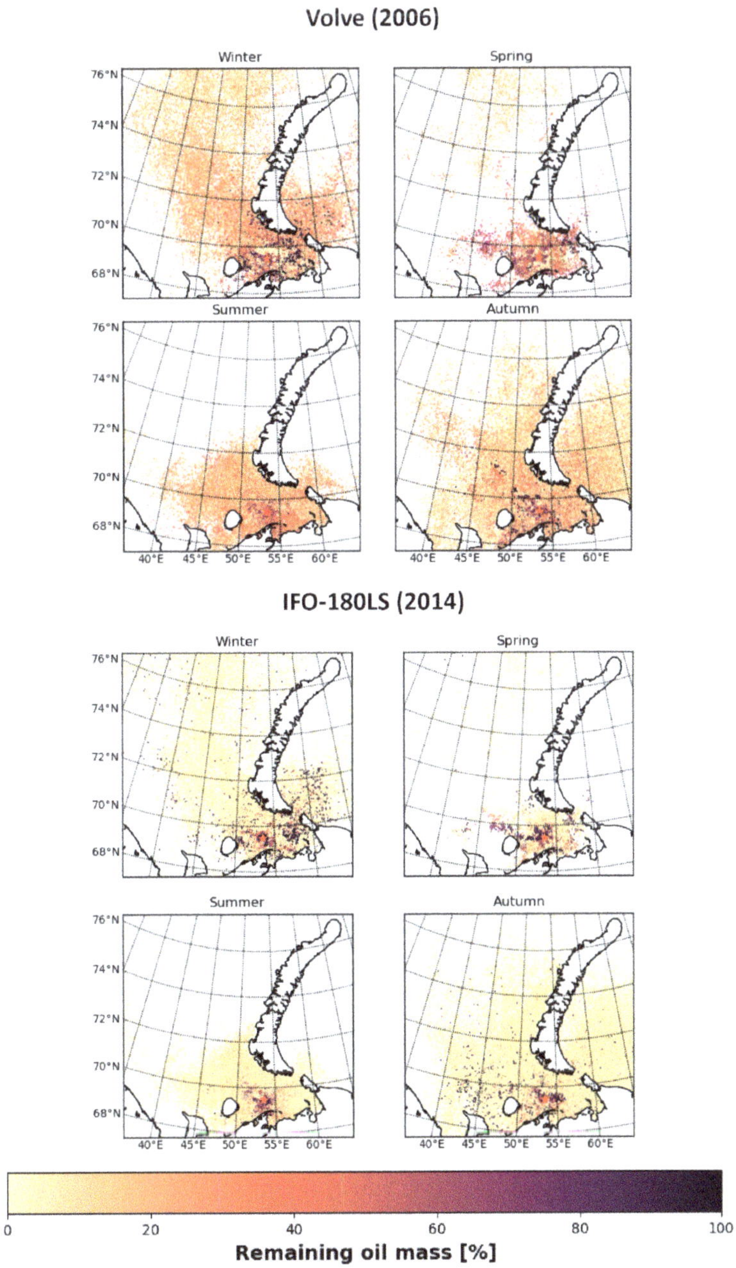

Figure 3. OpenDrift simulations for two studied continuous surface oil spills in the Barents Sea for annual seasons of 1988–2018: Volve (2006) and IFO-180LS (2014). Each presented probability map illustrates the initial spill location (marked with the red star) and seasonal patterns of remaining oil mass of each particle coloured by percentage relative to initial mass (shades of purple, as the colour palette).

3.3. Oil Characteristics

The simulations of oil spill accidents in four annual seasons included two oil types. One was a traditional heavy fuel oil–IFO-180 LS (2014), which is likely used by ships crossing the study area [74]. The Arctic risk map shows that such class of residual fuel oil is used in the area [72]. Volve (2006) was the other one, which is characterised as a medium oil type. It was additionally selected to give a broader picture when thinking about oil classes which might be transported from Europe and Asia, and to intercompare two different oil types in terms of their sea surface spreading. Regionally developed Arctic oils are mostly from light oil class, despite their higher viscosity. They are well studied because of local oil and gas industrial developments [75]. They also seem to be less impactful for biological species: from sensitivity and vulnerability perspectives [76]. Hence, this class was not chosen in our research.

OpenOil is based on a library of 1000 oils developed by the National Oceanic and Atmospheric Administration (NOAA). The library includes all necessary properties of various weathered oil types, showing changes in their physical and chemical state in time. Table 1 illustrates classification proposed by the World Petroleum Conference [77,78], with ranges of API gravity and oil density in relation to each category.

Table 1. Oil classification: according to its API gravity and density.

Oil Category	API, °	Oil Density (ρ), kg/m^3
Light	API ≥ 31.1	ρ ≤ 870
Medium	22.3 ≤ API < 31.1	870 > ρ ≥ 920
Heavy	10 ≤ API < 22.3	920 > ρ ≥ 1000
Extra-heavy	API < 10	ρ > 1000

API gravity (°) for the chosen oils can be calculated as follows [77]:

$$\text{API} = 141.5\text{d}\ (60/60) - 131.5, \tag{1}$$

where d is the relative density of the oil (kg/m^3) and 60/60 °F is the specific gravity of the oil at 60 °F compared with that of water at 60 °F.

The respective densities and API gravities of the oils were: IFO-180LS (2014) of 975.8 kg/m^3 and API ≅ 13.5°, and Volve (2006) of 892 kg/m^3 and API ≅ 27.13°. Distinct classes of oil, as the ones considered in this research, are differently impacted by weathering processes. Evaporation, for instance, affects more lighter refined oil products than heavy crude oil products since the former presents higher concentration of volatile components than the latter [79]. Therefore, oils can prevail in the environment for longer or shorter periods depending on their type and it is thus possible to evaluate which one could impact more severely a given region. The presence of sea ice adds new complexities to the oil modelling task. The ice cover dampens wave activities, thus decreasing dispersion, entrainment and emulsification [80]. Oil-in-ice studies conducted since around 1970 have shown that the oil surface drift is highly dependent on the sea ice concentration, C. As a rule-of-thumb, oil tends to drift as in open waters when C < 30% and with the ice field for high sea ice concentration values (C > 80%). No general consensus is yet established for the intermediate range. Nordam et al. (2019), for instance, suggested a linear transition between the two extremes whereas French-McCay et al. considered that oil drifts with the ice field already for C > 30% [81,82]. Overall, it is observed that the sea ice limits the oil drift and spreading, thus restricting their displacement.

4. Results and Discussion

4.1. Oil Behaviour and Spill Response Efficiency

Figure 3 shows scenarios of oil drifting trajectories under local oceanographic conditions and demonstrates most probable pathways of oil transport at sea surface. It displays differences in intensity of remaining oil mass on the sea surface: from lowest values of less than 20% to highest—at times more than 80%. The colour intensity shows how much oil likely remained on the surface: the darker the colour shade, the more oil remained on the surface. Findings from the conducted modelling showed that no distinctive elongated high-density oil trajectory patterns were observed in seasonal distribution of all simulations, which could occur, for example, in tropical regions caused by local oceanography. In contrast, oil had a rather even distribution of particles which tend to be carried in all directions from the initial spill location.

In a series of numerical experiments conducted in the Barents Sea by the RGC Risk Informatics, the Wild World Foundation showed that, in comparison to simulations performed during ice-free conditions, the area covered by the oil slick is more than 50% smaller when sea ice is present [83]. As can be noticed in Figure 3, particles corresponding to spring simulations were predominantly centred around the releasing point, similarly to the results presented in the aforementioned study. The smaller drift of particles in the spring simulations, which is also the period of maximum sea ice extent in south-east of the Barents Sea, might be seen as an indication of its role in the transport of oil.

4.1.1. Oil Types

The two selected scenarios, bunker oil IFO-180LS (2014) and medium oil Volve (2006), demonstrate that there is an observable difference between the two oil types. Volve has a more distinctive drifting pattern with wider surface coverage and potential to reach the nearby coasts of the archipelago, islands and mainland. IFO-180LS also has a wide polluted area, although less oil mass remained at sea. The share of weathering processes is noticeably higher. Both cases illustrated extensive trajectory movements and spill travel distances, at times exceeding 1000 km, when particles were found north of Novaya Zemlya. Pollution occurred in all water areas around the archipelago, except its north-eastern part. Small amounts reached Kola Peninsula in the west, 500 km away from the accident site. Some amount of the spilled oil drifted through the Kara Strait. The epicentre of the largest spill is within 150–200 km: with borders of the Kolguyev island in the west, Novaya Zemlya in the north, the Vaygach island in the east and the mainland in the south. Between two oil types, Volve has a much more noticeable presence in Figure 3 than IFO-180LS. This means more remaining oil mass presence, and, hence, medium oil types may be considered as a more hazardous oil type to the marine environment.

4.1.2. Summer

During summer months, most of the oil stayed within the radius of 300–400 km. This makes it a more localised emergency, which is a positive parameter from oil spill response preparedness perspective but a negative circumstance for the local marine and coastal ecosystems—it will be subjected to a higher oil concentration.

4.1.3. Autumn and Winter

In contrast, oceanographic and atmospheric conditions in autumn and winter seasons would result in a drift to the furthest distance in all directions. This may be explained by the presence of seasonal variability in sea and wind dynamics. A hypothetical oil spill accident may likely occur as a regional scale emergency—posing infrastructural difficulties for oil spill response operations. In both seasons, high oil particle density spots are also observed on the entire area of oil distribution. It is especially visible in the case of IFO-180LS (2014): dark spots in Figure 3, mostly west of the hypothetical spill accident site.

4.1.4. Spring

Spring simulations demonstrate that the highest oil remaining mass will concentrate in a small area producing most of the sea pollution in the land-surrounded region. This can be seen in the case of the Volve map where the remaining oil mass is mostly from the colour shade region of between 40 to 80% on the palette. Such oil distribution can provide a very noticeable impact within the accident area. The oil, however, is not localised similar to the summer pattern but is still likely to travel farther north in the Barents Sea. Much of the oil particles are observed west of the archipelago.

When we overlap the generated probability maps with already existing local sensitivity maps to oil spills, we can see where environmental impacts can be observed the most [84]. Figure 4 illustrates sensitivity areas based on spaced distribution of such biological species as phyto-, zoo-, ichthyoplankton, zoobenthos, fish, mammals and seabirds.

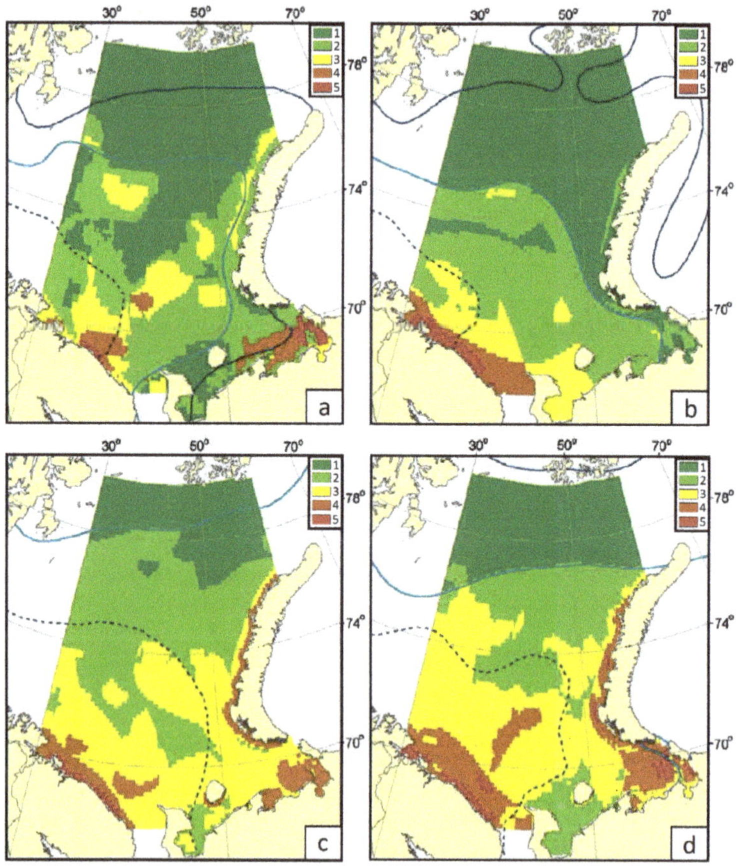

Figure 4. Sensitivity map of the Barents Sea: (**a**) Winter; (**b**) Spring; (**c**) Summer; (**d**) autumn [84]. Numbers in legend correspond to different sensitivity levels (1—very low, 2—low, 3—moderate, 4—high and 5—very high).

The presented data in Figure 4a for winter season shows higher susceptibility regions to the east from the initial accident site near Vaygach island. In case Figure 4b which is spring time, sensitivity levels near the epicentre are reasonably low: range of 1 to 2. Cases in Figure 4c,d, respectively for summer and autumn, illustrate similar trends: highly sensitive significant areas in the south-east of the sea and vast thin regions along the western coast of Novaya Zemlya: levels 4 and 5. Coastline of the Kola Peninsula has high susceptibility level in all four cases. All sensitive areas mostly correspond to seasonal species distribution [84].

Of the four seasons, the largest coverage for both studied oil types is observable during winter and autumn, whereas the coverage during spring and summer is smaller in size. In all seasons except summer, oil particles reached their highest latitudes of the Barents Sea, thus, drifting to remote areas of the high north. If hypothetical, summer oil spill accidents may be likely called local, in all other seasons, it can take a regional scale. Based on the 30-year simulations presented here, the oil spill may likely drift much further than the location of the nearest oil spill response centre in Naryan-Mar, thus, putting extra attention to accident preparedness in the study region.

In order to define efficiency of oil spill response operations in case of an accident in the study area in all seasons, the Circumpolar Oil Spill Response Viability Analysis tool was used. It has three primary OSR technologies with a few subdivisions in each: mechanical recovery, dispersants' use and in situ burning. They are all listed in Table 2. Numbers show how favourable the study site is for oil spill response operations. Here, the amount of percent means response conditions are either favourable (marked with plus), marginal (marked with plus-minus) or unfavourable (marked with minus) for a certain tool throughout a certain season. For example, if the tool has 0% in the plus column, 56%—in plus-minus and 44%—in minus, it stands for zero favourable conditions in the area for the season, 56%—marginal and 44%—unfavourable. It is 44% unlikely and 56% marginally likely that the tool will be an applicable abatement method for oil spill response.

Table 2. Mean seasonal operability ("+"—favourable conditions, "+/−"—marginal conditions, "−"—unfavourable conditions) for oil spill response methods in the Arctic marine environment.

OSR Tools		Seasonal OSR Operability, %											
		Autumn			Winter			Spring			Summer		
		+	+/−	−	+	+/−	−	+	+/−	−	+	+/−	−
Dispersants	vessel application	3	18	77	0	56	44	3	54	44	60	28	13
	helicopter application	19	68	14	5	5	90	19	21	61	54	19	27
	aircraft application	9	58	34	0	6	94	3	21	77	10	47	44
Mechanical recovery	3 vessels with boom	20	75	6	0	6	94	2	10	89	19	25	56
	2 vessels with boom	27	12	61	0	37	63	3	34	63	58	28	14
	1 vessel with outrigger	2	31	68	0	22	78	0	14	85	37	44	19
In situ burning	ignition from vessel	5	55	40	0	26	74	2	27	71	20	56	24
	ignition from helicopter	5	7	88	1	2	97	4	13	84	11	21	69

As seen from the table, the selected geographical location has the following estimates for successful oil spill response. In autumn, four out of eight available Arctic OSR tools are unlikely (−) to be applied, whereas the other four have marginal (+/−) local operational performance in the season. Winter and spring seasons are unfavourable (−) for almost all OSR tools. Summer conditions are best for vessel or helicopter application of dispersants or two-vessel boom containment, which is marked with higher percentages in "+" column of the table. However, it is important to consider limitations of the search and rescue helicopter flight distance of about 175 nm [85]. If we discuss the local context of Russian legislation, dispersant use in the Barents Sea is also subject to special permission: its application is assessed according to net environmental benefit analysis procedure [86]. In situ burning opportunity is available in summer or autumn with about 50% chance for its operability in the local environment, but still is put in the marginal segment. As a general conclusion, during the year, overall favourable conditions for oil spill abatement are rather unlikely: 20 "−" cases, 9—"+/−", and 3—"+". This greatly relates to the local environmental conditions, which limit performance of the OSR methods. This includes extremely low temperatures, excessive ice cover, high wind and wave activity, daylight unavailability, precipitation storms and low visibility. As a representative example, here are climatic parameters above which no oil spill response is possible: air temperature (−18 °C), wind (15 m/s), wave height (4 m), light availability (darkness) and visibility (air: 4.0 km; water: 0.3 km) [4].

In order to reduce the likelihood and severity of oil spill accidents to the Arctic environment and local communities, we recommend that shipping routes are selected seasonally, taking into consideration sensitivity maps, and avoiding areas of high environmental vulnerability. For example, in the studied region, summer seems to be the most suitable for shipping with the highest OSR operability and statistically least sea surface area polluted by a potential oil spill. However, oil spill preparedness should be as high as technologically possible, since summer has also several highly sensitive areas in the local marine environment. Here, most of the negative impacts will be localised within 300 km of the accident site—meaning exposure to higher oil concentrations. Out of four seasons, autumn seems to be one of the worst seasons for shipping due to mostly marginal/unfavourable oil spill response conditions and vast area of oil spill distribution. Winter, just as autumn, has a likely wide oil contamination area but also has unfavourable conditions for oil spill abatement. This season could be the second worst after autumn. The third worst is spring, having low chances for successful oil spill operations with existing OSR tools. As in cases of autumn and winter, a spring oil accident is likely to have regional scale.

This preliminary assessment aimed at drawing attention to the scale of hypothetical impact, and to provide a motivation for detailed risk assessment, in order to avoid possible environmental and socio-economic damages to this region of the Barents Sea. From our opinion, it is important to note that regional impacts and climate conditions similar to the 1988–2018 oceanographic and atmospheric data set may be expected in the near future, within the next 10 years. The short time span is due to possible unpredictability of future climate variations. As reported by IPCC in 2014 [87], the Earth's surface was successively warmer in each of the prior three decades than any preceding decade since 1850. Climate change is a self accelerating process, especially in the Arctic latitudes. However, there is certain unpredictability of how fast and severe those changes may be and how it will affect oceanographic and atmospheric phenomena. If the climate normal of the last century was 30 years as a standard, this may change during the course of the 21st century. There is already discussion in the National Oceanic and Atmospheric Administration (NOAA) to reconsider the 85-year-old concept of climate normal, due to an accelerating trend of ongoing climate change. NOAA suggested to work with shorter periods of 20, 15, 10 and 5 years [88]. As an illustration to this, we can compare two recent periods in temperature anomalies—1980–2018 and 2008–2018, which are shown in Figure 5.

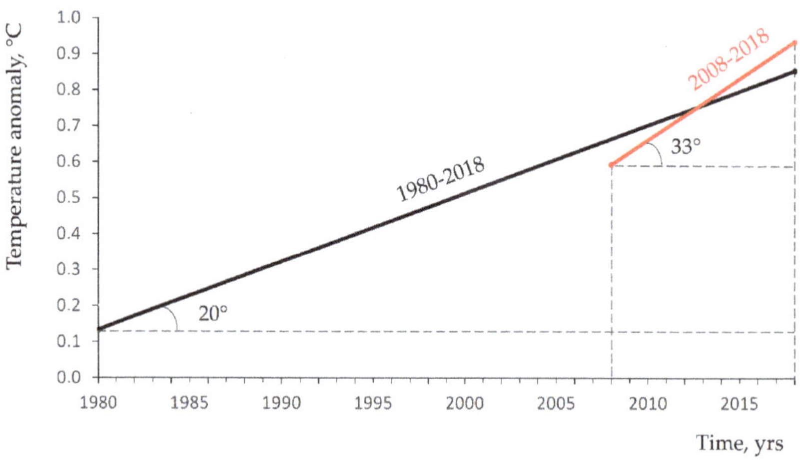

Figure 5. Acceleration of climate change: black line period of 1980–2018 and red line—2008–2018.

The trendlines illustrate the change in global surface temperatures relative to 1951–1980 average temperatures. The observation shows us that temperature rise accelerated in the last decade having a much more rapid tendency of deviation from climatic norm [89].

5. Conclusions

Simulations of surface oil spills were conducted in south-east of the Barents Sea to identify oil pollution trajectories. The study site is located near ecologically and economically important areas and lies along one of the most popular maritime routes. Two oil types were selected to demonstrate oil behaviour in the same Arctic environmental conditions in case of an oil spill: a popular bunker oil and a medium fuel oil. The study set provided statistical results for winter, spring, summer and autumn environmental conditions. Both cases demonstrated extensive spill travel distances, exceeding 1000 km and posing infrastructural and technological challenges for response operations. The selected geographical location has mostly unfavourable environmental conditions for oil spill abatement. Out of eight available Arctic oil spill response tools, only three showed potential for application in such climatic region and only in summer. Autumn, winter and spring seasons remain mostly difficult for organisation of abatement operations at sea. The area of highly probable impact is bordered by the Vaygach island and Novaya Zemlya in the east, the mainland of the Nenets autonomous region in the south, and the Kolguev Island in the west. In case an oil spill would occur at the location, all coastal areas within 300 km reach would be statistically vulnerable all year. From an environmental perspective, various ecosystem levels can be subjected to the risk of damage. The study area has vulnerable species during the whole year, but the autumn period seems to be the most dangerous. From a socio-economic perspective, impacts to both commercial and traditional fishing activities are likely to occur, however, this needs to be addressed in more detail in a separate research initiative.

Of the four seasons, the largest coverage for both studied oil types is observed during winter and autumn, whereas the coverage during spring and summer is smaller in size. In all seasons except summer, oil particles reached their highest latitudes of the Barents Sea, thus, drifting to remote areas of the high north. If hypothetical summer oil spill accidents may be likely called local, in all other seasons, it can take a regional scale. Hence, the accident consequences are most localised in summer in comparison with the other three seasons, where oil travel distances are large. Considering probable oil trajectories and possible impacts to the environment, summer would be the recommended shipping season along the studied route: e.g., oil and transportation of oil products. Autumn, in contrast, could be considered as a highly sensitive season for coastal offshore and onshore communities which are north from the initial spill site. Taking into account oil drift trajectories reaching the southern coast of Novaya Zemlya with its biodiversity levels, autumn shipping needs a higher level of oil spill response preparedness. The possibility of lowering the amount of shipping traffic in this season could be considered.

The produced probability maps indicated oil distribution patterns which may cause potential threats to local ecosystems. It is expected that the outcomes of this research are valuable to local and Arctic authorities, industrial actors and coastal communities. It may also contribute to creating a more extended picture for a hypothetical oil spill case in the considered local geographical point of the Barents Sea and improve local oil spill response preparedness. However, further research in environmental and socio-economic risk assessment is recommended to gain a detailed understanding of impacts to local flora and fauna as well as indigenous communities.

Author Contributions: Conceptualization, V.P.; methodology, V.P., V.C.M.d.A., L.R.H. and E.P.; software, V.C.M.d.A. and L.R.H.; validation, E.P.; data curation, V.C.M.d.A.; writing—original draft preparation, V.P. and V.C.M.d.A.; writing—review and editing, E.P.; visualization, V.P. and V.C.M.d.A.; supervision, L.R.H. and E.P. All authors have read and agreed to the published version of the manuscript.

Funding: This research was funded by Northern Periphery and Arctic Programme 2014–2020.

Informed Consent Statement: Not applicable.

Acknowledgments: The authors acknowledge the "Arctic Preparedness Platform for Oil Spill and other Environmental Accidents" (APP4SEA) project co-funded by the Northern Periphery and Arctic Programme 2014–2020. We also thank Henrikki Liimatainen for help with contribution to this work and Murmansk Marine Biological Institute of the Russian Academy of Sciences for providing sensitivity maps of the study region.

Conflicts of Interest: The authors declare no conflict of interest. The funders had no role in the design of the study; in the collection, analyses, or interpretation of data; in the writing of the manuscript, or in the decision to publish the results.

References

1. Wells, P.G. The iconic Torrey Canyon oil spill of 1967—Marking its legacy. *Mar. Pollut. Bull.* **2017**, *115*, 1–2. [CrossRef]
2. ITOPF, The International Tanker Owners Pollution Federation Limited. Oil Tanker Spill Statistics 2018. 2019. Available online: http://www.itopf.org/fileadmin/data/Documents/Company_Lit/Oil_Spill_Stats_2019.pdf (accessed on 15 October 2019).
3. Wenning, R.J.; Robinson, H.; Bock, M.; Rempel-Hester, M.A.; Gardiner, W. Current practices and knowledge supporting oil spill risk assessment in the Arctic. *Mar. Environ. Res.* **2018**, *141*, 289–304. [CrossRef]
4. Pavlov, V. Arctic Marine Oil Spill Response Methods: Environmental Challenges and Technological Limitations. In *Physics and Chemistry of the Arctic Atmosphere*; Pongrácz, E., Pavlov, V., Hänninen, N., Eds.; Springer: Cham, Switzerland, 2020; pp. 213–248.
5. Afenyo, M.; Khan, F.; Veitch, B.; Yang, M. Modeling oil weathering and transport in sea ice. *Mar. Pollut. Bull.* **2016**, *107*, 206–215. [CrossRef]
6. Oil Change International (OCI), Greenpeace. Frozen Future. The Gaps in Shell's Arctic Spill Response. 2015. Available online: http://priceofoil.org/content/uploads/2015/04/2015-The-gaps-in-Shells-Arctic-spill-response.pdf (accessed on 2 January 2016).
7. Panichkin, I. Arctic Oil and Gas Resource Development: Current Situation and Prospects. Available online: https://russiancouncil.ru/en/arcticoil (accessed on 19 November 2020).
8. Parviainen, T.; Lehikoinen, A.; Kuikka, S.; Haapasaari, P. Risk frames and multiple ways of knowing: Coping with ambiguity in oil spill risk governance in the Norwegian Barents Sea. *Environ. Sci. Policy* **2019**, *98*, 95–111. [CrossRef]
9. Arnarsson, S.; Dam, K.; Justus, D.; Latola, K.; Łuszczuk, M.; Sander, G.; Scheepstra, A.; Stępień, A.; Strahlendorff, M. Strategic Environmental Impact Assessment of Development of the Arctic. 2014. Available online: http://library.arcticportal.org/1792/1/SADA_report.pdf (accessed on 15 April 2017).
10. Kirby, M.F.; Law, R.J. Accidental Spills at Sea—Risk, Impact, Mitigation and the Need for co-Ordinated Post-Incident Monitoring. *Mar. Pollut. Bull.* **2010**, *60*, 797–803. [CrossRef]
11. Borch, O.J.; Andreassen, N.; Marchenko, N.; Ingimundarson, V.; Gunnarsdóttir, H.; Ludin, I.; Petrov, S.; Jakobsen, U.; Dali, B. Maritime Activity in the High North—Current and Estimated Level Up to 2025. MARPART Project Report 1. 2016. Available online: https://brage.bibsys.no/xmlui/bitstream/handle/11250/2413456/Utredning72016.pdf?sequence=5&isAllowed=y (accessed on 7 September 2017).
12. Kulakov, M.; Pogrebov, V.B.; Timofeyev, S.F.; Chernva, N.V.; Kiyko, O.A. Chapter 29. Ecosystem of the Barents and Kara Seas, Coastal Segment (22, P). In *The Sea*; Robinson, A.R., Brink, K.H., Eds.; President and Fellows of Harvard College: Cambridge, MA, USA, 2004; pp. 1135–1172.
13. Keramea, P.; Spanoudaki, K.; Zodiatis, G.; Gikas, G.; Sylaios, G. Oil Spill Modeling: A Critical Review on Current Trends, Perspectives, and Challenges. *J. Mar. Sci. Eng.* **2021**, *9*, 181. [CrossRef]
14. Alves, T.M.; Kokinou, E.; Zodiatis, G. A three-step model to assess shoreline and offshore susceptibility to oil spills: The South Aegean (Crete) as an analogue for confined marine basins. *Mar. Pollut. Bull.* **2014**, *86*, 443–457. [CrossRef]
15. Alves, T.M.; Kokinou, E.; Zodiatis, G.; Lardner, R.; Panagiotakis, C.; Radhakrishnan, H. Modelling of oil spills in confined maritime basins: The case for early response in the Eastern Mediterranean Sea. *Environ. Pollut.* **2015**, *206*, 390–399. [CrossRef]
16. Alves, T.M.; Kokinou, E.; Zodiatis, G.; Lardner, R. Hindcast, GIS and susceptibility modelling to assist oil spill clean-up and mitigation on the southern coast of Cyprus (Eastern Mediterranean). *Deep. Sea Res. Part II Top. Stud. Oceanogr.* **2016**, *133*, 159–175. [CrossRef]
17. Adler, E.; Inbar, M. Shoreline sensitivity to oil spills, the Mediterranean coast of Israel: Assessment and analysis. *Ocean Coast. Manag.* **2007**, *50*, 24–34. [CrossRef]
18. Beegle-Krause, C.J.; Simmons, H.; McPhee, M.; Lundmark Daae, R.; Reed, M. Literature Review: Fate of Dispersed Oil Under Ice. Final Report 1.4. 2013. Available online: http://www.arcticresponsetechnology.org/wp-content/uploads/2014/02/Report-1.4-Fate-of-Dispersed-Oil-under-Ice.pdf. (accessed on 9 December 2015).
19. Dobrovolsky, A.D.; Zalogin, B.C. USSR Seas (МоряСССР). 1982. Available online: http://ashipunov.info/shipunov/school/books/dobrovolskij1982_morja_sssr.pdf (accessed on 7 September 2017).
20. Emergency Prevention, Preparedness and Response (EPPR) Working Group of the Arctic Council. Guide to Oil Spill Response in Snow and Ice Conditions. 2015. Available online: https://oaarchive.arctic-council.org/handle/11374/403 (accessed on 2 January 2016).

21. Emergency Prevention, Preparedness and Response (EPPR) Working Group of the Arctic Council. Reykjavik, Iceland, 2–3 December 2015. Impact of Metocean Conditions on Oil Spill Response Viability in the Arctic (Previ-ously named Circumpolar Oil Spill Response GAP Analysis). 2015. Available online: http://arctic-council.org/eppr/wp-content/uploads/2015/06/2015_11_16_COSRVA_Analysis_Update.pdf (accessed on 30 December 2015).
22. Fingas, M. *Handbook of Oil Spill Science and Technology*; John Wiley & Sons: Hoboken, NJ, USA, 2015; p. 728.
23. Matishov, G.; Golubeva, N.; Titova, G.; Sydnes, A.; Voegele, B. Barents Sea, GIWA Regional Assessment 11. 2004. Available online: http://www.unep.org/dewa/giwa/areas/reports/r11/giwa_regional_assessment_11.pdf (accessed on 3 February 2017).
24. Sukhotin, A.; Denisenko, S.; Galaktionov, K. Pechora Sea ecosystems: Current state and future challenges. *Polar Biol.* **2019**, *42*, 1631–1645. [CrossRef]
25. Vorobiev, U.L.; Akimov, V.A.; Sokolov, U.I. *Предупреждение* и Ликвидация*Аварийных Разливов* Нефти и Нефтепродуктов*(Oil and oil Products Spill Prevention and Response)*; In-Oktavo: Moscow, Russia, 2005; p. 368.
26. RIA. Specially Protected Areas in the Russian Arctic. 2011. Available online: https://ria.ru/arctic_mm/20111226/527007298.html (accessed on 7 September 2017).
27. Conservation of Arctic Flora and Fauna (CAFF). *Protected Areas of the Arctic: Conserving a Full Range of Values*; CAFF: Ottawa, ON, Canada, 2002.
28. Pasvik State Nature Reserve (PSNR). Rare Species. 2019. Available online: http://www.pasvik51.ru/index.php/en/nature/rare-species (accessed on 12 November 2019).
29. UNEP-WCMC and IUCN. Protected Planet: The World Database on Protected Areas (WDPA). 2019. Available online: https://www.protectedplanet.net (accessed on 23 September 2019).
30. Mecklenburg, C.W.; Lynghammar, A.; Johannesen, E.; Byrkjedal, I.; Christiansen, J.S.; Dolgov, A.V.; Karamushko, O.V.; Mecklenburg, T.A.; Møller, P.R.; Steinke, D.; et al. *Marine Fishes of the Arctic Region*; Conservation of Arctic Flora and Fauna: Akureyri, Iceland, 2018; ISBN 978-9935-431-69-1.
31. Durif, C. Lumpfish. 2016. Available online: https://www.hi.no/en/hi/temasider/species/lumpfish (accessed on 15 October 2019).
32. Skaret, G. Polar Cod. 2013. Available online: https://www.hi.no/en/hi/temasider/species/polar-cod (accessed on 15 October 2019).
33. FishSource. Atlantic Cod: Barents Sea. 2019. Available online: https://www.fishsource.org/stock_page/1727 (accessed on 15 October 2019).
34. World Wildlife Fund (WWF). Arctic Geographical Information System. 2013. Available online: http://wwfarcticmaps.org/ (accessed on 11 June 2019).
35. Corbett, J.J.; Lack, D.A.; Winebrake, J.J.; Harder, S.; Silberman, J.A.; Gold, M. Arctic shipping emissions inventories and future scenarios. *Atmos. Chem. Phys. Discuss.* **2010**, *10*, 9689–9704. [CrossRef]
36. Mjelde, A.; Martinsen, K.; Eide, M.; Endresen, Ø. Environmental accounting for Arctic shipping—A framework building on ship tracking data from satellites. *Mar. Pollut. Bull.* **2014**, *87*, 22–28. [CrossRef] [PubMed]
37. Protection of the Arctic Marine Environment (PAME). The Increase in Arctic Shipping. 2020. Available online: https://pame.is/document-library/pame-reports-new/pame-ministerial-deliverables/2021-12th-arctic-council-ministerial-meeting-reykjavik-iceland/793-assr-1-the-increase-in-arctic-shipping-2013-2019/file (accessed on 5 July 2021).
38. Stobberup, K.; Erzini, K.; Lam, V.; Pauly, D. Fisheries Management and the Arctic in the Context of Climate Change. 2015. Available online: http://www.europarl.europa.eu/RegData/Etudes/STUD/2015/563380/IPOL_STU(2015)563380_EN.pdf (accessed on 3 September 2019).
39. Bogstad, B. Cod—Northeast Arctic. 2017. Available online: https://www.hi.no/en/hi/temasider/species/cod--northeast-arctic (accessed on 15 October 2019).
40. Carroll, J.; Vikebø, F.; Howell, D.; Broch, O.J.; Nepstad, R.; Augustine, S.; Skeie, G.M.; Bast, R.; Juselius, J. Assessing impacts of simulated oil spills on the Northeast Arctic cod fishery. *Mar. Pollut. Bull.* **2018**, *126*, 63–73. [CrossRef]
41. Ocean Trawlers. Decision for the 2016 Barents Sea Quota. 2015. Available online: http://www.oceantrawlers.com/decision-for-the-2016-quota-for-barents-sea/ (accessed on 7 September 2017).
42. Ocean Trawlers. Decision for the 2017 Barents Sea Quota. 2016. Available online: http://www.oceantrawlers.com/decision-for-the-2017-barents-sea-quota/ (accessed on 7 September 2017).
43. Staalesen, A. Less Cod in Barents Sea Must be Followed by Reduced Quotas, Researchers Say. 2017. Available online: https://thebarentsobserver.com/en/ecology/2017/06/less-cod-barents-sea-must-be-followed-reduced-quotas-researchers-say (accessed on 7 September 2017).
44. Bambulyak, A.; Frantzen, B.; Rautio, R. *Oil Transport from the Russian Part of the Barents Region. 2015 Status Report*; The Norwegian Barents Secretariat and Akvaplan-niva: Kirkenes, Norway, 2015; 105p.
45. Chernyaev, A.; Belova, L.; Prokhorova, N.; Pupova, Y. *Waters of Russia*; AQUA-PRESS: Rezh, Russia, 2000; pp. 92–94.
46. VEO Russia. Scientific Works of VEO Russia (ТрудыВЭС России). 2014. Available online: http://veorus.ru/upload/iblock/dc9/182_veor.pdf (accessed on 30 November 2017).
47. Matishov, G.G.; Brekhuntsov, A.M.; Dzhenyuk, S.L. Studies of the Kara Sea in the Modern Stage of Development of the Russian Arctic. 2013. Available online: http://eng.arctica-ac.ru/docs/1(9)/004_011_ARKTIKA_1(9)_03_2013.pdf (accessed on 7 September 2017).

48. Ports. Norway. Barents Sea Ports. 2017. Available online: http://ports.com/sea/barents-sea/ (accessed on 7 September 2017).
49. Marine Traffic. Density Maps: 2016. 2020. Available online: https://www.marinetraffic.com/en (accessed on 3 July 2020).
50. Nilsen, T. Norway Doubles Arctic Oil Estimates. 2017. Available online: https://thebarentsobserver.com/en/industry-and-energy/2017/04/norway-doubles-arctic-oil-estimates (accessed on 7 September 2017).
51. Staalesen, A. Norway Offers Oil Companies 93 New Blocks in Arctic Waters. 2017. Available online: https://thebarentsobserver.com/en/industry-and-energy/2017/06/norway-offers-oil-companies-93-new-blocks-arctic-waters (accessed on 7 September 2017).
52. Blanken, H.; Tremblay, L.B.; Gaskin, S.; Slavin, A. Modelling the long-term evolution of worst-case Arctic oil spills. *Mar. Pollut. Bull.* **2017**, *116*, 315–331. [CrossRef]
53. East and North Finland EU office (EU). Lapland. 2019. Available online: https://www.eastnorth.fi/regions/lapland (accessed on 20 November 2020).
54. U.S. Census Bureau. Alaska. 2019. Available online: https://www.census.gov/quickfacts/AK (accessed on 27 November 2020).
55. Norwegian Polar Institute & Association of Nenets People "Yasavey" (NPI & Yasavey). Monitoring of Development of Traditional Indigenous Land Use Areas in the Nenets Autonomous Okrug, Northwestern Russia. 2010. Available online: https://ipy-nenets.npolar.no/main%20pages/frame.html (accessed on 20 March 2020).
56. Rosstat: Federal State Statistics Service (Rosstat). Population. 2020. Available online: https://eng.gks.ru/population (accessed on 20 March 2020).
57. Tulaeva, S.A.; Tysiachniouk, M.S.; Henry, L.A.; Horowitz, L.S. Globalizing Extraction and Indigenous Rights in the Russian Arctic: The Enduring Role of the State in Natural Resource Governance. *Resources* **2019**, *8*, 179. [CrossRef]
58. Novikova, N. Pipeline Neighbors: How Can We Avoid Conflicts? *Resources* **2020**, *9*, 13. [CrossRef]
59. Nenets Autonomous Okrug: State Authority Portal (NAO Portal). Cooperative Societies. 2019. Available online: https://gkh.adm-nao.ru/ (accessed on 25 March 2020).
60. Department of Natural Resources, Ecology, Agricultural and Industrial Complex of Nenets Autonomous Region (NREAINAO). Environmental Report of Nenets Region: 2018. 2019. Available online: https://dprea.adm-nao.ru/ekologiya/doklady-o-sostoyanii-okruzhayushej-sredy-v-neneckom-avtonomnom-okruge/ (accessed on 20 March 2020).
61. Rusprofile. "Kolguev". 2020. Available online: https://www.rusprofile.ru/id/952639 (accessed on 25 March 2020).
62. Russian Association of Indigenous Peoples of the North (RAIPON). Nenets. 2014. Available online: http://www.raipon.info/peoples/nenets/nenets.php (accessed on 20 March 2020).
63. Varandej. Vaygach: Island of Mystery. 2017. Available online: https://varandej.livejournal.com/884648.html (accessed on 25 March 2020).
64. PLAN X. Adventure Team: "Sailing Half the Country". 2017. Available online: http://planx.media/devyataya-nedelya-polstrany-pod-parusom-beskrajnee-seroe-more/ (accessed on 25 March 2020).
65. Center of Strategic Development: North-West Russia (CSD). Strategy of Socio-Economic Development of the Nenets Autonomous Region by 2030. 2009. Available online: http://www.csr-nw.ru/files/csr/file_content_581.pdf (accessed on 20 March 2020).
66. Norwegian Coastal Administration (NCA). PAME's Arctic Ship Traffic Database: 2017. 2020. Available online: https://map.astd.is/ (accessed on 3 July 2020).
67. Protection of the Arctic Marine Environment (PAME). Arctic Shipping Status Reports. 2020. Available online: https://www.pame.is/projects/arctic-marine-shipping/arctic-shipping-status-reports (accessed on 3 July 2020).
68. Global Fishing Watch. Global Fishing Watch Map. 2019. Available online: https://globalfishingwatch.org/map/ (accessed on 3 September 2019).
69. Sea Around Us. Mapped data. Global Fishing in 2014. 2016. Available online: http://www.seaaroundus.org/data/#/spatial-catch (accessed on 3 September 2019).
70. Dagestad, K.-F.; Röhrs, J.; Breivik, Øyvind; Ådlandsvik, B. OpenDrift v1.0: A generic framework for trajectory modelling. *Geosci. Model Dev.* **2018**, *11*, 1405–1420. [CrossRef]
71. Hole, L.R.; Dagestad, K.-F.; Röhrs, J.; Wettre, C.; Kourafalou, V.H.; Androulidakis, Y.; Kang, H.; Le Hénaff, M.; Garcia-Pineda, O. The DeepWater Horizon Oil Slick: Simulations of River Front Effects and Oil Droplet Size Distribution. *J. Mar. Sci. Eng.* **2019**, *7*, 329. [CrossRef]
72. Det Norske Veritas (DNV). The Arctic RISK MAP. 2016. Available online: https://maps.dnvgl.com/arcticriskmap/ (accessed on 5 July 2021).
73. Det Norske Veritas (DNV). Circumpolar Oil Spill Response Viability Analysis. 2019. Available online: https://maps.dnvgl.com/cosrva/ (accessed on 5 July 2021).
74. Sørheim, K.R.; Daling, P.S.; Cooper, D.; Bust, I.; Faksness, L.G.; Altin, D.; Pettersen, T.A.; Bakken, O.M. Characterization of Low Sulfur Fuel Oils (LSFO)—A New Generation of Marine Fuel Oils. 2020. Available online: https://www.itopf.org/fileadmin/data/Documents/RDaward/Final_report_LSFO_Multipartner__2.0__May20.pdf (accessed on 5 July 2021).
75. Yashchenko, I.G.; Nesterova, G.V. Oil-Gas Resources of Arctic Sector. 2015. Available online: https://cyberleninka.ru/article/n/neftegazovye-resursy-arkticheskogo-sektora (accessed on 5 July 2021).
76. Nevalainen, M.; Helle, I.; Vanhatalo, J. Estimating the acute impacts of Arctic marine oil spills using expert elicitation. *Mar. Pollut. Bull.* **2018**, *131*, 782–792. [CrossRef]

77. Paliukaitė, M.; Vaitkus, A.; Zofka, A. Evaluation of bitumen fractional composition depending on the crude oil type and production technology. In Proceedings of the 9th International Conference Environmental Engineering, Vilnius, Lithuania, 22–23 May 2014.
78. Santos, R.G.; Loh, W.; Bannwart, A.C.; Trevisan, O.V. An overview of heavy oil properties and its recovery and transportation methods. *Braz. J. Chem. Eng.* **2014**, *31*, 571–590. [CrossRef]
79. Antigoni, Z. Oil Spill Dispersion Forecasting Models, Monitoring of Marine Pollution. 2018. Available online: https://www.intechopen.com/books/monitoring-of-marine-pollution/oil-spill-dispersion-forecasting-models (accessed on 10 July 2019).
80. Brandvik, P.; Daling, P.; Myrhaug, J. *Mapping Weathering Properties as a Function of Ice Conditions: A Combined Approach Using a Flume Basin Verified by Large—Scale Field Experiments*; U.S. Department of Energy: Washington, DC, USA, 2009.
81. Nordam, T.; Beegle-Krause, C.; Skancke, J.; Nepstad, R.; Reed, M. Improving oil spill trajectory modelling in the Arctic. *Mar. Pollut. Bull.* **2019**, *140*, 65–74. [CrossRef] [PubMed]
82. French-McCay, D.P.; Tajalli-Bakhsh, T.; Jayko, K.; Spaulding, M.L.; Li, Z. Validation of Oil Spill Transport and Fate Modeling in Arctic Ice. *Arct. Sci.* **2017**, *4*, 71–97. [CrossRef]
83. Wild World Foundation (WWF). Modeling of Potential Oil Spill Behavior When Operating Prirazlomnaya OIFP—Assessment of Possible Oil Spill Emergency Response. Technical Report, RGC Risk Informatics. 2012. Available online: https://wwf.ru/en/resources/publications/booklets/modeling-of-potential-oil-spill-behavior-when-operating-prirazlomnaya-oifp/ (accessed on 21 June 2020).
84. Shavykin, A.A.; Ilyin, G.V. *An Assessment of the Integral Vulnerability of the Barents Sea from Oil Contamination*; MMBI KSC RAS: Murmansk, Russia, 2010; 110p.
85. Jakobsen, M.; Milaković, A.-S.; Ehlers, S. Assessment of Helicopter Emergency Response Capacity in the Barents Sea. 2016. Available online: https://ntnuopen.ntnu.no/ntnu-xmlui/bitstream/handle/11250/2350647/13888_FULLTEXT.pdf?sequence=1&isAllowed=y (accessed on 5 July 2021).
86. Mokochunina, T.V.; Ossipov, K.; Maryutina, T.A. Regulatory Policy for Using Oil Dispersants to Eliminate Oil Emergency Spills in the Offshore Areas of the Russian Federation. 2021. Available online: https://www.cet-mipt.ru/storage/files/trzsPUSCbp5aYeg4Xq9G2kOvv2AOEYU8EhQc.pdf (accessed on 5 July 2021).
87. Intergovernmental Panel on Climate Change (IPCC). *Climate Change 2014: Synthesis Report. Contribution of Working Groups I, II and III to the Fifth Assessment Report of the Intergovernmental Panel on Climate Change*; Pachauri, R.K., Meyer, L.A., Eds.; IPCC: Geneva, Switzerland, 2014; 151p.
88. National Oceanic and Atmospheric Administration (NOAA). Defining Climate Normals in New Ways. 2020. Available online: https://www.ncdc.noaa.gov/news/defining-climate-normals-new-ways#:~{}:text=Scientists%20traditionally%20define%20a%20Climate,a%20recent%2030%2Dyear%20period.&text=So%2C%20our%20scientists%20have%20tradi (accessed on 11 December 2020).
89. Shaftel, H.; Callery, S.; Randal Jackson, R.; Bailey, D.; Callery, S. Global Temperature. 2021. Available online: https://climate.nasa.gov/vital-signs/global-temperature/ (accessed on 15 December 2021).

energies

MDPI

Article

Fly Ash-Based Geopolymers as Lower Carbon Footprint Alternatives to Portland Cement for Well Cementing Applications

Cameron Horan [1], Moneeb Genedy [1], Maria Juenger [2] and Eric van Oort [1,*]

[1] Hildebrand Department of Petroleum and Geosystems Engineering, University of Texas at Austin, Austin, TX 78712, USA
[2] Civil, Architectural, and Environmental Engineering, University of Texas at Austin, Austin, TX 78712, USA
* Correspondence: vanoort@austin.utexas.edu

Abstract: Ordinary Portland cement (OPC) is currently the preferred material for the creation of barriers in wells during their construction and abandonment globally. OPC, however, is a very carbon-intensive material with some inherent technical weaknesses. These include a low casing-to-cement bond strength which may allow for the formation of micro-annuli, which in turn can become a conduit for greenhouse gas transport (primarily of methane, a powerful greenhouse gas) to surface. Alkali-activated materials (AAMs), also known as geopolymers, have a much lower manufacturing carbon footprint than OPC and can be a good alternative to OPC for primary and remedial well cementing applications. This paper reports on a comprehensive study into the use of Class F fly ash-based geopolymers for a large variety of downhole well conditions, ranging from lower-temperature surface and intermediate casing cementing conditions to much higher temperature conditions (up to 204 °C (400 °F)) that can be encountered in high-pressure, high-temperature (HPHT) wells and geothermal wells. The rheological and mechanical properties of alkali-activated fly ash with six different sodium and potassium-based hydroxide and silicate activators were measured and compared to OPC. The results show that geopolymer formulation properties can be tuned to a variety of downhole cementing conditions. With the application of a suitable alkaline activator, geopolymers exhibit good compressive and tensile strength and an outstanding casing-to-cement bond strength of up to 8.8 MPa (1283 psi), which is more than an order of magnitude higher than OPC. This has important implications for preventing the creation of micro-annuli as a result of casing-to-cement interface debonding, thereby preventing the potential leakage of methane to the atmosphere on future wells that use geopolymers rather than OPC for barrier creation.

Keywords: geopolymer; fly ash; alkali-activated material; zonal isolation; cement replacement; carbon footprint reduction

Citation: Horan, C.; Genedy, M.; Juenger, M.; van Oort, E. Fly Ash-Based Geopolymers as Lower Carbon Footprint Alternatives to Portland Cement for Well Cementing Applications. *Energies* **2022**, *15*, 8819. https://doi.org/10.3390/en15238819

Academic Editor: Vijay Kumar Thakur

Received: 19 October 2022
Accepted: 21 November 2022
Published: 23 November 2022

1. Introduction

Almost every major energy company has announced plans to achieve net zero emissions before or by 2050 [1]. To achieve such goals, current practices and materials will need to be modified, and new ones will have to be adopted and implemented. As an example for the subsurface energy industry, shifting towards renewable energy sources such as geothermal energy and adopting carbon capture and underground storage are major steps toward reaching net zero goals [2,3]. Additionally, maintaining proper zonal isolation inside and outside wellbores is essential to prevent leakage incidents (see for instance [4]), which can release large quantities of greenhouse gasses (GHG) to the atmosphere. The concern revolves primarily around leakage of methane (CH_4), which has 86 times the GHG effect of carbon dioxide (CO_2) over a 20-year time period [5]. Ordinary Portland cement (OPC) is used extensively in the well construction industry for establishing cemented well barriers, but this comes at the cost of a large carbon footprint. The cement manufacturing industry is responsible for 5–8% of global CO_2 emissions [6]. OPC also has several technical weaknesses, which include low casing–cement bond strength, low tensile-to-compressive strength ratios, and failure behavior without significant re-healing effects when cement fractures under the influence of applied loads (e.g., cyclic loads, geomechanical and geochemical loads acting on the wells during their entire lifetime [7,8]). A low carbon footprint cementing material that can effectively address the weaknesses of OPC and provide zonal isolation and lost circulation control in a wide range of wellbore conditions is therefore highly desirable on the road towards reduced—and eventually net zero—emissions.

Several studies have been dedicated to the use of alkali-activated materials (AAMs), also known as geopolymers, as an alternative cementing material for wellbore cementing applications [7,9–15]. Geopolymers are formulated by mixing an aluminosilicate precursor powder (examples are fly ash, slag, or metakaolin) with an alkaline activating solution. The most common activators used in geopolymer formulations are sodium or potassium hydroxides and silicates. Class F fly ash-based geopolymers have shown desirable properties for cementing applications in previous investigations. They exhibit a high tolerance to contamination with drilling fluids [16–18], good resistance to acid attack [19], self-healing abilities [20–23], and lower shrinkage sensitivity at elevated pressure and temperature [24]. Moreover, high temperature stability up to 800 °C was observed for geopolymers activated with silicate activators [25,26], which is particularly useful when considering application in high-pressure/high temperature (HPHT) oil and gas wells as well as geothermal wells [27]. Most importantly for the present study, a well-cited study by Duxson et al. [28] indicates that geopolymer binders can deliver an 80%-or-greater reduction in CO_2 emissions compared to OPC. This is mainly because OPC manufacturing requires the high-temperature calcination (up to 1400 °C) of cement clinker, which not only consumes a large volume of fossil fuels to supply the necessary energy but also releases CO_2 as a reaction product. Such a calcination step is absent from geopolymer synthesis of ashes or slags [28]. Moreover, geopolymers yield additional environmental benefits compared to OPC associated with a lower water requirement (of particular importance in dry regions of the world where non-renewable water use may be a significant issue) and a limited requirement to use super-plasticizing agents which contribute to the cost and CO_2 footprint associated with the use of OPC [28,29].

Most of the studies conducted on the use of geopolymers as alternative materials for wellbore cementing applications have focused on relatively low-temperature applications (i.e., bottom hole static temperature (BHST) of less than 77 °C (170 °F)) or have used sodium-based activators [10,12,13,30–32]. This paper reports on a comprehensive investigation into the use of Class F fly ash-based geopolymers as cementing materials for various wellbore applications, including higher temperature conditions than previously studied. Potassium activators were also explored in addition to the more traditional sodium activators. Geopolymers formulated with six different activators, i.e., sodium and potassium-based hydroxide, as well as liquid and solid sodium and potassium silicates, were tested at different conditions, including relatively low-temperature conditions (BHST 77 °C (170 °F)), simulating surface and intermediate casing cementations; elevated temperature conditions (BHST 135 °C (275 °F)), representing deep intermediate and production casing cementations; and high-temperature conditions (BHST up to 204 °C (400 °F)), more representative of conditions in HPHT and geothermal wells. The viscosity, thickening time, and mechanical properties (i.e., compressive strength, tensile strength, and bond strength) of geopolymer slurries were measured and compared to OPC.

2. Materials

2.1. Cementitious Materials

Two types of cementitious materials were used in this study: Class H OPC and Class F fly ash (FA). The oxide contents of both OPC and FA are shown in Table 1. The chemical composition of the FA used in this study is representative of a Class F FA as defined by the

ASTM C618 standard, while the composition of the OPC is representative of a Class H OPC in accordance with the API 10A standard.

Table 1. Chemical compositions of Class H OPC and Class F FA used in this study.

Oxide	Content (wt%)	
	Class H OPC	Class F FA
SiO_2	20.9	44.7
Al_2O_3	3.6	23.2
Fe_2O_3	6.3	24.2
CaO	63.3	3.2
MgO	1.0	0.8
SO_3	3.0	0.6
Total alkali as sodium oxide, Na_2O_{eq}	0.55	1.0
Loss on ignition (LOI) at 750 °C	1.2	0.4
Insoluble residue/other	0.15	1.9

2.2. Mixtures Preparation

Table 2 shows the mixture design parameters for all OPC and geopolymer slurries. OPC slurries were prepared with a water-to-cement solids ratio by mass (W/S) of 0.385 following the procedure described in API RP 10B-2 using a high shear mixer. To overcome the strength retrogression reported in OPC slurries subjected to high temperature, an additional OPC mixture incorporating silica flour (OPC/SF) was investigated [33,34]. The OPC/SF mixture was prepared by replacing 30% of the cement weight with silica flour while maintaining the W/S ratio at 0.385.

Table 2. Mix design parameters of OPC and geopolymer slurries.

Mix ID	Activator	W/S	M_2O	SiO_2/M_2O	M_2O/FA
OPC	DI Water	0.385	—	—	—
OPC/SF	DI Water	0.385	—	—	—
LSH	8M Liquid Sodium Hydroxide	0.33	8M	—	—
LPH	8M Liquid Potassium Hydroxide	0.33	8M	—	—
LSS	Liquid Sodium Silicate	0.33	8M	0.12	0.1
SSS	Solid Sodium Silicate	0.33	8M	0.12	0.1
LPS	Liquid Potassium Silicate	0.33	8M	0.12	0.2
SPS	Solid Potassium Silicate	0.33	8M	0.12	0.2

Geopolymer slurries were formulated using six different activators, including liquid sodium hydroxide (LSH), liquid potassium hydroxide (LPH), liquid sodium silicate (LSS), solid sodium silicate (SSS), liquid potassium silicate (LPS), and solid potassium silicate (SPS). The slurries were prepared by initially hand-mixing dry components into the activating solution until a uniformly blended mixture was obtained, followed by mixing with a high-speed overhead paddle stirrer for 35 s. A constant W/S ratio of 0.33 was used for all geopolymer slurries.

For liquid hydroxide-based geopolymers, the activator solutions were prepared by dissolving appropriate amounts of sodium/potassium hydroxide pellets (98% purity) into deionized water (18 MΩ-cm resistivity) to reach an Na or K-based alkali concentration of 8 M.

Silicate-activated geopolymers had a constant SiO_2/M_2O weight ratio (where M represents either Na or K) of 0.12 and M_2O/FA weight ratio of 0.1 or 0.2, depending on the activator used, to achieve an 8M alkali concentration in the mixture. For the liquid silicate-based geopolymers, the solutions were prepared by dissolving solid sodium/potassium hydroxide pellets in deionized water, followed by mixing in commercially available sodium/potassium silicate solutions at the appropriate amounts. For the solid silicate-based geopolymers, the solutions were prepared by dissolving the required amount of solid sodium/potassium hydroxide pellets in deionized water, while sodium/potassium solid silicate powders were dry-mixed with the fly ash prior to mixing with the activating solution.

3. Experimental Methods

3.1. Viscosity

Viscosity testing was used to measure the rheological properties of geopolymer and OPC slurries. Following the API RP 10B-2 recommended practice, the slurries were tested immediately after preparation using a rotational viscometer. Three viscosity tests were conducted for each slurry at 23 °C using an Ofite Model 900 rotational viscometer equipped with an F1.0 spring and an R1B1 rotor and bob configuration. All geopolymer slurries were modeled using the Yield Power Law (YPL) rheology model shown in Equation (1) to determine the rheological constants:

$$\tau = \tau_y + K\dot{\gamma}^n \tag{1}$$

where τ is the shear stress (Pa), τ_y is the yield stress (Pa), K is the consistency index (Pa.s^n), n is the fluid behavior index (dimensionless), and $\dot{\gamma}$ is the shear rate (s^{-1}).

3.2. Mechanical Properties

3.2.1. Curing Conditions

A high-pressure high-temperature curing chamber (Model M5-700-8-2 by CTE with 371 °C (170 °F) and 5000 psi temperature and pressure limits) was used for three curing schemes, summarized in Table 3, to create all samples prepared for mechanical property determination. The schemes were chosen to reflect a wide range of wellbore conditions from a relatively low temperature of 77 °C (170 °F), representing conditions typically encountered during surface and intermediate casing cementations, to high temperature conditions of 204 °C (400 °F), more representative of HPHT and geothermal well cementation conditions. The bottomhole static temperature (BHST), bottomhole circulating temperature (BHCT), and bottomhole pressure (BHP) values for these curing schemes were selected following curing procedures and schedules given in recommended practice API RP 10B-2. All samples were cured for a period of 3 days, after which the temperature and pressure were ramped down to 32 °C (90 °F) and atmospheric pressure, respectively, in 30 min.

Table 3. Curing parameters for the three curing schemes used in this study.

Scheme	Initial Temp.	BHCT	Ramp Time	BHST	Ramp Time	BHP
1	23 °C (73.4 °F)	52 °C (125 °F)	120 min	77 °C (170 °F)	600 min	20.7 MPa (3000 psi)
2	23 °C (73.4 °F)	100 °C (212 °F)	150 min	135 °C (275 °F)	600 min	20.7 MPa (3000 psi)
3	23 °C (73.4 °F)	135 °C (275 °F)	180 min	204 °C (400 °F)	600 min	20.7 MPa (3000 psi)

3.2.2. Unconfined Compressive Strength (UCS)

Unconfined compressive strength (UCS) tests were carried out using cylinders (51 mm (2 inches) in diameter and 102 mm (4 inches) in height) created from geopolymer and OPC material following standard ASTM C39/C39M. Three cylinders were cured for each of the three curing schemes given in Table 3 for all mixtures discussed in this paper. After curing for 3 days, the cured materials were uniaxially loaded using a Forney-250 compression machine until failure with a load rate of 667 N/s (150 lbf /s) to determine UCS values.

3.2.3. Tensile Strength

The tensile strength of geopolymers and OPC samples was characterized using a split tension test following standard ASTM C496. Three cylinders of 51 mm (2 inches) in diameter and 102 mm (4 inches) in height were cured for each of the three curing schemes given in Table 3 for all mixtures discussed in this paper. Samples cured 3 days were then loaded using a Forney-250 compression machine to tensile failure with a load rate of 155 N/s (35 lbf/s).

3.2.4. Bond Strength

The shear bond strength of the geopolymer and OPC materials to steel was investigated using a pushout procedure, as described by [20,35]. A clean stainless-steel rod 12.7 mm (0.5 inches) in diameter was concentrically placed in a cylindrical mold with a 51 mm (2 inches) diameter. Geopolymer or OPC slurry was then cast around the rod to a height of 51 mm (2 inches). Three samples of each geopolymer or OPC mixture were cured following each of the three curing schemes in Table 3 and tested for bond strength after curing for 3 days. Bond strength was obtained by loading the steel rods using a Forney-250 compression machine at a load rate of 53 N/s (12 lbf/s) until debonding between the rod and the geopolymer or OPC material was observed. Bond strength is given by:

$$f_b = \frac{P}{\pi l d} \tag{2}$$

where f_b is bond strength (Pa), l is the height of the slurry (m), d is the diameter of the rod (m), and P is the peak load (N).

3.3. Thickening Time

An Ofite Model 2040 automated HPHT cement consistometer was used to determine thickening times of all slurries following the API standards RP 10A and 10B-2. During testing, temperature was raised from a surface temperature of 23 °C (73.4 °F) to the BHCT value in the corresponding ramp-up time for each wellbore condition as summarized in Table 3. Temperature was then kept constant at the BHCT value for the remainder of the test. Pressure was raised immediately from atmospheric pressure to 6.89 MPa (1000 psi) and then further increased to the BHP value in the corresponding ramp time as summarized in Table 3. Pressure was then kept constant at the BHP value for the remainder of the test. Note that thickening time is defined as the time required for the slurry to reach a consistency of 70 Bearden consistency (Bc) units, after which the slurry is considered to be unpumpable. The average value and range of the thickening time was obtained by testing each slurry three times.

4. Results and Discussion

4.1. Viscosity

The shear stress versus shear rate plots for OPC and geopolymer slurries are presented in Figure 1. The viscosity profiles were modeled using the YPL model (given in Equation (1)) and rheological constants (yield stress, fluid behavior index, and consistency index) were calculated. The average rheological constants and apparent viscosities at 300 rpm are summarized in Table 4. The results show that OPC (solid gray line) and OPC/SF (dashed gray line) slurries have similar viscosity profiles. The fluid behavior indices (n-values) of OPC and OPC/SF were found to be 0.46 and 0.61.

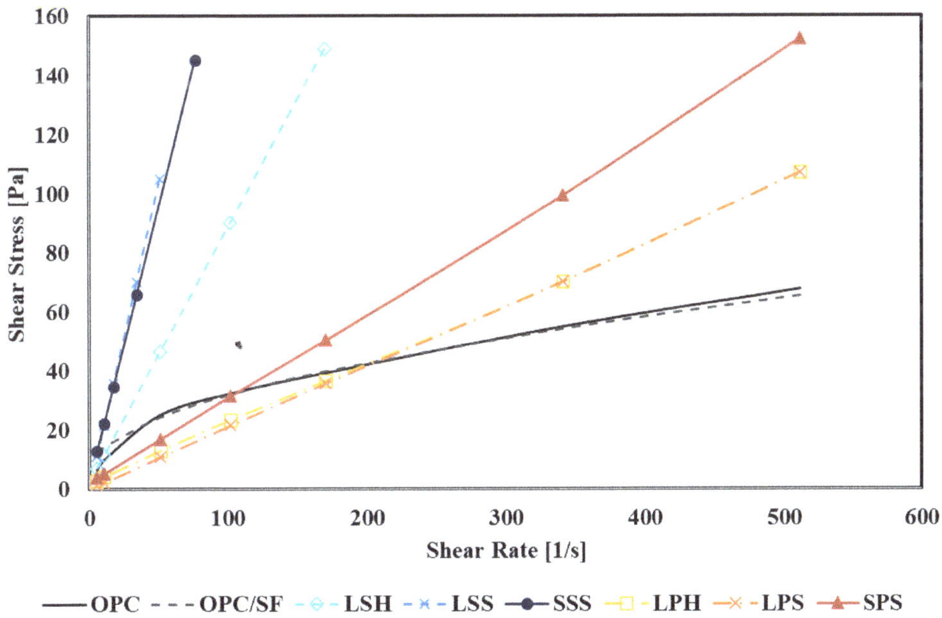

Figure 1. Viscosity profile for OPC and geopolymer slurries.

Table 4. Rheological model parameters for OPC and geopolymer slurries.

Mix ID	Apparent Viscosity at 300 rpm (Pa·s)	YPL Model Constants—See Equation (1)		
		Yield Stress τ_y (Pa)	Fluid Behavior Index n (Dimensionless)	Consistency Index K (Pa·s^n)
OPC	0.13 ± 0.03	6.70 ± 0.37	0.46 ± 0.06	3.87 ± 0.73
OPC/SF	0.12 ± 0.02	6.63 ± 2.91	0.61 ± 0.11	2.30 ± 0.72
LSH	>0.32	2.87 ± 0.21	1.00 ± 0.05	0.84 ± 0.30
LPH	0.21 ± 0.06	2.62 ± 0.98	1.02 ± 0.08	0.18 ± 0.09
LSS	>0.32	1.05 ± 0.72	1.00 ± 0.01	2.04 ± 0.48
SSS	>0.32	4.31 ± 2.40	1.02 ± 0.05	1.69 ± 0.51
LPS	0.21 ± 0.09	0.57 ± 0.93	1.01 ± 0.10	0.19 ± 0.05
SPS	0.30 ± 0.03	3.20 ± 1.05	1.04 ± 0.04	0.22 ± 0.04

For geopolymer slurries with all types of activators, it was found that their rheological behavior can be approximated as a Bingham Plastic because they have fluid behavior indices very close to 1.0 (i.e., in the range $0.95 < n < 1.05$). Moreover, the yield stresses of all geopolymer slurries were found to be lower than those of OPC slurries. All geopolymer slurries activated with sodium-based activators (represented by blue lines in Figure 1) were found to have high viscosities such that the apparent viscosity could not be measured by the viscometer (i.e., apparent viscosity values larger than 0.32 Pa·s). It was also found that the sodium hydroxide-based activator resulted in a slurry with a higher viscosity than slurries activated with sodium silicate-based activators. On the other hand, geopolymer slurries activated with potassium-based activators (represented by red, orange, and yellow lines in Figure 1) were found to have lower viscosities than those made with sodium-based activators. Liquid potassium activators were found to yield pumpable slurries (apparent viscosity ~0.2 Pa·s) with good rheological properties without the needs for thinners. In addition, the slurry with the liquid potassium activator showed lower viscosity than the solid potassium silicate-activated geopolymer.

4.2. Mechanical Properties

4.2.1. Unconfined Compressive Strength

The compressive strengths of OPC and geopolymer formulations cured under different curing schemes are summarized in Figure 2. At a curing temperature of 77 °C (170 °F), OPC was found to have a compressive strength at 3 days of 46.2 ± 2.3 MPa (6705 ± 337 psi). Incorporating SF in the OPC mixture reduced the compressive strength by 30%, which agrees with similar observations reported in the literature [36]. All the geopolymers had lower compressive strengths than OPC. In general, sodium-based geopolymers had higher compressive strengths than potassium-based geopolymers, with the highest geopolymer compressive strength at a value of 34.3 ± 1.2 MPa (4982 ± 173 psi) observed when using LSS as an activator. It was also found that silicate-based geopolymers showed higher compressive strengths than hydroxide-based geopolymers.

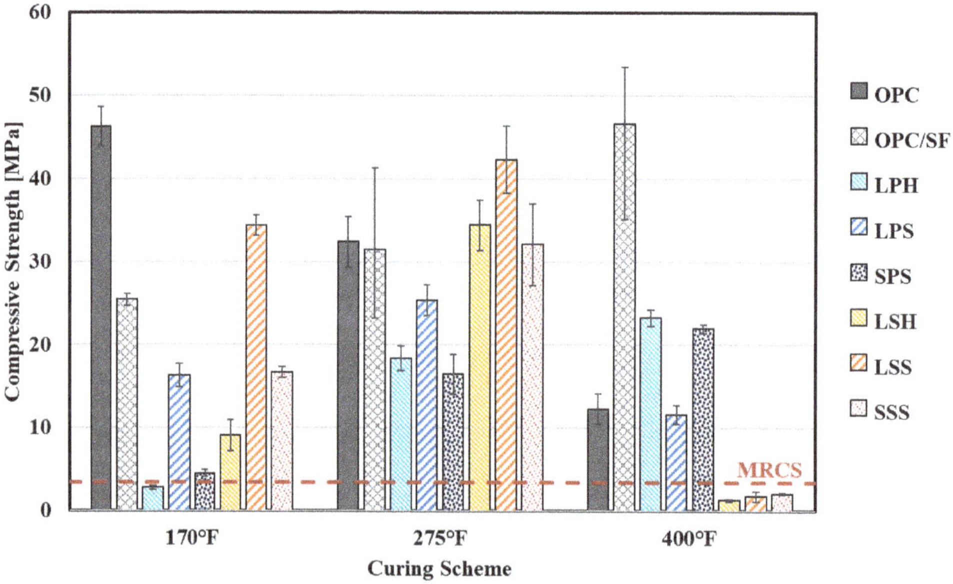

Figure 2. Three-day compressive strength for OPC and geopolymer formulations cured with different curing schemes (indicated by BHST value, see Table 3). Error bars represents the range of compressive strengths values measured for each mixture. Minimum Required Compressive Strength (MRCS) in accordance with standard API 10A is indicated by a dotted line.

When the curing temperature was increased from 77 °C (170 °F) to 135 °C (275 °F), the average compressive strength of OPC slurry decreased from 46.2 MPa (6705 psi) to 32.4 MPa (4693 psi). Incorporating silica flour (SF) into the OPC mixture cured at elevated temperature (135 °C (275 °F)) did not statistically significantly change its strength, resulting in a comparable strength to OPC alone. These results are consistent with expectations of strength retrogression from the literature [34,37]. Increasing the curing temperature from 77 °C (170 °F) to 135 °C (275 °F) resulted in an increase in the compressive strength for all geopolymer formulations. Although the potassium-based geopolymers had lower compressive strengths than OPC, LPS activation yielded a compressive strength up to 80% of the compressive strength of OPC. All geopolymers activated with sodium-based activators were found to have compressive strengths higher than geopolymers activated with potassium-based activators. Additionally, LSH and SSS were found to have compressive strengths comparable to OPC, while LSS had a compressive strength more than 30% higher than OPC.

Increasing curing temperature to the highest value tested (204 °C (400 °F)) reduced the average compressive strength of the neat OPC slurry to 12.3 MPa (1785 psi). However, the mixture incorporating silica flour (SF) in OPC showed a significant strength increase at this temperature, confirming the ability of silica flour addition to overcome strength retrogression [34]. In potassium-based geopolymer formulations, increasing the curing temperature to 204 °C (400 °F) resulted in increasing compressive strengths of geopolymers activated with LPH and SPS, while strength retrogression was observed in LPS. On the other hand, increasing the curing temperature to 204 °C (400 °F) resulted in a significant reduction of the compressive strengths of geopolymer formulations activated with sodium-based activators, such that they did not meet the minimum required compressive strength (MRCS) for cementing applications of 3.45 MPa (500 psi) according to the API 10A standard.

4.2.2. Tensile Strength

As expected, the tensile strengths of all formulations, as shown in Figure 3, followed trends similar to the compressive strengths. At 77 °C (170 °F), OPC was found to have a tensile strength of 1.8 ± 0.2 MPa (264 ± 30 psi). This result is consistent with the tensile strength of OPC cured at similar conditions reported in the literature [38]. In addition, no change in the tensile strength was observed when SF was added to the OPC mixture. Similar to compressive strength, the tensile strength values of potassium-based geopolymers were found to be lower than that of OPC. On the other hand, the tensile strength of geopolymers activated with sodium-based activators were found to be generally higher than those of geopolymers activated with potassium-based activators. The highest tensile strength amongst geopolymers at the low temperature condition was observed while using the LSS activator and was found to be similar to the tensile strength of OPC. In addition, it was observed that silicate based-geopolymers had higher tensile strengths than hydroxide-based geopolymers cured at 77 °C (170 °F).

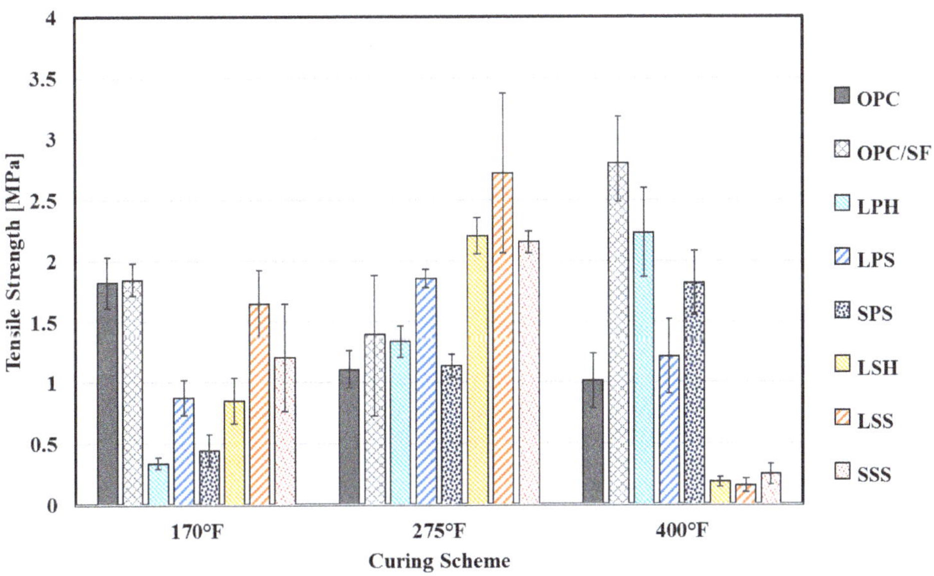

Figure 3. Three-day tensile strength for OPC and geopolymer formulations cured with different curing schemes (indicated by BHST value, see Table 3). Error bars represent the range of tensile strengths values measured for each mixture.

As the curing temperature increased to 135 °C (275 °F), a reduction of the tensile strength of OPC by 39% was observed compared to the lower curing temperature. Furthermore, incorporating SF had no significant impact on the tensile strength of OPC cured at the elevated temperature of 135 °C (275 °F). Increasing the curing temperature to 135 °C (275 °F) resulted in increasing the tensile strength of all geopolymer formulations. It was found that all sodium-based geopolymers had higher tensile strengths than potassium-based geopolymers. The highest tensile strength amongst geopolymers of 2.7 ± 0.7 MPa (394 ± 94 psi) was observed in LSS and was about 1.5 times higher than OPC cured at 135 °C (275 °F).

At the highest temperature tested of 204 °C (400 °F), no further reduction in the tensile strength of neat OPC compared to samples cured at 135 °C (275 °F) was observed. Furthermore, incorporating SF resulted in an increase of the tensile strength of OPC cured at the high temperature of 204 °C (400 °F) of 176%, which agrees with similar observations reported in the literature [39]. For geopolymer formulations, the increase of curing temperature from 135 °C (275 °F) to 204 °C (400 °F) resulted in a significant increase in the tensile strength of LPH and SPS-activated materials, while some strength reduction was observed for LPS-activated material. For all sodium-based geopolymers, high curing temperature resulted in low tensile strengths of less than 0.25 MPa (36 psi).

4.2.3. Bond Strength

The bond strength results for all OPC and geopolymer formulations cured at different curing conditions are summarized in Figure 4. For samples cured at 77 °C (170 °F), the bond strength of neat OPC was found to be 1.7 ± 0.08 MPa (247 ± 12 psi), which falls within the range of bond strengths reported in the literature [40,41]. Moreover, incorporating SF resulted in a 50% increase in the bond strength of OPC cured at 77 °C (170 °F). For geopolymer formulations, it was found that silicate-based geopolymers had higher bond strengths than hydroxide-based geopolymers. Moreover, the highest bond strengths amongst geopolymers at this temperature were observed in LSS and LPS and were found to be similar to the bond strength of OPC.

As the curing temperature increased to 135 °C (275 °F), a reduction in the bond strength was observed in OPC compared to the lower curing temperature condition (77 °C (170 °F). Moreover, incorporating SF had no significant impact on the bond strength of OPC cured at elevated temperature of 135 °C (275 °F). On the other hand, increasing the curing temperature from 77 °C (170 °F) to 135 °C (275 °F) resulted in a significant increase in the bond strength of geopolymer formulations, and all geopolymer formulations were found to have higher bond strengths than OPC. In general, silicate-based geopolymers were found to have higher bond strengths than hydroxide-based geopolymers. In contrast to the compressive and tensile strengths results, geopolymers activated with potassium-based activators and cured at 135 °C (275 °F) showed higher bond strengths than geopolymers activated with sodium-based activators. The highest bond strength amongst geopolymers was observed while using LPS as activator, which had an outstanding bond strength of 8.8 ± 0.35 MPa (1283 ± 51 psi), which is more than an order of magnitude higher than OPC.

Increasing the curing temperature to 204 °C (400 °F) had no statistically significant impact on the bond strength of OPC and OPC/SF compared to samples cured at 135 °C (275 °F). In addition, all geopolymer formulations activated with potassium-based activators cured at 204 °C (400 °F) were found to have similar bond strengths of more than 7.2 MPa (1044 psi). Such remarkably high bond strengths are more than an order of magnitude higher than OPC cured at the same conditions. On the other hand, similar to compressive and tensile strengths, increasing the curing temperature to high temperature of 204 °C (400 °F) resulted in negligible bond strengths (i.e., less than 0.1 MPa (15 psi)) for all geopolymers activated with sodium-based activators.

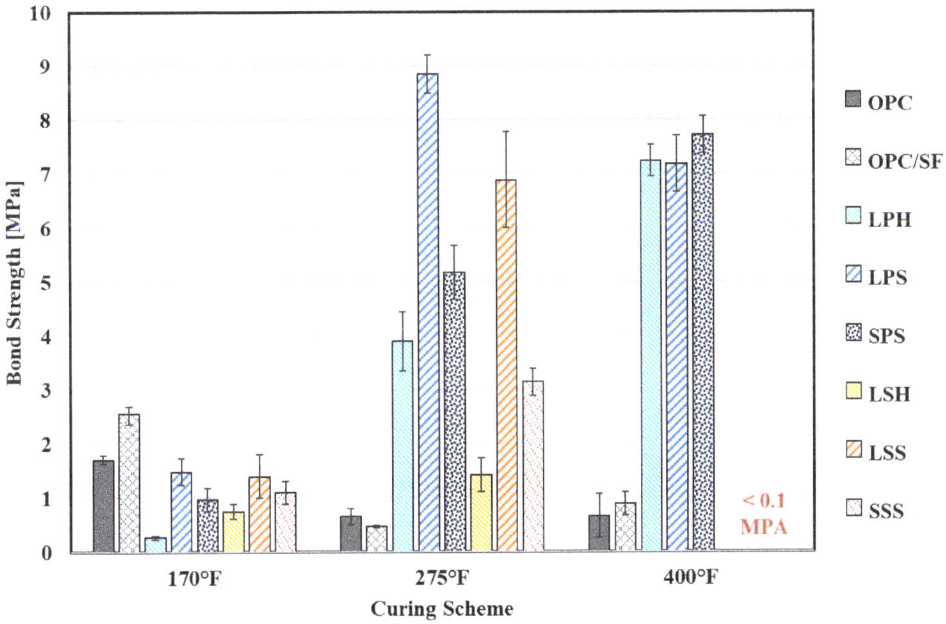

Figure 4. Three-day bond strength for OPC and geopolymer formulations cured with different curing schemes (indicated by BHST value, see Table 3). Error bars represents the range of bond strengths values measured for each mixture. A bond strength of less than 0.1 MPa was observed for geopolymers activated with LSS and SSS at 400 °F conditions.

4.3. Thickening Time

Cementing slurries are required to have thickening times longer than the planned placement time with some safety margins, while a very long thickening time is not desirable as it imposes an unnecessary waiting-on-cement time that translates into additional well cost [41,42]. The minimum required thickening time for most wellbore conditions is 2 h (see e.g., API 10-A and [41]). However, this minimum requirement increases for some applications such as deep HPHT and geothermal wells cementing applications [41,43]. The thickening times for all OPC and geopolymer slurries measured under different conditions are summarized in Table 5. At a BHCT of 52 °C (125 °F), OPC and OPC/SF were found to have similar thickening times a few minutes short of 2 h, which follows the expectations of Class H Cement without the use of retarders. At the same temperature, all potassium-based and sodium hydroxide-based geopolymers failed to reach a consistency of 70 Bc within 24 h, while LSS and SSS showed extended thickening times of 18.7 h and 13.9 h, respectively. With such long thickening times, accelerators will be necessary if geopolymers are to be used at such conditions.

As expected, an increase of BHCT resulted in a decrease in the thickening time of all slurries [41]. For potassium-based geopolymers, LPH showed an extended thickening time of 6.75 h, while LPS and SPS showed suitable thickening times of 3.25 and 3.5 h, respectively. All potassium-based geopolymers were found to have thickening times of more than 2.5 h at a BHCT of 135 °C (275 °F). A longer time might still be desirable for high-temperature applications, requiring the use of suitable retarders. In general, it was observed that sodium-based geopolymers have lower thickening times than potassium-based geopolymers. In addition, silicate-based geopolymers were found to have lower thickening times than hydroxide-based geopolymers tested at the same BHCT.

Table 5. Thickening time (formatted as hh:mm ± mm) for OPC and all geopolymer slurries. No retarders were used in any of the tests.

Mix ID	BHCT 52 °C (125 °F)	100 °C (212 °F)	135 °C (275 °F)
OPC	1:50 ± 2	1:22 ± 1	1:16 ± 4
OPC/SF	1:54 ± 4	1:28 ± 4	1:18 ± 3
LSH	>24:00	1:42 ± 3	N/A
LSS	18:41 ± 54 *	1:32 ± 3 *	N/A
SSS	13:52 ± 98 *	1:38 ± 7 *	N/A
LPH	>24:00	6:43 ± 14	2:35 ± 1
LPS	>24:00	3:14 ± 4	2:35 ± 14
SPS	>24:00	3:30 ± 9	2:32 ± 3

* denotes that the slurries' initial consistency was above 70 Bc.

5. Discussion and Summary

The experimental investigation shows that the type of activator and curing conditions affect the mechanical properties and thickening times of Class F fly ash-based geopolymer formulations. It also shows that such geopolymers can be straightforwardly engineered to provide primary and remedial cementing solutions for different wellbore conditions. For the relatively low temperature of 77 °C (170 °F) or lower, characteristic of conditions encountered when cementing surface and intermediate casing strings, an LSS geopolymer was found to be a viable cementing material. With a slightly lower compressive strength and similar tensile and bond strengths to OPC, LSS can provide similar performance with a lower carbon footprint when compared to OPC. Although it was found to have higher viscosity and extended thickening time at a corresponding BHCT of 52 °C (125 °F), additives such as potassium hydrogen phosphate, potassium chloride, and potassium nitrate may be able to adjust the rheological properties of the LSS formulation without negatively affecting the mechanical properties [6]. It is also possible to lower the viscosity of sodium-activated geopolymers in general by admixing non-aqueous fluids, such as residual oil-based mud or synthetic-based mud left over from the drilling operation, into the formulation, thereby creating a geopolymer-drilling fluid "hybrid" [17,18,20].

At an elevated temperature of 135 °C (275 °F), more common when cementing deeper intermediate and production strings, liquid silicate-activated geopolymers were found to have the best overall performance of all tested formulations. LPS and LSS-activated geopolymer materials were found to have twice the tensile strength and more than 13 times higher bond strength when compared to OPC. Moreover, the LPS geopolymer was found to have desirable rheological properties and thickening time at the corresponding BHCT of 100 °C (212 °F) without the use of additives, while super-plasticizing and retarding admixtures would be required for LSS formulations.

For a high temperature of 204 °C (400 °F), representing cementing conditions in HPHT and geothermal wells, potassium-based geopolymer materials showed high thermal stability. LPH and SPS-activated formulations showed increases in compressive, tensile, and bond strengths when compared to lower curing temperatures, without showing signs of strength retrogression over time. These formulations showed remarkable high casing-to-cement bond strengths that were more than an order of magnitude higher than OPC. Such excellent bonding will benefit zonal isolation and well integrity because the casing–cement bond of high-temperature wells will be able to much better resist the negative impact of cyclic pressure and temperature loading on micro-annulus formation at the cement–casing interface. This will reduce the associated risk of well leakage of methane to the atmosphere. The thickening times of LPH and SPS formulations were found to be on the order of 2.5 h without the use of any retarders. Retarders will be required to further extend the thickening time of geopolymer formulations at high temperature conditions.

6. Conclusions

This study shows the applicability of Class F fly-ash based geopolymers for a wide range of downhole well cementing conditions, ranging from low to high temperatures, making it a suitable, lower carbon footprint alternative to Portland cement. Fine-tuning of properties can be achieved primarily by changing the alkaline activator without changing the base material or having to add a significant number of admixtures, thereby also omitting the carbon footprint of the latter. An important discovery for potassium-activated geopolymer formulations is that they provide a significantly higher casing-to-cement bond strength than Portland cement at elevated and high temperature conditions. This will benefit well integrity and reduce the likelihood of methane leakage to the atmosphere through the de-bonding of the casing-to-cement interface (with the associated formation of a micro-annulus) in wells with compromised zonal isolation and integrity. This, then, adds to the overall carbon footprint reduction achieved by using geopolymers for reliable well barrier formation purposes.

Author Contributions: Conceptualization, M.J. and E.v.O.; Methodology, C.H. and M.G.; Validation, M.G. and E.v.O.; Formal analysis, M.G. and M.J.; Investigation, C.H. and M.G.; Resources, M.J. and E.v.O.; Writing – original draft, C.H. and M.G.; Writing – review & editing, M.J. and E.v.O.; Visualization, M.G.; Supervision, M.J. and E.v.O.; Project administration, M.J. and E.v.O.; Funding acquisition, M.J. and E.v.O. All authors have read and agreed to the published version of the manuscript.

Funding: This research was funded by the CODA Industry Affiliate Program (IAP) at the University of Texas at Austin.

Data Availability Statement: Not applicable.

Acknowledgments: We would like to thank Baker Hughes for their generous donation to establish the zonal isolation lab and fluids lab at UT Austin. A special thanks to PQ Corporation, SEFA Group, and Boral for providing materials and support. The work reported here was sponsored by current and former members of the CODA consortium at the University of Texas at Austin, including ConocoPhillips, Shell, BP, Total, Cenovus Energy, PQ Corporation, and Wellset, for which we would like to express our thanks. We would like to also thank Daryl Nygaard, the scientific instrument maker at the Hildebrand Petroleum and Geosystems Engineering Department at the University of Texas at Austin, for his help in this work.

Conflicts of Interest: The authors declare no conflict of interest.

References

1. Lin, A.C. Making Net Zero Matter. *Wash. Lee Law Rev.* **2022**, *79*, 679–768.
2. Hosseini, S.E.; Wahid, M.A. Hydrogen production from renewable and sustainable energy resources: Promising green energy carrier for clean development. *Renew. Sustain. Energy Rev.* **2016**, *57*, 850–866. [CrossRef]
3. Hastings, A.; Smith, P. Achieving Net Zero Emissions Requires the Knowledge and Skills of the Oil and Gas Industry. *Front. Clim.* **2020**, *2*, 601778. [CrossRef]
4. Kang, M. CO_2, Methane, and Brine Leakage through Subsurface Pathways: Exploring Modeling, Measurement, and Policy Options. Ph.D. Thesis, Princeton University, Princeton, MA, USA, 2014.
5. Kang, M.; Mauzerall, D.L.; Ma, D.Z.; Celia, M.A. Reducing methane emissions from abandoned oil and gas wells: Strategies and costs. *Energy Policy* **2019**, *132*, 594–601. [CrossRef]
6. Provis, J.L.; van Deventer, J.S. *Alkali Activated Materials: State-of-the-Art Report*; RILEM TC 224-AAM; Springer Science & Business Media: Dordrecht, The Netherlands, 2013; Volume 13.
7. Vrålstad, T.; Saasen, A.; Fjær, E.; Øia, T.; Ytrehus, J.D.; Khalifeh, M. Plug & abandonment of offshore wells: Ensuring long-term well integrity and cost-efficiency. *J. Pet. Sci. Eng.* **2019**, *173*, 478–491. [CrossRef]
8. Khalifeh, M.; Saasen, A. *Introduction to Permanent Plug and Abandonment of Wells*; Springer Nature: Berlin, Germany, 2020.
9. Khalifeh, M.; Hodne, H.; Saasen, A.; Integrity, O.; Eduok, E.I. Usability of Geopolymers for Oil Well Cementing Applications: Reaction Mechanisms, Pumpability, and Properties. In Proceedings of the SPE Asia Pacific Oil & Gas Conference and Exhibition, Perth, Australia, 10–12 October 2016. [CrossRef]
10. Khalifeh, M.; Saasen, A.; Hodne, H.; Godøy, R.; Vrålstad, T. Geopolymers as an Alternative for Oil Well Cementing Applications: A Review of Advantages and Concerns. *J. Energy Resour. Technol.* **2018**, *140*, 092801. [CrossRef]

11. Salehi, S.; Khattak, M.J.; Ali, N.; Rizvi, H.R. Development of Geopolymer-based Cement Slurries with Enhanced Thickening Time, Compressive and Shear Bond Strength and Durability. In Proceedings of the IADC/SPE Drilling Conference and Exhibition, Fort Worth, TX, USA, 1 March 2016. [CrossRef]
12. Salehi, S.; Ezeakacha, C.P.; Khattak, M.J. Geopolymer Cements: How Can You Plug and Abandon a Well with New Class of Cheap Efficient Sealing Materials. In Proceedings of the SPE Oklahoma City Oil and Gas Symposium, Oklahoma City, OK, USA, 27–30 March 2017. [CrossRef]
13. Salehi, S.; Khattak, J.; Saleh, F.K.; Igbojekwe, S. Investigation of mix design and properties of geopolymers for application as wellbore cement. *J. Pet. Sci. Eng.* **2019**, *178*, 133–139. [CrossRef]
14. van Oort, E.; Juenger, M.; Liu, X.; McDonald, M. Silicate-Activated Geopolymer Alternatives to Portland Cement for Thermal Well Integrity. Society of Petroleum Engineers. In Proceedings of the SPE Thermal Well Integrity and Design Symposium, Banff, AB, Canada, 17 November 2019. [CrossRef]
15. Genedy, M.; Juenger, M.; van Oort, E. Novel Cementing and Lost Circulation Solutions for Geothermal Wells based on Class F Fly Ash Geopolymers. *GRC Trans.* **2021**, *45*, 169–180. Available online: https://www.geothermal-library.org/index.php?mode=pubs&action=view&record=1034373 (accessed on 21 July 2022).
16. Ahdaya, M.; Imqam, A. Investigating geopolymer cement performance in presence of water-based drilling fluid. *J. Pet. Sci. Eng.* **2019**, *176*, 934–942. [CrossRef]
17. Liu, X.; Aughenbaugh, K.; Nair, S.; Shuck, M.; van Oort, E. Solidification of Synthetic-Based Drilling Mud Using Geopolymers. In Proceedings of the SPE Deepwater Drilling and Completions Conference, Galveston, TX, USA, 14–15 September 2016. [CrossRef]
18. Liu, X.; Nair, S.; Aughenbaugh, K.; van Oort, E. Mud-to-cement conversion of non-aqueous drilling fluids using alkali-activated fly ash. *J. Pet. Sci. Eng.* **2019**, *182*, 106242. [CrossRef]
19. Rostami, H.; Brendley, W. Alkali Ash Material: A Novel Fly Ash-Based Cement. *Environ. Sci. Technol.* **2003**, *37*, 3454–3457. [CrossRef] [PubMed]
20. Liu, X. Mud-to-Cement Conversion of Synthetic-Based Drilling Muds Using Geopolymers. Ph.D. Thesis, The University of Texas at Austin, Austin, TX, USA, 2017.
21. Liu, X.; Ramos, M.J.; Nair, S.D.; Lee, H.; Espinoza, D.N.; van Oort, E. True Self-Healing Geopolymer Cements for Improved Zonal Isolation and Well Abandonment. In Proceedings of the SPE/IADC Drilling Conference and Exhibition, The Hague, The Netherlands, 14–15 March 2017. [CrossRef]
22. Ross, J. An Evaluation of the Self-Healing Capabilities of Fly Ash-Based Geopolymers. Master's Thesis, The University of Texas at Austin, Austin, TX, USA, 2020.
23. Ross, J.H.; Genedy, M.; Juenger, M.C.; van Oort, E. Permeability Recovery by Self-Healing of Class F Fly Ash-Based Geopolymers. *Cement* **2022**, *10*, 100048. [CrossRef]
24. Panchmatia, P.; Olvera, R.; Genedy, M.; Juenger, M.C.G.; van Oort, E. Shrinkage behavior of Portland and geopolymer cements at elevated temperature and pressure. *J. Pet. Sci. Eng.* **2020**, *195*, 107884. [CrossRef]
25. Colangelo, F.; Cioffi, R.; Roviello, G.; Capasso, I.; Caputo, D.; Aprea, P.; Liguori, B.; Ferone, C. Thermal cycling stability of fly ash based geopolymer mortars. *Compos. Part Eng.* **2017**, *129*, 11–17. [CrossRef]
26. Kong, D.L.Y.; Sanjayan, J.G. Effect of elevated temperatures on geopolymer paste, mortar and concrete. *Cem. Concr. Res.* **2010**, *40*, 334–339. [CrossRef]
27. Denduluri, V.S.; Genedy, M.; Juenger, M.; McDonald, M.; van Oort, E. Using Geopolymers as a Sustainable Alternative Cementing Solution for Long-Term Zonal Isolation and Lost Circulation Challenges in Deep Geothermal Wells. *GRC Trans.* **2022**, *46*, 1438–1451. Available online: https://www.geothermal-library.org/index.php?mode=pubs&action=view&record=1034686 (accessed on 9 September 2022).
28. Duxson, P.; Provis, J.L.; Lukey, G.C.; van Deventer, J.S. The role of inorganic polymer technology in the development of 'green concrete'. *Cem. Concr. Res.* **2007**, *37*, 1590–1597. [CrossRef]
29. Juenger, M.C.G.; Winnefeld, F.; Provis, J.L.; Ideker, J.H. Advances in alternative cementitious binders. *Cem. Concr. Res.* **2011**, *41*, 1232–1243. [CrossRef]
30. Alvi, M.A.A.; Khalifeh, M.; Agonafir, M.B. Effect of nanoparticles on properties of geopolymers designed for well cementing applications. *J. Pet. Sci. Eng.* **2020**, *191*, 107128. [CrossRef]
31. Khalifeh, M.; Saasen, A.; Hodne, H.; Motra, H.B. Laboratory evaluation of rock-based geopolymers for zonal isolation and permanent P&A applications. *J. Pet. Sci. Eng.* **2019**, *175*, 352–362. [CrossRef]
32. Salehi, S.; Khattak, M.J.; Rizvi, H.; Karbalaei, S.F.; Kiran, R. Sensitivity analysis of fly ash geopolymer cement slurries: Implications for oil and gas wells cementing applications. *J. Nat. Gas Sci. Eng.* **2017**, *37*, 116–125. [CrossRef]
33. Grabowski, E.; Gillott, J.E. The effect of initial curing temperature on the performance of oilwell cements made with different types of silica. *Cem. Concr. Res.* **1989**, *19*, 703–714. [CrossRef]
34. Pyatina, T.; Sugama, T. Thermal-Shock Resistant Cement for Heat Storage. *GRC Trans.* **2019**, *43*, 36–59. Available online: https://www.geothermal-library.org/index.php?mode=pubs&action=view&record=1034102 (accessed on 2 October 2022).
35. Genedy, M.; Kandil, U.F.; Matteo, E.N.; Stormont, J.; Reda Taha, M.M. A new polymer nanocomposite repair material for restoring wellbore seal integrity. *Int. J. Greenh. Gas Control* **2017**, *58*, 290–298. [CrossRef]
36. Xie, Z.; Yao, X. Influence of the particle size distribution of silica flour on the mechanical and microstructural properties of oil well cement paste exposed to HTHP conditions. *Ceramics–Silikáty* **2019**, *63*, 239–247. [CrossRef]

37. Bezerra, U.T.; Martinelli, A.E.; Melo, D.M.A.; Melo, M.A.F.; Oliveira, V.G. The strength retrogression of special class Portland oilwell cement. *Cerâmica* **2011**, *57*, 150–154. [CrossRef]
38. Li, L.; Kellum, M.; Doan, A. In-situ Tensile Strength Testing: Awareness of Variations with Testing Environment. In Proceedings of the SPE Deepwater Drilling and Completions Conference, Galveston, TX, USA, 14–15 September 2016. [CrossRef]
39. Mahmoud, A.A.; Elkatatny, S.; Ahmed, S.A.; Mahmoud, M. Nanoclay Content Influence on Cement Strength for Oil Wells Subjected to Cyclic Steam Injection and High-Temperature Conditions. In Proceedings of the Abu Dhabi International Petroleum Exhibition & Conference, Abu Dhabi, UAE, 12 November 2018. [CrossRef]
40. Teodoriu, C.; Yi, M.C.; Ichim, A.; Salehi, S. A novel view of cement failure with application to geothermal well construction. In Proceedings of the 43rd Workshop on Geothermal Reservoir Engineering, Stanford, CA, USA, 12–14 February 2018.
41. Nelson, E.B.; Guillot, D. *Well Cementing*, 2nd ed.; Schlumberger: Sugarland, TX, USA, 2006.
42. Salam, K.K.; Arinkoola, A.O.; Ajagbe, B.; Sanni, O. Evaluation of Thickening Time of Oil Field Class G Cement Slurry at High Temperature and Pressure using Experimental Design. *Int. J. Eng. Sci.* **2013**, *2*, 361–367.
43. Fink, J. *Petroleum Engineer's Guide to Oil Field Chemicals and Fluids*, 2nd ed.; Gulf Professional Publishing: Amsterdam, The Netherlands, 2021.

energies

MDPI

Article

Fast Quantitative Modelling Method for Infrared Spectrum Gas Logging Based on Adaptive Step Sliding Partial Least Squares

Zhongbing Li [1], Wei Pang [1,*], Haibo Liang [2], Guihui Chen [3], Hongming Duan [1] and Chuandong Jiang [1]

[1] School of Electrical Engineering and Information, Southwest Petroleum University, Chengdu 610500, China; lzb@swpu.edu.cn (Z.L.); 202022000215@stu.swpu.edu.cn (H.D.); 202021000154@stu.swpu.edu.cn (C.J.)
[2] School of Mechatronic Engineering, Southwest Petroleum University, Chengdu 610500, China; secondbo@swpu.edu.cn
[3] School of Engineering, Southwest Petroleum University, Nanchong 637000, China; guihui528@swpu.edu.cn
* Correspondence: 201921000185@stu.swpu.edu.cn

Abstract: Infrared spectroscopy (IR) quantitative analysis technology has shown excellent development potential in the field of oil and gas logging. However, due to the high overlap of the IR absorption peaks of alkane molecules and the offset of the absorption peaks in complex environments, the quantitative analysis of IR spectroscopy applied in the field puts forward higher requirements for modelling speed and accuracy. In this paper, a new type of fast IR spectroscopy quantitative analysis method based on adaptive step-sliding partial least squares (ASS-PLS) is designed. A sliding step control function is designed to change the position of the local PLS analysis model in the full spectrum band adaptively based on the relative change of the current root mean square error and the global minimum root-mean-square error for rapid modelling. The study in this paper reveals the influence of the position and width of the local modelling window on the performance, and how to quickly determine the optimal modelling window in an uncertain sample environment. The performance of the proposed algorithm has been compared with three typical quantitative analysis methods by experiments on an IR spectrum dataset of 400 alkane samples. The results show that this method has a fast quantitative modelling speed with high analysis accuracy and stability. It has important practical value for promoting IR spectroscopy gas-logging technology.

Keywords: gas logging; infrared spectroscopy; fast quantitative modelling; adaptive step sliding; partial least squares

Citation: Li, Z.; Pang, W.; Liang, H.; Chen, G.; Duan, H.; Jiang, C. Fast Quantitative Modelling Method for Infrared Spectrum Gas Logging Based on Adaptive Step Sliding Partial Least Squares. *Energies* **2022**, *15*, 1325. https://doi.org/10.3390/en15041325

Academic Editors: Mofazzal Hossain, Hisham Khaled Ben Mahmud and Md Motiur Rahman

Received: 12 January 2022
Accepted: 2 February 2022
Published: 12 February 2022

1. Introduction

Gas logging refers to the process of measuring the type and content of hydrocarbon gas contained in the analysis of the drilling fluid generated during the oil and gas exploration process [1]. At first, it was mainly used as a pure safety system designed to monitor the content of toxic or flammable substances. With the development of technology, the measurement results of gas gushing from the well gradually transformed into a formation evaluation tool [2]. Gas logging has been an important means of oil and gas discovery and evaluation, and plays an irreplaceable role in natural gas exploration and development. It has important guiding significance to accurately analyse and evaluate the oil and gas content and type of downhole drilled formations in real time, for the timely discovery and interpretation of the reservoir environment, improving the efficiency of oil and gas exploration and ensuring the safety of drilling operations [3].

However, as oil and gas exploration move toward more efficient development in deeper formations, the natural environment of underground oil and gas reservoirs have become more and more complex. The requirements for gas-logging technology have become higher and higher [4]. Gas chromatography (GC), as one main traditional gas-logging technologies, uses chromatographic columns to separate the logging gas first, and then sends it to a flame ionization detector (FID) to detect hydrocarbon gases [5]. J. Breviere et al. [6] increased the detection components from C_1-C_5 (methane, ethane, propane, butane and pentane) to C8 (octane) and some acid gases through the analysis of GC combined with a mass spectrometer. A. Brumboiu et al. [7] introduced the gas-permeable membrane to high-speed GC in the analysis of drilling gas for the first time, so that the quantitative analysis of $C_1 \sim C_8$ and aromatic hydrocarbons can be completed within 50 s. C. A. Cramers et al. [8] gave the theoretical relationship between chromatographic analysis time and related pressure drop. Their research pointed out that it was the best way to increase the analysis speed by reducing the inner diameters of open-tube chromatographic columns and packed chromatographic columns, but the inlet pressure required would also increase correspondingly, resulting in high flow resistance inside the chromatographic column, which would severely restrict the application of the chromatographic column in fast GC [9].

Nevertheless, due to the long analysis period and poor ability to distinguish between hydrocarbons and non-hydrocarbons, GC is still difficult in terms of meeting the new requirements of real-time monitoring of drilling to find rare oil and gas reservoirs [10]. Besides improvements in quantitative methods, adding peripheral ancillary equipment and improving the structure of the chromatographic column are the main means to increase the detection rate, which will lead to many challenges regarding usage and ongoing maintenance cost [11]. Therefore, GC logging technology has gradually become more difficult in terms of meeting the actual needs and exploration requirements of complex oil and gas reservoirs.

With the development of spectroscopy technology and chemometrics, spectral technology has been a promising logging technology [12,13] due to its advantages of flexible, fast and nondestructive detection. Currently, Raman spectroscopy and IR spectroscopy are the two main spectral gas-logging technologies. Raman spectroscopy was first applied to natural gas detection in 1980 [14], and more and more analytical methods have emerged since then. I. Kosacki et al. [15] summarized the effective modelling spectral positions for $C_1 \sim C_4$ based on the principle of Raman spectroscopy. R. Sharma et al. [16] used the 64-channel concentric cavity unit operating enhanced Raman signal under high pressure (5bar), and the maximum error of repeated measurement of $C_1 \sim C_4$ reached 0.1%, but the scanning time of each spectrum reached 30 s. D. Chen et al. [17] introduced the template-oriented frog algorithm to extract useful Raman spectral features, finding the best combination of spectral features with the least overlap in a multicomponent mixture.

However, due to the weak Raman signal of gas, Raman spectroscopy requires a higher power laser (>4 W) or a high pressure (>50 bar) to achieve gas detection for industrial application. Generally, high-power lasers are bulky and need frequent maintenance, which makes their deployment in the field impractical and expensive. On the other hand, it is cumbersome and very unsafe to handle instruments operating under high pressure (especially in the case of natural gas). Therefore, the current Raman gas-logging technology has great difficulties in some extreme applications.

Relatively, IR spectroscopy has obvious advantages in detection speed and convenience [18,19]. K. Indo et al. [20] designed an artificial neural network IR spectroscopy analysis system to detect the content of pseudocomponents (not the true concentration) of $C_1 \sim C_6$ in the laboratory. R. Piazza et al. [21] explained the importance of instantaneous measurement of CO_2 in logging gas, and tested in five wells with nine formations, test pumps out of the station using infrared quantitative analysis technology, and found that the accuracy of carbon dioxide measurement was ±0.4% in the concentration range of 1.5~23%. Q. Ren et al. [22] used midinfrared spectroscopy to achieve high-precision detection of CO_2 content in the process of deep-sea natural gas exploration. Research by M. K. Moro [23] et al. pointed out that the vibration mode of alkane molecules exhibited complex differences in only the near-infrared band, and local modelling was one of the main methods to reduce the complexity of the model and the effects of nonlinear absorption, effectively.

Generally, the current application of IR spectroscopy in the field of gas logging depends on the quantitative analysis based on Lambert–Beer law. Combined with chemometrics, more and more models have been developed to achieve high-precision quantitative infrared spectroscopy analysis. U. Kamboj et al. [24] utilized partial least squares (PLS) analysis to establish a quantitative analysis model of near-infrared absorbance and concentration, and the root-mean-square *RMSE* error of prediction reached 0.04 on the sugar in milk dataset. Q. Gao et al. [25] compared the prediction performance of the PLS after different spectral preprocessing methods (Savitzky–Golay smoothing, multiplicative scattering correction, moving average, median filtering, normalization, standard normal variable transformation, baseline, detrending, and direct differential first-order derivatives), and applied data dimension-reduction methods (random frog [26], successive projections algorithm (SPA) [27], and principal component analysis) for variable selection to obtain an optimal fusion model with the best performance (*RMSE* is 0.271 on the Malus micromalus Makino dataset). Y. Yu et al. [28] introduced the instance-based transfer learning framework into extreme learning machine (ELM) [29] method combined with the two feature-extraction algorithms of synergy interval [30] and genetic algorithm to enhance the accuracy and stability of quantitative prediction models. H. Yu et al. [31] integrated machine learning and correlation analysis to analyse the content of petroleum naphtha, and the accuracy reached 90.48% on the petroleum naphtha dataset. T. Ouyang et al. [32] proposed the infrared spectral quantitative analysis method of Deep-ELM, and the RMSE accuracy of N_2O, NO_2 and NO reached 0.995, 0.984 and 0.985 respectively. L. Liu et al. [33] obtained good results by using wavelet transform to compress data and back-propagation neural network (BPNN) for quantitative analysis. Y. Li et al. [34] combined the two prescreening methods of competitive adaptive reweighted sampling (CARS) [35] and principal component analysis with the two quantitative analysis models of PLS and BPNN to improve the prediction accuracy for different applications.

Although types of artificial intelligence algorithms have been introduced into quantitative spectral analysis, the research on algorithm improvement based on PLS still has very important application value due to its high analysis accuracy and excellent robustness and modelling speed. Particularly, there are still many challenges in the popularization and application of infrared spectroscopy in logging gas measurement, due to the following difficulties:

1. Because of the high similarity of the molecular structure of alkane gas, the infrared absorption characteristic peaks are too overlapped to separate in multicomponent gas mixtures;
2. Because of the great uncertainties of the composition of the logging gas and the limited number of samples for modelling, fast modelling technology is becoming more and more urgent with the increment of required spectrum range and spectral resolution for analysis of more and more kinds of gases.

These two reasons make it important to establish an effective quantitative analysis model for IR spectroscopy gas logging. IR spectra possess complex and overlapping absorption bands, so mathematical procedures are needed for turning the spectra into meaningful information. For different types of oil and gas reservoirs, it is necessary to establish multiple quantitative analysis models with different applicability. This will undoubtedly take up more computing resources and increase the hardware burden of the analysis module, resulting in slow modelling speed.

In order to establish a rapid modelling method with good accuracy and to promote the application of infrared spectroscopy gas-logging technology on site, this paper proposes a fast modelling method based on adaptive step-sliding partial least squares (ASS-PLS). Three types of infrared spectrum datasets of logging gas are obtained by mixing six gases of methane (C_1), ethane (C_2), propane (C_3), n-butane (nC_4), n-pentane (nC_5) and carbon dioxide (CO_2). A sliding control function is designed to change the position of the local PLS analysis model in the full spectrum band adaptively, based on the relative change in the current root-mean-square error and the global minimum root-mean-square error for rapid modelling. The experiments are carried out to evaluate the influence of window position and window width on the quantitative analysis accuracy and the performance of the proposed method on the three types of datasets. The results show that the ASS-PLS method proposed in this paper can establish a rapid quantitative analysis model with good accuracy of logging gas, and can meet the actual needs of gas-logging operations.

2. Adaptive Step-Sliding Partial Least Squares

2.1. Infrared Quantitative Analysis Principle

Based on Lambert–Beer law [36], the absorption of a certain wavenumber of light by a substance is related to the concentration of the light-absorbing substance and the optical path length, as shown in Figure 1:

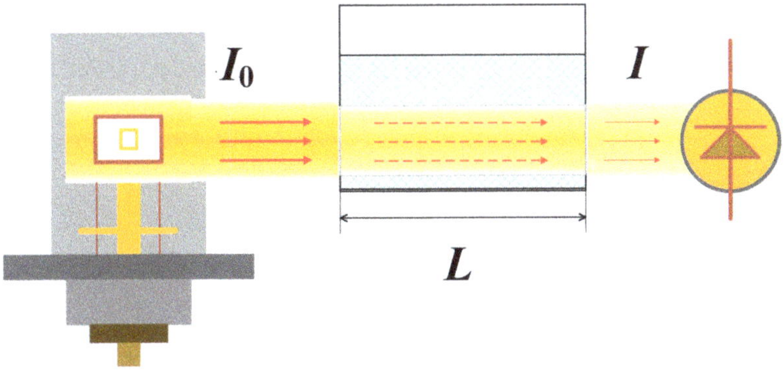

Figure 1. Schematic diagram of Lambert–Beer law.

The specific relational formula is shown in Equation (1), wherein the absorbance $A(\nu)$ of the substance with respect to infrared light of a specific frequency is defined as Equation (2), so the relation of Equation (1) can be further expressed as Equation (3):

$$I(\nu) = I_0(\nu)\exp[-\sigma(\nu)CL] \tag{1}$$

$$A(\nu) = -\lg\frac{I(\nu)}{I_0(\nu)} \tag{2}$$

$$A(\nu) = \sigma(\nu)CL \tag{3}$$

where ν is the frequency of incident infrared light, I_0 and I are the intensity of incident light and transmitted light, respectively, C is the concentration of the detected target gas, $\sigma(\nu)$ is the absorption cross-section of the IR spectrum at frequency ν, and L is the optical path length. For a multicomponent mixed gas with n components, the absorbance satisfies the linear additive property of Equation (4):

$$A(\nu) = \sum_1^n A_i(\nu) = \sum_1^n \sigma_i(\nu)LC_i \tag{4}$$

For the actual spectrum detection in engineering applications, L is known and determined because of the fixed optical path system, and $\sigma_i(\nu)$ is constant but unknown under one stable pressure, temperature and other physical or chemical environments. Therefore, the characteristic absorption coefficient $k(\nu)$ in the system can be defined as Equation (5):

$$k(\nu) = \sigma(\nu)L \tag{5}$$

Then, the absorbance matrix $\mathbf{A}$ and the component concentration matrix $\mathbf{C}$ in the mixed system have a linear relationship in Equation (6):

$$\mathbf{A}_{m\times n} = \mathbf{k}_{m\times p}\mathbf{C}_{p\times n} \tag{6}$$

where $\mathbf{A}_{m\times n}$ is the absorbance matrix of n samples at m wavenumber; $\mathbf{k}_{m\times p}$ is the absorption coefficients of p components at m wavenumber, and $\mathbf{C}_{p\times n}$ is the concentration of p components in n samples. Generally, the number of target components is much smaller than the number of wavenumber, that means $p << m$, so that the solution of the concentration in Equation (6) is a linear, overdetermined, equation-solving problem. Quantitative analysis modelling aims to establish the relationship between absorbance and the concentration of some known samples according to Equation (6), and quantitative analysis aims to obtain the concentration value of an unknown sample according to the established model.

2.2. Partial Least Squares Analysis

Partial least squares is one of the most effective methods to establish the linear relationship between multiple variables in Equation (6). Compared with traditional multiple linear regression, it combines the advantages of principal component regression and canonical correlation analysis, meanwhile it can overcome the problem of much larger sample size than variable dimensionality and the presence of multicollinearity within the variables [37]. Set the absorbance independent variable **A** and the substance concentration-dependent variable **C** to E_0 and F_0, respectively, after data standardization in Equation (7):

$$\begin{cases} E_0(i,j) = \frac{A(i,j)-mean(A(,j))}{std(A(,j))}, i = 1,2,\ldots,m; j = 1,2,\ldots,n \\ F_0(i,j) = \frac{C(i,j)-mean(C(,j))}{std(C(,j))}, i = 1,2,\ldots,p; j = 1,2,\ldots,n \end{cases} \tag{7}$$

where $mean()$ is an operator to calculate the average value, $std()$ is an operator to calculate the standard deviation, (i,j) denotes the element position in a matrix at i-th row j-th column, $(,j)$ denotes all elements in the j-th column. Suppose that the first principal components extracted are t_1 and u_1 respectively, and the weight coefficients are w_1 and m_1 respectively. So, t_1 and u_1 are the linear combination of E_0 and F_0, respectively, $t_1 = E_0 w_1$, $u_1 = F_0 m_1$. According to the principle of principal components, the variance of t_1 and u_1 is required to be maximized to make the principal components carry more information. According to the canonical correlation analysis, the correlation between t_1 and u_1 is required to be maximized for the best explanation from t_1 to u_1. Therefore, the covariance of t_1 and u_1 is required to be maximized and the weight coefficients must be unit vectors in Equation (8):

$$\begin{array}{l} \max\{Cov(t_1, u_1)\} = \max < E_0 w_1, F_0 m_1 > \\ s.t : {w_1}^T w_1 = 1, {m_1}^T m_1 = 1 \end{array} \tag{8}$$

This conditional extremum can be solved by the Lagrangian multiplier method to obtain w_1 and m_1 in Equation (9):

$$L = w_1^T {E_0}^T F_0 m_1 - \lambda(w_1^T w_1 - 1) - \theta(m_1^T m_1 - 1) \tag{9}$$

where λ and θ are both Lagrange multiplier. By solving Equation (9), the regression equation of E_0 and F_0 to t_1 can be established:

$$\begin{cases} E_0 = t_1 q_1^T + E_1 \\ F_0 = t_1 s_1^T + F_1 \end{cases} \tag{10}$$

where E_1 and F_1 are the residual matrix of the regression equation, q_1 and s_1 are the regression coefficient vectors calculated by the least-square method, $q_1 = {E_0}^T t_1/({t_1}^T t_1)$ and $s_1 = {F_0}^T t_1/({t_1}^T t_1)$.

Furthermore, the second principal components t_2 and u_2 can be obtained by replacing E_0 and F_0 with the residual matrices E_1 and F_1. If the rank of the **A** matrix is r, then the calculation will continue, to obtain:

$$F_0 = t_1 s_1^T + t_2 s_2^T + \ldots + t_r s_r^T + F_r = E_0 \left[\sum_{j=1}^{r} w_j^T s_j^T \right] + F_r \tag{11}$$

where F_r is the residual matrix when the maximum number of principal components r is extracted. The more principal components extracted, the better the linearity of the model, but the possible overfitting will lead to an unsatisfactory prediction effect.

A leave-one-out cross-validation test is adopted to determine the number of principal components h to be extracted from the spectral data. Each time the i-th spectrum observation data is discarded $(i = 1, 2, \ldots, n)$, the remaining $(n-1)$ spectrum observation data are used to fit a regression equation with h principal components, and then the discarded i-th spectrum observation data are incorporated into the regression equation to obtain the predicted concentration $\hat{c}_{(-i)j(h)}$. The above verification is repeated for $i = 1, 2, \ldots, n$. Then, we obtain the sum of squared prediction errors of the j-th concentration-dependent variable $(j = 1, 2, \ldots, p)$ in Equation (12):

$$S_j(h) = \sum_{i=1}^{n} \left(c_{ij} - \hat{c}_{(-i)j(h)}\right)^2, j = 1, 2, \ldots, p \tag{12}$$

then, the sum of squares of prediction errors of the concentration $\mathbf{C}$ is:

$$S(h) = \sum_{j=1}^{p} S_j(h) \tag{13}$$

In addition, all spectral data are used to fit the regression equation with h principal components again. Defining the predicted value of the i-th spectral data as $\hat{c}_{ij(h)}$, then the sum of the squared errors of the j-th concentration-dependent variable can be defined as Equation (14):

$$S_j^*(h) = \sum_{i=1}^{n} \left(c_{ij} - \hat{c}_{ij(h)}\right)^2, j = 1, 2, \ldots, p \tag{14}$$

then, the sum of the squared errors of the concentration $\mathbf{C}$ is:

$$S^*(h) = \sum_{j=1}^{p} SS_j(h) \tag{15}$$

When $S(h)$ reaches the minimum value, the corresponding h is the number of all the best principal components. Usually, there is always $S(h) < S^*(h)$, and $S^*(h) < S^*(h-1)$ at the same time. Define the cross-validity index Q_h^2:

$$Q_h^2 = 1 - S(h)/S^*(h-1) \tag{16}$$

After the principal component extraction process, the cross-validity index is hoped to be as small as possible. The threshold of Q_h is always set to 0.05, then $S(h)/S^*(h-1) \leq (1-0.05)^2 = 0.952$, and it is beneficial to the accuracy by increasing the number of the principal components. Therefore, for the process of extracting the principal components, if $Q_h^2 \geq 0.0975$ at the h-th principal component, the model meets the accuracy requirements and stops extracting. Otherwise, the $(h+1)$-th principal component extraction should be continued.

As shown in Equation (17), the root-mean-square error (*RMSE*, *RMSECV* for model calibration) is adopted as one of the most important evaluation indexes of quantitative methods.

$$RMSE(RMSECV) = \sqrt{\sum_{i=1}^{n} \left(C_i - \hat{C}_i\right)^2 / n} \tag{17}$$

where C_i is the real concentration of the sample, $\hat{C}_i$ is the predicted concentration, and n is the number of predicted samples.

2.3. Adaptive Step Sliding

Due to the uncertainty of the composition of the logging gas, the spectrometer is often required to scan with a wider frequency band. This means that it is necessary to abandon the irrelevant weak absorption and impurity absorption bands to establish a quantitative analysis model with strong robustness and high precision. Further, the similarity of the molecular structure of alkanes results in a continuous distribution of their strong characteristic absorption bands.

As shown in Figure 2b (blue box), the usual strategy is to traverse the whole spectrum with fixed sliding-step size in one search, and to establish the corresponding local model in the moving process to calculate the corresponding *RSMECV* value; then repeat the above process by changing the window width, and select the window with the minimum *RSMECV* value as the optimal window to establish the optimal model. This undoubtedly consumes a lot of computing resources and time [38].

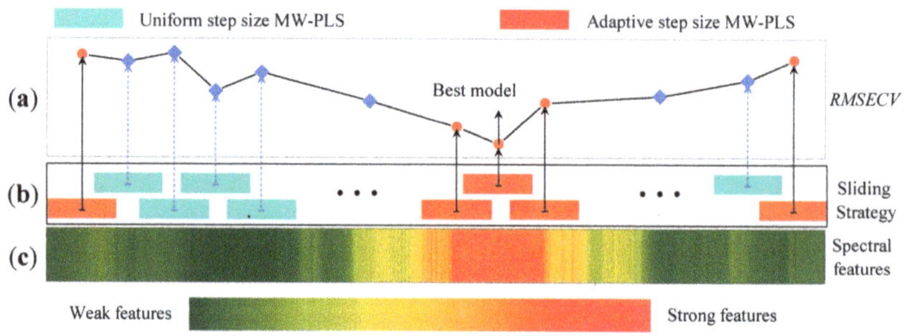

Figure 2. Process diagram of sliding window local modelling strategy: (**a**) *RMSECV* values during the sliding window. Red dots indicate ASS-PLS and blue dots indicate MW-PLS. (**b**) Window position for local modelling. (**c**) Analyse characteristic intensities of spectral bands.

Different from the uniform sliding strategy of MW-PLS, the ASS-PLS shown in Figure 2b (red box) can adaptively adjust the step size of each sliding according to the RMSECV value. This strategy can skip the spectral band information with more redundant information (blue dot) to achieve high-efficiency modelling. To this end, this paper designs an adaptive step-sliding control function of Equation (18) to determine the optimal local modelling interval with the best position of the moving window in only one search in the full spectrum. The real-time $\Delta RMSECV$ change value input by this function has an obvious nonlinear relationship with the output step size δ. In this way, the window can quickly slide on the weak feature band, and realize fine sliding on the strong feature band, so as to ensure that the window will not skip the ideal modelling interval during the sliding process.

$$\delta = \delta_0 + \beta(1 - \exp(-(\pi \cdot \Delta RMSECV/\xi)^2/\sqrt{2})) \tag{18}$$

where δ is the moving step size; δ_0 is the basic moving step size, which is a very small constant to avoid the function falling into dead loops when $\Delta RMSECV = 0$; β is the maximum offset of the moving step size; ξ is the kernel of the function, denoting the offset degree of $\Delta RMSECV$; $\Delta RMSECV$ is the offset value of real-time *RMSECV* relative to the minimum $RMSECV^*$, $\Delta RMSECV = RMSECV - RMSECV^*$, and $exp()$ is the exponential function.

In ASS-PLS shown by Algorithm 1, when the offset degree of *RMSECV* is not considered ($\xi \to \infty$), the ASS-PLS algorithm degenerates into moving window PLS (MW-PLS) with fixed moving speed.

Algorithm 1: ASS-PLS local modelling algorithm

Input : **A**: Absorbance matrix, **C**: Concentration matrix;
W_L^*: The origin point of spectral band to be analysed;
W_R^*: The end point of the spectrum to be analysed;
$[\beta, \xi]$: Sliding step size control factor;
$\mathbf{W}_d = [W_d^1, W_d^2, \ldots, W_d^m]$: Set m test window widths.
Output: $\mathbf{W}^*$: Global optimal local modelling interval;
$\mathbf{K}^*$: Global optimal local modelling interval.

$\begin{bmatrix} \mathbf{A}_{cal} & \mathbf{C}_{cal} \\ \mathbf{A}_{test} & \mathbf{C}_{test} \end{bmatrix} \xleftarrow[k]{Separation} [\mathbf{A}\ \mathbf{C}]$
for $j \leq m$ **do**
  $W_L = W_L^*$; $W_R = W_d^j$
  $RMSECV^* \xleftarrow{max} [\mathbf{C}_{cal}\ \mathbf{C}_{test}]$
  while $W_R \leq W_R^*$ **do**
    $\mathbf{W}_i = [W_L, W_R]$
    $\mathbf{K}_i \xleftarrow[\mathbf{A}_{cal}]{\mathbf{C}_{cal}}$ Establish local PLS model in window $\mathbf{W}_i$
    $\hat{\mathbf{C}} = \mathbf{K}_i \times \mathbf{C}_{test}$
    $RMSECV_i = \sqrt{\sum_{p=1}^{q} (\mathbf{C}_{test,p} - \hat{\mathbf{C}}_p)^2 / q}$
    $\Delta RMSECV = RMSECV_i - RMSECV^*$
    if $\Delta RMSECV < 0$ **then**
      $RMSECV^* = RMSECV_i$
      $\mathbf{W}^* = \mathbf{W}_i$, $\mathbf{K}^* = \mathbf{K}_i$
    end
    $\delta = \delta_0 + \beta(1 - \exp(-(\pi \cdot \Delta RMSECV/\xi)^2/\sqrt{2}))$
    $\mathbf{W}_i = \mathbf{W}_i + \delta$
  end
end

3. Experiment and Discussion

3.1. Experiment Dataset

Three types of spectral datasets are obtained by the designed infrared spectrum data acquisition system, shown in Figure 3. With methane (C_1, 99.999%), ethane (C_2, 99.999%), propane (C_3, 99.999%), n-butane (nC_4, 4.999%), n-pentane (nC_5, 3.999%), carbon dioxide (CO_2, 99.999%) and its mixture as the target gas, the test sample with nitrogen (N_2, 99.999%) as the carrier gas is fed into the gas-mixing system (LFIX-7000, Laifeng, Chengdu, China), the output error is ±1% of the input concentration), which realizes the mixed gas by controlling the intake flow. After being dehumidified by the drying tube (MD-070-24F-4091119-02, Perma Pure, US), the test sample is introduced into the airtight light path pool with a volume of 400 mL and an effective light range length of 4.8 m (PMG10030, YingSa, Shanghai, China). The internal temperature of the light path pool is kept constant at 27.5 °C by the heat preservation controlled by the temperature controller. Finally, the infrared spectrum data are acquired by the IR spectrometer (ALPHA II,Bruker, Germany) controlled by the computer.

The collection spectral range of the spectrometer is 2000~6000 cm^{-1}, the sampling interval is 1.03 cm^{-1}, and the number of collection points for a single sample is 3882.

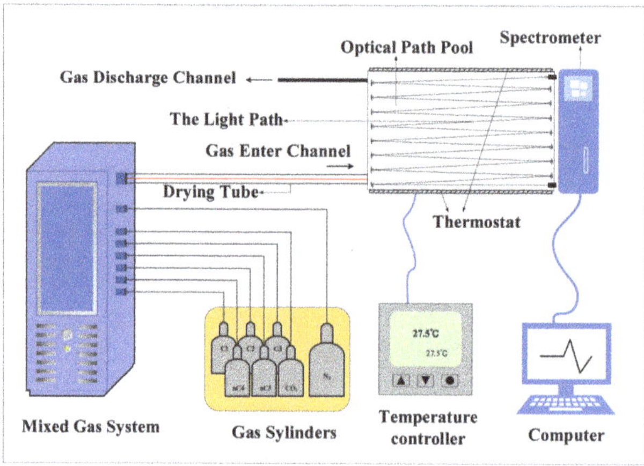

(a)

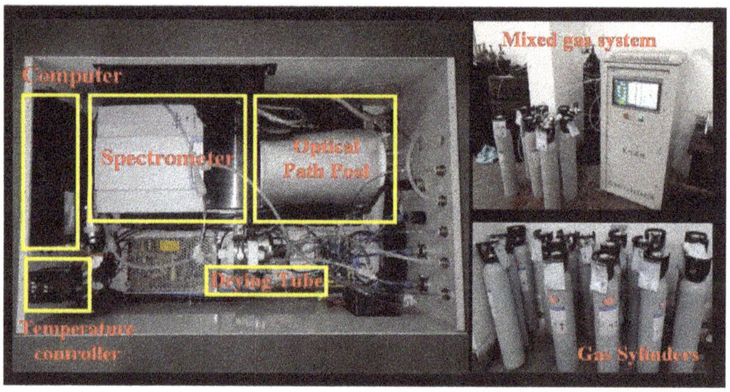

(b)

Figure 3. IR spectrum acquisition system. (**a**) Schematic diagram of collection process. (**b**) The actual hardware structure of gas-logging instrument.

The obtained IR spectrum dataset contains 400 samples. The datasets can be divided into three types. The first dataset (one-component dataset, sample serial number 1~100) consists of one component of six elementary gases C_1, C_2, C_3, nC_4, nC_5 and CO_2 with the concentration distribution in Table 1. As shown in Figure 4, the second dataset (three-component dataset, sample serial number 101~300) consists of three components (C_1, C_2, C_3) with the concentration of one component increasing and the other two random. As shown in Figure 5,the third dataset (six-component dataset, sample serial number 301~400) consists of six components (C_1, C_2, C_3, nC_4, nC_5, CO_2) with the concentration random.

Table 1. Standard concentration and sample concentration distribution of the one-component dataset.

Substance	Standard Gas Concentration	Concentration Gradient	Sample Index Number
C_1	99.999%	5%	1~20
C_2			21~40
C_3			41~60
CO_2			61~80
nC_4	4.999%	0.5%	81~90
nC_5	3.999%	0.4%	91~100

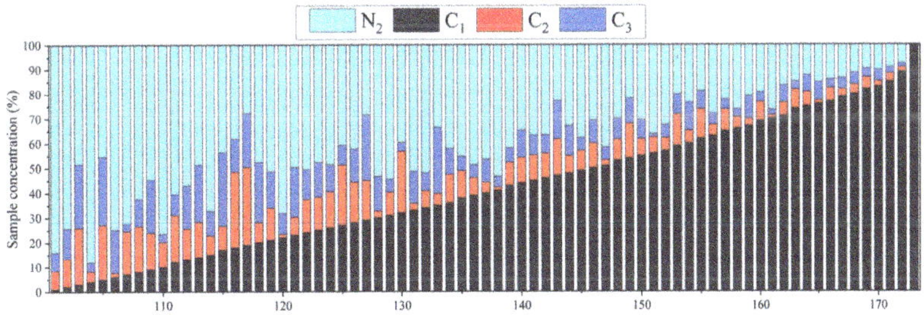

(**a**)

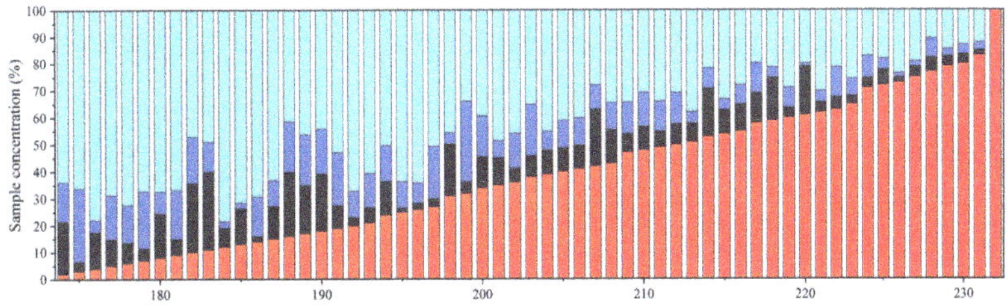

(**b**)

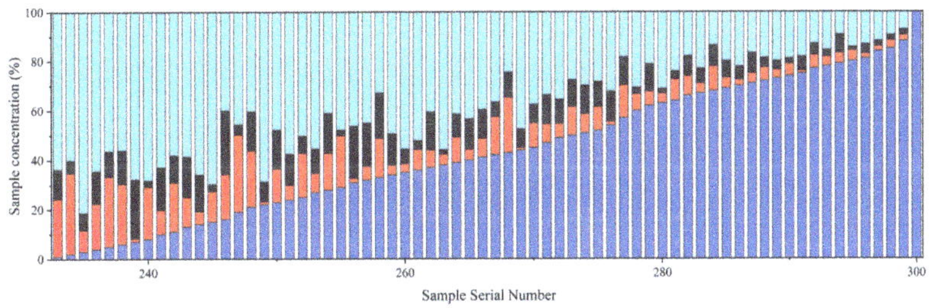

(**c**)

Figure 4. The concentration value of each sample in the three-component dataset. The horizontal axis represents the sample index number, and the vertical axis represents the sample concentration. (**a**) C_1 increments. (**b**) C_2 increments. (**c**) C_3 increments.

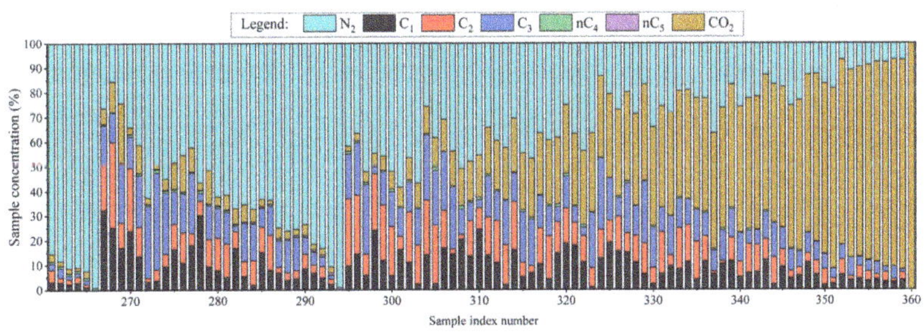

Figure 5. The concentration value of each sample in the six-component dataset.

Figure 6 shows the spectral curves of part samples (cutting the range of 2200~2700 cm^{-1}). It has a good absorption linearity with the increment of concentration of the samples in one-component dataset. As the number of components increases, the absorption peaks of different components are seriously covered up and overlapped in Figure 6d.

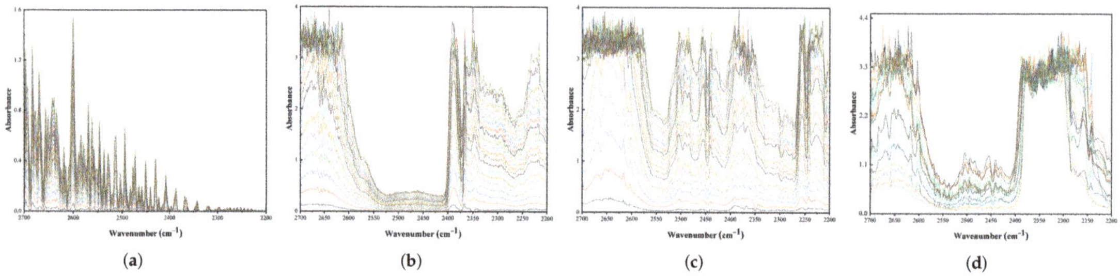

Figure 6. Spectral curve of the sample of the IR spectrum dataset. (**a**) Sample serial number 1~20 (C_1). (**b**) Sample serial number 21~40 (C_2). (**c**) Sample serial number 41~60 (C_3). (**d**) Sample serial number 101~120 (mixed dataset).

3.2. *Influencing Factors on Local Modelling*

In this paper, experiments were carried out firstly in the six-component dataset in order to explore the influence of the window width and window position (the centre position of the window is recorded as the position of the window) during the movement of the local modelling window for the six substances. Taking *RMSECV* as the model evaluation index, Figure 7 shows the effect of the local PLS model under different window positions (2050~4750 cm^{-1}, in increments of 100 cm^{-1}) and different window widths (5~300 cm^{-1}, in increments of 5 cm^{-1}).

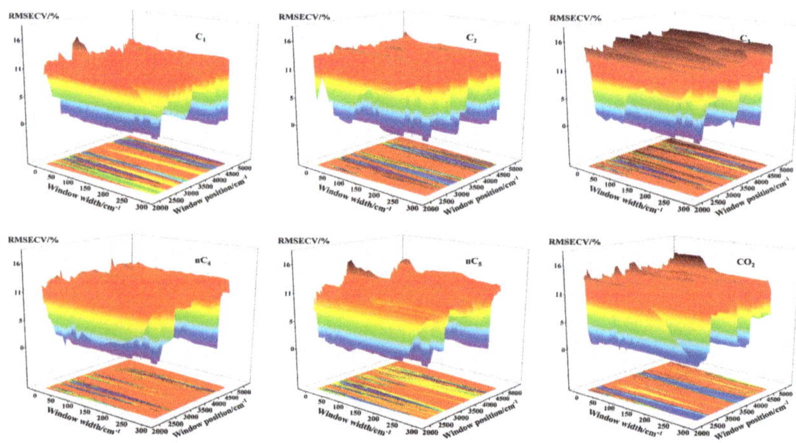

Figure 7. The accuracy of the local PLS model is affected by the width and position of the window.

Conclusions can be obtained from Figure 7 as follow:

1. The *RMSECV* of different substances are distributed in the range of 0~16%, and all of them show similar wavelike distribution in the direction of the window position;
2. The *RMSECV* of the same substance shows obvious stripe distribution on the window position;
3. The stripe distributions of C_1 ~C_5 are similar, but are obviously different from CO_2.

This indicates that the influence of the window position is much greater than the influence of the window width in the local PLS modelling; the influence of the window position to C_1 ~C_5 is alike, but is obviously different from CO_2, which is determined by the same C-H and C-C chemical bond in C_1 ~C_5 but the C=O chemical bond in CO_2.

3.2.1. Influence of the Window Position

In order to further confirm the influence of the window position, the *RMSECV* distributions of six substances with a window width of 5~300 cm^{-1} (in increments of 5 cm^{-1}) at 28 different positions (2050~4750 cm^{-1}, in increments of 100 cm^{-1}) were counted in Figure 8. Each box represents the *RMSECV* statistics of the local PLS modelling of all different window widths at the specified window position. The average value of each position is connected to observe the change trend (the red solid line in Figure 8).

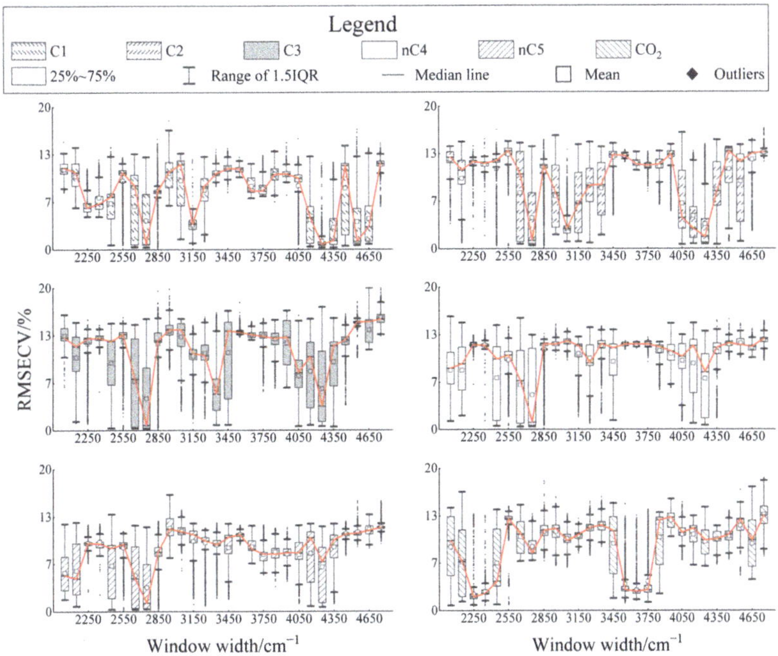

Figure 8. The influence of the local modelling position on the quantitative analysis model of six substances. The statistics include the range of 25% to 75%, the range of 1.5 times the interquartile range (1.5 IQR), the median, mean value and abnormal value (outliers).

It can be clearly seen that the influence of the window position fluctuates greatly, and the best window position for modelling of each hydrocarbon substance is around 2750 cm^{-1}, which is mainly determined by the symmetrical stretching vibration frequency of -CH_3 and -CH_2 [39]. A good quantitative analysis model can be established around 4250 cm^{-1} in the near infrared band or around 3350 cm^{-1} in the middle infrared band, but its analytical accuracy decreases greatly as the carbon chain grows from C_1 to C_5. It no longer has obvious modelling advantages when the carbon chain grows to C_4. Comparing the performance of the PLS models around 2750 cm^{-1}, 2650 cm^{-1} and 2850 cm^{-1}, it is found that the modelling window position is very important to build an optimal quantitative analysis model.

3.2.2. Influence of the Window Width

In order to further confirm the influence of the width, the *RMSECV* distributions of six substances in the range of 2050~4750 cm^{-1} and 30 different widths (10~300 cm^{-1}, in increments of 10 cm^{-1}) were counted in Figure 9. Similarly, each box represents the *RMSECV* statistics of the local PLS modelling of all different window positions under the specified width of the window.

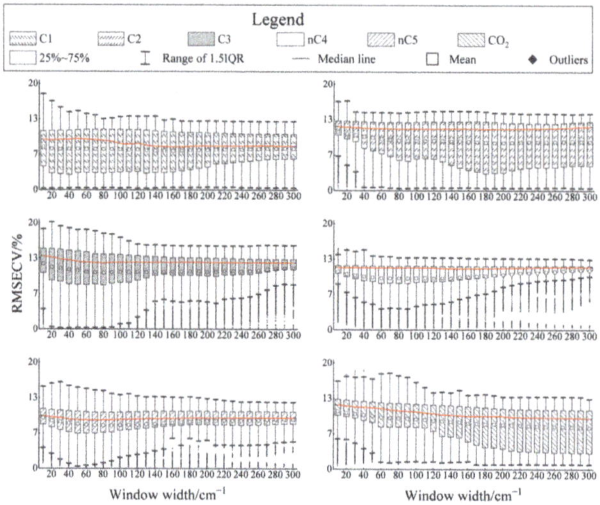

Figure 9. The influence of local modelling width on the quantitative analysis model of six substances.

Obviously, the mean and median values of each substance changed gently under different window widths. From the change in the range of 25∼75% and 1.5 IQR, the stability of the C_1∼C_5 models decreased, but it remained unchanged for CO_2, when the window width increased gradually. It can be explained that the performance of local models for hydrocarbons may be more easily affected by irrelevant interference information introduced due to too large a window because of the narrow band infrared absorption peak of hydrocarbons. As the length of the molecular carbon chain increases, the range of 1.5 IQR became narrower, further indicating that the search of the window position will become more important than the window width.

Figure 10 shows the IR spectra of three different samples with the same C_1 concentration of 15% and the selected modelling interval of ASS-PLS. Affected by other components, the absorption peaks (around 2750 cm^{-1}) of C_1 overlapped badly, which makes it very difficult to build a good quantitative analysis model based on absorption peaks directly. However, the ASS-PLS algorithm proposed in this paper can adaptively find the optimal modelling position (the red dashed box area) free from interference of other components, which is a great improvement for high-precision quantitative analysis of multicomponent mixtures.

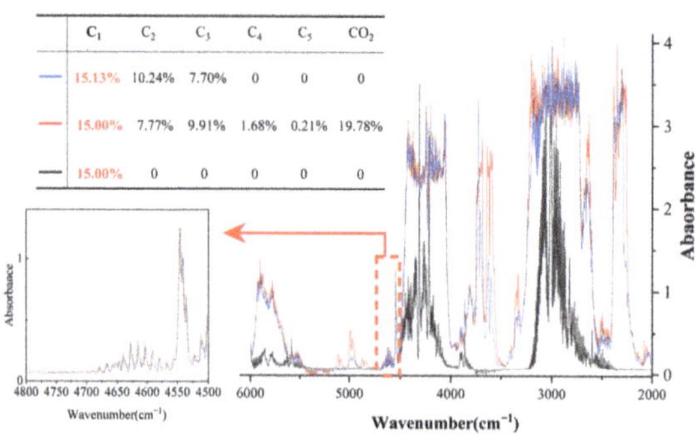

Figure 10. The IR spectra of three different samples with the same concentration of C_1 and the modelling interval selected by ASS-PLS (the red dotted box). The black line represents a sample in the one-component dataset. The red line represents a sample in the three-component dataset. The blue line represents a sample in the six-component dataset.

3.3. Quantitative Analysis Results

All the programs are implemented in MATLAB (Ver. R2021a) software with the computer processor Intel(R) Core (TM) i5-9600KF CPU @ 3.70 GHz. Quantitative tests were performed on the three types of datasets. Each dataset was divided into a training set and test set according to a random ratio of 7:3.

Figure 11 shows the correlation between the actual concentration and the predicted concentration of 30 samples in the one-component test set. It can be seen that the prediction accuracy of each substances is the best when the concentration is near 50%, but it starts to show a small degree of deviation toward the two ends gradually. This is probably caused by the instrument error of the mixed gas system in the preparation of experimental samples, which is ±1% of the concentration of the input gas source. However, the ASS-PLS model still has good analysis accuracy within ±0.4% of *RMSE* for one-component samples.

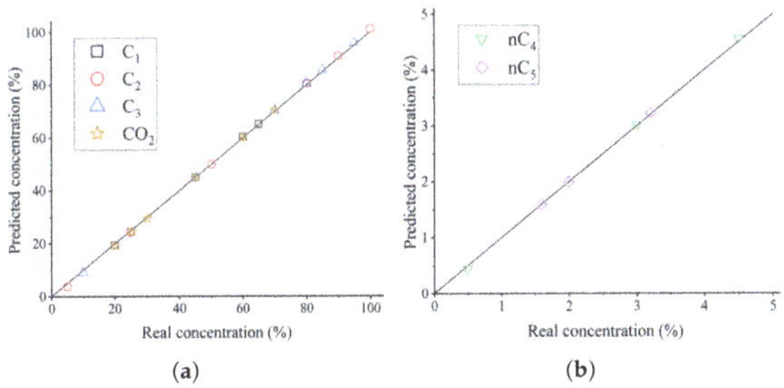

Figure 11. The correlation between the actual concentration and the predicted concentration of samples in the one-component test set. The horizontal axis is the real concentration, and the vertical axis is the predicted concentration of ASS-PLS model, the black line means predicted concentration equal to real concentration. (**a**) Prediction of C_1, C_2, C_3 and CO_2. (**b**) Prediction of nC_4 and nC_5.

Figure 12 shows the correlation between the real concentration and the predicted concentration of 60 samples in the three-component dataset and 30 samples the six-component dataset. Compared with the prediction of one-component samples, the quantitative analysis accuracy of mixed components reduced, especially in the low-concentration section. On the one hand, as the number of components increases, the IR characteristic absorption peaks of different substances overlap to varying degrees. On the other hand, the absorption intensity of low-concentration components is small in multicomponent samples, which makes the IR absorption characteristic peaks of low-concentration components more susceptible to noise and other components.

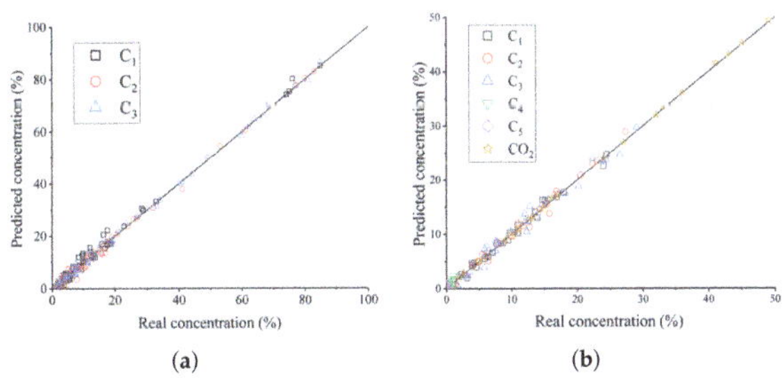

Figure 12. Prediction results in three-component dataset and six-component. (**a**) Prediction of the three-component dataset. (**b**) Prediction of the six-component dataset.

In addition, the actual concentration of CO_2 and the predicted concentration still maintain a good correlation away from the obvious overlap of the infrared characteristic absorption peaks because of the different molecular structure.

The performance of the proposed ASS-PLS method was further compared with full-spectrum PLS (F-PLS, blank contrast group), successive projections algorithm PLS (SPA-PLS), competitive adaptive reweighting sampling PLS (CARS-PLS) and moving window PLS (MW-PLS) by 10 repeated tests in the six-component dataset. Figure 13 shows the performance of the five models.

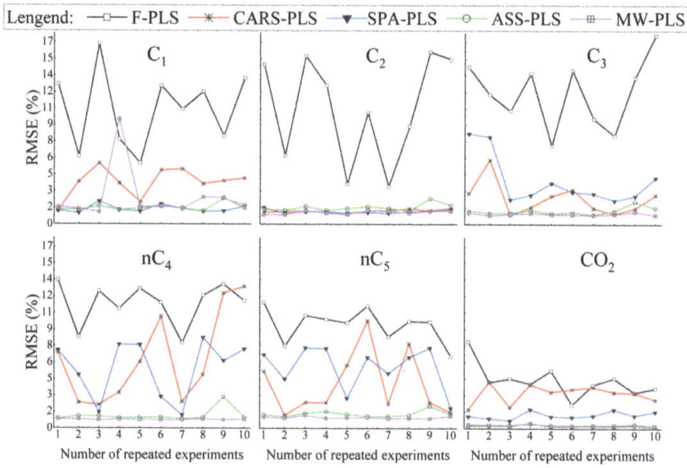

Figure 13. The performance of five different quantitative analysis methods for 10 repeated tests in the six-component dataset. The horizontal axis represents the number of repeated tests, and the vertical axis represents the model accuracy by *RMSE*.

It can be seen that the stability and accuracy of ASS-PLS and MW-PLS models are significantly better than the other three modelling methods, while the performance of F-PLS modelling is extremely poor. In addition, the prediction accuracy of CO_2 is always higher than that of the other five components, which also verifies that the prediction accuracy of hydrocarbons has been affected by the overlapping infrared absorption due to the similar molecular structure. The bolded text indicates the best prediction.

According to the *RSME* values in Table 2, the prediction accuracy in the one-component dataset is much higher than that in the multicomponent dataset for the same substance by the same quantitative analysis method, indicating that it is more difficult to establish a good quantitative analysis model for multicomponent mixtures. It is noticeable that the prediction accuracy of CO_2 has a smaller reduction in the multicomponent dataset because the characteristic infrared absorption band of CO_2 is not obviously affected with the increased number of hydrocarbon components. Furthermore, the prediction accuracy of ASS-PLS model is slightly better than the MW-PLS model, and significantly better than the CARS-PLS and SPA-PLS models.

It is extremely necessary to realize rapid modelling of quantitative analysis models in completely unknown application sites to promote the application of IR spectroscopy gas logging. Table 3 shows the number of wavenumber variables extracted by the modelling algorithm, analysis time and the modelling time (with the Q_h of PLS in (16) is 0.05) of the several algorithm models mentioned above in the six-component dataset.

Since the analysis time is small enough, the difference in the analysis time is almost negligible for practical applications. In addition to the higher model accuracy, the smaller modelling time will become more practical to update the model in time. Even CARS-PLS and SPA-PLS have a smaller modelling time, but the stabilities of the models are insufficient (shown in Figure 13) because of the less wavenumber variables extracted. Even MW-PLS and ASS-PLS have better prediction accuracy (shown in Table 2), but the modelling time of MW-PLS is nearly 349 times that of ASS-PLS due to lots of meaningless repetitions.

Experiments show that ASS-PLS has ideal modelling time with good prediction accuracy. Therefore, the robust accuracy and high-efficiency modelling of the ASS-PLS model is significantly more in line with the requirements of gas-logging applications for high stability, high precision and fast analysis.

Table 2. Comparison of the effects of five different analysis methods in three types of datasets.

Substance	Dataset	*RMSE* (%)				
		F-PLS	CARS-PLS	SPA-PLS	MW-PLS	ASS-PLS
C_1	A [1]	3.8690	0.8772	1.0151	0.4902	0.3586
	B [2]	7.5254	3.3254	1.3851	1.1743	1.1628
	C [3]	10.7304	3.8698	1.8750	1.9742	1.5020
C_2	A	3.6831	1.4582	1.3080	0.7736	0.5228
	B	7.2151	1.6526	1.8951	0.9414	1.4882
	C	10.7826	1.2329	2.1340	1.6683	1.3673
C_3	A	5.0227	1.5765	1.6862	0.5816	0.7659
	B	8.7754	1.7951	2.2651	1.1151	1.2864
	C	12.2141	2.3778	3.9925	1.6114	1.5182
nC_4	A	7.2256	0.7946	1.2773	0.2026	0.1930
	C	11.6557	6.4614	5.5224	1.7623	1.4150
nC_5	A	5.8746	0.2965	0.1137	0.1584	0.1874
	C	9.6520	4.1198	5.6541	1.4386	1.5524
CO_2	A	4.2137	0.2237	0.0963	0.1236	0.1731
	C	4.9388	0.7007	0.6676	0.3152	0.2897

[1] A is the one-component test set; [2] B is the three-component test set; [3] C is the six-component test set.

Table 3. Comparison of the number of wavenumber variables extracted, analysis time and the modelling time of several algorithm models in the six-component dataset.

Model (Q_h = 0.05)	Wavenumber Variables	Analysis Time (s)	Modelling Time (s)
F-PLS	3882	0.0208	41.2200
CARS-PLS	56	0.0056	32.6267
SPA-PLS	27	0.0033	27.2036
ASS-PLS	100	0.0047	32.7247
MW-PLS	100	0.0043	11,443.6870

4. Conclusions

This paper proposes a new ASS-PLS infrared spectroscopy gas-logging modelling method. The algorithm searches for the optimal quantitative analysis modelling interval of each component of mixtures in the full spectrum by sliding a window determined by an adaptive step-sliding control function. Three types of infrared datasets for the logging gas are constructed in this paper. Based on the datasets, it has been analysed that the ASS-PLS method can adaptively determine an optimal modelling interval within 2000~6000 cm^{-1} of infrared data. Comparative experiments show that the stability, speed and accuracy of ASS-PLS modelling are better than those of the other four modelling methods. In addition, the experimental results in different mixed types of samples also show that the gas-logging technology realized by infrared spectroscopy has the following distinct unique features compared with other industrial applications.

1. The infrared characteristic distribution of logging gas has an obvious concentration, and the local modelling strategy of continuous interception can effectively retain

the characteristic information and improve the prediction accuracy and stability of the model.

2. The accuracy of the local model under the continuous interception strategy is much more sensitive to the modelling position than the interception width.
3. The similarity of the alkane molecular structure can lead to a shift in the optimal modelling interval under different mixing types.

The proposed ASS-PLS modelling method can fit the above characteristics very well, and is helpful for improving IR spectroscopy gas-logging technology, and it will also have a great contribution to other molecular spectroscopy analyses with high similarity.

Author Contributions: Conceptualization, Z.L. and W.P.; methodology, Z.L.; software, W.P.; validation, Z.L., W.P., H.D. and C.J.; formal analysis, Z.L. and G.C.; investigation, Z.L. and H.L.; resources, H.L.; data curation, W.P.; writing—original draft preparation, W.P.; writing—review and editing, Z.L. and G.C.; visualization, W.P.; supervision, H.L.; project administration, H.L.; funding acquisition, H.L. All authors have read and agreed to the published version of the manuscript.

Funding: This work is supported in part by the National Natural Science Foundation of China (52074233), the Science and the Technology Cooperation Project of the China National Petroleum Corporation and Southwest Petroleum University Innovation Alliance (2020CX040000).

Institutional Review Board Statement: Not applicable.

Informed Consent Statement: Not applicable.

Data Availability Statement: Not applicable.

Acknowledgments: The authors would like to acknowledge the National Natural Science Foundation of China (52074233), the Science and Technology Cooperation Project of the China National Petroleum Corporation and Southwest Petroleum University Innovation Alliance (2020CX040000) and ChinaFrance Bohai Geoservices Co., Ltd. (CFB_TG_TI_2019_012). The contents of this paper are solely the authors' responsibility and not to be taken as the official views of any of the organizations listed above.

Conflicts of Interest: The authors declare no conflict of interest.

References

1. Ighodalo, E.; Davies, G.; D'Souza, S.A.; Ahmed, A. Increasing Certainty in Formation Evaluation Utilizing Advanced Mud Logging Gas Analysis. In Proceedings of the SPE Kingdom of Saudi Arabia Annual Technical Symposium and Exhibition, Dammam, Saudi Arabia, 25 April 2017. [CrossRef]
2. Ferroni, G.; Rivolta, F.; Schifano, R. Improved Formation Evaluation While Drilling With a New Heavy Gas Detector. In Proceedings of the Society of Professional Well Log Analysts Annual Logging Symposium, Cartagena, Colombia, 16–20 June 2012.
3. Zakrevskiy, K.E.; Gazizov, R.K.; Ryzhikov, E.A.; Freydin, K.V. Consistency evaluation technology for automatic well-log correlation using well logging data (Russian). *Neft. Khozyaystvo Oil Ind.* **2021**, *2021*, 22–26. [CrossRef]
4. Kandel, D.; Quagliaroli, R.; Segalini, G.; Barraud, B. Improved Integrated Reservoir Interpretation Using Gas While Drilling Data. *SPE Reserv. Eval. Eng.* **2001**, *4*, 489–501. [CrossRef]
5. Kriel, W.; Spence, A.; Kolodziej, E.; Hoolahan, S. Improved Gas Chromatographic Analysis of Reservoir Gas and Condensate Samples. In Proceedings of the SPE International Conference on Oilfield Chemistry, New Orleans, LA, USA, 2–5 March 1993. [CrossRef]
6. Breviere, J.; Herzaft, B.; Mueller, N. Gas Chromatography—Mass Spectrometry (Gcms)—A New Wellsite Tool For Continuous C1-C8 Gas Measurement In Drilling Mud—Including Original Gas Extractor And Gas Line Concepts. First Results And Potential. In Proceedings of the Society of Professional Well Log Analysts 43rd Annual Logging Symposium, Oiso, Japan, 2–5 June 2002.
7. Brumboiu, A.; Hawker, D.; Norquay, D.; Law, D. Advances in Chromatographic Analysis of Hydrocarbon Gases in Drilling Fluids? The Application of Semi-Permeable Membrane Technology to High Speed TCD Gas Chromatography. In Proceedings of the Society of Professional Well Log Analysts 46th Annual Logging Symposium, New Orleans, LA, USA, 22–26 June 2005.
8. Cramers, C.A.; Janssen, H.G.; van Deursen, M.M.; Leclercq, P.A. High-speed gas chromatography: An overview of various concepts. *J. Chromatogr. A* **1999**, *856*, 315–329. [CrossRef]
9. Agah, M.; Lambertus, G.; Sacks, R.; Wise, K. High-Speed MEMS-Based Gas Chromatography. *J. Microelectromech. Syst.* **2006**, *15*, 1371–1378. [CrossRef]
10. Rowe, M.; Splapikas, T. Jumping Mass Spectrometer. In Proceedings of the SPE Kingdom of Saudi Arabia Annual Technical Symposium and Exhibition, Dammam, Saudi Arabia, 23–26 April 2017. [CrossRef]

11. Rowe, M.; Muirhead, D. Mud-Gas Extractor and Detector Comparison. In Proceedings of the SPE Kingdom of Saudi Arabia Annual Technical Symposium and Exhibition, Dammam, Saudi Arabia, 23–26 April 2017. [CrossRef]
12. Sauer, C.; Lorén, A.; Schaefer, A.; Carlsson, P.A. On-Line Composition Analysis of Complex Hydrocarbon Streams by Time-Resolved Fourier Transform Infrared Spectroscopy and Ion–Molecule Reaction Mass Spectrometry. *Anal. Chem.* **2021**, *93*, 13187–13195. [CrossRef] [PubMed]
13. van der Veen, A.M.H.; Nieuwenkamp, G.; Zalewska, E.T.; Li, J.; de Krom, I.; Persijn, S.; Meuzelaar, H. Advances in metrology for energy-containing gases and emerging demands. *Metrologia* **2020**, *58*, 012001. [CrossRef]
14. Diller, D.E.; Chang, R.F. Composition of Mixtures of Natural Gas Components Determined by Raman Spectrometry. *Appl. Spectrosc.* **1980**, *34*, 411–414. [CrossRef]
15. Kosacki, I.; Srinivasan, S. Application of Raman Spectroscopy for Hydrocarbon Characterization. In Proceedings of the Nace Corrosion 2017, New Orleans, LA, USA, 26–30 March 2017.
16. Sharma, R.; Poonacha, S.; Bekal, A.; Vartak, S.; Weling, A.; Tilak, V.; Mitra, C. Raman analyzer for sensitive natural gas composition analysis. *Opt. Eng.* **2016**, *55*, 1–8. [CrossRef]
17. Rathmell, C. Data-Driven Raman Spectroscopy in Oil and Gas: Rapid Online Analysis of Complex Gas Mixtures. *Spectrosc. Suppl.* **2018**, *33*, 34–42.
18. Livanos, G.; Zervakis, M.; Pasadakis, N.; Karelioti, M.; Giakos, G. Deconvolution of petroleum mixtures using mid-FTIR analysis and non-negative matrix factorization. *Meas. Sci. Technol.* **2016**, *27*, 114005. [CrossRef]
19. Sell, J.K.; Jakoby, B. A simple mid-infrared measurement system based on a tunable filter for the analysis of ternary gas mixtures. *Meas. Sci. Technol.* **2013**, *24*, 084006. [CrossRef]
20. Indo, K.; Hsu, K.; Pop, J. Estimation of Fluid Composition From Downhole Optical Spectrometry. *SPE J.* **2015**, *20*, 1326–1338. [CrossRef]
21. Piazza, R.; Vieira, A.; Sacoragua, L.A.; Jones, C.; Dai, B.; Pearl, M.; Aguiar, H. Real-Time Downhole MID-IR Measurement of Carbon Dioxide Content. In Proceedings of the Society of Professional Well Log Analysts 60th Annual Logging Symposium, The Woodlands, TX, USA, 15–19 June 2019. [CrossRef]
22. Ren, Q.; Chen, C.; Wang, Y.; Li, C.; Wang, Y. A Prototype of ppbv-Level Midinfrared CO2 Sensor for Potential Application in Deep-Sea Natural Gas Hydrate Exploration. *IEEE Trans. Instrum. Meas.* **2020**, *69*, 7200–7208. [CrossRef]
23. Moro, M.K.; dos Santos, F.D.; Folli, G.S.; Romão, W.; Filgueiras, P.R. A review of chemometrics models to predict crude oil properties from nuclear magnetic resonance and infrared spectroscopy. *Fuel* **2021**, *303*, 121283. [CrossRef]
24. Kamboj, U.; Kaushal, N.; Jabeen, S. Near Infrared Spectroscopy as an efficient tool for the Qualitative and Quantitative Determination of Sugar Adulteration in Milk. *J. Phys. Conf. Ser.* **2020**, *1531*, 012024. [CrossRef]
25. Gao, Q.; Wang, M.; Guo, Y.; Zhao, X.; He, D. Comparative Analysis of Non-Destructive Prediction Model of Soluble Solids Content for Malus micromalus Makino Based on Near-Infrared Spectroscopy. *IEEE Access* **2019**, *7*, 128064–128075. [CrossRef]
26. Sun, J.; Yang, W.; Feng, M.; Liu, Q.; Kubar, M.S. An efficient variable selection method based on random frog for the multivariate calibration of NIR spectra. *RSC Adv.* **2020**, *10*, 16245–16253. [CrossRef]
27. Galvão, R.K.H.; Araújo, M.C.U.; Fragoso, W.D.; Silva, E.C.; José, G.E.; Soares, S.F.C.; Paiva, H.M. A variable elimination method to improve the parsimony of MLR models using the successive projections algorithm. *Chemom. Intell. Lab. Syst.* **2008**, *92*, 83–91. [CrossRef]
28. Yu, Y.; Huang, J.; Zhu, J.; Liang, S. An Accurate Noninvasive Blood Glucose Measurement System Using Portable Near-Infrared Spectrometer and Transfer Learning Framework. *IEEE Sens. J.* **2021**, *21*, 3506–3519. [CrossRef]
29. Wang, J.; Lu, S.; Wang, S.H.; Zhang, Y.D. A review on extreme learning machine. *Multimed. Tools Appl.* **2021**. [CrossRef]
30. Jiang, H.; Liu, G.; Mei, C.; Yu, S.; Xiao, X.; Ding, Y. Measurement of process variables in solid-state fermentation of wheat straw using FT-NIR spectroscopy and synergy interval PLS algorithm. *Spectrochim. Acta Part A Mol. Biomol. Spectrosc.* **2012**, *97*, 277–283. [CrossRef] [PubMed]
31. Yu, H.; Du, W.; Lang, Z.Q.; Wang, K.; Long, J. A Novel Integrated Approach to Characterization of Petroleum Naphtha Properties From Near-Infrared Spectroscopy. *IEEE Trans. Instrum. Meas.* **2021**, *70*, 1–13. [CrossRef]
32. Ouyang, T.; Wang, C.; Yu, Z.; Stach, R.; Mizaikoff, B.; Huang, G.B.; Wang, Q.J. NOx Measurements in Vehicle Exhaust Using Advanced Deep ELM Networks. *IEEE Trans. Instrum. Meas.* **2021**, *70*, 1–10. [CrossRef]
33. Liu, L.; Yan, L.; Xie, Y.; Xia, G. Content measurement of textile mixture by near infrared spectroscopy based on BP neural network. In Proceedings of the 2010 3rd International Congress on Image and Signal Processing, CISP 2010, Yantai, China, 16–18 October 2010; Volume 7, pp. 3354–3358. [CrossRef]
34. Li, Y.; Zhang, Y.; Zhang, W.B.; Xu, Y.Y.; Zhang, G.J. Comparative study of partial least squares and neural network models of near-infrared spectroscopy for aging condition assessment of insulating paper. *Meas. Sci. Technol.* **2020**, *31*, 045501. [CrossRef]
35. Kumar, K. Competitive adaptive reweighted sampling assisted partial least square analysis of excitation-emission matrix fluorescence spectroscopic data sets of certain polycyclic aromatic hydrocarbons. *Spectrochim. Acta Part A Mol. Biomol. Spectrosc.* **2021**, *244*, 118874. [CrossRef]
36. Huang, G.; He, J.; Zhang, X.; Feng, M.; Tan, Y.; Lv, C.; Huang, H.; Jin, Z. Applications of Lambert-Beer law in the preparation and performance evaluation of graphene modified asphalt. *Constr. Build. Mater.* **2021**, *273*, 121582. [CrossRef]
37. Silalahi, D.D.; Midi, H.; Arasan, J.; Mustafa, M.S.; Caliman, J.P. Kernel partial diagnostic robust potential to handle high-dimensional and irregular data space on near infrared spectral data. *Heliyon* **2020**, *6*, e03176. [CrossRef]

38. Jiang, W.; Lu, C.; Zhang, Y.; Ju, W.; Wang, J.; Hong, F.; Wang, T.; Ou, C. Moving-Window-Improved Monte Carlo Uninformative Variable Elimination Combining Successive Projections Algorithm for Near-Infrared Spectroscopy (NIRS). *J. Spectrosc.* **2020**, *6*, 3590301. [CrossRef]
39. Kazansky, V.B.; Subbotina, I.R.; Jentoft, F.C.; Schlögl, R. Intensities of C-H IR Stretching Bands of Ethane and Propane Adsorbed by Zeolites as a New Spectral Criterion of Their Chemical Activation via Polarization Resulting from Stretching of Chemical Bonds. *J. Phys. Chem. B* **2006**, *110*, 17468–17477. [CrossRef]

minerals

MDPI

Review

Catalysis of Minerals in Pyrolysis Experiments

Ming Zhong [1], Haiping Huang [1,2,3,*], Pengcheng Xu [4] and Jie Hu [4]

[1] School of Energy Resources, China University of Geosciences (Beijing), Beijing 100089, China
[2] College of Earth Sciences, Yangtze University, Wuhan 430100, China
[3] Department of Geoscience, University of Calgary, 2500 University Drive NW, Calgary, AB T2N 1N4, Canada
[4] Changqing Industry Group Co., Ltd., China Petroleum Changqing Oilfield Branch, Xi'an 710021, China
* Correspondence: hhp58@163.com

Abstract: Recent research in the field of oil and gas geochemistry has focused on the catalytic role of minerals in geological history. Thermal simulation experiments are considered a valuable means of studying the formation and transformation of hydrocarbons. In this paper, we review the catalytic mechanisms, processes, and various arguments for different types of minerals in thermal simulation experiments from the perspective of mineral additives. We focus on two categories: (1) minerals that provide direct catalysis, such as clay minerals, alkali metals, carbonate rocks, and some transition metal elements, and (2) minerals, such as serpentine, that promote aqueous hydrogen and act as the material basis, as well as the radioactive element uranium. We also discuss existing disputes and prospects for the development direction of mineral catalytic thermal simulation experiments.

Keywords: pyrolysis; kerogen; catalysis; geological process

Citation: Zhong, M.; Huang, H.; Xu, P.; Hu, J. Catalysis of Minerals in Pyrolysis Experiments. *Minerals* **2023**, *13*, 515. https://doi.org/10.3390/min13040515

Academic Editors: Mofazzal Hossain, Hisham Khaled Ben Mahmud and Md Motiur Rahman

Received: 1 March 2023
Revised: 4 April 2023
Accepted: 5 April 2023
Published: 6 April 2023

1. Introduction

Since the discovery that oil and natural gas mainly derive from kerogen, organic geochemistry has investigated the composition, types, hydrocarbon (gas) generation law, and sedimentary environment of kerogen using pyrolysis, chromatography, and other methods [1–3]. The aim of this study is to discuss the types and transformation of organic matter [4], kerogen oil generation [5], gas from crude oil cracking [6], and the safe utilization of coal and oil shale [7]. To date, three types of pyrolysis experiments exist, namely closed, semi-open, and open pyrolysis methods. The gold tube closed-system pyrolysis experiment is the most widely used. Gold's good ductility and stability enables the pyrolysis experiment to consider higher temperatures and pressure environments and thus more geological conditions. The open pyrolysis test uses different gas atmospheres (such as nitrogen) to purge the pyrolysis products into the analytical instrument, which has the advantages of a simple experiment and high analysis efficiency. However, the process of hydrocarbon formation and transport is not carried out in such an open-system experiment. So open systems are limited in what they can simulate. In the semi-open system, a device for discharging pyrolysis products is added on the basis of the closed system. Due to the limitation of equipment and technology, the internal pressure of the hydrocarbon expulsion process will change and affect the liquid hydrocarbon expulsion. In the design of pyrolysis experiments, the controlled factors of experimental purposes and samples should be fully considered to avoid systematic errors of experimental results caused by different pyrolysis systems [8].

The simulation method has gradually started to consider the pore throat characteristics of different types of compounds in various reservoirs and the features of hydrocarbon generation and expulsion under pressure [9,10]. Coal, oil, kerogen, carbonaceous mudstone, carbonate rock, and other organic matters can be the subjects of simulation [11–13]. Please refer to the review article by Peng et al. [14] for more information on the status of thermal simulation experiments. However, the geological history of crude oil and natural gas formation is much more complicated than previously thought, involving different surface interface behaviors and organic–inorganic interactions. Therefore, when designing experiments, geological factors, the selection of experimental equipment, and whether to add different minerals or water should be considered [15].

As far as current reports are concerned, different minerals provide varying catalytic effects in the process of thermal simulation experiments. The mechanism by which silicate clay minerals and carbonate minerals promote pyrolysis is still unclear, and the specific behavior of kerogen catalysis by some ions and minerals in geological processes is still controversial. Examples include coordinated metal ions in biomacromolecules from sedimentary environments, uranium captured during the formation of crude oil and natural gas, carbonate minerals such as calcite and dolomite, and potentially strongly catalytic metal ions. The catalytic mechanism of these metal ions during pyrolysis has not been thoroughly studied. This paper reviews the process of promoting kerogen to produce crude oil or natural gas from the perspectives of clay minerals, alkali metals, carbonate rocks, other metal minerals (mainly transition metals), uranium as thermal simulation additives, and serpentine olivine. This paper is mainly divided into two parts, one of which discusses how minerals directly play a catalytic role, while the other discusses how minerals and water coexist to generate hydrogen. The clear catalytic methods are summarized, while the controversial aspects are also reviewed.

2. Minerals Directly Promote Thermal Simulation Process

In modern crude oil and kerogen samples, a variety of minerals can be found. Prior to pyrolysis experiments, it is common practice to pretreat rocks with acidic fluids such as HCl and HF, which removes all minerals except kerogen. These minerals are influenced by the ancient environment during the earth's redox environment changes, and their unique characteristics can be used to infer sedimentary location environments, tectonic histories, and whether they have undergone epigenesis.

Moreover, these minerals can catalyze the formation of hydrocarbons in various forms. Typically, silico-aluminate dominated clay minerals promote the formation of petroleum in the form of unique solid acids, while various metallic elements enriched in other sedimentary environments are oxidation–reduction catalysts. In addition, contact between these minerals and liquids can result in secondary changes that promote the pyrolysis process.

2.1. *Effect of Clay Minerals on Pyrolysis*

The formation and structure of clay minerals in reservoirs are influenced by ancient water flow, environment, and provenance debris during the sedimentary period. Therefore, sedimentologists consider clay minerals as good indicators to reflect changes in sedimentary environments. Clay minerals are often used to divide sequences, construct and restore paleoenvironments, and judge major geo-historical events [16]. These silicate layered minerals contain metal elements such as potassium and aluminum, and their particle size is usually less than 3.9 μm. Researchers in the petroleum industry are highly interested in clay minerals [17].

The common types of clay minerals found in sedimentary environments are illite, montmorillonite, kaolinite, chlorite, and their mixed layers. Although clay minerals are not welcomed by reservoir engineers due to their special physical and chemical changes [18], such as water absorption and swelling, dispersed migration, and the release of metal ions including Fe and Al, they have gained attention in recent years among researchers exploring and developing unconventional reservoirs such as tight sandstone and shale. The nano-scale pores provided by clay minerals are excellent for preserving organic matter, and researchers are investigating this aspect [19].

Clay minerals were first used in the petrochemical field in the mid-twentieth century, and montmorillonite clay was first used as the cracking catalyst for crude oil in the chemical industry [20]. Since then, geologists and geochemists have been interested in the promotion of various mineral types in crude oil production. For instance, Eltantawy et al. [21] and Shimoyama [22] investigated the catalytic effects of clay minerals by heating natural fatty acids and alcohols in clay. The results showed that paraffin wax with a similar composition to crude oil and black carbonized residues attached to minerals was obtained, indicating that the catalysis of clay played a significant role in crude oil formation.

The catalytic effect of clay minerals, especially montmorillonite, in kerogen pyrolysis is widely recognized. However, the organic matter–water–rock reaction involved has not been fully explained. The catalytic effect of clay minerals in sedimentary layers mainly has the following two points: (1) the removal of heteroatoms from gum and asphaltene, promoting the formation of oil and the production of non-hydrocarbon substances such as CO_2 and H_2S, and (2) the condensation polymerization of kerogen, which forms C_1-C_{15} short-chain alkanes through condensation polymerization (Figure 1):

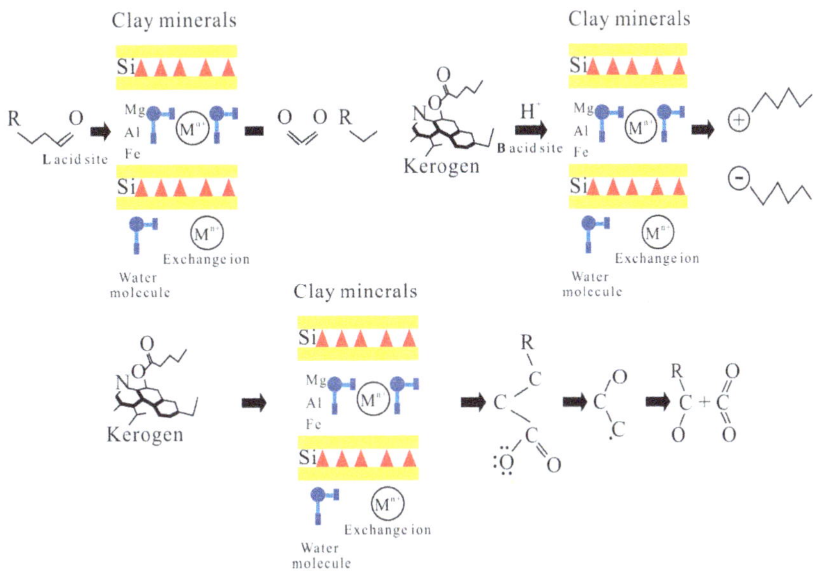

Figure 1. The main catalytic mechanism of clay minerals.

Currently, the accepted theory suggests that montmorillonite can promote the production of light hydrocarbons by facilitating the cracking of hydrocarbons and the isomerization of normal hydrocarbons via a mechanism involving carbocation intermediates [23]. The Brønsted acid sites in the interlayer space of montmorillonite are produced by interlayer dissociation water, which significantly promotes this process. In contrast, n-alkane isomerization and olefin alkane conversion through hydrogenation lead to high saturated hydrocarbon/n-alkane and olefin/alkane ratios in pyrolysis products [24]. Brønsted acid sites can promote the pyrolysis of nitrogen-containing organic compounds through Hoffman elimination [25], while Lewis acid sites participate in the decarboxylation reaction of aliphatic compounds during pyrolysis and CO_2 is generated [26].

Ma et al. [27] have proposed that the primary catalytic mechanism of montmorillonite for low-maturity lignite pyrolysis is the free radical mechanism at low temperatures. Specifically, in the decarboxylation of organic matter, clay minerals may receive electrons from it during the decarboxylation of organic matter, which leads to the formation of free radicals. Free radicals cause rearrangement, resulting in the cleavage of the C single bond and the formation of short chain free hydrocarbons (Figure 1). However, the presence of water during the pyrolysis experiment has complicated results. Montmorillonite has different isomerization effects on alkanes in the presence or absence of water. Water molecules are adsorbed on the surface of montmorillonite and in the interlayer space, providing an oxygen source for the generation of CO_2. It also inhibits the isomerization of n-alkanes in the pyrolysis process, leading to the dominant role of n-alkanes in the pyrolysis products. A lower interlayer water and surface adsorbed water content can greatly promote the isomerization of hydrocarbons and reduce the yield of CO_2 [28].

Due to the complexity of crude oil, we believe that the free radical mechanism and the carbocation intermediate mechanism will occur simultaneously in the transformation process of crude oil and kerogen. The clay minerals will also dehydrate and rearrange themselves when heated, resulting in a reduction in interlayer volume and the transformation to illite on a macro level. However, the addition of water will increase the solid acidity of minerals. During the pyrolysis process, the hydroxyl group in water is adsorbed on the surface of minerals and proton H is provided to the organic matter. Although the current report cannot distinguish the degree of contribution, future experiments could be designed to focus on this process.

When it comes to illite, while it does exhibit catalytic effects, it is still slightly inferior to montmorillonite [29,30]. Due to illite lacking interlayer water, catalysis can only occur on its surface through solid acid sites. Hence, Lewis acid sites mainly occur on its outer surface [31]. The addition of illite slightly increased the relative cycloalkane content of kerogen-generated hydrocarbons in the experiment [32]. In contrast, the effect of kaolinite on the anhydrous pyrolysis of bituminous coal is not consistent with its effect in the pyrolysis with water. The addition of kaolinite inhibited the generation of light hydrocarbons but increased the yield when water was added [33].

In a study conducted by Espitalie [34], the effects of different clay minerals on the pyrolysis of type II kerogen were investigated. Results indicated that calcium montmorillonite showed better catalytic effects, promoting the formation of light hydrocarbons, and had better catalytic effects than sodium montmorillonite. On the other hand, kaolinite, illite, and attapulgite showed poor catalytic effects, but the residual products of attapulgite had a higher H/C ratio compared to the other two. The study also suggested that pyrolysis experiments in open systems at high temperatures and without water can enhance the interactions between minerals and hydrocarbons.

Roets et al. [35] found that kaolinite and quartz have little influence on the pyrolysis products of coal. Öztaş [36] showed that the treatment of coal with HF and HCl increased the pyrolysis efficiency of the treated samples, suggesting that clay minerals in coal may reduce the pyrolysis yield. He et al. [37] used the relative concentrations of C_7 compounds to evaluate the difference in the catalytic effect of different clay mineral contents, with toluene, 1,2-dimethylcyclopentane, and n-heptane showing the most significant differences. According to Cai et al. [38], the catalytic effect of clay minerals is ranked as montmorillonite > kaolinite > illite, with kaolinite dominating in the high-mature to over-mature stage due to its free radical mechanism and amphiphilic nature.

In summary, the catalytic effect of clay minerals on the pyrolysis of hydrocarbons varies depending on the type of clay mineral and the experimental conditions. Montmorillonite has been found to be the most effective due to its interlayer water and solid acid sites in the interlayer space, while illite and kaolinite have limitations in their catalytic effect.

2.2. *Catalysis of Alkali Metals*

The deposition of organic matter, especially type III kerogen, is influenced by the terrestrial environment and contains significant amounts of alkali metals and alkaline earth metals, such as Na, K, Mg, and Ca. Among these elements, Ca is considered the most active alkaline earth metal [39,40]. Alkali metals in low maturity coal can exist in the form of hydrogen ions in oxygen functional groups or can be freely present in reservoirs such as various crystalline or amorphous minerals [41]. When heated, these alkali metals and alkaline earth metals decompose and generate basic oxides, such as MgO, CaO, Na_2O, and K_2O [42]. The basic oxides in coal promote the cracking of oxygen-containing functional groups, leading to a more compact coal structure. This process strengthens the reaction of tar macromolecules to form semi-coke, prevents tar macromolecules from escaping, and enhances the effect of demethanization and dehydrogenation [12]. For aromatic hydrocarbons, the catalysis of alkali metals can reduce the aromatic hydrocarbons with multi-side groups in pyrolysis products and increase the yield of small molecular aromatic hydrocarbons [43]. Generally speaking, alkali metals can accelerate the overall pyrolysis reaction rate, reduce the liquid yield, greatly improve the yield of gaseous hydrocarbons, and also reduce the contents of oxygen and sulfur in liquid products.

However, compared to marine carbonate reservoirs with vast storage capacity, the existing reports on the pyrolysis of kerogen III (coal) appear to be inconsistent with the reaction of crude oil, clay minerals, carbonate rocks, and kerogen. Karabakan and Yürüm [44] reported that carbonate minerals promote the pyrolysis of kerogen, and the total activation energy of kerogen decreases without HF treatment. They believe that alkaline earth metal oxides produced by carbonate minerals after heating promote the pyrolysis process. The reason why silicate minerals inhibit pyrolysis is not clear. Ballice et al. [45] came to a similar conclusion, stating that the inhibition of silicate minerals is greater than the promotion of carbonate in shale, which is attributed to the strong adsorption of clay minerals on hydrocarbons.

2.3. *Influence of Carbonate Rock on Pyrolysis*

Most carbonate rocks have a marine sedimentary environment and were formed during the early sedimentary period. Common carbonate minerals found in these rocks include aragonite ($CaCO_3$), calcite ($CaCO_3$), and dolomite ($CaMg(CO_3)_2$). Giant carbonate oil and gas reservoirs located in the Middle East have played a significant role in contributing to the world's energy reserves [46]. The secondary changes that carbonate rocks undergo during sedimentation and diagenesis greatly influence their reservoir properties. For example, the process of dolomitization causes a reduction in the volume of limestone, which has the potential to increase the reservoir area [47]. Furthermore, sedimentary processes and secondary changes contribute to the creation of various types of pores, providing carbonate reservoirs with ample storage space [48]. However, the uneven development of carbonate reservoirs and the high sulfur content in some of these reservoirs limit further exploration and development [49]. As such, carbonate reservoirs have been a focal point of research for petroleum geologists.

Petroleum geologists have always been interested in carbonate rocks, which they believe can serve as source rocks due to the discovery of both reservoirs and source rocks in these formations [50–52]. Scientists have also studied the oil generation process of carbonate rocks. Unfortunately, the content of organic carbon in modern carbonate reservoirs is often very low. Currently, there are several explanations for this phenomenon: (1) Diagenesis and dissolution during the formation of carbonate rocks can cause the loss of organic matter, leading to a loss of 15%–50% of organic carbon in clastic rocks and 80%–90% of organic carbon in carbonate rocks [53]. (2) Organic matter may not deposit with carbonate rocks during the deposition process [54]. (3) The organic matter may have reached the threshold of hydrocarbon generation and been expelled. After hydrocarbon expulsion, the total organic carbon (TOC) of type I, type II, and type III kerogen decreased by 80%, 50%, and 20% [55], respectively, and carbonate rocks with high organic carbon content are not conducive to organic matter preservation after hydrocarbon expulsion [56]. (4) Organic matter may transform during carbonate rock deposition processes (e.g., running water, oxidation, and recrystallization), but this process appears to have little effect on organic carbon [57]. Reservoir geochemists find it challenging to understand carbonate reservoirs with conventional petroleum geology methods, and the indicators used to evaluate source rocks (e.g., vitrinite reflectance, organic matter deterioration degree) seem to have no effect on carbonate rocks [58]. This has prevented the development of a deep understanding of carbonate oil generation.

Calcite in carbonate rocks was first observed by Shimoyama and Johns [59] and has very close contact with organic matter during the pyrolysis of calcite fatty acids. They suggest that calcite provides catalysis in the pyrolysis process. However, it can be seen from the surface/interface relationship that the interface between calcite and polar macromolecules in crude oil [60] is more easily adsorbed. Therefore, this phenomenon is not enough to conclude that calcite has catalytic effect. Tannenbaum and Aizenshtat et al. [61] suggested that the heteroatoms present in carbonate rocks provide the essential catalysis for hydrocarbon generation from organic matter and kerogen. Jones et al. [62] concluded that there is no significant difference in the oil production process between carbonate rocks and shale, whereas Huizinga et al. [23] pointed out that $CaCO_3$ has little impact on the thermal evolution and petroleum cracking behavior of kerogen. Pan et al. [63] further observed that the addition of calcite during crude oil cracking experiments had little effect on the cracking of main components in crude oil, but the formation rate of methane and total hydrocarbon gas was relatively low compared to other experimental groups. They postulated that calcite inhibits the secondary cracking of moisture (C_2–C_5) into methane.

Chen et al. [64] combined Rock–Eval pyrolysis data and hydrocarbon generation kinetics to conduct pyrolysis experiments on middle-upper Ordovician argillaceous carbonate rocks in Tarim Basin, China. The kerogen conversion rate reached 90%, indicating a high original potential. The argillaceous component has a relatively strong ability to preserve organic matter [65], but on the whole, carbonate rocks contain few argillaceous components. Therefore, whether well-preserved argillaceous carbonate rock experiments can accurately represent the hydrocarbon generation process in geological processes remains to be discussed.

Dolomite, unlike calcite, is currently regarded as a cheap and excellent catalyst for biomass gasification in the industry. Islam et al. [66] suggested that its catalytic effect comes from decomposition products, such as MgO and CaO, that form when the temperature exceeds 600 °C. However, the temperature of pyrolysis typically does not reach this limit. Thus, the catalytic mechanism of dolomite pyrolysis remains unclear.

Wu et al. [67] studied the effect of dolomite and calcite on the pyrolysis of crude oil using closed flexible gold capsules. They found that the presence of dolomite slightly reduced the yield of C_1–C_5 but significantly increased the yield of C_6–C_{13}. Moreover, they conducted pyrolysis experiments of $MgCO_3$ and $MgCO_3$ + $CaCO_3$, which produced similar results to those of petroleum and dolomite. Thus, it is evident that the influence of dolomite on petroleum pyrolysis is closely related to the content of Mg in dolomite. In addition, the lower content of dolomite can promote the TSR (Thermochemical Sulfate Reduction) of crude oil. Specifically, dolomite -$MgCO_3$ promotes the formation of $SO_4{}^{2-}$. The coexistence of $SO_4{}^{2-}$, water, and H_2S then initiates the sulfate thermochemical reduction (TSR) reaction, which destroys hydrocarbons and produces CO_2 and H_2S [68]. However, the high content of dolomite can absorb or convert some H_2S into other forms [69]. Moreover, the reaction of $MgCO_3$ with water can generate $Mg(OH)_2$ (R2), which further reduces the amount of H_2S generated.

$$Mg(OH)_2 + H_2S \rightarrow MgS + 2H_2O \; MgS + 2O_2 \rightarrow MgSO_4 \tag{1}$$

Maimaiti et al. [70] conducted a pyrolysis study using the same method as Wu et al. [67] to investigate the effect of carbonate rocks on crude oil cracking. They observed that different mineral types, such as calcite and dolomite, had varying degrees of inhibition on methane production at high maturity stages (EasyRo > 3.5%), but had no effect on C_2–C_5 hydrocarbons, while simultaneously inhibiting C_{14+}. Additionally, the yield of CO_2 in the pyrolysis products with carbonate minerals increased significantly, and unique changes in carbon isotopes were observed. Unlike typical thermally driven isotope fractionation, the addition of dolomite promoted the shift of carbon isotopes of methane, whereas calcite inhibited it. In crude oil + dolomite, crude oil + calcite, and simple crude oil pyrolysis experiments, $\delta 13C_1$ was −49.89, −54.08, and −51.77%, respectively (EasyRo = 0.96%). However, the carbon isotope fractionation process of ethane and propane remained unaffected by minerals. They speculate that this phenomenon may be due to dolomite increasing the diversity of methane sources during pyrolysis and the differential adsorption of various organic gases by different minerals. Nevertheless, there is still insufficient evidence to support this claim.

Furthermore, Wang et al. [71] observed that increasing the carbonate concentration in source rocks led to a corresponding increase in the yield of CO_2 and heavier carbon isotopes in CO_2. They believe that carbonate rocks significantly contribute to CO_2 in pyrolysis products, which is derived from the dissolution of carbonate minerals by organic acids during pyrolysis, despite the lower pyrolysis temperature compared to the decomposition temperature of carbonate rocks.

2.4. Effect of Metal Minerals on Pyrolysis Experiments

The precursors of crude oil and kerogen contain many metal elements that exist in organisms in the form of metalloenzymes, playing an important role in the evolution of life and the maintenance of biological functions [72]. These metal elements can be captured at any stage of deposition, hydrocarbon generation, and migration, and are relatively enriched in macromolecules such as colloid and asphaltene, and even complexed or bonded into kerogen [73]. For instance, porphyrin, which is widely found in chlorophyll and heme, is also present in petroleum and kerogen, with the metal elements coordinated being different depending on the living body and its environmental conditions [74]. The proportions of vanadoporphyrin and nickel porphyrin in kerogen, crude oil, and rock represent marine and continental sedimentary environments, respectively. Additionally, these bio-lipids are also considered important catalysts for transforming into rearranged alkanes in crude oil, and the relative content of rearranged alkanes in crude oil produced in some geological periods lacking chlorophyll is lower [75]. Furthermore, these metals and their oxides are important catalysts for the transformation of the Earth's environment from simple to complex [76]. Based on chemical principles, active intermediates are formed by the coordination between ligands and receptors due to the d orbitals around the metal elements not being fully occupied. This reduces the activation energy of the reaction and promotes the formation of organic matter [77]. Moreover, the larger internal pores generated after the organic–inorganic interaction between minerals and organic matter also promote this process [78]. However, Lewan et al. [79] presented different opinions, reporting no obvious difference in gas type, yield, dryness, and formation rate of low-cooked kerogen after 72 h of aqueous pyrolysis at 330 and 355 °C. They argued that various transition metals with catalytic properties were not in an active state during the geological maturation of kerogen and exhibited suitable catalytic activity.

Gao and colleagues [80] conducted anhydrous pyrolysis experiments on low-rank coal with the transition metal Ni. The addition of the transition metal Ni catalyzed the decomposition of organic matter and the synthesis of abiogenic methane during the maturity of source rocks. With the addition of Ni, the content of CH_4 in pyrolysis products increased, and the yields of ethane and propane decreased. They concluded that Ni played a catalytic role in the whole process of kerogen, crude oil, and heavy hydrocarbon gas cracking, and Fischer–Tropsch Type (FTT) reaction occurred in the pyrolysis experiment, which catalyzed together with heavy hydrocarbon gas cracking, significantly improving methane yield and greatly improving CH_4 yield at all temperatures (see Figure 2). In all the experimental groups, δD became smaller, but δ13C gradually became positive with the increase in temperature.

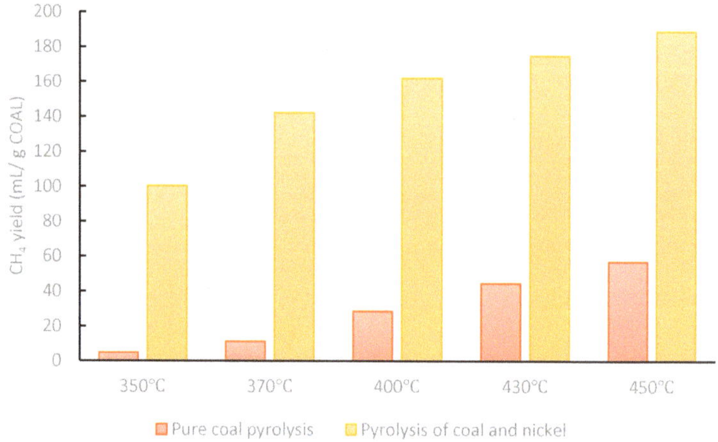

Figure 2. Variation diagram of pyrolysis yield of increasing Ni (Adapted from [80]).

He et al. [81] investigated the impact of iron-containing minerals, specifically $FeCO_3$, FeS_2, and Fe_2O_3, on the pyrolysis of 1-methylnaphthalene under high pressure. They explored different catalytic mechanisms by utilizing isotope differences at different temperature ranges. The coexistence of water and minerals greatly increased the ratio of i-C4/n-C_4 isomers, which is consistent with the pyrolysis trend observed in water-bearing minerals by Cai et al. [82]. The reactions at different mineral interfaces also varied. The presence of FeS_2 and Fe_2O_3 minerals promoted the reaction between water and hydrocarbons, leading to a significant increase in CH_4 yield. In contrast, the experimental group containing $FeCO_3$ showed little difference in CH_4 yield but exhibited positive carbon isotope. The authors hypothesized that the FTT reaction occurs at the interface between minerals and hydrocarbons at higher temperatures (400–450 °C), contributing to abiotic CH_4 production.

The catalytic action of transition metals in the pyrolysis process is related to their chemical forms. For instance, in the low-temperature stage, Fe^{3+} combines with oxygen-containing functional groups in coal or π bonds in unsaturated hydrocarbons using its own d empty orbital, resulting in chemical adsorption. The cations in coal have a positive effect on the organic structure of coal, making the coal molecules more polar and increasing their pyrolysis activity. As the temperature increases, the form of iron changes into crystalline oxide, which can catalyze the transformation of heavy components into primary pyrolysis products, reducing tar yield and increasing the gaseous yield [82]. However, He et al. [83] argued that transition metals, including iron, have a catalytic effect only in the elemental state, which can explain the differences observed in the experiment by Lewan et al. [79]. Guan et al. [84] reported that the different occurrence states of the same element have a significant influence on the results of thermal simulation experiments. In thermal simulation experiments conducted at low temperatures for an extended period (150 °C, 30 days), the Mn solution promoted crude oil yield, whereas Mn elemental substance showed obvious inhibition. Interestingly, the final crude oil yield of low-concentration Mn solution was 18.19 kg/t TOC, much lower than that of high-concentration Mn solution, which yielded 5.4 kg/t TOC (Figure 3).

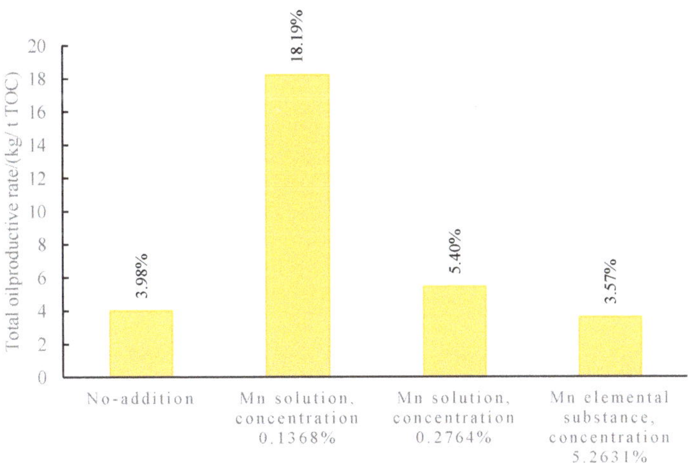

Figure 3. Comparison diagram of different forms of Mn element addition on oil production rate in thermal simulation experiment (adapted from [84]).

3. Thermal Simulation Process Promoted by Catalytic Hydrogenation of Minerals

Hydrogen plays a crucial role in the transformation of organic matter, as it not only provides the material basis for kerogen but also enhances the hydrogenation reaction during oil and gas generation and promotes the thermal evolution of source rocks. Specifically, hydrogen atoms can combine with carbon free radicals during the thermal simulation process, leading to the formation of smaller molecular hydrocarbons and ultimately increasing the final hydrocarbon yield [85,86]. Surprisingly, even a highly evolved solid pitch can generate hydrocarbons again under the influence of hydrogen [87].

Several sources of exogenous hydrogen are present in sedimentary basins, including water in sedimentary environments, hydrogen-containing fluids resulting from volcanic hydrothermal activity in the mantle, serpentine in ultrabasic volcanic rocks, and enrichment and radiation of radioactive minerals.

3.1. *Water Is an Important Source of Hydrogen for Hydrocarbon in Natural Basins*

Water has been reported to increase the production of gaseous hydrocarbons in pyrolysis experiments [88,89]. This is because water undergoes a disproportionation reaction during pyrolysis reaction, releasing hydrogen that promotes the cracking of kerogen oil or crude oil. This reduces cross-linking reactions and promotes thermal cracking, thereby altering the original pyrolysis law [90].

Water is transported to the deep part of the earth in the subduction zone of the subducting plate [91], where it reacts with iron under different temperature and pressure conditions to form a unique high-magnesium iron hydride [92,93]. Finally, dehydrogenation takes place and hydrogen-rich fluids in the mantle return to the surface through volcanic hydrothermal activity [94,95]. Meanwhile, hydrogen plays various roles in the Earth's C cycle [96–99]. Mantle-derived gas also has a profound impact on the transformation of organic matter and the formation of hydrocarbons. The hydrocarbons that may be affected by mantle-derived gas can be determined by the difference in hydrogen isotopes [95,100–103].

3.2. *The Role of Hydrogen Generation from Serpentine in Pyrolysis*

During the early stages of Earth's evolution, catalytic metals carried up from the mantle by volcanic activity were believed to play a significant role in the formation of primitive life and small organic molecules(Table 1). Transition metals such as iron, manganese, cobalt, zinc, and vanadium were transported into the ancient ocean via hydrothermal vents. Similar to their modern-day counterparts, the hydrothermal vents in the deep sea support a variety of complex biological and chemical processes, including sulfate reduction, iron oxidation and reduction, serpentine formation, and methane production, which contribute to the largest primary productivity in the deep ocean [104–108].

Table 1. Primary metabolic reactions of various microorganisms near hydrothermal solution (cited from Martin et al. [76]).

	Metabolism	Reaction	Examples in Vent Environments
Anaerobic	Methanogenesis	$4H_2 + CO_2 \rightarrow CH_4 + 2H_2O$ $CH_3CO_2^- + H_2O \rightarrow CH_4 + HCO_3^-$ $4HCOO^- + H^+ \rightarrow 3HCO_3^- + CH_4$	*Methanococcus* spp. common in magma-hosted vents; Methanosarcinales at Lost City
	S° reduction	$S^0 + H_2 \rightarrow H_2S$	Lithotrophic and heterotrophic; hyperthermophilic archaea
	Anaerobic CH4 oxidation	$CH_4 + SO_4^{2-} \rightarrow HS^- + HCO_3^- + 2H_2O$	*Methanosarcina* spp. and epsilonproteobacteria at mudvolcanoes and methane seeps
	Sulfate reduction	$SO_4^{2-} + H^+ + 4H_2 \rightarrow HS^- + 4H_2O$	Deltaproteobacteria
	Fe reduction	$8Fe^{3+} + CH_3CO_2^- + 4H_2O \rightarrow 2HCO_3^- + 8Fe^{2+} + 9H^+$	Epsilonproteobacteria, thermophilic bacteria and hyperthermophilic Crenarchaeota
	Fermentation	$C_6H_{12}O_6 \rightarrow 2C_6H_6O + 2CO_2$	Many genera of bacteria and archaea
Aerobic	Sulfide oxidation	$HS^- + 2O_2 \rightarrow SO_4^{2-} + H^+$	Many genera of bacteria; common vent animal symbionts
	CH4 oxidation	$CH_4 + 2O_2 \rightarrow HCO_3^- + H^+ + H_2O$	Common in hydrothermal systems; vent animal symbionts
	H2 oxidation	$H_2 + 0.5O_2 \rightarrow H_2O$	Common in hydrothermal systems; vent animal symbionts
	Fe oxidation	$Fe^{2+} + 0.5O_2 + H^+ \rightarrow Fe^{3+} + 0.5H_2O$	Common in low-temperature vent fluids; rock-hosted microbial mats
	Mn oxidation	$Mn^{2+} + 0.5O_2 + H_2O \rightarrow MnO_2 + 2H^+$	Common in low-temperature vent fluids; rock-hosted microbial mats; hydrothermal plumes
	Respiration	$C_6H_{12}O_6 + 6O_2 \rightarrow 6CO_2 + 6H_2O$	Many genera of bacteria

The process of igneous oxidation of olivine and pyroxene in the lower oceanic crust and upper mantle of the earth is referred to as serpentinization. In the presence of water, the Fe(II) present in these minerals is oxidized by Fe(III), leading to the precipitation of Fe_3O_4 and the release of diatomic H_2, which can react with carbon to produce methane (as shown in Formula (2)). This process generates a substantial amount of heat and significantly affects the pH around the minerals, resulting in the deposition of carbonate rocks that are widespread in the hydrothermal vents of modern mid-ocean ridges.

$$(Fe,Mg)_2SiO_4 + H_2O + CO_2 \rightarrow Mg_3Si_2O_5(OH)_4 + Fe_3O_4 + CH_4 \quad (2)$$

Hydrocarbons, primarily C_1–C_4 and formic acid, are the most commonly detected compounds in modern hydrothermal activities on the ocean floor. Although only a small portion of these compounds are believed to be produced by the thermal cracking of other hydrocarbons, recent studies have detected some macromolecular hydrocarbons and lipid compounds with the participation of microorganisms near the deep sea [109,110]. However, the source of macromolecular compounds near hydrothermal solutions in the marine environment is still controversial [111], and man-made organic pollution is one possible factor. It is also challenging to distinguish the contribution of organic and inorganic actions to hydrocarbons. Therefore, only a small portion of CH_4 is currently considered to be generated by the cracking of other hydrocarbons [112].

When organic matter is affected by volcanoes, it can also be impacted [113–115]. Different stages of volcanic hydrothermal activity in sedimentary basins have different catalytic behaviors on hydrocarbon behavior. The upward movement of magma is selective, and without other favorable channels, sedimentary basins that lack rigidity are more susceptible to invasion [116]. Once organic matter accumulates, the heat brought by the magmatic hydrothermal solution accelerates its maturity, promotes the thermal metamorphism of mudstone and sandstone, squeezes clastic material, and produces structural cracks. If this effect occurs during the sedimentary period, it can change the original sedimentary environment [117]. Metal minerals and other inorganic gases with catalytic properties brought by hydrothermal solutions will provide all the necessary elements for primary productivity, resulting in explosive growth of aquatic organisms. The addition of volcanic ash leads to water stratification, which enhances the reduction and preservation conditions of organic matter in the basin, improves the primary biological productivity of the basin, and easily forms high-quality source rocks, as shown in Figure 4 [84].

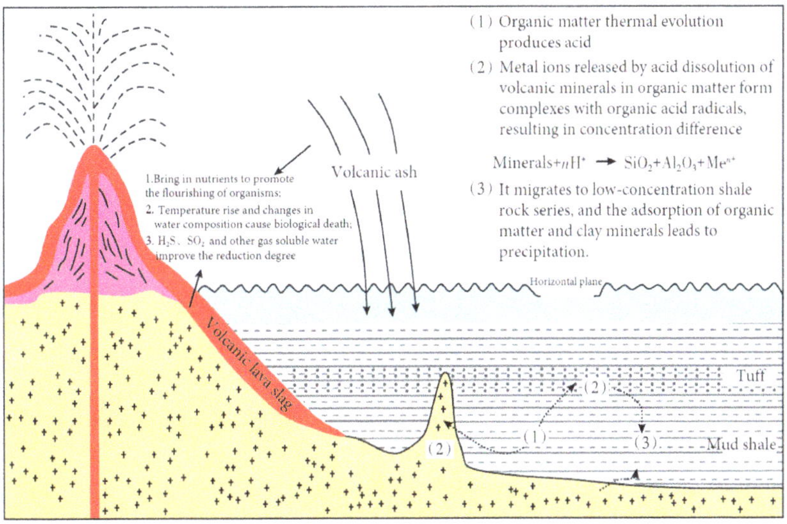

Figure 4. Influence model of volcanic activity on sedimentary basins (Ref. [85]).

3.3. The Role of Radioactive Hydrogen Production from Uranium in Pyrolysis

Uranium, a special transition metal element, is the cornerstone of the nuclear industry and has garnered significant attention for its potential in energy transformation [118]. Geologists are particularly interested in the unique sandstone-type pitchblende and its associated relationships and formation mechanisms. This phenomenon has been reported in most sedimentary basins worldwide [119–124].

In general, the deposition of uranium in sedimentary basins is closely related to the transportation, reduction, enrichment, and preservation of organic matter (Figure 5) [119,125,126]. Uranium is highly reactive and can react with most elements, especially in terms of its strong reduction ability. Under reduction conditions, U(VI) is converted to an immobile tetravalent state, U(IV) (see Formula (3)), which predominantly occurs in shale that is rich in organic matter [127]. The reduction and mineralization of uranium may occur simultaneously with the bacterial sulfate reduction in crude oil [128,129].

$$UO_2{}^{2+} + CH_4 + 3H_2O \rightarrow 4UO_2 + HCO_3^- + 9H^+ \quad (3)$$

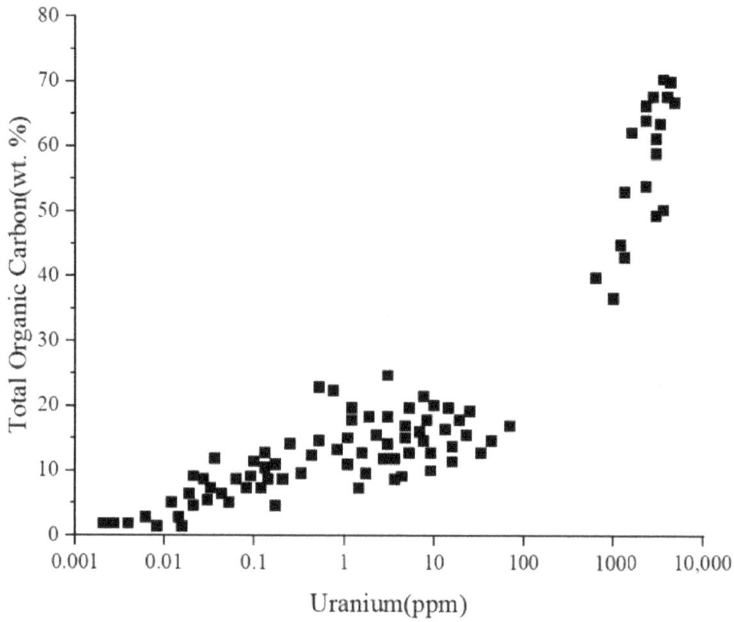

Figure 5. Correlation between uranium and TOC (quoted from Schulz et al. [125]).

When uranium is enriched to a certain extent, high radioactive anomalies can also be found in the formation. The decay of uranium can release α, β, and γ rays, and the particles in these rays can decompose water to provide ${}_{1}^{3}H_{4}$ for hydrogen and oxygen, resulting in the generation of active hydrogen. The radioactive decay and the provision of hydrogen are also considered to promote the pyrolysis process of organic matter in natural strata. The reaction process is as follows:

$$H_2O + (\alpha, \gamma) \rightarrow H\cdot + OH\cdot \tag{4}$$

$$2OH\cdot \rightarrow HOOH \tag{5}$$

$$2HOOH \rightarrow 2H_2O + O_2 \uparrow \tag{6}$$

$$2H\cdot \rightarrow H_2 \uparrow \tag{7}$$

Uranium enrichment can result in high radioactive anomalies in formations, releasing α, β, and γ rays that can decompose water to generate foreign hydrogen and promote the pyrolysis process of organic matter in natural strata. This foreign hydrogen compensates for the hydrogen required for the thermal degradation of organic matter and enhances the methane yield and total hydrocarbon yield [86]. In sedimentary basins, compounds can reach metastable thermodynamic equilibrium during hydrocarbon generation, leading to the formation of olefin-alkane equilibrium when the H concentration is sufficient [130]. Uranium ore can catalyze the pyrolysis of immature kerogen II, as discussed by Cai et al. [131], with liquid and gas products increasing significantly during gold tube closed-system pyrolysis. Deuterium in hydrocarbon gas was consumed, indicating that water and uranium ore can participate in the thermal evolution of organic matter and provide a hydrogen source for hydrocarbon generation. However, under this condition, the carbocation mechanism is inhibited to some extent, as the i-C4/n-C4 ratio in the experimental group containing uranium ore and water is low.

According to Yang et al. [132], the catalysis of uranium can cause an exponential change in kerogen structure. With an increase in uranium content in kerogen, the gas/oil ratio and aromatic hydrocarbon content of pyrolysis products continuously increase compared to the hydrocarbon generation law of conventional marine shale. FT-ICR-MS results show that macromolecular compounds in uranium-rich samples are more alkylated and oxygenated compounds are more abundant, which can be attributed to free radicals generated by the in situ reaction of highly active OH produced by the radioactive cracking of water with organic matter during diagenesis. However, the more complex cross-linked structure of uranium-rich samples can hinder further cracking.

4. Conclusions and Outlook

Previous studies have shown that different minerals exhibit different forms of catalysis in laboratory-scale experiments. However, detailed consideration of the geological background and the corresponding design of thermal simulation experiments remain a primary concern, particularly with respect to shale and deep carbonate rocks that are currently the focus of research in China. Future research should focus on the development and application of combined measuring equipment, such as TG-IR, TG-MS, and TG-XRD, which can more accurately reflect the catalytic process and mechanism by monitoring changes in molecular markers, minerals, and parameters during the experimental process. Furthermore, with the continuous development of computational chemistry, there is a need to explore more accurate and precise modeling of thermal simulation experiments on computers. Numerical simulations can easily change experimental parameters and provide a more intuitive molecular reaction mechanism. They are also more environmentally friendly and faster than thermal simulation experiments.

Author Contributions: Conceptualization and funding acquisition, H.H.; methodology and writing—review and editing, M.Z.; data curation, P.X.; writing—original draft preparation, visualization, J.H. All authors have read and agreed to the published version of the manuscript.

Funding: This work was supported by National Natural Science Foundation of China (Grant number 41873049).

Data Availability Statement: Data available on request from the authors.

Conflicts of Interest: We declare that we have no known competitive financial interests or personal relationships.

References

1. Hunt, J.M. Distribution of carbon in crust of earth. *AAPG Bull.* **1972**, *56*, 2273–2277.
2. Han, Q.; Li, M.; Liu, X.; Xiao, H.; Ren, J.; Guo, C. A maturation scale for molecular simulation of kerogen thermal degradation. *Org. Geochem.* **2023**, *175*, 104507. [CrossRef]
3. Wang, Z.; Tang, Y.; Wang, Y.; Zheng, Y.; Chen, F.; Wu, S.; Fu, D. Kinetics of shale oil generation from kerogen in saline basin and its exploration significance: An example from the Eocene Qianjiang Formation, Jianghan Basin, China. *J. Anal. Appl. Pyrolysis* **2020**, *150*, 104885. [CrossRef]
4. Lai, D.; Zhan, J.-H.; Tian, Y.; Gao, S.; Xu, G. Mechanism of kerogen pyrolysis in terms of chemical structure transformation. *Fuel* **2017**, *199*, 504–511. [CrossRef]
5. Spigolon, A.L.; Lewan, M.D.; de Barros Penteado, H.L.; Coutinho, L.F.C.; Mendonça Filho, J.G. Evaluation of the petroleum composition and quality with increasing thermal maturity as simulated by hydrous pyrolysis: A case study using a Brazilian source rock with Type I kerogen. *Org. Geochem.* **2015**, *83*, 27–53. [CrossRef]
6. Tian, H.; Wang, Z.; Xiao, Z.; Li, X.; Xiao, X. Oil cracking to gases: Kinetic modeling and geological significance. *Chin. Sci. Bull.* **2006**, *51*, 2763–2770. [CrossRef]
7. Lu, Y.; Wang, Y.; Zhang, J.; Wang, Q.; Zhao, Y.; Zhang, Y. Investigation on the characteristics of pyrolysates during co-pyrolysis of Zhundong coal and Changji oil shale and its kinetics. *Energy* **2020**, *200*, 117529. [CrossRef]
8. Behar, F.; Vandenbroucke, M.; Tang, Y.; Marquis, F.; Espitalie, J. Thermal cracking of kerogen in open and closed systems: Determination of kinetic parameters and stoichiometric coefficients for oil and gas generation. *Org. Geochem.* **1997**, *26*, 321–339. [CrossRef]

9. Shao, D.; Ellis, G.S.; Li, Y.; Zhang, T. Experimental investigation of the role of rock fabric in gas generation and expulsion during thermal maturation: Anhydrous closed-system pyrolysis of a bitumen-rich Eagle Ford Shale. *Org. Geochem.* **2018**, *119*, 22–35. [CrossRef]
10. Zhao, Z.; Feng, Q.; Liu, X.; Lu, H.; Peng, P.A.; Liu, J.; Hsu, C.S. Petroleum maturation processes simulated by high-pressure pyrolysis and kinetic modeling of low-maturity type I Kerogen. *Energy Fuels* **2022**, *36*, 1882–1893. [CrossRef]
11. Nelson, P.F.; Smith, I.W.; Tyler, R.J.; Mackie, J.C. Pyrolysis of coal at high temperatures. *Energy Fuels* **1988**, *2*, 391–400. [CrossRef]
12. Sathe, C.; Pang, Y.; Li, C.-Z. Effects of heating rate and ion-exchangeable cations on the pyrolysis yields from a Victorian brown coal. *Energy Fuels* **1999**, *13*, 748–755. [CrossRef]
13. Chen, J.; Zhang, X.; Chen, Z.; Pang, X.; Yang, H.; Zhao, Z.; Pang, B.; Ma, K. Hydrocarbon expulsion evaluation based on pyrolysis Rock-Eval data: Implications for Ordovician carbonates exploration in the Tabei Uplift, Tarim. *J. Pet. Sci. Eng.* **2021**, *196*, 107614. [CrossRef]
14. Peng, W.; Hu, G.; Liu, Q.; Jia, N.; Fang, C.; Gong, D.; Yu, C.; Lyu, Y.; Wang, P.; Feng, Z. Research status on thermal simulation experiment and several issues of concern. *J. Nat. Gas Geosci.* **2018**, *3*, 283–293. [CrossRef]
15. Ma, W.; Hou, L.; Luo, X.; Liu, J.; Tao, S.; Guan, P.; Cai, Y. Generation and expulsion process of the Chang 7 oil shale in the Ordos Basin based on temperature-based semi-open pyrolysis: Implications for in-situ conversion process. *J. Pet. Sci. Eng.* **2020**, *190*, 107035. [CrossRef]
16. Merriman, R.J. Clay minerals and sedimentary basin history. *Eur. J. Mineral.* **2005**, *17*, 7–20. [CrossRef]
17. Sposito, G.; Skipper, N.T.; Sutton, R.; Park, S.-H.; Soper, A.K.; Greathouse, J.A. Surface geochemistry of the clay minerals. *Proc. Natl. Acad. Sci. USA* **1999**, *96*, 3358–3364. [CrossRef] [PubMed]
18. Wilson, M.; Wilson, L.; Patey, I.; Shaw, H. The influence of individual clay minerals on formation damage of reservoir sandstones: A critical review with some new insights. *Clay Miner.* **2014**, *49*, 147–164. [CrossRef]
19. Chen, S.; Han, Y.; Fu, C.; Zhu, Y.; Zuo, Z. Micro and nano-size pores of clay minerals in shale reservoirs: Implication for the accumulation of shale gas. *Sediment. Geol.* **2016**, *342*, 180–190. [CrossRef]
20. Thomas, C.L.; Hickey, J.; Stecker, G. Chemistry of clay cracking catalysts. *Ind. Eng. Chem.* **1950**, *42*, 866–871. [CrossRef]
21. Eltantawy, I.; Arnold, P. Catalytic decomposition of organic molecules by clays. *Nat. Phys. Sci.* **1973**, *244*, 144. [CrossRef]
22. Shimoyama, A.; Johns, W.D. Catalytic conversion of fatty acids to petroleum-like paraffins and their maturation. *Nat. Phys. Sci.* **1971**, *232*, 140–144. [CrossRef]
23. Huizinga, B.J.; Tannenbaum, E.; Kaplan, I.R. The role of minerals in the thermal alteration of organic matter—IV. Generation of n-alkanes, acyclic isoprenoids, and alkenes in laboratory experiments. *Geochim. Et Cosmochim. Acta* **1987**, *51*, 1083–1097. [CrossRef] [PubMed]
24. Yuan, P.; Liu, H.; Liu, D.; Tan, D.; Yan, W.; He, H. Role of the interlayer space of montmorillonite in hydrocarbon generation: An experimental study based on high temperature–pressure pyrolysis. *Appl. Clay Sci.* **2013**, *75*, 82–91. [CrossRef]
25. Liu, H.; Yuan, P.; Liu, D.; Bu, H.; Song, H.; Qin, Z.; He, H. Pyrolysis behaviors of organic matter (OM) with the same alkyl main chain but different functional groups in the presence of clay minerals. *Appl. Clay Sci.* **2018**, *153*, 205–216. [CrossRef]
26. Bu, H.; Yuan, P.; Liu, H.; Liu, D.; Liu, J.; He, H.; Zhou, J.; Song, H.; Li, Z. Effects of complexation between organic matter (OM) and clay mineral on OM pyrolysis. *Geochim. Et Cosmochim. Acta* **2017**, *212*, 1–15. [CrossRef]
27. Ma, X.; Zheng, G.; Sajjad, W.; Xu, W.; Fan, Q.; Zheng, J.; Xia, Y. Influence of minerals and iron on natural gases generation during pyrolysis of type-III kerogen. *Mar. Pet. Geol.* **2018**, *89*, 216–224. [CrossRef]
28. Song, H.; Liu, H.; Bu, H.; Liu, D.; Li, Y.; Du, P. Effects of montmorillonite charge reduction on the high-temperature/high-pressure pyrolysis of organic matter. *Appl. Clay Sci.* **2021**, *213*, 106243. [CrossRef]
29. Espitalie, J.; Madec, M.; Tissot, B. Role of mineral matrix in kerogen pyrolysis: Influence on petroleum generation and migration. *AAPG Bull.* **1980**, *64*, 59–66.
30. Horsfield, B.; Douglas, A. The influence of minerals on the pyrolysis of kerogens. *Geochim. Et Cosmochim. Acta* **1980**, *44*, 1119–1131. [CrossRef]
31. Wu, L.M.; Zhou, C.H.; Keeling, J.; Tong, D.S.; Yu, W.H. Towards an understanding of the role of clay minerals in crude oil formation, migration and accumulation. *Earth-Sci. Rev.* **2012**, *115*, 373–386. [CrossRef]
32. Thompson, K.F. Mechanisms controlling gas and light end composition in pyrolysates and petroleum: Applications in the interpretation of reservoir fluid analyses. *Org. Geochem.* **2006**, *37*, 798–817. [CrossRef]
33. Franklin, H.D.; Peters, W.A.; Howard, J.B. Mineral matter effects on the rapid pyrolysis and hydropyrolysis of a bituminous coal. 1. Effects on yields of char, tar and light gaseous volatiles. *Fuel* **1982**, *61*, 155–160. [CrossRef]
34. Espitalié, J.; Makadi, K.S.; Trichet, J. Role of the mineral matrix during kerogen pyrolysis. *Org. Geochem.* **1984**, *6*, 365–382. [CrossRef]
35. Roets, L.; Bunt, J.R.; Neomagus, H.W.; Strydom, C.A.; Van Niekerk, D. The effect of added minerals on the pyrolysis products derived from a vitrinite-rich demineralised South African coal. *J. Anal. Appl. Pyrolysis* **2016**, *121*, 41–49. [CrossRef]
36. Öztaş, N.; Yürüm, Y. Pyrolysis of Turkish Zonguldak bituminous coal. Part 1. Effect of mineral matter. *Fuel* **2000**, *79*, 1221–1227. [CrossRef]
37. He, M.; Wang, Z.; Moldowan, M.J.; Peters, K. Insights into catalytic effects of clay minerals on hydrocarbon composition of generated liquid products during oil cracking from laboratory pyrolysis experiments. *Org. Geochem.* **2022**, *163*, 104331. [CrossRef]

38. Cai, J.; Du, J.; Song, M.; Lei, T.; Wang, X.; Li, Y. Control of clay mineral properties on hydrocarbon generation of organo-clay complexes: Evidence from high-temperature pyrolysis experiments. *Appl. Clay Sci.* **2022**, *216*, 106368. [CrossRef]
39. Linares-Solano, A.; Hippo, E.J.; Walker, P.L., Jr. Catalytic activity of calcium for lignite char gasification in various atmospheres. *Fuel* **1986**, *65*, 776–779. [CrossRef]
40. Skodras, G.; Sakellaropoulos, G. Mineral matter effects in lignite gasification. *Fuel Process. Technol.* **2002**, *77*, 151–158. [CrossRef]
41. Walker, P.L., Jr. Catalysis of lignite char gasification by exchangeable calcium and magnesium. *Fuel* **1984**, *63*, 1214–1220.
42. Li, C.-Z.; Sathe, C.; Kershaw, J.R.; Pang, Y. Fates and roles of alkali and alkaline earth metals during the pyrolysis of a Victorian brown coal. *Fuel* **2000**, *79*, 427–438. [CrossRef]
43. Qiao, L.; Deng, C.; Lu, B.; Wang, Y.; Wang, X.; Deng, H.; Zhang, X. Study on calcium catalyzes coal spontaneous combustion. *Fuel* **2022**, *307*, 121884. [CrossRef]
44. Karabakan, A.; Yürüm, Y. Effect of the mineral matrix in the reactions of oil shales: 1. Pyrolysis reactions of Turkish Göynük and US Green River oil shales. *Fuel* **1998**, *77*, 1303–1309. [CrossRef]
45. Ballice, L. Stepwise chemical demineralization of Göynük (Turkey) oil shale and pyrolysis of demineralization products. *Ind. Eng. Chem. Res.* **2006**, *45*, 906–912. [CrossRef]
46. Trice, R. Challenges and insights in optimizing oil production from Middle East mega karst reservoirs. In Proceedings of the SPE Middle East Oil and Gas Show and Conference, Manama, Bahrain, 12 March 2005; Volume 3.
47. Weber, J.; Cheshire, M.C.; Bleuel, M.; Mildner, D.; Chang, Y.-J.; Ievlev, A.; Littrell, K.C.; Ilavsky, J.; Stack, A.G.; Anovitz, L.M. Influence of microstructure on replacement and porosity generation during experimental dolomitization of limestones. *Geochim. Et Cosmochim. Acta* **2021**, *303*, 137–158. [CrossRef]
48. Ibrahem, Y.; Morozov, V.; Sudakov, V.; Idrisov, I.; Kolchugin, A. Sedimentary diagenesis and pore characteristics for the reservoir evaluation of Domanik formations (Semiluksk and Mendymsk) in the central part of Volga-Ural petroleum province. *Pet. Res.* **2022**, *7*, 32–46. [CrossRef]
49. Gürgey, K. An attempt to recognise oil populations and potential source rock types in Paleozoic sub-and Mesozoic-Cenozoic supra-salt strata in the southern margin of the Pre-Caspian Basin, Kazakhstan Republic. *Org. Geochem.* **2002**, *33*, 723–741. [CrossRef]
50. Klemme, H.; Ulmishek, G.F. Effective petroleum source rocks of the world: Stratigraphic distribution and controlling depositional factors. *AAPG Bull.* **1991**, *75*, 1809–1851.
51. Liu, W.; Xu, Y.; Chen, J. Comprehensive geochemical identification of highly evolved marine carbonate rocks as hydrocarbon-source rocks as exemplified by the Ordos Basin. *Sci. China Ser. D* **2006**, *49*, 384–396.
52. Zhu, G.; Yan, H.; Chen, W.; Yan, L.; Zhang, K.; Li, T.; Chen, Z.; Wu, G.; Santosh, M. Discovery of Cryogenian interglacial source rocks in the northern Tarim, NW China: Implications for Neoproterozoic paleoclimatic reconstructions and hydrocarbon exploration. *Gondwana Res.* **2020**, *80*, 370–384. [CrossRef]
53. Morad, S. Carbonate cementation in sandstones: Distribution patterns and geochemical evolution. In *Carbonate Cementation in Sandstones: Distribution Patterns and Geochemical Evolution*; Wiley-Blackwell: Oxford, UK, 1998; pp. 1–26.
54. Hunt, J.M. The Origin of Petroleum in Carbonate Rocksk1. In *Developments in Sedimentology*; Elsevier: Amsterdam, The Netherlands, 1967; Volume 9, pp. 225–251.
55. Daly, A.R.; Edman, J.D. *EALoss of Organic Carbon from Source Rocks during Thermal Maturation*; AAPG Bulletin: Tulsa, OK, USA, 1987.
56. Xia, L.-W.; Cao, J.; Wang, M.; Mi, J.-L.; Wang, T.-T. A review of carbonates as hydrocarbon source rocks: Basic geochemistry and oil–gas generation. *Pet. Sci.* **2019**, *16*, 713–728. [CrossRef]
57. Tissot, B.; du Petrole, E.N.S.; Welte, D. Petroleum Formation and Occurrence. In *A New Approach to Oil and Gas Exploration*; Springer-Verlag: Berlin, Germany, 1978.
58. Huo, Z.; Pang, X.; Chen, J.; Zhang, J.; Song, M.; Guo, K.; Li, P.; Li, W.; Liang, Y. Carbonate source rock with low total organic carbon content and high maturity as effective source rock in China: A review. *J. Asian Earth Sci.* **2019**, *176*, 8–26. [CrossRef]
59. Shimoyama, A.; Johns, W.D. Formation of alkanes from fatty acids in the presence of CaCO3. *Geochim. Et Cosmochim. Acta* **1972**, *36*, 87–91. [CrossRef]
60. Subramanian, S.; Sørland, G.H.; Simon, S.; Xu, Z.; Sjöblom, J. Asphaltene fractionation based on adsorption onto calcium carbonate: Part 2. Self-association and aggregation properties. *Colloids Surf. A Physicochem. Eng. Asp.* **2017**, *514*, 79–90. [CrossRef]
61. Tannenbaum, E.; Aizenshtat, Z. Formation of immature asphalt from organic-rich carbonate rocks—I. Geochemical correlation. *Org. Geochem.* **1985**, *8*, 181–192. [CrossRef]
62. Jones, R. *Comparison of Carbonate and Shale Source Rocks*; AAPG Bulletin: Tulsa, OK, USA, 1984.
63. Pan, C.; Jiang, L.; Liu, J.; Zhang, S.; Zhu, G. The effects of calcite and montmorillonite on oil cracking in confined pyrolysis experiments. *Org. Geochem.* **2010**, *41*, 611–626. [CrossRef]
64. Chen, Z.; Chai, Z.; Cheng, B.; Liu, H.; Cao, Y.; Cao, Z.; Qu, J. Geochemistry of high-maturity crude oil and gas from deep reservoirs and their geological significance: A case study on Shuntuoguole low uplift, Tarim Basin, western China. *AAPG Bull.* **2021**, *105*, 65–107. [CrossRef]
65. Lanyu, W.; Dongfeng, H.; Yongchao, L.; Ruobing, L.; Xiaofeng, L. Advantageous shale lithofacies of Wufeng Formation-Longmaxi Formation in Fuling gas field of Sichuan Basin, SW China. *Pet. Explor. Dev.* **2016**, *43*, 208–217.
66. Islam, M.W. A review of dolomite catalyst for biomass gasification tar removal. *Fuel* **2020**, *267*, 117095. [CrossRef]

67. Wu, L.; Fang, X.; Geng, A. Influence of carbonate minerals on the pyrolysis and phase behavior of oil. *Mar. Pet. Geol.* **2021**, *134*, 105374. [CrossRef]
68. Machel, H.G.; Krouse, H.R.; Sassen, R. Products and distinguishing criteria of bacterial and thermochemical sulfate reduction. *Appl. Geochem.* **1995**, *10*, 373–389. [CrossRef]
69. Shand, M.A. *The Chemistry and Technology of Magnesia*; John Wiley & Sons: Hoboken, NJ, USA, 2006.
70. Maimaiti, A.; Wang, Q.; Hao, F.; Yang, X.; Zhang, H.; Tian, J. Oil pyrolysis with carbonate minerals: Implications for the thermal stability of deep crude oil. *Mar. Pet. Geol.* **2022**, *146*, 105929. [CrossRef]
71. Wang, W.; Ji, L.; Song, D.; Zhang, D.; Lü, C.; Su, L. Origin of inorganic carbon dioxide associated with hydrocarbon generation: Evidence from hydrous pyrolysis experiments and natural and shale gases. *J. Asian Earth Sci. X* **2022**, *7*, 100079. [CrossRef]
72. Bullock, R.M.; Chen, J.G.; Gagliardi, L.; Chirik, P.J.; Farha, O.K.; Hendon, C.H.; Jones, C.W.; Keith, J.A.; Klosin, J.; Minteer, S.D. Using nature's blueprint to expand catalysis with Earth-abundant metals. *Science* **2020**, *369*, eabc3183. [CrossRef]
73. Curiale, J. Distribution and Occurrence of Metals in Heavy Crude Oils and Solid Bitumens–Implications for Petroleum Exploration: Section II. In *Characterization, Maturation, and Degradation*; AAPG Bulletin: Tulsa, OK, USA, 1987.
74. Treibs, A. Chlorophyll-und Häminderivate in bituminösen Gesteinen, Erdölen, Erdwachsen und Asphalten. Ein Beitrag zur Entstehung des Erdöls. *Justus Liebigs Ann. Der Chem.* **1934**, *510*, 42–62. [CrossRef]
75. Mango, F.D. Transition metal catalysis in the generation of petroleum: A genetic anomaly in Ordovician oils. *Geochim. Et Cosmochim. Acta* **1992**, *56*, 3851–3854. [CrossRef]
76. Martin, W.; Baross, J.; Kelley, D.; Russell, M.J. Hydrothermal vents and the origin of life. *Nat. Rev. Microbiol.* **2008**, *6*, 805–814. [CrossRef]
77. Miller, D.M.; Buettner, G.R.; Aust, S.D. Transition metals as catalysts of "autoxidation" reactions. *Free. Radic. Biol. Med.* **1990**, *8*, 95–108. [CrossRef]
78. Yan, J.; Jiang, X.; Han, X.; Liu, J. A TG–FTIR investigation to the catalytic effect of mineral matrix in oil shale on the pyrolysis and combustion of kerogen. *Fuel* **2013**, *104*, 307–317. [CrossRef]
79. Lewan, M.; Kotarba, M.; Więcław, D.; Piestrzyński, A. Evaluating transition-metal catalysis in gas generation from the Permian Kupferschiefer by hydrous pyrolysis. *Geochim. Et Cosmochim. Acta* **2008**, *72*, 4069–4093. [CrossRef]
80. Gao, J.; Zou, C.; Li, W.; Ni, Y.; Yuan, Y. Transition metal catalysis in natural gas generation: Evidence from nonhydrous pyrolysis experiment. *Mar. Pet. Geol.* **2020**, *115*, 104280. [CrossRef]
81. He, K.; Zhang, S.; Wang, X.; Mi, J.; Zhang, W.; Guo, J.; Zhang, W. Pyrolysis of 1-methylnaphthalene involving water: Effects of Fe-bearing minerals on the generation, C and H isotope fractionation of methane from H2O-hydrocarbon reaction. *Org. Geochem.* **2021**, *153*, 104151. [CrossRef]
82. He, L.; Li, S.; Lin, W. Catalytic cracking of pyrolytic vapors of low-rank coal over limonite ore. *Energy Fuels* **2016**, *30*, 6984–6990. [CrossRef]
83. He, L.; Hui, H.; Li, S.; Lin, W. Production of light aromatic hydrocarbons by catalytic cracking of coal pyrolysis vapors over natural iron ores. *Fuel* **2018**, *216*, 227–232. [CrossRef]
84. Guan, D.; Ma, Z.; Zhao, Y.; Wang, H.; Wang, Q.; Liu, Q. *Oil Generation Theory and Simulation Experiment Technology of Continental Source Rocks in China*; Petroleum Industry Press in China: Xi'an, China, 2022; pp. 199–245. (In Chinese)
85. Jin, Z.; Zhang, J. Exploration strategy for deep basin gas reservoirs. *Pet. Explor. Dev.* **1999**, *26*, 4–5.
86. Mao, G.; Liu, C.; Zhang, D.; Qiu, X.; Wang, J.; Liu, B.; Liu, J.; Qu, S.; Deng, Y.; Wang, F. Effects of uranium on hydrocarbon generation of hydrocarbon source rocks with type-III kerogen. *Sci. China Earth Sci.* **2014**, *57*, 1168–1179. [CrossRef]
87. Liao, Y.; Fang, Y.; Wu, L.; Geng, A.; Hsu, C.S. The characteristics of the biomarkers and δ13C of n-alkanes released from thermally altered solid bitumens at various maturities by catalytic hydropyrolysis. *Org. Geochem.* **2012**, *46*, 56–65. [CrossRef]
88. Waples, D.W. Time and temperature in petroleum formation: Application of Lopatin's method to petroleum exploration. *AAPG Bull.* **1980**, *64*, 916–926.
89. Lewan, M.D.; Roy, S. Role of water in hydrocarbon generation from Type-I kerogen in Mahogany oil shale of the Green River Formation. *Org. Geochem.* **2011**, *42*, 31–41. [CrossRef]
90. Shuai, Y.; Zhang, S.; Luo, P.; Liu, J.; Hu, G. Experimental evidence for formation water promoting crude oil cracking to gas. *Chin. Sci. Bull.* **2012**, *57*, 4587–4593. [CrossRef]
91. Wyllie, P. Plate tectonics and magma genesis. *Geol. Rundsch.* **1981**, *70*, 128–153. [CrossRef]
92. Mao, H.-K.; Hu, Q.; Yang, L.; Liu, J.; Kim, D.Y.; Meng, Y.; Zhang, L.; Prakapenka, V.B.; Yang, W.; Mao, W.L. When water meets iron at Earth's core–mantle boundary. *Natl. Sci. Rev.* **2017**, *4*, 870–878. [CrossRef]
93. Ohtani, E. The role of water in Earth's mantle. *Natl. Sci. Rev.* **2020**, *7*, 224–232. [CrossRef]
94. Jin, Z.; Zhang, L.; Yang, L.; Hu, W. A preliminary study of mantle-derived fluids and their effects on oil/gas generation in sedimentary basins. *J. Pet. Sci. Eng.* **2004**, *41*, 45–55. [CrossRef]
95. Bali, E.; Audétat, A.; Keppler, H. Water and hydrogen are immiscible in Earth's mantle. *Nature* **2013**, *495*, 220–222. [CrossRef] [PubMed]
96. Guélard, J.; Beaumont, V.; Rouchon, V.; Guyot, F.; Pillot, D.; Jézéquel, D.; Ader, M.; Newell, K.; Deville, E. Natural H2 in K ansas: Deep or shallow origin? *Geochem. Geophys. Geosystems* **2017**, *18*, 1841–1865. [CrossRef]
97. Holloway, J.R.; O'DAY, P.A. Production of CO2 and H2 by diking-eruptive events at mid-ocean ridges: Implications for abiotic organic synthesis and global geochemical cycling. *Int. Geol. Rev.* **2000**, *42*, 673–683. [CrossRef]

98. Zgonnik, V. The occurrence and geoscience of natural hydrogen: A comprehensive review. *Earth-Sci. Rev.* **2020**, *203*, 103140. [CrossRef]
99. Catling, D.C.; Zahnle, K.J.; McKay, C.P. Biogenic methane, hydrogen escape, and the irreversible oxidation of early Earth. *Science* **2001**, *293*, 839–843. [CrossRef]
100. Andersen, T.; Neumann, E.-R. Fluid inclusions in mantle xenoliths. *Lithos* **2001**, *55*, 301–320. [CrossRef]
101. Liu, X.; Fu, X.; Liu, D.; Wei, W.; Lu, X.; Liu, C.; Wang, W.; Gao, H. Distribution of mantle-derived CO2 gas reservoir and its relationship with basement faults in Songliao Basin, China. *J. Nat. Gas Sci. Eng.* **2018**, *56*, 593–607. [CrossRef]
102. Walter, B.F.; Giebel, R.J.; Steele-MacInnis, M.; Marks, M.A.; Kolb, J.; Markl, G. Fluids associated with carbonatitic magmatism: A critical review and implications for carbonatite magma ascent. *Earth-Sci. Rev.* **2021**, *215*, 103509. [CrossRef]
103. Hiett, C.D.; Newell, D.L.; Jessup, M.J.; Grambling, T.A.; Scott, B.E.; Upin, H.E. Deep CO2 and N2 emissions from Peruvian hot springs: Stable isotopic constraints on volatile cycling in a flat-slab subduction zone. *Chem. Geol.* **2022**, *595*, 120787. [CrossRef]
104. Jørgensen, B.B.; Isaksen, M.F.; Jannasch, H.W. Bacterial sulfate reduction above 100 C in deep-sea hydrothermal vent sediments. *Science* **1992**, *258*, 1756–1757. [CrossRef] [PubMed]
105. Fisher, A.T. Marine hydrogeology: Recent accomplishments and future opportunities. *Hydrogeol. J.* **2005**, *13*, 69–97. [CrossRef]
106. Russell, M. First life: Billions of years ago, deep under the ocean, the pores and pockets in minerals that surrounded warm, alkaline springs catalyzed the beginning of life. *Am. Sci.* **2006**, *94*, 32–39. [CrossRef]
107. Lartaud, F.; Little, C.T.; De Rafelis, M.; Bayon, G.; Dyment, J.; Ildefonse, B.; Gressier, V.; Fouquet, Y.; Gaill, F.; Le Bris, N. Fossil evidence for serpentinization fluids fueling chemosynthetic assemblages. *Proc. Natl. Acad. Sci. USA* **2011**, *108*, 7698–7703. [CrossRef] [PubMed]
108. Konn, C.; Charlou, J.-L.; Holm, N.G.; Mousis, O. The production of methane, hydrogen, and organic compounds in ultramafic-hosted hydrothermal vents of the Mid-Atlantic Ridge. *Astrobiology* **2015**, *15*, 381–399. [CrossRef]
109. Baross, J.A.; Hoffman, S.E. Submarine hydrothermal vents and associated gradient environments as sites for the origin and evolution of life. *Orig. Life Evol. Biosph.* **1985**, *15*, 327–345. [CrossRef]
110. Ménez, B.; Pisapia, C.; Andreani, M.; Jamme, F.; Vanbellingen, Q.P.; Brunelle, A.; Richard, L.; Dumas, P.; Réfrégiers, M. Abiotic synthesis of amino acids in the recesses of the oceanic lithosphere. *Nature* **2018**, *564*, 59–63. [CrossRef]
111. Reeves, E.P.; Fiebig, J. Abiotic synthesis of methane and organic compounds in Earth's lithosphere. *Elem. Int. Mag. Mineral. Geochem. Petrol.* **2020**, *16*, 25–31. [CrossRef]
112. Beaudry, P.; Stefánsson, A.; Fiebig, J.; Rhim, J.H.; Ono, S. High temperature generation and equilibration of methane in terrestrial geothermal systems: Evidence from clumped isotopologues. *Geochim. Et Cosmochim. Acta* **2021**, *309*, 209–234. [CrossRef]
113. Coveney, R.M., Jr.; Goebel, E.D.; Zeller, E.J.; Dreschhoff, G.A.; Angino, E.E. Serpentinization and the origin of hydrogen gas in Kansas. *AAPG Bull.* **1987**, *71*, 39–48.
114. Lollar, B.S.; Frape, S.; Weise, S.; Fritz, P.; Macko, S.; Welhan, J. Abiogenic methanogenesis in crystalline rocks. *Geochim. Et Cosmochim. Acta* **1993**, *57*, 5087–5097. [CrossRef]
115. Klein, F.; Grozeva, N.G.; Seewald, J.S. Abiotic methane synthesis and serpentinization in olivine-hosted fluid inclusions. *Proc. Natl. Acad. Sci. USA* **2019**, *116*, 17666–17672. [CrossRef]
116. Galland, O.; Holohan, E.; Van Wyk de Vries, B.; Burchardt, S. Laboratory modelling of volcano plumbing systems: A review. In *Physical Geology of Shallow Magmatic Systems: Dykes, Sills and Laccoliths*; Springer-Verlag: Berlin, Germany, 2018; pp. 147–214.
117. Zahedi, M.K.; MacDonald, S. Volcanics: A Commonly Underestimated part of Petroleum Exploration. *ASEG Ext. Abstr.* **2018**, *2018*, 1–6. [CrossRef]
118. Dincer, I. Green methods for hydrogen production. *Int. J. Hydrog. Energy* **2012**, *37*, 1954–1971. [CrossRef]
119. Landais, P. Organic geochemistry of sedimentary uranium ore deposits. *Ore Geol. Rev.* **1996**, *11*, 33–51. [CrossRef]
120. Fisher, W.L.; Galloway, W.; Proctor, C.V., Jr.; Nagle, J. *Depositional Systems in the Jackson Group of Texas their Relationship to Oil Gas, and Uranium*; The University of Texas at Austin, Bureau of Economic Geology: Austin, TX, USA, 1970.
121. Lecomte, A.; Michels, R.; Cathelineau, M.; Morlot, C.; Brouand, M.; Flotté, N. Uranium deposits of Franceville basin (Gabon): Role of organic matter and oil cracking on uranium mineralization. *Ore Geol. Rev.* **2020**, *123*, 103579. [CrossRef]
122. Gauthier-Lafaye, F.; Weber, F. Natural nuclear fission reactors: Time constraints for occurrence, and their relation to uranium and manganese deposits and to the evolution of the atmosphere. *Precambrian Res.* **2003**, *120*, 81–100. [CrossRef]
123. Robertson, J.; Hendry, M.J.; Kotzer, T.; Hughes, K.À. Geochemistry of uranium mill tailings in the Athabasca Basin, Saskatchewan, Canada: A review. *Crit. Rev. Environ. Sci. Technol.* **2019**, *49*, 1237–1293. [CrossRef]
124. Kang, H.; Chen, Y.-l.; Li, D.-p.; Zhao, J.-x.; Cui, F.-r.; Xu, Y.-l. Deep-penetrating geochemistry for concealed sandstone-type uranium deposits: A case study of Hadatu uranium deposit in the Erenhot Basin, North China. *J. Geochem. Explor.* **2020**, *211*, 106464. [CrossRef]
125. Schulz, H.-M.; Yang, S.; Schovsbo, N.H.; Rybacki, E.; Ghanizadeh, A.; Bernard, S.; Mahlstedt, N.; Krüger, M.; Amann-Hildebrandt, A.; Krooss, B.M. The Furongian to Lower Ordovician Alum Shale Formation in conventional and unconventional petroleum systems in the Baltic Basin–A review. *Earth-Sci. Rev.* **2021**, *218*, 103674. [CrossRef]
126. Nakashima, S.; Disnar, J.; Perruchot, A.; Trichet, J. Experimental study of mechanisms of fixation and reduction of uranium by sedimentary organic matter under diagenetic or hydrothermal conditions. *Geochim. Et Cosmochim. Acta* **1984**, *48*, 2321–2329. [CrossRef]

127. Partin, C.A.; Bekker, A.; Planavsky, N.J.; Scott, C.; Gill, B.C.; Li, C.; Podkovyrov, V.; Maslov, A.; Konhauser, K.O.; Lalonde, S.V. Large-scale fluctuations in Precambrian atmospheric and oceanic oxygen levels from the record of U in shales. *Earth Planet. Sci. Lett.* **2013**, *369*, 284–293. [CrossRef]
128. Lovley, D.R.; Phillips, E.J. Reduction of uranium by Desulfovibrio desulfuricans. *Appl. Environ. Microbiol.* **1992**, *58*, 850–856. [CrossRef] [PubMed]
129. Cai, C.; Dong, H.; Li, H.; Xiao, X.; Ou, G.; Zhang, C. Mineralogical and geochemical evidence for coupled bacterial uranium mineralization and hydrocarbon oxidation in the Shashagetai deposit, NW China. *Chem. Geol.* **2007**, *236*, 167–179. [CrossRef]
130. Seewald, J.S. Organic–inorganic interactions in petroleum-producing sedimentary basins. *Nature* **2003**, *426*, 327–333. [CrossRef]
131. Cai, Y.-W.; Zhang, S.-C.; He, K.; Mi, J.-K.; Zhang, W.-L.; Wang, X.-M.; Wang, H.-J.; Wu, C.-D. Effects of U-ore on the chemical and isotopic composition of products of hydrous pyrolysis of organic matter. *Pet. Sci.* **2017**, *14*, 315–329. [CrossRef]
132. Yang, S.; Schulz, H.-M.; Horsfield, B.; Schovsbo, N.H.; Noah, M.; Panova, E.; Rothe, H.; Hahne, K. On the changing petroleum generation properties of Alum Shale over geological time caused by uranium irradiation. *Geochim. Et Cosmochim. Acta* **2018**, *229*, 20–35. [CrossRef]

energies

MDPI

Article

Research on the Application Performance of the Helmholtz High-Efficiency Wave-Elimination Chamber in Gas Wave Tubes

Wutekuer Nuermaimaiti, Xuewu Liu *, Pengze Yan, Jiupeng Zou and Dapeng Hu

School of Chemical Engineering, Dalian University of Technology, Dalian 116024, China; wutekuer@mail.dlut.edu.cn (W.N.); pengzeyolo@163.com (P.Y.); zoujiupeng@dlut.edu.cn (J.Z.); Hudapeng@dlut.edu.cn (D.H.)

* Correspondence: liuxuewu@163.com

Abstract: A gas wave refrigerator is a device that uses gas pressure to expand and refrigerate; the moving shock wave in the gas wave tube is the key to cooling. If the shock wave is reflected to the open mouth of the gas wave tube, it will heat the expanded refrigeration gas at the open end, severely reducing the refrigeration efficiency. In order to reduce this effect, a Helmholtz middle wave chamber-type gas wave tube is presented here. This study describes the performance of this structure in the gas wave refrigeration process. The refrigeration performance of the structure's gas wave tube use was evaluated experimentally and numerically using computational fluid dynamics (CFD). The function of the middle wave elimination chamber was studied and explained: the chamber turns the reflected shock wave back to the end section of the tube and converts part of the incoming shock wave to an expansion wave, which may counteract the reflected shock wave by escaping from wave elimination chamber. The experimental results showed that the structure increased refrigeration efficiency by about 2.9% (50 Hz), 2% (40 Hz), and 2.2% (25 Hz) at the corresponding ejection frequencies, and reduced the reflected shock wave intensity by 11%, which proves the high-efficiency energy dissipation performance and the significant increase in refrigeration efficiency of the Helmholtz-type gas wave tube.

Keywords: gas wave tube; wave-eliminating chamber; Helmholtz chamber; refrigeration efficiency

Citation: Nuermaimaiti, W.; Liu, X.; Yan, P.; Zou, J.; Hu, D. Research on the Application Performance of the Helmholtz High-Efficiency Wave-Elimination Chamber in Gas Wave Tubes. *Energies* **2022**, *15*, 882. https://doi.org/10.3390/en15030882

Academic Editor: Dmitry Eskin

Received: 3 December 2021
Accepted: 7 January 2022
Published: 26 January 2022

1. Introduction

Due to the high storage and transportation risk factors, the efficient use of pressure energy has long plagued researchers. Whether the pressure energy is converted from other energy or associated pressure energy, it must be used or converted in time to efficiently use its resources. A gas wave refrigerator (also called a thermal separator) is a piece of equipment that uses gas pressure to expand and refrigerate [1,2]. Its advantages include simple structure, low cost [3], convenient operation and maintenance, high pressure resistance, operation with liquid, low power consumption, etc. The basic principle is to inject high pressure gas from a rotating distributor (rotary type) or a wall jet oscillator (static type) into several gas wave tubes with initial openings in turn. The pulse jet formed will perform expansion work on the stagnant gas in the tube, and the enthalpy of the jet gas itself will be lowered and cooled [4].

Research on static gas wave refrigerators is still in the development stage. The air wave tube (also called the receiving tube) is closed at the end to eject air intermittently. A "contact surface" formed between the ejected air and the trapped air in the tube moves forward, and multiple compression waves are continuously generated until a positive shock wave is formed [2,5]. The shock wave sweeps the gas along the tube, increasing the temperature and pressure of the gas; the energy of the shock wave dissipates as it is transmitted to the gas through the tube wall, thereby greatly reducing the enthalpy and temperature of the incident gas (refrigeration) [6,7]. The forward movement of the shock wave continuously transfers energy to the original gas in the tube, thereby attenuating the shock wave. However, when the shock wave is not completely attenuated, it can move to the end of the tube and reflect back to the outlet of the gas wave tube. The reflected shock wave will have two adverse effects on gas refrigeration. The first is to heat the cold air that was refrigerated to be discharged, such that the temperature of the cold air rises again. The second is to increase the pressure of the jet inlet, increase the oscillator load, and reduce the incident pressure ratio. Both losses will directly reduce the efficiency of gas wave refrigeration.

Elimination the reflected shock wave is an important research direction to improve the efficiency of gas wave refrigeration. The conventional method is to lengthen the gas wave tube to reduce the shock wave such that it is unable to reflect; however, this wastes a significant amount of materials, causes vibration and condensate accumulation, and blocks the gas wave tube, making the effective tube length heavy and short. Fang et al. [8] proposed the addition of a shock absorption chamber with a limited volume at the end of each gas wave tube to dissipate and absorb the shock wave; the refrigeration efficiency was greatly improved, yet the fluctuation in the ejection frequency was still large. Hu et al. [9] proposed using multiple shock absorption cavities in series to further attenuate the reflected shock waves. Zou et al. [10,11] experimentally studied the damping dissipation of wire mesh in the end of the gas wave tube, and Li et al. [12]. experimentally studied the shock wave with an orifice composite structure and achieved certain results. Dai et al. [13]. used CFD calculations to show that two composite structures can attenuate the shock wave to an acceptable level. In addition, Zhu [14], Li [15], and Liu et al. [16] studied the matching of the length of the gas wave tube with the ejection frequency, such that the gas wave tube can operate at a high-efficiency ejection frequency and reduce the impact of reflected shock waves. It is also possible to increase the heat exchange by changing the tube wall structure, increasing the energy dissipation in the tube, and reducing the reflected shock wave. For example, in the research of Al-Obaidi et al. [17–19], different dimple geometrical configurations with a combination of corrugated tubes and twisted tape were numerically investigated. The study found that vortices and rotational flow originated behind and near the dimple, twisted tape, and corrugation surfaces. The rotational flow and vortices promote mixing in flow between the thermal boundary layer and velocity boundary flow layer, thus increasing the heat transfer enhancement. This method can also improve refrigeration efficiency. Zou et al. [10,11] proposed moving the shock wave absorption chamber forward to a proper position behind the middle of the gas wave tube, creating a middle shock wave absorption chamber that was used to reduce the return of the reflected shock wave and fold the reflected shock wave back to the back section of the gas wave tube to accelerate its dissipation. The refrigeration efficiency was improved; however, there was significant incident energy loss in the fluid.

In a uniform straight gas wave tube, the reflected shock wave will maintain a high intensity and return to the exit position. If a wave-eliminating cavity is installed, the reflected shock wave can be attenuated by multiple reflections and energy dissipation in the cavity, and the intensity of the reflected returning shock wave will be significantly reduced, improving the cooling efficiency of a gas wave tube with the same tube length. In order to efficiently use the ejection pressure energy and reduce the reflected shock wave, one should also try to reconstruct the distribution and reflection of the wave system in the gas wave tube, improve the shock wave forward path, block the shock wave reflection, and increase the dissipation in the back section of the gas wave tube, thus improving shock wave attenuation. According to this idea, we proposed and studied the Helmholtz-type wave-elimination chamber structure, which was used to enhance the incident shock wave, prevent the return of the reflected shock wave, and fold the reflected shock wave back to the back section of the gas wave tube to accelerate energy dissipation and reduce the impact on the exit position of the reflected shock wave. Based on this, a new and efficient Helmholtz gas wave tube was developed.

2. Using CFD to Model the Chamber Geometry and Simulate the Flow Field of a Gas Wave Tube

2.1. Creating a Chamber Geometry Model

According to the previous analysis, to improve the efficiency of a static gas wave refrigerator it is necessary to increase the shock wave forward path, block the shock wave reflection, and increase the dissipation of the back section of the gas wave tube to promote shock wave attenuation. The proposed Helmholtz wave-elimination chamber structure is strictly symmetrical along the axis in geometry, and the three-dimensional model can be simplified to two-dimensional in the calculation. If the ejection frequency is controlled well, this cavity in the middle can make a small part of the shock wave that escapes from the left side of the cavity reflect the shock wave, which will meet the incident rightward expansion wave caused by the next ejection jet and the two will cancel out. Thus, the harm of the reflected shock waves is basically eliminated. Figure 1 shows the schematic diagram of the Helmholtz wave-elimination chamber.

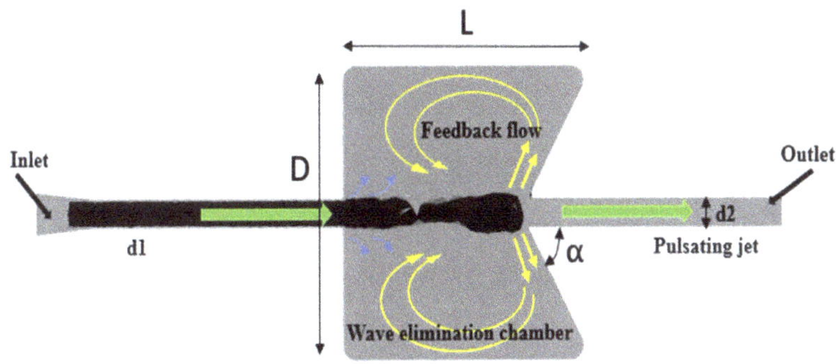

Figure 1. Schematic diagram of a Helmholtz wave-elimination chamber.

After the jet enters the chamber, the boundary layer disturbance is amplified by the resonance of the oscillating chamber to form an annular feedback flow vortex. The vortex ring further develops to form a large-area cavitation, blocking the continuity of the jet and forming a self-oscillating pulse jet.

Figure 2 shows the schematic diagram of a traditional wave-elimination chamber. Generally, after the jet enters the chamber, the boundary layer disturbance is amplified by the wave-elimination chamber and directly hits the wall of the vertical chamber. After a large amount of kinetic energy is lost, the feedback flow vortex is formed; however, the regular vortex ring cannot be completely formed, causing the jet to lose a significant amount of pressure.

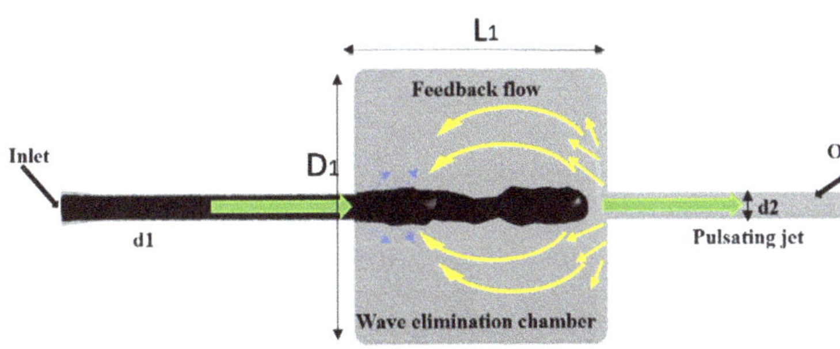

Figure 2. Schematic diagram of a traditional wave-elimination chamber.

This geometric model adopts the Helmholtz wave-elimination chamber and traditional wave-elimination chamber comparison model. The dimensionless parameters of Helmholtz geometry were obtained from the related literature [20–22], and they can be used as a comparison with the traditional chamber. For comparison, these parameters are shown in Table 1.

Table 1. Comparison chamber parameters.

Structure	Helmholtz Wave-Elimination Chamber (mm)	Traditional Wave-Elimination Chamber (mm)
Entrance diameter d_1	10	10
Outlet diameter d_2	10	10
Chamber length L	100	100
Chamber diameter D	100	100
Inclination angle of posterior chamber wall α	60°	90°

2.2. Calculation Models and Boundary Conditions

CFD was used to simulate a double-chamber gas wave tube with the traditional wave-elimination chamber and the Helmholtz wave-elimination chamber, designed in the middle of the tube; an absorption chamber was also added at the ends of the two kinds of gas wave tubes. The two kinds of tube structures and the calculation mesh division are shown in Figure 3. Using a two-dimensional calculation model, a maximum mesh size of 0.2 mm was determined by independence analysis, and the mesh at the opening of the tube was densified.

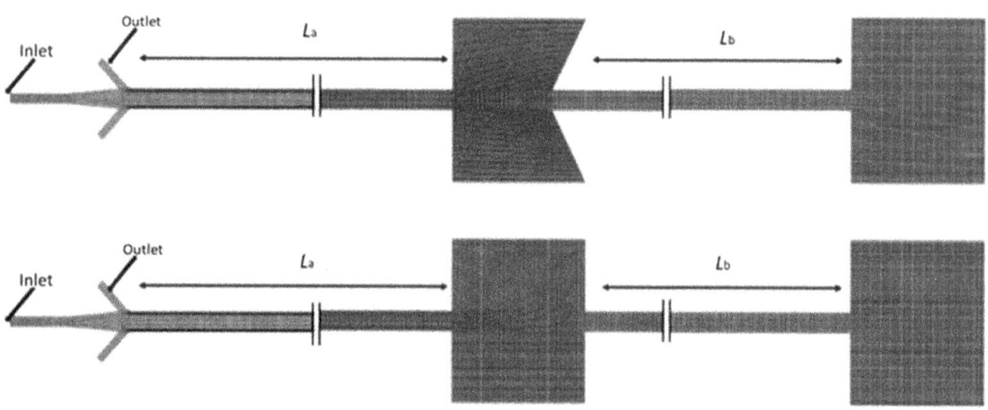

Figure 3. Structures and mesh of each type of gas wave tube.

The end absorption chamber was closed on the same side as the wave-elimination chamber; the front section of the middle wave-elimination chamber was L_a = 2000 mm, the rear section was L_b = 1000 mm, and the total length of the two kinds of tubes with the chamber was 3200 mm. Since only the efficiency of the two gas wave tubes were compared, the pulse flow setting was realized by programming the pulsating intake function (UDF) in the simulation process. We used a trapezoidal wave to simulate the incident wave. As shown in Figure 4, the time for the rising and falling of the leading edge was 1/40 s of the period T.

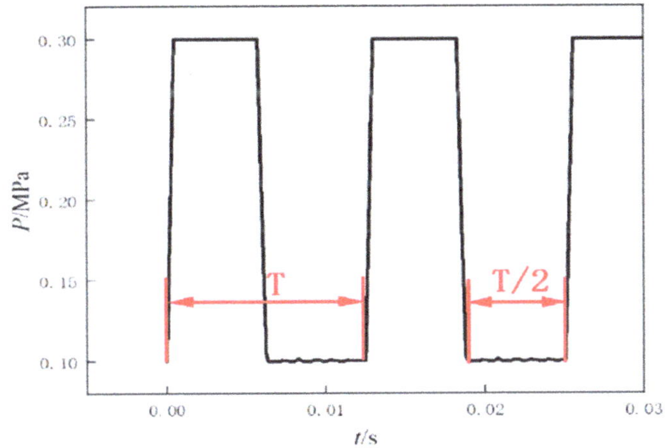

Figure 4. Trapezoidal total pressure waveform of external stimulation flow.

The pulse jet function (UDF) shown in Figure 4 is as follows:

$$P_A = \begin{cases} P_L + [(P_h - P_L)/0.0005]t & (0 < t < T/40) \\ P_h & (T/40 < t < 19T/40) \\ P_h - [(P_h - P_L)/0.0005](t - 0.0095) & (19T/40 < t < T/2) \\ P_L & (T/2 < t < T) \end{cases} \quad (1)$$

where t represents any time, T is the oscillation period, P_A is the certain instantaneous excitation pressure value, P_L is the minimum pressure, and P_h is the maximum pressure.

The total pressure of the gas wave tube port in the ejection stage was 0.2 MPa, 0.3 MPa, and 0.4 MPa, and the temperature was 300 K, the cold air exhaust stage was subsonic flow, the static pressure value of the fixed nozzle was 0.1 MPa, and the corresponding expansion ratios ε (the ratio of the total inlet pressure to the static outlet pressure) were 2, 3, and 4, respectively, at fixed wall boundaries. The gas hypothesis conformed to the ideal gas state equation. The gas injection frequency f was 25 Hz~80 Hz (in 5 Hz increments); the gas was assumed to conform to the ideal gas equation of state, and the viscosity was 1.789×10^{-5} kg/m·s.

2.3. Numerical Calculation Method

The inlet of the gas wave tube pulsating jet is transonic, and it produces compressible and strong turbulent flow. The Reynolds average method for solving the time-averaged Navier–Stokes equation should be adopted, considering that the nozzle jet is turbulent and the turbulent flow is a highly non-linear and complex flow. The simulation method adopts the two-equation realizable k-ε turbulence model, and the finite volume method is used to discretize the control equations [23]. The diffusion term selects a central difference scheme with high computational efficiency and second-order accuracy, which converges quickly. The convection term is anisotropic. In order to avoid numerical oscillations, the MUSCL (Monotonic Upstream-centered Scheme for Conservation Laws) format of the Roe flux differential splitting in the upside-down style is used for discretization [24], and the second-order fully implicit time step is used for iteration [25].

2.4. Refrigeration Efficiency of the Gas Wave Tube

The refrigeration efficiency of the gas wave tube is the ratio of the actual enthalpy drop of the jet gas entering and exiting the gas wave tube to the ideal isentropic enthalpy drop. The specific heat of the ideal gas does not change with pressure, as shown in the following formula:

$$\eta = \frac{C_p m (T_{in} - T_{out})}{C_p m T_{in}\left[1 - \left(\frac{P_{out}}{P_{in}}\right)^{\frac{k-1}{k}}\right]} = \frac{T_{in} - T_{out}}{T_{in}\left[1 - \left(\frac{1}{\varepsilon}\right)^{\frac{k-1}{k}}\right]} \quad (2)$$

where k is the isentropic process index of the gas, T_{in} is the total temperature at the inlet of the jet gas, and T_{out} is the time-average total temperature at the outlet of the gas wave tube.

During the simulation process, the mass-weighted average total temperature T_{av} of the outlet section of the gas wave tube is monitored over time. When the change with time is stable (we selected the seventh period), the outlet section of the gas wave tube outlet section T_{av} and Δt, the product of the mass flux Δm, are accumulated and then divided by the integral value m of the total exhaust mass at the outlet of the gas wave tube; finally, the time-average total temperature T_{out} of the gas at the outlet of the gas wave tube can be obtained.

3. Flow Field Simulation and Results Analysis of the Two Kinds of Chambers

For the three-dimensional simulation of the two cavities, the simulation conditions were as follows: the ejection frequency was 80 Hz, the total pressure of pulsating ejection was 0.3 MPa, the temperature was 300 K, the static pressure at the outlet stage was 0.1 MPa, and the corresponding expansion ratio ε was 3. Figure 5 shows a cloud diagram of the velocity vortex field changes inside two kinds of chambers. It can be seen from the figure that the jet in the Helmholtz-wave-elimination chamber begins to spread from the edge of the jet with time after ejection. When the time of reaching the back chamber wall is t, the edge of the jet collides with the back wall of the wave-elimination chamber, forming a tiny backflow. The backflow forms a vortex in the wave-elimination chamber, and the formation of this vortex will in turn disturb the edge shear layer of the jet, forming positive feedback, and finally strengthening the vortex flow inside the oscillation chamber and intercepting the continuous jet to form a self-oscillating jet. In the traditional wave-elimination chamber, the time it takes for the jet to enter the chamber, the wave-elimination chamber to amplify the boundary layer disturbance, and the jet to directly hit the vertical chamber wall, is t. After a large amount of kinetic energy is lost, the feedback flow vortex is formed; however, the regular vortex ring cannot be completely formed, causing the jet to lose a significant amount of pressure.

Figure 6 shows the pressure changes at the outlet of the two chambers. The Helmholtz chamber excited a stable self-oscillating jet under a condition of small pressure loss. The total pressure of the Helmholtz wave-elimination chamber was much higher than the traditional wave-breaking chamber. The total pressure retention rate increased by 7%. This shows that the forward collision of the jet in the traditional chamber consumes a great deal of energy. Reducing the jet pressure front jump speed and the slow pressure drop causes a large amount of energy loss when the jet collides with the back chamber wall and cannot be avoided. Therefore, the total outlet pressure is lower than in the Helmholtz chamber, which affects the refrigeration efficiency.

Figure 7 shows the outlet velocities of the two chambers. The Helmholtz wave-elimination chamber jet velocity loss was significantly lower than that of the traditional wave-elimination chamber, and the velocity increased by about 50 m/s, which also doubled the amplitude of the jet's self-oscillation pulsation and increased the jet turbulence at the end of the gas wave tube, improving the energy dissipation efficiency.

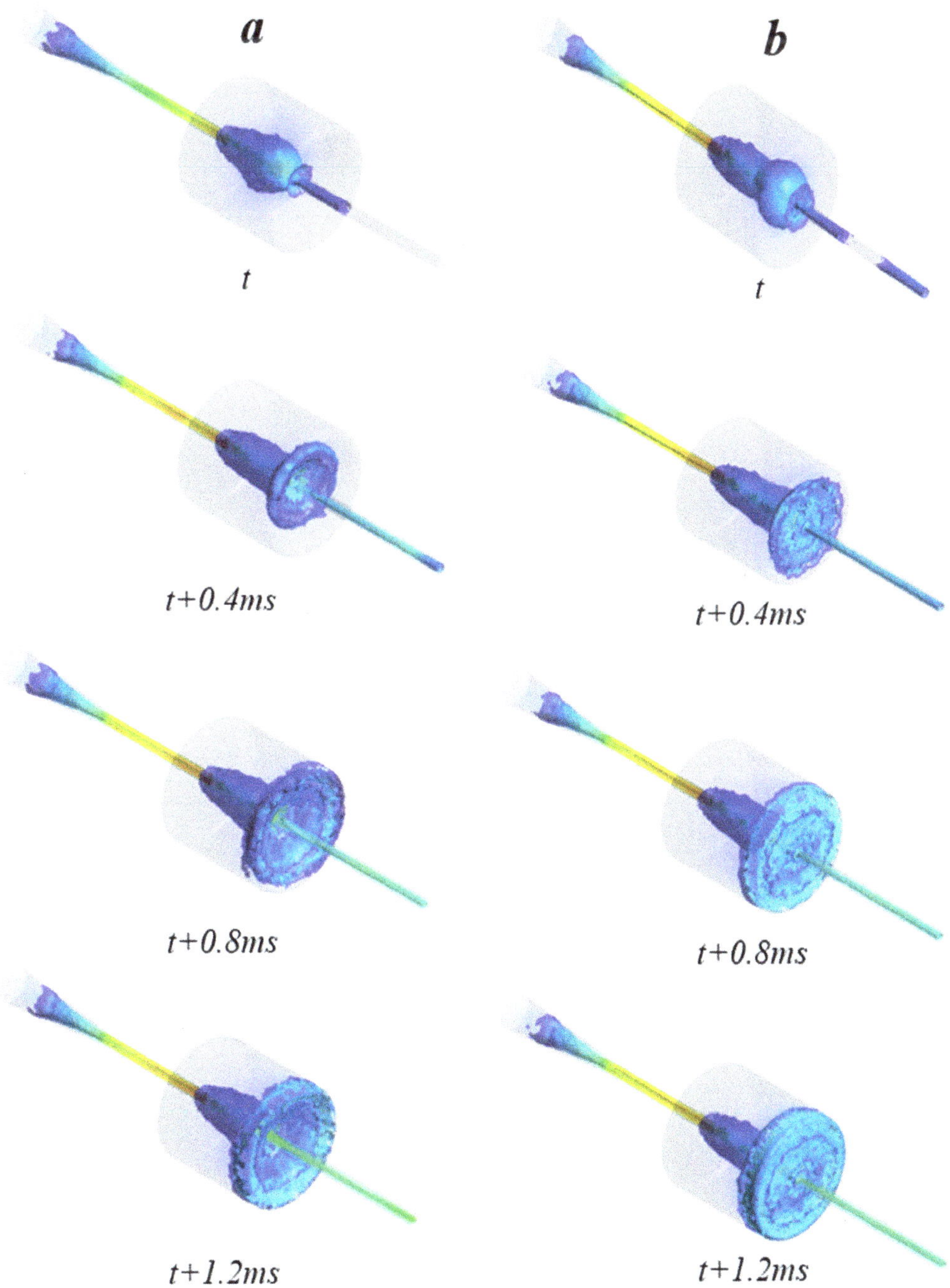

Figure 5. Variations of the velocity vortex field inside the two chambers: (**a**) Helmholtz wave-elimination chamber; (**b**) traditional wave-elimination chamber.

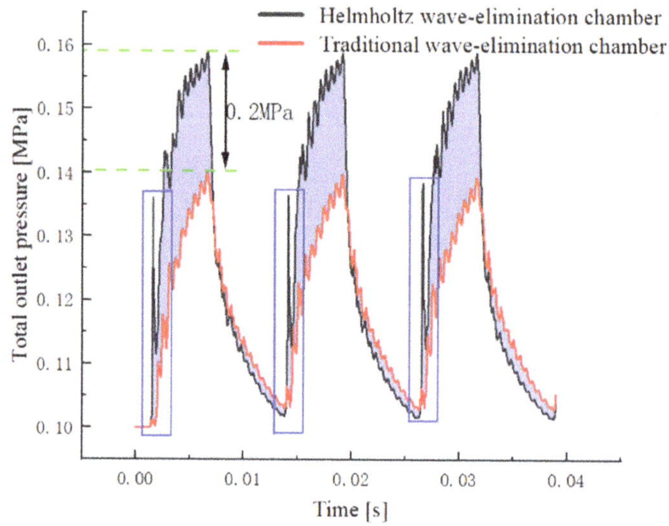

Figure 6. Outlet pressure of the two chambers.

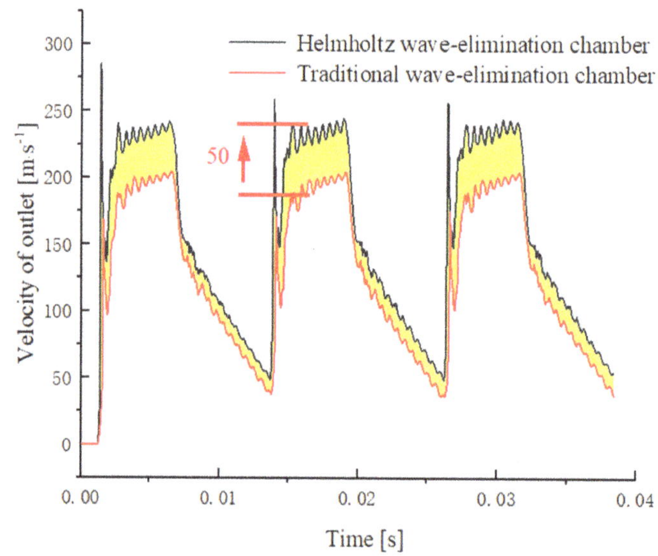

Figure 7. Velocity of the two chambers.

We observed the effect of the change in the expansion ratio on the outlet pressure of the Helmholtz wave-elimination chamber. When the pulsating ejection frequency was 80 Hz, the total pressure of pulsating ejection was 0.2 MPa, 0.3 MPa, and 0.4 MPa, and the temperature was 300 K, the cold air exhaust stage was subsonic flow, the static pressure value of the fixed nozzle was 0.1 MPa, and the corresponding expansion ratios ε were 2, 3, and 4, respectively. It can be seen from Figure 8 that reducing the expansion ratio slows down the pressure front jump speed and pressure drop speed. Therefore, the Helmholtz chamber would have a better effect on the end jet at a high expansion ratio.

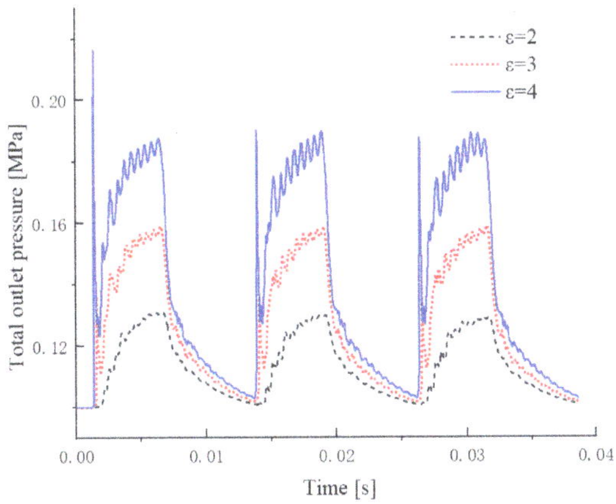

Figure 8. Outlet pressure change under different expansion ratios.

4. Comparison and Analysis of the Efficiency of the Two Kinds of Gas Wave Tube Refrigeration

Through the method of two-dimensional numerical simulation, the calculation was carried out by combining different frequencies of pulsating jets with a closed gas wave tube structure. The intensity of the shock wave is defined as the ratio of the total gas pressure in the gas wave tube to the static pressure at the outlet. The changes in the shock wave intensity in the two gas wave tubes with time are shown in Figure 9. Under the action of the wave elimination chamber, the reflected shock wave in the gas wave tube was significantly weakened; it can be seen that the Helmholtz gas wave tube P/P_{os} value was only 1.7, and the P/P_{os} value of the traditional gas wave tube was 1.9. The reflected shock wave intensity was reduced by about 11%.

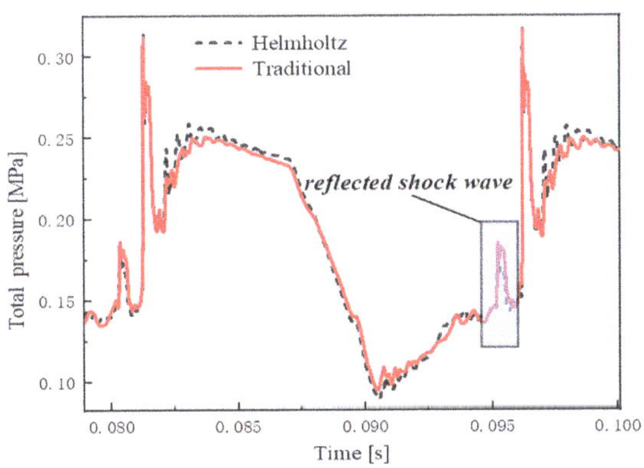

Figure 9. The curve of the shock wave intensity in the two gas wave tubes at the front end of the chamber versus time.

Figure 10 shows the gas mass weighted average total temperature T_{av} with time for the two gas wave tube structures in a cycle where the jet duty ratio was 1/2, the expansion ratios ε were 2, 3, and 4, and the jet frequencies f were 25 Hz and 80 Hz. It can be seen from the figure that the jet gas in the exhaust stage is expanded and cooled by the original gas in the compression tube and then discharged from the outlet to obtain a lower-temperature refrigerant gas. From Figure 10a,b, it can be seen that the cooling temperature of the

Helmholtz central wave-elimination chamber was significantly lower than that of the gas wave tube of the traditional wave-elimination chamber.

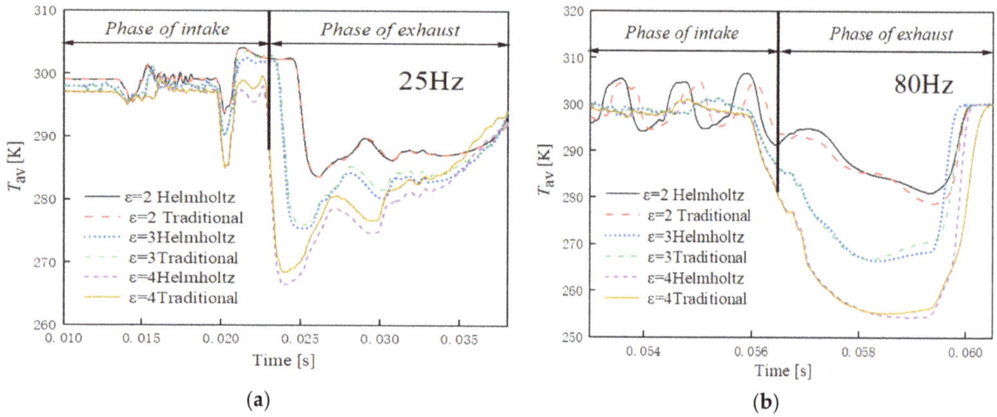

Figure 10. The variation in the total temperature T_{av} of the two gas wave tube structures with time in one cycle: (**a**) 25 Hz; (**b**) 80 Hz.

Figure 11 shows the numerical simulation of the refrigeration efficiency of the Helmholtz wave-elimination chamber and the traditional wave-elimination chamber as a function of the ejection frequency. Under the condition where the jet duty ratio was 1/2 and the expansion ratios ε were 2, 3, and 4, the refrigeration efficiency increased by about 11–18% with the frequency conversion; the refrigeration efficiency showed a relative increase of about 2.9% (50 Hz), 2% (40 Hz), and 2.2% (25 Hz). In addition, due to the increase in the expansion ratio, the turbulence intensity, the temperature difference, and the pressure gradient in the tube, the efficiency was generally slightly improved. Furthermore, because the temperature in the tube was higher, the wave speed increased. The corresponding efficiency extremes were also improved with the increase in the ejection frequency.

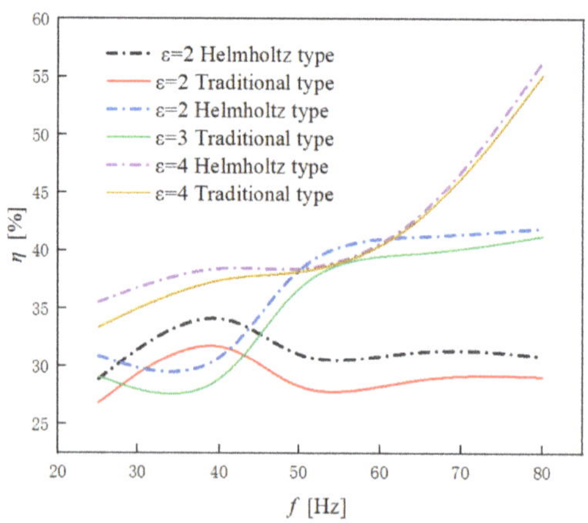

Figure 11. The refrigeration efficiency of the two types of gas wave tubes numerically simulated with the change of the ejection frequency.

Through the numerical simulation method, combined with the structural calculation of the pulsating jet and the gas wave tube, it was shown that the Helmholtz-type gas wave tube reduces the loss of incident kinetic energy and increases the consumption of reflected shock waves. The reflected shock wave is reflected multiple times between the middle chamber and the end of the gas wave tube, which weakens the intensity of the reflected shock wave and improves the refrigeration efficiency of the gas wave tube.

5. Experimental Study and Comparison of Two Kinds of Gas Wave Tube Refrigeration Efficiency

We constructed a single-tube gas wave refrigerator to verify the influence on the refrigeration efficiency of the two gas wave tube structures. The exhaust gas of the single-tube small-volume gas wave refrigerator is easily affected by the temperature of the surrounding environment, making it difficult to accurately measure the exhaust temperature. However, if the changing law of environmental influence factors is observed, the contrasting values of different air wave tube structure refrigeration efficiencies are easy to measure and distinguish; hence, it is a more economical and effective experimental research method.

5.1. Experimental Facilities

The experimental device is shown in Figure 12. The single-tube experimental machine was built by accurately aligning the jet nozzle with the opening of the single air wave tube and then fixing it. Between the nozzle and the gas wave tube orifice, a jet pulse modulation disc with two symmetrical and transparent arc openings was inserted. When the arc opening turns to the nozzle and the opening position of the gas wave tube, air is ejected into the gas wave tube. During the exhaust phase, the gas wave tube entrance is blocked by the pulse modulation disc, and the air flows out from the front end of the gas wave tube toward the outlet. According to the requirements of intake duty ratio 1/2 and duty ratio 1/4, the ratio of the transparent arc length to the closed arc length of the pulse modulation disc was 1:1 and 1:2, respectively, as shown in Figure 13.

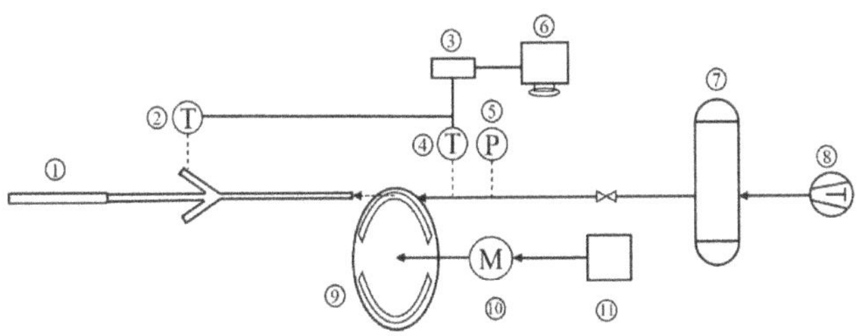

Figure 12. Schematic diagram of the experimental process. 1—gas wave tube; 2, 4—temperature sensor; 3—PLC (Programmable Logic Controller); 5—pressure gauge; 6—computer; 7—air tank; 8—air compressor; 9—pulse jet modulation disc; 10—motor; 11—inverter.

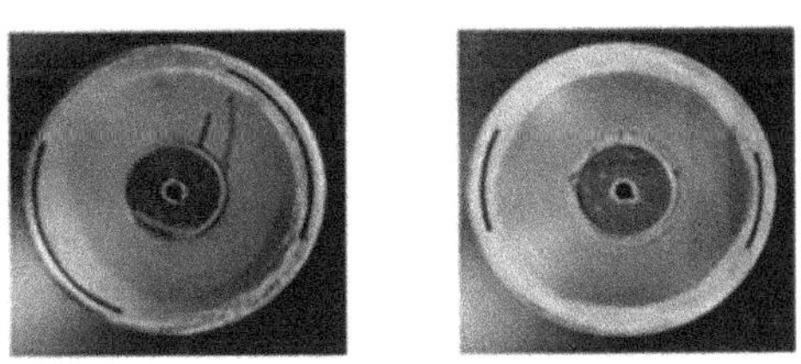

(a) (b)

Figure 13. Pulse modulation discs with different duty cycles: (**a**) 1/2 duty cycle; (**b**) 1/4 duty cycle.

The pulse modulation disc was driven by an AC motor whose speed was regulated by a frequency converter. The frequency f was changed by adjusting the motor speed n (r/min). Because there were two arc-shaped openings on the disc, f = 2n/60 Hz. The outlet structure adopted an acrylic plate to engrave the flow channel; the gas wave tube was copper tube, the front tube length was 2000 mm, and the inner diameter was 10 mm. The distance between the two chambers of the two airwave tubes was 1000 mm, and the inner diameter of the tube was 10 mm. The inner diameter of the chamber was about 96 mm, and the length of the inner chamber was about 100 mm. There are three types of wave-elimination chamber. The first two are hollow cylinders made of aluminum. The difference between the two is whether both ends are open. These three structures are shown in Figure 14. The Helmholtz wave-elimination chamber proposed in this paper was made of Plexiglass.

(a) (b) (c)

Figure 14. Different types of chambers: (**a**) traditional chamber; (**b**) end wave chamber; (**c**) Helmholtz chamber.

5.2. Experimental Results and Analysis

In the actual experiment, the ambient temperature was uncontrollable and was different from the idealized instantaneous ejection into the gas wave tube and instantaneous exhaust used in the simulation. For a certain distance between the nozzle and the gas wave tube opening, the jet caused a certain degree of loss before entering the gas wave tube, which, along with other complications, made the refrigeration efficiency obtained by the experiment lower than the value obtained by the simulation.

The pressure sensor (5) was inserted into the air supply pipe close to the nozzle to measure the total pressure P_{in} at the inlet of the air wave pipe. The valve was used to adjust the pressure to 0.1 and 0.2 MPa. The exhaust static pressure was the atmospheric pressure, and the corresponding expansion ratios ε were close to 2 and 3, respectively. A temperature sensor (4) was also inserted into the air supply pipe close to the nozzle for measurement, which was used as the total intake temperature T_{out}. For the measurement of the cooling temperature of the outlet air, because the modulation disc 9 was attached to the arc-shaped groove on the end of the gas wave tube, the pulse exhaust and reflection caused by the speed change made it difficult to accurately measure the exhaust temperature in the edge area of the pulse jet modulation disc. Therefore, this experiment adopted an indirect method that involved setting a flow channel near the opening of the gas wave tube, inserting a temperature sensor (2), measuring the time-average temperature T_{a} near the opening, and then approximately converting to the exhaust temperature T_{out}. Although the accurate refrigeration temperature cannot be measured in this way, it is clear enough that the measurement results of uniform conditions can be used to compare the refrigeration efficiency of various gas wave tubes.

Because the room temperature injection gas and the refrigerating gas alternately pass through the opening of the gas wave tube, and the ambient gas outside the metal gas wave tube heats the small amount of refrigerating gas, the time-average temperature T_{a} measured at the opening of the gas wave tube is much higher than the actual cooling discharge temperature T_{out}. Regardless of the influence of environmental heating, it is assumed that the heat transfer $Q = mc_p\Delta T$ of the ejection and exhaust gas to the sensor are equal. Assuming that the specific heat capacity c_{p} is constant, and the mass m of injection and exhaust gas is equal, the following equations are given according to the relationship of equal heat transfer: $T_{in} - T_a = T_a - T_{out}$ and $T_{out} = 2T_a - T_{in}$. The refrigeration efficiency values of different gas wave tube structures under different ejection conditions and frequency conditions can be calculated by substituting the calculated exhaust temperature T_{out}, the measured total intake temperature T_{in}, and the expansion ratio ε into Equation (2).

In fact, corresponding to the improved efficiency of the new gas wave tube, the ambient gas outside the tube had a greater impact on the refrigerant gas inside the tube, and the heating was also stronger. However, since the above simplified formula is not considered, the calculated efficiency curve gap between the two gas wave tubes is smaller than the actual value. In addition, the heating coefficient of the nozzle with the temperature difference is difficult to estimate, so it was not considered. This leads to a certain error in the results between the actual experiment and the numerical simulation under ideal boundary conditions; however, it is still accurate enough to compare the refrigeration efficiency of the two gas wave tubes.

Figure 15a shows the cooling efficiency obtained from the measured time-average temperature at the exit section of the gas wave tube of the two structures under the condition where the jet duty ratio was 1/2 and the expansion ratios ε were 2 and 3. With the increase in frequency, the cooling efficiency also increased by 2% to 6%; the Helmholtz chopper chamber increased by about 3.5% (50 Hz) and 3.7% (90 Hz) compared with the gas wave tube of the traditional middle chopper chamber. In addition, the efficiency extreme value increased with the fluctuation of the ejection frequency. Figure 15b shows the cooling efficiency obtained by the measured time-average temperature at the exit section of the gas wave tube of the two structures under the condition where the jet duty ratio was 1/4 and the expansion ratios ε were 2 and 3. The refrigeration efficiency also increased by 2% to 5%. The Helmholtz central wave-elimination chamber and the gas wave tube of the traditional central wave-elimination chamber were relatively improved by about 5% (50 Hz) and 2% (90 Hz), respectively.

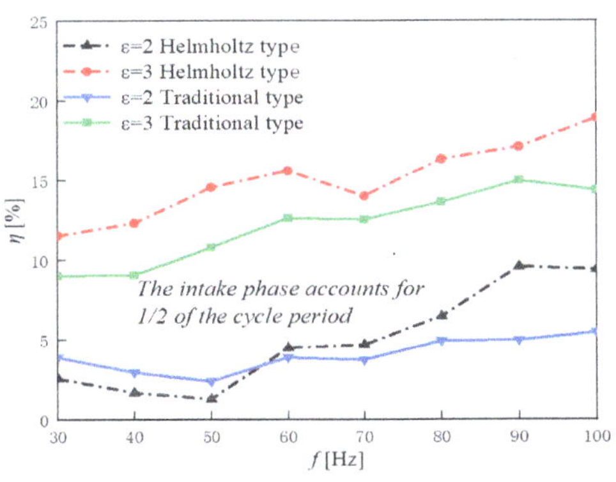

(a)

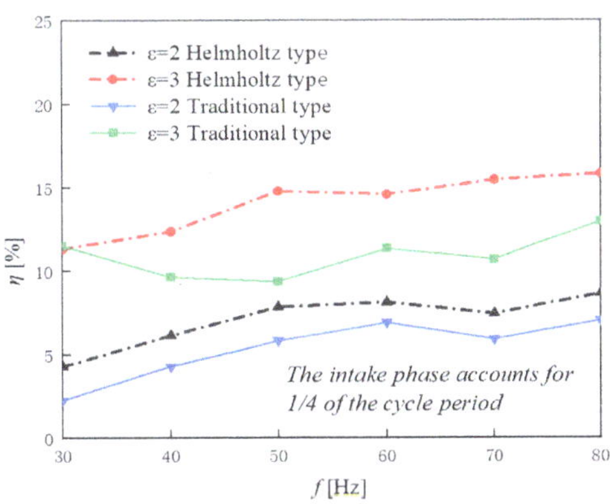

(b)

Figure 15. Refrigeration efficiency change diagram with frequency: (**a**) 1/2 duty cycle; (**b**) 1/4 duty cycle.

It can be seen from the experimental results that the refrigeration efficiency and flatness of the Helmholtz wave-elimination chamber gas wave tube were significantly higher than that of the existing traditional wave-elimination chamber gas wave tube. The ejection frequencies corresponding to the peak and trough efficiency of the two kinds of gas wave tubes in the experiment were also basically consistent with the simulation results.

6. Conclusions

The gas wave tube structure of the static gas wave refrigerator was optimized by the method of combining CFD numerical simulation and experimental studies. The shock wave absorption chamber was changed from a traditional chamber to a Helmholtz chamber. The refrigeration efficiencies of the two gas wave tube structures with varying jet frequencies when the expansion ratio was 2, 3, and 4 were compared and analyzed. The results showed that:

(1) Through chamber flow field analysis and CFD three-dimensional numerical simulation, it was shown that Helmholtz-type wave-elimination chamber reduced forward shock wave dissipation. The collision between the incident airflow and the chamber wall of a traditional wave-breaking chamber consumes a great deal of kinetic energy, resulting in a total pressure retention rate much lower than that of the Helmholtz wave-breaking chamber. The use of Helmholtz-type chamber increased the total pressure retention rate by 7%.

(2) Through the two-dimensional numerical simulation method, combining the pulsating ejection gas and the gas wave tube structure, it was shown that the Helmholtz gas wave tube reduced the incident kinetic energy loss, weakened the reflected shock wave, and improved the cooling efficiency. The intensity of the reflected shock wave was reduced by about 11%. In the three expansion ratios, the relative increase in refrigeration efficiency was about 2.9% (50 Hz), 2% (40 Hz), and 2.2% (25 Hz).

(3) A single-tube gas wave machine model was used to conduct a comparative experimental study on the efficiency of the two gas wave tube structures. Although the experimental model structure itself had uncontrollable ambient temperature and the loss of gas at the inlet, the overall efficiency was very low. It can be clearly seen from the comparison of the gas wave tube structures that the refrigeration efficiency of the Helmholtz-type gas wave tube increased significantly, and the difference in the trend of the refrigeration efficiency range of the two kinds of gas wave tubes corresponded to the peak and valley values of the efficiency. Although the ejection frequency was more consistent with the simulation results, the refrigeration efficiency of the gas wave tube was significantly increased.

Author Contributions: Conceptualization, W.N.; methodology, W.N.; software, W.N.; validation, W.N., X.L. and P.Y.; formal analysis, J.Z.; investigation, W.N.; resources, D.H.; data curation, W.N.; writing—original draft preparation, W.N.; writing—review and editing, X.L.; supervision, D.H.; project administration, X.L.; funding acquisition, D.H. All authors have read and agreed to the published version of the manuscript.

Funding: This work was supported by the National Key Research and Development Program of China (2018YFA0704601) and the Natural Science Foundation of China (21776029).

Institutional Review Board Statement: Not applicable.

Informed Consent Statement: Not applicable.

Data Availability Statement: Not applicable.

Acknowledgments: The authors thank the opportunity and financial support to carry on this project from the Natural Science Foundation of China.

Conflicts of Interest: The authors declare no conflict of interest.

References

1. Marchal, P.; Malek, S.; Vitard, J.C. Skid-mounted rotating thermal separator. *Oil Gas* **1984**, *11*, 55–58.
2. Yu, H.-R. Flow in heat separator. *J. Dalian Univ. Technol.* **1984**, *23*, 1–7.
3. Peng, X.-H.; Zhong, H.-G. Application of gas wave refrigerator in simulated altitude test. *Gas Turbine Exp. Res.* **2012**, *25*, 27–30.
4. Liu, P.-Q.; Xu, S.-Y.; Wang, Z.-W.; Liu, S.; Hu, D.; College of Chemical Engineering, Dalian University of Technolog. Influence of offset angle on refrigeration efficiency of gas wave refrigerator and prediction for optimal offset angle. *CIESC J.* **2014**, *65*, 4271–4277.

Zheng, M.-F.; Liu, X.; Huang, C.; Lin, Y.; Lei, X.; Li, X.; College of Chemistry and Chemical Engineering, Fuzhou University. Incident shock wave attenuation in oscillatory tube and influence on performance of pressure wave refrigerator. *CIESC J.* **2014**, *65*, 3410–3417.

Li, Z.-C.; Xu, L. Research and application of gas wave refrigerator. *Cryogenics* **2002**, *126*, 22–27.

Yu, Y.; Liu, P.-Q.; Wang, Y.-L.; Leng, C.; Zhao, Y.; Wang, J.; Hu, D.; School of Chemical Machinery, Dalian University of Technology. Flow and thermodynamic properties of efficient gas wave refrigeration plant. *CIESC J.* **2017**, *68*, 3039–3048.

Fang, Y.-Q.; Zheng, J.; Liu, R.-J.; Zhu, C.; Fan, J.; Hu, D.-P. Experimental investigation of influence factor of gas wave refrigeration efficiency. *Aerodyn. Exp. Meas. Control.* **1993**, *7*, 15–18.

Hu, D.-P.; Liu, P.-Q.; Zhu, C. A weakening reflected shock wave method in oscillating tube. *CIESC J.* **2008**, *59*, 562–566.

10. Zou, J.-P.; Liu, X.-W.; Chen, S.-H. Experiment about weakening the energy of the reflecting shock waves in the incepting tube. *Cryogenics* **2001**, *121*, 48–53.
11. Liu, X.-W.; Zou, J.-P.; Chen, S.-H.; Jin, L.-A. Study on the influence of gas wave tube end boundary condition to refrigeration efficiency. *Fluid Mach.* **2001**, *29*, 55–57.
12. Li, X.-L.; Zhu, C. Elimination of reflect shock wave in oscillatory tube. *CIESC J.* **2001**, *52*, 379–380.
13. Dai, Y.-Q.; Hu, D.-P.; Liu, W.; Zhu, C. The CFD analysis of oscillatory tube equipped with compound damps. *Low Temp. Spec. Gases* **2003**, *21*, 23–24.
14. Zhu, X.-Q. Study on the optimum length of the accept tube of gas wave refrigerator. *J. Wuxi Inst. Light Ind.* **1993**, *12*, 135–141.
15. Li, X.-L. Experimental study on influences of the tube length on cooling effects of the oscillating tube. *Refrigeration* **1993**, *2*, 15–18.
16. Liu, P.-Q.; Hu, D.-P.; Zhao, W.-J.; Zhu, C.; Dai, Y.; Liu, R.; Zou, J. Numerical prediction for extreme jet flow frequency of rotary gas wave refrigerator. *CIESC J.* **2009**, *60*, 1–5.
17. Al-Obaidi, A.R. Study the influence of concavity shapes on augmentation of heat-transfer performance, pressure field, and fluid pattern in three-dimensional pipe. *Heat Transf.* **2021**, *50*, 4354–4381. [CrossRef]
18. Al-Obaidi, A.R.; Alhamid, J.; Hamad, F. Flow felid and heat transfer enhancement investigations by using a combination of corrugated tubes with a twisted tape within 3D circular tube based on different dimple configurations. *Heat Transf.* **2021**, *50*, 6868–6885. [CrossRef]
19. Al-Obaidi, A.R. Investigation of the flow, pressure drop characteristics, and augmentation of heat performance in a 3D flow pipe based on different inserts of twisted tape configurations. *Heat Transf.* **2021**, *50*, 5049–5079. [CrossRef]
20. Li, D.; Kang, Y.; Ding, X.-L.; Wang, X.-C.; Fang, Z.-L. Effects of area discontinuity at nozzle inlet on the characteristics of high speed self-excited oscillation pulsed waterjets. *Exp. Therm. Fluid Sci.* **2016**, *79*, 254–265. [CrossRef]
21. Fang, Z.-L.; Zheng, F.-D.; Xiong, T.; Wei, W.; Jiang, P.; Wu, Q.; Wang, Y.; Fei, Y. Large eddy simulation of self-excited oscillation inside Helmholtz oscillator. *Int. J. Multiph. Flow* **2020**, *126*, 103253. [CrossRef]
22. Xue, X.; Yu, Y.; Zhao, J. Study on draining off water mechanism and interaction characteristic of high-temperature and high-pressure combustion-gas jets with the water. *Appl. Therm. Eng.* **2018**, *143*, 570–581. [CrossRef]
23. Wang, F.-J. *Computational Fluid Dynamics Analysis: The Principle and Application of CFD Software*; Tsinghua University Press Co., Ltd.: Beijing, China, 2014.
24. Deconinck, H.; Roe, P.L.; Struijs, R.A. multidimensional generalization of Roe's flux difference splitter for the Euler equations. *Comput. Fluids* **1993**, *22*, 215–222. [CrossRef]
25. Saito, T.; Voinovich, P.; Zhao, W.; Shibasaki, K.; Shibasaki, S.; Takayama, K. Experimental and numerical study of pressure wave refrigerator performance. *Shock. Waves* **2003**, *13*, 253–259. [CrossRef]

MDPI
St. Alban-Anlage 66
4052 Basel
Switzerland
www.mdpi.com

MDPI Books Editorial Office
E-mail: books@mdpi.com
www.mdpi.com/books

www.ingramcontent.com/pod-product-compliance
Lightning Source LLC
Chambersburg PA
CBHW041935260926
51850CB00018B/175
* 9 7 8 3 7 2 5 8 1 1 2 1 2 *